AN INTRODUCTION TO QUANTUM OPTICS AND QUANTUM FLUCTUATIONS

An Introduction to Quantum Optics and Quantum Fluctuations

Peter W. Milonni

Los Alamos National Laboratory and University of Rochester

OXFORD
UNIVERSITY PRESS

OXFORD
UNIVERSITY PRESS

Great Clarendon Street, Oxford, OX2 6DP,
United Kingdom

Oxford University Press is a department of the University of Oxford.
It furthers the University's objective of excellence in research, scholarship,
and education by publishing worldwide. Oxford is a registered trade mark of
Oxford University Press in the UK and in certain other countries

First published 2019
First published in paperback 2023

Published in the United States of America by Oxford University Press
198 Madison Avenue, New York, NY 10016, United States of America

British Library Cataloguing in Publication Data
Data available

Library of Congress Cataloging in Publication Data
Data available

ISBN 978–0–19–921561–4 (Hbk.)
ISBN 978–0–19–889268–7 (Pbk.)

DOI: 10.1093/oso/9780199215614.001.0001

To my mother, Antoinette Marie Milonni

and

To the memory of my mother-in-law, Xiu-Lan Feng

Preface

The quantum theory of light and its fluctuations are applied in areas as diverse as the conceptual foundations of quantum theory, nanotechnology, communications, and gravitational wave detection. The primary purpose of this book is to introduce some of the most basic theory for scientists who have studied quantum mechanics and classical electrodynamics at a graduate or advanced undergraduate level. Perhaps it might also offer some different perspectives and some material that are not presented in much detail elsewhere.

Any book purporting to be a serious introduction to quantum optics and fluctuations should include field quantization and some of its consequences. It is not so easy to decide which other aspects of this broad field are most apt or instructive. I have for the most part written about matters of fundamental and presumably long-lasting significance. These include spontaneous emission and its role as a source of quantum noise; field fluctuations and fluctuation-induced forces; fluctuation–dissipation relations; and some distinctly quantum aspects of light. I have tried to focus on the essential physics, and in calculations have favored the Heisenberg picture, as it often suggests interpretations along classically familiar lines. Some historical notes that might be of interest to some readers are included; these and other digressions appear in small type. Also included are exercises for readers wishing to delve further into some of the material.

I am grateful to my longtime friends Paul R. Berman, Richard J. Cook, Joseph H. Eberly, James D. Louck, and G. Jordan Maclay for discussions over many years about much of the material in this book. Jordan read most of the book in its nearly final version and made insightful comments and suggestions. Thanks also go to Sönke Adlung and Harriet Konishi of Oxford University Press for their patience and encouragement.

Peter W. Milonni
Los Alamos, New Mexico

Contents

Contents

1
Elements of Classical Electrodynamics

This chapter is a brief refresher in some aspects of (mostly) classical electromagnetic theory. It is mainly background and accompaniment for the rest of the book, with a few small conceptual points not always found in standard treatises.

1.1 Electric and Magnetic Fields

Maxwell's equations for the electric field $\mathbf{E}$ and the magnetic induction field $\mathbf{B}$ are:

$$\nabla \cdot \mathbf{E} = \rho/\epsilon_0, \quad \text{or} \quad \oint_S \mathbf{E} \cdot \mathbf{n}\, dS = \frac{1}{\epsilon_0} \int_V \rho\, dV = \frac{1}{\epsilon_0} Q, \tag{1.1.1}$$

$$\nabla \cdot \mathbf{B} = 0, \quad \text{or} \quad \oint_S \mathbf{B} \cdot \mathbf{n}\, dS = 0, \tag{1.1.2}$$

$$\nabla \times \mathbf{E} = -\frac{\partial \mathbf{B}}{\partial t}, \quad \text{or} \quad \oint_C \mathbf{E} \cdot d\mathbf{r} = -\frac{\partial}{\partial t} \int_S \mathbf{B} \cdot \mathbf{n}\, dS, \tag{1.1.3}$$

$$\nabla \times \mathbf{B} = \mu_0 \mathbf{J} + \frac{1}{c^2}\frac{\partial \mathbf{E}}{\partial t}, \quad \text{or} \quad \oint_C \mathbf{B} \cdot d\mathbf{r} = \mu_0 I + \frac{1}{c^2}\frac{\partial}{\partial t} \oint_S \mathbf{E} \cdot \mathbf{n}\, dS, \tag{1.1.4}$$

where ρ is the electric charge density, $\mathbf{J}$ is the electric current density, and $c = 1/\sqrt{\epsilon_0 \mu_0}$ is the speed of light in vacuum. The fields $\mathbf{E}$ and $\mathbf{B}$ are defined such that the force on a point charge q moving with velocity $\mathbf{v}$ is

$$\mathbf{F} = q(\mathbf{E} + \mathbf{v} \times \mathbf{B}). \tag{1.1.5}$$

Newton's second law ($\mathbf{F} = m\mathbf{a}$) describes the (non-relativistic) motion of a charge q of mass m in the $\mathbf{E}$ and $\mathbf{B}$ fields.

Equation (1.1.1) is Gauss's law: the flux of $\mathbf{E}$ through any closed surface S is proportional to the net charge Q in the volume V enclosed by S. Equation (1.1.2) implies there is no magnetic charge analogous to Q. Equation (1.1.3) is Faraday's law of induction: the line integral of the electric field around any closed curve C—the electromotive force (emf) in a wire loop, for example, or just a loop in free space—is minus the rate of change with time of the magnetic flux through the loop; the minus sign enforces Lenz's law, the (experimental) fact that the emf induced in a coil when a pole of a magnet is pushed into it produces a current acting to repel the magnet.

An Introduction to Quantum Optics and Quantum Fluctuations. Peter W. Milonni.
© Peter W. Milonni 2019. Published in 2019 by Oxford University Press.
DOI:10.1093/oso/9780199215614.001.0001

Equation (1.1.4) relates the integral of $\mathbf{B}$ around a loop C to the current I in C and the flux of $\mathbf{E}$ through C; the first term expresses Oersted's law (an electric current can deflect a compass needle), while the second term corresponds to the *displacement current* that Maxwell, relying on mechanical analogies, added to the current density $\mathbf{J}$. With this additional term (1.1.1) and (1.1.4), together with the identity $\nabla\cdot(\nabla\times\mathbf{B}) = 0$, imply the continuity equation

$$\nabla \cdot \mathbf{J} + \frac{\partial \rho}{\partial t} = 0, \tag{1.1.6}$$

which says, in particular, that electric charge is conserved. (The additional term also implied wave equations for the electric and magnetic fields and therefore the possibility of nearly instantaneous communication between any two points on Earth!) Maxwell's equations express all the laws of electromagnetism discovered experimentally by the pioneers (Ampère, Cavendish, Coulomb, Faraday, Lenz, Oersted, etc.) in a wonderfully compact form.

If the charge density ρ does not change with time, it follows that $\nabla \cdot \mathbf{J} = 0$ and, from Maxwell's equations, that the electric and magnetic fields do not change with time and are uncoupled:

$$\nabla \cdot \mathbf{E} = \rho/\epsilon_0 \quad \text{and} \quad \nabla \times \mathbf{E} = 0, \tag{1.1.7}$$

$$\nabla \cdot \mathbf{B} = 0 \quad \text{and} \quad \nabla \times \mathbf{B} = \mu_0 \mathbf{J}. \tag{1.1.8}$$

According to Ampère's law ($\nabla \times \mathbf{B} = \mu_0\mathbf{J}$), the magnetic field produced by a steady current I in a straight wire has the magnitude

$$B = \frac{\mu_0}{2\pi}\frac{I}{r} = \frac{1}{4\pi\epsilon_0}\frac{2I}{c^2 r} \tag{1.1.9}$$

at a distance r from the wire and points in directions specified by the right-hand rule.[1] It then follows from (1.1.5) that the (attractive) force f per unit length between two long, parallel wires separated by a distance r and carrying currents I and I' is

$$f = \frac{\mu_0}{2\pi}\frac{II'}{r} = \frac{1}{4\pi\epsilon_0}\frac{2II'}{c^2 r}. \tag{1.1.10}$$

Until recently this was used to define the ampere (A) as the current $I = I'$ in two long parallel wires that results in a force of 2×10^{-7} N/m when the wires are separated by 1 m. This definition of the ampere implied the definition $\mu_0 = 4\pi \times 10^{-7}$ Wb/A·m, the weber (Wb) being the unit of magnetic flux. With this definition of the ampere,

---

[1] The fact that a wire carrying an electric current generates what Faraday would later identify as a magnetic field was discovered by Oersted. While lecturing to students in the spring of 1820, Oersted noticed that when the circuit of a "voltaic pile" was closed, there was a deflection of the needle of a magnetic compass that happened to be nearby. Ampère, at the time a mathematics professor in Paris, performed and analyzed further experiments on the magnetic effects of electric currents.

the coulomb (C) was defined as the charge transported in 1 s by a steady current of 1 A. Then, in the Coulomb law,

$$\mathbf{F}_2 = \frac{1}{4\pi\epsilon_0} \frac{q_1 q_2}{r_{12}^3} \mathbf{r}_{12} \tag{1.1.11}$$

for the force on a point charge q_2 due to a point charge q_1, with $\mathbf{r}_{12}$ the vector pointing from q_1 to q_2, ϵ_0 is inferred from the *defined* values of μ_0 and c: $\epsilon_0 = 8.854 \times 10^{-12}$ $C^2/N\cdot m^2$, or $1/4\pi\epsilon_0 = 8.9874 \times 10^9$ $N\cdot m^2/C^2$.

In the revised International System of Units (SI), the ampere is defined, based on a fixed value for the electron charge, as the current corresponding to $1/(1.602176634 \times 10^{-19})$ electrons per second. The free-space permittivity ϵ_0 and permeability μ_0 in the revised system are experimentally determined rather than exactly defined quantities; the relation $\epsilon_0\mu_0 = 1/c^2$, with c defined as 299792458 m/s, remains exact.

Equation (1.1.11) implies that the Coulomb interaction energy of two equal charges q separated by a distance r is

$$U(r) = \frac{1}{4\pi\epsilon_0} \frac{q^2}{r}. \tag{1.1.12}$$

We can use this formula to make rough estimates of binding energies. Consider, for example, the H_2^+ ion. The total energy is $E_{tot} = E_{nn} + E_{en} + E_{kin}$, where E_{nn} is the proton–proton Coulomb energy, E_{en} is the Coulomb interaction energy of the electron with the two protons, and E_{kin} is the kinetic energy. According to the virial theorem of classical mechanics, $E_{tot} = -E_{kin}$, implying

$$E_{tot} = \frac{1}{2}(E_{nn} + E_{en}). \tag{1.1.13}$$

$E_{nn} = e^2/(4\pi\epsilon_0 r)$, where $e = 1.602 \times 10^{-19}$ C and $r \cong 0.106$ nm is the internuclear separation. A rough estimate of E_{en} is obtained by assuming that the electron sits at the midpoint between the two protons:

$$E_{en} \approx -\frac{1}{4\pi\epsilon_0} \times 2 \times \frac{e^2}{\frac{1}{2}r} = -4E_{nn}. \tag{1.1.14}$$

Then,

$$E_{tot} = -\frac{3}{2} \times \frac{1}{4\pi\epsilon_0} \frac{e^2}{r} \approx -20.4 \text{ eV}. \tag{1.1.15}$$

Since the binding (ionization) energy of the hydrogen atom is 13.6 eV, the binding energy of H_2^+, defined as the binding energy between a hydrogen atom and a proton, is estimated to be (20.4 - 13.6) eV = 6.8 eV. Quantum-mechanical calculations yield 2.7 eV for this binding energy. Chemical binding energies on the order of a few electron volts are typical.

Consider as another example the energy released in the fission of a U^{235} nucleus. Since there are 92 protons, the Coulomb interaction energy of the protons is

$$U_1 \approx \frac{1}{4\pi\epsilon_0} \frac{(92e)^2}{R}, \tag{1.1.16}$$

where R is the nuclear radius. If the nucleus is split in two, the volume decreases be a factor of 2, and the radius therefore decreases to $(1/2)^{1/3}R$, since the volume is proportional to the radius cubed. The sum of the Coulomb interaction energies of the daughter nuclei is therefore

$$U_2 \approx 2 \times \frac{1}{4\pi\epsilon_0} \times (46e)^2/[(1/2)^{1/3}R] = 0.63U_1. \tag{1.1.17}$$

The energy released in fission is $U_f = U_1 - U_2 = 0.37U_1$. Taking $R = 10^{-14}$ m for the nuclear radius, we obtain $U_f = 4.8 \times 10^8$ eV $= 480$ MeV, compared with the actual value of about 170 MeV per nucleus. Thus we obtain the correct order of magnitude with only electrostatic interactions, without accounting for the strong force between nucleons and without having to know that $E = mc^2$.[2] The physical origin of the energy released in this simple model is the Coulomb interaction of charged particles, just as in a chemical combustion reaction. But the energy released in chemical reactions typically amounts to just a few electron volts per atom; the enormously larger energy released per nucleus in the fission of U^{235} is due to the small size of the nucleus compared with an atom and to the large number of charges (protons) involved.

1.2 Earnshaw's Theorem

Electrostatics is based on (1.1.7). We introduce a scalar potential $\phi(\mathbf{r})$ such that $\mathbf{E}(\mathbf{r}) = -\nabla\phi(\mathbf{r})$, so that $\nabla \times \mathbf{E} = 0$ is satisfied identically. Then, $\nabla \cdot \mathbf{E} = \rho/\epsilon_0$ implies the Poisson equation

$$\nabla^2\phi = -\rho/\epsilon_0, \tag{1.2.1}$$

or the Laplace equation

$$\nabla^2\phi = 0 \tag{1.2.2}$$

in a region free of charges. The reader has probably enjoyed solving these equations in homework problems for various symmetrical configurations of charge distributions and conductors subject to boundary conditions. Here, we will recall only one implication of the electrostatic Maxwell equations, *Earnshaw's theorem*: a charged particle cannot be held at a point of stable equilibrium by any electrostatic field. This follows simply from Gauss's law (see Figure 1.1). The theorem is easily generalized to any number of charges: no arrangement of positive and negative charges in free space can be in stable equilibrium under electrostatic forces alone.

---

[2] "Somehow the popular notion took hold long ago that Einstein's theory of relativity, in particular his famous equation $E = mc^2$, plays some essential role in the theory of fission . . . but relativity is not required in discussing fission." — R. Serber, *The Los Alamos Primer*, University of California Press, Berkeley, 1992, p. 7.

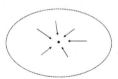

Fig. 1.1 A point inside some imagined closed surface in free space. For that point to be one of stable equilibrium for a positive point charge, for example, the electric field must point everywhere toward it, which would imply a negative flux of electric field through the surface. This would violate Gauss's law, because $\nabla \cdot \mathbf{E} = 0$ in free space.

A more formal proof of Earnshaw's theorem starts from the force $\mathbf{F} = q\mathbf{E} = -q\nabla\phi$ on a point charge q, or, equivalently, the potential energy $U(\mathbf{r}) = q\phi(\mathbf{r})$; $\nabla \cdot \mathbf{E} = 0$ in free space implies Laplace's equation,

$$\nabla^2 U = \frac{\partial^2 U}{\partial x^2} + \frac{\partial^2 U}{\partial y^2} + \frac{\partial^2 U}{\partial z^2} = 0, \tag{1.2.3}$$

which means that the potential energy has no local maximum or minimum inside the surface of Figure 1.1; a local maximum or minimum would require that all three second derivatives in Laplace's equation have the same sign, which would contradict the equation. It is only possible at any point to have a maximum along one direction and a minimum along another (saddle points). In particular, no combination of forces involving $1/r$ potential energies, such as, for example, electrostatic plus gravitational interactions, can result in points of stable equilibrium, since the sum of the Laplacians over all the potentials is zero.

The Reverend Samuel Earnshaw presented his theorem in 1842 in the context of the "luminiferous ether" and elasticity theory. He showed that forces varying as the inverse square of the distance between particles could not produce a stable equilibrium, and concluded that the ether must be held together by non-inverse-square forces. Maxwell stated the theorem as "a charged body placed in a field of electric force cannot be in stable equilibrium," and proved it for electrostatics.[3]

Earnshaw's theorem in electrostatics only says that stable equilibrium cannot occur with electrostatic forces *alone*. If other forces act to hold negative charges in place, a positive charge can, of course, be kept in stable equilibrium by a suitable distribution of the negative charges. Similarly, a charge can be in stable equilibrium in electric fields that vary in time, or in a dielectric medium (held together by non-electrostatic forces!) in which any displacement of the charge results in a restoring force acting back on it, as occurs for a charge at the center of a dielectric sphere with permittivity $\epsilon < \epsilon_0$.[4]

Things are a little more complicated in magnetostatics. There are no magnetic monopoles, and the potential energy of interest is $U(\mathbf{r}) = -\mathbf{m} \cdot \mathbf{B}$ for a magnetic dipole $\mathbf{m}$ in a magnetic field $\mathbf{B}$. For induced magnetic dipoles, $\mathbf{m} = \alpha_m \mathbf{B}$, where $\alpha_m > 0$ for a paramagnetic material

[3] J. C. Maxwell, *Treatise on Electricity and Magnetism*, Volume 1, Dover Publications, New York, 1954, p. 174.

[4] See, for instance, D. F. V. James, P. W. Milonni, and H. Fearn, Phys. Rev. Lett. **75**, 3194 (1995).

(the dipole tends to align with the $\mathbf{B}$ field), $\alpha_m < 0$ for a diamagnetic material (the dipole tends to "anti-align" with the field), the potential energy is

$$U(\mathbf{r}) = -\int_0^{\mathbf{B}(\mathbf{r})} \alpha_m \mathbf{B} \cdot d\mathbf{B} = -\frac{1}{2}\alpha_m B^2(\mathbf{r}), \qquad (1.2.4)$$

and $\nabla^2 U = -(1/2)\alpha_m \nabla^2 B^2$. For there to be a point of stable equilibrium the flux of the force $\mathbf{F}$ through any surface surrounding the point in free space must be negative, which, from the divergence theorem, requires that $\nabla \cdot \mathbf{F} = -\nabla^2 U < 0$, or $\alpha_m \nabla^2 B^2 < 0$ at that point. Now in free space $\nabla \times \mathbf{B} = 0$, and, consequently, $\nabla \times (\nabla \times \mathbf{B}) = \nabla(\nabla \cdot \mathbf{B}) - \nabla^2 \mathbf{B} = 0$, so $\nabla^2 \mathbf{B} = 0$ and

$$\nabla^2(\mathbf{B} \cdot \mathbf{B}) = 2\mathbf{B} \cdot \nabla^2 \mathbf{B} + 2|\nabla \mathbf{B}|^2 = 2|\nabla \mathbf{B}|^2 \geq 0. \qquad (1.2.5)$$

Therefore, we cannot have $\alpha_m \nabla^2 B^2 < 0$ in the paramagnetic case, that is, a paramagnetic particle cannot be held in stable equilibrium in a magnetostatic field. But it is possible for a diamagnetic particle to be in stable equilibrium in a magnetostatic field: this is simply because B^2, unlike any of the three components of $\mathbf{B}$ itself, does *not* satisfy Laplace's equation and *can* have a local minimum. Ordinary diamagnetic materials (wood, water, proteins, etc.) are only very weakly diamagnetic, but levitation is possible in sufficiently strong magnetic fields. The most spectacular practical application at present of magnetic levitation—"maglev" trains—is based on the levitation of superconductors ($\alpha_m \to -\infty$) in magnetic fields.

1.3 Gauges and the Relativity of Fields

The electric and magnetic fields of interest in optical physics are far from static and must, of course, be described by the coupled, time-dependent Maxwell equations. In this section, we briefly review some gauge and Lorentz transformation properties implied by these equations.

We introduce a vector potential $\mathbf{A}$ such that $\mathbf{B} = \nabla \times \mathbf{A}$, consistent with $\nabla \cdot \mathbf{B} = 0$. From (1.1.3), it follows that we can write $\mathbf{E} = -\nabla\phi - \partial\mathbf{A}/\partial t$, and, from (1.1.4) and the identity $\nabla \times (\nabla \times \mathbf{A}) = \nabla(\nabla \cdot \mathbf{A}) - \nabla^2\mathbf{A}$,

$$\nabla(\nabla \cdot \mathbf{A} + \frac{1}{c^2}\frac{\partial\phi}{\partial t}) - \nabla^2\mathbf{A} + \frac{1}{c^2}\frac{\partial^2\mathbf{A}}{\partial t^2} = \mu_0\mathbf{J}. \qquad (1.3.1)$$

In terms of ϕ and $\mathbf{A}$, (1.1.1) becomes

$$\nabla^2\phi + \frac{\partial}{\partial t}(\nabla \cdot \mathbf{A}) = -\rho/\epsilon_0. \qquad (1.3.2)$$

These last two equations for the potentials ϕ and $\mathbf{A}$ are equivalent to the Maxwell equations (1.1.3) and (1.1.1), and the definitions of ϕ and $\mathbf{A}$ ensure that the remaining two Maxwell equations are satisfied. But ϕ and $\mathbf{A}$ are not uniquely specified by $\mathbf{B} = \nabla \times \mathbf{A}$ and $\mathbf{E} = -\nabla\phi - \partial\mathbf{A}/\partial t$: we can satisfy Maxwell's equations with different potentials $\mathbf{A}'$ and ϕ' obtained from the *gauge transformations* $\mathbf{A} = \mathbf{A}' + \nabla\chi$, and $\phi = \phi' - \partial\chi/\partial t$ with $\mathbf{B} = \nabla \times \mathbf{A} = \nabla \times \mathbf{A}'$, and $\mathbf{E} = -\nabla\phi - \partial\mathbf{A}/\partial t = -\nabla\phi' - \partial\mathbf{A}'/\partial t$.[5]

---

[5] The word "gauge" in this context was introduced by Hermann Weyl in 1929.

1.3.1 Lorentz Gauge

We can, for example, choose the Lorentz gauge in which the scalar and vector potentials are chosen such that we obtain the following equation:[6]

$$\nabla \cdot \mathbf{A} + \frac{1}{c^2}\frac{\partial \phi}{\partial t} = 0. \tag{1.3.3}$$

Then, from (1.3.1) and (1.3.2),

$$\nabla^2 \mathbf{A} - \frac{1}{c^2}\frac{\partial^2 \mathbf{A}}{\partial t^2} = -\mu_0 \mathbf{J}, \tag{1.3.4}$$

$$\nabla^2 \phi - \frac{1}{c^2}\frac{\partial^2 \phi}{\partial t^2} = -\rho/\epsilon_0. \tag{1.3.5}$$

The advantage of the Lorentz gauge, as the name suggests, comes when the equations of electrodynamics are formulated so as to be "manifestly" invariant under the Lorentz transformations of relativity theory, as discussed below.

Recall a solution of the scalar wave equation

$$\nabla^2 \psi - \frac{1}{c^2}\frac{\partial^2 \psi}{\partial t^2} = f(\mathbf{r}, t) \tag{1.3.6}$$

using the Green function G satisfying

$$\left(\nabla^2 - \frac{1}{c^2}\frac{\partial^2}{\partial t^2}\right) G(\mathbf{r}, t; \mathbf{r}', t') = \delta^3(\mathbf{r} - \mathbf{r}')\delta(t - t'). \tag{1.3.7}$$

From a standard representation for the delta function,

$$\delta^3(\mathbf{r} - \mathbf{r}')\delta(t - t') = \left(\frac{1}{2\pi}\right)^4 \int d^3k \int_{-\infty}^{\infty} d\omega\, e^{i(\mathbf{k}\cdot\mathbf{R} - \omega T)}, \tag{1.3.8}$$

with $\mathbf{R} = \mathbf{r} - \mathbf{r}'$, and $T = t - t'$. The corresponding Fourier decomposition of the Green function,

$$G(\mathbf{r}, t; \mathbf{r}', t') = \int d^3k \int_{-\infty}^{\infty} d\omega\, g(\mathbf{k}, \omega) e^{i(\mathbf{k}\cdot\mathbf{R} - \omega T)}, \tag{1.3.9}$$

and its defining equation (1.3.7), imply that

---

[6] Recall that the "Lorentz gauge" is really a class of gauges, as we can replace $\mathbf{A}$ by $\mathbf{A} + \nabla\psi$, and ϕ by $\phi - \partial\psi/\partial t$, and still satisfy (1.3.3) as long as $\nabla^2\psi - (1/c^2)\,\partial^2\psi/\partial t^2 = 0$. The Coulomb gauge condition, similarly, remains satisfied under such "restricted" gauge transformations with $\nabla^2\psi = 0$, but for potentials that fall off at least as fast as $1/r$, r being the distance from the center of a localized charge distribution, $\psi = 0$. What is generally called the Lorentz gauge condition was actually proposed a quarter-century before H. A. Lorentz by L. V. Lorenz, who also formulated equations equivalent to Maxwell's, independently of Maxwell but a few years later. See J. D. Jackson and L. B. Okun, Rev. Mod. Phys. **73**, 663 (2001).

$$G(\mathbf{r}, t; \mathbf{r}', t') = -\left(\frac{1}{2\pi}\right)^4 \int d^3k \int_{-\infty}^{\infty} d\omega \, \frac{e^{i(\mathbf{k}\cdot\mathbf{R}-\omega T)}}{k^2 - \omega^2/c^2}$$

$$= -\left(\frac{1}{2\pi}\right)^4 \int_0^{\infty} dk\, k^2 \int_0^{\pi} d\theta \sin\theta \int_0^{2\pi} d\phi \int_{-\infty}^{\infty} d\omega \frac{e^{ikR\cos\theta} e^{-i\omega T}}{k^2 - \omega^2/c^2}$$

$$= \left(\frac{1}{2\pi}\right)^3 \frac{c}{2iR} \int_{-\infty}^{\infty} dk\, e^{ikR} \int_{-\infty}^{\infty} d\omega \left(\frac{1}{\omega - kc} - \frac{1}{\omega + kc}\right) e^{-i\omega T}. \quad (1.3.10)$$

How can we deal with the singularities at $\omega = \pm kc$ in the integration over ω? A physically reasonable assumption is that $G(\mathbf{r}, t; \mathbf{r}', t')$ is 0 for $T = t - t' < 0$, that is, for times before the delta function "source" is turned on. We can satisfy this condition by introducing the positive infinitesimal ϵ and defining the *retarded* Green function:

$$G(\mathbf{r}, t; \mathbf{r}', t') = \left(\frac{1}{2\pi}\right)^3 \frac{c}{2iR} \int_{-\infty}^{\infty} dk e^{ikR} \int_{-\infty}^{\infty} d\omega \left(\frac{1}{\omega - kc + i\epsilon}\right.$$

$$\left. - \frac{1}{\omega + kc + i\epsilon}\right) e^{-i\omega T}. \quad (1.3.11)$$

Now the poles lie not on the real axis but in the lower half of the complex plane. Since $e^{-i\omega T} \to 0$ for $T < 0$ and large, positive imaginary parts of ω, we can replace the integration path in (1.3.11) by one along the real axis and closed in a large (radius $\to \infty$) semicircle in the upper half-plane. And since there are no poles inside this closed path, we have the desired property that

$$G(\mathbf{r}, t; \mathbf{r}', t') = 0 \quad (t < t'). \quad (1.3.12)$$

For $T = t - t' > 0$, similarly, we can close the integration path with an infinitely large semicircle in the lower half of the complex plane. The integration path now encloses the poles at $\omega = \pm kc - i\epsilon$, and the residue theorem gives

$$G(\mathbf{r}, t; \mathbf{r}', t') = \left(\frac{1}{2\pi}\right)^3 \frac{c}{2iR} \int_{-\infty}^{\infty} dk\, e^{ikR}(-2\pi i)[e^{-ikcT} - e^{ikcT}]$$

$$= -\frac{c}{4\pi R}[\delta(R - cT) - \delta(R + cT)] = \frac{c}{4\pi R}\delta(R - cT)$$

$$= -\frac{c}{4\pi|\mathbf{r} - \mathbf{r}'|}\delta[|\mathbf{r} - \mathbf{r}'| - c(t - t')] \quad (t > t'). \quad (1.3.13)$$

The solution of (1.3.5), for example, is then

$$\phi(\mathbf{r}, t) = \frac{-1}{\epsilon_0} \int d^3r' \int_{-\infty}^{\infty} dt'\, G(\mathbf{r}, t; \mathbf{r}', t')\rho(\mathbf{r}', t')$$

$$= \frac{c}{4\pi\epsilon_0} \int d^3r' \int_{-\infty}^{\infty} dt'\, \frac{\rho(\mathbf{r}', t')\delta[|\mathbf{r} - \mathbf{r}'| - c(t - t')]}{|\mathbf{r} - \mathbf{r}'|}$$

$$= \frac{1}{4\pi\epsilon_0} \int d^3r'\, \frac{\rho(\mathbf{r}', t - |\mathbf{r} - \mathbf{r}'|/c)}{|\mathbf{r} - \mathbf{r}'|} \quad (1.3.14)$$

under the assumption that it is the retarded Green function that is physically meaningful, rather than the "advanced" Green function or some linear combination of advanced and retarded Green functions.[7] The contribution of the charge density at $\mathbf{r}'$ to the scalar potential at $\mathbf{r}$ at time t depends on the value of the charge density at the retarded time $t - |\mathbf{r} - \mathbf{r}'|/c$, and likewise for the vector potential. Evaluation of these potentials gives expressions that are more complicated than trivially retarded versions of their static forms, as we now recall for a simple but important example.

For a point charge q moving such that its position at time t is $\mathbf{u}(t)$, $\rho(\mathbf{r}', t') = q\delta^3[\mathbf{r}' - \mathbf{u}(t')]$, and the scalar potential is

$$\phi(\mathbf{r}, t) = \frac{q}{4\pi\epsilon_0} \int d^3r' \int_{-\infty}^{\infty} dt' \frac{\delta^3[\mathbf{r}' - \mathbf{u}(t')]\delta(t' - t + |\mathbf{r} - \mathbf{r}'|/c)}{|\mathbf{r} - \mathbf{r}'|}. \tag{1.3.15}$$

To perform the integration, we change variables from x', y', z', t' to $y_1 = x' - u_x(t')$, $y_2 = y' - u_y(t')$, $y_3 = z' - u_z(t')$ and $y_4 = t' - t + |\mathbf{r} - \mathbf{r}'|/c$:

$$\phi(\mathbf{r}, t) = \frac{q}{4\pi\epsilon_0} \frac{1}{|\mathbf{r} - \mathbf{r}'|} \int\int\int\int dy_1\, dy_2\, dy_3\, dy_4\, J^{-1}\delta(y_1)\, \delta(y_2)\, \delta(y_3)\, \delta(y_4), \tag{1.3.16}$$

where now $\mathbf{r}' = \mathbf{u}(t')$, $t' = t - |\mathbf{r} - \mathbf{r}'|/c$, and J is the 4×4 Jacobian determinant,

$$J = \frac{\partial(y_1, y_2, y_3, y_4)}{\partial(x', y', z', t')}, \tag{1.3.17}$$

which is found by straightforward algebra to be

$$J = 1 - [\dot{\mathbf{u}}(t')/c] \cdot \frac{\mathbf{r} - \mathbf{r}'}{|\mathbf{r} - \mathbf{r}'|}. \tag{1.3.18}$$

Therefore,

$$\phi(\mathbf{r}, t) = \frac{q}{4\pi\epsilon_0} \frac{1}{|\mathbf{r} - \mathbf{r}'| - [\dot{\mathbf{u}}(t')/c] \cdot (\mathbf{r} - \mathbf{r}')/|\mathbf{r} - \mathbf{r}'|}, \tag{1.3.19}$$

or, in more compact notation,

$$\phi(\mathbf{r}, t) = \frac{1}{4\pi\epsilon_0} \left[\frac{q}{R(1 - \mathbf{v} \cdot \hat{\mathbf{n}}/c)} \right]_{\text{ret}}, \tag{1.3.20}$$

where R is the distance from the charge to the observation point $\mathbf{r}$, $\hat{\mathbf{n}}$ is the unit vector pointing from the point charge to the point of observation, $\mathbf{v} = \dot{\mathbf{u}}$ is the velocity of the charge, and the subscript "ret" means that all the quantities in brackets are evaluated at the retarded time $t' = t - |\mathbf{r} - \mathbf{r}'|/c$. Likewise, the solution of (1.3.4) for the retarded vector potential is

$$\mathbf{A}(\mathbf{r}, t) = \frac{1}{4\pi\epsilon_0 c^2} \left[\frac{q\mathbf{v}}{R(1 - \mathbf{v} \cdot \hat{\mathbf{n}}/c)} \right]_{\text{ret}}, \tag{1.3.21}$$

since the current density associated with the point charge is $\mathbf{J} = q\mathbf{v}\delta^3[\mathbf{r} - \mathbf{u}(t)]$.

---

[7] We follow here the nearly universal practice in classical electrodynamics of setting to 0 the (zero-temperature) solutions of the homogeneous Maxwell equations, that is, we presume there are no "source-free" fields. In *quantum* electrodynamics, however, there are fluctuating fields, with observable physical consequences, even at zero temperature. Nontrivial solutions of the homogeneous Maxwell equations also appear in the classical theory called *stochastic electrodynamics*. See Section 7.4.1.

These *Liénard–Wiechert potentials* are complicated. For one thing, $\phi(\mathbf{r}, t)$, for instance, is *not* simply $q/4\pi\epsilon_0[R]_{\text{ret}}$, which "almost everyone would, at first, think."[8] Instead, $\phi(\mathbf{r}, t)$ depends not only on the position of the charge at the retarded time t', but also on what the velocity was at t'. For a charge moving with constant velocity v along the x axis, for example,

$$t' = t - \frac{1}{c}|\mathbf{r} - \mathbf{u}(t')| = t - \frac{1}{c}\sqrt{(x - vt')^2 + y^2 + z^2} \qquad (1.3.22)$$

if we define our coordinates such that, at $t = 0$, the charge is at $(x = 0, y = 0, z = 0)$. The solution of this equation for t' $(< t)$ is

$$t' = \left(1 - \frac{v^2}{c^2}\right)^{-1}\left[t - \frac{xv}{c^2} - \frac{1}{c}\sqrt{(x - vt)^2 + \left(1 - \frac{v^2}{c^2}\right)(y^2 + z^2)}\right]. \qquad (1.3.23)$$

Since $R = c(t - t')$ and the component of velocity along $\mathbf{r}'$ at the retarded time t' is $v \times (x - vt')/|\mathbf{r}'|$, it follows from (1.3.22) and (1.3.23) that

$$[R - R\mathbf{v} \cdot \hat{\mathbf{n}}/c]_{\text{ret}} = c(t - t') - \frac{v}{c}(x - vt') = \sqrt{(x - vt)^2 + (1 - \frac{v^2}{c^2})(y^2 + z^2)}, \quad (1.3.24)$$

and therefore that

$$\phi(x, y, z, t) = \frac{q}{4\pi\epsilon_0}\frac{1}{\sqrt{(x - vt)^2 + (1 - v^2/c^2)(y^2 + z^2)}}$$

$$= \frac{q}{4\pi\epsilon_0}\frac{1}{\sqrt{1 - v^2/c^2}}\frac{1}{\sqrt{(x - vt)^2/(1 - v^2/c^2) + y^2 + z^2}} \qquad (1.3.25)$$

and

$$A_x(x, y, z, t) = \frac{qv}{4\pi\epsilon_0 c^2}\frac{1}{\sqrt{1 - v^2/c^2}}\frac{1}{\sqrt{(x - vt)^2/(1 - v^2/c^2) + y^2 + z^2}} \qquad (1.3.26)$$

for a charged particle moving with constant velocity v along the x direction.

We can derive these results more simply using the fact that, in special relativity theory, ϕ and $\mathbf{A}$ transform as the components of a four-vector $(\phi/c, \mathbf{A})$. In a space-time coordinate system (x', y', z', t') in which a charge q is at rest,

$$\phi'(x', y', z', t') = \frac{q}{4\pi\epsilon_0}\frac{1}{\sqrt{x'^2 + y'^2 + z'^2}}, \quad \mathbf{A}'(x', y', z', t') = 0. \qquad (1.3.27)$$

The coordinates (x, y, z, t) in the "lab" frame, in which the charge is moving in the positive x direction with constant velocity v, are related to the rest-frame coordinates by the Lorentz transformations:

---

[8] *Feynman, Leighton, and Sands*, Volume II, p. 21–9. (We refer to books in the Bibliography using their authors' italicized surnames.)

$$x' = \frac{x - vt}{\sqrt{1 - v^2/c^2}}, \quad t' = \frac{t - vx/c^2}{\sqrt{1 - v^2/c^2}}, \quad y' = y, \quad z' = z. \tag{1.3.28}$$

The potential $\phi(x, y, x, t)$, for instance, is obtained by transforming $\phi'(x', y', z', t')$ from the rest frame of the charge to a frame moving with velocity $-v$ along the x axis:

$$\phi(x, y, z, t) = \frac{\phi'(x', y', z', t') + vA'_x(x', y', z', t')/c^2}{\sqrt{1 - v^2/c^2}} = \frac{1}{\sqrt{1 - v^2/c^2}} \frac{q/4\pi\epsilon_0}{\sqrt{x'^2 + y'^2 + z'^2}}$$

$$= \frac{q}{4\pi\epsilon_0} \frac{1}{\sqrt{1 - v^2/c^2}} \frac{1}{\sqrt{(x - vt)^2/(1 - (v^2/c^2)) + y^2 + z^2}}, \tag{1.3.29}$$

which is just (1.3.25). That we obtained (1.3.25) directly from the solution of the wave equation for ϕ without making any Lorentz transformations is not surprising, of course, because the Maxwell equations are the correct equations of electromagnetic theory in special relativity; they are correct in any inertial frame. Indeed, the Liénard–Wiechert potentials were obtained before the development of the theory of special relativity. What special relativity shows is that v can be regarded as the relative velocity between the coordinate system in which the charge is at rest and the system in which it is moving with velocity v.

Once we have ϕ and $\mathbf{A}$, we can obtain the electric and magnetic fields using $\mathbf{E} = -\nabla\phi - \partial\mathbf{A}/\partial t$ and $\mathbf{B} = \nabla \times \mathbf{A}$. From (1.3.25) and the corresponding formulas for $\mathbf{A}$,

$$E_x = \frac{q}{4\pi\epsilon_0} \frac{1}{\sqrt{1 - v^2/c^2}} \frac{(x - vt)}{[(x - vt)^2/(1 - v^2/c^2) + y^2 + z^2]^{3/2}},$$

$$E_y = \frac{q}{4\pi\epsilon_0} \frac{1}{\sqrt{1 - v^2/c^2}} \frac{y}{[(x - vt)^2/(1 - v^2/c^2) + y^2 + z^2]^{3/2}},$$

$$E_z = \frac{q}{4\pi\epsilon_0} \frac{1}{\sqrt{1 - v^2/c^2}} \frac{z}{[(x - vt)^2/(1 - v^2/c^2) + y^2 + z^2]^{3/2}}, \tag{1.3.30}$$

and

$$\mathbf{B} = \frac{1}{c^2}\mathbf{v} \times \mathbf{E}. \tag{1.3.31}$$

More generally, the electric and magnetic fields transform as

$$E'_x = E_x,$$

$$E'_y = \frac{E_y - vB_z}{\sqrt{1 - v^2/c^2}},$$

$$E'_z = \frac{E_z + vB_y}{\sqrt{1 - v^2/c^2}},$$

$$B'_x = B_x,$$

$$B'_y = \frac{B_y + vE_z/c^2}{\sqrt{1 - v^2/c^2}},$$

$$B'_z = \frac{B_z - vE_y/c^2}{\sqrt{1 - v^2/c^2}}, \qquad (1.3.32)$$

when the primed frame moves with respect to the unprimed frame at a constant velocity v in the x direction.[9]

The result (1.3.31), for example, can be obtained from the Coulomb field in a frame in which the charge is at rest, using these transformation laws to relate the fields in the two inertial frames. In particular, a purely electric field in one frame implies a magnetic field in another, and vice versa.[10]

In the case of a charged particle moving with a velocity that varies in time, the electric and magnetic fields can be calculated from the Liénard–Wiechert potentials, as is done in standard texts. Here, we only recall the formulas for the (retarded) fields in the radiation zone when the particle motion is non-relativistic ($v \ll c$):

$$\mathbf{E}(\mathbf{r}, t) = \frac{q}{4\pi\epsilon_0} \frac{1}{c^2 r^3} \mathbf{r} \times (\mathbf{r} \times \dot{\mathbf{v}}), \qquad (1.3.33)$$

$$\mathbf{B}(\mathbf{r}, t) = \frac{q}{4\pi\epsilon_0} \frac{1}{c^3 r^2} \dot{\mathbf{v}} \times \mathbf{r}. \qquad (1.3.34)$$

The power radiated per solid angle is calculated using these fields and the Poynting vector:

$$\frac{dP}{d\Omega} = \frac{1}{4\pi\epsilon_0} \frac{q^2}{4\pi c^3} |\dot{\mathbf{v}}|^2 \sin^2\theta, \qquad (1.3.35)$$

where θ is the angle between $\mathbf{r}$ and the acceleration $\dot{\mathbf{v}}$. Integration over all solid angles results in the (non-relativistic) Larmor formula for the radiated power:

$$P = \int_0^{2\pi} d\phi \int_0^{\pi} d\theta \sin\theta \frac{dP}{d\Omega} = \frac{1}{4\pi\epsilon_0} \frac{2q^2\dot{v}^2}{3c^3}. \qquad (1.3.36)$$

---

[9] These transformations are applied in Section 2.8 to blackbody radiation fields.

[10] "What led me more or less directly to the special theory of relativity was the conviction that the electromotive force acting on a body in motion in a magnetic field was nothing else but an electric field." — Einstein, quoted in R. S. Shankland, Am. J. Phys. **32**, 16 (1964), p. 35.

1.3.2   Coulomb Gauge

In the Coulomb gauge we choose χ such that $\nabla \cdot \mathbf{A} = 0$.[11] In this gauge,

$$\nabla^2 \phi = -\rho/\epsilon_0 \tag{1.3.37}$$

and

$$\nabla^2 \mathbf{A} - \frac{1}{c^2}\frac{\partial^2 \mathbf{A}}{\partial t^2} = -\mu_0 \mathbf{J} + \frac{1}{c^2}\nabla\frac{\partial \phi}{\partial t}. \tag{1.3.38}$$

The scalar potential satisfies the Poisson equation (1.3.37) and is given in terms of the charge density $\rho(\mathbf{r}, t)$ by the instantaneous Coulomb potential,

$$\phi(\mathbf{r}, t) = \frac{1}{4\pi\epsilon_0}\int \frac{\rho(\mathbf{r}', t)}{|\mathbf{r} - \mathbf{r}'|}d^3 r', \tag{1.3.39}$$

if the charge distribution is specified throughout all space. (Of course, this is not always the case; in many examples in electrostatics, for example, the potentials are specified on conductors, and surface charge distributions are deduced *after* solving Laplace's equation with boundary conditions.) Equation (1.3.38) can be rewritten using *Helmholtz's theorem*: any vector field $\mathbf{F}(\mathbf{r}, t)$ can be uniquely decomposed in transverse and longitudinal parts defined respectively by[12]

$$\mathbf{F}^{\perp}(\mathbf{r}, t) = \frac{1}{4\pi}\nabla \times \nabla \times \int \frac{\mathbf{F}(\mathbf{r}', t)}{|\mathbf{r} - \mathbf{r}'|}d^3 r', \tag{1.3.40}$$

$$\mathbf{F}^{\|}(\mathbf{r}, t) = -\frac{1}{4\pi}\nabla \int \frac{\nabla' \cdot \mathbf{F}(\mathbf{r}', t)}{|\mathbf{r} - \mathbf{r}'|}d^3 r'. \tag{1.3.41}$$

In other words, $\mathbf{F} = \mathbf{F}^{\perp} + \mathbf{F}^{\|}$, with $\nabla \cdot \mathbf{F}^{\perp} = \nabla \times \mathbf{F}^{\|} = 0$. In the Coulomb gauge the vector potential $\mathbf{A}$ is a transverse vector field ($\nabla \cdot \mathbf{A} = 0$); writing $\mathbf{J} = \mathbf{J}^{\perp} + \mathbf{J}^{\|}$ in (1.3.38), we have

$$\begin{aligned}
\nabla^2 \mathbf{A} - \frac{1}{c^2}\frac{\partial^2 \mathbf{A}}{\partial t^2} &= -\mu_0 \mathbf{J}^{\perp} - \mu_0 \mathbf{J}^{\|} + \frac{1}{c^2}\nabla\frac{\partial \phi}{\partial t} \\
&= -\mu_0 \mathbf{J}^{\perp} + \frac{\mu_0}{4\pi}\nabla \int \frac{\nabla' \cdot \mathbf{J}(\mathbf{r}', t)}{|\mathbf{r} - \mathbf{r}'|}d^3 r' + \frac{1}{4\pi\epsilon_0 c^2}\nabla\frac{\partial}{\partial t}\int \frac{\rho(\mathbf{r}', t)}{|\mathbf{r} - \mathbf{r}'|}d^3 r' \\
&= -\mu_0 \mathbf{J}^{\perp},
\end{aligned} \tag{1.3.42}$$

where we have used the charge conservation condition (1.1.6).

Although the Lorentz gauge is perfectly suited for relativistic theory, the Coulomb gauge also offers some advantages, and is almost always used in quantum optics. In the Coulomb gauge, the longitudinal field $\mathbf{E}^{\|} = -\nabla\phi$ is effectively eliminated and replaced by Coulomb interactions of the charges, and quantization of the field then involves

---

[11] For explicit forms of the χ's that effect the gauge transformations, see J. D. Jackson, Am. J. Phys. **70**, 917 (2002).

[12] A proof of Helmholtz's theorem is outlined in Appendix B.

only the transverse fields $\mathbf{A}$, $\mathbf{E}^\perp$, and $\mathbf{B}$. ($\mathbf{B}^\parallel = 0$ in any gauge.) But the Coulomb interactions in the Coulomb gauge are instantaneous, not retarded (see (1.3.39)). In the Lorentz gauge, in contrast, the potentials (and therefore the electric and magnetic fields) do not propagate instantaneously and are retarded as long as we choose the retarded Green function for the wave equation:

$$\phi(\mathbf{r},t) = \frac{1}{4\pi\epsilon_0} \int d^3r' \, \frac{\rho(\mathbf{r}',t-|\mathbf{r}-\mathbf{r}'|/c)}{|\mathbf{r}-\mathbf{r}'|}, \tag{1.3.43}$$

$$\mathbf{A}(\mathbf{r},t) = \frac{\mu_0}{4\pi} \int d^3r' \, \frac{\mathbf{J}(\mathbf{r}',t-|\mathbf{r}-\mathbf{r}'|/c)}{|\mathbf{r}-\mathbf{r}'|}, \tag{1.3.44}$$

and

$$\mathbf{E}(\mathbf{r},t) = -\nabla\phi(\mathbf{r},t) - \frac{\partial \mathbf{A}}{\partial t} = -\nabla\left[\frac{1}{4\pi\epsilon_0} \int d^3r' \, \frac{\rho(\mathbf{r}',t-|\mathbf{r}-\mathbf{r}'|/c)}{|\mathbf{r}-\mathbf{r}'|}\right]$$
$$- \frac{\partial}{\partial t}\left[\frac{\mu_0}{4\pi} \int d^3r' \, \frac{\mathbf{J}(\mathbf{r}',t-|\mathbf{r}-\mathbf{r}'|/c)}{|\mathbf{r}-\mathbf{r}'|}\right]. \tag{1.3.45}$$

The expression for the same electric field when the Coulomb gauge is used is

$$\mathbf{E}(\mathbf{r},t) = -\nabla\left[\frac{1}{4\pi\epsilon_0} \int d^3r' \, \frac{\rho(\mathbf{r}',t)}{|\mathbf{r}-\mathbf{r}'|}\right]$$
$$- \frac{\partial}{\partial t}\left[\frac{\mu_0}{4\pi} \int d^3r' \, \frac{\mathbf{J}^\perp(\mathbf{r}',t-|\mathbf{r}-\mathbf{r}'|/c)}{|\mathbf{r}-\mathbf{r}'|}\right] \tag{1.3.46}$$

when we use the retarded Green function for the solution of the wave equation (1.3.42). Of course, $\mathbf{E}$ cannot depend on the choice of gauge, and so the expressions (1.3.45) and (1.3.46) must be equivalent, and, in particular, (1.3.46) must be a retarded field, even though the *instantaneous* Coulomb field appears in the first term. We show in Appendix A that this is so.

We can express the electric field in other forms. First, write (1.3.45) more compactly as

$$\mathbf{E}(\mathbf{r},t) = -\frac{1}{4\pi\epsilon_0} \int d^3r' \left(\nabla\frac{[\rho]}{R} + \frac{[\dot{\mathbf{J}}]}{c^2 R}\right) \tag{1.3.47}$$

by defining $[f] = f(\mathbf{r}',t-|\mathbf{r}-\mathbf{r}'|/c)$, $\dot{f} = (\partial/\partial t)f(\mathbf{r}',t-|\mathbf{r}-\mathbf{r}'|/c)$, and $\mathbf{R} = \mathbf{r} - \mathbf{r}'$. Using

$$\nabla\frac{[\rho]}{R} = \frac{1}{R}\nabla[\rho] + [\rho]\nabla\left(\frac{1}{R}\right) = \frac{1}{R}\nabla\rho(\mathbf{r}',t-|\mathbf{r}-\mathbf{r}'|/c) - [\rho]\hat{\mathbf{R}}/R^2$$
$$= -\frac{\hat{\mathbf{R}}}{cR}[\dot{\rho}] - [\rho]\hat{\mathbf{R}}/R^2, \tag{1.3.48}$$

where the unit vector $\hat{\mathbf{R}} = \mathbf{R}/R$, we write the electric field as

$$\mathbf{E}(\mathbf{r},t) = \frac{1}{4\pi\epsilon_0} \int d^3r' \left(\frac{[\rho]}{R^2}\hat{\mathbf{R}} + \frac{[\dot{\rho}]}{cR}\hat{\mathbf{R}} - \frac{[\dot{\mathbf{J}}]}{c^2 R}\right). \tag{1.3.49}$$

Similarly,

$$\mathbf{B}(\mathbf{r}, t) = \frac{\mu_0}{4\pi} \int d^3 r' \left[\left(\frac{[\mathbf{J}]}{R^2} + \frac{[\dot{\mathbf{J}}]}{cR} \right) \times \hat{\mathbf{R}} \right]. \tag{1.3.50}$$

Expressions (1.3.49) and (1.3.50), which may be regarded as time-dependent generalizations of the Coulomb and Biot-Savart laws, are the *Jefimenko equations* for the electric and magnetic fields produced by a charge density $\rho(\mathbf{r}, t)$ and a current density $\mathbf{J}(\mathbf{r}, t)$.[13]

1.4 Dipole Radiators

Radiation by accelerated charges is, in one way or another, responsible for all light. In optical physics, we are particularly concerned with charge acceleration in the form of oscillations of bound electrons. In the crudest description, the radiation from an excited atom, for example, can be regarded as radiation from an oscillating electric dipole formed by the negatively charged electrons and the positively charged nucleus. (This will be clarified in the following chapters.) In fact, the radiation resulting from an electric dipole transition in an atom is very similar in some ways to that from a dipole antenna. We begin our discussion of dipole radiation by considering the simple antenna sketched in Figure 1.2.

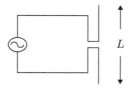

Fig. 1.2 An antenna wire of length L center-fed by an AC current.

The current I in the wire oscillates in time at the frequency ω and vanishes at the end points $z = \pm L/2$. It takes the form of a standing wave:

$$I(z', t) = I_m \frac{\sin(\frac{1}{2}kL - k|z'|)}{\sin \frac{1}{2}kL} e^{-i\omega t} \quad (I(\pm L/2, t) = 0), \tag{1.4.1}$$

with $k = \omega/c$, and I_m the peak current. The vector potential (1.3.44) in this example is

$$\mathbf{A}(\mathbf{r}, t) = \hat{\mathbf{z}} \frac{\mu_0 I_m}{4\pi} e^{-i\omega t} \int_{-L/2}^{L/2} dz' \frac{\sin[k(L/2 - |z'|)]}{\sin \frac{1}{2}kL} \frac{e^{ik|\mathbf{r} - \hat{\mathbf{z}}z'|}}{|\mathbf{r} - \hat{\mathbf{z}}z'|}, \tag{1.4.2}$$

where, as usual, it is implied that we must take the real part of the right side.

---

[13] See K. T. McDonald, Am. J. Phys. **65**, 1074 (1997), and references therein.

Fig. 1.3 The vector **r** from the middle of the antenna wire to the point of observation.

For large distances from the antenna, we can approximate $|\mathbf{r} - \mathbf{r}'|$ by r in the denominator of the integrand and use (see Figure 1.3)

$$|\mathbf{r} - \hat{\mathbf{z}}z'| = (r^2 + z'^2 - 2\mathbf{r} \cdot \hat{\mathbf{z}}z')^{1/2} \cong r - z' \cos\theta \qquad (1.4.3)$$

in the exponent in the numerator:

$$
\begin{aligned}
\mathbf{A}(\mathbf{r}, t) &\cong \hat{\mathbf{z}} \frac{\mu_0 I_m}{4\pi r} e^{-i(\omega t - kr)} \frac{1}{\sin\frac{1}{2}kL} \int_{-L/2}^{L/2} dz' \sin\frac{1}{2}k(L - |z'|) e^{-ikz'\cos\theta} \\
&= \hat{\mathbf{z}} \frac{\mu_0 I_m}{2\pi r} e^{-i(\omega t - kr)} \frac{1}{\sin\frac{1}{2}kL} \int_0^{L/2} dz' \sin\frac{1}{2}k(L - z') \cos(kz'\cos\theta) \\
&= \hat{\mathbf{z}} \frac{\mu_0 I_m}{2\pi kr} e^{-i(\omega t - kr)} \frac{\cos(\frac{1}{2}kL\cos\theta) - \cos\frac{1}{2}kL}{\sin\theta \sin\frac{1}{2}kL} \qquad (1.4.4)
\end{aligned}
$$

and (after taking the real part)

$$
\begin{aligned}
\mathbf{B}(\mathbf{r}, t) = \nabla \times \mathbf{A} &\cong \left[\frac{y\hat{\mathbf{x}} - x\hat{\mathbf{y}}}{r} \right] \frac{\mu_0 I_m}{2\pi r} \sin(\omega t - kr) \frac{\cos(\frac{1}{2}kL\cos\theta) - \cos\frac{1}{2}kL}{\sin\theta \sin\frac{1}{2}kL} \\
&= -\mathbf{e}_\phi \frac{\mu_0 I_m}{2\pi r} \sin(\omega t - kr) \frac{\cos(\frac{1}{2}kL\cos\theta) - \cos\frac{1}{2}kL}{\sin\theta \sin\frac{1}{2}kL}, \qquad (1.4.5)
\end{aligned}
$$

where $\mathbf{e}_\phi = -\hat{\mathbf{x}}\sin\phi + \hat{\mathbf{y}}\cos\phi$ is the azimuthal-angle unit vector at x, y in spherical coordinates. Similarly,

$$\mathbf{E}(\mathbf{r}, t) = -\mathbf{e}_\theta \frac{I_m}{2\pi r} \sqrt{\frac{\mu_0}{\epsilon_0}} \sin(\omega t - kr) \frac{\cos(\frac{1}{2}kL\cos\theta) - \cos\frac{1}{2}kL}{\sin\theta \sin\frac{1}{2}kL}, \qquad (1.4.6)$$

where $\mathbf{e}_\theta = \hat{\mathbf{x}}\cos\theta\cos\phi + \hat{\mathbf{y}}\cos\theta\sin\phi - \hat{\mathbf{z}}\sin\theta$ is the polar-angle unit vector at x, y, z in spherical coordinates. The cycle-averaged Poynting vector,

$$\mathbf{S}(\mathbf{r}) = \mathbf{E} \times \mathbf{H} = \frac{1}{\mu_0} \mathbf{E} \times \mathbf{B} = \hat{\mathbf{r}} \frac{I_m^2}{8\pi^2 r^2} \sqrt{\frac{\mu_0}{\epsilon_0}} \left(\frac{\cos[\frac{1}{2}kL\cos\theta] - \cos\frac{1}{2}kL}{\sin\theta \sin\frac{1}{2}kL} \right)^2, \qquad (1.4.7)$$

follows by simple algebra and the identity $\mathbf{e}_\theta \times \mathbf{e}_\phi = \hat{\mathbf{r}}$. The radiated power is therefore

$$P = r^2 \int_0^{2\pi} d\phi \int_0^{\pi} d\theta \sin\theta |\mathbf{S}|$$

$$= \frac{I_m^2}{4\pi} \sqrt{\frac{\mu_0}{\epsilon_0}} \int_0^{\pi} d\theta \sin\theta \left[\frac{\cos(\frac{1}{2}kL\cos\theta) - \cos\frac{1}{2}kL}{\sin\theta \sin\frac{1}{2}kL} \right]^2 . \tag{1.4.8}$$

The integral can be evaluated in terms of sine (Si) and cosine (Ci) integrals:

$$P = \frac{I_m^2}{4\pi} \sqrt{\frac{\mu_0}{\epsilon_0}} \Big\{ (\gamma + \log(kL) - \mathrm{Ci}(kL) + \frac{1}{2}\sin(kL)[\mathrm{Si}(2kL) - 2\mathrm{Si}(kL)]$$

$$+ \frac{1}{2}\cos(kL)[\gamma + \log(\frac{1}{2}kL) + \mathrm{Ci}(2kL) - 2\mathrm{Ci}(kL)] \Big\}, \tag{1.4.9}$$

where $\gamma = 0.57721$ is Euler's constant. This is plotted versus $kL/2\pi = L/\lambda$ in Figure 1.4.

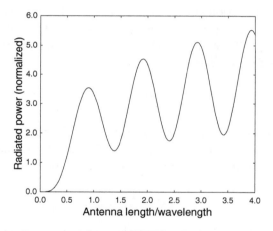

Fig. 1.4 The normalized power $P/[(I_m^2/4\pi)\sqrt{\mu_0/\epsilon_0}]$ (see (1.4.9)) versus (antenna length L)/(radiation wavelength λ).

For reasons given below, the *half-wave antenna* defined by $\frac{1}{2}kL = \pi/2$ (that is, $L = \lambda/2$, with $\lambda = \omega/2\pi c = k/2\pi$ the wavelength of the radiated field) is of particular interest.[14] The power in this case is

$$P = \frac{I_m^2}{4\pi} \sqrt{\frac{\mu_0}{\epsilon_0}} \int_0^{\pi} d\theta \sin\theta \frac{\cos^2[(\pi/2)\cos\theta]}{\sin^2\theta} = \frac{I_m^2}{4\pi} \sqrt{\frac{\mu_0}{\epsilon_0}} (1.22) = \frac{1}{2} I_m^2 R_{\mathrm{rad}}. \tag{1.4.10}$$

[14] There are, of course, many different types of antennas! Half-wave (dipole) and quarter-wave ("monopole") antennas have frequently been employed in wireless communications because of their "omni-directional" radiation patterns. The quarter-wave antenna consists basically of a single conducting rod of length $\lambda/4$ mounted on a conducting surface which might, for instance, be a copper foil on a printed circuit board. The single end-fed element and its "image" produce a radiation pattern and other properties similar to those of the dipole antenna.

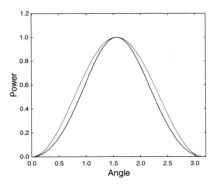

Fig. 1.5 Normalized radiation pattern, as a function of angle θ, of the half-wave antenna (solid curve) compared with that of an oscillating Hertzian dipole (dotted curve).

R_{rad} is the *radiation resistance* of the half-wave antenna: the antenna loses power in the form of radiation, and R_{rad} is the electrical resistance that causes the same average power loss when it passes a current $I_m \cos \omega t$. If the antenna is perfectly conducting, as we have implicitly assumed, R_{rad} is its equivalent circuit resistance. For the half-wave antenna it follows from (1.4.10) that

$$R_{\text{rad}} = \frac{1.22}{2\pi} \sqrt{\frac{\mu_0}{\epsilon_0}} \cong 73 \ \Omega. \tag{1.4.11}$$

If I_{in} is the current feed to the antenna, and R_{in} the input resistance, and if the dipole itself has nearly zero ohmic resistance, then ideally all the power loss from the antenna is from radiation, and, therefore, the (time-averaged) input and the radiated powers are equal:

$$\frac{1}{2} I_{\text{in}}^2 R_{\text{in}} = \frac{1}{2} I_m^2 R_{\text{rad}}. \tag{1.4.12}$$

Since $I_{\text{in}} = I_m \sin \frac{1}{2} kL = I_m$ for the half-wave antenna (see (1.4.1)), $R_{\text{in}} = R_{\text{rad}}$. In order for all the power from the current generator to end up as power transmitted by the antenna, the impedance must be purely real, that is, there are no capacitive or inductive effects that cause energy to be stored in the near field of the antenna and reflected back along the transmission line to the generator rather than radiated away. This is one advantage of the half-wave antenna: the imaginary part of its impedance can be small. Another is that its radiation pattern is relatively broad compared with that of a longer antenna. In fact its radiation pattern is quite similar to that of a Hertzian dipole (see Figure 1.5).

According to (1.4.8), the directional dependence with respect to the antenna axis ($\hat{\mathbf{z}}$) is given by the following equation:[15]

---

[15] An important fact about (passive) antennas is *reciprocity*: the transmission and reception characteristics, including directionality, are the same. For example, a single antenna on a cell phone serves for both reception and transmission.

$$F(\theta) = \left[\frac{\cos(\frac{1}{2}kL\cos\theta) - \cos\frac{1}{2}kL}{\sin\theta\sin\frac{1}{2}kL}\right]^2. \tag{1.4.13}$$

In the case of the half-wave antenna, this reduces to

$$F_{1/2}(\theta) = \frac{\cos^2(\frac{\pi}{2}\cos\theta)}{\sin^2\theta}. \tag{1.4.14}$$

This is compared in Figure 1.5 with the radiation pattern $F_d(\theta) = \sin^2\theta$ (see (1.4.24)) of an oscillating *Hertzian* electric dipole, and in Figure 1.6 with that of an antenna with $L = 3\lambda/2$. The antenna *gain* G is defined as the ratio of the magnitude of the Poynting vector in the direction in which it is largest to the power per unit area of a hypothetical *isotropic* radiator of the same total power. Thus, for the half-wave antenna,

$$G = |\mathbf{S}(\theta = \pi/2)| \times \frac{4\pi r^2}{P} = \frac{4\pi r^2}{P}\frac{1}{r^2}\frac{I_m^2}{4\pi}\sqrt{\frac{\mu_0}{\epsilon_0}}F_{1/2}(\theta = \pi/2) = \frac{2}{1.22} = 1.64, \tag{1.4.15}$$

or $G_{dB} = 10\log_{10}G = 2.15$. For the Hertzian dipole reviewed in Section 1.4.1, in contrast, $G_{dB} = 1.5$, as the reader can easily verify.

The field from a half-wave antenna is similar to that of a Hertzian electric dipole not only in its angular distribution but also in its dependence on r; the expression for the electric field, for instance, has terms that go as $1/r$, $1/r^2$, and $1/r^3$. This leads to concepts like "storage fields," "reactive fields," and "radiation fields." Since the oscillating electric dipole is the basic radiation source of primary interest in optical physics, we will discuss these concepts for this particular example.

1.4.1 The Hertzian Electric Dipole

Half-wave antennas, or any transmitters made from conducting elements, are not very practical at optical wavelengths; the radiators of primary interest at optical wavelengths are excited atoms and molecules. The closest classical model for these radiators

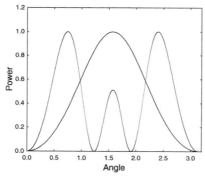

Fig. 1.6 Normalized radiation pattern, as a function of angle θ, of a center-fed linear antenna with $kL = \pi$ (half-wave antenna, solid curve) and $kL = 3\pi$ (dotted curve).

is an electric dipole effectively localized at a point $\mathbf{r} = 0$.[16] For such a *Hertzian* dipole it is simpler to calculate the fields by introducing the Hertz vector $\mathbf{Z}$ instead of the vector and scalar potentials $\mathbf{A}$ and ϕ, respectively. We write

$$\mathbf{A} = \frac{1}{c^2}\frac{\partial \mathbf{Z}}{\partial t}, \qquad \phi = -\nabla \cdot \mathbf{Z}, \tag{1.4.16}$$

which satisfy the Lorentz gauge condition (1.3.3). In terms of $\mathbf{Z}$ the wave equation (1.3.4) becomes

$$\frac{\partial}{\partial t}\left[\nabla^2 \mathbf{Z} - \frac{1}{c^2}\frac{\partial^2 \mathbf{Z}}{\partial t^2}\right] = -\mu_0 c^2 \mathbf{J}. \tag{1.4.17}$$

If we take the divergence of both sides of this equation and then use the continuity equation (1.1.6), it follows that the wave equation (1.3.5) is satisfied by ϕ. For a dipole $\mathbf{p}(t)$ at $\mathbf{r} = 0$, $\mathbf{J}(\mathbf{r}, t) = \dot{\mathbf{p}}(t)\delta^3(\mathbf{r})$ and therefore

$$\nabla^2 \mathbf{Z} - \frac{1}{c^2}\frac{\partial^2 \mathbf{Z}}{\partial t^2} = -\frac{1}{\epsilon_0}\mathbf{p}(t)\delta^3(\mathbf{r}). \tag{1.4.18}$$

This shows the advantage of using the Hertz vector: it satisfies the wave equation with a source proportional to the dipole density.[17] The solution of this wave equation based on the retarded Green function (1.3.13) is simply

$$\mathbf{Z}(\mathbf{r}, t) = \frac{1}{4\pi\epsilon_0}\int d^3 r' \frac{\mathbf{p}(t - |\mathbf{r} - \mathbf{r}'|/c)}{|\mathbf{r} - \mathbf{r}'|}\delta^3(\mathbf{r}') = \frac{1}{4\pi\epsilon_0}\frac{\mathbf{p}(t - r/c)}{r}. \tag{1.4.19}$$

The vector and scalar potentials then follow from (1.4.16), and from them one easily calculates the electric and magnetic fields: for a linear dipole $\mathbf{p}(t) = \hat{\mathbf{z}}p(t)$,

$$\begin{aligned}
\mathbf{E}(\mathbf{r}, t) &= -\frac{\partial \mathbf{A}}{\partial t} - \nabla\phi = \frac{1}{c^2}\frac{\partial^2 \mathbf{Z}}{\partial t^2} + \nabla(\nabla \cdot \mathbf{Z}) \\
&= \frac{1}{4\pi\epsilon_0}[3(\hat{\mathbf{z}} \cdot \hat{\mathbf{r}})\hat{\mathbf{r}} - \hat{\mathbf{z}}]\left[\frac{1}{r^3}p(t - r/c) + \frac{1}{cr^2}\dot{p}(t - r/c)\right] \\
&\quad + \frac{1}{4\pi\epsilon_0}[(\hat{\mathbf{z}} \cdot \hat{\mathbf{r}})\hat{\mathbf{r}} - \hat{\mathbf{z}}]\left[\frac{1}{c^2 r}\ddot{p}(t - r/c)\right],
\end{aligned} \tag{1.4.20}$$

$$\begin{aligned}
\mathbf{B}(\mathbf{r}, t) &= \nabla \times \mathbf{A} = \frac{1}{c^2}\nabla \times \frac{\partial \mathbf{Z}}{\partial t} \\
&= \frac{1}{4\pi\epsilon_0 c}(\hat{\mathbf{z}} \times \hat{\mathbf{r}})\left[\frac{1}{cr^2}\dot{p}(t - r/c) + \frac{1}{c^2 r}\ddot{p}(t - r/c)\right].
\end{aligned} \tag{1.4.21}$$

The integral of the normal component of the Poynting vector $\mathbf{S} = \epsilon_0 c^2(\mathbf{E} \times \mathbf{B})$ over a sphere of radius r centered at the dipole is

---

[16] Electric quadrupole and magnetic dipole transitions are generally much weaker than electric dipole transitions but are very important in astrophysics.

[17] In Section 1.5, the definition of the Hertz vector is generalized to the case of a dielectric medium.

$$P(r,t) = r^2 \int_0^{2\pi} d\phi \int_0^\pi d\theta \sin\theta\, \mathbf{S}\cdot\hat{\mathbf{r}}$$

$$= \frac{1}{6\pi\epsilon_0}\left[\frac{1}{2r^3}\frac{d}{dt}(p^2) + \frac{1}{cr^2}\frac{d}{dt}(p\dot{p}) + \frac{1}{c^2r}\frac{d}{dt}(\dot{p})^2 + \frac{\ddot{p}^2}{c^3}\right]$$

$$= P_s(r,t) + \frac{\ddot{p}^2}{6\pi\epsilon_0 c^3}, \tag{1.4.22}$$

as is easily verified; p and its derivatives are, of course, to be evaluated at the retarded time $t - r/c$. In the limit $r \to \infty$, only the last term in brackets survives, and it gives the radiated power

$$P = P(r \to \infty, t) = \frac{1}{4\pi\epsilon_0}\frac{2\ddot{p}^2}{3c^3}. \tag{1.4.23}$$

This has the same form as the Larmor formula (1.3.36) for the power radiated by an accelerated charge when we replace the charge times the acceleration $(q\dot{v})$ in that formula by $\ddot{p}$. For a dipole oscillating at frequency ω, $\ddot{p} = -\omega^2 p$, and, therefore, $P \propto \omega^4$ (see (1.4.31)). This is the origin of the dependence of Rayleigh scattering on the fourth power of the frequency (see Section 1.10). The angular distribution function describing the radiation is seen from (1.4.29) and (1.4.30) to be simply

$$F_d(\theta) = \sin^2\theta. \tag{1.4.24}$$

Exercise 1.1: (a) Verify (1.4.22). (b) Consider a sinusoidally oscillating dipole moment $p(t)$. Show that (1.4.22), when averaged over a cycle of oscillation, gives (1.4.23) for *all* distances r from the dipole.

1.4.2 Storage Fields and Radiation Fields

What is the physical significance of the contribution

$$P_s(r,t) = \frac{1}{6\pi\epsilon_0}\frac{d}{dt}\left[\frac{p^2}{2r^3} + \frac{p\dot{p}}{cr^2} + \frac{\dot{p}^2}{c^2r}\right] \tag{1.4.25}$$

to $P(r,t)$ in (1.4.22)? One thing that is clear is that $P_s(r,t)$ is power flowing out of a sphere of radius r *in the near field of the dipole* as opposed to the far field, where $P_s(r,t) \to 0$. It is also seen that, for a dipole that is nonzero for a finite period of time,

$$\int_{t_1}^{t_2} P_s(r,t)\,dt = 0, \tag{1.4.26}$$

where t_1 is a time before the dipole moment is excited to a nonzero value, and t_2 is a time after which it is again zero. The power $P_s(r,t)$ out of the spherical surface in close proximity to the dipole is then positive while the dipole moment is being turned, on and negative while it is turning off: energy flows out of the sphere and then back into it. We can, therefore, regard the near fields of the dipole as *storage* fields. Unlike the far fields, which propagate as radiation "forever" —or until the radiant energy is absorbed—the storage fields vanish when the source of excitation of the dipole is switched off.

The stored field energy here is analogous to that in the field of an ideal inductor connected to an AC power supply. There is an electromagnetic field in the neighborhood of the inductor, but no energy is taken from the power supply because the inductor current and the voltage are $\pi/2$ out of phase; the impedance is purely imaginary—there is reactance but no resistance. No energy is lost to radiation, and the field produced by the inductor is entirely of the storage type. In the case of the ideal (nearly) half-wave antenna, the impedance is purely real (resistive), and energy delivered by the transmission line goes entirely into radiation, with no energy remaining stored in the near field of the antenna. The Hertzian electric dipole, in contrast, has no resonant length that would permit a purely real impedance.

The storage field in the immediate neighborhood of an antenna is "reactive" in the sense that measurements involving the near field can affect the driving circuit. The same is true for the electric dipole. If, for example, we have an electric dipole radiator A and we put a second dipole B in its near (reactive) field, the rate of radiation and even the spectrum of radiation from A can be significantly altered. If B is in the far (radiation) field of A, however, the radiation from A is essentially unaffected. Similarly, atoms in close proximity can radiate slower or faster than they do in free space, depending on their separation and other factors, whereas atoms separated by large distances generally radiate independently (see Section 7.9).

Reactive fields have more complicated spatial characteristics than radiation fields, for which the intensity falls off as the inverse square of the distance from the source, and the electric and magnetic fields are in phase and orthogonal. At low frequencies, reactive fields extend over large distances. At 60 Hz, $kr \approx 1$ when $r \approx 800$ km, and electrical interference effects are nearly always associated with reactive fields. At optical frequencies, in contrast, we sense in everyday life only far fields, because we are always much more than an optical wavelength from the Sun, light bulbs, and other sources.

For the familiar sources we have considered, the intensity of radiation obviously decreases as the inverse square of the distance from the source, consistent with the conservation of energy. We note that, in spite of occasional claims to the contrary, the intensity of radiation from *any* source that varies smoothly with space and time, such that the source function and its first derivatives are bounded, cannot fall off more slowly than the inverse square of the distance from the source (*Hannay's theorem*).

A difference between storage (reactive) fields and radiation fields is nicely illustrated by the following example.[18] Suppose you live near high-voltage power lines and find a way to extract energy from the surrounding 60 Hz field. Are you stealing from the utility company?

---

[18] R. Schmitt, EDN (March, 2000), 77.

It might be argued that you are only using electromagnetic energy that has been lost by radiation from the lines and of no use to anyone else. But you would be stealing, because the (near) field at your residence is a storage field, not a radiation field.

1.4.3 Sinusoidally Oscillating Electric Dipoles

An important special case for us is an electric dipole moment that oscillates with frequency ω ($p(t) = p_0 \cos \omega t$). The fields (1.4.20) and (1.4.21) of such a source are

$$
\mathbf{E}(\mathbf{r}, t) = \frac{p_0}{4\pi\epsilon_0} [3(\hat{\mathbf{z}} \cdot \hat{\mathbf{r}})\hat{\mathbf{r}} - \hat{\mathbf{z}}] \left(\frac{1}{r^3} - \frac{i\omega}{cr^2} \right) e^{-i\omega(t-r/c)}
$$
$$
- \frac{p_0}{4\pi\epsilon_0} [(\hat{\mathbf{z}} \cdot \hat{\mathbf{r}})\hat{\mathbf{r}} - \hat{\mathbf{z}}] \frac{\omega^2}{c^2 r} e^{-i\omega(t-r/c)} \tag{1.4.27}
$$

and

$$
\mathbf{B}(\mathbf{r}, t) = \frac{p_0}{4\pi\epsilon_0} \frac{1}{c} [\hat{\mathbf{z}} \times \hat{\mathbf{r}}] \left(-\frac{i\omega}{cr^2} - \frac{\omega^2}{c^2 r} \right) e^{-i\omega(t-r/c)}. \tag{1.4.28}
$$

In terms of the unit vectors $\hat{\mathbf{r}}, \mathbf{e}_\theta, \mathbf{e}_\phi$ in spherical coordinates, the electric and magnetic field vectors in the far field are

$$
\mathbf{E}(\mathbf{r}, t) \cong -\mathbf{e}_\theta \frac{p_0}{4\pi\epsilon_0} \frac{k_0^2}{r} \sin\theta e^{-i\omega(t-r/c)}, \quad k_0 = \omega/c, \quad \text{(radiation zone)}, \tag{1.4.29}
$$

and

$$
\mathbf{B}(\mathbf{r}, t) \cong -\mathbf{e}_\phi \frac{p_0}{4\pi\epsilon_0} \frac{1}{c} \frac{k_0^2}{r} \sin\theta e^{-i\omega(t-r/c)} \quad \text{(radiation zone)}. \tag{1.4.30}
$$

In the radiation zone $\mathbf{E}$ and $\mathbf{B}$ are in phase and orthogonal, and the field takes on the character of a plane wave, with the Poynting vector pointing in the $\hat{\mathbf{r}}$ direction. The cycle-averaged power is given by (1.4.23):

$$
P = \frac{1}{4\pi\epsilon_0} \frac{\omega^4 p_0^2}{3c^3}. \tag{1.4.31}
$$

Very close to the dipole, $\mathbf{E}$ and $\mathbf{B}$ are out of phase by $\pi/2$:

$$
\mathbf{E}(\mathbf{r}, t) \cong (2\hat{\mathbf{r}} \cos\theta + \mathbf{e}_\theta \sin\theta) \frac{p_0}{4\pi\epsilon_0} \frac{1}{r^3} e^{-i\omega(t-r/c)} \quad \text{(near field)}, \tag{1.4.32}
$$

$$
\mathbf{B}(\mathbf{r}, t) \cong -\mathbf{e}_\phi \frac{i p_0}{4\pi\epsilon_0} \frac{1}{c} \frac{k_0}{r^2} \sin\theta e^{-i\omega(t-r/c)} \quad \text{(near field)}. \tag{1.4.33}
$$

The field is predominantly electric and has the characteristics of the familiar electric dipole of electrostatics.

1.5   Dielectrics and the Refractive Index

So far, we have only reviewed some aspects of radiation from a localized source into a vacuum. We now turn our attention to electromagnetic fields in a dielectric medium. The "medium," of course, must for many purposes be treated according to quantum mechanics, which we will do in later chapters, but for now we take a purely classical approach and assume that a dielectric medium consists of "atoms" that acquire electric dipole moments when exposed to electric fields. The current density at point $\mathbf{r}$ in the medium is taken to be

$$\mathbf{J}(\mathbf{r}, t) = \mathbf{J}_f(\mathbf{r}, t) + \sum_j q\dot{\mathbf{x}}_j(t)\delta^3(\mathbf{r} - \mathbf{r}_j) = \mathbf{J}_f(\mathbf{r}, t) + \sum_j \dot{\mathbf{p}}_j(t)\delta^3(\mathbf{r} - \mathbf{r}_j). \quad (1.5.1)$$

Here, $\mathbf{J}_f$ is the current density due to "free" charges, in addition to the current density due to the dipole moments of the atoms. The jth atom is located at $\mathbf{r}_j$ and has an electric dipole moment $\mathbf{p}_j$ due to a charge $-q$ separated from a charge q ($q > 0$) by the vector $\mathbf{x}_j$ pointing from $-q$ to q: $\mathbf{p}_j(t) = q\mathbf{x}_j$. We assume further that the medium is well described as a uniform continuum of N atoms per unit volume, and replace (1.5.1) by

$$\mathbf{J}(\mathbf{r}, t) = \mathbf{J}_f(\mathbf{r}, t) + N\int d^3r'\,\dot{\mathbf{p}}(\mathbf{r}', t)\delta^3(\mathbf{r} - \mathbf{r}')$$

$$= \mathbf{J}_f(\mathbf{r}, t) + N\dot{\mathbf{p}}(\mathbf{r}, t) = \mathbf{J}_f(\mathbf{r}, t) + \frac{\partial \mathbf{P}(\mathbf{r}, t)}{\partial t}. \quad (1.5.2)$$

$\mathbf{P}$ is the polarization density, usually called the *polarization*. The continuity equation (1.1.6) then implies that, in addition to the free charge density ρ_f, there is a charge density $\rho_p = -\nabla \cdot \mathbf{P}$ at a point $\mathbf{r}$: a spatially varying density of dipoles can produce an imbalance of otherwise equally distributed charge. In the Maxwell equation (1.1.1), therefore, the charge density ρ in the case of a dielectric must be $\rho_f + \rho_p$:

$$\nabla \cdot \mathbf{E} = \frac{1}{\epsilon_0}(\rho_f + \rho_p) = \frac{1}{\epsilon_0}(\rho_f - \nabla \cdot \mathbf{P}), \quad (1.5.3)$$

or

$$\nabla \cdot \mathbf{D} = \rho_f, \quad (1.5.4)$$

where the electric displacement field is

$$\mathbf{D} = \epsilon_0\mathbf{E} + \mathbf{P}. \quad (1.5.5)$$

Using (1.5.2) and (1.5.5) in (1.1.4), we obtain the differential Maxwell equations for fields in a dielectric medium:

$$\nabla \cdot \mathbf{D} = \rho_f, \quad (1.5.6)$$

$$\nabla \cdot \mathbf{B} = 0, \quad (1.5.7)$$

$$\nabla \times \mathbf{E} = -\frac{\partial \mathbf{B}}{\partial t}, \quad (1.5.8)$$

$$\nabla \times \mathbf{H} = \mathbf{J}_f + \frac{\partial \mathbf{D}}{\partial t}. \quad (1.5.9)$$

We have introduced the magnetic field strength $\mathbf{H}$ by writing the magnetic induction field $\mathbf{B}$ (also known as the magnetic flux density) as $\mathbf{B} = \mu_0 \mathbf{H}$.

The fields $\mathbf{E}$ and $\mathbf{B}$, the fields acting on charges ($\mathbf{F} = q[\mathbf{E} + \mathbf{v} \times \mathbf{B}]$), are the fundamental fields of electromagnetism; $\mathbf{D}$ and $\mathbf{H}$ in material media are defined in terms of these fields. To relate $\mathbf{D}$ to $\mathbf{E}$, we need to relate the polarization $\mathbf{P}$ to $\mathbf{E}$. At this point in classical theory, we resort to a model for the dipole moments of the individual atoms. We imagine something like the "plum pudding" model of an atom, but, to make things as simple as we can, we imagine that the dipole moment of an atom results from the displacement of an electron of charge e ($e < 0$) from a nucleus of charge $-e$. The nucleus is assumed to be much heavier than the electron, so that as a first approximation, we need only be concerned with the motion of the electron. In the absence of a field or other perturbation, the electron is assumed to be at a position of stable equilibrium within the atom, that is, a displacement from its equilibrium position ($\mathbf{x} = 0$) results in a restoring force $-k_s\mathbf{x}$, where the "spring constant" $k_s > 0$. The equation of motion for the electron (mass m) when the atom is in an electromagnetic field is then

$$\mathbf{F} = m\ddot{\mathbf{x}} = -k_s\mathbf{x} + e\mathbf{E} + e\mathbf{v} \times \mathbf{B}. \tag{1.5.10}$$

For the interaction of light with matter, we are mostly concerned with the electric component rather than the magnetic component of an optical field. When light is incident on an atom, for example, its effect is mainly on the atomic electrons, since the nuclei are so much more massive than the electrons. For a monochromatic plane wave with a linearly polarized electric field,

$$\mathbf{E} = \hat{\mathbf{z}}E_0 \cos(\omega t - kz), \tag{1.5.11}$$

we have $\nabla \times \mathbf{E} = \hat{\mathbf{y}}kE_0 \sin(\omega t - kz)$, and, from $\partial\mathbf{B}/\partial t = -\nabla \times \mathbf{E}$,

$$\mathbf{B} = \hat{\mathbf{y}}\frac{k}{\omega} \cos(\omega t - kz) = \hat{\mathbf{y}}\frac{1}{c}E_0 \cos(\omega t - kz). \tag{1.5.12}$$

The ratio of the electric force $q\mathbf{E}$ to the magnetic force $q\mathbf{v} \times \mathbf{B}$ is therefore on the order of v/c, which, based on the Bohr model, is $e^2/(4\pi\epsilon_0 n\hbar c) \cong 1/(137n)$ for an electron in a hydrogen atom in a state with principal quantum number n. We therefore approximate (1.5.10) by

$$\ddot{\mathbf{x}} + \omega_0^2\mathbf{x} = \frac{e}{m}\mathbf{E}(t) \quad (\omega_0 = \sqrt{k_s/m}). \tag{1.5.13}$$

For optical fields, the electric field $\mathbf{E}$ does not vary significantly over the dimensions of an atom; the field wavelengths are much larger than an atom. So, $\mathbf{E}$ in (1.5.10) can be taken to be the field at the center ($\mathbf{r}$) of the atom, or, what is approximately equivalent, the position of the electron. For a (linearly polarized) monochromatic field applied to the atom, therefore, the equation of motion for the electron is

$$\ddot{\mathbf{x}} + \omega_0^2 \mathbf{x} = \frac{e}{m} \mathbf{E}_0 \cos[\omega t + \phi(\mathbf{r})], \qquad (1.5.14)$$

where $\phi(\mathbf{r})$ is a phase at the position of the atom. The electron oscillates about its equilibrium position at the frequency of the field:

$$\mathbf{x} = \frac{(e/m)\mathbf{E}_0}{\omega_0^2 - \omega^2} \cos[\omega t + \phi(\mathbf{r})]. \qquad (1.5.15)$$

We ignore the solution of the homogeneous equation ($\ddot{x} + \omega_0^2 \mathbf{x} = 0$) on the grounds that its contribution to $\mathbf{x}$ will damp out due to dissipative effects (e.g. radiation) on the electron's oscillation, which we have not included in writing (1.5.13). Then

$$\mathbf{P} = N\mathbf{p} = Ne\mathbf{x} = \frac{Ne^2/m}{\omega_0^2 - \omega^2} \mathbf{E}_0 \cos[\omega t + \phi(\mathbf{r})] = N\alpha(\omega)\mathbf{E}_0 \cos[\omega t + \phi(\mathbf{r})] \quad (1.5.16)$$

and

$$\mathbf{D} = \epsilon_0 \mathbf{E} + \mathbf{P} = \epsilon_0 \left[1 + \frac{N\alpha(\omega)}{\epsilon_0}\right] \mathbf{E} = \epsilon(\omega)\mathbf{E}, \qquad (1.5.17)$$

$$\epsilon(\omega) = \epsilon_0 + N\alpha(\omega), \qquad (1.5.18)$$

where we have introduced the permittivity $\epsilon(\omega)$ as well as the polarizability $\alpha(\omega)$ relating the induced dipole moment in each atom to the field acting on the atom. Taking the curl of both sides of (1.5.8) and using (1.5.9) for $\nabla \times \mathbf{H}$ as well as the identity $\nabla \times (\nabla \times \mathbf{E}) = \nabla(\nabla \cdot \mathbf{E}) - \nabla^2 \mathbf{E}$, we find that the electric field satisfies the equation

$$\nabla(\nabla \cdot \mathbf{E}) - \nabla^2 \mathbf{E} = -\frac{1}{c^2} \frac{\partial^2 \mathbf{D}}{\partial t^2} = -\frac{1}{c^2} \frac{\epsilon}{\epsilon_0} \frac{\partial^2 \mathbf{E}}{\partial t^2} \qquad (1.5.19)$$

if $\mathbf{J}_f = 0$. For the uniform distribution of atoms we are assuming, ϵ is independent of position in the dielectric, and

$$\nabla \cdot \mathbf{E} = \nabla \cdot [\frac{\epsilon_0}{\epsilon}\mathbf{D}] = \frac{\epsilon_0}{\epsilon} \nabla \cdot \mathbf{D} = 0 \qquad (1.5.20)$$

if $\rho_f = 0$. Therefore, our monochromatic electric field satisfies the homogeneous wave equation,

$$\nabla^2 \mathbf{E} - \frac{n^2}{c^2} \frac{\partial^2 \mathbf{E}}{\partial t^2} = 0, \qquad (1.5.21)$$

which identifies $c/n(\omega)$ as the phase velocity for frequency ω, and $n(\omega)$ as the refractive index:

$$n^2(\omega) = \epsilon(\omega)/\epsilon_0 = 1 + N\alpha(\omega)/\epsilon_0 = 1 + \frac{Ne^2/m\epsilon_0}{\omega_0^2 - \omega^2}. \qquad (1.5.22)$$

The first two equalities are general relations, but the third is specific to our classical model for an atom (or, more precisely, our classical model for how an atom responds to an electromagnetic field). We haven't said anything about how ω_0 relates to a real atom, or about what in our model distinguishes between different types of atoms. As the reader might guess, or already know, ω_0 should be identified with a transition frequency ($\nu_0 = \omega_0/2\pi$) of the atom. It is often a reasonably good approximation for ground-state atoms to take ω_0 to be the Bohr transition frequency between the ground state and the first excited state. More generally, we have to sum over contributions to $n^2(\omega)-1$ from all transitions between the ground state (or whatever state(s) the atoms are assumed to occupy) and all the other states of the atom (including continuum states) that can make transitions to and from the ground state. This is discussed further in Section 2.5.

Exercise 1.2: (a) Consider a sinusoidally oscillating "guest" electric dipole inside a "host" dielectric medium with (real) refractive index $n(\omega)$ at the dipole oscillation frequency ω. Show that the $\mathbf{E}$ and $\mathbf{B}$ fields from the dipole can be obtained from (1.4.27) and (1.4.28) by simply replacing ϵ_0 by $\epsilon(\omega) = \epsilon_0 n^2(\omega)$, and c by $c/n(\omega)$. (b) Assuming that local field corrections are negligible, show that the power radiated by the guest dipole is $n(\omega)$ times the power that would be radiated into free space, that is,

$$P = n(\omega)\frac{1}{4\pi\epsilon_0}\frac{p_0^2\omega^4}{3c^3}.$$

(More generally, when the magnetic permeability μ differs significantly from μ_0, the correct expression for the radiated power is $[\mu(\omega)/\mu_0]P$.)

Transition frequencies of many familiar dielectrics typically exceed optical frequencies, in which case (1.5.22) can be approximated by the *Cauchy dispersion formula*:

$$n(\omega) \cong 1 + \frac{Ne^2}{2m\epsilon_0\omega_0^2}\left(1 + \frac{\omega^2}{\omega_0^2}\right) \tag{1.5.23}$$

for $\omega_0 \gg \omega$, and $n^2 \cong 2n - 1$, or, in terms of the field wavelength $\lambda = 2\pi c/\omega$,

$$n(\lambda) \cong 1 + \frac{Ne^2\lambda_0^2}{8\pi^2mc^2\epsilon_0}\left(1 + \frac{\lambda_0^2}{\lambda^2}\right) = A\left(1 + \frac{B}{\lambda^2}\right). \tag{1.5.24}$$

Measured refractive indices are commonly fit to this formula or to simple extensions of it that follow from (1.5.22) or its generalization that allows for more than one resonance frequency of the medium. For air, for example, $A \cong 28.79 \times 10^{-5}$ and $B \cong 5.67 \times 10^{-7}$ m^2 at optical wavelengths. This illustrates the fact that the classical *Lorentz model* leading to the equation of motion (1.5.13) for bound electrons is often in good qualitative agreement with experiment. To bring it into quantitative agreement with reality for certain phenomena, all we have to do is modify its predictions to include oscillator strengths (Section 2.5). Needless to say, classical theory can account for neither these oscillator strengths nor the resonance frequencies of atoms.

1.5.1   The Superposition Principle and the Extinction Theorem

According to the superposition principle, the total electric field at any point in a medium is the sum of the field incident on the medium and the fields from all the atoms constituting the medium. The electric field of light of frequency ω induces in each atom of a dielectric medium an electric dipole moment that oscillates and radiates at frequency ω. The dipole field at $\mathbf{r}$ at time t from a single atom at $\mathbf{r}_j$ depends on the dipole moment of the atom at the retarded time $t - |\mathbf{r} - \mathbf{r}_j|/c$. So if a light wave is incident on a dielectric from a vacuum, the total field at any point inside or outside the medium is a superposition of fields *that all propagate with the phase velocity c.* But we know this total field inside the medium propagates with a phase velocity that is not c but $c/n(\omega)$: the dipole fields from all the atoms must evidently add with the incident field to give a net field having a phase velocity $c/n(\omega)$. We will try to elucidate this remarkable result with a simple model.

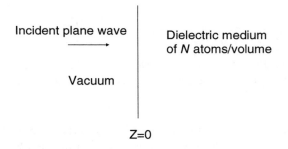

Fig. 1.7 A plane wave incident on a dielectric medium. The total field at any point is the sum of the incident plane wave and the fields from all the atoms of the medium.

Suppose a monochromatic, linearly polarized field of frequency ω is normally incident from vacuum onto a dielectric medium occupying the half-space $z \geq 0$ (see Figure 1.7). The medium consists of N electrically polarizable particles per unit volume, each characterized by a polarizability $\alpha(\omega)$; for simplicity, we will assume that all the particles have the same polarizability. At any point $\mathbf{r}$, inside or outside the medium, the total electric field $\mathbf{E}_T(\mathbf{r}, t)$ is the sum of the incident field $\mathbf{E}_i(\mathbf{r}, t)$ and the field $\mathbf{E}_d(\mathbf{r}, t)$ radiated by all the dipoles of the medium:

$$\mathbf{E}_T(\mathbf{r}, t) = \mathbf{E}_i(\mathbf{r}, t) + \mathbf{E}_d(\mathbf{r}, t). \tag{1.5.25}$$

We write the incident field as

$$\mathbf{E}_i(z, t) = \hat{\mathbf{x}} E_i(z) e^{-i\omega t} = \hat{\mathbf{x}} E_{i0} e^{ik_0 z} e^{-i\omega t} \quad (k_0 = \omega/c), \tag{1.5.26}$$

and, similarly,

$$\mathbf{E}_T(z,t) = \hat{\mathbf{x}} E_T(z) e^{-i\omega t} \quad \text{and} \quad \mathbf{E}_d(z,t) = \hat{\mathbf{x}} E_d(z) e^{-i\omega t}; \tag{1.5.27}$$

$E_T(z)$ and $E_d(z)$ are to be determined.

The dipole moment $p(\mathbf{r}',t)$ of an atom at a point $\mathbf{r}' = (x', y', z')$ inside the medium is induced by the total field $\mathbf{E}_T(\mathbf{r}',t)$ at $\mathbf{r}'$. In our model,

$$p(z',t) = \alpha(\omega) E_T(z') e^{-i\omega t}.^{19} \tag{1.5.28}$$

The direction of the dipole moment will be along the direction $\hat{\mathbf{x}}$ of the total field. We obtain from (1.4.27) the electric field at $\mathbf{r}$ due to the induced dipole at $\mathbf{r}'$:

$$\mathbf{E}(\mathbf{r},t) = \frac{1}{4\pi\epsilon_0} \Big\{ [3(\hat{\mathbf{x}} \cdot \hat{\mathbf{n}})\hat{\mathbf{n}} - \hat{\mathbf{x}}] [\frac{1}{R^3} - \frac{ik_0}{R^2}]$$
$$- [(\hat{\mathbf{x}} \cdot \hat{\mathbf{n}})\hat{\mathbf{n}} - \hat{\mathbf{x}}] \frac{k_0^2}{R} \Big\} \alpha(\omega) E_T(z') e^{ik_0 R} e^{-i\omega t}. \tag{1.5.29}$$

Here, $R = |\mathbf{r} - \mathbf{r}'|$, and $\hat{\mathbf{n}}$ is the unit vector pointing from $\mathbf{r}'$ to $\mathbf{r}$. Then,

$$E_d(z) = \frac{1}{4\pi\epsilon_0} N\alpha(\omega) \int_{-\infty}^{\infty} dx' \int_{-\infty}^{\infty} dy' \int_0^{\infty} dz' E_T(z') e^{ik_0 R}$$
$$\times \Big\{ [3(\hat{\mathbf{x}} \cdot \hat{\mathbf{n}})^2 - 1](\frac{1}{R^3} - \frac{ik_0}{R^2}) - [(\hat{\mathbf{x}} \cdot \hat{\mathbf{n}})^2 - 1] \frac{k_0^2}{R} \Big\}. \tag{1.5.30}$$

The integration over x' and y' is straightforward but a little messy; here, we just state the result:[20]

$$E_d(z) = \frac{ik_0}{2\epsilon_0} N\alpha(\omega) \int_0^{\infty} dz' \, E_T(z') e^{ik_0|z-z'|}$$
$$= \frac{ik_0}{2} [n^2(\omega) - 1] \int_0^{\infty} dz' \, E_T(z') e^{ik_0|z-z'|}. \tag{1.5.31}$$

The superposition principle, previously given in the form shown in (1.5.25), therefore takes the form of an integral equation for the total field amplitude $E_T(z)$ in any plane of constant z:

$$E_T(z) = E_{i0} e^{ik_0 z} + \frac{ik_0}{2} [n^2(\omega) - 1] \int_0^{\infty} dz' \, E_T(z') e^{ik_0|z-z'|}. \tag{1.5.32}$$

This integral equation obviously allows for all multiple scatterings by the electrically polarizable particles constituting the medium.

---

[19] This assumes that the difference between the local electric field experienced by the dipole and the "macroscopic," average field $E_T(z')$ is negligible. When this difference is taken into account it leads under the most simplifying assumptions to the Lorentz–Lorenz formula; see *Born and Wolf*. It does not affect the reflection and transmission coefficients.

[20] H. Fearn, D. F. V. James, and P. W. Milonni, Am. J. Phys. **64**, 986 (1996). This paper includes a discussion of different interpretations of the Ewald–Oseen extinction theorem.

Inside the dielectric: For $z > 0$, we try a solution of (1.5.32) of the form

$$E_T(z) = \text{constant} \times e^{ikz} = E_{T0}e^{ikz}. \tag{1.5.33}$$

Plugging this into (1.5.32), we obtain the condition for this solution to be valid:

$$E_{T0}e^{ikz} = \left[E_{i0} + \frac{k_0}{2(k_0 - k)}(n^2 - 1)]E_{T0} \right] e^{ik_0 z} - \frac{k_0^2}{k_0^2 - k^2}(n^2 - 1)E_{T0}e^{ikz}, \tag{1.5.34}$$

or

$$E_{i0} + \frac{k_0}{2(k_0 - k)}E_{T0} = 0 \tag{1.5.35}$$

and

$$-\frac{k_0^2}{k_0^2 - k^2}(n^2 - 1) = 1. \tag{1.5.36}$$

The second condition gives $k = n(\omega)k_0$; with this, the first condition becomes

$$E_{T0} = \frac{2}{n(\omega) + 1}E_{i0}, \tag{1.5.37}$$

that is,

$$E_T(z, t) = \frac{2}{n(\omega) + 1}E_{i0}e^{-i\omega[t - n(\omega)z/c]} \tag{1.5.38}$$

in the medium. These results for the particular case of normal incidence are familiar from the solution of Maxwell's equations with the appropriate boundary conditions for the fields: the field inside the medium propagates with the phase velocity $\omega/k = c/n(\omega)$ and the amplitude transmission (Fresnel) coefficient is $2/(n(\omega) + 1)$.

Outside the dielectric: For $z < 0$ we use the results for the field inside the medium in (1.5.32) and obtain

$$E_T(z) = E_{i0}e^{ik_0 z} - \frac{n(\omega) - 1}{n(\omega) + 1}E_{i0}e^{-ik_0 z}, \tag{1.5.39}$$

which is familiar from the solution of Maxwell's equations with the appropriate boundary conditions: the amplitude reflection (Fresnel) coefficient is $-[n(\omega) - 1]/[n(\omega) + 1]$.

We have shown, therefore, that the Fresnel formulas for normal incidence are consistent with the superposition principle. In particular, although *all* the fields we have added up to obtain the total field propagate at the speed of light in vacuum, the total field inside the medium propagates with the phase velocity $c/n(\omega)$. The incident field propagating with phase velocity c is "extinguished" inside the medium by the fields from the induced dipoles of the medium (see (1.5.35)) and replaced by a field with phase velocity $c/n(\omega)$. This is the upshot of the *Ewald–Oseen extinction theorem*.[21] Note that, in our model, there is no "extinction distance" defining how far the incident field must penetrate into the medium before it is replaced by the field (1.5.38).

We have assumed that the medium consists of a uniform continuum of N atoms per unit volume. This is a valid approximation if there are many atoms within a volume λ^3 (wavelength cubed). For an ideal gas at standard temperature and pressure, there are roughly 6×10^6 molecules in a volume of $(600 \text{ nm})^3$, and so our assumption is well justified at optical wavelengths. In the case of a very dilute medium, or if the scatterers are arranged periodically in space and relatively far apart, as in x-ray diffraction, there will be scattering of radiation in directions other than those defined by the $\mathbf{k}$ vectors of the incident and refracted fields. This also occurs whenever the number of particles in any given small volume fluctuates about its mean value (Section 1.10).

The Ewald–Oseen extinction theorem provides a satisfying physical description of reflection and refraction, but it is generally very much easier to work directly with the differential Maxwell equations and the boundary conditions they imply. The differential Maxwell equations in a sense "automatically" perform the summation over the dipole fields. In the plane-wave model we have considered, for example, (1.5.21) reduces to

$$\left(\frac{d^2}{dz^2} + k_0^2 \right) E_T(z) = -k_0^2(n^2 - 1)E_T(z), \tag{1.5.40}$$

with the formal solution

$$E_T(z) = E_{i0}e^{ik_0 z} - k_0^2[n^2 - 1]\int_0^\infty dz' \, G(z - z')E_T(z'). \tag{1.5.41}$$

The appropriate Green function $G(z - z')$, defined by

$$\left(\frac{\partial^2}{\partial z^2} + k_0^2 \right) G(z - z') = \delta(z - z'), \tag{1.5.42}$$

is

$$G(z - z') = \frac{1}{2ik_0}e^{ik_0|z-z'|}. \tag{1.5.43}$$

Thus, (1.5.41) is equivalent to (1.5.32) but is obtained without the complicated summation over the fields of all the induced dipoles of the medium.

In the case of non-normal incidence, the induced dipoles do not all have the same phase of oscillation in planes of constant z; their relative phasing in this case gives

---

[21] Our results are simplified special cases of the more general extinction theorem first obtained by P. P. Ewald in 1912 for crystalline media and by C. W. Oseen in 1915 for isotropic media. Ewald's work turned out to play a significant role in Max von Laue's discovery of x-ray diffraction by crystals. In his Nobel lecture, von Laue recalled that "during [a conversation with Ewald] I was suddenly struck by the obvious question of the behavior of waves which are short by comparison with the lattice-constants of the space-lattice. And it was at that point that my intuition for optics suddenly gave me the answer: lattice spectra would have to ensue." (M. von Laue, "Concerning the Detection of X-ray Interferences," http://www.nobelprize.org/prizes/physics/1914/laue/lecture, accessed October 25, 2018.)

the direction of propagation of the reflected wave, much like the directional emission by phased antenna arrays. Note, in particular, the obvious but interesting fact that, although all the individual induced dipoles radiate with the characteristic dipole pattern (see (1.4.27)), the total field consists of refracted and reflected fields propagating in well-defined directions, with no radiation "scattered" in other directions when there is a uniform and fixed distribution of dipoles.

The expression (1.5.32) for the total electric field is obtained by adding to the incident field the field produced by all the dipoles of the medium. We can approach it in a different way—a way, however, that presupposes the conclusions that follow from the extinction theorem it implies—as follows. A monochromatic plane wave in a medium of refractive index n undergoes a phase shift $(n-1)k_0 z$ over that accrued in propagation over the same distance z in a vacuum. After propagation through an infinitesimally thin slab occupying the space from $z = 0$ to Δz, the field amplitude at z is therefore

$$E_{i0} e^{ik_0[z+(n-1)\Delta z]} \cong E_{i0} e^{ik_0 z} + ik_0(n-1)E_{0i}e^{ik_0 z}\Delta z, \qquad (1.5.44)$$

which is the field $E_T(z)$ given by (1.5.32) when $n \cong 1$ (and, therefore, $n-1 \cong \frac{1}{2}(n^2-1)$) and the medium is taken to be infinitesimally thin.

1.5.2 Hertz Vector for Dielectrics

In Section 1.4, we introduced the Hertz vector field $\mathbf{Z}(\mathbf{r}, t)$ and used it to obtain the field from an electric dipole in free space. The electric and magnetic fields at frequency ω $(\mathbf{Z}(\mathbf{r}, t) = \mathbf{Z}(\mathbf{r}, \omega)e^{-i\omega t})$ are given by

$$\mathbf{E}(\mathbf{r}, \omega) = -k_0^2 \mathbf{Z}(\mathbf{r}, \omega) + \nabla[\nabla \cdot \mathbf{Z}(\mathbf{r}, \omega)] \quad (k_0 = \omega/c), \qquad (1.5.45)$$

$$\mathbf{H}(\mathbf{r}, \omega) = -i\omega\epsilon_0 \nabla \times \mathbf{Z}(\mathbf{r}, \omega). \qquad (1.5.46)$$

For an electric dipole in a dielectric medium, we replace k_0^2 in (1.5.45) by $k_0^2 \epsilon_b(\omega)/\epsilon_0$, and ϵ_0 in (1.5.46) by $\epsilon_b(\omega)$:

$$\mathbf{E}(\mathbf{r}, \omega) = k_0^2[\epsilon_b(\omega)/\epsilon_0]\mathbf{Z}(\mathbf{r}, \omega) + \nabla[\nabla \cdot \mathbf{Z}(\mathbf{r}, \omega)] \quad (k_0 = \omega/c), \qquad (1.5.47)$$

$$\mathbf{H}(\mathbf{r}, \omega) = -i\omega\epsilon_b(\omega)\nabla \times \mathbf{Z}(\mathbf{r}, \omega). \qquad (1.5.48)$$

We denote by $\epsilon_b(\omega)$ the permittivity at frequency ω of the dielectric. We will be interested here (and in Section 1.10) in a source inside a "background" dielectric medium, whence the subscript b on $\epsilon_b(\omega)$. To simplify things, we will assume that the medium is transparent (no absorption) at frequency ω, so that $\epsilon_b(\omega)$ may be taken to be purely real. The identifications (1.5.47) and (1.5.48) are consistent with the propagation of a wave of frequency ω with the phase velocity $c/n_b(\omega)$ in the medium, $n_b(\omega) = \sqrt{\epsilon_b(\omega)/\epsilon_0}$, as will be clear in the following.

The curl of $\mathbf{E}(\mathbf{r},\omega)$ in (1.5.47) is simply

$$\nabla \times \mathbf{E}(\mathbf{r},\omega) = k_0^2[\epsilon_b(\omega)/\epsilon_0]\nabla \times \mathbf{Z}(\mathbf{r},\omega), \qquad (1.5.49)$$

since the curl of a gradient is zero. Now, apply the curl operation to this equation, assuming no free currents and therefore $\nabla \times \mathbf{H}(\mathbf{r},\omega) = -i\omega\mathbf{D}(\mathbf{r},\omega)$:

$$\nabla \times (\nabla \times \mathbf{E}) = i\omega\mu_0\nabla \times \mathbf{H} = \omega^2\mu_0\mathbf{D} = k_0^2[\epsilon_b(\omega)/\epsilon_0]\nabla \times (\nabla \times \mathbf{Z})$$
$$= k_0^2[\epsilon_b(\omega)/\epsilon_0][\nabla(\nabla \cdot \mathbf{Z}) - \nabla^2\mathbf{Z}], \qquad (1.5.50)$$

implying

$$\nabla^2\mathbf{Z} = \frac{\epsilon_0}{\epsilon_b}\frac{\omega^2}{k_0^2}\mu_0\mathbf{D} + \nabla(\nabla \cdot \mathbf{Z}) = -\frac{1}{\epsilon_b}\mathbf{D} + (\mathbf{E} - \frac{\epsilon_b}{\epsilon_0}k_0^2\mathbf{Z}), \qquad (1.5.51)$$

$$\nabla^2\mathbf{Z} + k^2\mathbf{Z} = \mathbf{E} - \frac{1}{\epsilon_b}\mathbf{D}, \qquad k^2 = k_0^2\epsilon_b(\omega)/\epsilon_0 = n_b^2(\omega)\omega^2/c^2. \qquad (1.5.52)$$

If $\mathbf{D}(\mathbf{r},\omega) = \epsilon_b(\omega)\mathbf{E}(\mathbf{r},\omega)$, the right side is zero, and all we have done is rederived what we already know: the field propagates with phase velocity $\omega/k(\omega) = c/n_b(\omega)$. Suppose, however, that within the medium there is a localized "source" characterized by a dipole moment density $\mathbf{P}_s(\mathbf{r},\omega) = \mathbf{p}_0(\omega)\delta^3(\mathbf{r})$. Then, $\mathbf{D} = \epsilon_b\mathbf{E} + \mathbf{P}_s$, and

$$\nabla^2\mathbf{Z} + k^2\mathbf{Z} = -\frac{1}{\epsilon_b}\mathbf{p}_0(\omega)\,\delta^3(\mathbf{r}). \qquad (1.5.53)$$

The solution for $\mathbf{Z}(\mathbf{r},\omega)$ is simply (see (1.4.18) and (1.4.19))

$$\mathbf{Z}(\mathbf{r},\omega) = \frac{1}{4\pi\epsilon_b(\omega)}\mathbf{p}_0(\omega)\frac{e^{ikr}}{r}, \qquad (1.5.54)$$

and, from this, one obtains the electric and magnetic fields due to the source in the medium.

Imagine, somewhat more generally, that the source within the medium occupies a volume V and is characterized by a permittivity $\epsilon_s(\omega)$. Then, $\mathbf{D}(\mathbf{r},\omega) = \epsilon(\omega)\mathbf{E}(\mathbf{r},\omega)$, where $\epsilon = \epsilon_s$ inside the source and $\epsilon = \epsilon_b$ outside, and

$$\nabla^2\mathbf{Z} + k^2\mathbf{Z} = (1 - \epsilon/\epsilon_b)\mathbf{E}. \qquad (1.5.55)$$

The solution of this equation is

$$\mathbf{Z}(\mathbf{r},\omega) = -\frac{1}{4\pi}\int d^3r'\left[1 - \frac{\epsilon(\mathbf{r}',\omega)}{\epsilon_b(\omega)}\right]\mathbf{E}(\mathbf{r}',\omega)\frac{e^{ik|\mathbf{r}-\mathbf{r}'|}}{|\mathbf{r}-\mathbf{r}'|}$$
$$= -\frac{1}{4\pi}\left[1 - \frac{\epsilon_s(\omega)}{\epsilon_b(\omega)}\right]\int_V d^3r'\,\mathbf{E}(\mathbf{r}',\omega)\frac{e^{ik|\mathbf{r}-\mathbf{r}'|}}{|\mathbf{r}-\mathbf{r}'|}. \qquad (1.5.56)$$

In the last step, we assume that $\epsilon_s(\mathbf{r},\omega)$ is nonzero only inside the volume V, where it has a constant value $\epsilon_s(\omega)$. Imagine further that the extent of the volume V is sufficiently small, compared with a wavelength, that we can approximate (1.5.56) by

$$\mathbf{Z}(\mathbf{r},\omega) = -\frac{1}{4\pi}\left[1 - \frac{\epsilon_s(\omega)}{\epsilon_b(\omega)}\right] V\mathbf{E}_{\text{ins}}(\omega)\frac{e^{ikr}}{r}, \qquad (1.5.57)$$

with r the distance from the center of the source (at $\mathbf{r} = 0$) to the observation point, and $\mathbf{E}_{\text{ins}}(\omega)$ the (approximately constant) electric field in the source volume V. This has the same form as (1.5.54), with $\mathbf{p}_0(\omega) = \epsilon_b(\omega)[\epsilon_s(\omega)/\epsilon_b(\omega) - 1]V\mathbf{E}_{\text{ins}}(\omega)$. In other words, $\mathbf{Z}(\mathbf{r},\omega)$ has the same form as the Hertz vector for an electric dipole moment

$$\mathbf{p}_0(\omega) = [\epsilon_s(\omega) - \epsilon_b(\omega)]V\mathbf{E}_{\text{ins}}(\omega). \qquad (1.5.58)$$

Consider, for example, a small dielectric sphere of radius a: $V = 4\pi a^3/3$. The field inside such a sphere is $\mathbf{E}_{\text{ins}}(\omega) = [3\epsilon_b/(\epsilon_s + 2\epsilon_b)]\mathbf{E}_{\text{out}}(\omega)$, where $\mathbf{E}_{\text{out}}(\omega)$ is the electric field outside the sphere. The dipole moment (1.5.58) in this case is therefore related to $\mathbf{E}_{\text{out}}(\omega)$ by $\mathbf{p}_0(\omega) = \alpha(\omega)\mathbf{E}_{\text{out}}(\omega)$, where we define the polarizability

$$\alpha(\omega) = 4\pi\epsilon_b\left(\frac{\epsilon_s - \epsilon_b}{\epsilon_s + 2\epsilon_b}\right)a^3 \quad \text{(small dielectric sphere)}. \qquad (1.5.59)$$

1.5.3   Why Sine Waves Are Special

We have considered solutions of the equation of motion (1.5.13) for the electron displacement only for the special case of a monochromatic field. Suppose we do not specify $\mathbf{E}(t)$ but require that the time dependence of the electron displacement follows exactly that of $\mathbf{E}(t)$, that is, $\mathbf{x}(t) = A\mathbf{E}(t)$, where A is constant in time. Then,

$$A[\ddot{\mathbf{E}}(t) + \omega_0^2\mathbf{E}(t)] = \mathbf{E}(t). \qquad (1.5.60)$$

The only nontrivial solutions of this equation for the forced oscillation of the electron are of the form

$$\mathbf{E}(t) = \frac{\mathbf{E}_0}{\omega_0^2 - \omega^2}e^{-i\omega t}, \qquad (1.5.61)$$

where $\mathbf{E}_0$ and ω are arbitrary. In other words, monochromatic fields have the *unique* time dependence for which the forced oscillations of bound electrons vary in time in exactly the same way as the forcing field. The same argument applies anytime the medium, whatever it may be, responds (linearly) to a field in the manner of a driven harmonic oscillator, and explains why "the sinusoid is the only wave that travels without change of shape" in its temporal variation at a fixed point in space.[22]

---

[22] M. V. Berry and D. A. Greenwood, Am. J. Phys. **43**, 91 (1975).

Exercise 1.3: Suppose that $\mathbf{x}$ responds to a derivative of $\mathbf{E}(t)$, or that there is some time delay τ in its response, and that we require

$$\mathbf{x}(t) = A\frac{d^n}{dt^n}\mathbf{E}(t - \tau).$$

Is a monochromatic field still the unique driving field for which this more general condition holds?

In our elementary introduction to dielectrics, we have mostly assumed that the atoms form a continuum. More generally (and realistically), the space between atoms leads us to distinguish between the *local* electric field at any point and the average electric field implicit in our discussion thus far. Some other aspects of dielectric media are addressed in later chapters. At this point, we turn our attention to the energy and the linear momentum of optical fields in dielectrics.

1.6 Electromagnetic Energy and Intensity in Dielectrics

Consider a wave of frequency ω (electric and magnetic fields equal to the real parts of $\mathbf{E} = \mathbf{E}_\omega(\mathbf{r})\exp(-i\omega t)$ and $\mathbf{H} = \mathbf{H}_\omega(\mathbf{r})\exp(-i\omega t)$) propagating in a medium with permittivity $\epsilon(\omega)$. Assume that the magnetic permeability $\mu(\omega) \cong \mu_0$, which is an excellent approximation at optical frequencies. For field frequencies far removed from any absorption resonance of the medium, the cycle-averaged energy density u is

$$u = \frac{1}{4}\left[\frac{d}{d\omega}(\epsilon\omega)|\mathbf{E}_\omega|^2 + \mu_0|\mathbf{H}_\omega|^2\right]. \tag{1.6.1}$$

We will show that, in spite of the approximations made in the standard derivation of this energy density (as in *Landau, Lifshitz, and Pitaevskii* or *Jackson*, for example—see also Section 3.4), (1.6.1) is *exact* provided that (1) the response of the atoms to the field is well described by the oscillator (Lorentz) model, and (2) there is no absorption at frequency ω.

We again model the dielectric as a collection of N electric dipole oscillators per unit volume, each having the natural oscillation frequency ω_0 and satisfying the equation of motion (1.5.13). The field is assumed for definiteness to be linearly polarized, and the components of $\mathbf{x}$ and $\mathbf{E}$ along the polarization direction are denoted by x and E:

$$\ddot{x} + \omega_0^2 x = \frac{e}{m}E. \tag{1.6.2}$$

The polarization density and the permittivity are, respectively,

$$P = Nex = \frac{Ne^2/m}{\omega_0^2 - \omega^2} \tag{1.6.3}$$

and

$$\epsilon(\omega) = \epsilon_0 \left(1 + \frac{Ne^2/m\epsilon_0}{\omega_0^2 - \omega^2}\right) = \epsilon_0 \left(1 + \frac{\omega_p^2}{\omega_0^2 - \omega^2}\right). \tag{1.6.4}$$

Here, $\omega_p = (Ne^2/m\epsilon_0)^{1/2}$ is the plasma frequency, so called because the refractive index for free (unbound) electrons is given (approximately) by (1.5.22), with $\omega_0 = 0$: $n^2(\omega) \cong 1 - \omega_p^2/\omega^2$.

Recall Poynting's theorem:

$$\oint \mathbf{S} \cdot \hat{n} da = -\int_V \left[\mathbf{E} \cdot \frac{\partial \mathbf{D}}{\partial t} + \mu_0 \mathbf{H} \cdot \frac{\partial \mathbf{H}}{\partial t}\right] dV$$

$$= -\int_V \left[\frac{1}{2}\frac{\partial}{\partial t}(\epsilon_0 \mathbf{E}^2 + \mu_0 \mathbf{H}^2) + \mathbf{E} \cdot \frac{\partial \mathbf{P}}{\partial t}\right] dV. \tag{1.6.5}$$

The integral of the normal component of $\mathbf{S} = \mathbf{E} \times \mathbf{H}$ on the left side is, as usual, over a surface enclosing a volume V. From (1.6.2),

$$\mathbf{E} \cdot \frac{\partial \mathbf{P}}{\partial t} = \frac{m}{e}(\ddot{x} + \omega_0^2 x) \cdot Ne\dot{x} = N\frac{\partial}{\partial t}\left(\frac{1}{2}m\dot{x} + \frac{1}{2}m\omega_0^2 x^2\right), \tag{1.6.6}$$

and therefore

$$\oint \mathbf{S} \cdot \hat{n} da = -\int \dot{u} dV, \tag{1.6.7}$$

$$u = \frac{1}{2}\epsilon_0 E^2 + \frac{1}{2}\mu_0 H^2 + N\left(\frac{1}{2}m\dot{x}^2 + \frac{1}{2}m\omega_0^2 x^2\right). \tag{1.6.8}$$

Here, u is obviously the *total* energy density, that of the field plus that of the medium. Using

$$x = \frac{e/m}{\omega_0^2 - \omega^2} E_\omega \cos \omega t, \quad \dot{x} = -\frac{\omega e/m}{\omega_0^2 - \omega^2} E_\omega \sin \omega t, \tag{1.6.9}$$

and (1.6.4) for a monochromatic field $E_\omega \cos \omega t$, we find after cycle averaging that

$$u = \frac{1}{4}\epsilon_0 E_\omega^2 + \frac{1}{4}\mu_0 H_\omega^2 + \frac{Ne^2}{4m}\frac{\omega_0^2 + \omega^2}{(\omega_0^2 - \omega^2)^2} E_\omega^2$$

$$= \frac{1}{4}\epsilon_0 \left[1 + \frac{\omega_p^2}{\omega_0^2 - \omega^2} + \frac{2\omega^2\omega_p^2}{(\omega_0^2 - \omega^2)^2}\right] E_\omega^2 + \frac{1}{4}\mu_0 H_\omega^2$$

$$= \frac{1}{4}\left[\epsilon(\omega) + \omega\frac{d\epsilon}{d\omega}\right] E_\omega^2 + \frac{1}{4}\mu_0 H_\omega^2, \tag{1.6.10}$$

confirming that (1.6.1) defines the total energy density, that of the field *and* the medium.[23]

---

[23] The parts of the solutions for $x(t)$ and $\dot{x}(t)$ that oscillate at frequency ω_0 and are not driven by the applied field may be ignored on the grounds that they are eventually wiped out by dissipative effects that we have assumed to be otherwise negligible.

From $H_\omega^2 = (\epsilon/\mu_0)E_\omega^2$, we obtain

$$u = \frac{1}{2}\left(\epsilon + \frac{1}{2}\omega\frac{d\epsilon}{d\omega}\right)E_\omega^2 = \frac{1}{2}n(n + \omega\frac{dn}{d\omega})\epsilon_0 E_\omega^2 = \frac{1}{2}nn_g\epsilon_0 E_\omega^2 \tag{1.6.11}$$

for the energy density, and

$$S = \frac{1}{2}\sqrt{\frac{\epsilon}{\mu_0}}E_\omega^2 \tag{1.6.12}$$

for the magnitude of the (cycle-averaged) Poynting vector. According to the latter equation, the field intensity I is related to the square of the electric field strength by

$$I = S = \frac{\epsilon_0}{2}\sqrt{\frac{\epsilon}{\epsilon_0}}\sqrt{\frac{1}{\epsilon_0\mu_0}}E_\omega^2 = \frac{1}{2}nc\epsilon_0 E_\omega^2. \tag{1.6.13}$$

Together with (1.6.11), this implies

$$I = \frac{cu}{n + \omega dn/d\omega} = v_g u, \tag{1.6.14}$$

where

$$v_g = \frac{c}{n + \omega dn/d\omega} = c/n_g \tag{1.6.15}$$

is the *group velocity* and n_g is the *group index*.

Subject to the conditions (1) and (2) stated earlier, these results are exact. We can take a more general approach, which will be valid for *non-monochromatic* fields but still restricted to a linear, homogeneous, isotropic, and non-absorbing medium. In this approach, we do not rely on any particular model for the medium, and simply use the definition of the (real) permittivity $\epsilon(\omega)$ to write

$$\mathbf{D}(\mathbf{r}, t) = \int_{-\infty}^{\infty} d\omega' \mathbf{D}(\mathbf{r}, \omega')e^{-i\omega't} = \int_{-\infty}^{\infty} d\omega' \epsilon(\omega')\mathbf{E}(\mathbf{r}, \omega')e^{-i\omega't} \tag{1.6.16}$$

and

$$\mathbf{E} \cdot \frac{\partial \mathbf{D}}{\partial t} = -i\int_{-\infty}^{\infty} d\omega' \int_{-\infty}^{\infty} d\omega'' \omega' \epsilon(\omega')\mathbf{E}(\mathbf{r}, \omega') \cdot \mathbf{E}^*(\mathbf{r}, \omega'')e^{-i(\omega'-\omega'')t}. \tag{1.6.17}$$

We have used the relation $\mathbf{E}(\mathbf{r}, -\omega'') = \mathbf{E}^*(\mathbf{r}, \omega'')$, as required for $\mathbf{E}(\mathbf{r}, t)$ to be real. From (1.6.5) and (1.6.7), the rate of change of the total energy density is

$$\begin{aligned}
\frac{\partial u}{\partial t} &= \mathbf{E} \cdot \frac{\partial \mathbf{D}}{\partial t} + \mu_0\mathbf{H} \cdot \frac{\partial \mathbf{H}}{\partial t} \\
&= -i\int_{-\infty}^{\infty} d\omega' \int_{-\infty}^{\infty} d\omega'' \omega' \epsilon(\omega')\mathbf{E}(\mathbf{r}, \omega') \cdot \mathbf{E}^*(\mathbf{r}, \omega'')e^{-i(\omega'-\omega'')t} \\
&\quad + \frac{1}{2}\mu_0\frac{\partial}{\partial t}(\mathbf{H}^2).
\end{aligned} \tag{1.6.18}$$

Integrating, we identify

$$u(\mathbf{r}) = \int_{-\infty}^{\infty} d\omega' \int_{-\infty}^{\infty} d\omega'' \frac{\omega' \epsilon(\omega')}{\omega' - \omega''} \mathbf{E}(\mathbf{r}, \omega') \cdot \mathbf{E}^*(\mathbf{r}, \omega'') e^{-i(\omega' - \omega'')t} + \frac{1}{2}\mu_0 \mathbf{H}^2. \quad (1.6.19)$$

We are only interested in the general form of u, so we ignore the constant of integration; if the field vanishes at $t = -\infty$, we can take this constant to be 0. Next, interchange the integration variables ω' and ω'' and make the change of variables $\omega' \to -\omega'$, $\omega'' \to -\omega''$. In the process, we can take $\epsilon(-\omega'') = \epsilon(\omega'')$; this follows because (a) $\mathbf{D}(\mathbf{r}, t)$ is real and therefore $\epsilon^*(\omega) = \epsilon(-\omega)$ (see (1.6.16)), and (b) we are assuming that there is no absorption and therefore that $\epsilon(\omega'')$ is purely real. Since u is real and therefore equal to its complex conjugate,

$$u(\mathbf{r}) = -\int_{-\infty}^{\infty} d\omega' \int_{-\infty}^{\infty} d\omega'' \frac{\omega'' \epsilon(\omega'')}{\omega' - \omega''} \mathbf{E}(\mathbf{r}, \omega') \cdot \mathbf{E}^*(\mathbf{r}, \omega'') e^{-i(\omega' - \omega'')t} + \frac{1}{2}\mu_0 \mathbf{H}^2. \quad (1.6.20)$$

We now write u as $1/2$ the sum of (1.6.19) and (1.6.20) to arrive at a general expression for the total energy density when absorption is negligible:

$$u(\mathbf{r}) = \frac{1}{2} \int_{-\infty}^{\infty} d\omega' \int_{-\infty}^{\infty} d\omega'' \frac{\omega' \epsilon(\omega') - \omega'' \epsilon(\omega'')}{\omega' - \omega''}$$
$$\times \mathbf{E}(\mathbf{r}, \omega') \cdot \mathbf{E}^*(\mathbf{r}, \omega'') e^{-i(\omega' - \omega'')t} + \frac{1}{2}\mu_0 \mathbf{H}^2. \quad (1.6.21)$$

This shows, incidentally, that the expression (1.6.19) for u is not divergent, as might appear to be so at first glance because of the term $1/(\omega' - \omega'')$ in the integrand.

For a monochromatic field,

$$\mathbf{E}(\mathbf{r}, \omega') = \frac{1}{2}\mathbf{E}_\omega(\mathbf{r})[\delta(\omega' - \omega) + \delta(\omega' + \omega)]. \quad (1.6.22)$$

($\mathbf{E}_\omega(\mathbf{r})$ is assumed to be real, just for simplicity.) Putting this in (1.6.21), cycle averaging it, and using

$$\lim_{\omega' \to \omega} \frac{\omega \epsilon(\omega) - \omega' \epsilon(\omega')}{\omega - \omega'} = \frac{d}{d\omega}(\epsilon\omega), \quad (1.6.23)$$

we obtain again exactly (1.6.11).

The expression for the energy density when absorption is negligible, shown in (1.6.11), is generally valid only when the field is nearly monochromatic. But if different frequency components of the field are uncorrelated, we can effectively take

$$\mathbf{E}(\mathbf{r}, \omega') \cdot \mathbf{E}^*(\mathbf{r}, \omega'') \to \frac{1}{2}|\tilde{\mathbf{E}}(\mathbf{r}, \omega')|^2 [\delta(\omega' - \omega'') + \delta(\omega' + \omega'')] \quad (1.6.24)$$

in (1.6.21). Then, if we write $\mathbf{H}(\mathbf{r}, t)$ as a Fourier integral analogous to (1.6.16), and use (1.6.23) again, we can replace (1.6.21) by

$$u(\mathbf{r}) = \frac{1}{4} \int_{-\infty}^{\infty} d\omega \left[\frac{d}{d\omega}(\epsilon\omega)|\tilde{\mathbf{E}}(\mathbf{r}, \omega)|^2 + \mu_0|\tilde{\mathbf{H}}(\mathbf{r}, \omega)|^2 \right], \quad (1.6.25)$$

which is to be regarded as an average over field fluctuations such that different frequency components are uncorrelated and

$$[\mathbf{E}^2(\mathbf{r},t)]_{\text{avg}} = \int_{-\infty}^{\infty} d\omega |\tilde{\mathbf{E}}(\mathbf{r},\omega)|^2. \tag{1.6.26}$$

For such fluctuations, (1.6.25) is exact, to the extent that absorption is negligible.

Since $\mu_0|\mathbf{H}(\mathbf{r},\omega)|^2 = \epsilon(\omega)|\mathbf{E}(\mathbf{r},\omega)|^2$ for a dielectric medium, (1.6.25) is equivalent to

$$
\begin{aligned}
u(\mathbf{r}) &= \frac{1}{4} \int_{-\infty}^{\infty} d\omega \left[2\epsilon(\omega) + \omega\frac{d\epsilon}{d\omega}\right] |\tilde{\mathbf{E}}(\mathbf{r},\omega)|^2 \\
&= \frac{1}{2} \int_{0}^{\infty} d\omega \left[2\epsilon(\omega) + \omega\frac{d\epsilon}{d\omega}\right] |\tilde{\mathbf{E}}(\mathbf{r},\omega)|^2,
\end{aligned}
\tag{1.6.27}
$$

where we have used the fact that $\epsilon(-\omega) = \epsilon(\omega)$ when there is no absorption.

To allow for a magnetic susceptibility $\mu(\omega)$ that differs from μ_0, we simply replace μ_0 by $(d/d\omega)(\mu\omega)$ in the expression (1.6.25) for the energy density.

The group index n_g can be negative. But under our assumption that absorption is negligible at frequencies ω of interest, n_g is, in fact, positive for any passive medium. This follows from the identity

$$n_g = \frac{1}{2}\sqrt{\frac{\epsilon}{\epsilon_0}} + \frac{1}{2}\sqrt{\frac{\epsilon_0}{\epsilon}}\frac{d}{d\omega}\left(\frac{\omega\epsilon}{\epsilon_0}\right). \tag{1.6.28}$$

According to the Kramers–Kronig dispersion relation between the real (ϵ_R) and imaginary (ϵ_I) parts of the permittivity (Section 6.6.1),

$$\epsilon_R(\omega) - \epsilon_0 = \frac{2}{\pi} \int_0^{\infty} \frac{\omega'\epsilon_I(\omega')}{\omega'^2 - \omega^2} d\omega'. \tag{1.6.29}$$

In the usual form of this relation, the integral on the right is the Cauchy principal part, but since $\epsilon_I(\omega) = 0$ by our assumption of no absorption at frequency ω, what appears in (1.6.29) is an ordinary integral. Therefore,

$$\frac{d\epsilon_R}{d\omega} = \frac{4\omega}{\pi} \int_0^{\infty} \frac{\omega'\epsilon_I(\omega')}{(\omega'^2 - \omega^2)^2} d\omega', \tag{1.6.30}$$

which is positive for any passive (non-amplifying) medium, that is, for any medium for which $\epsilon_I(\omega) > 0$ at all frequencies. It then follows from (1.6.28) with $\epsilon = \epsilon_R$ that $n_g > 0$. A similar proof can be given for negative-index media.

1.7 Electromagnetic Momentum

The momentum density of the electromagnetic field in free space is $(1/c^2)\mathbf{E} \times \mathbf{H}$, that is, $1/c^2$ times the Poynting vector. What is the momentum density when a field propagates in a dielectric medium? This question has been the subject of quite a large number of papers over quite a long period of time, much of the work focusing on the formal symmetry properties of the energy-momentum tensor. The two most common expressions by far for the momentum density in a dielectric are those of Minkowski and Abraham.[24] The Abraham momentum density $\mathbf{g}_A$ has the same form in terms of $\mathbf{E}$ and $\mathbf{H}$ as the momentum density of the field in free space, whereas the Minkowski momentum density $\mathbf{g}_M$ involves $\mathbf{D}$ and $\mathbf{B}$:

$$\mathbf{g}_A = \frac{1}{c^2}\mathbf{E} \times \mathbf{H} \quad \text{and} \quad \mathbf{g}_M = \mathbf{D} \times \mathbf{B}. \tag{1.7.1}$$

(We will continue to take the magnetic permeability μ to be equal to its vacuum value μ_0, so that $\mathbf{B} = \mu_0\mathbf{H}$.)

It is interesting to see what these momentum densities are for a single-photon field. For this purpose, we first return to (1.6.11) for the energy density:

$$u = \frac{1}{2}nn_g\epsilon_0 E_\omega^2. \tag{1.7.2}$$

In a volume V of uniform energy density, a single photon corresponds to an energy density $\hbar\omega/V$; equating this to (1.7.2) implies that, in effect,

$$E_\omega^2 = \frac{2\hbar\omega}{nn_g\epsilon_0 V} \tag{1.7.3}$$

for a single-photon field. This expression and $H_\omega^2 = (\epsilon/\mu_0)E_\omega^2 = n^2(\epsilon_0/\mu_0)E_\omega^2$ suggests, for instance, that the (cycle-averaged) Abraham momentum density for a single-photon field is

$$p_A = \frac{1}{2}\frac{n}{c}\sqrt{\frac{\epsilon_0}{\mu_0}}E_\omega^2 = \frac{1}{n_g}\frac{\hbar\omega}{c}\frac{1}{V}. \tag{1.7.4}$$

Photon momentum densities are frequently considered without regard to dispersion, that is, with $dn/d\omega = 0$ and, therefore, $n_g = n$. Then, the expressions above for E_ω^2 and H_ω^2 imply

$$p_A = \frac{1}{n}\frac{\hbar\omega}{c}\frac{1}{V} \quad \text{(no dispersion)}, \tag{1.7.5}$$

$$p_M = n\frac{\hbar\omega}{c}\frac{1}{V} \quad \text{(no dispersion)}. \tag{1.7.6}$$

Two examples of momentum exchange between light and matter are instructive. The first is based on an argument of Fermi's that the Doppler effect is a consequence of this momentum exchange. Consider an atom of mass M inside a dielectric medium with refractive index $n(\omega)$. The atom has a sharply defined transition frequency ω_0 and is initially moving with velocity v away from a source of light of frequency ω

[24] H. Minkowski, Nachr. Ges. Wiss. Goettingen, Math. Phys. Kl. 53–111 (1908) and Math. Ann. **68**, 472–525 (1910); M. Abraham, Rend. Circ. Mat. Palermo **30**, 33–46 (1910).

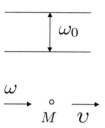

Fig. 1.8 An atom of mass M with a transition frequency ω_0 is moving with velocity v away from a source of light of frequency ω. The atom is inside a dielectric medium with refractive index $n(\omega)$. Because of the Doppler effect, absorption by the atom requires that $\omega \cong \omega_0(1 + nv/c)$. What does this say about photon momentum?

(see Figure 1.8). Because the light in the atom's reference frame has a Doppler-shifted frequency $\omega(1 - nv/c)$ determined by the phase velocity (c/n) of light in the medium, the atom can absorb a photon if $\omega(1 - nv/c) = \omega_0$, or

$$\omega \cong \omega_0(1 + nv/c). \tag{1.7.7}$$

We denote the momentum of a photon in the medium by $\wp$ and consider the implications of (non-relativistic) energy and momentum conservation. The initial energy is $E_i = \hbar\omega + \frac{1}{2}Mv^2$, and the final energy, after the atom has absorbed a photon, is $\frac{1}{2}Mv'^2 + \hbar\omega_0$, where v' is the velocity of the atom after absorption. The initial momentum is $\wp + Mv$, and the final momentum is just Mv'. Therefore,

$$\frac{1}{2}M(v'^2 - v^2) \cong Mv(v' - v) = Mv(\wp/M) = \hbar(\omega - \omega_0), \tag{1.7.8}$$

or

$$\omega \cong \omega_0 + \frac{\wp v}{\hbar}. \tag{1.7.9}$$

From (1.7.7) and $\omega \cong \omega_0$, we conclude that

$$\wp = n\frac{\hbar\omega}{c}. \tag{1.7.10}$$

In our second example, we consider a block of mass M, refractive index n, and thickness a, initially sitting at rest on a frictionless surface (see Figure 1.9). A single-photon pulse of frequency ω passes through the block, which is assumed to be non-absorbing at frequency ω and to have anti-reflection coatings on its front and back surfaces. If the photon momentum is $\wp_{\text{in}}$ inside the block and $\wp_{\text{out}}$ outside, the block picks up a momentum $MV = \wp_{\text{out}} - \wp_{\text{in}}$ when the pulse enters. If the space outside the block is vacuum, $\wp_{\text{out}} = mc$, where we define a mass $m = E/c^2 = \hbar\omega/c^2$. Similarly, $\wp_{\text{in}} = mv$, where v is the velocity of light in the block. Without dispersion, $v = c/n$

Fig. 1.9 A block with refractive index n on a frictionless surface and initially at rest is displaced when it transmits a photon.

and the momentum of the photon in the block is evidently $\wp_{in} = mc/n = \hbar\omega/nc$. The effect of dispersion is to replace $v = c/n$ by $v_g = c/n_g$ and $\wp_{in} = \hbar\omega/nc$ by $\wp_{in} = \hbar\omega/n_g c$. With or without dispersion, the photon momentum in this example has the Abraham form. An essential feature of this simplified argument is that the velocity of light in the medium is the phase velocity $v = c/n$ (or, more generally, the group velocity v_g). This, together with momentum conservation, is what leads to the conclusion that the field momentum has the Abraham form.[25]

It is important to distinguish between the *field* momentum and the momentum p_{med} imparted to the medium. In the example shown in Figure 1.9, the total initial momentum is $\hbar\omega/c$, and, after the single-photon field propagates into the block, the total momentum, neglecting dispersion, is $(\hbar\omega/c)(1/n + p_{med})$, and so the block recoils with momentum $p_{med} = (\hbar\omega/c)(1 - 1/n)$ in the direction of propagation of the field.[26] After the field has propagated out of the block, the total momentum must again be $\hbar\omega/c$ by momentum conservation, and so the block must again be at rest.

This can, in principle, be tested experimentally. Conservation of momentum requires that $MV = m(c - v)$. When the pulse exits the block, the block recoils and comes to rest, and is left with the net displacement

$$\Delta x = V\Delta t = \frac{m}{M}(c - v)\frac{a}{v} = \frac{\hbar\omega}{Mc^2}(n_g - 1)a \qquad (1.7.11)$$

as a result of the light having passed through it. If, however, the photon momentum inside the block were assumed to be $n\hbar\omega/c$, for example, the displacement of the block would in similar fashion be predicted to be

$$\Delta x = \frac{\hbar\omega}{Mc^2}n_g(1 - n)a. \qquad (1.7.12)$$

These different assumptions about the photon momentum can lead to different predictions not only for the magnitude of the block displacement but also for its direction.

---

[25] N. L. Balazs, Phys. Rev. **91**, 408 (1953).

[26] The momentum p_{med} has been calculated directly and fully quantum mechanically, based on the Lorentz force, by R. Loudon, J. Mod. Opt. **49**, 821 (2002); see (1.8.25). If $n > 1$, the incoming field pushes the block in the direction of propagation of the field, a conclusion opposite to that reached by J. H. Poynting, in Phil. Mag. **9**, 393 (1905). See also R. Loudon and C. Baxter, in Proc. R. Soc. **468**, 1825 (2012).

The displacement (1.7.11) (with $n = n_g$) appears in a delightful article by Frisch entitled "Take a Photon . . ."[27] Balazs used this example earlier to argue that the momentum of a photon in a dielectric is $\hbar\omega/nc$.[28] His analysis, based on the conservation of momentum and the center-of-gravity theorem, is reminiscent of Einstein's argument in his paper entitled "The Principle of Conservation of Motion of the Center of Gravity and the Inertia of Energy."[29] Suppose we replace the block in Figure 1.9 by an empty box $(n = 1)$ of mass M and length a and that some source at the left side of the box emits at time $t = 0$ a pulse of light of energy E and momentum $p = E/c$. Conservation of momentum implies that the initially stationary box recoils with a velocity $v = -E/Mc$. The light pulse is assumed to be completely absorbed when it reaches the right side of the box at time $t = a/c$. The box is then again at rest, but has been displaced by $\Delta x = vt = -Ea/Mc^2$ from its initial position. Since the system is isolated, the center-of-gravity theorem requires that its center of mass has not moved. This suggests that the light has an inertial mass m such that the change $ma + M\Delta x$ of the center of mass of the system is zero, or, in other words, that $E = mc^2$.

Our first example suggests at first thought that the momentum of the photon is $n\hbar\omega/c$, while the second example seems to show that it is $\hbar\omega/n_g c$. Consider more carefully the first example. There is little doubt that the Doppler shift is $nv\omega/c$, regardless of dispersion, as we have assumed, but does this imply that the momentum of a photon in a dielectric is $n\hbar\omega/c$? We will show below that the forces exerted by a plane monochromatic wave on the polarizable particles of a dielectric result in a momentum density of magnitude

$$p_{\text{med}} = \frac{\epsilon_0}{2c}n(nn_g - 1)E_\omega^2 = (n - \frac{1}{n_g})\frac{\hbar\omega}{c}\frac{1}{V}; \qquad (1.7.13)$$

the second equality applies to a single photon, and follows from (1.7.3). Now, since the Doppler shift implies that an absorber (or emitter) inside a dielectric recoils with momentum $n\hbar\omega/c$, all we can safely conclude from momentum conservation is that a momentum $n\hbar\omega/c$ is taken from (or given to) the *combined system of field and dielectric*. Given that the medium has a momentum density (1.7.13) due to the force exerted on it by the propagating field, we can attribute to the field (by conservation of momentum) a momentum density

$$n\frac{\hbar\omega}{c}\frac{1}{V} - p_{\text{med}} = \frac{1}{n_g}\frac{\hbar\omega}{c}\frac{1}{V} = p_A. \qquad (1.7.14)$$

That is, the momentum of the field in this interpretation is given by the Abraham formula, consistent with the conclusion of the thought experiment shown in Figure 1.9. The momentum $n\hbar\omega/c$ evidently gives the momentum not of the field as such but of the combined system of field plus dielectric. It is the momentum density equal to the *total* energy density $u = \hbar\omega/V$ for a monochromatic field divided by the phase velocity c/n of the propagating wave. Experiments on the recoil of objects immersed

[27] O. R. Frisch, Contemp. Phys. **7**, 45 (1965).

[28] N. L. Balazs, Phys. Rev. **91**, 408 (1953).

[29] A. Einstein, Ann. Physik. **20**, 627 (1906).

in dielectric media have, in fact, shown that the recoil momentum is $n\hbar\omega/c$ per unit of energy $\hbar\omega$ of the field and is independent of the group index n_g, just as in the Doppler effect. But this does not necessarily imply that $n\hbar\omega/c$ is the momentum of a photon of light in the medium. Regardless of how this momentum is apportioned between the field and the medium, the important thing for the theory is that it correctly predicts the *observable forces* exerted by electromagnetic fields. We next turn our attention specifically to forces on polarizable particles in applied electromagnetic fields.

Radiation pressure was first demonstrated by Lebedev in experiments in which small vanes with blackened and polished surfaces were hung from a torsion fiber in an evacuated jar and were deflected when they reflected light.[30] The effect had been predicted earlier by Maxwell, and independently by A. Bartoli. Independently of Lebedev, Nichols and Hull carried out a series of experiments in which they carefully subtracted out the force on the mirror due to effects of the ambient gas and quantitatively confirmed the Maxwell–Bartoli theory "within the probable errors of observation."[31]

1.8   Forces and Momenta

It is an excellent approximation for many purposes in optical physics to regard an atom as an electrically polarizable particle such that the induced electric dipole moment in a field of frequency ω is $\mathbf{d} = \alpha(\omega)\mathbf{E}_\omega \exp(-i\omega t)$. The polarizability $\alpha(\omega)$ may be assumed to be independent of the applied field and real if the applied field is not too strong and if absorption and scattering processes by the atom are negligible, as is typically the case when the field frequency ω is far from any resonance frequency of the atom. With these assumptions, we now consider the forces acting on atoms in quasi-monochromatic fields.

We begin with the force

$$\mathbf{F} = (\mathbf{d} \cdot \nabla)\mathbf{E} + \dot{\mathbf{d}} \times \mathbf{B} = (\mathbf{d} \cdot \nabla)\mathbf{E} + \mathbf{d} \times (\nabla \times \mathbf{E}) + \frac{\partial}{\partial t}(\mathbf{d} \times \mathbf{B}) = \mathbf{F}_1 + \mathbf{F}_2 \quad (1.8.1)$$

on an electric dipole in the plane-wave electric field

$$\mathbf{E} = \mathbf{E}_0(t)e^{-i(\omega t - \mathbf{k}\cdot\mathbf{r})}, \quad (1.8.2)$$

where we have defined $\mathbf{F}_1 = (\mathbf{d} \cdot \nabla)\mathbf{E} + \mathbf{d} \times (\nabla \times \mathbf{E})$, and $\mathbf{F}_2 = \partial(\mathbf{d} \times \mathbf{B})/\partial t$. Changing notation slightly and writing

$$\mathbf{E} = \mathbf{E}_0(\mathbf{r}, t)e^{-i\omega t} = e^{-i\omega t}\int_{-\infty}^{\infty} d\Omega\, \tilde{\mathbf{E}}_0(\mathbf{r}, \Omega)e^{-i\Omega t}, \quad (1.8.3)$$

in which $|\partial\mathbf{E}_0/\partial t| \ll \omega|\mathbf{E}_0|$ for a quasi-monochromatic field, we can approximate $\mathbf{d}$ as follows:

[30] P. N. Lebedev, Ann. Physik. **6**, 433 (1901).

[31] E. F. Nichols and G. F. Hull, Phys. Rev. **17**, 26 (1903); Astrophys. J. **57**, 315 (1903). Nichols and Hull reported that their experiments confirmed the theory to an accuracy of 1%, but further experiments and analyses showed that it was more like 10%. See M. Bell and S. E. Green, Proc. Phys. Soc. **45**, 320 (1933), and G. F. Hull, M. Bell, and S. E. Green, Proc. Phys. Soc. **46**, 589 (1934). I thank Dr. S. G. Lukishova for bringing this point to my attention.

$$\mathbf{d}(\mathbf{r},t) = \int_{-\infty}^{\infty} d\Omega \, \alpha(\omega+\Omega)\tilde{\mathbf{E}}_0(\mathbf{r},\Omega)e^{-i(\omega+\Omega)t}$$

$$\cong \int_{-\infty}^{\infty} d\Omega \, [\alpha(\omega) + \Omega\alpha'(\omega)]\tilde{\mathbf{E}}_0(\mathbf{r},\Omega)e^{-i(\omega+\Omega)t}$$

$$= \alpha(\omega)\mathbf{E}(\mathbf{r},t) + i\alpha'(\omega)\frac{\partial\mathbf{E}_0}{\partial t}e^{-i\omega t}. \tag{1.8.4}$$

Here, $\alpha' = d\alpha/d\omega$, and it is assumed that dispersion is sufficiently weak that terms $d^m\alpha/d\omega^m$ can be neglected for $m \geq 2$. Putting (1.8.4) into (1.8.1), we obtain, after some straightforward manipulations and cycle averaging,

$$\mathbf{F}_1 = (\mathbf{d}\cdot\nabla)\mathbf{E} + \mathbf{d}\times(\nabla\times\mathbf{E}) = \nabla\left[\frac{1}{4}\alpha(\omega)|\mathbf{E}_0|^2\right] + \frac{1}{4}\alpha'(\omega)\mathbf{k}\frac{\partial}{\partial t}|\mathbf{E}_0|^2. \tag{1.8.5}$$

Since the refractive index n for a sufficiently dilute medium is given by $n^2-1 = N\alpha/\epsilon_0$, N being the density of dipoles in the dielectric, $\alpha' = (2n\epsilon_0/N)(dn/d\omega)$, and

$$\mathbf{F}_1 = \nabla\left[\frac{1}{4}\alpha(\omega)|\mathbf{E}_0|^2\right] + \frac{\epsilon_0}{2N}\mathbf{k}n\frac{dn}{d\omega}\frac{\partial}{\partial t}|\mathbf{E}_0|^2. \tag{1.8.6}$$

The first term is the *dipole force* associated with the energy $W = -\frac{1}{2}\alpha(\omega)\mathbf{E}^2$ involved in inducing an electric dipole moment in an electric field:

$$W = -\int_0^{\mathbf{E}} \mathbf{d}\cdot d\mathbf{E}' = -\alpha(\omega)\int_0^{\mathbf{E}} \mathbf{E}'\cdot d\mathbf{E}' = -\frac{1}{2}\alpha(\omega)\mathbf{E}^2. \tag{1.8.7}$$

The second term in (1.8.6) is non-vanishing only because of dispersion ($dn/d\omega \neq 0$), is in the direction of propagation of the field, and implies for a (uniform) density N a momentum density of magnitude

$$g_D = \frac{1}{2}\epsilon_0 n^2 \frac{dn}{d\omega}\frac{\omega}{c}|\mathbf{E}_0|^2 = \frac{1}{2}\frac{\epsilon_0}{c}n^2(n_g-n)|\mathbf{E}_0|^2, \tag{1.8.8}$$

since $k = n(\omega)\omega/c$. The force $\mathbf{F}_2 = (\partial/\partial t)(\mathbf{d}\times\mathbf{B})$ in (1.8.1) may be evaluated similarly:

$$\mathbf{F}_2 = \frac{1}{2}\alpha(\omega)\frac{\mathbf{k}}{\omega}\frac{\partial}{\partial t}|\mathbf{E}_0|^2, \tag{1.8.9}$$

when we use $\mathbf{k}\cdot\mathbf{E}_0 = 0$ and the assumption that $|\dot{\mathbf{E}}_0| \ll \omega|\mathbf{E}_0|$; $\mathbf{F}_2$ implies that the following momentum density is imparted to the medium:

$$\mathbf{g}^A = N\mathbf{d}\times\mathbf{B} = \frac{1}{2}\epsilon_0(n^2-1)\frac{\mathbf{k}}{\omega}|\mathbf{E}_0|^2. \tag{1.8.10}$$

The magnitude of the total momentum density in the medium due to the force on the individual dipoles is therefore

$$g_D + g^A = \frac{\epsilon_0}{2c}\left[n^2(n_g-n) + n(n^2-1)\right]|\mathbf{E}_0|^2 = \frac{\epsilon_0}{2c}n(nn_g-1)|\mathbf{E}_0|^2 \tag{1.8.11}$$

in the approximation that the field is sufficiently uniform that we can ignore the dipole force $\nabla(\frac{1}{4}\alpha|\mathbf{E}_0|^2)$.

The complete momentum density for the field and the medium is obtained by adding to (1.8.11) the Abraham momentum density g_A of the field. According to (1.7.1), $g_A = (\epsilon_0/2c)n|\mathbf{E}_0|^2$, so that the total momentum density is

$$g_A + g_D + g^A = \frac{\epsilon_0}{2c}[n + n(nn_g - 1)]|\mathbf{E}_0|^2 = \frac{\epsilon_0}{c}n^2 n_g|\mathbf{E}_0|^2. \qquad (1.8.12)$$

To express these results in terms of single photons, we again replace $|\mathbf{E}_0|^2$ by its photon counterpart, $2\hbar\omega/(\epsilon_0 nn_g V)$; then, (1.8.12) takes the form

$$p_A + p_D + p^A = n\frac{\hbar\omega}{c}\frac{1}{V}, \qquad (1.8.13)$$

consistent with the discussion in Section 1.7. This is the total momentum density per photon, assuming that the dipole force is negligible. The momentum density of the medium per photon of the field follows from (1.8.11):

$$p_{\text{med}} = p_D + p^A = \frac{\epsilon_0}{2c}n(nn_g - 1)\frac{2\hbar\omega}{nn_g\epsilon_0 V} = (n - \frac{1}{n_g})\frac{\hbar\omega}{c}\frac{1}{V}, \qquad (1.8.14)$$

as stated earlier (see (1.7.13)).

Consider, for example, spontaneous emission by a guest atom in a host dielectric medium: the atom loses energy $\hbar\omega$ and recoils with a linear momentum of magnitude $n\hbar\omega/c$, independent of the group index n_g and consistent with (1.8.13) and momentum conservation.[32]

The rate of change of the momentum density $\mathbf{g}^A$ is the force density

$$\mathbf{f}^A = N\frac{\partial}{\partial t}(\mathbf{d} \times \mathbf{B}) = N\alpha(\omega)\mu_0\frac{\partial}{\partial t}(\mathbf{E} \times \mathbf{H}) = \frac{1}{c^2}[n^2(\omega) - 1]\frac{\partial}{\partial t}(\mathbf{E} \times \mathbf{H}), \quad (1.8.15)$$

when dispersion is ignored; this is the so-called *Abraham force density*. The difference between the Minkowski and Abraham momentum densities is $\mathbf{g}^A$: $\mathbf{g}_M = \mathbf{g}_A + \mathbf{g}^A$.

The momentum density (1.8.8) is attributable directly to the second term on the right side of (1.8.4), that is, to the part of the induced dipole moment that arises from dispersion. This dispersive contribution is obviously a general property of induced dipole moments in applied fields.

The expression (1.8.1) for the force on a single polarizable particle can be generalized to allow for absorption by the particle, in which case $\alpha(\omega)$ is complex. One obtains

$$F_i = \frac{1}{2}\text{Re}\sum_{j=1}^{3}\left[\alpha(\omega)\mathbf{E}_{0j}\frac{\partial \mathbf{E}_{0j}^*}{\partial x_i}\right] \qquad (1.8.16)$$

for the ith component of the cycle-averaged force in a field of frequency ω.[33] The absorption and particle recoil associated with the imaginary part of the polarizability,

[32] P. W. Milonni and R. W. Boyd, Laser Phys. **15**, 1 (2005).

[33] P. C. Chaumet and M. Nieto-Vesperinas, Opt. Lett. **25**, 1065 (2000).

$\alpha_I(\omega)$, is the basis for laser cooling and trapping of atoms.[34] In Section 1.10, we relate the force proportional to $\alpha_I(\omega)$ in (1.8.16) to the "scattering force" that accompanies Rayleigh scattering.

Exercise 1.4: (a) Derive (1.8.16). What assumptions are made in your derivation? (b) Consider electrons in a plane monochromatic wave of frequency ω. Under what approximations can we associate with the electrons a polarizability $\alpha(\omega) = -\omega_p^2/\omega^2$, where ω_p is the plasma frequency (see (1.6.4))? (c) How are the expressions for the polarizability and the force (1.8.16) changed if we include a dissipative force $-b\dot{\mathbf{x}}$ in the Newton equation of motion for each electron?

We have mostly ignored the (cycle-averaged) dipole force,

$$\mathbf{F}_{\text{dipole}} = \nabla\left[\frac{1}{4}\alpha(\omega)|\mathbf{E}_0|^2\right], \tag{1.8.17}$$

because it vanishes in our idealized model in which the electric field is defined by (1.8.2). When the field is not well approximated by (1.8.2), the dipole force on atoms must of course be taken into account. It implies, for example, that, in a laser beam with frequency ω and with a Gaussian transverse spatial profile, atoms with $\alpha(\omega) > 0$ will move toward the center of the beam, where the intensity is greatest. This occurs when ω is detuned to the "red" side of an atomic resonance (e.g. $\omega < \omega_0$ in (1.5.22)), as has been demonstrated experimentally.

The dipole force also comes into play when we consider a pulse of radiation. Consider, for example, a plane-wave pulse propagating with a group velocity v_g:

$$\mathbf{E}(z,t) = \mathbf{E}_0(t - z/v_g)e^{-i(\omega t - kz)}. \tag{1.8.18}$$

Then,

$$\mathbf{F}_{\text{dipole}} = \frac{\partial}{\partial z}\left[\frac{1}{4}\alpha(\omega)|\mathbf{E}_0|^2\right]\hat{\mathbf{z}} = \frac{-1}{v_g}\frac{\partial}{\partial t}\left[\frac{1}{4}\alpha(\omega)|\mathbf{E}_0|^2\right]\hat{\mathbf{z}}. \tag{1.8.19}$$

Multiplying this by the number density N, and using again the relation $n^2 - 1 = N\alpha/\epsilon_0$, we infer that a momentum density

$$g_d = -\frac{1}{4v_g}\epsilon_0[n^2(\omega) - 1]|\mathbf{E}_0|^2 \tag{1.8.20}$$

is taken up by the medium. On a per-photon basis, for a pulse of volume V, this is

$$p_d = -\frac{\epsilon_0}{4c}n_g(n^2 - 1)\frac{2\hbar\omega}{\epsilon_0 n n_g}\frac{1}{V} = -\frac{1}{2}\left(n - \frac{1}{n}\right)\frac{\hbar\omega}{c}\frac{1}{V}. \tag{1.8.21}$$

---

[34] For an elementary introduction to laser cooling and trapping, see, for instance, *Milonni and Eberly*.

Adding this to (1.8.14), which does not include any effect of the dipole force, gives the total momentum of the medium per photon:

$$p_{\text{med}} = (p_D + p^A + p_d)V = \left[\frac{1}{2}\left(n + \frac{1}{n}\right) - \frac{1}{n_g}\right]\frac{\hbar\omega}{c}. \qquad (1.8.22)$$

If dispersion is negligible ($n_g \cong n$),

$$p_{\text{med}} = \frac{1}{2}\left(n - \frac{1}{n}\right)\frac{\hbar\omega}{c}, \qquad (1.8.23)$$

as obtained by other approaches.[35] If we add to (1.8.22) the Abraham momentum density $\hbar\omega/cn_g V$ for the propagating field, we obtain a total momentum per photon

$$p_{\text{total}} = \frac{1}{2}\left(n + \frac{1}{n}\right)\frac{\hbar\omega}{c}. \qquad (1.8.24)$$

In a detailed, quantum-mechanical calculation that allows for absorption and an interface between two dielectrics, Loudon showed that, for a non-dispersive, transparent dielectric block, a single-photon pulse normally incident from a vacuum imparts the momentum

$$p = \frac{\hbar\omega}{c}(1 + R) = nT\left[\frac{1}{2}\left(n + \frac{1}{n}\right)\frac{\hbar\omega}{c}\right], \qquad (1.8.25)$$

where $R = (n - 1)^2/(n + 1)^2$ is the power reflection coefficient at the interface between the vacuum and the block of refractive index n.[36] Since $nT = 4n/(n + 1)^2$ is the fraction of the incident field intensity that goes into the medium, (1.8.25) might be expected from (1.8.24). However, Loudon's calculation requires no assumptions about the form of the photon momentum in a dielectric, and is based directly on the Lorentz force.

1.9   Stress Tensors

There are 500 papers arguing about which energy-momentum tensor is the right one to use inside a dielectric medium; 490 of them are idiotic. It's like if someone passes you a plate of cookies and you start arguing about which [cookie] is #1 and which is #2. They're all edible![37]

Electromagnetic forces on material media can be calculated using the Maxwell stress tensor. The force on a volume V of a material medium is defined as the integral over

[35] W. Shockley, Proc. Natl. Acad. Sci. USA **60**, 807 (1968); H. A. Haus, Physica (Amsterdam) **43**, 77 (1969). Haus allows for a magnetic permeability $\mu \neq \mu_0$. When expressed on a per-photon basis, his result for $\mu = \mu_0$ is equivalent to (1.8.23). He refers to this as the momentum "assigned to the material in the presence of a plane wave packet," consistent with our interpretation. He also identifies the sum of (1.8.23) and the Abraham field momentum as the total conserved momentum, consistent with (1.8.24).

[36] R. Loudon, J. Mod. Opt. **49**, 821 (2002).

[37] S. Coleman, http://arxiv.org/abs/1110.5013, p. 46, accessed October 23, 2018.

this volume of the Lorentz force density $\rho_f \mathbf{E} + \mathbf{J}_f \times \mathbf{B}$, and, from the macroscopic Maxwell equations (1.5.6)–(1.5.9), it follows that this integral, equal to the rate of change of the "mechanical" momentum $\mathbf{g}_m$ of the material in the volume V, is

$$\frac{d\mathbf{g}_m}{dt} = \int_V \frac{d\mathbf{p}_m}{dt}\, dV = \int_V [\rho_f \mathbf{E} + \mathbf{J}_f \times \mathbf{B}]dV$$

$$= \int_V \left[(\nabla \cdot \mathbf{D})\mathbf{E} + (\nabla \times \mathbf{H}) \times \mathbf{B} - \frac{\partial \mathbf{D}}{\partial t} \times \mathbf{B} \right] dV$$

$$= \int_V \left[(\nabla \cdot \mathbf{D})\mathbf{E} + (\nabla \times \mathbf{H}) \times \mathbf{B} + \mathbf{D} \times \frac{\partial \mathbf{B}}{\partial t} - \frac{\partial}{\partial t}(\mathbf{D} \times \mathbf{B}) \right] dV$$

$$= \int_V \left[(\nabla \cdot \mathbf{D})\mathbf{E} + (\nabla \cdot \mathbf{H})\mathbf{B} - \mathbf{D} \times (\nabla \times \mathbf{E}) - \mathbf{B} \times (\nabla \times \mathbf{H}) \right] dV$$

$$- \frac{d}{dt} \int_V (\mathbf{D} \times \mathbf{B})\, dV, \tag{1.9.1}$$

and therefore that

$$\int_V \left[\frac{d}{dt}(\mathbf{p}_m + \mathbf{D} \times \mathbf{B}) \right] dV = \int_V \left[(\nabla \cdot \mathbf{D})\mathbf{E} + (\nabla \cdot \mathbf{H})\mathbf{B} - \mathbf{D} \times (\nabla \times \mathbf{E}) \right.$$

$$\left. - \mathbf{B} \times (\nabla \times \mathbf{H}) \right] dV. \tag{1.9.2}$$

We have added 0 in the guise of $(\nabla \cdot \mathbf{H})\mathbf{B}$ to the right sides in order to put these equations in a form that is symmetrical in the electric and magnetic fields. If there is a linear relation between $\mathbf{D}$ and $\mathbf{E}$, Maxwell's equations imply that we can write the ith Cartesian component of a force density as

$$\tilde{f}_i = \frac{d}{dt}(\mathbf{p}_m + \mathbf{D} \times \mathbf{B})_i = \sum_{j=1}^{3} \frac{\partial T_{ij}^M}{\partial x_j}, \tag{1.9.3}$$

where the stress tensor T^M has components

$$T_{ij}^M = E_i D_j + H_i B_j - \frac{1}{2}(\mathbf{E} \cdot \mathbf{D} + \mathbf{H} \cdot \mathbf{B})\delta_{ij} \quad (i, j = 1, 2, 3). \tag{1.9.4}$$

Equation (1.9.4) defines the Minkowski form of the stress tensor, as indicated by the superscript M. For fields propagating in a vacuum, the stress tensor is obtained by taking $D_j = \epsilon_0 E_j$ and $B_j = \mu_0 H_j$ in (1.9.4):

$$T_{ij} = \epsilon_0 E_i E_j + \mu_0 H_i H_j - \frac{1}{2}(\epsilon_0 \mathbf{E} \cdot \mathbf{E} + \mu_0 \mathbf{H} \cdot \mathbf{H})\delta_{ij}. \tag{1.9.5}$$

The force transmitted through a surface element $d\mathbf{S}$ is

$$dF_i = \sum_{j=1}^{3} T_{ij}\, dS_j, \tag{1.9.6}$$

where the direction of $d\mathbf{S}$ is defined to be that of a vector normal to it and pointing outward from it. This formal definition of dF_i does not require that any matter be present.

Consider the simplest nontrivial example, a plane wave with electric and magnetic fields

$$\mathbf{E}(\mathbf{r}, t) = \hat{\mathbf{x}} E_0 \cos \omega(t - z/c), \quad \mathbf{H}(\mathbf{r}, t) = \hat{\mathbf{y}} \sqrt{\frac{\epsilon_0}{\mu_0}} E_0 \cos \omega(t - z/c), \quad (1.9.7)$$

normally incident from $z < 0$ on a perfectly conducting plate at $z = 0$. Equation (1.9.6) gives, for the force per unit area on the plate,

$$F/A = -T_{zz} = -\left(\epsilon_0 E_z^2 - \frac{1}{2}(\epsilon_0 E_x^2 + \mu_0 H_y^2) \right)_{z=0} \quad (1.9.8)$$

with the minus sign appearing because $d\mathbf{S}$ in (1.9.6) points in the $-z$ direction. We have used the fact that the tangential component of $\mathbf{E}$ and the normal component of $\mathbf{B}$ both vanish at $z = 0$. With our assumption of normal incidence, the pressure (1.9.8) on the plate reduces to

$$F/A = \frac{1}{2}\mu_0 (H_y)_{z=0}^2 = \frac{1}{2}\mu_0 \left(2\sqrt{\frac{\epsilon_0}{\mu_0}} E_0 \cos \omega t \right)^2, \quad (1.9.9)$$

since the electric field E_x at $z = 0$ vanishes and the magnetic field amplitude H_y at $z = 0$, the sum of the incident and reflected field amplitudes, is just twice that of the incident field. Replacing $\cos^2 \omega t$ by its average value, we obtain

$$F/A = \epsilon_0 E_0^2, \quad (1.9.10)$$

which, of course, is the expected result: the pressure on the mirror is $2I/c$, where the (time-averaged) incident intensity $I = (c/2)\epsilon_0 E_0^2$.

> **Exercise 1.5:** (a) Use the stress tensor to show that the pressure exerted by a plane wave of intensity I, incident from a vacuum onto a perfectly conducting plate with angle of incidence θ, is $F/A = (2I/c) \cos^2 \theta$, regardless of the field polarization. (b) How is (1.9.10) changed if the field is incident from a transparent dielectric medium of refractive index n?

Recall the differential form of Poynting's theorem that follows from the macroscopic Maxwell equations:

$$\nabla \cdot \mathbf{S} = -\mathbf{J} \cdot \mathbf{E} - \mathbf{E} \cdot \frac{\partial \mathbf{D}}{\partial t} - \mathbf{H} \cdot \frac{\partial \mathbf{B}}{\partial t}. \quad (1.9.11)$$

The Poynting vector $\mathbf{S} = \mathbf{E} \times \mathbf{H}$ gives the flux of electromagnetic energy, and the right side of this equation is the rate of change of total energy, that of the field plus that of the material medium. In the absence of any elastic forces that can cause mechanical energy to be transported within a material medium, only energy attributable to the field propagates out of any given volume element of the medium. This suggests that $\mathbf{S} = \mathbf{E} \times \mathbf{H}$ gives the energy flux of the field in the medium as well as in free space. From this assumption, and the relation $\mathbf{g} = \mathbf{S}/c^2$ of special relativity theory for the momentum density associated with any process, electromagnetic or otherwise, by which energy is transported with a flux $\mathbf{S}$, we are led to assign to the field the momentum density

$$\mathbf{g}_A = \mathbf{E} \times \mathbf{H}/c^2. \tag{1.9.12}$$

This defines the Abraham momentum density (see (1.7.1)). Using $\mathbf{D} \times \mathbf{B} = (1/c^2)(1 + n^2 - 1)\mathbf{E} \times \mathbf{H}$ for an effectively non-dispersive and isotropic linear medium with refractive index n, we can write (1.9.3) as

$$\left(\rho\mathbf{E} + \mathbf{J} \times \mathbf{B} + \mathbf{f}^A + \frac{\partial \mathbf{g}_A}{\partial t} \right)_i = \sum_{j=1}^{3} \frac{\partial T_{ij}^M}{\partial x_j}, \tag{1.9.13}$$

where

$$\mathbf{f}^A = \frac{1}{c^2}(n^2 - 1)\frac{\partial}{\partial t}(\mathbf{E} \times \mathbf{H}) \tag{1.9.14}$$

is again the Abraham force density.

Equation (1.9.3) suggests that the force density acting on the material medium is

$$(\mathbf{f}_M)_i = \sum_{j=1}^{3} \frac{\partial T_{ij}^M}{\partial x_j} - \left(\frac{\partial \mathbf{g}_M}{\partial t} \right)_i, \tag{1.9.15}$$

whereas, according to (1.9.13), the force density acting on the medium is

$$(\mathbf{f}_A)_i = \sum_{j=1}^{3} \frac{\partial T_{ij}^M}{\partial x_j} - \left(\frac{\partial \mathbf{g}_A}{\partial t} \right)_i, \tag{1.9.16}$$

that is,

$$\mathbf{f}_A = \mathbf{f}_M + \frac{\partial \mathbf{g}_M}{\partial t} - \frac{\partial \mathbf{g}_A}{\partial t} = \mathbf{f}_M + \mathbf{f}^A. \tag{1.9.17}$$

Thus, in the Minkowski formulation, the force acting on the particles of the dielectric medium is obtained by subtracting $\partial \mathbf{g}_M/\partial t$ from $\sum_j \partial T_{ij}^M/\partial x_j$, suggesting that the field momentum density is $\mathbf{g}_M$. In the Abraham interpretation as just described, however, the force on the medium is obtained by subtracting $\partial \mathbf{g}_A/\partial t$ from $\sum_j \partial T_{ij}^M/\partial x_j$ under the assumption that the momentum density of the field is $\mathbf{g}_A$. In this interpretation, there appears the force density $\mathbf{f}^A$ that, together with the Lorentz force density $\rho\mathbf{E} + \mathbf{J} \times \mathbf{H}$, gives the total force on the material medium, as is clear from (1.9.13).

Of course, both the Minkowski and the Abraham momentum densities, $\mathbf{g}_M$ and $\mathbf{g}_A$, respectively, are defined in terms of measurable quantities and are themselves therefore measurable in principle. Either momentum density will comport with conservation of linear momentum, the only difference being in how the total momentum is apportioned between the field and the material medium. This is obvious from the equation

$$\rho \mathbf{E} + \mathbf{J} \times \mathbf{B} + \frac{\partial \mathbf{g}_M}{\partial t} = \rho \mathbf{E} + \mathbf{J} \times \mathbf{B} + \mathbf{f}^A + \frac{\partial \mathbf{g}_A}{\partial t}. \qquad (1.9.18)$$

On the left side we could interpret $\mathbf{g}_M$ as field momentum density; on the right side, we could interpret $\mathbf{g}_A$ as field momentum density, but then, compared with the left side, we have an "additional" (Abraham) force and momentum (density) associated with the medium. The generally accepted view, which we advocate here, is that $\mathbf{g}_A$ is the momentum density of the field.

The Abraham force $\mathbf{f}^A = (1/c^2)(n^2 - 1)(\partial/\partial t)(\mathbf{E} \times \mathbf{H})$ obviously plays an important role in comparisons of the Abraham and Minkowski formulations of the stress tensor. It will average to zero over times long compared with an optical period, but it can be measured if the electric and magnetic fields are applied continuously *and* if they vary slowly enough. Such measurements were reported by James and by Walker et al.[38] In one of the experiments, an annular disk of high permittivity ($\epsilon \approx 3620$), serving as a torsion pendulum, was subjected to a static, vertical magnetic field and a slowly varying (0.4 Hz) radial electric field between its inner and outer surfaces. Then, the Abraham force is azimuthal, and the oscillations of the electric field should cause the disk to oscillate about the direction of the magnetic field. Such oscillations were observed, and the measurements were said to confirm the existence of the Abraham force to an accuracy of about 5%.

The stress tensor T_{ij}^M is the spatial part of a four-dimensional energy-momentum tensor employed by Minkowksi in the context of the electrodynamics of moving bodies. For an isotropic medium that is effectively dispersionless at frequencies of interest and that has no free charges or currents ($\rho = \mathbf{J} = 0$), the Minkowski force density (1.9.15) reduces to

$$\mathbf{f}_M = -\frac{1}{2} \mathbf{E}^2 \nabla \epsilon. \qquad (1.9.19)$$

Minkowski's energy-momentum tensor (1.9.4) is not symmetric. Even the 3×3 matrix T_{ij}^M defined here is not symmetric in the general case of an anisotropic medium. This lack of symmetry led Abraham to introduce a different, symmetric energy-momentum tensor, the 3×3 (spatial) part of which is

$$T_{ij}^A = \frac{1}{2}(E_i D_j + E_j D_i) + \frac{1}{2}(H_i B_j + H_j B_i) - \frac{1}{2}(\mathbf{E} \cdot \mathbf{D} + \mathbf{H} \cdot \mathbf{H})\delta_{ij}. \qquad (1.9.20)$$

In terms of this tensor,

$$(\mathbf{f}_A)_i = \sum_{j=1}^{3} \frac{\partial}{\partial x_j} T_{ij}^A - \left(\frac{\partial \mathbf{g}_A}{\partial t}\right)_i = (\mathbf{f}_M)_i + (\mathbf{f}^A)_i, \qquad (1.9.21)$$

[38] R. P. James, Proc. Natl. Acad. Sci. USA **61**, 1149 (1968); G. B. Walker, D. G. Lahoz, and G. Walker, Can J. Phys. **53**, 2577 (1975).

or

$$\mathbf{f}_A = -\frac{1}{2}\mathbf{E}^2\nabla\epsilon + \mathbf{f}^A \tag{1.9.22}$$

for an isotropic and dispersionless medium. Here, $\mathbf{f}_A$ and $\mathbf{f}^A$ must not be confused: $\mathbf{f}_A$ is the force density acting on the medium in the Abraham formulation of the stress tensor, whereas $\mathbf{f}^A$ is the "Abraham force density" that appears in addition to the Lorentz force density in this formulation.

A more general formulation of the stress tensor describing electromagnetic forces in a non-dispersive dielectric fluid leads to the force density

$$\mathbf{f} = -\nabla P + \nabla\left(\rho\frac{\partial\epsilon}{\partial\rho}\frac{1}{2}\mathbf{E}^2\right) + \mathbf{f}_A$$
$$= -\nabla P + \nabla\left(\rho\frac{\partial\epsilon}{\partial\rho}\frac{1}{2}\mathbf{E}^2\right) - \frac{1}{2}\mathbf{E}^2\nabla\epsilon + \frac{1}{c^2}(n^2-1)\frac{\partial}{\partial t}(\mathbf{E}\times\mathbf{H}), \tag{1.9.23}$$

where ρ is the mass density, and P is the pressure in the fluid, which depends on the field only to the extent that the field can affect the density and temperature.[39] The second term on the right side is associated with electrostriction. Under conditions of frequent interest, such as when there is mechanical equilibrium, it is cancelled by the first term, and the net force density reduces to the Abraham force density (1.9.22):

$$\mathbf{f} = -\frac{1}{2}\mathbf{E}^2\nabla\epsilon + \frac{1}{c^2}(n^2-1)\frac{\partial}{\partial t}(\mathbf{E}\times\mathbf{H}) = \mathbf{f}_A. \tag{1.9.24}$$

To relate electrostriction to pressure, consider the change du in electromagnetic energy density when the mass density ρ undergoes an infinitesimal change $d\rho$:

$$du = d(\frac{1}{2}\epsilon|\mathbf{E}|^2) = \frac{1}{2}\frac{d\epsilon}{d\rho}|\mathbf{E}|^2\,d\rho. \tag{1.9.25}$$

Equating this to the "PdV" work per unit volume in an isothermal process, $P\,dV/V = -P\,d\rho/\rho$ (since ρV is constant), implies the pressure

$$P_{\text{el}} = -\frac{1}{2}\rho\frac{d\epsilon}{d\rho}|\mathbf{E}|^2, \tag{1.9.26}$$

so that the first two terms on the right side of (1.9.23) can be expressed as $-\nabla(P+P_{\text{el}})$ and can be assumed to vanish if there is a mechanical equilibrium.

It is interesting to compare the force density (1.9.24) to that obtained by starting from the microscopic perspective in which the force on an electric dipole moment $\mathbf{d}$ in an electromagnetic field is defined by (1.8.1):

$$\mathbf{F} = (\mathbf{d}\cdot\nabla)\mathbf{E} + \dot{\mathbf{d}}\times\mathbf{B}. \tag{1.9.27}$$

Introducing the polarizability α, we can express this force as

---

[39] *Landau, Lifshitz, and Pitaevskii*, (75.17).

$$\mathbf{F} = \alpha\left[(\mathbf{E} \cdot \nabla)\mathbf{E} + \frac{\partial \mathbf{E}}{\partial t} \times \mathbf{B}\right] = \alpha\left[\nabla(\frac{1}{2}\mathbf{E}^2) - \mathbf{E} \times (\nabla \times \mathbf{E})\right.$$
$$\left. + \mu_0 \frac{\partial}{\partial t}(\mathbf{E} \times \mathbf{H}) - \mu_0 \mathbf{E} \times \frac{\partial \mathbf{H}}{\partial t}\right]$$
$$= \alpha\left[\nabla(\frac{1}{2}\mathbf{E}^2) + \mu_0 \frac{\partial}{\partial t}(\mathbf{E} \times \mathbf{H})\right]. \tag{1.9.28}$$

Using $N\alpha = \epsilon - \epsilon_0$, we obtain the force density for a medium of N dipoles per unit volume:

$$\mathbf{f}_{\text{dipoles}} = (\epsilon - \epsilon_0)\left[\nabla(\frac{1}{2}\mathbf{E}^2) + \mu_0 \frac{\partial}{\partial t}(\mathbf{E} \times \mathbf{H})\right]$$
$$= \epsilon_0(n^2 - 1)\nabla(\frac{1}{2}\mathbf{E}^2) + \frac{1}{c^2}(n^2 - 1)\frac{\partial}{\partial t}(\mathbf{E} \times \mathbf{H})$$
$$= \epsilon_0(n^2 - 1)\nabla(\frac{1}{2}\mathbf{E}^2) + \mathbf{f}^A, \tag{1.9.29}$$

which differs from both $\mathbf{f}_M$ and $\mathbf{f}_A$ as defined by (1.9.19) and (1.9.22), respectively.

This difference is related to the difference between the local field acting on an electric dipole and the macroscopic field: write the force density (1.9.23) equivalently as[40]

$$\mathbf{f} = -\nabla P + \mathbf{f}_{\text{dipoles}} + \nabla\{[\rho \frac{d\epsilon}{d\rho} - (\epsilon - \epsilon_0)]\frac{1}{2}\mathbf{E}^2\}. \tag{1.9.30}$$

The last term on the right vanishes if $\epsilon - \epsilon_0$ is proportional to the density ρ, that is, if the local field and the macroscopic field are taken to be the same; this assumption was made in the derivation of (1.9.29). Therefore, aside from a force density associated with a pressure gradient, the difference between $\mathbf{f}$ and $\mathbf{f}_{\text{dipoles}}$ is attributable to the difference between the macroscopic and local fields. If we assume, for instance, the Lorentz–Lorenz local field that results in the Clausius–Mossotti relation,

$$\frac{\epsilon - \epsilon_0}{\epsilon + 2\epsilon_0} = A\rho, \tag{1.9.31}$$

the difference between $\mathbf{f}$ and $\mathbf{f}_{\text{dipoles}}$ is second order in $\epsilon - \epsilon_0$, that is,

$$\rho \frac{d\epsilon}{d\rho} - (\epsilon - \epsilon_0) = \frac{1}{3\epsilon_0}(\epsilon - \epsilon_0)^2, \tag{1.9.32}$$

and is negligible for a sufficiently dilute medium.

1.10   Rayleigh Scattering

So far, we have not considered in much detail any scattering or absorption processes that can take energy out of a light beam and put it into radiation in other directions, heat, or some other form of energy. We take up the quantum theory of absorption

[40] J. P. Gordon, Phys. Rev. A **8**, 14 (1973).

by atoms in Chapter 2, and restrict ourselves in the remainder of this chapter to scattering. One of the simplest scattering processes is the *Rayleigh scattering* of light by electrically polarizable particles (e.g., atoms) whose size is much smaller than a wavelength. There is no net loss of energy by the field; energy is only transferred from an incident field to radiation in other directions.

1.10.1   Attenuation Coefficient Due to Rayleigh Scattering

The cross section for Rayleigh scattering can be deduced as follows. An electric field $\mathbf{E}_0 \cos \omega t$ induces an electric dipole moment $\mathbf{p}(t) = \alpha(\omega)\mathbf{E}_0 \cos \omega t$ in each of N identical polarizable particles per unit volume, each particle having a spatial extent small compared with a wavelength $\lambda = 2\pi c/\omega$. This oscillating dipole generates electric and magnetic fields, and the power radiated by the dipole is given by the following equation (see Exercise 1.2):

$$\frac{dW_{\rm rad}}{dt} = n(\omega)\frac{\omega^4}{12\pi\epsilon_0 c^3}\alpha^2(\omega)\mathbf{E}_0^2 = \sigma_{\rm R}(\omega)I, \qquad (1.10.1)$$

where $W_{\rm rad}$ denotes energy of the radiated field, $I = \frac{1}{2}n(\omega)c\epsilon_0\mathbf{E}_0^2$ is the intensity of the field incident on the dipole, and

$$\sigma_{\rm R}(\omega) = \frac{1}{6\pi N^2}\left(\frac{\omega}{c}\right)^4 [n^2(\omega) - 1]^2 \qquad (1.10.2)$$

is the (Rayleigh) scattering cross section. We have used the formula $n^2(\omega) - 1 = N\alpha(\omega)/\epsilon_0$ to express $\sigma_{\rm R}(\omega)$ in terms of the refractive index $n(\omega)$. A plane wave of frequency ω propagating in the z direction in a medium of N scatterers per unit volume will diminish in intensity according to $dI/dz = -N\sigma_{\rm R}(\omega)I$, or

$$I(z) = I(0)e^{-a_{\rm R}(\omega)z}, \qquad (1.10.3)$$

where

$$a_{\rm R}(\omega) = N\sigma_{\rm R}(\omega) = \frac{1}{6\pi N}\left(\frac{\omega}{c}\right)^4 [n^2(\omega) - 1]^2 \qquad (1.10.4)$$

is the attenuation coefficient for Rayleigh scattering. Its ω^4 dependence means that the amount of scattering increases sharply with increasing frequency (n generally varies much more slowly with ω than ω^4). That's one reason why the sky is blue.

   More generally we can replace the polarizability in (1.10.1) by the polarizability of whatever particles are doing the scattering, provided the particles are small compared with the wavelength. For example, we can use the polarizability (1.5.59) for scattering by dielectric spheres. Rayleigh scattering can occur due to the presence of small dust particles, for example. John W. Strutt (Lord Rayleigh) deduced that scattering by air molecules alone is sufficient to cause the "heavenly azure."

1.10.2 Density Fluctuations

How does this simple derivation of a_R comport with the discussion in Section 1.5 indicating that a light beam propagating in a continuous medium characterized by a (real) refractive index suffers no attenuation? The answer can be traced to our assumption there that the density N of scatterers is constant throughout the medium. If the density is truly constant, the scattered fields from different dipoles, with their uniformly distributed phases, cancel out in every direction except the forward (z) direction, with the result that the total field propagates without any "side scattering." Rayleigh scattering is actually a consequence of density *fluctuations*, which were ignored in our simplistic derivation of a_R.

The role of density fluctuations in Rayleigh scattering may be understood roughly as follows. The electric dipole moment induced in a particle at $\mathbf{r}_i$ in an electric field $\mathbf{E}_0 \cos(\omega t - \mathbf{k}_0 \cdot \mathbf{r})$, is

$$\mathbf{p}_i(t) = \alpha(\omega)\mathbf{E}_0 e^{-i\omega t} e^{i\mathbf{k}_0 \cdot \mathbf{r}_i} \qquad (k_0 = \omega/c). \tag{1.10.5}$$

The electric field from this dipole at a point $\mathbf{r}$ in the radiation zone is, therefore, proportional to

$$\ddot{p}(t - |\mathbf{r} - \mathbf{r}_i|/c) = -\omega^2 \alpha(\omega)E_0 e^{-i\omega(t - |\mathbf{r} - \mathbf{r}_i|/c)} e^{i\mathbf{k}_0 \cdot \mathbf{r}_i}. \tag{1.10.6}$$

For large distances from the dipole ($r \gg r_i$),

$$k_0|\mathbf{r} - \mathbf{r}_i| = k_0\sqrt{r^2 - 2\mathbf{r} \cdot \mathbf{r}_i + r_i^2} \cong k_0 r - k_0 \hat{\mathbf{r}} \cdot \mathbf{r}_i, \tag{1.10.7}$$

and therefore

$$\begin{aligned}\ddot{p}(t - |\mathbf{r} - \mathbf{r}_i|/c) &\cong -\omega^2 \alpha(\omega)E_0 e^{-i\omega(t - r/c)} e^{i\mathbf{k}_0 \cdot \mathbf{r}_i} e^{-ik_0 \hat{\mathbf{r}} \cdot \mathbf{r}_i} \\ &= -\omega^2 \alpha(\omega)E_0 e^{-i\omega(t - r/c)} e^{i\mathbf{K} \cdot \mathbf{r}_i}.\end{aligned} \tag{1.10.8}$$

$\mathbf{K} = \mathbf{k}_0 - k_0\hat{\mathbf{r}}$ is the difference between the $\mathbf{k}$ vectors of incident ($\mathbf{k}_0$) and scattered ($k_0\hat{\mathbf{r}}$) plane waves. Assuming that the medium is sufficiently dilute that multiple scattering is negligible, the total scattered field at large distances from $\mathcal{N}$ dipoles at positions $\mathbf{r}_1, \mathbf{r}_2, \ldots, \mathbf{r}_\mathcal{N}$ is proportional to the average over particle positions of

$$\begin{aligned}-\omega^2 \alpha(\omega)E_0 e^{-i\omega(t-r/c)} \sum_{i=1}^{\mathcal{N}} e^{i\mathbf{K}\cdot\mathbf{r}_i} &= -\omega^2 \alpha(\omega)E_0 e^{-i\omega(t-r/c)} \sum_{i=1}^{\mathcal{N}} e^{i\Phi_i} \\ &= -\omega^2 \alpha(\omega)E_0 e^{-i\omega(t-r/c)} F(\mathbf{K}),\end{aligned} \tag{1.10.9}$$

where the phase $\Phi_i = \mathbf{K} \cdot \mathbf{r}_i$ and the *structure factor* is

$$F(\mathbf{K}) = \sum_{i=1}^{\mathcal{N}} e^{i\Phi_i}. \tag{1.10.10}$$

Now *if*, for example, the scatterers are continuously and uniformly distributed, $F(\mathbf{K}) = 0$ *except* for the forward scattering direction, the direction of propagation of the incident field, for which $\mathbf{K} = 0$; all the scattering is in the forward direction and, as shown in Section 1.5, this forward scattering is responsible for the index of refraction. In the forward direction, the dipole fields effectively add in phase, and, in this sense, the forward scattering is *coherent*. If $\mathbf{K} \neq 0$, there is, for every Φ_i, a Φ_j such that $\exp(i\Phi_j) = -\exp(i\Phi_i)$. So if the scatterers are continuously and uniformly distributed, there is no side scattering. But if the scatterers are *randomly* distributed, as is the case in an ideal gas and approximately the case in a gas of weakly interacting particles, it is the *average* of the scattered field over the particle positions that vanishes. The average of the scattered *intensity*, being proportional to

$$|F(\mathbf{K})|^2 = \left| \sum_{i=1}^{\mathcal{N}} e^{i\Phi_i} \right|^2 = \mathcal{N} + \sum_{i \neq j}^{\mathcal{N}} \sum_{j=1}^{\mathcal{N}} e^{i(\Phi_i - \Phi_j)}, \qquad (1.10.11)$$

does not vanish. The average of the last term on the right side is zero, and so the average intensity is just proportional to the number $\mathcal{N}$ of scatterers. This averaging justifies the implicit assumption made in our derivation of a_R that the particles scatter independently, with no interference among fields scattered by different particles. In that case, a_R is then simply proportional to the number of particles in the scattering volume, that is, to the number density N.

These simplistic arguments can be made a bit more rigorous. Density fluctuations in a gas, for example, can be accounted for if, instead of working with (1.10.11), we work with

$$|F(\mathbf{K})|^2 = \left| \sum_{i=1}^{\mathcal{M}} \mathcal{M}_i e^{i\Phi_i} \right|^2, \qquad (1.10.12)$$

where the sum is now over $\mathcal{M}$ small (compared with a wavelength) volume elements ("cells"), there being $\mathcal{M}_i$ ($\gg 1$) particles in the ith cell, which contributes a phase Φ_i to the scattered field. If $\overline{\mathcal{M}}$ is the average value of $\mathcal{M}_i$, independent of i, and we write $\mathcal{M}_i = \overline{\mathcal{M}} + \Delta\mathcal{M}_i$, the average

$$\overline{|F(\mathbf{K})|^2} = \overline{\left| \sum_{i=1}^{\mathcal{M}} \Delta\mathcal{M}_i e^{i\Phi_i} \right|^2} = \sum_{i=1}^{\mathcal{M}} \sum_{j=1}^{\mathcal{M}} \overline{\Delta\mathcal{M}_i \Delta\mathcal{M}_j} e^{i(\Phi_i - \Phi_j)}$$

$$= \sum_{i=1}^{\mathcal{M}} \overline{(\Delta\mathcal{M}_i)^2} = \sum_{i=1}^{\mathcal{M}} \overline{\mathcal{M}_i} = \mathcal{M}\overline{\mathcal{M}} = \mathcal{N} \qquad (1.10.13)$$

if the number fluctuations in different cells are uncorrelated and if, as in an ideal gas, these fluctuations obey Poisson statistics, for which $\overline{(\Delta\mathcal{M}_i)^2} = \overline{\mathcal{M}_i}$. Under these assumptions, we reach again the conclusion that a_R is proportional to N. Of course, if there are correlations in the density fluctuations, as in a liquid, we cannot assume Poisson statistics. The strong light scattering in critical opalescence, for example, is a consequence of correlated density fluctuations.

1.10.3   Polarization by Rayleigh Scattering

The polarization of light by Rayleigh scattering observed in skylight, for example, follows simply from the electric field in the radiation zone of an oscillating dipole $\hat{\mathbf{x}}p(t)$, where $\hat{\mathbf{x}}$ is the unit vector in the direction of the electric field inducing the dipole moment (see (1.4.20)):

$$\mathbf{E}(\mathbf{r},t) = \frac{1}{4\pi\epsilon_0}[(\hat{\mathbf{x}}\cdot\hat{\mathbf{r}})\hat{\mathbf{r}} - \hat{\mathbf{x}}]\frac{1}{c^2 r}\frac{d^2}{dt^2}p(t - r/c). \tag{1.10.14}$$

Here, $\hat{\mathbf{r}}$ is the unit vector pointing from the dipole to the point of observation ($\mathbf{r} = r\hat{\mathbf{r}}$). Thus, the scattered light observed at right angles to the plane defined by the directions of polarization and propagation is polarized in the x direction, since

$$(\hat{\mathbf{x}}\cdot\hat{\mathbf{r}})\hat{\mathbf{r}} - \hat{\mathbf{x}} = (\hat{\mathbf{x}}\cdot\hat{\mathbf{y}})\hat{\mathbf{y}} - \hat{\mathbf{x}} = -\hat{\mathbf{x}} \tag{1.10.15}$$

when $\hat{\mathbf{r}} = \hat{\mathbf{y}}$ (see Figure 1.10). If, however, the incident field is polarized in the y direction, there is no scattered field along y direction, since

$$(\hat{\mathbf{y}}\cdot\hat{\mathbf{r}})\hat{\mathbf{r}} - \hat{\mathbf{y}} = (\hat{\mathbf{y}}\cdot\hat{\mathbf{y}})\hat{\mathbf{y}} - \hat{\mathbf{y}} = 0. \tag{1.10.16}$$

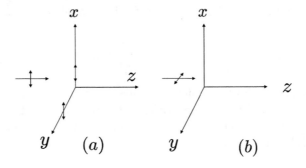

Fig. 1.10 An incident field propagating in the z direction with polarization (a) along x, and (b) along y. In case (a), the field scattered in the y direction is also polarized along x. In (b), there is no scattering in the y direction.

For unpolarized incident light propagating in the z direction, the induced dipole radiates in the y direction only when its oscillation has a nonzero component in the x direction (see Figure 1.10), in which case the scattered radiation is polarized along x. Rayleigh-scattered light is, therefore, polarized.

We have assumed that the scatterers are spherically symmetric, meaning that their induced electric dipole moments are directly proportional to the electric field incident upon them. In other words, we have assumed that the polarizability $\alpha(\omega)$ is a scalar.

This is the case for (unperturbed) atoms, but, for molecules, the polarizability is generally a tensor (Section 2.3); this anisotropy is understandable from a classical model in which a molecule rotates while it scatters radiation. The resulting depolarization in Rayleigh scattering by a gas of molecules is usually small, on the order of a few per cent, but can be much larger in the case of liquids. Direct measurements of Rayleigh scattering cross sections have been reported for Ar as well as for various molecular scatterers. The measured cross sections were found to be within a few per cent of the values predicted by the formula (1.10.2) after small depolarization corrections are included for the molecular scatterers.[41]

Exercise 1.6: Show that the differential cross section for Rayleigh scattering for unpolarized light is

$$\frac{d\sigma}{d\Omega} = \frac{3}{16\pi}\sigma_{R}(\omega)(1+\cos^2\theta) = \frac{\pi^2[n^2(\lambda)-1]^2}{\lambda^4 N^2}\frac{1}{2}(1+\cos^2\theta),$$

where $\sigma_R(\omega)$ is the total Rayleigh cross section defined by (1.10.2), and θ is the observation angle with respect to the propagation direction of the incident light.

1.10.4   Sundry Remarks

In his classic book, Rosenfeld obtains, instead of (1.10.4),

$$a_R = N\sigma_R = \frac{1}{6\pi n N}\left(\frac{\omega}{c}\right)^4 [n^2(\omega)-1]^2. \tag{1.10.17}$$

The difference between Rosenfeld's result and (1.10.4) lies in his ignoring the proportionality of the emission rate of an oscillating dipole in a dielectric to n (see, for example, Exercise 1.2). He therefore ends up with an "extra" factor of n in the denominator of a_R.[42]

Rayleigh, similarly, did not have the factor n in the radiation rate of the induced dipoles, but, in addition, did not have an n dependence of the field intensity (or energy) in his 1899 derivation of (1.10.4); such things were not well established at the time. It is interesting that he had returned to the problem after nearly 20 years as a result of a letter he received from Maxwell many years earlier. Maxwell asked about Rayleigh's formula for the attenuation coefficient, hoping to use it to estimate molecular diameters. Rayleigh quotes from Maxwell's letter: "I have left your papers on the light of the sky . . . at Cambridge, and it would take me, even if I had them, some time to get them assimilated sufficiently to answer the following question, which I think will involve less expense to the energy of the race if you stick the data into your formula and send me the result . . . Suppose that there are N spheres of density ρ and diameter S in unit volume of the medium. Find the refractive index of the compound

[41] M. Sneep and W. Ubachs, J. Quant. Spectrosc. Rad. Trans. **92**, 293 (2005).
[42] L. Rosenfeld, *Theory of Electrons*, Dover Publications, New York, 1965.

medium and the coefficient of extinction of light passing through it. The object of the enquiry is, of course, to obtain data about the size of the molecules of air."[43]

Rayleigh's analysis, in present notation, may be summarized as follows. (We include n and the group index n_g, neither of which appear in Rayleigh's derivation, to show why he arrived at the (presumably) correct result.) The rate at which energy density u of the incident field is lost (Rayleigh: "there is no waste of energy on the whole, this represents the loss of energy in the [incident] wave"[44]) is

$$\frac{du}{dt} = -N \left[\frac{dW_{\text{rad}}}{dt} \right] = Nn \frac{\omega^4}{12\pi\epsilon_0 c^3} \alpha^2(\omega) \mathbf{E}_0^2. \tag{1.10.18}$$

Since $u = \frac{1}{2}\epsilon_0 n n_g \mathbf{E}_0^2$ (see (1.6.11)),

$$\frac{du}{dz} = \frac{n_g}{c}\frac{du}{dt} = -n n_g \frac{N\omega^4\alpha^2}{6\pi\epsilon_0^2 c^3} \frac{u}{n n_g} = -a_R u, \tag{1.10.19}$$

in agreement with (1.10.4).

Initially, Rayleigh thought that skylight might be "similar in composition to that of dilute solutions of copper, which acquire their blue tint by a partial suppression of the extreme red."[45] Experiments by John Tyndall convinced him that the blue sky was due to scattering. He concluded from calculations based on $n \cong 1.0003$ for air at $\lambda = 600$ nm, and Maxwell's estimate of 1.9×10^{19} molecules/cm^3 for air at standard temperature and pressure, that "foreign" particles are not required to make the sky blue. He calculated from (1.10.4) that the intensity of light decreases by $1/e$ after a propagation distance of 83 km. He then compared this estimate with data reported by astronomers: "Perhaps the best data for a comparison are those afforded by the varying brightness of stars at various altitudes. Bouguer and others estimate about 0.8 for the transmission of light through the entire atmosphere for a star at the zenith. This corresponds to 8.3 kilometres of air at standard pressure. At this rate the transmission through 83 kilometres would be $(.8)^{10}$ or .11 instead of $1/e$ or 0.37. It appears then that the actual transmission through 83 kilometres is only about 3 times less than that calculated (with the above value for n) from molecular [scattering] without any allowance for foreign matter at all. And we may conclude that the light scattered from the molecules would suffice to give us a blue sky, not so very greatly darker than that actually enjoyed."[46]

We remarked earlier that forward scattering is coherent in that the fields from the different dipoles add in phase, or "coherently." Rayleigh knew this: "The combination of the [scattered] waves which travel in the [forward] direction in question has this peculiarity, that the phases are no more distributed at random. The intensity of the

---

[43] Lord Rayleigh, Phil. Mag. **47**, 375 (1899), p. 375.
[44] Ibid., p. 399.
[45] Lord Rayleigh, Nature **3**, 234 (1871), p. 236.
[46] Lord Rayleigh, Phil. Mag. **47**, 375 (1899), p. 382.

[scattered] light is no longer to be arrived at by addition of individual intensities, but must be calculated with consideration of the particular phases involved. If we consider a number of particles which all lie upon a [forward] ray, we see that the phases of the [scattered] vibrations which issue along this line are all the same."[47]

Rayleigh scattering refers to scattering by particles small compared with the wavelength of the incident light. If a dielectric sphere consists of $\mathcal{N}$ atoms, the electric field of the scattered wave will be approximately $\mathcal{N}$ times as large as the electric field scattered by one atom, and the scattered intensity will be $\mathcal{N}^2$ times as large and therefore proportional to the sixth power of the diameter. As the diameter increases, therefore, the Rayleigh scattering cross section increases. When the diameter is comparable to or exceeds the wavelength, however, the rate of increase of scattered intensity with diameter diminishes because different atoms are no longer driven with the same phase of the incident field; the fields scattered by different atomic dipoles are no longer all in phase. Then we are in the regime of *Mie scattering*, in which case the scattering cross section is much less sensitive to the wavelength than the λ^{-4} dependence of the Rayleigh scattering cross section; it is even possible for red light to be more strongly scattered than blue when the particle size is comparable to the wavelength. The scattering is strongest in the forward direction, and the light scattered at 90° is not completely polarized. This diminution of the scattering cross section with increasing particle size is obviously most pronounced for shorter wavelengths and results in the scattered light being "whiter" (less blue, more red) than it is for Rayleigh scattering when the incident light is white. Thus, in the absence of absorption, the light scattered by particles that are larger than a wavelength of light looks white when the incident light is white. This is why clouds, if they are not thick rain clouds, are white; the diameters of water droplets and ice grains in clouds are typically on the order of 0.01 mm.

We have ignored multiple scattering, which acts not only to reduce the degree of polarization of Rayleigh-scattered light but also to weaken the wavelength dependence of the scattering, even for particle dimensions small compared with a wavelength. When white light is incident on a medium of such particles, the blue component of the *singly* scattered light is stronger than the red component because of Rayleigh scattering. When this scattered light propagates in the medium, it loses blue more than red, so that, by the time it is incident on a secondary scatterer, its red component has grown relatively larger. But the secondary scattering again enhances the blue over the red; the net result of multiple scattering in an optically thick medium is to give the scattered light approximately the same white color as the incident light, provided that there is no selective absorption to alter the spectrum. (In the present context, "optically thick" at frequency ω refers to a thickness large compared with the inverse of the attenuation coefficient $a_R(\omega)$ for Rayleigh scattering.) In other words, white light scattered or reflected by an optically thick but weakly absorbing medium of

---

[47] Ibid., p. 378.

particles that are small compared with an optical wavelength is white. This is why milk, consisting predominantly of scatterers (fat globules) that are small compared with optical wavelengths, is white, whereas a small amount of milk in water gives a bluish hue. Multiple scattering also explains why skylight appears whiter at the horizon than at the zenith: shorter wavelengths scatter more light to our eyes, but longer wavelengths suffer less scattering out of the line of sight, and the net effect for the long path from the horizon to our eyes is that the light is white.

Exercise 1.7: (a) Verify Rayleigh's computations for the attenuation coefficient as cited above, assuming Maxwell's estimate for the molecular number density of air at standard temperature and pressure. (b) Consider the propagation of light in the atmosphere, assuming a refractive index $n \cong 1.0003$ for air, and conditions of standard temperature and pressure, for which the density N of air molecules is 2.69×10^{19} cm^{-3}. Estimate the propagation distance for which the light decreases in intensity by 50%.

1.11   Scattering Force and the Optical Theorem

When a non-absorbing polarizable particle scatters radiation from a beam of incident light, there is no loss of total field energy; energy lost by the incident beam goes into energy of the scattered radiation. A loss of energy and momentum along the "forward" direction of propagation of the incident beam results in a force acting to push the scattering particle in the forward direction. We will calculate this force as a special case of the rather general formula (1.8.16). We will assume that the (Rayleigh) scatterer is a small, non-absorbing dielectric sphere whose polarizability for radiation of frequency ω is given by (1.5.59), and that it is surrounded by a non-absorbing medium of refractive index $n_b(\omega)$.

We will focus our attention on the term proportional to $\alpha_I(\omega)$ in the formula (1.8.16). Although we associated this contribution to the force with absorption, it occurs even if the sphere does not absorb any radiation of frequency ω. In other words, the polarizability of the particle must have an imaginary part even if there is no absorption as such. This is simply a consequence of energy conservation: a monochromatic plane wave propagating in the z direction varies with z as $\exp(ikz) = \exp(ik_R z)\exp(-k_I z)$, and so the (intensity) attenuation coefficient is

$$a(\omega) = 2k_I = 2\frac{\omega}{c}n_I(\omega). \tag{1.11.1}$$

The complex refractive index $n(\omega)$ is given by

$$n^2(\omega) = n_R^2 - n_I^2 + 2in_R n_I = 1 + N_b\alpha_b(\omega)/\epsilon_0 + N\alpha(\omega)/\epsilon_0$$
$$= n_b^2(\omega) + N\alpha(\omega)/\epsilon_0, \tag{1.11.2}$$

where N_b and $\alpha_b(\omega)$ are, respectively, the number density and the polarizability that determine the (real) refractive index of the "background" medium in which the Rayleigh scatterers are embedded, these having a polarizability $\alpha(\omega) = \alpha_R(\omega) + i\alpha_I(\omega)$. It follows that

$$n_I(\omega) = N\alpha_I(\omega)/2\epsilon_0 n_R(\omega) \cong N\alpha_I(\omega)/2\epsilon_0 n_b(\omega). \tag{1.11.3}$$

We are assuming that the density of Rayleigh scatterers is sufficiently small that $n_R^2(\omega) \cong n_b^2(\omega)$, as we want to obtain a relation involving the polarizability of a single scatterer. Then

$$a(\omega) \cong 2\frac{\omega}{c}[N\alpha_I(\omega)/2\epsilon_0 n_b(\omega)]. \tag{1.11.4}$$

In this same approximation, the attenuation coefficient (1.10.4) due to Rayleigh scattering is

$$a_R(\omega) \cong \frac{2}{3\pi N}\left(\frac{\omega}{c}\right)^4 |n(\omega) - 1|^2 = \frac{N}{6\pi\epsilon_0^2}\left(\frac{\omega}{c}\right)^4 |\alpha(\omega)|^2, \tag{1.11.5}$$

where we allow for $\alpha(\omega)$ to be complex by taking its squared modulus. By conservation of energy, assuming there are no loss processes for the incident field aside from scattering, $a(\omega)$ must be equal to the attenuation coefficient for Rayleigh scattering; equating (1.11.4) and (1.11.5), therefore, we obtain a relation between the imaginary part of the polarizability and the squared modulus of the complex polarizability:

$$\alpha_I(\omega) = \frac{1}{4\pi\epsilon_0}n_b(\omega)\frac{2\omega^3}{3c^3}|\alpha(\omega)|^2. \tag{1.11.6}$$

Then, from (1.5.59) and the term in the total force (1.8.16) proportional to $\alpha_I(\omega)$, we obtain the scattering force on the dielectric sphere in terms of the incident beam intensity I, the radius a of the sphere, and the refractive indices n_b and n of the materials constituting the background medium and the sphere, respectively:

$$F_{\text{scat}} = \frac{8\pi}{3}\left(\frac{\omega}{c}\right)^4 \frac{n_b^5(\omega)I}{c}\left[\frac{n^2(\omega) - n_b^2(\omega)}{n^2(\omega) + 2n_b^2(\omega)}\right]^2 a^6. \tag{1.11.7}$$

This expression for the scattering force appears in the theory of laser tweezers. In that application, the scattering force must be kept small compared with the force acting to confine a particle within a laser beam.

The relation (1.11.6) is a special case of the *optical theorem*, an exact and general relation between the total cross section $\sigma_{\text{tot}}(\omega)$ and the imaginary part of the forward scattering amplitude $f(\omega)$. In the present context, the forward ($\theta = \pi/2$) scattering amplitude may be read off from the electric field (1.4.29) in the radiation zone and related to the incoming field $\mathbf{E}_{\text{inc}}$ that induces the dipole moment $p_0\hat{\mathbf{z}}$:

$$\mathbf{E}_{\text{scat}} = \frac{p_0\hat{\mathbf{z}}}{4\pi\epsilon_0}\frac{\omega^2}{c^2}\frac{e^{-i\omega(t-r/c)}}{r} = \frac{\alpha(\omega)}{4\pi\epsilon_0}\mathbf{E}_{\text{inc}}\frac{\omega^2}{c^2}\frac{e^{i\omega r/c}}{r} = f(\omega)\mathbf{E}_{\text{inc}}\frac{e^{i\omega r/c}}{r}, \tag{1.11.8}$$

where we define the forward scattering amplitude

$$f(\omega) = \frac{\alpha(\omega)}{4\pi\epsilon_0}\frac{\omega^2}{c^2}.$$
(1.11.9)

Now, the total cross section $\sigma_{\text{tot}}(\omega)$, which accounts for attenuation of the incident field due to absorption as well as scattering, is defined by writing the attenuation coefficient as

$$a(\omega) = N\sigma_{\text{tot}}(\omega) = 2k_I = 2\frac{\omega}{c}n_I(\omega) = 2\frac{\omega}{c}\left[\frac{N\alpha_I(\omega)}{2\epsilon_0}\right]$$
(1.11.10)

in the dilute-medium limit needed to deduce a relation appropriate to the absorption and scattering by a single particle in vacuum ($n_b = 1$). From (1.11.9) and (1.11.10), therefore,

$$\sigma_{\text{tot}}(\omega) = 4\pi\frac{c}{\omega}f_I(\omega),$$
(1.11.11)

which is the general statement of the optical theorem. We have deduced it in the classical electromagnetic context and related it, in the particular case of Rayleigh scattering, to the conservation of energy. In the quantum theory of scattering, the optical theorem is a statement of conservation of probability, or unitarity.

Exercise 1.8: Derive the relation between the refractive index and the forward scattering amplitude for a dilute medium.

In our derivation of the Rayleigh scattering cross section and attenuation coefficient, we took the polarizability $\alpha(\omega)$ to be real, whereas the optical theorem requires that it be complex. To resolve this difference, we return to the classical electron oscillator model and recall first that a force $\mathbf{F}(t)$ acting on an electron causes its energy to change at the rate

$$\frac{dW}{dt} = \mathbf{F}\cdot\dot{\mathbf{x}}.$$
(1.11.12)

If this force is to result in a radiated power (1.4.23), it must evidently satisfy

$$\int_{t_1}^{t_2} dt\,\mathbf{F}(t)\cdot\dot{\mathbf{x}}(t) = -\frac{1}{4\pi\epsilon_0}\frac{2}{3c^3}\int_{t_1}^{t_2} dt\,\ddot{p}^2(t)$$

$$= -\frac{e^2}{6\pi\epsilon_0 c^3}\left[\ddot{\mathbf{x}}(t)\cdot\dot{\mathbf{x}}(t)\,\big|_{t_1}^{t_2} - \int_{t_1}^{t_2} dt\,\dddot{\mathbf{x}}(t)\cdot\dot{\mathbf{x}}(t)\right],$$
(1.11.13)

where the second equality follows from an integration by parts and the definition $\mathbf{p} = e\mathbf{x}$ of the electric dipole moment. We assume that the electron undergoes periodic (or nearly periodic) motion and that the time difference $t_2 - t_1$ is equal to some integral number of periods. Then, we conclude that $\mathbf{F}(t) = (e^2/6\pi\epsilon_0 c^3)\,\dddot{\mathbf{x}}\,(t)$, or

$$\mathbf{E}_{\mathrm{RR}}(t) = \frac{e}{6\pi\epsilon_0 c^3}\,\dddot{\mathbf{x}}\,(t), \tag{1.11.14}$$

where $\mathbf{E}_{\mathrm{RR}}(t) = \mathbf{F}(t)/e$ is the *radiation reaction field*. We can obtain $\mathbf{E}_{\mathrm{RR}}$ by calculating directly the electric field of the electron at the position of the electron, but our simplistic derivation here will suffice for present purposes.[48]

The electron oscillator equation of motion that includes the radiation reaction field $\mathbf{E}_{\mathrm{RR}}$, as well as the applied field $\mathbf{E}$, is (cf. (1.6.2))

$$\ddot{\mathbf{x}} + \omega_0^2 \mathbf{x} = \frac{e}{m}[\mathbf{E} + \mathbf{E}_{\mathrm{RR}}] = \frac{e}{m}\mathbf{E} + \frac{e^2}{6\pi\epsilon_0 mc^3}\,\dddot{\mathbf{x}}\,.^{49} \tag{1.11.15}$$

This modifies the solution given in (1.5.15) for an electron in a field $\mathbf{E}(t) = \mathbf{E}_0 \exp(-i\omega t)$, and implies the polarizability

$$\alpha(\omega) = \frac{e^2/m}{\omega_0^2 - \omega^2 - i\tau_e\omega^3}, \qquad \tau_e = e^2/6\pi\epsilon_0 mc^3 = 6.3 \times 10^{-24}\ \mathrm{s}. \tag{1.11.16}$$

It is easily checked that this polarizability satisfies the optical theorem (1.11.6) for a single scatterer in vacuum ($n_b = 1$), that is, the optical theorem for Rayleigh scattering. For a medium with $n_b \neq 1$ we must account for the fact that the dipole radiation rate is proportional to n_b (see Exercise 1.2) in order to satisfy the optical theorem.

Now, from $\mathbf{p} = \alpha(\omega)\mathbf{E}_0 \exp(-i\omega t)$, we have

$$\ddot{\mathbf{p}}^2 = \frac{1}{2}\omega^4[\alpha_R^2(\omega) + \alpha_I^2(\omega)]\mathbf{E}_0^2 = \frac{1}{2}\omega^4|\alpha(\omega)|^2\mathbf{E}_0^2 \tag{1.11.17}$$

when we average over a period of oscillation of the applied field. The approximation that $\alpha(\omega)$ is purely real in our derivation of the Rayleigh scattering cross section is therefore justified if $|\alpha(\omega)|^2 \cong \alpha_R^2(\omega)$, or $\tau_e\omega^3 \ll |\omega_0^2 - \omega^2|$.

1.12   Thomson Scattering

For N_e free, approximately collisionless, and independently scattering electrons per unit volume (cf. (1.6.4)),

$$n^2(\omega) - 1 = -\omega_p^2/\omega^2 = -N_e e^2/m\epsilon_0\omega^2, \tag{1.12.1}$$

and the differential cross section obtained in Exercise 1.6 for unpolarized incident light reduces to

---

[48] In a direct calculation (Section 4.3) one also obtains a term proportional to $\ddot{\mathbf{x}}$ that is associated with the "electromagnetic mass" of the electron. This mass diverges for a point electron, and the divergence persists in a fully relativistic quantum-electrodynamical calculation. It is removed by *mass renormalization*.

[49] This equation of motion presents some well-known peculiarities. For one thing, the polarizability (1.11.16) does not satisfy the Kramers–Kronig dispersion relations. See Section 6.8.

$$\frac{d\sigma}{d\Omega} = \Big(\frac{e^2}{4\pi\epsilon_0 mc^2}\Big)^2 \frac{1}{2}(1 + \cos^2\theta) = \frac{1}{2}r_0^2(1 + \cos^2\theta), \qquad (1.12.2)$$

where r_0 is the "classical electron radius."[50] The total cross section is obtained by integrating over all solid angles:

$$\sigma_T = \frac{8\pi}{3}r_0^2 = 6.65 \times 10^{-29}\ \mathrm{m}^2, \qquad (1.12.3)$$

which is the total cross section for *Thomson scattering*. The attenuation coefficient due to Thomson scattering by N_e electrons per unit volume, $N_e\sigma_T$, is independent of the field frequency.

The differential cross section for Thomson scattering can also be derived as the classical limit of the Klein–Nishina formula for the differential cross section for Compton scattering:

$$\frac{d\sigma_C}{d\Omega} = \frac{1}{2}\Big(\frac{e^2}{4\pi\epsilon_0 mc^2}\Big)^2 \Big(\frac{\omega_f}{\omega_i}\Big)^2 \Big(\frac{\omega_f}{\omega_i} + \frac{\omega_i}{\omega_f} - \sin^2\theta\Big), \qquad (1.12.4)$$

where ω_i and ω_f are the frequencies of the incident and scattered photons, respectively, and are related by the Compton formula

$$\omega_f = \frac{\omega_i}{1 + (\hbar\omega_i/mc^2)(1 - \cos\theta)}, \qquad (1.12.5)$$

or, expressed in wavelengths, $\lambda_f - \lambda_i = (\hbar/mc)(1 - \cos\theta)$. In the classical limit, $\hbar\omega_i/mc^2 \to 0$, $\omega_f \to \omega_i$ and (1.12.4) reduces to (1.12.2).

Exercise 1.9: Suppose that the molecules in the air around you were somehow completely ionized. Show that the propagation distance over which radiation is nearly fully attenuated due to Thomson scattering would be a mere few meters.

According to the "standard model" of cosmology, the universe was originally so hot ($\sim 10^{11}$ K) that electrons could not be bound, and the Thomson scattering caused by unbound electrons made matter opaque. Matter and radiation were strongly interacting and in thermal equilibrium. But as the universe expanded and cooled to around 3000 K after $\sim 10^5$ years, the electrons became bound to protons, the scattering of radiation became much weaker, and the universe became transparent. The cosmic microwave background radiation (Figure 2.5) started out as thermal equilibrium ("blackbody") radiation at approximately 3000 K and became the presently observed background radiation at 2.7 K with the expansion and cooling of the universe.

A simplistic estimate can be made, based on Thomson scattering, of the time it takes for light to make its way from the center of the Sun to the surface. The Sun has a density $\rho \sim 10^3$ kg/m^3, and, if we assume it consists in large measure of ionized hydrogen, this implies

---

[50] For incident and scattered fields characterized by the unit linear polarization vectors $\mathbf{e}_i$ and $\mathbf{e}_s$, $d\sigma/d\Omega = r_0^2(\mathbf{e}_i \cdot \mathbf{e}_s)^2$. Equation (1.12.2) for unpolarized light is obtained by averaging over two (orthogonal) polarizations of the incident field and summing over (unobserved) polarizations of the scattered field.

an electron number density $N_e \sim \rho/m_p \sim 10^{30}$ m^{-3} (m_p denotes the proton mass). The attenuation coefficient due to Thomson scattering is therefore $N_e \sigma_T \sim 66.5$ m^{-1}, implying a mean free path between Thomson scatterings of $d \sim 1/(N_e \sigma_T) \sim 0.01$ m for an "average photon." Now if the average photon is created in a nuclear reaction at the center of the Sun and reaches the "surface" of the Sun after $\mathcal{N}$ Thomson scatterings, then $R^2 \sim \mathcal{N} d^2$, where $R \sim 7 \times 10^8$ m is the solar radius; this assumes a photon diffusion process in which the average displacement of a photon in a collision with an electron is zero, and the mean-square displacement is d^2.[51] Then, the total distance traversed by an average photon as it makes its way to the surface is $\sim \mathcal{N} d = (R^2/d^2)d$ and the time it takes for it to diffuse out to the surface from the center is $t_d \sim (R^2/d)/c \sim 10^{11}$ s $\sim 10^4$ years: "sunlight that we receive today was created by nuclear reactions at the centre of the Sun at a time when our ancestors were fighting woolly mammoths and sabre-toothed tigers."[52]

References and Suggested Additional Reading

For further analysis of the storage and radiation fields of an electric dipole, see, for instance, L. Mandel, J. Opt. Soc. Am. **62**, 1011 (1972). See also P. R. Berman, Am. J. Phys. **76**, 48 (2008). The Hannay theorem mentioned in our discussion is proved in J. H. Hannay, Proc. R. Soc. Lond., A**452**, 2351 (1996).

A rigorous treatment of the Ewald–Oseen extinction theorem may be found in *Born and Wolf*, Section 2.4.2, "The Ewald–Oseen extinction theorem and a rigorous derivation of the Lorentz–Lorenz formula," pp. 105–110.

The discussion of the energy density for field frequencies where absorption is negligible follows parts of R. Loudon, J. Phys. A: Math. Gen. **3**, 233 (1970), Yu. S. Barash and V. L. Ginzburg, Sov. Phys. Usp. **19**, 263 (1976), and *Ginzburg*.

The experiments of Lebedev and of Nichols and Hull are reviewed by A. V. Masalov and E. Garmire, respectively, in *Quantum Photonics: Pioneering Advances and Emerging Applications*, eds. R. W. Boyd, S. G. Lukishova, and V. N. Zadkov, Springer, Berlin, 2019.

There is extensive literature on the different expressions for electromagnetic momenta in material media, especially the Abraham and Minkowski forms. Reviews of theory and experiment include, among others, F. N. H. Robinson, Phys. Rep. **16**, 313 (1975); I. Brevik, Phys. Rep. **52**, 133 (1979); R. N. Pfeifer, T. A. Nieminen, N. R. Heckenberg, and H. Rubinsztein-Dunlop, Rev. Mod. Phys. **79**, 1197 (2007); S. M. Barnett and R. Loudon, Phil. Trans. R. Soc. A **368**, 927 (2010); C. Baxter and R. Loudon, J. Mod. Opt. **57**, 830 (2010); P. W. Milonni and R. W. Boyd, Adv. Optics and Photonics **2**,

---

[51] Diffusion implies $R \propto \sqrt{\mathcal{N}} d$ (see Section 6.1).

[52] A. R. Choudhuri, *Astrophysics for Physicists*, Cambridge University Press, Cambridge, 2010, p. 56.

519 (2010) and B. A. Kemp, *Progress in Optics* **60**, ed. E. Wolf, Elsevier, Amsterdam, 2015, pp. 437–488. The Abraham and Minkowski momenta are kinetic and canonical momenta, respectively, as shown by S. M. Barnett, Phys. Rev. Lett. **104**, 070401 (2010). The dispersive contribution to the force (1.8.6) can be understood from work of D. F. Nelson, Phys. Rev. A **44**, 3985 (1991), and H. Washimi and V. I. Karpman, Sov. Phys. JETP **44**, 528 (1977). Among many other papers on forces in and on dielectric media are R. Loudon, L. Allen and D. F. Nelson, Phys. Rev. E **55**, 1071 (1997); S. M. Barnett and R. Loudon, J. Phys. B: At. Mol. Opt. Phys. **39**, S671 (2006); and several other papers by these authors. For a succinct survey of various aspects of electromagnetic momentum, including many references, see D. J. Griffiths, Am. J. Phys. **80**, 7 (2012).

The classic experiments confirming that the recoil momentum of objects in a dielectric medium of refractive index n is consistent with a momentum $n\hbar\omega/c$ per unit of energy $\hbar\omega$ of the field were carried out by R. V. Jones and J. C. S. Richards, Proc. R. Soc. Lond. A **221**, 480 (1954), and R. V. Jones and B. Leslie, Proc. R. Soc. Lond. A **360**, 347 (1978). These experiments measured the recoil of irradiated mirrors suspended in different fluids. More recently, this momentum was observed in the recoil of irradiated atoms in a Bose–Einstein condensate: G. K. Campbell, A. E. Leanhardt, J. Mun, M. Boyd, E. W. Streed, W. Ketterle, and D. E. Pritchard, Phys. Rev. Lett. **94**, 170403 (2005). In all these experiments, the recoil momentum was found to depend on n but not n_g, as expected from the analyses in Sections 1.7 and 1.8.

Various aspects of light scattering by atoms and gases are treated from a quantum-theoretical perspective by P. R. Berman, Contemp. Phys. **49**, 313 (2008). For a treatment of the role of interparticle correlations in the suppression of Rayleigh scattering, based on simplified models, see A. G. Rojo and P. R. Berman, Am. J. Phys. **78**, 94 (2010).

Bohren and Huffman is recommended for readers interested in pursuing further the physics of light scattering in the atmosphere. See also C. F. Bohren, "Atmospheric Optics," in *The Optics Encyclopedia*, T. G. Brown et al., eds., Wiley-VCH, Weinheim, 2004, pp. 53–91; C. F. Bohren, *Clouds in a Glass of Beer. Simple Experiments in Atmospheric Physics*, Dover Publications, Mineola, New York, 1987 and C. F. Bohren, *What Light through Yonder Window Breaks? More Experiments in Atmospheric Physics*, Dover Publications, Mineola, New York, 1991.

2

Atoms in Light: Semiclassical Theory

In this chapter, we formulate the interaction of an atom with an optical field in the electric dipole approximation and in semiclassical radiation theory, the theory in which matter is treated quantum mechanically while the field is described by the *classical* Maxwell equations. We show that the most familiar experimental facts about the photoelectric effect are well described without having to invoke the quantum (photon) nature of light. We derive the Kramers–Heisenberg dispersion formula and review the two-state model of an atom and the optical Bloch equations, all within semiclassical theory. As a prelude to the quantization of the field in Chapter 3, we review some salient features of blackbody radiation and its beautiful analysis by Einstein, who thus gave some of the most compelling evidence for the quantum nature of light. Einstein's fluctuation formula is used to interpret the "photon bunching" measured in the Hanbury Brown–Twiss experiment, which marked the beginning of modern quantum optics.

2.1 Atom–Field Interaction

Consider, for simplicity, a one-electron atom. The electron, with negative charge e, and the nucleus, with charge $-e$, define an electric dipole moment $\mathbf{d} = e\mathbf{x}$, where $\mathbf{x}$ is the vector pointing from the nucleus to the electron. We will, for now, assume that the nucleus is fixed at a point $\mathbf{R} = 0$ and that the electron displacement $\mathbf{x}$ is small compared with field wavelengths of interest, which is an excellent approximation for optical wavelengths. Then, the interaction Hamiltonian for the atom in an applied electric field $\mathbf{E}$ may be taken to have the electric dipole form

$$H_{\mathrm{I}}(t) = -\mathbf{d} \cdot \mathbf{E}(t) \tag{2.1.1}$$

familiar from classical electromagnetism. $\mathbf{d}$ is a quantum-mechanical operator, whereas the electric field $\mathbf{E}(t)$ in the semiclassical theory is a prescribed, ordinary function of time, not an operator.

In quantum optics, it is usually sufficient to characterize an atom simply in terms of its energy levels and transition electric dipole moments. Let H_0 be the Hamiltonian for the electron in the Coulomb field of the nucleus, and denote the eigenstates and eigenvalues of H_0 by $|i\rangle$ and E_i, respectively: $H_0|i\rangle = E_i|i\rangle$. The states $|i\rangle$ form a complete set, $\sum_i |i\rangle\langle i| = 1$, so we can write

An Introduction to Quantum Optics and Quantum Fluctuations. Peter W. Milonni.
© Peter W. Milonni 2019. Published in 2019 by Oxford University Press.
DOI:10.1093/oso/9780199215614.001.0001

$$H_0 = \left(\sum_i |i\rangle\langle i| \right) H_0 \left(\sum_j |j\rangle\langle j| \right) = \sum_i \sum_j |i\rangle E_j \langle i|j\rangle\langle j| = \sum_i E_i |i\rangle\langle i|, \qquad (2.1.2)$$

where we have used the theorem that the eigenstates corresponding to different eigenvalues of the Hermitian operator H_0 are orthogonal (and assumed to be normalized to unity): $\langle i|j\rangle = \delta_{ij}$, the Kronecker delta. Similarly,

$$\mathbf{d} = \left(\sum_i |i\rangle\langle i| \right) \mathbf{d} \left(\sum_j |j\rangle\langle j| \right) = \sum_i \sum_j \mathbf{d}_{ij} |i\rangle\langle j|. \qquad (2.1.3)$$

$\mathbf{d}_{ij} = \langle i|\mathbf{d}|j\rangle$ is the electric dipole matrix element between states $|i\rangle$ and $|j\rangle$:

$$\mathbf{d}_{ij} = e \int d^3x \, \phi_i^*(\mathbf{x}) \mathbf{x} \phi_j(\mathbf{x}), \qquad (2.1.4)$$

where $\phi_i(\mathbf{x}) = \langle \mathbf{x}|i\rangle$ is the Schrödinger wave function for the electron state with energy eigenvalue E_i. Thus, the Hamiltonian describing the atom in a prescribed electric field $\mathbf{E}(t)$ is

$$H = H_0 + H_{\mathrm{I}} = \sum_i E_i |i\rangle\langle i| - \sum_{i,j} \mathbf{d}_{ij} \cdot \mathbf{E}(t) |i\rangle\langle j|. \qquad (2.1.5)$$

The variation in time of a quantum system can be described in different ways, or "pictures," all of which involve the time evolution operator $U(t)$ satisfying

$$i\hbar \frac{\partial U}{\partial t} = HU. \qquad (2.1.6)$$

$U(t)$ is a unitary operator: $U^\dagger(t)U(t) = U(t)U^\dagger(t) = 1$, which follows from the fact that H is Hermitian ($H^\dagger = H$).

The assumption that the set of all eigenvectors of the Hamiltonian comprises a complete set is well justified for our purposes: completeness is a property of the eigenvectors of any Hermitian operator that is bounded from below but not from above.[1] An operator H is said to be bounded from below if, for any state vector $|\psi\rangle$, $\langle \psi|H|\psi\rangle/\langle \psi|\psi\rangle$ is larger than some fixed number. It is not bounded from above if, for any real number R, there exists a $|\psi\rangle$ such that $\langle \psi|H|\psi\rangle/\langle \psi|\psi\rangle > R$. In the case of a Hermitian operator H acting in a space of *finite* dimension d, completeness follows from the fact that there are d orthogonal and therefore linearly independent eigenvectors of H. It will not be necessary for us to distinguish between Hermitian operators and self-adjoint operators.

We now review briefly and in general terms the Schrödinger, interaction, and Heisenberg pictures.

---

[1] For a proof of this completeness theorem, see T. D. Lee, *Particle Physics and Field Theory*, Volume I, Harwood Academic Publishers, Chur, 1981, pp. 12–13.

2.1.1 Schrödinger Picture

In the Schrödinger picture, the state vector $|\psi(t)\rangle$ of the system evolves in time from $t = 0$ as

$$|\psi(t)\rangle = U(t)|\psi(0)\rangle, \qquad (2.1.7)$$

and (2.1.6) implies the time-dependent Schrödinger equation

$$i\hbar\frac{\partial}{\partial t}|\psi\rangle = H|\psi\rangle = [H_0 + H_{\mathrm{I}}(t)]|\psi\rangle. \qquad (2.1.8)$$

Unitary time evolution implies $\langle\psi(t)|\psi(t)\rangle = \langle\psi(0)|\psi(0)\rangle$, which is equal to 1 when $|\psi(0)\rangle$ is appropriately normalized. Unitary time evolution thus ensures "conservation of probability."

It is very often convenient to expand $|\psi(t)\rangle$ in terms of the (complete) set of eigenstates of the unperturbed Hamiltonian H_0:

$$|\psi(t)\rangle = \sum_i c_i(t)|i\rangle. \qquad (2.1.9)$$

Then, (2.1.8) and the orthonormality of the states $|i\rangle$ result in the ordinary differential equations satisfied by the expansion coefficients $c_i(t)$:

$$i\hbar\dot{c}_i(t) = E_i c_i(t) + \sum_j \langle i|H_{\mathrm{I}}(t)|j\rangle c_j(t). \qquad (2.1.10)$$

$|c_i(t)|^2$ is interpreted as the probability that a single measurement at time t will find the system in the state $|i\rangle$, and $\sum_i |c_i(t)|^2 = \sum_i |c_i(0)|^2 = 1$ is guaranteed by the unitarity of $U(t)$. The form (2.1.10) of the Schrödinger equation is especially useful when only a small number of states $|i\rangle$ have significant occupation probabilities, for then we can approximate it by a small number of coupled, ordinary differential equations.

2.1.2 Interaction Picture

If we write the time evolution operator as

$$U(t) = U_0(t)u(t), \qquad (2.1.11)$$

where $U_0(t)$, satisfying $i\hbar\partial U_0/\partial t = H_0 U_0$, is the (unitary) time evolution operator for the unperturbed system, we obtain from (2.1.6) an equation for $u(t)$:

$$i\hbar\frac{\partial u}{\partial t} = U_0^\dagger(t)H_{\mathrm{I}}(t)U_0(t)u(t) = h_{\mathrm{I}}(t)u(t). \qquad (2.1.12)$$

$h_{\mathrm{I}}(t)$ is the interaction Hamiltonian in the interaction picture. If the interaction described by $h_{\mathrm{I}}(t)$ is assumed to begin at a time $t = 0$, then (2.1.11) implies $u(0) = 1$, and (2.1.12) has the formal solution

$$u(t) = 1 + \frac{1}{i\hbar} \int_0^t dt' h_I(t') + \left(\frac{1}{i\hbar}\right)^2 \int_0^t dt' \int_0^{t'} dt'' h_I(t') h_I(t'') + \ldots \quad (2.1.13)$$

The state vector $|\psi_I(t)\rangle$ in the interaction picture is related to the state vector $|\psi(t)\rangle$ in the Schrödinger picture by $|\psi_I(t)\rangle = U_0^\dagger(t)|\psi(t)\rangle = u(t)|\psi(0)\rangle = u(t)|\psi_I(0)\rangle$ and satisfies

$$i\hbar \frac{\partial}{\partial t} |\psi_I(t)\rangle = h_I(t)|\psi_I(t)\rangle. \quad (2.1.14)$$

The interaction picture is convenient for the perturbative calculation of transition probabilities. If a system is initially in the state $|i\rangle$, an eigenstate of the unperturbed Hamiltonian, then the probability amplitude at time t that it is in the eigenstate $|f\rangle$ of the unperturbed Hamiltonian is

$$\langle f|\psi(t)\rangle = \langle f|U(t)|i\rangle = \langle f|U_0(t)u(t)|i\rangle = e^{-iE_f t/\hbar} \langle f|u(t)|i\rangle, \quad (2.1.15)$$

and the probability for the transition from the initial state to the final state is

$$p_{i \to f}(t) = |\langle f|\psi(t)\rangle|^2 = |\langle f|u(t)|i\rangle|^2 = |\langle f|\psi_I(t)\rangle|^2. \quad (2.1.16)$$

Thus, to lowest order in the perturbation series (2.1.13),

$$\begin{aligned}
p_{i \to f}(t) &\cong \frac{1}{\hbar^2} \left| \int_0^t dt' \langle f|h_I(t')|i\rangle \right|^2 \\
&= \frac{1}{\hbar^2} \left| \int_0^t dt' \langle f|U_0^\dagger(t')H_I(t')U_0(t')|i\rangle \right|^2 \\
&= \frac{1}{\hbar^2} \left| \int_0^t dt' e^{i(E_f - E_i)t'/\hbar} \langle f|H_I(t')|i\rangle \right|^2.
\end{aligned} \quad (2.1.17)$$

2.1.3   Heisenberg Picture

The expectation value at time t of a dynamical variable corresponding to the Hermitian operator A is

$$\langle \psi(t)|A|\psi(t)\rangle = \langle \psi(0)|U^\dagger(t)AU(t)|\psi(0)\rangle = \langle \psi(0)|A_H(t)|\psi(0)\rangle. \quad (2.1.18)$$

$A_H(t) = U^\dagger(t)AU(t) = U^\dagger(t)A_H(0)U(t)$ is an operator in the Heisenberg picture. In this picture, operators, not state vectors, evolve in time. From (2.1.6), it follows that the time evolution of $A_H(t)$ is governed by the *Heisenberg equation of motion*

$$i\hbar \frac{dA_H}{dt} = A_H H - H A_H = [A_H, H], \quad (2.1.19)$$

where it has been assumed that A has no explicit, prescribed dependence on time. If it does have an explicit time dependence, (2.1.19) is replaced by the more general time evolution equation

$$i\hbar \frac{dA_H}{dt} = [A_H, H] + i\hbar \frac{\partial A_H}{\partial t}. \qquad (2.1.20)$$

The Heisenberg picture has the nice feature that the Heisenberg equations have a formal similarity to the corresponding classical equations when there is a corresponding classical system. This is an especially attractive feature when it comes to describing the propagation of light, for then the Heisenberg equations for field operators are formally similar to equations based on the classical Maxwell equations and therefore lend themselves to physical, and in many cases classical-like, interpretations.

2.2 Why the Electric Dipole Interaction?

The interaction Hamiltonian for a charged particle in an electromagnetic field that is fully characterized by the vector potential $\mathbf{A}$ is generally taken to be

$$H_{\mathrm{I}} = -\frac{e}{m}\mathbf{A} \cdot \mathbf{p} + \frac{e^2}{2m}\mathbf{A}^2, \qquad (2.2.1)$$

where e and m are, respectively, the charge and mass of the particle, and $\mathbf{A}$ is the vector potential at its position $\mathbf{x}$. For reasons discussed in Section 4.1, this is often called the minimal coupling interaction. But very often the electric dipole form (2.1.1) is assumed to describe the atom–field interaction. Calculations with it tend to be easier than they are with the minimal coupling form, at least when the dipole approximation is appropriate; we postpone discussion of this point to Section 4.2. Here, we show how the electric dipole form of the interaction can be derived from the minimal coupling form in the semiclassical theory.

Recall first that (2.2.1) follows from the Lagrangian

$$L = \frac{1}{2}m\mathbf{v}^2 + e\mathbf{A} \cdot \mathbf{v} \qquad (2.2.2)$$

for a charged particle in an electric field $\mathbf{E}$ ($\mathbf{E} = -\partial\mathbf{A}/\partial t$, $\mathbf{B} = \nabla\times\mathbf{A}$). This Lagrangian gives the correct equation of motion for the particle. That is, the principle of least action,

$$\frac{d}{dt}\left(\frac{\partial L}{\partial \dot{x}_k}\right) - \frac{\partial L}{\partial x_k} = 0, \qquad (2.2.3)$$

and the Lagrangian (2.2.2) imply

$$m\ddot{\mathbf{x}} = e\mathbf{E} + e\mathbf{v} \times \mathbf{B}. \qquad (2.2.4)$$

The interaction Hamiltonian (2.2.1) follows from the identification

$$H = \sum_k \dot{x}_k \frac{\partial L}{\partial \dot{x}_k} - L = \sum_k p_k \dot{x}_k - L, \qquad (2.2.5)$$

where $p_k = \partial L/\partial \dot{x}_k$ is the momentum conjugate to x_k. Explicitly, $\mathbf{p} = m\mathbf{v} + e\mathbf{A}$, and

$$H = (m\mathbf{v} + e\mathbf{A}) \cdot \mathbf{v} - (\tfrac{1}{2}m\mathbf{v}^2 + e\mathbf{A} \cdot \mathbf{v}) = \tfrac{1}{2}m\mathbf{v}^2 = \frac{1}{2m}(\mathbf{p} - e\mathbf{A})^2 = \frac{\mathbf{p}^2}{2m} + H_{\mathrm{I}}. \quad (2.2.6)$$

For particle displacements small compared with distances over which there is any significant variation of the field, it is often a good approximation to evaluate $\mathbf{A}$ at the center of the region over which the particle moves. Then, we replace $\mathbf{A}(\mathbf{x}, t)$ by $\mathbf{A}(t)$, a prescribed function of time only:

$$L = \frac{1}{2}m\mathbf{v}^2 + e\mathbf{A}(t) \cdot \mathbf{v}, \quad (2.2.7)$$

$$\mathbf{p} = m\mathbf{v} + e\mathbf{A}(t), \quad (2.2.8)$$

$$H = \frac{\mathbf{p}^2}{2m} - \frac{e}{m}\mathbf{A}(t) \cdot \mathbf{p} + \frac{e^2}{2m}\mathbf{A}^2(t), \quad (2.2.9)$$

$$m\ddot{\mathbf{x}} = -e\dot{\mathbf{A}}(t) = e\mathbf{E}(t). \quad (2.2.10)$$

Next, we recall that, according to the principle of least action, we can add a time derivative $(d/dt)S(\mathbf{x}, t)$ to L without affecting any physics. (More precisely, the condition that $\int_{t_1}^{t_2} dt\, L(\mathbf{x}, \dot{\mathbf{x}}, t)$ is stationary—Hamilton's principle—is unaffected when $L \to L + dS/dt$.) Let $S(\mathbf{x}, t) = -e\mathbf{A}(t) \cdot \mathbf{x}$, and consider, instead of (2.2.7), the equally valid Lagrangian

$$L' = \frac{1}{2}m\mathbf{v}^2 + e\mathbf{A}(t) \cdot \mathbf{v} + \frac{d}{dt}[-e\mathbf{A}(t) \cdot \mathbf{x}] = \frac{1}{2}m\mathbf{v}^2 - e\dot{\mathbf{A}}(t) \cdot \mathbf{x}$$

$$= \frac{1}{2}m\mathbf{v}^2 + e\mathbf{x} \cdot \mathbf{E}(t). \quad (2.2.11)$$

With this Lagrangian, $p_k = \partial L'/\partial \dot{x}_k = mv_k$, or

$$\mathbf{p} = m\mathbf{v}, \quad (2.2.12)$$

and

$$H' = m\mathbf{v} \cdot \mathbf{v} - L' = m\mathbf{v}^2 - [\tfrac{1}{2}m\mathbf{v}^2 + e\mathbf{x} \cdot \mathbf{E}(t)] = \frac{\mathbf{p}^2}{2m} - e\mathbf{x} \cdot \mathbf{E}(t), \quad (2.2.13)$$

which implies an electric dipole interaction of the form (2.1.1). Note that $\mathbf{p}$ in this Hamiltonian is the same as the mechanical or "kinetic" momentum $m\mathbf{v}$ defined by (2.2.12). In the original Hamiltonian H, however, $\mathbf{p} = m\mathbf{v} + e\mathbf{A}(t)$ (see (2.2.8)) is not the kinetic momentum but rather the kinetic momentum plus a momentum $e\mathbf{A}(t)$.

This argument justifies the electric dipole form of the interaction for a classically described particle in an electromagnetic field. To extend it to the semiclassical theory, we start from the Schrödinger equation for a charged particle in a prescribed vector potential $\mathbf{A}(t)$, and we also include now a scalar potential $V(\mathbf{x})$:

$$i\hbar \frac{\partial}{\partial t}|\psi\rangle = \left[\frac{\mathbf{p}^2}{2m} + V(\mathbf{x}) - \frac{e}{m}\mathbf{A}(t) \cdot \mathbf{p} + \frac{e^2}{2m}\mathbf{A}^2(t) \right] |\psi\rangle = H|\psi\rangle. \quad (2.2.14)$$

We define a new state vector $|\psi'\rangle$ by writing

$$|\psi\rangle = e^{-iS(\mathbf{x},t)/\hbar}|\psi'\rangle = e^{ie\mathbf{A}(t)\cdot\mathbf{x}/\hbar}|\psi'\rangle = \mathcal{U}(\mathbf{x},t)|\psi'\rangle. \qquad (2.2.15)$$

Then, (2.2.14) becomes

$$i\hbar\mathcal{U}(\mathbf{x},t)\left[\frac{ie}{\hbar}\dot{\mathbf{A}}(t)\cdot\mathbf{x} + \frac{\partial}{\partial t}\right]|\psi'\rangle = \left[\frac{\mathbf{p}^2}{2m} + V(\mathbf{x}) - \frac{e}{m}\mathbf{A}(t)\cdot\mathbf{p}\right.$$
$$\left. + \frac{e^2}{2m}\mathbf{A}^2(t)\right]\mathcal{U}(\mathbf{x},t)|\psi'\rangle, \qquad (2.2.16)$$

or

$$e\mathbf{x}\cdot\mathbf{E}(t)|\psi'\rangle + i\hbar\frac{\partial}{\partial t}|\psi'\rangle = \mathcal{U}^\dagger(\mathbf{x},t)\left[\frac{\mathbf{p}^2}{2m} + V(\mathbf{x}) - \frac{e}{m}\mathbf{A}(t)\cdot\mathbf{p}\right.$$
$$\left. + \frac{e^2}{2m}\mathbf{A}^2(t)\right]\mathcal{U}(\mathbf{x},t)|\psi'\rangle, \qquad (2.2.17)$$

which follows from the unitarity of $\mathcal{U}: \mathcal{U}^\dagger(\mathbf{x},t)\mathcal{U}(\mathbf{x},t) = \mathcal{U}(\mathbf{x},t)\mathcal{U}^\dagger(\mathbf{x},t) = 1$. From the general operator identity

$$e^B C e^{-B} = C + [B,C] + \frac{1}{2!}[B,[B,C]] + \dots \qquad (2.2.18)$$

we obtain

$$\mathcal{U}^\dagger(\mathbf{x},t)\mathbf{p}\,\mathcal{U}(\mathbf{x},t) = \mathbf{p} + e\mathbf{A}(t) \qquad (2.2.19)$$

and

$$i\hbar\frac{\partial}{\partial t}|\psi'\rangle = \left[\frac{\mathbf{p}^2}{2m} + V(\mathbf{x}) - e\mathbf{x}\cdot\mathbf{E}(t)\right]|\psi'\rangle = H'|\psi'\rangle, \qquad (2.2.20)$$

which is the Schrödinger equation with the electric dipole interaction Hamiltonian (2.1.1).[2] A similar transformation can be carried out when the field is quantized, except that an additional term appears in the transformed Hamiltonian (Section 4.1).

The operator $\mathbf{p}$ appearing in the Hamiltonian (2.2.20) is not only a canonical momentum, which satisfies $[x_i, p_j] = i\hbar\delta_{ij}$ and takes the form $(\hbar/i)\nabla$ in the coordinate representation of state vectors; it is also the "kinetic" (or "mechanical") momentum $m\mathbf{v} = m\dot{\mathbf{x}}$, that is, $m\dot{\mathbf{x}} = m\partial H'/\partial\mathbf{p} = \mathbf{p}$.

The operator $\mathbf{p}$ in the Hamiltonian (2.2.14), however, while it is likewise a canonical momentum, is not the kinetic momentum, since, in that Hamiltonian, $\mathbf{p} = m\mathbf{v} + e\mathbf{A}(t)$ (see (2.2.8)).

The question naturally arises concerning the initial ($t = 0$) state of the atom, which might, for instance, be the hydrogen ground-state wave function $\phi_0(\mathbf{r}) = \langle\mathbf{r}|\phi_0\rangle = (\pi a_0^3)^{-1/2}e^{-r/a_0}$, where a_0 is the Bohr radius: should we associate this with the initial

---

[2] The identity (2.2.18) can be proved by defining $f(x) = e^{xB}Ce^{-xB}$ and expanding $f(x)$ in a Taylor series about $x = 0$. Here, B and C are operators, and x is an ordinary number (sometimes called a *c-number*, c denoting "classical").

$|\psi\rangle$ of (2.2.14) or the initial $|\psi'\rangle$ of (2.2.20)? Recall again that it is the canonical momentum $\mathbf{p}$ (such that $[x_i, p_j] = i\hbar\delta_{ij}$) that takes the form $(\hbar/i)\nabla$ in the coordinate representation. When we formulate the atom–field interaction according to (2.2.20), the canonical momentum is $\mathbf{p} = m\mathbf{v}$ (see (2.2.12)) and

$$\langle\psi'(0)|m\mathbf{v}|\psi'(0)\rangle = \int d^3r\phi_0^*(\mathbf{r})m\frac{\hbar}{i}\nabla\phi_0(\mathbf{r}) = 0, \qquad (2.2.21)$$

if we take the initial state $|\psi'(0)\rangle$ to be $|\phi_0\rangle$. This is as expected: the average electron velocity for the ground state of the atom is zero. But, in the Hamiltonian (2.2.14), the canonical momentum $\mathbf{p}$ is equal to $m\mathbf{v} + e\mathbf{A}(t)$ (see (2.2.8)), which would, in general, imply a non-vanishing (and possibly gauge-dependent) average electron velocity $\langle\psi(0)|m\mathbf{v}|\psi(0)\rangle$ if we take $|\psi(0)\rangle = |\phi_0\rangle$. To obtain the desired result $\langle m\mathbf{v}\rangle = 0$ when (2.2.14) is used, we must instead take the initial state vector to be the transformed state $|\psi(0)\rangle = e^{ie\mathbf{A}(t)\cdot\mathbf{x}/\hbar}|\psi'(0)\rangle = e^{ie\mathbf{A}(t)\cdot\mathbf{x}/\hbar}|\phi_0\rangle$, as required by (2.2.15).[3] Then,

$$\begin{aligned}
\langle\psi(0)|m\mathbf{v}|\psi(0)\rangle &= \langle|\psi'(0)|e^{-ie\mathbf{A}(t)\cdot\mathbf{x}/\hbar}[\mathbf{p} - e\mathbf{A}(t)]e^{ie\mathbf{A}(t)\cdot\mathbf{x}/\hbar}|\psi'(0)\rangle \\
&= \langle\psi'(0)|\mathbf{p}|\psi'(0)\rangle = \langle\phi_0|\mathbf{p}|\phi_0\rangle \\
&= \int d^3r\phi_0^*(\mathbf{r})m\frac{\hbar}{i}\nabla\phi_0(\mathbf{r}) = 0, \qquad (2.2.22)
\end{aligned}$$

where we have used (2.2.19). This example suggests that, when the dipole approximation is valid, the electric dipole form of the Hamiltonian is more convenient than the minimal coupling form. And because it involves the electric field rather than the vector potential, the electric dipole form of the atom–field interaction is now almost always used in quantum optics. This is discussed further in Section 4.2.

More generally, a time-dependent transformation $|\psi\rangle = \mathcal{U}|\psi'\rangle$ transforms the Schrödinger equation $i\hbar(\partial/\partial t)|\psi\rangle = H|\psi\rangle$ to

$$i\hbar\frac{\partial}{\partial t}|\psi'\rangle = \left(\mathcal{U}^\dagger H\mathcal{U} - i\mathcal{U}^\dagger\frac{\partial\mathcal{U}}{\partial t}\right)|\psi'\rangle = H'|\psi'\rangle. \qquad (2.2.23)$$

The Hamiltonian H' is obviously not simply a unitary transformation of H.[4] For example,

$$\langle\psi|H|\psi\rangle = \langle\psi'|\mathcal{U}^\dagger H\mathcal{U}|\psi'\rangle \neq \langle\psi'|H'|\psi'\rangle. \qquad (2.2.24)$$

For the transformation we have considered, where the minimal coupling Hamiltonian is replaced by the electric dipole Hamiltonian,

[3] K. Rzążewski and R. W. Boyd, J. Mod. Opt. **51**, 1137 (2004). See also W. E. Lamb, Jr., R. R. Schlicher, and M. O. Scully, Phys. Rev. A **36**, 2763 (1987).

[4] This aspect of such time-dependent transformations is discussed by M. M. Nieto, Phys. Rev. Lett. **38**, 1042 (1977). He remarks that "an external-field single-particle Hamiltonian conceals the fact that physically it is the *total* coupled-field Hamiltonian which need be invariant. The coupled Hamiltonian only implies that the energy of the entire system is conserved under a time-dependent transformation, not necessarily a piece of it (the external-field problem)" (p. 1043). In our semiclassical radiation theory, we do not include the electromagnetic field energy as part of the Hamiltonian; the field is treated only as an "external" perturbation with a prescribed time dependence that is unaffected by the atom.

$$\langle\psi|H|\psi\rangle = \langle\psi|\big[\frac{1}{2m}(\mathbf{p}-e\mathbf{A})^2 + V(\mathbf{x})\big]|\psi\rangle = \langle\psi'|\big[\frac{\mathbf{p}^2}{2m}+V(\mathbf{x})\big]|\psi'\rangle, \qquad (2.2.25)$$

whereas

$$\langle\psi'|H'|\psi'\rangle = \langle\psi'|\big[\frac{\mathbf{p}^2}{2m}+V(\mathbf{x})-e\mathbf{x}\cdot\mathbf{E}\big]|\psi'\rangle. \qquad (2.2.26)$$

Although the minimal coupling and electric dipole Hamiltonians are equivalent, they may not appear so in approximate calculations. The transformation to the electric dipole form involves the commutation relation $[x_i, p_j] = i\hbar\delta_{ij}$, which requires an infinite-dimensional Hilbert space. In *approximate* calculations involving a restricted set of states, therefore, it should not be surprising that the two forms of interaction can yield different results. Recall the remarks following (2.1.10).

If $[x, p] = i\hbar$ were to hold in an n-dimensional Hilbert space, we would have $\text{Tr}[x, p] = ni\hbar$, where "Tr" denotes the trace operation, that is, summation over the diagonal elements of a matrix. But it is also easy to show, for any two finite-dimensional matrices A and B, that $\text{Tr}(AB - BA) = 0$. Since $ni\hbar \neq 0$, we conclude that the commutation relation $[x, p] = i\hbar$ cannot hold in a finite-dimensional Hilbert space.

The commutation relation $[x, p] = i\hbar$ was arrived at historically by what has been called "systematic guesswork." Let us briefly recall the sort of guesswork that led to matrix mechanics and the representation of observables by operators in an abstract vector space. Heisenberg considered a particle in periodic motion with a frequency ω_n, in which case its coordinate $x(t)$ can be expressed classically as the Fourier series

$$x(t) = \sum_{m=-\infty}^{\infty} x(n, m)e^{im\omega_n t}. \qquad (2.2.27)$$

The experimental observations indicating that an electron in an atom is actually characterized by Bohr *transition* frequencies suggested to Heisenberg that (2.2.27) should be replaced in the atomic domain by

$$x_Q(t) = \sum_{m=-\infty}^{\infty} x_Q(n, n-m)e^{i\omega_{n,n-m}t}. \qquad (2.2.28)$$

Things get more interesting when one considers $x^2(t)$, which classically is given by

$$x^2(t) = \sum_m \sum_j x(n, m)x(n, j)e^{i(m+j)\omega_n t} = \sum_k \sum_m x(n, m)x(n, k-m)e^{ik\omega_n t}$$

$$\equiv \sum_k (x^2)(n, k)e^{ik\omega_n t}. \qquad (2.2.29)$$

If it is again supposed that $k\omega_n$ should be replaced by $\omega_{n,n-k}$, and $(x^2)(n, k)$ by $(x_Q^2)(n, n-k)$ in the atomic domain, then (2.2.29) should be replaced by

$$(x_Q^2)(t) = \sum_k (x_Q^2)(n, n-k)e^{i\omega_{n,n-k}t}, \qquad (2.2.30)$$

and the question arises, how might $(x_Q^2)(n, n-k)$ be related to the coefficients $x_Q(n, n-k)$ appearing in (2.2.28)?

Since $\omega_{n,n-k} = \omega_{n,m} + \omega_{m,n-k}$,[5] it is consistent with (2.2.30) and the association of $x_Q(n,m)$ with $e^{i\omega_{n,m}t}$ to suppose that

$$(x_Q^2)(n, n-k)e^{i\omega_{n,n-k}t} = \sum_m x_Q(n,m)e^{i\omega_{n,m}t}x_Q(m, n-k)e^{i\omega_{m,n-k}t} \tag{2.2.31}$$

or

$$(x_Q^2)(n, n-k) = \sum_m x_Q(n,m)x_Q(m, n-k), \tag{2.2.32}$$

which will be recognized (as Max Born first did) as a matrix multiplication.[6] Heisenberg remarked that "this type of combination is an almost necessary consequence of the frequency combination rules."[7] This reflected his philosophy of founding quantum mechanics "exclusively upon relationships between quantities which in principle are observable."[8] It was by systematic guesswork of this sort that matrix mechanics came to be. The commutation rule $[x_i, p_j] = i\hbar\delta_{ij}$ was first deduced by Born and Jordan in 1925.

2.2.1 Beyond the Electric Dipole Approximation

Although the electric dipole approximation is an excellent one for most purposes in quantum optics, it is by no means universally valid. We will now digress briefly and present the leading corrections to the electric dipole approximation, first using the minimal coupling Hamiltonian. For definiteness, we take the vector potential to point in the x direction and to vary spatially only with z:

$$H = \mathbf{p}^2/2m + V(\mathbf{x}) - \frac{e}{m}A_x(z,t)p_x + \frac{e^2}{2m}A_x^2(z,t). \tag{2.2.33}$$

In the first correction to the dipole approximation,

$$A_x(z,t) \cong A_x(0,t) + A_x'(0,t)z \quad (A_x' \equiv \partial A_x/\partial z), \tag{2.2.34}$$

and

$$H \cong \mathbf{p}^2/2m + V(\mathbf{x}) - \frac{e}{m}\big[A_x(0,t) + A_x'(0,t)z\big]p_x + \frac{e^2}{2m}\big[A_x^2(0,t)$$
$$+ 2A_x'(0,t)A_x(0,t)z\big], \tag{2.2.35}$$

assuming the term proportional to $[A_x'(0,t)z]^2$ can be neglected. The Heisenberg equations of motion for x and p_x, for example, are then

---

[5] This is the Ritz "combination principle," which states that spectral lines of atoms are differences of "terms" as defined by spectroscopists at the time. Originally deduced from spectroscopic observations, it is consistent with the relation $\omega_{n,m} = (E_n - E_m)/\hbar$, and, not surprisingly, it played an important role in considerations that led to the Bohr model. Bohr's original statement of the correspondence principle related atomic transition frequencies to frequency components of the motion of the corresponding classical system.

[6] Heisenberg recognized that it is not generally true in quantum theory that $A(t)B(t) = B(t)A(t)$ for two quantities $A(t)$ and $B(t)$.

[7] Quoted in *Van der Waerden*, p. 265.

[8] Ibid., p. 261.

$$\dot{x} = p_x/m - \frac{e}{m}A_x(0,t) - \frac{e}{m}A'_x(0,t)z, \tag{2.2.36}$$

$$\dot{p}_x = -\frac{\partial V}{\partial x}, \tag{2.2.37}$$

and therefore

$$m\ddot{x} = -\frac{\partial V}{\partial x} - e\dot{A}_x(0,t) - e\frac{\partial A'_x(0,t)}{\partial t}z - eA'_x(0,t)\dot{z}, \tag{2.2.38}$$

or, since $E_x = -\dot{A}_x$, and $B_y = \partial A_x/\partial z = A'_x$,

$$m\ddot{x} = -\frac{\partial V}{\partial x} + eE_x(0,t) + eE'_x(0,t)z + e(\mathbf{v}\times\mathbf{B})_x. \tag{2.2.39}$$

We can still make the transformation (2.2.15), which takes us from the minimal coupling form of the Hamiltonian to a "multipolar" form, but now we must allow for the spatial variation of the vector potential. We skip the trivial details and only state the result for the transformed Hamiltonian:

$$H' = \mathbf{p}^2/2m + V(\mathbf{x}) - exE_x(0,t) - Q_{xz}E'_x(0,t) - \mu_y B_y(0,t), \tag{2.2.40}$$

which is a little more complicated than (2.2.35). Here,

$$Q_{xz} = \frac{e}{2}xz \tag{2.2.41}$$

and

$$\mu_y = \frac{e}{2m}(zp_x - xp_z) \tag{2.2.42}$$

are associated with electric quadrupole and magnetic dipole contributions, respectively. The Heisenberg equations of motion for x and p_x that follow from H' are

$$\dot{x} = p_x/m - \frac{e}{2m}zB_y, \tag{2.2.43}$$

and

$$\dot{p}_x = -\frac{\partial V}{\partial x} + eE_x(0,t) + \frac{e}{2}zE'_x(0,t) - \frac{e}{2m}p_z B_y, \tag{2.2.44}$$

or

$$m\ddot{x} = -\frac{\partial V}{\partial x} + eE_x(0,t) + \frac{e}{2}zE'_x(0,t) - \frac{e}{2m}p_z B_y - \frac{e}{2}\dot{z}B_y - \frac{e}{2}z\dot{B}_y. \tag{2.2.45}$$

From $\nabla \times \mathbf{E} = -\dot{\mathbf{B}}$,

$$\frac{e}{2}z\dot{B}_y = -\frac{e}{2}zE'_x(0,t) \tag{2.2.46}$$

and

$$m\ddot{x} = -\frac{\partial V}{\partial x} + eE_x(0,t) + eE'_x(0,t)z + e(\mathbf{v}\times\mathbf{B})_x, \tag{2.2.47}$$

as obtained from the minimal coupling Hamiltonian.

2.3   Semiclassical Radiation Theory

The Hamiltonian (2.1.5) provides a basis for *semiclassical radiation theory*, in which atoms are treated according to quantum theory, and the field according to the classical Maxwell equations. While fundamentally inconsistent, semiclassical radiation theory is a very useful—and very often entirely adequate—approach to the interaction of light with matter. For instance, as we now discuss, it correctly describes the main features of the photoelectric effect, features that are usually thought to be inexplicable without photons.

Suppose that the electric field applied to an atom is

$$\mathbf{E}(t) = \frac{1}{2}\left(\mathbf{E}_0 e^{-i\omega t} + \mathbf{E}_0^* e^{i\omega t}\right), \tag{2.3.1}$$

so that

$$H_{\mathrm{I}}(t) = -\frac{1}{2}\mathbf{d}\cdot\left(\mathbf{E}_0 e^{-i\omega t} + \mathbf{E}_0^* e^{i\omega t}\right).^9 \tag{2.3.2}$$

Then, the transition probability (2.1.17) is

$$p_{i\to f}(t) \cong \frac{1}{4\hbar^2}\left|\mathbf{d}_{fi}\cdot\mathbf{E}_0\int_0^t dt' e^{i(\omega_{fi}-\omega)t'} + \mathbf{d}_{fi}\cdot\mathbf{E}_0^*\int_0^t dt' e^{i(\omega_{fi}+\omega)t'}\right|^2, \tag{2.3.3}$$

where $\omega_{fi} = (E_f - E_i)/\hbar$ is the (angular) Bohr transition frequency between the initial and final states of the atom. Now,

$$\int_0^t dt' e^{i(\omega_{fi}\pm\omega)t'} = e^{i(\omega_{fi}\pm\omega)t/2}\frac{\sin[\frac{1}{2}(\omega_{fi}\pm\omega)t]}{\frac{1}{2}(\omega_{fi}\pm\omega)}, \tag{2.3.4}$$

and, for a transition to a state of higher energy ($\omega_{fi} > 0$) with $\omega \approx \omega_{fi}$, the term with the $+$ sign is negligible compared to the term with the $-$ sign:

$$p_{i\to f}(t) \cong \frac{1}{4\hbar^2}|\mathbf{d}_{fi}\cdot\mathbf{E}_0|^2\frac{\sin^2[\frac{1}{2}(\omega_{fi}-\omega)t]}{[\frac{1}{2}(\omega_{fi}-\omega)]^2}. \tag{2.3.5}$$

It has been assumed that the initial state i is approximately undepleted.

There is often a continuous distribution of possible final states; this occurs, for instance, in photoionization. Then, the total transition probability from state i, that is, the sum over the transition probabilities from i to all possible final states f, is

$$p_{if}(t) \cong \frac{1}{4\hbar^2}\int dE_f \rho_E(E_f)|\mathbf{d}_{fi}\cdot\mathbf{E}_0|^2\frac{\sin^2[\frac{1}{2}(\omega_{fi}-\omega)t]}{[\frac{1}{2}(\omega_{fi}-\omega)]^2}, \tag{2.3.6}$$

where $\rho_E(E_f)dE_f$ is the number of final states in the energy interval $[E_f, E_f + dE_f]$. For times t large enough that the sinc function in (2.3.6) is sharply peaked compared with the variation of $\rho_E(E_f)|\mathbf{d}_{fi}\cdot\mathbf{E}_0|^2$ around $E_f = E_i + \hbar\omega$,

---

[9] We now denote the electric dipole Hamiltonian by H rather than H'.

$$p_{if}(t) \cong \frac{1}{4\hbar^2} |\mathbf{d}_{fi} \cdot \mathbf{E}_0|^2 \rho_E(E_i + \hbar\omega) \hbar \int d\omega_{fi} \frac{\sin^2[\frac{1}{2}(\omega_{fi} - \omega)t]}{[\frac{1}{2}(\omega_{fi} - \omega)]^2}. \tag{2.3.7}$$

Using

$$\int_0^\infty d\omega_{fi} \frac{\sin^2[\frac{1}{2}(\omega_{fi} - \omega)t]}{[\frac{1}{2}(\omega_{fi} - \omega)]^2} \cong \int_{-\infty}^\infty d\omega_{fi} \frac{\sin^2[\frac{1}{2}(\omega_{fi} - \omega)t]}{[\frac{1}{2}(\omega_{fi} - \omega)]^2} = 2\pi t, \tag{2.3.8}$$

we obtain the transition *rate*

$$R_{if} = \frac{d}{dt} p_{if}(t) = \frac{\pi}{2\hbar} |\mathbf{d}_{fi} \cdot \mathbf{E}_0|^2 \rho_E(E_i + \hbar\omega), \tag{2.3.9}$$

which, of course, is just Fermi's golden rule for the example under consideration.

Exercise 2.1: Derive (2.3.9) using the Schrödinger picture.

This result can be applied to the photoelectric effect. Recall that the most significant observations in the photoelectric effect are that (1) the maximum kinetic energy E_f of the photoelectrons is a linear function of the light frequency ω: $E_f = \hbar\omega - W$, where W is the "work function"; (2) the photocurrent or rate of ejection of electrons is proportional to the light intensity; and (3) electrons are ejected immediately upon exposure of the surface to light. These observations are traditionally explained in terms of photons of energy $\hbar\omega$, but they can all be explained using (2.3.9), *which was derived assuming that the field is a classical wave*. Thus, the linear dependence of the photoelectron kinetic energy on the light frequency follows from the evaluation of the final density of states at $E_f = \hbar\omega + E_i$ in (2.3.9): our semiclassical derivation of (2.3.9) leads automatically to an energy conservation condition such that the change in the electron energy is $\hbar\omega$. Observation (2) follows from the proportionality of the rate (2.3.9) to E_0^2 and therefore to the light intensity: as the light intensity increases, the rate of ejection of electrons increases proportionately. And, finally, observation (3) is just a consequence of (2.3.9) being a *rate*. We did require the time t in our derivation of (2.3.9) to be "long enough," but only long enough to enforce energy conservation; this aspect of the calculation is not peculiar to our semiclassical calculation, but appears also when we quantize the field.

There is nothing very subtle about this semiclassical explanation of the photoelectric effect, so why has the photoelectric effect for so long been invoked as evidence for the existence of photons? The answer lies in the history of the photoelectric effect. Noting that blackbody radiation and other phenomena suggest a "heuristic point of view" in which "the energy of light is discontinuously distributed in space," Einstein in 1905 *predicted* the linear relation between radiation frequency and stopping power in the photoelectric effect.[10] This prediction was verified in 1916 by Millikan, who,

[10] Einstein, quoted in the English translation of his paper by A. B. Arons and M. B. Peppard, Am. J. Phys. **33**, 367 (1965), p. 368.

however, for some time considered Einstein's heuristic picture "reckless" and "pretty generally abandoned."[11] The success (and simplicity) of Einstein's heuristic picture, and the failure of any *purely classical* theory to explain the photoelectric effect, suggested that light quanta (named *photons* by the physical chemist G. N. Lewis in 1926) are *necessary* for the explanation of the observations (1)–(3) above. The preceding discussion shows, as was emphasized in particular by P. A. Franken and W. E. Lamb, Jr., that these observations are explained by semiclassical radiation theory—quantum theory for matter, and classical theory for radiation.

Semiclassical radiation theory cannot, of course, account for all aspects of the interaction of light with matter. In the case of the photoelectric effect, for example, it is inconsistent with energy conservation, simply because the classical expression for energy in the field is not identically equal to some integer q times the energy $\hbar\omega$:

$$\frac{1}{2}\epsilon_0 \int d^3r (\mathbf{E}^2 + \mathbf{H}^2) \neq q\hbar\omega. \qquad (2.3.10)$$

Failure of semiclassical theory is especially evident in the case of spontaneous emission, the process in which an atom, without any external cause, makes a transition to a state of lower energy, and the electromagnetic field gains an energy $\hbar\omega$. Spontaneous emission is the origin of most of the light around us. The reason it cannot be described by semiclassical radiation theory is simple: in semiclassical radiation theory, the sources of the field, as well as the field itself, are ordinary "c-numbers," rather than operators in a Hilbert space. Thus, it is the oscillating *expectation value* of the electric dipole moment of an excited atom, for example, that is taken to be the source of spontaneous radiation in semiclassical radiation theory. But this expectation value vanishes for any free atom in an energy eigenstate, because there is no preferred direction for the electric dipole moment formed by the electron and the nucleus. Therefore, according to semiclassical radiation theory, an excited atom should not radiate spontaneously! This failure of semiclassical radiation theory is discussed further in Section 4.7.

Similarly, semiclassical radiation theory cannot account for recoil in spontaneous emission, an effect that plays an important role in the physics of ultracold gases. For, according to semiclassical radiation theory, the radiated field is a continuous electromagnetic wave emanating from the source atom; for an electric dipole transition, this wave has the familiar dipole radiation pattern with inversion symmetry about the atom. There is no preferred direction of emission that could "tell" the atom in which direction it should recoil in order to conserve total momentum. But as discussed in Section 2.8, the Planck spectrum implies that an atom must recoil when it radiates.

Figure 2.1 illustrates another failure of semiclassical radiation theory. An atom spontaneously radiates a field corresponding to a single photon of energy $\hbar\omega$. There is a non-vanishing probability that this process results in the detection of a photon at

---

[11] R. A. Millikan, Phys. Rev. **7**, 355 (1916), p. 355.

D_1, as well as a non-vanishing probability for the detection of a photon at D_2. But the probability of detecting photons at *both* D_1 and D_2 is zero. This is just what is predicted by the quantum theory of the field. In semiclassical radiation theory, however, the field emitted by the atom is described classically as a continuous wave. Part of this wave can be measured at D_1, and part of it can be measured at D_2, resulting in a non-vanishing probability of photodetection at both D_1 and D_2. Experiments of this type have been performed, and they show that this prediction of semiclassical radiation theory is incorrect (see Section 5.7).

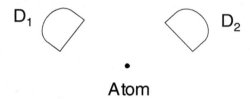

Fig. 2.1 An atom undergoes a single spontaneous emission event in the presence of detectors D_1 and D_2. There is a finite probability for a photon count to occur at D_1, and a finite probability for a photon count at D_2, but zero probability for counts at *both* D_1 and D_2.

2.3.1 Semiclassical Radiation Theory in the Heisenberg Picture

Heisenberg equations of motion for $\mathbf{x}$ and $\mathbf{p}$ are not very convenient for calculating effects of fields on atoms, mainly because they are usually too difficult to solve. Consider instead the operators σ_{ij}, which are defined by writing a Schrödinger-picture operator A acting in the Hilbert space of an atom as

$$A = 1 \times A \times 1 = \left(\sum_i |i\rangle\langle i| \right) A \left(\sum_j |j\rangle\langle j| \right) = \sum_{i,j} \langle i|A|j\rangle |i\rangle\langle j| = \sum_{i,j} A_{ij}\sigma_{ij}. \quad (2.3.11)$$

We have again used the completeness property of the eigenstates $|i\rangle$ of the unperturbed atomic Hamiltonian $H_0 = \mathbf{p}^2/2m + V(\mathbf{x})$, and have defined the matrix elements $A_{ij} = \langle i|A|j\rangle$ and the operators $\sigma_{ij} = |i\rangle\langle j|$. In the Heisenberg picture

$$A_H(t) = \sum_{i,j} A_{ij}\sigma_{ij}(t), \quad (2.3.12)$$

$\sigma_{ij}(t) = U^\dagger(t)\sigma_{ij}U(t)$. The Schrödinger-picture operators satisfy the commutation relation

$$[\sigma_{ij}, \sigma_{k\ell}] = |i\rangle\langle j|k\rangle\langle \ell| - |k\rangle\langle \ell|i\rangle\langle j| = \delta_{jk}\sigma_{i\ell} - \delta_{i\ell}\sigma_{kj} \quad (2.3.13)$$

that follows from the orthonormality of the different eigenstates of H_0, and, of course, the Heisenberg-picture operators $\sigma_{ij}(t)$ satisfy the same commutation relation as a consequence of the unitarity of the time evolution operator $U(t)$.

Write the electric dipole Hamiltonian in terms of the σ operators:

$$H = H_0 - \mathbf{d} \cdot \mathbf{E}(t) = \sum_{i,j} \langle i|H_0|j\rangle \sigma_{ij} - \sum_{i,j} \langle i|\mathbf{d}|j\rangle \cdot \mathbf{E}(t)\sigma_{ij}$$

$$= \sum_i E_i \sigma_{ii} - \sum_{i,j} \mathbf{d}_{ij} \cdot \mathbf{E}(t)\sigma_{ij}. \tag{2.3.14}$$

The Heisenberg equations of motion for the σ operators, $i\hbar\dot{\sigma}_{ij} = [\sigma_{ij}, H]$, take the form

$$i\hbar\dot{\sigma}_{ij}(t) = \sum_k E_k[\sigma_{ij}, \sigma_{kk}] - \sum_{k,\ell} \mathbf{d}_{k\ell} \cdot \mathbf{E}(t)[\sigma_{ij}, \sigma_{k\ell}]$$

$$= \sum_k E_k(\delta_{jk}\sigma_{ik} - \delta_{ik}\sigma_{kj}) - \sum_{k,\ell} \mathbf{d}_{k\ell} \cdot \mathbf{E}(t)(\delta_{jk}\sigma_{i\ell} - \delta_{i\ell}\sigma_{kj})$$

$$= (E_j - E_i)\sigma_{ij} - \sum_k [\mathbf{d}_{jk} \cdot \mathbf{E}(t)\sigma_{ik} - \mathbf{d}_{ki} \cdot \mathbf{E}(t)\sigma_{kj}], \tag{2.3.15}$$

or

$$\dot{\sigma}_{ij}(t) = -i\omega_{ji}\sigma_{ij}(t) + \frac{i}{\hbar}\sum_k [\mathbf{d}_{jk} \cdot \mathbf{E}(t)\sigma_{ik}(t) - \mathbf{d}_{ki} \cdot \mathbf{E}(t)\sigma_{kj}(t)], \tag{2.3.16}$$

where again $\omega_{ji} = (E_j - E_i)/\hbar$. Within semiclassical radiation theory and the electric dipole approximation, these equations are an exact formulation of the time evolution of an atom in an applied electric field $\mathbf{E}(t)$. Solutions of these equations give us, via (2.3.12), any Heisenberg-picture operator acting in the Hilbert space of the atom.

To appreciate the physical significance of $\sigma_{ij}(t)$ itself, suppose that the atom is in the state $|g\rangle$ at the time $t = 0$ at which the field is applied, and consider the expectation value

$$\langle\sigma_{ij}(t)\rangle = \langle g|\sigma_{ij}(t)|g\rangle = \langle g|U^\dagger(t)\sigma_{ij}(0)U(t)|g\rangle = \langle g|U^\dagger(t)|i\rangle\langle j|U(t)|g\rangle. \tag{2.3.17}$$

$|\psi(t)\rangle = U(t)|g\rangle$ is just the Schrödinger-picture state vector at time t, and so

$$\langle\sigma_{ij}(t)\rangle = \langle\psi(t)|i\rangle\langle j|\psi(t)\rangle = a_i^*(t)a_j(t), \tag{2.3.18}$$

where $a_i(t) = \langle i|\psi(t)\rangle$ is the probability amplitude at time t for the atom to be found in the energy eigenstate $|i\rangle$ of its unperturbed Hamiltonian. In particular, $\langle\sigma_{ii}(t)\rangle = \langle g|\sigma_{ii}(t)|g\rangle = |a_i(t)|^2$ is the probability at time t that the atom will be found in the eigenstate $|i\rangle$ of energy E_i, given that it started out in the eigenstate $|g\rangle$ of energy E_g when the field was turned on.

2.3.2 Density-Matrix Equations

In much of the literature, the atom–field interaction is treated using the density matrix

$$\rho = \sum_{\psi} p_{\psi} |\psi\rangle\langle\psi|, \tag{2.3.19}$$

where, as reviewed in Chapter 3, Section 3.7, p_{ψ} is the probability assigned to state $|\psi\rangle$. From the Schrödinger equation $i\hbar|\dot{\psi}\rangle = H|\psi\rangle$,

$$i\hbar\dot{\rho} = \sum_{\psi} p_{\psi}\left(H|\psi\rangle\langle\psi| - |\psi\rangle\langle\psi|H\right) = [H, \rho]. \tag{2.3.20}$$

From this equation, we find that the density-matrix elements $\rho_{ij} = \langle i|\rho|j\rangle$, where $|i\rangle$ and $|j\rangle$ are again eigenstates of the unperturbed atom Hamiltonian H_0, satisfy

$$
\begin{aligned}
i\hbar\dot{\rho}_{ij} &= \sum_{\psi} p_{\psi}\left[\langle i|H|\psi\rangle\langle\psi|j\rangle - \langle i|\psi\rangle\langle\psi|H|j\rangle\right] \\
&= \sum_{\psi} p_{\psi} \sum_{k}\left[\langle i|H|k\rangle\langle k|\psi\rangle\langle\psi|j\rangle - \langle i|\psi\rangle\langle\psi|k\rangle\langle k|H|j\rangle\right] \\
&= \sum_{k}\left[\langle i|H|k\rangle\rho_{kj} - \rho_{ik}\langle k|H|j\rangle\right],
\end{aligned} \tag{2.3.21}
$$

where we have used the completeness relation $\sum_{k}|k\rangle\langle k| = 1$. Since

$$\langle i|H|k\rangle = \langle i|H_0|k\rangle - \langle i|\mathbf{d}\cdot\mathbf{E}(t)|k\rangle = E_k\delta_{ik} - \mathbf{d}_{ik}\cdot\mathbf{E}(t), \tag{2.3.22}$$

and, likewise, $\langle k|H|j\rangle = E_j\delta_{jk} - \mathbf{d}_{kj}\cdot\mathbf{E}(t)$,

$$\dot{\rho}_{ij} = i\omega_{ji}\rho_{ij} + \frac{i}{\hbar}\sum_{k}\left[\mathbf{d}_{ik}\cdot\mathbf{E}(t)\rho_{kj} - \rho_{ik}\mathbf{d}_{kj}\cdot\mathbf{E}(t)\right]. \tag{2.3.23}$$

This differs from (2.3.16), but it obviously has the same general form. Damping effects, due to spontaneous emission or collisions, for example, are not included in (2.3.16) and (2.3.23).[12] We include effects of damping in the two-state model of an atom in Section 2.7, and, in Section 4.8, we consider in particular the effects of spontaneous emission in the more general case of a "multistate atom." Equation (2.3.16) is applicable if, for example, the applied field is far off resonance with any transition from the initial state of the atom, or if it is in the form of a pulse whose duration is short compared with relaxation times characterizing any relevant damping effects. Even under such conditions, however, these equations can generally be solved analytically only under various approximations. We consider next an example of such an approximate solution.

---

[12] Some examples of numerical solutions of the more general equations, including damping, Doppler broadening, different types of applied electric fields, the Earth's magnetic field, and other effects are presented by P. W. Milonni, R. Q. Fugate, and J. M. Telle, J. Opt. Soc. Am. A **15**, 217 (1998), and P. W. Milonni, H. Fearn, R. Q. Fugate, and J. M. Telle, J. Opt. Soc. Am. A **16**, 2555 (1999).

2.3.3   The Kramers–Heisenberg Dispersion Formula

Take expectation values of both sides of (2.3.16):

$$\langle \dot{\sigma}_{ij}(t) \rangle = -i\omega_{ji}\langle \sigma_{ij}(t) \rangle + \frac{i}{\hbar}\sum_{k}[\mathbf{d}_{jk}\cdot\mathbf{E}(t)\langle \sigma_{ik}(t) \rangle - \mathbf{d}_{ki}\cdot\mathbf{E}(t)\langle \sigma_{kj}(t) \rangle], \qquad (2.3.24)$$

or

$$\langle \dot{\sigma}_{ij}(t) \rangle = -i\omega_{ji}\langle \sigma_{ij}(t) \rangle + \frac{i}{\hbar}\mathbf{d}_{ji}\cdot\mathbf{E}(t)[\langle \sigma_{ii}(t) \rangle - \langle \sigma_{jj}(t) \rangle]$$

$$+ \frac{i}{\hbar}\sum_{k\neq i}\mathbf{d}_{jk}\cdot\mathbf{E}(t)\langle \sigma_{ik}(t) \rangle - \frac{i}{\hbar}\sum_{k\neq j}\mathbf{d}_{ki}\cdot\mathbf{E}(t)\langle \sigma_{kj}(t) \rangle. \qquad (2.3.25)$$

Suppose that the initial state to which these expectation values refer is the state $|i\rangle$ of the atom, and that the applied field frequency is sufficiently far removed from any absorption resonance that the atom remains with high probability in this state, so that $\langle \sigma_{kk}(t) \rangle = |a_k(t)|^2 \cong \delta_{ik}$ Then, to calculate $\langle \sigma_{ij}(t) \rangle$ to lowest order in the applied field, we retain only the first two terms on the right side of (2.3.25) and write, for $i \neq j$,

$$\langle \dot{\sigma}_{ij}(t) \rangle \cong -i\omega_{ji}\langle \sigma_{ij}(t) \rangle + \frac{i}{\hbar}\mathbf{d}_{ji}\cdot\mathbf{E}(t). \qquad (2.3.26)$$

We assume a monochromatic field, and, to simplify slightly, we take it to have the linearly polarized form $\mathbf{E}(t) = \mathbf{E}_0\cos\omega t$. Then,

$$\langle \dot{\sigma}_{ij}(t) \rangle \cong -i\omega_{ji}\langle \sigma_{ij}(t) \rangle + \frac{i}{\hbar}\mathbf{d}_{ji}\cdot\mathbf{E}_0\cos\omega t, \qquad (2.3.27)$$

which has the solution

$$\langle \sigma_{ij}(t) \rangle = \frac{1}{2\hbar}\mathbf{d}_{ji}\cdot\mathbf{E}_0\left(\frac{e^{-i\omega t}}{\omega_{ji}-\omega} + \frac{e^{i\omega t}}{\omega_{ji}+\omega}\right). \qquad (2.3.28)$$

Consider the expectation value $\langle \mathbf{d}(t) \rangle$ of the electric dipole moment of the atom in state $|i\rangle$. According to (2.3.12),

$$\langle \mathbf{d}(t) \rangle = \sum_{k,\ell}\mathbf{d}_{k\ell}\langle \sigma_{k\ell}(t) \rangle = \sum_{k,\ell}\mathbf{d}_{k\ell}\langle i|U^{\dagger}(t)|k\rangle\langle \ell|U(t)|i\rangle. \qquad (2.3.29)$$

Note that the diagonal matrix elements $\mathbf{d}_{kk}$ of $\mathbf{d}$ are zero: since the Coulomb potential and the Hamiltonian have inversion symmetry, the expectation value of the electric dipole moment of an unperturbed atom in a stationary state must vanish. If, as we have assumed, the atom remains with high probability in the initial state $|i\rangle$, we can approximate (2.3.29) by

$$\langle \mathbf{d}(t) \rangle = \sum_{k}\mathbf{d}_{ki}\langle i|U^{\dagger}(t)|k\rangle\langle i|U(t)|i\rangle + \sum_{\ell}\mathbf{d}_{i\ell}\langle i|U^{\dagger}(t)|i\rangle\langle \ell|U(t)|i\rangle$$

$$= \sum_{k}\mathbf{d}_{ki}\langle \sigma_{ki}(t) \rangle + \sum_{k}\mathbf{d}_{ik}\langle \sigma_{ik}(t) \rangle = 2\mathrm{Re}\left(\sum_{j}\mathbf{d}_{ij}\langle \sigma_{ij}(t) \rangle\right), \qquad (2.3.30)$$

and, from (2.3.28),

$$\langle d_x(t) \rangle = \frac{1}{\hbar} \sum_j |(\mathbf{d}_{ij})_x|^2 \left(\frac{1}{\omega_{ji} - \omega} + \frac{1}{\omega_{ji} + \omega} \right) E_0 \cos \omega t$$

$$= \frac{1}{3\hbar} \sum_j |\mathbf{d}_{ij}|^2 \left(\frac{1}{\omega_{ji} - \omega} + \frac{1}{\omega_{ji} + \omega} \right) E_0 \cos \omega t \qquad (2.3.31)$$

for the x component of the *induced* dipole moment expectation value in a field polarized along the x direction; we have again invoked spherical symmetry to replace $|(\mathbf{d}_{ij})_x|^2$ by $|\mathbf{d}_{ij}|^2/3$. Finally, we identify the polarizability $\alpha_i(\omega)$ associated with the state $|i\rangle$ as the coefficient of proportionality between the induced dipole moment expectation value and the applied field ($\langle \mathbf{d}(t) \rangle = \langle i | \mathbf{d}(t) | i \rangle = \alpha_i(\omega) \mathbf{E}_0 \cos \omega t$):

$$\alpha_i(\omega) = \frac{1}{3\hbar} \sum_j |\mathbf{d}_{ij}|^2 \left(\frac{1}{\omega_{ji} - \omega} + \frac{1}{\omega_{ji} + \omega} \right) = \frac{2}{3\hbar} \sum_j |\mathbf{d}_{ij}|^2 \frac{\omega_{ji}}{\omega_{ji}^2 - \omega^2}. \qquad (2.3.32)$$

This is the *Kramers–Heisenberg dispersion formula* for the (linear) polarizability of an atom in state $|i\rangle$. It is related to dispersion by the formula $n^2(\omega) - 1 = N\alpha_i(\omega)/\epsilon_0$ for the refractive index $n(\omega)$ of a dilute medium of N atoms per unit volume, all in the state i. In most cases of practical interest, $|i\rangle$ is the ground state. We have ignored any effects of spontaneous emission or collisions or other line-broadening effects, which have the effect of adding imaginary parts to the frequency denominators $\omega_{ij} \pm \omega$; if the field frequency is far enough removed from any resonance frequency ω_{ji} of the atom, these imaginary parts have a negligible effect on the dispersive (real) part of the refractive index. In general, of course, the polarizability and the refractive index are complex, their imaginary parts being associated with field attenuation or amplification.

If we do not assume spherical symmetry, we obtain a polarizability *tensor* $\tilde{\alpha}(\omega)$ such that the μth Cartesian component of $\langle \mathbf{d}(t) \rangle$, when the incident field has Cartesian components $E_{0\nu}$, is

$$\langle d_\mu(t) \rangle = \mathrm{Re}\left[\tilde{\alpha}_{\mu\nu}(\omega) E_{0\nu} e^{-i\omega t} \right] \qquad (2.3.33)$$

for a molecule in state $|i\rangle$; we follow the convention here of summing over repeated indices ($\nu = 1, 2, 3$). If we again ignore line-broadening effects, the polarizability tensor is found by proceeding as above to be

$$\tilde{\alpha}_{i,\mu\nu}(\omega) = \tilde{\alpha}_{i,\nu\mu}(-\omega) = \frac{1}{\hbar} \sum_j \left[\frac{(d_\mu)_{ij}(d_\nu)_{ji}}{\omega_{ji} - \omega} + \frac{(d_\nu)_{ij}(d_\mu)_{ji}}{\omega_{ji} + \omega} \right]. \qquad (2.3.34)$$

When the scatterer is spherically symmetric, as is the case for isolated atoms, $\tilde{\alpha}(\omega)$ is diagonal, and $\tilde{\alpha}_{i,xx}(\omega) = \tilde{\alpha}_{i,yy}(\omega) = \tilde{\alpha}_{i,zz}(\omega) = \alpha_i(\omega)$, as defined by (2.3.32).

The classical model for a bound electron in a monochromatic field, based on the equation

$$\ddot{\mathbf{x}} + \omega_0^2 \mathbf{x} = \frac{e}{m} \mathbf{E}_0 \cos \omega t \qquad (2.3.35)$$

and the induced dipole moment $ex = \alpha_c(\omega)\mathbf{E}_0 \cos \omega t$, leads trivially to the formula

$$\alpha_c(\omega) = \frac{e^2/m}{\omega_0^2 - \omega^2} \cong -\frac{e^2}{m\omega^2} \qquad (2.3.36)$$

for the polarizability in the limit $\omega \gg \omega_0$ for Thomson scattering (Section 1.12). The presumption that the polarizability (2.3.32) reduces to this classical result in the high-frequency limit in which the discreteness of the atom's energy levels is unimportant leads to

$$-\frac{2}{3\hbar\omega^2} \sum_j |\mathbf{d}_{ij}|^2 \omega_{ji} = -\frac{e^2}{m\omega^2}, \quad \text{or} \quad \frac{2m}{3\hbar} \sum_j \omega_{ji} |\mathbf{x}_{ij}|^2 = 1, \qquad (2.3.37)$$

which is the *Thomas–Reiche–Kuhn sum rule* (for a one-electron atom) obtained in early work on the quantum theory of dispersion.[13]

This sum rule played an important role in the inductive reasoning that led to the postulate $[x, p] = i\hbar$. To gain some appreciation for this, consider the diagonal matrix element

$$\langle i|[x, p]|i\rangle = \langle i|x1p|i\rangle - \langle i|p1x|i\rangle = \sum_j \langle i|x|j\rangle\langle j|p|i\rangle - \sum_j \langle i|p|j\rangle\langle j|x|i\rangle$$

$$= \sum_j x_{ij}p_{ji} - \sum_j p_{ij}x_{ji} \qquad (2.3.38)$$

for some bound state $|i\rangle$. The matrix elements of p can be related to those of x as follows: from $p = m\dot{x} = (m/i\hbar)[x, H_0]$,

$$p_{ij} = -\frac{im}{\hbar}\langle i|xH_0 - H_0x|j\rangle = -\frac{im}{\hbar}(E_j - E_i)\langle i|x|j\rangle = -im\omega_{ji}x_{ij}. \qquad (2.3.39)$$

Therefore,

$$\langle i|[x, p]|i\rangle = -im\left(\sum_j x_{ij}\omega_{ij}x_{ji} - \sum_j \omega_{ji}x_{ij}x_{ji}\right) = 2im\sum_j \omega_{ji}|x_{ij}|^2$$

$$= \frac{2im}{3}\sum_j \omega_{ji}|\mathbf{x}_{ij}|^2 = i\hbar, \qquad (2.3.40)$$

where we have used $\omega_{ij} = -\omega_{ji}$ and the Thomas–Reiche–Kuhn sum rule. This demonstrates a connection between the canonical commutation rule and the Thomas–Reiche–Kuhn sum rule, a connection that, together with Heisenberg's introduction of matrix elements discussed in Section 2.3.2, provides a faint flavor of the kind of considerations that led to the hypothesis $[x, p] = i\hbar$ for *all* quantum systems and for *all* canonical coordinates and momenta.

2.3.4   The AC Stark Shift and the Ponderomotive Potential

Consider the energy of an induced electric dipole moment d_i of an atom initially in an eigenstate of energy E_i:

---

[13] It is the most notable of various sum rules relating to the optical properties of material media. See M. Altarelli, D. L. Dexter, H. M. Nussenzveig, and D. Y. Smith, Phys. Rev. B **6**, 4502 (1972).

$$\delta E_i = -\int_0^{\mathbf{E}} \mathbf{d}_i \cdot d\mathbf{E}' = -\alpha_i \int_0^{\mathbf{E}} \mathbf{E}' \cdot d\mathbf{E}' = -\frac{1}{2}\alpha_i \mathbf{E}^2$$

$$= -\frac{1}{2}\mathbf{d}_i \cdot \mathbf{E} \to -\frac{1}{4}\alpha_i(\omega)E_0^2, \tag{2.3.41}$$

where, in the last step, we have taken the cycle average of the square of an applied electric field $\mathbf{E}(t) = \mathbf{E}_0 \cos\omega t$. δE_i is the *AC Stark shift* of the energy level E_i. The factor $1/2$ in the energy $-(1/2)\mathbf{d}_i \cdot \mathbf{E}$ appears because the dipole moment is *induced* by the field; the energy of a *permanent* electric dipole moment $\mathbf{d}$ in an applied field $\mathbf{E}$, in contrast, is the familiar $-\mathbf{d} \cdot \mathbf{E}$. Using (2.3.32) for the polarizability, we can write the AC Stark shift as

$$\delta E_i = -\Big[\frac{1}{6\hbar}\sum_j |\mathbf{d}_{ij}|^2 \frac{\omega_{ji}}{\omega_{ji}^2 - \omega^2}\Big]E_0^2. \tag{2.3.42}$$

If the discreteness of the atomic electron's energy levels can be ignored, that is, if $\omega \gg |\omega_{ji}|$ for frequencies $|\omega_{ji}|$ of all electric dipole transitions starting or ending at level i,

$$\delta E_i \cong E_P = \frac{1}{\omega^2}\Big(\frac{1}{6\hbar}\sum_j |\mathbf{d}_{ij}|^2\omega_{ji}\Big)E_0^2 = \frac{e^2 E_0^2}{4m\omega^2}, \tag{2.3.43}$$

where we have used the Thomas–Reiche–Kuhn sum rule (2.3.37). E_P is the *ponderomotive potential*, or "quiver energy," of an unbound electron in an electric field $\mathbf{E}_0 \cos\omega t$. It is the kinetic energy $(1/2)m\dot{\mathbf{x}}^2$ obtained by solving the equation of motion $m\ddot{\mathbf{x}} = e\mathbf{E}_0 \cos\omega t$ for $\dot{\mathbf{x}}$ and then replacing $\sin^2 \omega t$ by its average. We can regard it alternatively as the cycle average of the $\mathbf{A}^2$ term in the minimal coupling Hamiltonian:

$$\frac{e}{2m}\mathbf{A}^2 = \frac{e}{2m}[(\mathbf{E}_0/\omega)\sin\omega t]^2 \to \frac{e^2 E_0^2}{4m\omega^2} = E_P, \tag{2.3.44}$$

where we have taken $\mathbf{A} = \mathbf{A}_0 \sin\omega t = -(\mathbf{E}_0/\omega)\sin\omega t$ ($\mathbf{E} = -\dot{\mathbf{A}} = \mathbf{E}_0 \cos\omega t$).

We can also derive the AC Stark shift (and the Kramers–Heisenberg formula for the polarizability) using the minimal coupling Hamiltonian.[14] The energy from the term $(e^2/2m)\mathbf{A}^2$ in the minimal coupling Hamiltonian is just

$$\delta E_i^{(1)} = \frac{e^2 E_0^2}{4m\omega^2}\frac{2m}{3\hbar}\sum_j \omega_{ji}|\mathbf{x}_{ij}|^2, \tag{2.3.45}$$

where we have once again used the Thomas–Reiche–Kuhn sum rule. The level shift of state i due to $-(e/m)\mathbf{A} \cdot \mathbf{p}$ in the minimal coupling Hamiltonian can be obtained by simply replacing $|\mathbf{d}_{ij}|^2 E_0^2$ in (2.3.42) by $[(e/m)|\mathbf{p}_{ij}|\mathbf{A}_0]^2 = (e^2/m^2\omega^2)|\mathbf{p}_{ij}|^2 E_0^2$:

$$\delta E_i^{(2)} = -\Big(\frac{1}{6\hbar}\sum_j |\mathbf{p}_{ij}|^2 \frac{\omega_{ji}}{\omega_{ji}^2 - \omega^2}\Big)\frac{e^2}{m^2\omega^2}E_0^2, \tag{2.3.46}$$

[14] See, for instance, K. Rzążewski and R. W. Boyd, J. Mod. Opt. **51**, 1137 (2004).

or, since $|\mathbf{p}_{ij}|^2 = m^2\omega_{ji}^2|\mathbf{x}_{ij}|^2 = m^2\omega_{ji}^2|\mathbf{d}_{ij}|^2/e^2$ (see (2.3.39)),

$$\delta E_i^{(2)} = -\frac{E_0^2}{6\hbar\omega^2}\sum_j |\mathbf{d}_{ij}|^2 \frac{\omega_{ji}^3}{\omega_{ji}^2 - \omega^2}. \tag{2.3.47}$$

The complete level shift calculated with the minimal coupling Hamiltonian, to second order in E_0, is therefore

$$\delta E_i^{(1)} + \delta E_i^{(2)} = \frac{E_0^2}{6\hbar\omega^2}\sum_j |\mathbf{d}_{ij}|^2\omega_{ji} - \frac{E_0^2}{6\hbar\omega^2}\sum_j |\mathbf{d}_{ij}|^2 \frac{\omega_{ji}^3}{\omega_{ji}^2 - \omega^2}$$

$$= -\frac{E_0^2}{6\hbar\omega^2}\sum_j |\mathbf{d}_{ij}|^2 \frac{\omega^2\omega_{ji}}{\omega_{ji}^2 - \omega^2}$$

$$= \delta E_i. \tag{2.3.48}$$

δE_i can, of course, be derived more conventionally, and with either Hamiltonian, using standard second-order perturbation theory for the level shift.

The AC Stark effect is generally small for optical field frequencies far removed from atomic resonances, but it has been observed in many experiments. It has been measured, for example, as a shift in the field frequency at which there is a two-photon absorption resonance.[15] The formula (2.3.48) obviously does not apply when ω equals one of the transition frequencies ω_{ji}. In that case, it is necessary to account for damping effects, which typically add an imaginary—and generally frequency-dependent—part to $(\omega_{ji}^2 - \omega^2)$ in the formula for the polarizability.[16] For sufficiently strong, near-resonance fields, the quadratic dependence of the AC Stark shift on the electric field no longer holds, and effects *linear* in the applied field strength are observed (Section 5.6).

The Stark effect occurs in any electric field, including a static (DC) field. To derive the DC ($\omega = 0$) Stark shift, we can use (2.3.48), but without the factor $1/2$ that arose from averaging over the rapid oscillations of $\cos^2\omega t$:

$$\delta E_i = -\frac{E_0^2}{\hbar}\sum_j \frac{|\langle j|ez|i\rangle|^2}{\omega_{ji}} = -E_0^2\sum_j \frac{|\langle j|ez|i\rangle|^2}{E_j - E_i} \tag{2.3.49}$$

for an electric field along the z direction. This turns out to scale with the principal quantum number n as n^7; high-lying (Rydberg) states consequently have large polarizabilities and can have relatively large Stark shifts in microwave fields. Here again, the formula (2.3.49) for the Stark shift cannot apply when the energy denominator approaches zero, in which case the DC Stark shift becomes linear rather than quadratic in the electric field.

---

[15] P. F. Liao and J. E. Bjorkholm, Phys. Rev. Lett. **34**, 1 (1975).

[16] See, for instance, P. W. Milonni, R. Loudon, P. R. Berman, and S. M. Barnett, Phys. Rev. A **77**, 043835 (2008).

How this linear dependence comes about can be understood from the two-state equations (2.5.12) and (2.5.13) for the static-field ($\omega = 0$) case:

$$\ddot{\sigma}_x + \omega_0^2 \sigma_x = -2\omega_0 \Omega \sigma_z, \tag{2.3.50}$$

$$\dot{\sigma}_z = \frac{2\Omega}{\omega_0} \dot{\sigma}_x, \tag{2.3.51}$$

and therefore

$$\ddot{\sigma}_x(t) + (\omega_0^2 + 4\Omega^2)\sigma_x(t) = -2\omega_0 \Omega \sigma_z(0) - 4\Omega^2 \sigma_x(0), \tag{2.3.52}$$

implying a shifted resonance frequency

$$\omega_0 + \delta\omega_0 = (\omega_0^2 + 4\Omega^2)^{1/2}. \tag{2.3.53}$$

For $\omega_0^2 \gg 4\Omega^2$, $\delta\omega_0 \cong 2\Omega^2/\omega_0$. We equate $\delta\omega_0$ to $(\delta E_2 - \delta E_1)/\hbar$, where δE_2 and δE_1 are the energy shifts of the unperturbed upper- and lower states, respectively, of the two-state atom; this is consistent with (2.3.49), which implies

$$\delta E_2 = -\delta E_1 = \frac{E_0^2}{\hbar\omega_0}|\langle j|ez|i\rangle|^2 = \hbar\Omega^2/\omega_0. \tag{2.3.54}$$

If $\omega_0^2 \ll 4\Omega^2$, however, $\delta\omega_0 \cong 2\Omega = 2|\langle j|ez|i\rangle|E_0/\hbar$ and

$$\delta E_2 = -\delta E_1 = |\langle j|ez|i\rangle|E_0/\hbar. \tag{2.3.55}$$

In this case, there is, *in effect* a permanent dipole moment that can be parallel or anti-parallel to the applied electric field. Such a linear Stark effect occurs for sufficiently strong electric fields in the hydrogen atom, where the $2s$ and $2p$ levels are separated by the ~ 1050 MHz Lamb shift. A linear Stark shift and "permanent" electric dipole moments arising from small energy differences also occur in the rotational (microwave) spectroscopy of polar molecules. These dipole moments are not really permanent, as they vanish in the absence of an applied field.

Exercise 2.2: Estimate the electric field strength needed to observe the linear Stark effect in the $2s$ and $2p$ levels of the hydrogen atom ($|j\rangle = |n = 2, \ell = 1, m = 0\rangle$, $|i\rangle = |n = 2, \ell = 0, m = 0\rangle$).

The formula (2.3.48) for the AC Stark shift has been used to calculate the energy-level shift due to blackbody radiation of a highly excited atom. The argument is essentially as follows. Write the cycle-averaged field energy density $(1/4)\epsilon_0 E_0^2$ as $\rho(\omega)d\omega$, the energy density of blackbody radiation in the frequency interval $[\omega, \omega + d\omega]$, and replace (2.3.42) by

$$\delta E_i = -\frac{1}{6\hbar} \sum_j |\mathbf{d}_{ij}|^2 \int_0^\infty d\omega \frac{\omega_{ji}}{\omega_{ji}^2 - \omega^2} \frac{1}{2} \frac{4}{\epsilon_0} \rho(\omega)$$

$$= -\frac{1}{4\pi\epsilon_0} \frac{4}{3\pi c^3} \sum_j |\mathbf{d}_{ij}|^2 \int_0^\infty \frac{d\omega \omega^3}{e^{\hbar\omega/k_B T} - 1} \frac{\omega_{ji}}{\omega_{ji}^2 - \omega^2} \qquad (2.3.56)$$

when we use the Planck formula for $\rho(\omega)$.[17] We have inserted a factor $1/2$, since (2.3.42) was derived under the assumption of a linearly polarized field, whereas the blackbody radiation now being considered is unpolarized. Level shifts in blackbody radiation have been studied with Rydberg atoms (principal quantum numbers $n \approx 15$), where the transition frequencies $|\omega_{ji}|$ are such that $\hbar|\omega_{ji}|/k_B T \ll 1$, and (2.3.56) may be approximated by

$$\delta E_i \cong \frac{1}{4\pi\epsilon_0} \frac{4}{3\pi c^3} \sum_j \omega_{ji} |\mathbf{d}_{ij}|^2 \int_0^\infty \frac{d\omega \omega}{e^{\hbar\omega/k_B T} - 1} = \frac{1}{4\pi\epsilon_0} \frac{\pi e^2}{3m\hbar c^3} (k_B T)^2, \qquad (2.3.57)$$

where we have used the Thomas–Reiche–Kuhn sum rule (2.3.37). Shifts consistent with this expression have been measured.[18] But see Section 6.8.1.

Another type of Stark shift is that due to the electric fields of charged particles. This is an important source of line broadening in plasmas.[19]

2.4   Electric Dipole Matrix Elements

The electric dipole matrix elements

$$\mathbf{d}_{ij} = \langle \psi_i | e\mathbf{x} | \psi_j \rangle = e \int d^3 r \psi_i^*(\mathbf{r}) \mathbf{r} \psi_j(\mathbf{r}) \qquad (2.4.1)$$

for the hydrogen atom can be evaluated exactly and analytically from its stationary-state wave functions $\psi_i(\mathbf{r})$. Evaluation of these matrix elements for other atoms generally relies on results of spectroscopic measurements together with formulas from the theory of the addition of angular momenta. The relevant angular momentum theory is discussed in detail in specialized treatises.[20] Here, we will only summarize a few pertinent points.

---

[17] The integral here should be interpreted as the Cauchy principal value, so that there is no singularity at $\omega = \omega_{ji}$.

[18] L. Hollberg and J. L. Hall, Phys. Rev. Lett. **53**, 230 (1984).

[19] See, for instance, H. R. Griem, *Principles of Plasma Spectroscopy*, Cambridge University Press, Cambridge, 1997.

[20] See, for instance, *Cowan*, whose phase conventions we follow.

In terms of the polar and azimuthal angles θ and ϕ with respect to a set of orthogonal axes x, y, z, a vector $\mathbf{V}$ has Cartesian components $V_x = V \sin\theta \cos\phi, V_y = V \sin\theta \sin\phi, V_z = V \cos\theta$. Define a different set of components V_q, $q = 0, \pm 1$:

$$V_0 = V_z = V\cos\theta,$$

$$V_1 = -\frac{d}{\sqrt{2}}\sin\theta e^{i\phi} = -\frac{1}{\sqrt{2}}(V_x + iV_y),$$

$$V_{-1} = \frac{d}{\sqrt{2}}\sin\theta e^{-i\phi} = \frac{1}{\sqrt{2}}(V_x - iV_y), \tag{2.4.2}$$

which are proportional, respectively, to the spherical harmonics $Y_{10}(\theta, \phi)$, $Y_{11}(\theta, \phi)$, and $Y_{1-1}(\theta, \phi)$. This relation to spherical harmonics, their transformation properties under rotations, and their role in the quantum theory of angular momentum is the motivation for the definitions given in (2.4.2). (The V_q are said to be the components of an irreducible tensor operator of rank 1, which is just a vector in the ordinary sense.) It follows from these definitions that the scalar product of two vectors $\mathbf{V}$ and $\mathbf{W}$ is

$$\mathbf{V} \cdot \mathbf{W} = V_x W_x + V_y W_y + V_z W_z = \sum_{q=-1}^{1} (-1)^q V_q W_{-q}, \tag{2.4.3}$$

and that

$$|\mathbf{V}|^2 = |V_x|^2 + |V_y|^2 + |V_z|^2 = \sum_{q=-1}^{1} |V_q|^2. \tag{2.4.4}$$

Let the stationary state $|i\rangle$ of an atom be characterized by the angular momentum quantum numbers J, M, where the "magnetic" quantum number M can have one of the $2J + 1$ values $-J, -J + 1, \ldots, J - 1, J$, and, similarly, let the state $|j\rangle$ be characterized by angular momentum quantum numbers J', M'. We write the matrix elements of the components $(d_{ij})_q$ of $\mathbf{d}_{ij}$, as defined by (2.4.2), in terms of quantum numbers characterizing the states $|i\rangle$ and $|j\rangle$:

$$(d_{ij})_q = \langle \alpha J M | d_q | \alpha' J' M' \rangle, \tag{2.4.5}$$

where α and α' are additional quantum numbers, besides J, M and J', M', characterizing the stationary states $|i\rangle$ and $|j\rangle$, respectively. According to the Wigner–Eckart theorem,

$$\langle \alpha J M | d_q | \alpha' J' M' \rangle = (-1)^{J-M} \begin{pmatrix} J & 1 & J' \\ -M & q & M' \end{pmatrix} \langle \alpha J \| d \| \alpha' J' \rangle, \tag{2.4.6}$$

where $\langle \alpha J \| d \| \alpha' J' \rangle$ is a *reduced matrix element*, and the quantity in large parentheses is the $3j$ symbol, numerical values of which can be found on the web or in books such as *Cowan*. The reduced matrix element is independent of M and M'. Selection rules for the angular momentum quantum numbers in dipole transitions follow from the properties of the $3j$ symbols: transitions between states with angular momentum quantum numbers J, M, and J', M', can only occur (that is, the $3j$ symbol is not 0) if $\Delta J = J - J' = 0, \pm 1$, and J and J' are not both 0. In addition, the $3j$ symbol is zero unless $q = M - M' = 0, \pm 1$. Other types of transitions (magnetic dipole, electric quadrupole, etc.) are generally very much weaker than "allowed" (electric dipole) transitions.

In some applications, we must allow not only for the fine structure of atomic spectra due to spin-orbit coupling, but also for the (much finer!) "hyperfine structure" due to the magnetic coupling of electron angular momentum ($\mathbf{J}$) and nuclear spin angular momentum ($\mathbf{I}$); this generally occurs in atoms with an odd number of protons or an odd number of neutrons, for example, hydrogen and the alkalis.[21] The reduced matrix element $\langle \alpha F || d || \alpha' F' \rangle$ ($\mathbf{F} = \mathbf{J} + \mathbf{I}$) in this case depends on the electron angular momentum quantum numbers J, J', as well as the nuclear spin angular momentum quantum number I:

$$\langle \alpha F || d || \alpha' F' \rangle = (-1)^{F'} \sqrt{(2F+1)(2F'+1)} \begin{Bmatrix} J & I & F \\ F' & 1 & J' \end{Bmatrix} \langle \alpha J || d || \alpha' J' \rangle. \qquad (2.4.7)$$

The curly brackets define the *6j symbol*, numerical values for which can again be found on the web or in *Cowan*, for example.

For an example of the application of these formulas, consider the radiative decay process in which an atom in free space makes a transition from a state $|j\rangle = |\alpha J M\rangle$ of energy $E(\alpha J)$ to a state $|i\rangle = |\alpha' J' M'\rangle$ of energy $E(\alpha' J')$ $(E(\alpha J) > E(\alpha' J'))$. The rate for this process is

$$A_{ji} = A(\alpha J M \to \alpha' J' M') = \frac{\omega_0^3}{3\pi\epsilon_0\hbar c^3} |\langle \alpha J M | \mathbf{d} | \alpha' J' M' \rangle|^2, \qquad (2.4.8)$$

$\omega_0 = (E_{\alpha J} - E_{\alpha' J'})/\hbar$.[22] The states $|\alpha J M\rangle$ and $|\alpha' J' M'\rangle$ might, for instance, be magnetic substates of the $3P_{3/2}$ and $3S_{1/2}$ levels of the sodium atom, in which case $J = 3/2$ and $J' = 1/2$. The total radiative decay rate of the state $|\alpha J M\rangle$ is obtained by summing (2.4.8) over all the lower states to which a transition from $|\alpha J M\rangle$ can occur:

$$A(\alpha J M) = \frac{\omega_0^3}{3\pi\epsilon_0\hbar c^3} \sum_{M'q} |\langle \alpha J M | d_q | \alpha' J' M' \rangle|^2$$

$$= \frac{\omega_0^3}{3\pi\epsilon_0\hbar c^3} |\langle \alpha J || d || \alpha' J' \rangle|^2 \sum_{M'q} \begin{pmatrix} J & 1 & J' \\ -M & q & M' \end{pmatrix}^2$$

$$= \frac{\omega_0^3}{3\pi\epsilon_0\hbar c^3} |\langle \alpha J || d || \alpha' J' \rangle|^2 \frac{1}{2J+1}, \qquad (2.4.9)$$

---

[21] A very useful compilation of reference data for alkali atoms has been made available by D. Steck at http://steck.us/alkalidata, accessed October 25, 2018.

[22] This formula for the spontaneous emission rate is derived in Chapter 4. See, for example, (4.4.21).

where we have used a sum rule satisfied by the $3j$ symbols.[23] $A(\alpha J M)$ is independent of M, and so the different magnetic substates of the upper level have the same radiative decay rate. For the sodium $3P_{3/2}$ level, for example, the radiative lifetime is $1/A(\alpha J M) = 16$ ns. Knowing this, we can calculate the reduced matrix element $\langle \alpha J M \| d \| \alpha' J' \rangle$ from (2.4.9), and, from that, the electric dipole matrix elements in (2.4.6).

We can account for hyperfine structure in the same fashion. The rate for the radiative decay from a state $|\alpha F M_F\rangle$ to a state $|\alpha' F' M_{F'}\rangle$ ($E(\alpha F) > E(\alpha', F')$; $M_F = -F, -F+1, \ldots, F-1, F$; $M_{F'} = -F', -F'+1, \ldots, F'-1, F'$), is

$$A(\alpha F M_F \to \alpha' F' M_{F'}) = \frac{\omega_0^3}{3\pi\epsilon_0\hbar c^3} |\alpha F M_F|d|\alpha' F' M_{F'}\rangle|^2, \qquad (2.4.10)$$

$\omega_0 = (E_{\alpha F} - E_{\alpha' F'})/\hbar$. Consider transitions between levels with electron angular momentum quantum numbers J and J' that have hyperfine splittings into states with quantum numbers F, M_F and $F', M_{F'}$. Since the hyperfine splittings are very small compared with the fine structure splittings, we will assume that all the hyperfine transitions $|\alpha F M\rangle \to |\alpha' F' M_{F'}\rangle$ have approximately the same transition frequency ω_0. Then, the total radiative decay rate of the state $|\alpha F M_F\rangle$ is

$$\begin{aligned}
A(\alpha F M_F) &\cong \frac{\omega_0^3}{3\pi\epsilon_0\hbar c^3} \sum_{F'} \sum_{M_{F'} q} |\langle \alpha F M_F|d_q|\alpha' F' M_{F'}\rangle|^2 \\
&= \frac{\omega_0^3}{3\pi\epsilon_0\hbar c^3} \sum_{F'} \sum_{M_{F'}} \begin{pmatrix} F & 1 & F' \\ -M_F & q & M_{F'} \end{pmatrix}^2 |\langle \alpha F\|d\|\alpha' F'\rangle|^2 \\
&= \frac{\omega_0^3}{3\pi\epsilon_0\hbar c^3} |\langle \alpha J\|d\|\alpha' J'\rangle|^2 (2F+1) \sum_{F'} (2F'+1) \begin{Bmatrix} J & I & F \\ F' & 1 & J' \end{Bmatrix}^2 \\
&\quad \times \sum_{M_{F'} q} \begin{pmatrix} F & 1 & F' \\ -M_F & q & M_{F'} \end{pmatrix}^2 \\
&= \frac{\omega_0^3}{3\pi\epsilon_0\hbar c^3} |\langle \alpha J\|d\|\alpha' J'\rangle|^2 \frac{1}{2J+1}, \qquad (2.4.11)
\end{aligned}$$

where we have used a sum rule for the $6j$ symbols as well as the sum rule used above for the $3j$ symbols.[24] This is identical with (2.4.9), meaning that each hyperfine state belonging to a given J multiplet has (approximately) the same radiative decay rate. For the $3P_{3/2}$ level of sodium, for example, $J = 3/2$, the nuclear spin $I = 3/2$, $F = |J - I|, |J - I| + 1, \ldots, |J + I| = 0, 1, 2, 3$, and there are $2F + 1$ values of M_F for each F for a total of 16 states F, M_F, each of which has a radiative lifetime $\cong 16$ ns (see Figure 2.2). For the $3S_{1/2}$ level, $J' = 1/2$ and $F' = 1, 2$ for a total of eight states

[23] See, for instance, *Cowan*, p. 145.
[24] Ibid., p. 149.

$F', M_{F'}$. All the electric dipole matrix elements between the states $|3P_{3/2}FM_F\rangle$ and $|3S_{2/2}F'M_{F'}\rangle$ are given in terms of $3j$ and $6j$ symbols by

$$\langle 3P_{3/2}FM_F|d_q|3S_{1/2}F'M_{F'}\rangle = (-1)^{F+F'-M_F}\sqrt{(2F+1)(2F'+1)}\begin{pmatrix} F & 1 & F' \\ -M_F & q & M_{F'} \end{pmatrix}$$
$$\times \begin{Bmatrix} \frac{3}{2} & \frac{3}{2} & F \\ F' & 1 & \frac{1}{2} \end{Bmatrix}\langle \alpha J||d||\alpha'J'\rangle, \qquad (2.4.12)$$

$\alpha = 3P_{3/2}$, $\alpha' = 3S_{1/2}$, $J = 3/2$, $J' = 1/2$.

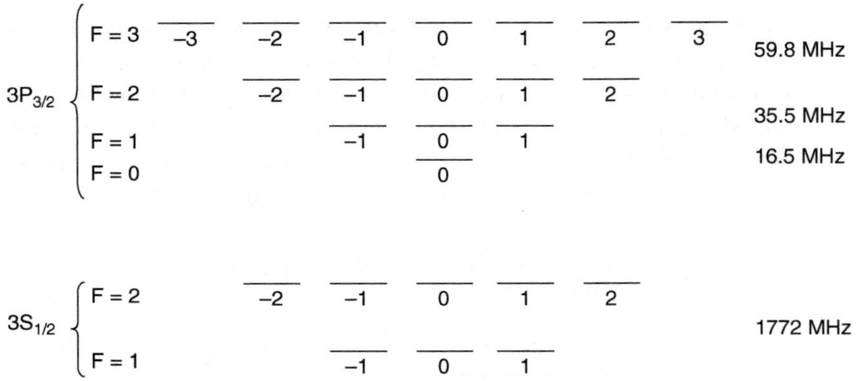

Fig. 2.2 Hyperfine splittings of the 589 nm sodium D$_2$ line. (The energy splittings indicated are obviously not to scale.)

Consider now an atom in the applied electric field

$$\mathbf{E}(t) = \frac{1}{2}\left(\mathbf{E}_0 e^{-i\omega t} + \mathbf{E}_0^* e^{i\omega t}\right). \qquad (2.4.13)$$

For an absorptive transition in which the atom goes from state $|i\rangle$ to state $|j\rangle$, such that $E_j - E_i = \hbar\omega$, the relevant matrix element is

$$\mathbf{d}_{ji} \cdot \mathbf{E}_0 = \sum_{q=-1}^{1}(-1)^q(d_{ji})_q(E_0)_{-q}, \qquad (2.4.14)$$

as can be inferred from (2.3.3). Here, as above, we identify $|j\rangle$ with $|\alpha JM\rangle$, and $|i\rangle$ with $|\alpha'J'M'\rangle$. The selection rule $\Delta J = 0, \pm 1$ ($J = 0 \leftrightarrow J' = 0$ forbidden) applies regardless of the polarization of the field. For a field having an amplitude $\mathcal{E}_0$ and linearly polarized along a direction $\hat{z}$, for example,

$$\mathbf{E}(t) = \mathcal{E}_0 \hat{z} \cos \omega t, \tag{2.4.15}$$

$(E_0)_1 = (E_0)_{-1} = 0$, and

$$\mathbf{d}_{ji} \cdot \mathbf{E}_0 = (d_{ji})_0 (E_0)_0 = \langle \alpha J M | d_0 | \alpha' J' M' \rangle \mathcal{E}_0. \tag{2.4.16}$$

In this case, only the $q = 0$ component of $\mathbf{d}_{ji}$ contributes, and the condition $q = M - M'$ therefore implies the selection rule $\Delta M = 0$ for absorption in a linearly polarized field. For a circularly polarized field

$$\mathbf{E}(t) = \frac{1}{\sqrt{2}} \mathcal{E}_0 \big[(\hat{x} \pm i\hat{y}) e^{-i\omega t} + (\hat{x} \mp i\hat{y}) e^{i\omega t} \big], \tag{2.4.17}$$

$(E_0)_0 = 0$, $(E_0)_1 = -\mathcal{E}_0$, $(E_0)_{-1} = \mathcal{E}_0$, and only the $q = +1$ or $q = -1$ component of $\mathbf{d}_{ji}$ contributes, depending on whether the field is right-circularly polarized ($q = -1$) or left-circularly polarized ($q = 1$).[25] The condition $q = M - M'$ therefore implies the selection rules $\Delta M = -1$, and $\Delta M = 1$, for right-circularly polarized and left-circularly polarized fields, respectively. The signs of the ΔM's are reversed in the case of induced (stimulated) emission in polarized fields. Similar selection rules apply to hyperfine transitions, in which case the J's and M's are replaced by F's and M_F's.

2.5 Two-State Atoms

The atom–field interactions of particular interest in quantum optics are those in which field frequencies are close to atomic transition frequencies. In the simplest example, a single field frequency is near resonance with a single atomic transition, and only the two states of this transition have significant occupation probabilities. Then, the atom–field interaction is well described by the approximation in which the atom is regarded as a two-state system characterized completely by a transition frequency and a dipole matrix element (see Figure 2.3).

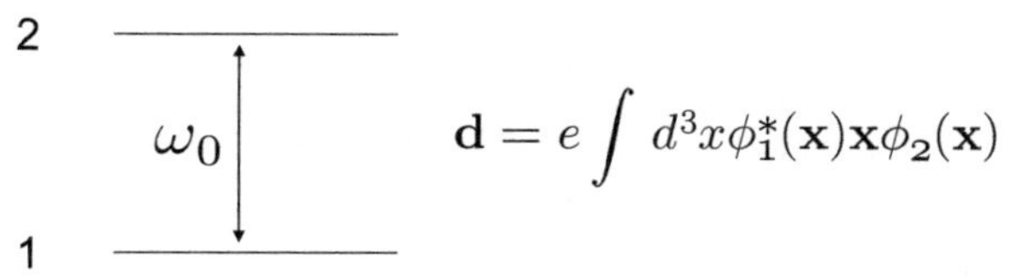

Fig. 2.3 A two-state atom with transition frequency ω_0 and transition electric dipole moment **d**.

---

[25] The factor $1/\sqrt{2}$ is chosen such that the fields (2.4.15) and (2.4.17) have the same time-averaged intensity.

For such a "two-state atom," the Hamiltonian (2.3.14) reduces to

$$H = E_1\sigma_{11} + E_2\sigma_{22} - \mathbf{d}_{12} \cdot \mathbf{E}(t)\sigma_{12} - \mathbf{d}_{21} \cdot \mathbf{E}(t)\sigma_{21}. \tag{2.5.1}$$

Now,

$$E_1\sigma_{11} + E_2\sigma_{22} = \frac{1}{2}(E_2 - E_1)(\sigma_{22} - \sigma_{11}) + \frac{1}{2}(E_2 + E_1)(\sigma_{22} + \sigma_{11}). \tag{2.5.2}$$

$\sigma_{22} + \sigma_{11}$ is the constant unit operator in the Hilbert space of the two-state atom, and so the last term may be dropped because it commutes with every operator in the Hamiltonian (and therefore does not affect Heisenberg equations of motion). We simplify further by assuming again a purely monochromatic, linearly polarized applied field $\mathbf{E}(t) = \mathbf{E}_0 \cos \omega t$. By an appropriate choice of the (arbitrary) constant phases of the wave functions $\phi_1(\mathbf{x})$ and $\phi_2(\mathbf{x})$ (see (2.1.4)) we can make $\mathbf{d}_{12} \cdot \mathbf{E}_0 = \mathbf{d}_{21} \cdot \mathbf{E}_0 = \mathbf{d} \cdot \mathbf{E}_0$ real and define the (real) *Rabi frequency*

$$\Omega = \frac{1}{\hbar}\mathbf{d} \cdot \mathbf{E}_0 = \frac{d}{\hbar}E_0. \tag{2.5.3}$$

Then the Hamiltonian for the interaction of our two-state atom with an applied monochromatic field becomes

$$H = \frac{1}{2}\hbar\omega_0\sigma_z - \hbar\Omega(\sigma + \sigma^\dagger)\cos \omega t, \tag{2.5.4}$$

where $\omega_0 = (E_2 - E_1)/\hbar$ is the two-state-atom transition frequency, and $\sigma_z = \sigma_{22} - \sigma_{11}$, $\sigma = \sigma_{12}$, $\sigma^\dagger = \sigma_{21}$. The following commutation relations are easily verified:

$$[\sigma_z, \sigma] = -2\sigma, \quad \text{and} \quad [\sigma, \sigma^\dagger] = -\sigma_z, \tag{2.5.5}$$

and, together with (2.5.4), they imply the Heisenberg equations of motion

$$\dot{\sigma} = -i\omega_0\sigma - i\Omega\sigma_z \cos \omega t, \tag{2.5.6}$$

$$\dot{\sigma}_z = -2i\Omega(\sigma - \sigma^\dagger)\cos \omega t. \tag{2.5.7}$$

We also introduce the operators

$$\sigma_x = \sigma + \sigma^\dagger \quad \text{and} \quad \sigma_y = i(\sigma - \sigma^\dagger). \tag{2.5.8}$$

In terms of these operators, the Heisenberg equations (2.5.6) and (2.5.7) are

$$\dot{\sigma}_x = -\omega_0\sigma_y, \tag{2.5.9}$$

$$\dot{\sigma}_y = \omega_0\sigma_x + 2\Omega\sigma_z \cos \omega t, \tag{2.5.10}$$

$$\dot{\sigma}_z = -2\Omega\sigma_y \cos \omega t, \tag{2.5.11}$$

or, equivalently,

$$\ddot{\sigma}_x + \omega_0^2\sigma_x = -2\omega_0\Omega\sigma_z \cos \omega t, \tag{2.5.12}$$

$$\dot{\sigma}_z = \frac{2\Omega}{\omega_0}\dot{\sigma}_x \cos \omega t. \tag{2.5.13}$$

In the absence of an applied field, the operators σ and $\sigma^\dagger$ are "lowering" and "raising" operators, respectively, for the two-state atom: $\sigma|1\rangle = |1\rangle\langle 2|1\rangle = 0$, $\sigma|2\rangle = |1\rangle\langle 2|2\rangle =$

$|1\rangle$ and, similarly, $\sigma^\dagger|1\rangle = |2\rangle$, $\sigma^\dagger|2\rangle = 0$. The operators σ_x, σ_y, and σ_z are the usual Pauli operators for a two-state system, most familiar historically in the context of spin-1/2 systems (the two states there being "spin down" and "spin up"). Thus, $[\sigma_x, \sigma_y] = 2i\sigma_z$, and so on. Equation (2.5.13) gives the rate of change of the two-state-atom occupation probabilities induced by the applied field, that is, it accounts for absorption and induced (stimulated) emission processes. Writing it as

$$\frac{1}{2}\hbar\omega_0\dot{\sigma}_z = -E(t)[d\dot{\sigma}_x(t)], \qquad (2.5.14)$$

we see that it has the same form as the classical equation $\dot{W}(t) = -E(t)\dot{p}(t)$ for the rate of change of the energy of an electric dipole $p(t)$ in an electric field $E(t)$ along the direction of the dipole.

2.5.1 How to Make a Two-State Atom

The fact that the frequency of an applied field is nearly resonant with a transition frequency of the atom might make it reasonable to treat the atom as a two-*level* system, but, in the absence of a magnetic field, there will be degenerate magnetic states belonging to each of these energy levels. But a two-*state* atom can be realized by optically pumping with polarized light.[26] This is illustrated in Figure 2.4, where we indicate transitions among degenerate magnetic substates of a higher energy level with total angular momentum quantum number $F = 2$ and a lower energy level with total angular momentum quantum number $F = 1$. The two energy levels might, for example, be the $3S_{1/2}(F = 1)$ and $3P_{3/2}(F = 2)$ levels of the sodium atom, with F the total (electron-plus-nucleus) angular momentum quantum number, and the different F's labeling different energy levels (hyperfine structure). If the atom is irradiated with σ_+ circularly polarized light, which can only induce $\Delta M = +1$ absorptive transitions, the effect is to put the atom in states of larger M. Together with the selection rule $\Delta M = 0, \pm 1$ for spontaneous emission, this results in a "drift" of the atom to higher values of M, and, eventually, it can only be in one of the two states $(F = 1, M = +1)$ or $(F = 2, M = +2)$. For simplicity, but without affecting the basic explanation of the making of a two-state atom in this example, we have ignored stimulated emission.

2.5.2 The Oscillator Model of an Atom

You may think that this is a funny model of an atom if you have heard about electrons whirling around in orbits. But that is just an oversimplified picture. The correct picture of an atom, which is given by [quantum mechanics], says that, *so far as problems involving light are concerned*, the electrons behave as though they were held by springs. — Richard Feynman[27]

From the discussion of the physical significance of $\langle\sigma_{ij}(t)\rangle$ in Section 2.5.1, it is clear that $\langle\sigma_z(t)\rangle = \langle\sigma_{22}(t)\rangle - \langle\sigma_{11}(t)\rangle$ is the two-state-atom "population difference," the

---

[26] J. A. Abate, Opt. Commun. **10**, 269 (1974).

[27] *Feynman, Leighton, and Sands*, Volume I, p. 31–4.

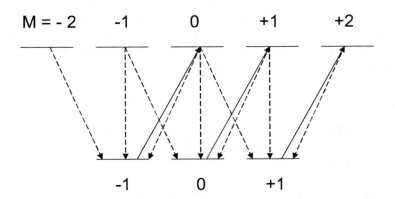

Fig. 2.4 Optical pumping with circularly polarized light that induces $\Delta M = +1$ transitions (solid arrows). Spontaneous emission (dashed arrows) can occur between states with $\Delta M = 0$ or ± 1. After a few absorption and spontaneous emission events, the atom can only be in one of the states $(F = 1, M = +1)$ or $(F = 2, M = +2)$. It has become a two-state atom.

occupation probability of the higher-energy state minus the occupation probability of the lower-energy state. If the two-state atom remains in the lower state with probability $\cong 1$, then $\langle \sigma_z(t) \rangle \cong -1$, and (2.5.12) implies the approximate equation

$$\langle \ddot{\sigma}_x \rangle + \omega_0^2 \langle \sigma_x \rangle = 2\omega_0 \Omega \cos \omega t. \tag{2.5.15}$$

Let $\mathbf{E}_0$ point in the x direction, and consider the expectation value

$$\langle x(t) \rangle = x_{12} \langle \sigma_{12}(t) \rangle + x_{21} \langle \sigma_{21}(t) \rangle = x_{12} \langle \sigma_x(t) \rangle \tag{2.5.16}$$

of the electron coordinate along the same direction, assuming again that the phases of the wave functions have been chosen such that $d_{12} = d_{21} = ex_{12}$ is real. From (2.5.15),

$$\langle \ddot{x} \rangle + \omega_0^2 \langle x \rangle = \frac{2\omega_0 e}{\hbar} x_{12}^2 E_0 \cos \omega t = f_{12} \frac{e}{m} E_0 \cos \omega t, \tag{2.5.17}$$

where we define the *oscillator strength*

$$f_{12} = \frac{2m\omega_0}{\hbar} x_{12}^2 = \frac{2m\omega_0}{3\hbar} |\mathbf{x}_{12}|^2. \tag{2.5.18}$$

We see, therefore, that the expectation value of the electron coordinate in the two-state-atom model satisfies an equation of the same form as the classical equation of motion (2.3.35) if the atom remains with high probability in the lower state; the only difference is that we must introduce the dimensionless factor f_{12} in the coupling to the field. In other words, once we introduce the oscillator strength, which comes from quantum theory, a two-state atom that remains with high probability in the lower state responds to an electric field in just the way described by the classical oscillator model of a bound electron. The natural resonance frequency in the classical model is the two-state-atom transition frequency ω_0.

We have assumed that the two-state atom remains in the lower state with high probability. If, instead, it is the upper state that remains populated with high probability, we find that $\langle x(t) \rangle$ satisfies the classical oscillator equation but with a minus sign on the right side. In other words, in this case, the oscillator strength is $f_{21} = -f_{12}$: the two-state atom in its upper state behaves as a "negative oscillator." More generally, we define the oscillator strength of the $i \to j$ transition of an atom to be

$$f_{ij} = \frac{2m\omega_{ji}}{3\hbar} |\mathbf{x}_{ij}|^2, \tag{2.5.19}$$

which is positive for upward transitions ($\omega_{ji} = (E_j - E_i)/\hbar > 0$) and negative for downward transitions. In terms of the transition oscillator strengths, the Thomas–Reiche–Kuhn sum rule (2.3.37) is

$$\sum_j f_{ij} = 1 \tag{2.5.20}$$

for any state i of a one-electron atom. (For an N-electron atom, the 1 on the right side is replaced by N.)

Exercise 2.3: Show that the refractive index $n(\omega)$ of a dilute monatomic gas is given by the formula

$$n^2(\omega) - 1 = \frac{e^2}{m\epsilon_0} \sum_i \sum_j \frac{N_i f_{ij}}{\omega_{ji}^2 - \omega^2},$$

with N_i the number density of atoms in state $|i\rangle$. Note that, if excited states are populated, there are contributions to the refractive index from "negative oscillators." This means that, if excited-state populations are produced by applying an electric current, for instance, the refractive index should change as the current, and therefore the density of "negative oscillators," is changed. This was observed by H. Kopfermann and R. W. Ladenburg, Nature **122**, 438 (1928).

2.5.3 The Rotating-Wave Approximation

Now take a slightly more general approach to the two-state atom in an applied field, starting with the Hamiltonian (see (2.5.1))

$$H = \frac{1}{2}\hbar\omega_0\sigma_z - (\mathbf{d}_{12}\sigma_{12} + \mathbf{d}_{21}\sigma_{21}^\dagger) \cdot \mathbf{E}(t), \tag{2.5.21}$$

with $\mathbf{E}(t) = \frac{1}{2}(\mathbf{E}_0 e^{-i\omega t} + \mathbf{E}_0^* e^{i\omega t})$, $\mathbf{E}_0$ now not necessarily real. Using again the commutation relations (2.5.5), we obtain the Heisenberg equations of motion

$$\dot{\sigma} = -i\omega_0 \sigma - \frac{i}{\hbar} \mathbf{d}_{21} \cdot \mathbf{E}(t)\sigma_z, \tag{2.5.22}$$

$$\dot{\sigma}_z = -\frac{2i}{\hbar}[\mathbf{d}_{12} \cdot \mathbf{E}(t)\sigma - \mathbf{d}_{21} \cdot \mathbf{E}(t)\sigma^\dagger]. \tag{2.5.23}$$

For $\omega \cong \omega_0$, it is seen from (2.5.22) that $\sigma(t)$ varies *approximately* as $e^{-i\omega t}$, prompting us to write

$$\sigma(t) = S(t)e^{-i\omega t}, \tag{2.5.24}$$

with $S(t)$ assumed to be slowly varying compared with the field oscillation at frequency ω. In this *rotating-wave approximation*, it follows from (2.5.22)–(2.5.24) that the expectation values $\langle S(t) \rangle$ and $w(t) = \langle \sigma_z(t) \rangle$ satisfy

$$\langle \dot{S}(t) \rangle = -i\Delta\langle S(t) \rangle - \frac{i}{2\hbar} \mathbf{d}_{21} \cdot \mathbf{E}_0 w(t), \tag{2.5.25}$$

$$\dot{w}(t) = -\frac{i}{\hbar}\big[\mathbf{d}_{12} \cdot \mathbf{E}_0^* \langle S(t) \rangle - \mathbf{d}_{21} \cdot \mathbf{E}_0 \langle S^\dagger(t) \rangle\big], \tag{2.5.26}$$

where we define the atom–field *detuning* $\Delta = \omega_0 - \omega$. To simplify slightly we will assume again that the (arbitrary) phases of the wave functions $\phi_1(\mathbf{x})$ and $\phi_2(\mathbf{x})$ have been conveniently chosen, now such that the *Rabi frequency*

$$\Omega = \frac{1}{\hbar} \mathbf{d}_{21} \cdot \mathbf{E}_0 \tag{2.5.27}$$

is real. Then, in terms of the real variables $u(t)$ and $v(t)$ defined by writing $\langle S(t) \rangle = \frac{1}{2}[u(t) - iv(t)]$, we obtain from (2.5.25) and (2.5.26)

$$\dot{u} = -\Delta v, \tag{2.5.28}$$
$$\dot{v} = \Delta u + \Omega w, \tag{2.5.29}$$
$$\dot{w} = -\Omega v, \tag{2.5.30}$$

with $w(t) = \langle \sigma_z(t) \rangle$. If the detuning Δ and the Rabi frequency Ω are small compared with ω, the assumption of slowly varying quantities underlying the rotating-wave approximation is justified, and the rotating-wave approximation will accurately describe the dynamics of a two-state atom in a field whose frequency is resonant (or nearly resonant) with the two-state-atom transition frequency. This is almost always the case in situations where the two-state model is sensible.

Equations (2.5.28)–(2.5.30) imply that $u^2(t) + v^2(t) + w^2(t)$ is a constant. Using (2.3.18) and some simple algebra, we find that this constant is 1:

$$u^2(t) + v^2(t) + w^2(t) = |a_1(t)|^2 + |a_2(t)|^2 = 1, \tag{2.5.31}$$

which is just a statement of "conservation of probability," that is, the probability is 1 that the two-state atom is either in the upper state or the lower state at any time t.

Exercise 2.4: Estimate the field intensity necessary for the Rabi frequency to be comparable to an optical frequency ω. For this purpose, you will need to assume a numerical value for the transition dipole moment $|\mathbf{d}_{12}|$. You can use the formula $A_{21} = |\mathbf{d}_{21}|^2 \omega_0^3 / 3\pi\epsilon_0 \hbar c^3$ for the rate of spontaneous photon emission from the upper state, a typical value being about 10^8 s^{-1}. This formula is derived in Section 4.4.

To better appreciate the rotating-wave approximation, consider, as in our derivation of the Kramers–Heisenberg formula, the expectation value of the electric dipole moment induced in an atom by a linearly polarized field $\mathbf{E}_0 \cos \omega t$, with $\mathbf{E}_0$ in the x direction. For a two-state atom, this induced dipole moment along the x direction has the magnitude

$$\langle d(t) \rangle = (d_{12})_x [\langle \sigma(t) \rangle + \langle \sigma^\dagger(t) \rangle] = (d_{12})_x (u \cos \omega t - v \sin \omega t), \qquad (2.5.32)$$

with $(d_{12})_x$ real. Using the particular solutions $v = 0$ and $u = \Omega/\Delta$ of (2.5.28) and (2.5.29) with $w = -1$ for a two-state atom in its lower state, we obtain $\langle d(t) \rangle = ([(d_{12})_x]^2/\hbar\Delta)E_0 \cos \omega t$, implying in the rotating-wave approximation (indicated by the superscript) the polarizability

$$\alpha^{\mathrm{RWA}}(\omega) = \frac{[(d_{12})_x]^2}{\hbar} \frac{1}{\omega_0 - \omega} = \frac{|\mathbf{d}_{12}|^2}{3\hbar} \frac{1}{\omega_0 - \omega} \qquad (2.5.33)$$

for a two-state atom. Comparison with the two-state approximation to the Kramers–Heisenberg formula (2.3.32),

$$\alpha(\omega) = \frac{2|\mathbf{d}_{12}|^2}{3\hbar} \frac{\omega_0}{\omega_0^2 - \omega^2} = \frac{|\mathbf{d}_{12}|^2}{3\hbar} \left(\frac{1}{\omega_0 - \omega} + \frac{1}{\omega_0 + \omega} \right), \qquad (2.5.34)$$

shows that the rotating-wave approximation omits the "antiresonant" term $1/(\omega_0 + \omega)$, that is, we are assuming in the rotating-wave approximation that $|\omega_0 - \omega| \ll \omega_0 + \omega$, consistent with the remarks following (2.5.30). For a two-state atom in its upper state, the polarizability is found similarly to simply have the opposite sign to the lower-state polarizability:

$$\alpha^{\mathrm{RWA}}(\omega) = -\frac{[(d_{12})_x]^2}{\hbar} \frac{1}{\omega_0 - \omega} = -\frac{|\mathbf{d}_{12}|^2}{3\hbar} \frac{1}{\omega_0 - \omega}, \qquad (2.5.35)$$

$$\alpha(\omega) = -\frac{2|\mathbf{d}_{12}|^2}{3\hbar} \frac{\omega_0}{\omega_0^2 - \omega^2} = -\frac{|\mathbf{d}_{12}|^2}{3\hbar} \left(\frac{1}{\omega_0 - \omega} + \frac{1}{\omega_0 + \omega} \right). \qquad (2.5.36)$$

The rotating-wave-approximation equations (2.5.28)–(2.5.30) imply

$$\ddot{v} + \Delta^2 v = \Omega \dot{w} = -\Omega^2 v \qquad (2.5.37)$$

and therefore a shifted resonance frequency ω_0' given by

$$\Delta' = \omega_0' - \omega = \left(\Delta^2 + \Omega^2\right)^{1/2} \cong \Delta\left(1 + \frac{\Omega^2}{2\Delta^2}\right) = \Delta + \frac{\Omega^2}{2\Delta}, \tag{2.5.38}$$

or

$$\omega_0' = \omega_0 + \frac{\Omega^2}{2\Delta} = \omega_0 + \delta\omega_0 \tag{2.5.39}$$

for $\Delta^2 \gg \Omega^2$. Now, from (2.3.42), we obtain the AC Stark shifts δE_2 and δE_1 of the upper and lower states, respectively:

$$\delta E_2 = -\frac{|\mathbf{d}_{12}|^2 E_0^2}{6\hbar}\left(\frac{-\omega_0}{\omega_0^2 - \omega^2}\right) \cong \frac{|\mathbf{d}_{12}|^2 E_0^2}{12\hbar\Delta} = \frac{[(d_{12})_x]^2 E_0^2}{4\hbar\Delta}, \tag{2.5.40}$$

$$\delta E_1 = -\frac{\mathbf{d}^2 E_0^2}{6\hbar}\left(\frac{\omega_0}{\omega_0^2 - \omega^2}\right) \cong -\frac{\mathbf{d}^2 E_0^2}{12\hbar\Delta} \cong -\frac{[(d_{12})_x]^2 E_0^2}{4\hbar\Delta}, \tag{2.5.41}$$

for $\omega_0 + \omega \cong 2\omega_0$. Therefore,

$$\delta\omega_0 = \frac{\Omega^2}{2\Delta} = \frac{1}{\hbar}(\delta E_2 - \delta E_1), \tag{2.5.42}$$

which is just the Stark-shifted *frequency*—in the rotating-wave approximation—of the two-state atom.

As remarked earlier, much of the theory for two-state atoms in an applied electric field follows the theory of two-state (spin-1/2) systems in magnetic fields. In magnetic resonance theory, it is not necessary to make the rotating-wave approximation in deriving equations of the form shown in (2.5.30) when the applied magnetic field is circularly polarized. For two-level atoms in circularly polarized light, however, we cannot derive these equations without making the rotating-wave approximation. This is because real atoms, even when they may be treated very accurately as two-(energy-)level systems, are never *exactly* two-*state* systems: even with circularly polarized light, there are virtual, "non-energy-conserving" transitions, both spontaneous and stimulated, to states other than the two of interest. These result in small energy-level shifts analogous to the Bloch–Siegert shifts of spin-1/2 systems in magnetic fields.[28]

We have ignored damping effects that result in an imaginary part of the polarizability, which is associated with attenuation of field energy (see Sections 1.11 and 2.7). They can very often be taken into account by replacing the near-resonance polarizability (2.5.33) by

$$\alpha(\omega) = \frac{|\mathbf{d}_{12}|^2}{3\hbar}\frac{1}{\omega_0 - \omega - i\beta} = \frac{|\mathbf{d}_{12}|^2}{3\hbar}\frac{\omega_0 - \omega + i\beta}{(\omega_0 - \omega)^2 + \beta^2}, \tag{2.5.43}$$

in which case the real (α_R) and imaginary (α_I) parts of the near-resonance polarizability are related by

[28] For a discussion of this point, see R. J. C. Spreeuw and J. P. Woerdman, Phys. Rev. A**44**, 4765 (1991).

$$\alpha_R(\omega) = \frac{\omega_0 - \omega}{\beta}\alpha_I(\omega). \qquad (2.5.44)$$

For a dilute medium of N atoms per unit volume, the real part of the refractive index near the resonance frequency ω_0 is (see (1.5.22))

$$n_R(\omega) = [1 + N\alpha_R(\omega)/\epsilon_0]^{1/2} \cong 1 + \frac{1}{2\epsilon_0}N\alpha_R(\omega), \qquad (2.5.45)$$

and the attenuation (absorption) coefficient is (see (1.11.1))

$$a(\omega) = 2\frac{\omega}{c}\frac{1}{2\epsilon_0}N\alpha_I(\omega). \qquad (2.5.46)$$

Therefore,

$$n_R(\omega) \cong 1 + \frac{1}{2\epsilon_0}N\frac{\omega_0 - \omega}{\beta}\alpha_I(\omega) = 1 + \frac{c}{2\omega}a(\omega) = 1 + \frac{\lambda}{2\pi}\frac{\omega_0 - \omega}{\gamma}a(\omega), \qquad (2.5.47)$$

where λ is the wavelength and $\gamma = 2\beta$ is the full width at half-maximum of the (Lorentzian) absorption lineshape function. This implies that the (real) refractive index close to a resonance frequency *decreases* with increasing frequency, so-called *anomalous* dispersion because the index of refraction ordinarily increases with frequency (recall, for instance, (1.5.24)). In the case of an amplifying medium, $a(\omega) = -g(\omega) < 0$ and

$$n_R(\omega) \cong 1 + \frac{\lambda}{2\pi}\frac{\omega - \omega_0}{\gamma}g(\omega), \qquad (2.5.48)$$

where $g(\omega)$ is the amplification (gain) coefficient.

2.6 Pulsed Excitation and Rabi Oscillations

Suppose the field frequency is exactly equal to the transition frequency ω_0 and that the atom is initially in the lower state. Then, $w(0) = \langle\sigma_{22}(0)\rangle - \langle\sigma_{11}(0)\rangle = |a_2(0)|^2 - |a_1(0)|^2 = -1$, and, from (2.5.31), $u(0) = v(0) = 0$. The solution of (2.5.28)–(2.5.30) is then

$$u(t) = 0, \qquad v(t) = -\sin\Omega t, \qquad w(t) = -\cos\Omega t. \qquad (2.6.1)$$

Since $w(t) = |a_2(t)|^2 - |a_1(t)|^2 = 2|a_2(t)|^2 - 1$, the probability that the atom will be found in the upper state at time t is

$$|a_2(t)|^2 = \frac{1}{2}(1 - \cos\Omega t), \qquad (2.6.2)$$

and, similarly, the lower-state probability is $|a_1(t)|^2 = (1/2)(1 + \cos\Omega t)$. If the atom is initially in the lower state but $\Delta \neq 0$,

$$|a_1(t)|^2 = \frac{1}{2}\left(1 + \frac{\Delta^2}{\Omega^2 + \Delta^2} + \frac{\Omega^2}{\Omega^2 + \Delta^2}\cos\sqrt{\Omega^2 + \Delta^2}t\right), \qquad (2.6.3)$$

$$|a_2(t)|^2 = \frac{1}{2}\frac{\Omega^2}{\Omega^2 + \Delta^2}\left(1 - \cos\sqrt{\Omega^2 + \Delta^2}t\right). \qquad (2.6.4)$$

According to these solutions, a two-state atom will oscillate between the two states at the Rabi frequency Ω if $\Delta = 0$, and at the "generalized Rabi frequency" $(\Omega^2 + \Delta^2)^{1/2}$

if $\Delta \neq 0$. The atom is said to undergo *Rabi oscillations* between the two states of the resonant transition.

These solutions for a monochromatic applied field do not account for processes such as collisions or spontaneous emission. Suppose the field is in the form of a pulse that is on for a time τ_p short compared with the time τ_c over which such processes can significantly affect the atom. In this case, we write

$$E(t) = \mathcal{E}(t) \cos \omega t, \tag{2.6.5}$$

where $\mathcal{E}(t)$ is the pulse "envelope," which is assumed to be slowly varying compared with the "carrier" $\cos \omega t$. Then we obtain, instead of (2.5.28)–(2.5.30),

$$\dot{u} = -\Delta v, \tag{2.6.6}$$

$$\dot{v} = \Delta u + \frac{d}{\hbar}\mathcal{E}(t)w, \tag{2.6.7}$$

$$\dot{w} = -\frac{d}{\hbar}\mathcal{E}(t)v. \tag{2.6.8}$$

These are the *optical Bloch equations* in the absence of damping. The solutions for $\Delta = 0$, and for the atom in the lower state at times $t \to -\infty$ before the pulse has any appreciable amplitude, are $u(t) = 0$, and

$$v(t) = -\sin\left[\frac{d}{\hbar}\int_{-\infty}^{t} dt'\mathcal{E}(t')\right], \qquad w(t) = -\cos\left[\frac{d}{\hbar}\int_{-\infty}^{t} dt'\mathcal{E}(t')\right]. \tag{2.6.9}$$

After the pulse is off, but before collisions and spontaneous emission have any appreciable effect, the atom is left in a state for which

$$v = -\sin\theta, \qquad w = -\cos\theta, \tag{2.6.10}$$

where

$$\theta = \frac{d}{\hbar}\int_{-\infty}^{\infty} dt'\mathcal{E}(t') \tag{2.6.11}$$

is called the *pulse area*.

Thus, a short pulse whose carrier frequency is resonant with the two-state-atom transition frequency will leave the atom in a state that depends only on the pulse area and the initial state of the atom. For $\theta = \pi$, a *π pulse*, the atom is left in a state with $v = 0$ and $w = 1$, that is, an atom that was originally in the lower state is left with probability 1 in the excited state by the action of a π pulse. The two-state dynamics is often described in terms of the motion of a three-vector with components (u, v, w). Since $u^2(t) + v^2(t) + w^2(t) = 1$, this vector points from the origin to a point (u, v, w) on the surface of a unit sphere, the *Bloch sphere*. The south pole of this sphere at $w = -1$ corresponds to the lower state of the two-state atom, whereas the north pole at $w = +1$ corresponds to the upper state. Thus, a π pulse rotates the *Bloch vector* (u, v, w) from the south pole to the north pole in the case of a two-state atom initially in its lower state (south pole). A $\pi/2$ pulse rotates the Bloch vector from the south pole to the uv plane, leaving the atom with equal probabilities of being in the upper and lower states, that is, $w = 0$, and therefore $|a_1|^2 = |a_2|^2$, after the application of a $\pi/2$ pulse. A 2π pulse rotates the Bloch sphere by $360°$, so that the atom goes from the lower state to the upper state and then back to the lower state.

We can go further and consider solutions of (2.6.6)–(2.6.8) with $\Delta \neq 0$ and different initial conditions. We must, in general, also include damping terms in the optical Bloch equations, as is done in Section 2.7; such damping terms account for *homogeneous broadening*. It is also necessary, more generally, to consider ensembles of two-state atoms having different transition frequencies and therefore different detunings from the applied field, the case of so-called *inhomogeneous broadening*. Since all this is fairly straightforward, and the reader can find discussions of theory and experiments relating to optical Bloch equations in books and on the web, we will not at this point delve further into the effects of short pulses on two-state atoms.

We have treated the field as prescribed and unaffected by the atom. For a field propagating in a medium of effectively two-state atoms, we must, of course, include the effect of the atoms on the field. In semiclassical radiation theory, the density N of atoms times the expectation value of the dipole moment induced in an atom by the field is the polarization density **P** in Maxwell's equations. The propagation equation for the slowly varying field amplitude in the plane-wave approximation is found straightforwardly to be[29]

$$\frac{\partial \mathcal{E}}{\partial z} + \frac{1}{c}\frac{\partial \mathcal{E}}{\partial t} = \frac{i\omega}{2\epsilon_0 c}Nd(u - iv). \tag{2.6.12}$$

The coupled equations (2.6.6)–(2.6.8) and (2.6.12) then describe the effect of the field on the atoms, and vice versa, as the field propagates in the medium. More generally, we must include damping effects on the atoms as well as the field in the coupled *Maxwell–Bloch equations*. As with the optical Bloch equations, solutions of the Maxwell–Bloch equations are discussed in detail in the literature.[30]

2.7 Transition Rates and the Golden Rule

If the time τ_p over which the atom is exposed to the field is *not* short compared with the time for collisions, spontaneous emission and other atomic "relaxation" processes, we must include the effects of relaxation in (2.6.6)–(2.6.8). Inelastic collisions and spontaneous emission from the upper state produce a change in the population difference w. In most cases of interest, the effect of these processes on (2.6.8) is to add

---

[29] See, for instance, *Allen and Eberly* or *Milonni and Eberly*.

[30] Semiclassical, coupled Maxwell and Bloch equations, in essentially the form in which they are now widely used to describe the propagation of radiation in resonant media, were evidently first introduced and solved for particular cases by F. T. Arecchi and R. Bonifacio, IEEE J. Quantum Electron. **1**, 169 (1965), and S. L. McCall and E. L. Hahn, Phys. Rev. Lett. **18**, 908 (1967).

a term $-(1/T_1)(w + 1)$ to the right side, where T_1 is the relaxation time associated with processes that cause an atom to drop from the upper state to the lower state, that is, that cause $|a_2|^2$ to decrease and $|a_1|^2$ to increase. In other words, because of these processes, $w \to -1$ in the absence of an applied field or any other perturbation that can excite the atom. Since the "off-diagonal" components u and v of the Bloch vector involve $a_1^* a_2$ and $a_1 a_2^*$, T_1-type processes also cause u and v to relax (to 0) in the absence of an applied field.

But we must also account for *elastic* collisions and other "dephasing" processes that do not affect the state occupation probabilities $|a_2|^2$ and $|a_1|^2$ but, acting alone, cause $a_1^* a_2$ and $a_1 a_2^*$ and therefore u and v to relax to 0. In other words, both inelastic and dephasing processes act to dampen u and v. Letting β denote the total off-diagonal relaxation rate, we account for all relaxation processes by modifying (2.6.6)–(2.6.8) as follows:

$$\dot{u} = -\beta u - \Delta v, \tag{2.7.1}$$

$$\dot{v} = -\beta v + \Delta u + \frac{d}{\hbar}\mathcal{E}(t)w, \tag{2.7.2}$$

$$\dot{w} = -\frac{1}{T_1}(w + 1) - \frac{d}{\hbar}\mathcal{E}(t)v. \tag{2.7.3}$$

It may be shown from the definitions of u, v, and w that $\beta \geq 1/2T_1$. The notation T_1 follows that used for the two-state (spin-1/2) dynamics in magnetic resonance theory, where one conventionally writes $1/T_2$ instead of β. Concepts such as the Bloch vector and the Bloch sphere in two-state-atom theory are carryovers from magnetic resonance theory. In the present context of a two-state atom in an optical field, (2.7.1)–(2.7.3) are the optical Bloch equations with damping.

The optical Bloch equations with damping are applicable for long as well as short pulses. It often happens that the dephasing rate $\beta \gg 1/T_1$. This occurs, for instance, in a gaseous medium when the elastic collision rate is much larger than the rate of any inelastic process, in which case β is approximately equal to the collision rate. Then, if the pulse duration τ_p is large compared with $1/\beta$, we can approximate u and v by the quasi-steady-state solutions of (2.7.1) and (2.7.2) and use these solutions in (2.7.3):

$$\dot{w} = -\frac{1}{T_1}(w + 1) - \frac{\beta}{\Delta^2 + \beta^2}\frac{d^2}{\hbar^2}\mathcal{E}^2(t)w, \tag{2.7.4}$$

or, in terms of the lower- and upper-state probabilities $p_1(t) = |a_1(t)|^2$ and $p_2(t) = |a_2(t)|^2$, respectively,

$$\dot{p}_1 = \frac{1}{T_1}p_2 + \frac{\sigma(\Delta)}{\hbar\omega}I(p_2 - p_1), \tag{2.7.5}$$

$$\dot{p}_2 = -\frac{1}{T_1}p_2 - \frac{\sigma(\Delta)}{\hbar\omega}I(p_2 - p_1). \tag{2.7.6}$$

$I(t) = (c\epsilon_0/2)\mathcal{E}^2(t)$ is the (cycle-averaged) field intensity (in vacuum), and the cross section

$$\sigma(\Delta) = \frac{1}{4\pi\epsilon_0} \frac{4\pi\omega d^2}{\hbar c} \frac{\beta}{\Delta^2 + \beta^2} = \frac{1}{4\pi\epsilon_0} \frac{4\pi^2\omega d^2}{\hbar c} S(\Delta), \qquad (2.7.7)$$

where the lineshape function

$$S(\Delta) = \frac{\beta/\pi}{\Delta^2 + \beta^2}, \qquad \int_0^\infty d\omega S(\Delta) = 1. \qquad (2.7.8)$$

From the form of these *rate equations* for the occupation probabilities, it is seen that

$$R(\Delta) = \frac{\sigma(\Delta)}{\hbar\omega} I \qquad (2.7.9)$$

is the rate of absorption from the lower state, and stimulated emission from the upper state, and $\sigma(\Delta)$ is the cross section for absorption and stimulated emission. The damping (dephasing) of off-diagonal "coherence" at the damping rate β results in a broadening of the transition. The line broadening in this case of homogeneous broadening is described by a Lorentzian function of width $\propto \beta$.

Equations (2.7.5) and (2.7.6) imply that a field in a medium modeled as a collection of two-state atoms will gain or lose energy, depending on whether $p_2 > p_1$ (gain) or $p_1 > p_2$ (loss). If there are N atoms per unit volume, the variation of the intensity of a traveling wave with propagation distance z is described in the rate-equation approximation by

$$\frac{dI}{dz} = R(\Delta)\hbar\omega(N_2 - N_1) = g(\Delta)I, \qquad (2.7.10)$$

where $N_2 = Np_1$, $N_1 = Np_1$, and the gain (per unit length) coefficient is

$$g(\Delta) = \sigma(\Delta)(N_2 - N_1). \qquad (2.7.11)$$

With (2.7.9), and the formula (4.4.21) for the spontaneous emission rate A_{21}, the gain coefficient for a field of frequency $\nu = \omega/2\pi$ can be expressed in the oft-quoted form:

$$g(\nu) = \frac{\lambda^2 A_{21}}{8\pi}(N_2 - N_1)\tilde{S}(\nu). \qquad (2.7.12)$$

Here $\lambda = c/\nu$, $\nu_0 = \omega_0/2\pi$, and the lineshape function in terms of $\nu = \omega/2\pi$ is

$$\tilde{S}(\nu) = \frac{\delta\nu/\pi}{(\nu - \nu_0)^2 + \delta\nu^2}, \qquad \delta\nu = \beta/2\pi, \qquad (2.7.13)$$

for a homogeneously broadened transition; for transition frequencies and decay rates of practical physical interest ($\nu_0 \gg \delta\nu$), $\tilde{S}(\nu)$ may be assumed to be normalized, since

$$\int_0^\infty d\nu \tilde{S}(\nu) = \frac{1}{2} + \frac{2}{\pi}\tan^{-1}\frac{\nu_0}{\delta\nu} \cong 1. \qquad (2.7.14)$$

$\delta\nu$ is the width (half-width at half-maximum) of the Lorentzian lineshape function $\tilde{S}(\nu)$. In writing (2.7.12), we have, as before, invoked spherical symmetry and replaced d^2 by $|\mathbf{d}|^2/3$, since d in (2.7.7) is the transition moment along a particular direction.

For constant intensity I, (2.7.5) and (2.7.6) have the steady-state solutions

$$p_1 = \frac{1 + \frac{1}{2}I/I_{\text{sat}}}{1 + I/I_{\text{sat}}}, \qquad (2.7.15)$$

$$p_2 = \frac{\frac{1}{2}I/I_{\text{sat}}}{1 + I/I_{\text{sat}}}, \qquad (2.7.16)$$

respectively, where the *saturation intensity*

$$I_{\text{sat}} = \hbar\omega/2T_1\sigma(\Delta). \qquad (2.7.17)$$

For intensities $I \gg I_{\text{sat}}$, the transition becomes "saturated," that is, the upper- and lower-state occupation probabilities are approximately equal. This means that a sufficiently intense, long pulse in a resonant medium can propagate without absorption or amplification; absorption from the lower state, and stimulated emission from the upper state, are equally probable. The saturated medium in this case is sometimes said to be *bleached*.

Exercise 2.5: In the case of a transition for which the only relaxation process is spontaneous emission, $\beta = 1/2T_1 = A/2$, half the rate A of spontaneous emission. (This result is derived in Section 5.1.) Show that, in this case, (2.7.7) can be expressed as $\sigma(\Delta = 0) = 3\lambda^2/2\pi$, where λ is the transition wavelength. Assuming the typical order of magnitude $A = 10^8$ s^{-1} for an optical transition, estimate the saturation intensity I_{sat} when the field is tuned exactly to the atomic transition frequency.

These derivations of transition rates and the rate equations for the occupation probabilities show that the rate equations are in general *approximations* to the optical Bloch equations, valid in particular when the field acting on the atom is on for a time long compared with the off-diagonal dephasing time $1/\beta$.

It is also instructive to consider the rate equations when the applied field can be described completely by an intensity distribution $\mathcal{I}(\omega)$ such that $\mathcal{I}(\omega)d\omega$ is the intensity in the frequency interval $[\omega, \omega + d\omega]$. In this case, we generalize (2.7.9) by defining an absorption (and stimulated emission) rate

$$R = \int_0^\infty d\omega \frac{\sigma(\omega_0 - \omega)}{\hbar\omega}\mathcal{I}(\omega) = \int_0^\infty d\omega \frac{1}{\hbar\omega}\frac{\omega d^2}{\hbar c\epsilon_0}\frac{\beta}{(\omega_0 - \omega)^2 + \beta^2}\mathcal{I}(\omega)$$

$$\cong \frac{d^2}{\hbar^2 c\epsilon_0}\mathcal{I}(\omega_0)\int_0^\infty d\omega \frac{\beta}{(\omega_0 - \omega)^2 + \beta^2} \qquad (2.7.18)$$

if we assume $\mathcal{I}(\omega)$ is sufficiently "broadband" that it varies negligibly with ω compared with the Lorentzian function $\beta/[(\omega_0 - \omega)^2 + \beta^2]$. Evaluating the integral over the latter function in (2.7.18), we obtain

$$R = \frac{\pi d^2}{\hbar^2 c \epsilon_0} \mathcal{I}(\omega_0) = \frac{\pi |\mathbf{d}|^2}{3\hbar^2 c \epsilon_0} \mathcal{I}(\omega_0). \tag{2.7.19}$$

If we write $\mathcal{I}(\omega) = c\rho(\omega)$, where $\rho(\omega)d\omega$ is a field energy density in the frequency interval $[\omega, \omega + d\omega]$, then

$$R = \frac{\pi |\mathbf{d}|^2}{3\hbar^2 \epsilon_0} \rho(\omega_0) = B\rho(\omega_0). \tag{2.7.20}$$

In this broadband case, the absorption rate is independent of the atom's "lineshape" function $\beta/(\Delta^2 + \beta^2)$ and the atom is specified by its B coefficient. Using the formula $A = |\mathbf{d}|^2 \omega_0^3/3\pi\epsilon_0\hbar c^3$ for the spontaneous emission transition rate (Section 4.4), we obtain

$$A/B = \hbar\omega_0^3/\pi^2 c^3, \tag{2.7.21}$$

a famous relation that will play a role in Section 2.8.

The rates (2.7.9) and (2.7.20) are special cases of the "golden-rule" rate (2.3.9). Thus, if we take $\rho_E(E_i + \hbar\omega) = S(\Delta)/\hbar$, where $S(\Delta) = \beta/\pi(\Delta^2 + \beta^2)$ is the normalized $[\int_{-\infty}^{\infty} d\Delta S(\Delta) = 1]$ absorption lineshape function, then (2.3.9) is equivalent to (2.7.9). If, instead, we go back to the derivation of (2.3.9) and let the *field* have a broad distribution of frequencies compared with $S(\Delta)$, then we arrive at (2.7.20). In other words, both rates (2.7.9) and (2.7.20) are golden-rule rates, the latter applicable when the field has a broad distribution of frequencies compared with the atomic lineshape, and the former when the field has a narrow frequency distribution. But, in the "narrowband" case, for pulses whose duration is short compared with the atomic dipole dephasing time $1/\beta$, there are Rabi oscillations that are not described by the golden-rule or rate equations.

These rates can also be derived from the classical oscillator model, provided that we put in the oscillator strength and modify (2.3.35) to include a damping rate β:

$$\ddot{\mathbf{x}} + 2\beta\dot{\mathbf{x}} + \omega_0^2 \mathbf{x} = \frac{e}{m}\mathbf{E}_0 \cos\omega t, \tag{2.7.22}$$

and therefore

$$\mathbf{x} = \mathbf{x}_\omega e^{-i\omega t} + \mathbf{x}_\omega^* e^{i\omega t} \tag{2.7.23}$$

for $t \gg 1/\beta$, where

$$\mathbf{x}_\omega = \frac{(e/2m)\mathbf{E}_0}{\omega_0^2 - \omega^2 - 2i\beta\omega}. \tag{2.7.24}$$

The rate at which the field does work on the atom is

$$\frac{dW}{dt} = \mathbf{F} \cdot \dot{\mathbf{x}} = (-i\omega e\mathbf{E}_0 \cos\omega t) \cdot (\mathbf{x}_\omega e^{-i\omega t} - \mathbf{x}_\omega^* e^{i\omega t}) = \frac{\omega^2 e^2}{m}\mathbf{E}_0^2 \frac{\beta}{(\omega_0^2 - \omega^2)^2 + 4\beta^2\omega^2}$$

$$\cong \frac{e^2}{2mc\epsilon_0} I \frac{\beta}{(\omega_0 - \omega)^2 + \beta^2} \tag{2.7.25}$$

if we assume $\omega \cong \omega_0$. Introducing the oscillator strength by replacing $e^2/2mc\epsilon_0$ by $e^2 f_{12}/2mc\epsilon_0$, where f_{12} is defined by (2.5.18), we obtain

$$\frac{dW}{dt} = \sigma(\Delta)I, \tag{2.7.26}$$

with $\sigma(\Delta)$ given by (2.7.7). Finally, we make the connection to a quantum-mechanical transition rate by dividing by $\hbar\omega$:

$$R(\Delta) = \frac{1}{\hbar\omega}\frac{dW}{dt}, \tag{2.7.27}$$

which is identical to (2.7.9).

In the case of a broadband field with spectral energy density $\rho(\omega)$, we obtain, similarly,

$$\frac{dW}{dt} = \frac{\pi e^2}{2\epsilon_0 m}\rho(\omega_0) \tag{2.7.28}$$

for the rate at which energy is absorbed, and when we include the oscillator strength and divide by $\hbar\omega_0$, we obtain (2.7.20).

One thing we cannot do with the classical electron oscillator model is derive the observed (Planck) spectrum of blackbody radiation. From the Larmor formula (1.3.36), we have the radiation rate of an atom according to the classical oscillator model:

$$\frac{dW_{\text{rad}}}{dt} = \frac{1}{4\pi\epsilon_0}\frac{2e^2\ddot{\mathbf{x}}^2}{3c^3} = \frac{1}{4\pi\epsilon_0}\frac{2e^2}{3c^3}(-\omega_0^2\mathbf{x})^2 = \frac{1}{4\pi\epsilon_0}\frac{4e^2\omega_0^2}{3mc^3}(\frac{1}{2}m\omega_0^2\mathbf{x}^2)$$

$$\rightarrow \frac{1}{4\pi\epsilon_0}\frac{4e^2\omega_0^2}{3mc^3}(\frac{3}{2}k_BT), \tag{2.7.29}$$

where we have replaced $\frac{1}{2}m\omega_0^2\mathbf{x}^2$ by the thermal-equilibrium value $\frac{3}{2}k_BT$ required by classical statistical mechanics. The condition that, in thermal equilibrium, this radiation rate should equal the absorption rate (2.7.28) implies the Rayleigh–Jeans spectrum:

$$\rho(\omega_0) = \rho_{\text{RJ}}(\omega_0) = \frac{\omega_0^2}{\pi^2c^3}k_BT, \tag{2.7.30}$$

which, besides disagreeing with observation, suffers from the infamous "ultraviolet catastrophe." The Planck spectrum (2.8.3), of course, gives a finite electromagnetic energy density: $\int_0^\infty d\omega\rho(\omega) = (\pi^2 k_B^4/15\hbar^3c^3)T^4$.

2.8 Blackbody Radiation and Fluctuations

Semiclassical radiation theory is in excellent agreement with many, but not all, observations and experiments on the interaction of light with matter. From a historical perspective, the most significant failure of *any* theory based on classical electromagnetic theory is found in the spectrum of blackbody radiation. We now discuss Einstein's analysis of the energy and momentum fluctuations characteristic of blackbody radiation. This preceded the full development of the quantum theory of atoms and electromagnetic radiation, but is forever relevant to the foundations of the quantum theory of radiation because it reveals among other things that (1) light has both wave and particle properties, (2) atoms must recoil when they absorb or emit light and, in particular, when they radiate spontaneously, and (3) radiation from thermal (and, more generally, "chaotic") sources is characterized by what is now called *photon bunching*.

We start with a quick review of Einstein's last but simplest derivation, in 1916, of the Planck spectrum. We consider two-state atoms in thermal equilibrium at temperature T with radiation of spectral energy density $\rho(\omega)$. We denote by N_1 and N_2 the number densities of atoms in the lower and upper states, respectively, so that $N_2/N_1 = e^{-\hbar\omega_0/k_BT}$, where k_B is Boltzmann's constant. Atoms in the upper state can drop to the lower state by spontaneous emission, characterized by the rate A, and by stimulated emission, characterized by a rate $B_{21}\rho(\omega_0)$. They can also go from the lower state to the upper state by absorption, characterized by a rate $B_{12}\rho(\omega_0)$. In equilibrium, the absorption and emission processes balance on average, giving

$$AN_2 + B_{21}N_2\rho(\omega_0) = B_{12}N_1\rho(\omega_0). \tag{2.8.1}$$

In the limit $T \to \infty$, $N_1 = N_2$ and, assuming $\rho(\omega_0) \to \infty$ in this limit, we conclude from (2.8.1) that $B_{12} = B_{21} = B$ and therefore that

$$\rho(\omega_0) = \frac{A/B}{N_1/N_2 - 1} = \frac{A/B}{e^{\hbar\omega_0/k_BT} - 1}. \tag{2.8.2}$$

If it is to be valid at all temperatures, (2.8.2) must, in particular, give the correct spectral energy density in the "classical limit" $\hbar\omega_0/k_BT \ll 1$, that is, it must reduce to the classical Rayleigh–Jeans spectrum in this limit. For isotropic and unpolarized radiation, there are $(\omega^2/\pi^2c^3)d\omega$ modes of the field in the frequency interval $[\omega, \omega+d\omega]$ (see Section 4.5) and, according to the equipartition theorem of classical physics, an energy k_BT is associated with each frequency. For (2.8.2) to reduce to the Rayleigh–Jeans spectrum $(\rho(\omega_0) = (\omega_0^2/\pi^2c^3)k_BT)$ in the classical limit, therefore, we must have $(A/B)(k_BT/\hbar\omega_0) = \omega_0^2 k_BT/\pi^2c^3$, or $A/B = \hbar\omega_0^3/\pi^2c^3$, which is just (2.7.21). Then,

$$\rho(\omega_0) = \frac{\hbar\omega_0^3/\pi^2c^3}{e^{\hbar\omega_0/k_BT} - 1}. \tag{2.8.3}$$

Equivalently, the energy per unit volume in a wavelength interval $[\lambda, \lambda + d\lambda]$ is $\rho_\lambda(\lambda)d\lambda$, where

$$\rho_\lambda(\lambda) = \frac{8\pi hc/\lambda^5}{e^{hc/k_BT\lambda} - 1}. \tag{2.8.4}$$

The spectrum of the cosmic microwave background radiation is very well described by the Planck formula (see Figure 2.5). In fact, it appears to be the most accurate confirmation ever of the Planck formula!

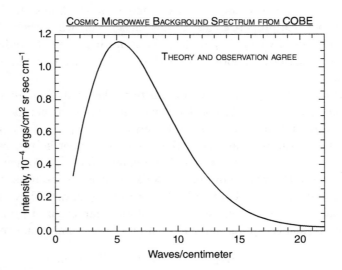

Fig. 2.5 Measured cosmic microwave background spectrum compared with the Planck spectrum for $T = 2.726$ K. Theory and data agree so well that their differences in the figure are imperceptible. From G. F. Smoot, Rev. Mod. Phys. **79**, 1349 (2007), with permission.

Exercise 2.6: (a) We have assumed that the upper and lower energy levels of the atoms are non-degenerate. Show that the Planck spectrum (2.8.3) is independent of any atomic level degeneracies. (b) Consider blackbody radiation in an expanding universe in which wavelengths λ change to $\lambda' = f\lambda$ when the universe expands by a factor f. Assuming no photons are created or lost in the expansion, show that the Planck formula still applies after the expansion, but with a smaller temperature $T' = T/f$, where T is the temperature before the expansion. (See S. Weinberg, *The First Three Minutes*, Bantam Books, New York, 1977, p. 162.)

Einstein deduced the Planck spectrum from physical considerations; he did not have or require explicit expressions for his A and B coefficients., but was nevertheless able to derive the Planck spectrum (2.8.3) in an "astonishingly simple and general way."[31] In a letter to M. Besso, he wrote that "A splendid light has dawned on me about the absorption and emission of light."[32]

---

[31] Einstein, quoted in *Van der Waerden*, p. 64.
[32] Quoted in *Pais (1982)*, p. 405.

2.8.1 Recoil of Atoms in Absorption and Emission

A theory [of blackbody radiation] can only be regarded as justified when it is able to show that the impulses transmitted by the radiation field to matter lead to motions that are in accordance with the theory of heat. – Einstein[33]

Less well known is Einstein's derivation of the Planck spectrum from the perspective of momentum fluctuations. Suppose that an atom undergoes n emission and absorption events in a time δt, and that each such event imparts a recoil momentum P_i in, say, the x direction. Then, the momentum change along x of the atom during the time δt is $P = \sum_{i=1}^{n} P_i$, and, assuming the P_i to be statistically independent, we obtain the mean-square momentum

$$\overline{P^2} = \sum_i^n \overline{P_i^2} = \frac{1}{3} n \left(\frac{\hbar \omega_0}{c} \right)^2 \tag{2.8.5}$$

if we associate a recoil momentum $|P_i| = \hbar \omega_0 / c$ with each absorption and emission event occurring at the transition frequency ω_0; the factor $1/3$ appears because we are considering the motion of the atom along the x axis in an isotropic field. The average number of such events occurring in the time δt is

$$n = A p_2 \delta t + (p_1 + p_2) B \rho(\omega_0) \, \delta t, \tag{2.8.6}$$

where p_1 and p_2 are the lower- and upper-state occupation probabilities, respectively. Therefore, using (2.8.1) expressed in terms of thermal-equilibrium occupation probabilities,

$$\overline{P^2}/\delta t = \frac{1}{3} \left(\frac{\hbar \omega_0}{c} \right)^2 [A p_2 + (p_1 + p_2) B \rho(\omega_0)] = \frac{2}{3} \left(\frac{\hbar \omega_0}{c} \right)^2 B p_1 \rho(\omega_0). \tag{2.8.7}$$

As discussed below, there is another effect on an atom in the field with which it is in thermal equilibrium: an atom with velocity v experiences a retarding force

$$F = - \left(\frac{\hbar \omega_0}{c^2} \right) (p_1 - p_2) B \left[\rho(\omega_0) - \frac{\omega_0}{3} \frac{d\rho}{d\omega_0} \right] v = -R_{\text{ret}} v. \tag{2.8.8}$$

This retarding force, according to Einstein, acts in thermal equilibrium to balance the mean-square momentum increase given by (2.8.7). Thus, an atom of mass m experiences a total momentum change of

$$m v(t + \delta t) - m v(t) = P - R_{\text{ret}} v(t) \delta t \tag{2.8.9}$$

in a time interval $[t, t + \delta t]$. We now square both sides and average, using $\frac{1}{2} m \overline{v^2}(t) = \frac{1}{2} k_B T$ for the average kinetic energy per degree of freedom in thermal equilibrium.

---

[33] Quoted in *Van der Waerden*, p. 77.

Letting δt be so small that we need only retain terms linear in it, and assuming $\overline{v(t)P} = 0$, since the momentum change P is equally likely to be positive or negative, we obtain

$$\overline{P^2}/\delta t = 2R_{\mathrm{ret}}k_B T, \tag{2.8.10}$$

or, from (2.8.7) and (2.8.8),

$$\rho(\omega_0) - \frac{\omega_0}{3}\frac{d\rho}{\omega_0} = \frac{\hbar\omega_0}{3k_B T}\frac{p_1}{p_1-p_2}\rho(\omega_0) = \frac{\pi^2 c^3}{3\omega_0^2 k_B T}\left[\rho^2(\omega_0) + \frac{\hbar\omega_0^3}{\pi^2 c^3}\rho(\omega_0)\right], \tag{2.8.11}$$

where we have used the relation $A/B = \hbar\omega_0^3/\pi^2 c^3$ and also $p_1/(p_1-p_2) = B\rho(\omega_0)/A + 1$, which follows from (2.8.1) and $p_1 + p_2 = 1$. The solution of the differential equation (2.8.11) with $\rho(0) = 0$ is the Planck spectrum (2.8.3).

We can understand the form of the retarding force (2.8.8) as follows. The field energy density in the frequency interval $[\omega, \omega + d\omega]$ and attributable to radiation within a solid angle $d\Omega$ is $\rho(\omega)d\omega d\Omega/4\pi$. Radiation directed at an angle θ with respect to the velocity v of a moving atom appears to the atom to have the Doppler-shifted frequency

$$\omega' \cong \omega\left(1 - \frac{v}{c}\cos\theta\right) \quad (v/c \ll 1), \tag{2.8.12}$$

and, because of the "aberration effect," this radiation appears to the atom to be directed at the angle θ', satisfying

$$\cos\theta' \cong \cos\theta - \frac{v}{c}\sin^2\theta. \tag{2.8.13}$$

The moving atom finds itself in a transformed field described by

$$\rho'(\omega',\theta')d\omega'd\Omega' \cong \left(1 - \frac{2v}{c}\cos\theta\right)\rho(\omega)d\omega d\Omega, \tag{2.8.14}$$

as shown below. From (2.8.12) and (2.8.13),

$$\begin{aligned}
\rho'(\omega',\theta') &\cong \left(1 - \frac{2v}{c}\cos\theta\right)\rho(\omega)\frac{d\omega}{d\omega'}\frac{d(\cos\theta)}{d(\cos\theta')} \cong \left(1 - \frac{3v}{c}\cos\theta'\right)\rho(\omega) \\
&\cong \left(1 - \frac{3v}{c}\cos\theta'\right)\rho\left(\omega' + \frac{v\omega'}{c}\cos\theta'\right) \\
&\cong \left(1 - \frac{3v}{c}\cos\theta'\right)\left[\rho(\omega') + \frac{d\rho(\omega')}{d\omega'}\left(\frac{v}{c}\right)\omega'\cos\theta'\right]. \tag{2.8.15}
\end{aligned}$$

The net rate of momentum transfer to the moving atom from absorption and stimulated emission due to radiation in the solid angle $d\Omega'$ is

$$F = (p_1 - p_2)B\rho'(\omega',\theta')\frac{d\Omega'}{4\pi}\frac{\hbar\omega'}{c}\cos\theta', \tag{2.8.16}$$

since absorption ($\propto p_1$) causes the atom to recoil in the direction of the field whereas stimulated emission ($\propto p_2$) causes it to recoil in the opposite direction. Finally, we add the forces associated with all propagation directions of radiation of frequency ω:

$$\begin{aligned}
F &= \frac{\hbar\omega'}{c}\frac{B}{4\pi}(p_1-p_2)\int_0^{2\pi}d\phi'\int_0^{\pi}d\theta'\sin\theta'\rho'(\omega',\theta')\cos\theta' \\
&\cong -\left(\frac{\hbar\omega}{c^2}\right)(p_1-p_2)B\left[\rho(\omega) - \frac{\omega}{3}\frac{d\rho}{d\omega}\right]v, \tag{2.8.17}
\end{aligned}$$

where, to lowest order in v/c, we have dropped all the primes.

To derive (2.8.14), consider first a plane wave propagating in a direction making an angle θ with respect to the direction (x) of an atom's velocity, and denote the field components of this wave by

$$E_x = 0, \quad E_y = E_0 \cos(\omega t - \mathbf{k} \cdot \mathbf{r}), \quad E_z = 0,$$

$$B_x = -\frac{k_z}{\omega} E_0 \cos(\omega t - \mathbf{k} \cdot \mathbf{r}), \quad B_y = 0, \quad B_z = \frac{k_x}{\omega} E_0 \cos(\omega t - \mathbf{k} \cdot \mathbf{r}), \quad (2.8.18)$$

$\mathbf{k} = k_x \hat{\mathbf{x}} + k_z \hat{\mathbf{z}} = k(\hat{\mathbf{x}} \cos\theta + \hat{\mathbf{z}} \sin\theta)$, $k = \omega/c$. The electric field components in the moving atom's frame are given by (1.3.32):

$$E'_x = 0, \quad E'_y = \gamma\left(1 - \frac{v}{c}\cos\theta\right)E_y, \quad E'_z = 0, \quad (\gamma = 1/\sqrt{1 - v^2/c^2}), \quad (2.8.19)$$

or

$$\mathbf{E}'^2 = \gamma^2\left(1 - \frac{v}{c}\cos\theta\right)^2\mathbf{E}^2 \cong \left(1 - 2\frac{v}{c}\cos\theta\right)\mathbf{E}^2 \quad (2.8.20)$$

for $v \ll c$. Now, $\rho(\omega)d\omega d\Omega \propto \rho(\omega)d\omega d(\cos\theta)$ is the energy density in the infinitesimal frequency interval $[\omega, \omega + d\omega]$ and the infinitesimal solid angle interval $[\Omega, \Omega + d\Omega]$ for the field in the rest frame of the atom, and, as such, it transforms in the same way as $\mathbf{E}^2$ (or $\mathbf{B}^2$) for a plane wave with $k_x = k\cos\theta$. In other words, (2.8.14) follows simply from (2.8.20).

The same argument, without the small-velocity approximation, can be used to derive the general form of the Lorentz-transformed blackbody spectrum:

$$\rho'(\omega', \theta') = \frac{1}{4\pi}\frac{\hbar\omega'^3/\pi^2 c^3}{e^{\hbar\omega/k_B T} - 1}, \quad (2.8.21)$$

where $\omega = \gamma(1 + \frac{v}{c}\cos\theta')\omega'$ and $\rho'(\omega', \theta')$ is now defined such that $\rho'(\omega', \theta')d\omega' d\Omega'$ is the energy density in the frequency interval $[\omega', \omega' + d\omega']$ and the solid angle interval $[\Omega', \Omega' + d\Omega']$.[34]

In Einstein's analysis, it is assumed that an atom recoils with a linear momentum $\hbar\omega_0/c$ when it makes a transition of frequency ω_0. In particular, an atom recoils with this momentum when it emits radiation *spontaneously*. Recoil in *stimulated* emission and absorption can be understood using semiclassical radiation theory, but semiclassical theory, for reasons already mentioned in Section 2.3, does not correctly account for recoil in spontaneous emission. Recoil in spontaneous emission is, *on average*, zero, for the same reason it is zero in semiclassical radiation theory: there is no preferred direction of emission. It is for this reason that spontaneous emission does not contribute to the (average) retarding force (2.8.8), which originates solely from absorption and stimulated emission events. But there is a *mean-square* recoil momentum of an atom in spontaneous emission, and Einstein's analysis reveals that the recoil accompanying spontaneous emission is required if blackbody radiation is to be described by the Planck spectrum. For this reason, Einstein concluded that "outgoing radiation in the form of spherical waves does not exist."[35] In fact, no outgoing radiation from an atom

---

[34] G. W. Ford and R. F. O'Connell, Phys. Rev. E **88**, 044101 (2013). These authors include the contribution to ρ from zero-point energy, which retains its form ($\rho(\omega) \propto \omega^3$) under Lorentz transformations. See Section 7.4.

[35] Quoted in *Van der Waerden*, p. 76.

can exist that does not carry linear momentum, and this linear momentum must, of course, be equal in magnitude and opposite in direction to the recoil momentum of the atom.

The impossibility of outgoing spherical waves can be deduced more generally, quite apart from Einstein's considerations about radiation from atoms, based on a theorem developed by Brouwer in 1909 and which, for our purposes, says that it is impossible to have a continuous vector field that is constant on a sphere and tangent everywhere to the surface of the sphere. (This is often called the "hairy ball theorem": it is impossible to have a combed hairy ball without a cowlick or bald spot.) Thus, a linearly polarized, transverse electric field in the far zone of an antenna, for example, cannot have an amplitude that depends only on the distance from the antenna, and so the radiated power cannot be isotropic.[36] Bouwkamp and Casimir deduced from a multipole expansion of the radiation field that "a system of currents isotropically radiating in all directions of free space is physically impossible."[37]

An isotropic radiator is possible, however, with an incoherent superposition of currents. For example, the power radiated by a point source consisting of three identical electric dipoles, one oscillating along a z axis, and two oscillating $\pi/2$ out of phase along orthogonal x and y axes, will be isotropic if the fields from these dipoles are mutually incoherent.

The angular distribution of spontaneous emission from an atom is formally the same as that of a classical radiator having the same multipole character, except, of course, that it describes a probability for detection of a single photon. For electric dipole transitions, for example, the probability of detecting a photon resulting from a $\Delta M = 0$ transition is proportional to $\sin^2 \theta$, where θ is the angle between the quantization (z) axis and the point of observation. The normalized probability is

$$f_0(\theta) = \frac{\sin^2 \theta}{\int_0^{2\pi} d\phi \int_0^\pi d\theta \sin \theta \sin^2 \theta} = \frac{3}{8\pi} \sin^2 \theta. \tag{2.8.22}$$

For $\Delta M = \pm 1$ transitions, similarly,

$$f_{\pm 1}(\theta) = \frac{3}{16\pi}(1 + \cos^2 \theta). \tag{2.8.23}$$

These expressions can be deduced from classical radiation theory, from the fact that an electric dipole moment $\mathbf{p}(t) = \mathbf{p}e^{-i\omega t}$ produces a far-field radiation pattern proportional to $1 - |\hat{\mathbf{p}} \cdot \hat{\mathbf{r}}|^2$, where $\hat{\mathbf{p}}$ is in the direction of $\mathbf{p}$, and $\hat{\mathbf{r}}$ is the unit vector pointing from the dipole to the point of observation. For $\hat{\mathbf{p}} = \hat{z}$, $1 - |\hat{\mathbf{p}} \cdot \hat{\mathbf{r}}|^2 = 1 - \cos^2 \theta = \sin^2 \theta$, and the angular distribution of the radiated power is given by (2.8.22). For $\hat{\mathbf{p}} = (1\sqrt{2})(\hat{x} \pm i\hat{y})$, $1 - |\hat{\mathbf{p}} \cdot \hat{\mathbf{r}}|^2 = 1 - |(1/\sqrt{2})e^{\pm i\phi} \sin \theta|^2 = (1/2)(1 + \cos^2 \theta)$, and the angular distribution of the radiated power is given by (2.8.23).

2.8.2 Einstein's Fluctuation Formula: Wave–Particle Duality

Recall the formula $\overline{\delta E^2} = \overline{E^2} - \overline{E}^2 = k_B T^2 \partial \overline{E}/\partial T$ for the variance in energy of a system in thermal equilibrium at temperature T.[38] Using $\overline{E} = \rho(\omega)V d\omega$ for the (statistical-ensemble) average energy of blackbody radiation in a volume V and a

---

[36] H. F. Mathis, Proc. IRE **39**, 970 (1951).

[37] C. J. Bouwkamp and H. B. G. Casimir, Physica **20**, 539 (1954), p. 553.

[38] This formula, which was derived by Einstein in 1904, follows easily from the definition $\overline{E} = \sum_j E_j e^{-E_j/k_B T} / \sum_j e^{-E_j/k_B T}$, the E_j's being the possible energies of the system.

frequency interval $[\omega, \omega + d\omega]$, we obtain from this formula and (2.8.3) the *Einstein fluctuation formula*

$$\overline{\delta E^2} = [\hbar\omega\rho(\omega) + \frac{\pi^2 c^3}{\omega^2}\rho^2(\omega)]V d\omega. \tag{2.8.24}$$

This formula, obtained by Einstein in 1909, was important in particular as an early indicator of wave–particle duality and Bose–Einstein statistics, as we now discuss.[39]

If, instead of using the Planck spectrum in the aforementioned energy variance formula, we use the Rayleigh–Jeans formula for $\rho(\omega)$, we obtain

$$\overline{\delta E^2} = \frac{\omega^2}{\pi^2 c^3}(k_B T)^2 V d\omega = \frac{\pi^2 c^3}{\omega^2}\rho^2(\omega)V d\omega \tag{2.8.25}$$

instead of (2.8.24). In other words, the second term in brackets in (2.8.24) can be associated with a *classical wave* approach to blackbody radiation.

Now, take a different point of view, and regard blackbody radiation as a collection of *particles of energy* $\hbar\omega$. Since there are $(\omega^2/\pi^2 c^3)V d\omega$ modes of the field (or states of the particles) in the volume V and the frequency interval $[\omega, \omega + d\omega]$, we assume

$$\overline{E} = \overline{n}(\omega)\hbar\omega \times \frac{\omega^2}{\pi^2 c^3}V d\omega = \rho(\omega)V d\omega. \tag{2.8.26}$$

If we assume also that the particle numbers belonging to different modes fluctuate independently according to a Poisson distribution, so that

$$\overline{\Delta n^2} = \overline{n^2} - \overline{n}^2 = \overline{n}, \tag{2.8.27}$$

we deduce that

$$\overline{\delta E^2} = \overline{\Delta n^2}(\hbar\omega)^2\frac{\omega^2}{\pi^2 c^3}V d\omega = \overline{n}(\omega)(\hbar\omega)^2\frac{\omega^2}{\pi^2 c^3}V d\omega = \hbar\omega\rho(\omega)V d\omega, \tag{2.8.28}$$

which corresponds to the first term in (2.8.24). It is worth noting that this result can also be derived using $\overline{\delta E^2} = k_B T^2 \partial\overline{E}/\partial T$, $\overline{E} = \rho(\omega)V d\omega$, and the Wien formula

$$\rho(\omega) = \frac{\hbar\omega^3}{\pi^2 c^3}e^{-\hbar\omega/k_B T}, \tag{2.8.29}$$

which, incidentally, follows from Einstein's reasoning based on (2.8.1) but without including stimulated emission. (Einstein had shown earlier that the entropy of radiation

---

[39] Einstein's derivation was criticized by Ornstein and Zernike and also by Ehrenfest, in essence because the derivation implicitly ignored interference between different modes of the field. But Einstein's formula was derived in a different way by Born, Heisenberg, and Jordan (Z. Phys. **35**, 557 (1926)), who accounted for the fact that the position and momentum of a harmonic oscillator in the new matrix mechanics were represented by non-commuting matrices; the term Einstein associated with "particles" in the fluctuation formula was closely connected in their derivation with the zero-point energy of a harmonic oscillator. But this wasn't the end of the story. Some aspects of the history of the Einstein fluctuation formula and its role in the development of the theory of quantized fields are reviewed by A. S. Wightman, Fortschr. Phys. **44**, 143 (1996).

obeying the Wien distribution has the same form as the entropy of an ideal gas of particles of energy $\hbar\omega$: radiation satisfying (2.8.29) appears thermodynamically to consist of independent quanta of energy $\hbar\omega$. This was done in his photoelectric effect paper of 1905, and was the basis for his "heuristic point of view" based on energy quanta.)

From the fluctuation formula, (2.8.24), Einstein thus inferred that wave and particle characteristics of light were, as he said, "not to be considered as mutually incompatible."[40] His fluctuation formula contains at once a "wave term" that may be deduced from classical electromagnetic wave theory and a "particle term" derivable from the concept of quanta of energy $\hbar\omega$. (His conclusion that these quanta also carry a linear momentum $\hbar\omega/c$ came seven years later, in the work discussed in Section 2.8.2; with the acuity of hindsight, it is surprising it took so long.) As discussed in Chapter 3, this wave–particle duality—and much more!—is contained in the quantum field theory of light, in which we cannot take too literally Einstein's remark that "the energy of light is discontinuously distributed in space."[41]

> **Exercise 2.7**: But suppose we do take very literally the notion of photons as particles. What would you say is the average distance between these particles in blackbody radiation? Express your answer in terms of a "typical wavelength" of blackbody radiation, for example, the wavelength where the Planck spectrum for a given temperature T has its largest value.

Finally, we again write the Planck spectrum as $\rho(\omega) = (\hbar\omega^3/\pi^2 c^3)\overline{n}(\omega)$ and use (2.8.24):

$$\overline{\delta E^2} = \frac{\hbar^2 \omega^4}{\pi^2 c^3}[\overline{n}(\omega) + \overline{n}^2(\omega)]V\,d\omega. \tag{2.8.30}$$

If we write this, in turn, in a way that relates the energy fluctuations to the fluctuations in $n(\omega)$,

$$\overline{\delta E^2} = (\hbar\omega)^2 \overline{\Delta n^2} \times \frac{\omega^2}{\pi^2 c^3} V\,d\omega, \tag{2.8.31}$$

we see that

$$\overline{\Delta n^2} = \overline{n} + \overline{n}^2, \tag{2.8.32}$$

or, equivalently,

$$\overline{n^2} = \overline{n} + 2\overline{n}^2, \tag{2.8.33}$$

the second term on the right being associated with the wave term in the Einstein fluctuation formula. The variance (2.8.32) in particle (photon) number is exactly that given

---

[40] Quoted in *Pais (1982)*, p. 404.

[41] Quoted in the English translation of Einstein's paper by A. B. Arons and M. B. Peppard, Am. J. Phys. **33**, 367 (1965), p. 368.

by the Bose–Einstein probability distribution (also known as the discrete-exponential distribution):

$$p_n = \frac{\overline{n}^n}{(\overline{n}+1)^{n+1}}. \tag{2.8.34}$$

2.8.3 Fluctuations and Dissipation

As part of his work relating to the energy fluctuations of blackbody radiation, Einstein considered the fluctuations in the momentum of a mirror which was free to move perpendicular to its planar surface, due to the radiation pressure of blackbody radiation. He assumed that the radiation pressure also caused the mirror to experience a drag force, such that, in a short time interval δt, its momentum would change from Mv to $Mv - \gamma v \delta t + \Delta$, where γ is a constant, and M is the mirror mass. Then, if terms of order $(\delta t)^2$ are neglected, and v and Δ are assumed to be uncorrelated,

$$\frac{1}{2} M \overline{v^2} = \frac{1}{2} M \overline{v^2} - 2\gamma M \overline{v^2} \delta t + \overline{\Delta^2}. \tag{2.8.35}$$

The equilibrium condition is

$$\overline{\Delta^2}/\delta t = 2\gamma M \overline{v^2} = 2\gamma k_B T \tag{2.8.36}$$

when the equipartition theorem is used for the mirror's kinetic energy ($\frac{1}{2} M \overline{v^2} = \frac{1}{2} k_B T$). In 1910, Einstein calculated γ for a mirror of area S that is assumed to transmit all frequencies except those in a narrow range $[\omega, \omega + d\omega]$, which are perfectly reflected, and used (2.8.36) to obtain

$$\overline{\Delta^2}/\delta t = \frac{\pi^2 c^2}{\omega^2} \left[\rho^2(\omega) + \frac{\hbar \omega^3}{\pi^2 c^3} \rho(\omega) \right] S d\omega \tag{2.8.37}$$

when the Planck spectrum is assumed for $\rho(\omega)$. The factor in brackets has exactly the same form as that in (2.8.11), which Einstein derived later, in 1916. Once again, we can identify wave and particle contributions to fluctuations relating to blackbody radiation, in this case, the mean-square fluctuations in the momentum of a mirror bathed in blackbody radiation.

But there is something else here of great importance: inspired no doubt by his earlier work on Brownian motion (Section 6.1)—where he obtained "the first fluctuation–dissipation theorem ever noted"[42]—Einstein, in his analysis, connected the momentum fluctuations of the mirror ($\overline{\Delta}^2$) to a dissipative force ($-\gamma v$). If there were no fluctuations, the dissipative force would bring the mirror to rest, in contradiction to the equipartition theorem. Blackbody radiation involves both fluctuations and dissipation, as well as a particular relation (2.8.36) between them. In his later analysis of the momentum fluctuations of atoms in thermal equilibrium with radiation (see Section 2.8.1), Einstein again established a relation between (momentum) fluctuations and dissipation (see (2.8.10)) and used it to derive the Planck spectrum.

---

[42] *Pais (1982)*, p. 56.

2.9   Photon Bunching

In the experiment shown in Figure 2.6, quasi-monochromatic, collimated, and spectrally filtered light from a source in thermal equilibrium, or nearly so, is incident on a beam splitter and results in intensities $I_1(t)$ and $I_2(t)$ at photomultipliers PM1 and PM2. The photomultipliers work by the photoelectric effect to produce electric currents proportional to the intensities, and correlator electronics record the average over t of the product of $I_1(t)$ and $I_2(t+\tau)$; because of the "stationary" property of the light in the experiment, the averages $\langle I_1(t)\rangle$, $\langle I_2(t)\rangle$, and $\langle I_1(t)I_2(t+\tau)\rangle$ are independent of t.[43] What is found is a positive correlation for small τ (see Figure 2.7): there is a statistical tendency for photons (or, more precisely, photoelectrons) to be "bunched" for small τ.

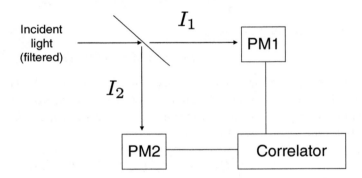

Fig. 2.6 Experiment to measure the average of $I_1(t)I_2(t+\tau)$.

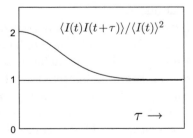

Fig. 2.7 Photon bunching near $\tau = 0$.

---

[43] Henceforth, we denote the average of a quantity Q by $\langle Q\rangle$. We will regard this as an ensemble average, but, for our purposes here, it can also be thought of as a time average.

Such photon bunching was observed in intensity correlation experiments by Hanbury Brown and Twiss in the 1950s. These experiments actually measured correlations in the fluctuations of the outputs (photoelectron counts) at the two detectors. We will ignore the realities that made the experiments difficult and just focus on the observed photon bunching effect.

Since the fields at PM1 and PM2 in Figure 2.6 are just samples of the field incident on the beam splitter, consider simply $\langle I(t)I(t+\tau)\rangle$, where $I(t)$ is the intensity of the field incident on the (50/50) beam splitter. The observed effect implies that the average over t of this quantity for $\tau = 0$ is larger than the squared average of I itself: $\langle I(t)I(t+\tau)\rangle > \langle I(t)\rangle^2$.

How can we understand this in terms of photons, each of which ideally yields a single photoelectron? Since intensity I is proportional to the average number $\langle n\rangle$ of photons in the field, the simplest thing that comes to mind is to equate (time-independent) $\langle I^2(t)\rangle$ to the quantum-mechanical expectation value $\langle n^2\rangle$ and try to explain the bunching effect with $\langle n^2\rangle$. But this cannot be quite correct because, if we imagine a *one*-photon field, for example, we cannot "split the photon" and record photoelectron counts at both PM1 and PM2. (Recall Figure 2.1.) And if we detect a single photon (photoelectron) from an n-photon field, there are $n-1$ photons left to detect and so, if the probability of counting a single photon is proportional to n, we expect the probability of counting two photons to be proportional to $n(n-1)$. We might therefore suspect that the quantity relevant to our idealized photon bunching experiment is not $\langle n^2\rangle$ but $\langle n(n-1)\rangle = \langle n^2\rangle - \langle n\rangle$. As discussed in Chapter 3, this surmise is correct: the photon coincidence counting rate in our idealized model for the Hanbury Brown–Twiss experiment is

$$\langle n(n-1)\rangle = 2\langle n\rangle^2, \tag{2.9.1}$$

where we have used (2.8.33) for chaotic radiation. The factor 2 explains the photon bunching indicated in Figure 2.7 for $\tau = 0$, and our derivation shows that *photon bunching of thermal radiation is attributable to the wave term*, $\propto \langle n\rangle^2$, *in the Einstein fluctuation formula*.[44] Actually, the bunching effect can be understood from a purely *classical* approach to the intensity fluctuations of thermal radiation or some other type of "chaotic" radiation (Section 3.6.5, (3.6.110)); this, in effect, supports our discussion relating to Figures 2.6 and 2.7 in terms of intensities.

The novel conceptual aspect of the Hanbury Brown–Twiss experiments was the measurement of *intensity* correlations (or intensity "interference"), in contrast to nearly all previously studied optical interference effects involving the interference of

---

[44] Thermal radiation, strictly speaking, should be distinguished from blackbody radiation. Blackbody radiation is radiation in thermal equilibrium with a reservoir whose heat capacity is so large that its temperature stays the same even as it gains or loses energy by heat transfer. It is spatially homogeneous, isotropic, and unpolarized. Thermal radiation refers to radiation from a source that is in thermal equilibrium, or nearly so, but it might not be homogeneous, isotropic, or unpolarized because of its having passed through spatial filters, lenses, polarizers, or other optical elements.

fields; interference effects were shown to be possible without coherent sources.[45] Some people at first viewed the experiments with skepticism, suggesting that a revision of quantum theory would be required if the reported results were correct. E. T. Jaynes recalled that there was even a suspicion that a hoax was being perpetrated![46]

The confusion was rooted in large part in Dirac's famous statement that "each photon . . . interferes only with itself. Interference between two different photons never occurs," whereas different photons in the experiments *did* seem to be interfering in some sense in that they tended to arrive in pairs.[47] From the present perspective, Dirac's statement needs to be revised: we should not think in terms of photons interfering but rather probability amplitudes. But, aside from any of that, we can appreciate the results of Hanbury Brown and Twiss based on the *classical* wave contribution to Einstein's fluctuation formula.

Bunching is *not* simply a consequence of the bosonic character of photons, for it is possible for light to exhibit *anti-bunching* or neither bunching nor anti-bunching. A complete description of the statistical properties of light evidently requires the quantum field theory of light, to which we turn in Chapter 3.

An example of the skepticism expressed by some physicists when they first heard of the Hanbury Brown–Twiss effect has been described by V. Radhakrishnan, who recalled that, during a colloquium by Hanbury Brown on the subject of the intensity correlations, and the stellar interferometer based on these correlations, "Richard Feynman jumped up and said 'It can't work!' In his inimitable style, Hanbury responded, 'Yes, I know. We were told so. But we built it anyway, and it did work'. Late that night, Feynman phoned and woke Hanbury up to say 'you are right'. He also wrote a letter in which he magnanimously admitted his mistake and acknowledged the importance of this phenomenon that, at first sight, appears counterintuitive, even to quantum theorists."[48]

As already noted, the Hanbury Brown–Twiss bunching effect can be understood from purely classical arguments. For this purpose, we assume sources with uncorrelated random phases. To illustrate this, we will briefly describe the basic idea behind the stellar interferometer invented by Hanbury Brown and Twiss. Referring to Figure 2.8, we assume for utmost simplicity that the electric fields from two point sources "a" and "b" have the same frequency ω and the same (real) amplitude A_0, but that their phases ϕ_a and ϕ_b vary randomly and independently. Aside from the factor $\exp(-i\omega t)$, the total electric field at a point $\mathbf{r}$ is

$$\mathcal{E}(\mathbf{r}) = A_0 e^{i\phi_a} \frac{e^{ik|\mathbf{r}-\mathbf{r}_a|}}{|\mathbf{r}-\mathbf{r}_a|} + A_0 e^{i\phi_b} \frac{e^{ik|\mathbf{r}-\mathbf{r}_b|}}{|\mathbf{r}-\mathbf{r}_b|}, \qquad (2.9.2)$$

---

[45] Early attempts to measure coincidences of the type found by Hanbury Brown and Twiss were unsuccessful. See A. Ádám, L. Jánossy, and P. Varga, Acta Phys. Hung. **4**, 301 (1955), and E. Brannen and H. I. S. Ferguson, Nature **178**, 481 (1956).

[46] E. T. Jaynes, "Quantum Beats," in A. O. Barut, ed., *Foundations of Radiation Theory and Quantum Electrodynamics*, Plenum Press, New York, 1980, p. 37.

[47] P. A. M. Dirac, *The Principles of Quantum Mechanics*, fourth edition, Oxford University Press, Oxford, 1958, p. 9.

[48] V. Radhakrishnan, Physics Today, **55**, 75 (2002), p. 75.

where $\mathbf{r}_a$ and $\mathbf{r}_b$ specify the positions of the two sources. The electric field at detector D1, for example, is

$$\mathcal{E}_1 \cong \frac{1}{r} A_0 \Big[e^{i\phi_a} e^{ikr_{1a}} + e^{i\phi_b} e^{ikr_{1b}} \Big], \tag{2.9.3}$$

where r_{1a} and r_{1b} are the distances from "a" to D1 and from "b" to D1, respectively, and it is assumed that the source–detector distance r in the simplified model shown in Figure 2.8 is much greater than both d and D. Similarly, the field at detector D2 is

$$\mathcal{E}_2 \cong \frac{1}{r} A_0 \Big[e^{i\phi_a} e^{ikr_{2a}} + e^{i\phi_b} e^{ikr_{2b}} \Big], \tag{2.9.4}$$

where r_{2a} and r_{2b} are the distances from "a" to D2 and from "b" to D2, respectively.

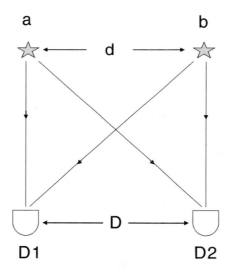

Fig. 2.8 Hanbury Brown–Twiss intensity interferometry. The intensity of light from two sources (a and b) is measured at two detectors (D1 and D2), and the average of the product of the intensities at D1 and D2 as a function of the detector separation D provides information about the distance between the sources. The arrangement shown is, of course, only schematic, as the scale of the source and detector distances and positions will, in practice, be very different.

The intensity at D1 is proportional to

$$|\mathcal{E}_1|^2 = \frac{A_0^2}{r^2} \Big[2 + e^{ik(r_{1b}-r_{1a})} e^{i(\phi_b-\phi_a)} + e^{-ik(r_{1b}-r_{1a})} e^{-i(\phi_b-\phi_a)} \Big], \tag{2.9.5}$$

and, since ϕ_a and ϕ_b are assumed to vary randomly and independently, the average value of $|\mathcal{E}_1|^2$ is just

$$\langle |\mathcal{E}_1|^2 \rangle = \frac{2A_0^2}{r^2}. \tag{2.9.6}$$

Similarly, for detector D2,

$$|\mathcal{E}_2|^2 = \frac{A_0^2}{r^2}\left[2 + e^{ik(r_{2b}-r_{2a})}e^{i(\phi_b-\phi_a)} + e^{-ik(r_{2b}-r_{2a})}e^{-i(\phi_b-\phi_a)}\right], \qquad (2.9.7)$$

and

$$\langle|\mathcal{E}_2|^2\rangle = \frac{2A_0^2}{r^2}, \qquad (2.9.8)$$

and the product of the average intensities at D1 and D2 is proportional to

$$\langle|\mathcal{E}_1|^2\rangle\langle|\mathcal{E}_2|^2\rangle = \frac{4A_0^4}{r^4}. \qquad (2.9.9)$$

The average of the product of the intensities at the two detectors is found by multiplying $|\mathcal{E}_1|^2$ by $|\mathcal{E}_2|^2$ and then averaging over the fluctuating independent random variables ϕ_a and ϕ_b. But there are two terms in this product that are independent of ϕ_a and ϕ_b; they arise from the multiplication of the second term in (2.9.5) by the third term in (2.9.7) and from the multiplication of the third term in (2.9.5) by the second term in (2.9.7). Terms in the product that do depend on ϕ_a and ϕ_b give zero when averaged. Thus,

$$\langle|\mathcal{E}_1|^2|\mathcal{E}_2|^2\rangle = \frac{A_0^4}{r^4}\left\{4 + 2\cos[k(r_{1b} - r_{2b} - r_{1a} + r_{2a})]\right\}, \qquad (2.9.10)$$

or, in terms of the average intensities at the two detectors, which are assumed in our model to be equal,

$$\frac{\langle I_1 I_2\rangle}{\langle I_1\rangle\langle I_2\rangle} = 1 + \frac{1}{2}\cos[k(r_{1b} - r_{2b} - r_{1a} + r_{2a})]. \qquad (2.9.11)$$

Writing $I_1 = \langle I_1\rangle + \Delta I_1$, and $I_2 = \langle I_2\rangle + \Delta I_2$, we can write this equivalently as

$$\frac{\langle I_1 I_2\rangle}{\langle I_1\rangle\langle I_2\rangle} = 1 + \frac{\langle \Delta I_1 \Delta I_2\rangle}{\langle I_1\rangle\langle I_2\rangle} = 1 + \frac{1}{2}\cos[k(r_{1b} - r_{2b} - r_{1a} + r_{2a})]. \qquad (2.9.12)$$

Exercise 2.8: Show that the dependence of $\langle \Delta I_1 \Delta I_2 \rangle$ on the detector separation D can be used to infer the angular separation $\Delta\theta = d/r$ of the sources a and b. (Note: for $r \gg d$, $k(r_{1b} - r_{2b} - r_{1a} + r_{2a}) \cong \mathbf{d} \cdot (\mathbf{k}_2 - \mathbf{k}_1)$, where $\mathbf{k}_i$ is the wave vector of the field incident on detector i.) In other words, angular diameters can be determined by measuring the correlations in the fluctuations of the electric currents of photoelectric detectors at D1 and D2. (In 1950, Hanbury Brown and Twiss demonstrated their intensity interferometry by measuring the diameter of the Sun using two radio telescopes with λ ($= 2\pi/k$) $= 2.4$ m, and thereafter measured the angular diameters of 32 other stars; in the case of a single source, the correlation function $\langle I_1 I_2 \rangle / \langle I_1 \rangle \langle I_2 \rangle$ involves a Fourier transform of the source distribution. A significant shortcoming of intensity interferometry is that the signal-to-noise ratio limits it in practice to relatively bright stars. Since the mid-1970s, stellar interferometry has been done exclusively by amplitude (Michelson-type) interferometry.)

We noted earlier, in connection with Dirac's remark that "each photon . . . interferes only with itself. Interference between two different photons never occurs," that what are interfering are not photons as such but probability amplitudes.[49] Thus, in Figure 2.8, there are two indistinguishable ways by which photons can be found at both D1 and D2, and therefore two probability amplitudes that must be added to give the total probability amplitude, whose absolute square is proportional to the predicted rate for counting photons in coincidence at D1 and D2: (1) a photon from "a" can be detected at D1 with a photon from "b" detected at D2, or (2) a photon from "a" can be detected at D2 with a photon from "b" detected at D1. It is the interference of probability amplitudes for these two indistinguishable processes that explains the Hanbury Brown–Twiss correlations.

References and Suggested Additional Reading

The brief discussion of Heisenberg's path to matrix mechanics in Section 2.2 follows in part that of I. J. R. Aitchison, D. A. MacManus, and T. M. Snyder, Am. J. Phys. **72**, 1370 (2004).

M. Göppert-Mayer (Ann. Physik **401**, 273 (1931)) employed the electric dipole form of the Hamiltonian, which she derived using the classical Lagrangian approach reviewed in Section 2.2, in a treatment of two-photon absorption. Since that original work, there have been many discussions of the transformation relating the minimal coupling and

[49] P. A. M. Dirac, *The Principles of Quantum Mechanics*, fourth edition, Oxford University Press, Oxford, 1958, p. 9.

electric dipole forms of the Hamiltonian in semiclassical radiation theory. See, for instance, *Scully and Zubairy* or *Mandel and Wolf (1995)*. Different approaches to the transformation are discussed by J. R. Ackerhalt and P. W. Milonni, J. Opt. Soc. Am. B **1**, 116 (1984).

Among reviews of the photon concept and semiclassical radiation theory are M. O. Scully and M. Sargent, Physics Today **25**, 38 (March, 1972); P. W. Milonni, Phys. Rep. **25**, 1 (1976); and L. Mandel, *Progress in Optics* **13**, ed. E. Wolf, North-Holland, Amsterdam, 1976, pp. 27–68.

The role of the ponderomotive potential and the $\mathbf{A}^2$ term in the minimal coupling Hamiltonian is discussed in the context of strong-field ionization by P. W. Milonni and J. R. Ackerhalt, Phys. Rev. A **39**, 1139 (1989).

A standard reference for two-state atoms in resonant fields is *Allen and Eberly*. Extensive treatments of atoms in laser fields may be found in works such as *Shore* or *Berman and Malinovsky*. Another excellent source is the "exploration through problems and solutions" of *Budker, Kimball, and DeMille*. Homogeneous and inhomogeneous line broadening are discussed at an introductory level in *Milonni and Eberly*.

Einstein's work on the blackbody problem has, of course, attracted much commentary. See, for instance, *Milonni (1994)*, Chapter 1, and references therein.

More detailed treatments of the Hanbury Brown–Twiss effect can be found in *Loudon* and *Mandel and Wolf (1995)*. For a review of the application of Hanbury Brown–Twiss intensity correlations "from stars to nuclear collisions," see G. Baym, http://arxiv.org/abs/nucl-th/9804026, accessed October 19, 2018.

3

Quantum Theory of the Electromagnetic Field

The Hamiltonian for the electromagnetic field is the Hamiltonian for a (generally infinite) set of uncoupled harmonic oscillators. To "quantize" the field in the simplest way, we regard the coordinates and momenta of these oscillators as operators obeying the canonical commutation relations. The states of definite field energy—the eigenstates of the field Hamiltonian—are states of definite *photon* number. Their basic importance notwithstanding, these states do not characterize the light from stars, lamps, lasers, or any other familiar source.

Different quantum states of the field oscillators imply different quantum states of light with different interference and statistical properties. In this chapter, we quantize the field and set the stage for the fully quantum-theoretical treatment of matter–field interactions in the following chapters. We consider in particular the most important states of light for this purpose—thermal, coherent, and vacuum states—and derive some of their statistical properties. We also address some foundational matters relating to the consistency of quantum electrodynamics and its conceptual framework.

3.1 The Harmonic Oscillator

We first review briefly the quantum mechanics of the harmonic oscillator, with the Hamiltonian

$$H = \frac{1}{2}(p^2 + \omega^2 q^2) \tag{3.1.1}$$

for frequency ω and unit mass. q and p are, respectively, the (Hermitian) coordinate and momentum operators satisfying the commutation relation $[q, p] = i\hbar$.

It follows directly from the Heisenberg equations of motion, without requiring their solutions, that the eigenvalues of H form a ladder of equally spaced energies: denote the eigenvalues of H by $E_0, E_1, E_2, \ldots$, the eigenstates by $|E_n\rangle$, and the matrix elements $\langle E_m|q|E_n\rangle$ and $\langle E_m|p|E_n\rangle$ by q_{mn} and p_{mn}, respectively. Then, the Heisenberg equation $p = \dot{q} = -(i/\hbar)(qH - Hq)$ implies

$$p_{mn} = -(i/\hbar)(E_n - E_m)q_{mn}, \tag{3.1.2}$$

and, similarly, the Heisenberg equation $-\omega^2 q = \dot{p} = -(i/\hbar)(pH - Hp)$ implies

An Introduction to Quantum Optics and Quantum Fluctuations. Peter W. Milonni.
© Peter W. Milonni 2019. Published in 2019 by Oxford University Press.
DOI:10.1093/oso/9780199215614.001.0001

$$q_{mn} = (i/\hbar\omega^2)(E_n - E_m)p_{mn}. \tag{3.1.3}$$

From (3.1.2) and (3.1.3),

$$p^2_{mn} = \left(\frac{E_n - E_m}{\hbar\omega}\right)^2 p^2_{mn}. \tag{3.1.4}$$

For non-vanishing matrix elements p_{mn} and q_{mn}, we must have $(E_n - E_m)/\hbar\omega = \pm 1$, from which it follows that[1]

$$E_n = n\hbar\omega + E_0, \quad n = 0, 1, 2, \ldots . \tag{3.1.5}$$

We can exclude negative values of n since, from (3.1.1), the eigenvalues E_n cannot be negative. We also conclude from (3.1.4) that matrix elements q_{mn} and p_{mn} for which $m \neq n \pm 1$ must vanish.

We now review the basic theory of the harmonic oscillator based on the (non-Hermitian) operators

$$a = \sqrt{\frac{1}{2\hbar\omega}}(p - i\omega q), \quad a^\dagger = \sqrt{\frac{1}{2\hbar\omega}}(p + i\omega q), \tag{3.1.6}$$

in terms of which

$$q = i\sqrt{\frac{\hbar}{2\omega}}(a - a^\dagger), \quad p = \sqrt{\frac{\hbar\omega}{2}}(a + a^\dagger). \tag{3.1.7}$$

These definitions and $[q, p] = i\hbar$ imply

$$[a, a^\dagger] = 1 \tag{3.1.8}$$

and

$$H = \frac{1}{2}\hbar\omega(aa^\dagger + a^\dagger a) = \hbar\omega(a^\dagger a + \frac{1}{2}). \tag{3.1.9}$$

The energy spectrum of the harmonic oscillator is therefore determined by the eigenvalues of the operator $a^\dagger a$. We now denote the eigenvalues and (presumed normalized) eigenkets of $a^\dagger a$ by n and $|n\rangle$, respectively:

$$a^\dagger a|n\rangle = n|n\rangle. \tag{3.1.10}$$

Since $\langle n|a^\dagger a|n\rangle$ is the scalar product of $a|n\rangle$ with itself, it follows from (3.1.10) that n is real and positive. H being Hermitian means not only that the eigenvalues n must be real, but that eigenstates belonging to different eigenvalues are orthogonal: $\langle n|m\rangle = \delta_{mn}$.

Now, $a^\dagger aa|n\rangle = (aa^\dagger a + [a^\dagger a, a])|n\rangle$, and (3.1.8) implies $[a^\dagger a, a] = -a$, $a^\dagger aa|n\rangle = (n-1)a|n\rangle$. Thus, if $|n\rangle$ is an eigenket of $a^\dagger a$ with eigenvalue n, then $a|n\rangle$ is an eigenket

---

[1] E. P. Wigner, Phys. Rev. **77**, 711 (1950).

of $a^\dagger a$ with eigenvalue $n-1$: $a|n\rangle = C|n-1\rangle$. Taking the norm of both sides of this equation, we deduce that $|C|^2 = \langle n|a^\dagger a|n\rangle = n$, and so we can take

$$a|n\rangle = \sqrt{n}|n-1\rangle. \tag{3.1.11}$$

In similar fashion, we deduce that

$$a^\dagger|n\rangle = \sqrt{n+1}|n+1\rangle. \tag{3.1.12}$$

Because of these relations, a and $a^\dagger$ are called *lowering* (or *annihilation*) and *raising* (or *creation*) operators, respectively. From (3.1.11) we see that we can generate eigenstates with lower and lower eigenvalues by repeated applications of the lowering operator a. But since $n \geq 0$, we must have $a|n\rangle = 0$ for $n < 1$, and (3.1.11) implies that this holds only for $n = 0$. We conclude then that the eigenvalues n of $a^\dagger a$ are all the positive integers (including zero) and that the energy levels of the harmonic oscillator are

$$E_n = \left(n + \frac{1}{2}\right)\hbar\omega, \quad n = 0, 1, 2, \ldots. \tag{3.1.13}$$

It is easy to relate this approach using raising and lowering operators to the more traditional Schrödinger-equation approach to the harmonic oscillator. For instance, from (3.1.11) we have $a|0\rangle = 0$, or $\langle q|(p - i\omega q)|0\rangle = 0$ for the ground state $|0\rangle$. $\langle q|0\rangle$, by definition, is just the wave function $\psi_0(q)$ for the ground state, and, likewise, $\langle q|p|0\rangle = (\hbar/i)d\psi_0/dq$, so

$$\left(\frac{\hbar}{i}\frac{d}{dq} - i\omega q\right)\psi_0(q) = 0, \tag{3.1.14}$$

or $\psi_0(q) = (\omega/\pi\hbar)^{1/4}e^{-\omega q^2/2\hbar}$, after we normalize such that $\int_{-\infty}^{\infty} dq|\psi_0(q)|^2 = 1$; this is the familiar ground-state wave function for the harmonic oscillator. The excited-state wave functions $\psi_n(q) = \langle q|n\rangle$ may be generated by application of $a^\dagger$ according to (3.1.12):

$$|n\rangle = (n!)^{-1/2}(a^\dagger)^n|0\rangle \tag{3.1.15}$$

and

$$\begin{aligned}
\psi_n(q) &= (n!)^{-1/2}(2\hbar\omega)^{-n/2}\langle q|(p + i\omega q)^n|0\rangle \\
&= [(2\hbar\omega)^n n!]^{-1/2}\left(\frac{\hbar}{i}\frac{d}{dq} + i\omega q\right)^n \psi_0(q) \\
&= i^n(2^n n!)^{-1/2}\left(\frac{\omega}{\pi\hbar}\right)^{1/4}\left(\xi - \frac{d}{d\xi}\right)^n e^{-\xi^2/2}, \quad \xi = (\omega/\hbar)^{1/2}q. \tag{3.1.16}
\end{aligned}$$

These are the excited-state eigenfunctions proportional to $e^{-\xi^2/2}H_n(\xi)$, where H_n is the Hermite polynomial of degree n.

We can use a and $a^\dagger$ to work out familiar properties of the harmonic oscillator. For instance, $\langle n|q|n\rangle = \langle n|p|n\rangle = 0$ and

$$\langle n|q^2|n\rangle = -\frac{\hbar}{2\omega}\langle n|(a - a^\dagger)^2|n\rangle = \frac{\hbar}{\omega}\left(n + \frac{1}{2}\right), \tag{3.1.17}$$

$$\langle n|p^2|n\rangle = \hbar\omega\left(n + \frac{1}{2}\right). \tag{3.1.18}$$

Therefore, $\Delta q_n \Delta p_n = (n + \frac{1}{2})\hbar$ for state $|n\rangle$, consistent with the uncertainty relation $\Delta q \Delta p \geq \hbar/2$ (where $\Delta q = (\langle q^2\rangle - \langle q\rangle^2)^{1/2}$, $\Delta p = (\langle p^2\rangle - \langle p\rangle^2)^{1/2}$).

The eigenstates $|n\rangle$ form a complete set: $\sum_{n=0}^{\infty} |n\rangle\langle n| = 1$.

Exercise 3.1: (a) Using the completeness property for a Hermitian operator that is bounded from below but not from above (see Section 2.1), prove that the Hermite polynomials form a complete set. (b) Why is it necessary for the eigenvalues of a Hamiltonian operator in quantum theory to be bounded from below? (c) Does the completeness property apply for Hermitian operators that are bounded from above but not from below?

3.1.1   Zero-Point Energy

The lowest possible energy of the harmonic oscillator—the energy it would have even at the absolute zero of temperature—is $\frac{1}{2}\hbar\omega$. This "zero-point energy" is closely related to the position-momentum uncertainty relation, and may even be regarded as a consequence of it: from $\langle\Delta q^2\rangle\langle\Delta p^2\rangle \geq \hbar^2/4$, it follows that

$$\langle H\rangle = \frac{1}{2}(\langle p^2\rangle + \omega^2\langle q^2\rangle) \geq \hbar^2/8\langle q^2\rangle + \frac{1}{2}\omega^2\langle q^2\rangle \tag{3.1.19}$$

for any state of the oscillator. (We assume without loss of generality that $\langle q\rangle = \langle p\rangle = 0$, and therefore that $\langle\Delta q^2\rangle = \langle q^2\rangle$, $\langle\Delta p^2\rangle = \langle p^2\rangle$.) The right-hand side has its smallest value when $\langle q^2\rangle = \hbar/2\omega$, and therefore

$$\langle H\rangle \geq \frac{1}{2}\hbar\omega \tag{3.1.20}$$

for any state of the oscillator.

Exercise 3.2: The rotational energy levels of a diatomic molecule are given approximately by $CJ(J+1)$, $J = 0, 1, 2, \ldots$, with C a constant depending on the particular molecule. The zero-point rotational energy is therefore zero. Why is this not a violation of the uncertainty relation?

3.2   Field Hamiltonian

At optical frequencies, the magnetic permeability is generally very close to its free-space value μ_0, so we will put $\mu = \mu_0$ ($\mathbf{B} = \mu_0 \mathbf{H}$) and work with Maxwell's equations in the form

$$\nabla \cdot \mathbf{D} = \rho, \tag{3.2.1}$$

$$\nabla \cdot \mathbf{B} = 0, \tag{3.2.2}$$

$$\nabla \times \mathbf{E} = -\frac{\partial \mathbf{B}}{\partial t}, \tag{3.2.3}$$

$$\nabla \times \mathbf{B} = \mu_0 \mathbf{J} + \mu_0 \frac{\partial \mathbf{D}}{\partial t}. \tag{3.2.4}$$

For now, we describe the field classically: each of the components $\mathbf{D}$, $\mathbf{E}$, and $\mathbf{B}$ are ordinary functions of $\mathbf{r}$ and t. $E_x(\mathbf{r}, t)$, for instance, is a real number at any point $\mathbf{r}$ at any time t. In a region devoid of any free charges and currents, $\mathbf{D} = \epsilon_0 \mathbf{E}$, $\rho = \mathbf{J} = 0$, and, in the Coulomb gauge (see Section 1.3), the scalar potential ϕ can be set to zero, whereas the vector potential satisfies

$$\nabla^2 \mathbf{A} - \frac{1}{c^2} \frac{\partial^2 \mathbf{A}}{\partial t^2} = 0, \quad \nabla \cdot \mathbf{A} = 0, \tag{3.2.5}$$

where we have used $c = 1/\sqrt{\epsilon_0 \mu_0}$ for the speed of light in vacuum.

The simplest nontrivial type of solution of (3.2.5) oscillates sinusoidally in time:

$$\mathbf{A}(\mathbf{r}, t) = \alpha(t) \mathbf{A}_0(\mathbf{r}) + \alpha^*(t) \mathbf{A}_0^*(\mathbf{r}), \tag{3.2.6}$$

with

$$\dot{\alpha}(t) = -i\omega \alpha(t), \tag{3.2.7}$$

and $\mathbf{A}_0(\mathbf{r})$ is a solution of the Helmholtz equation

$$\nabla^2 \mathbf{A}_0(\mathbf{r}) + \frac{\omega^2}{c^2} \mathbf{A}_0(\mathbf{r}) = 0, \tag{3.2.8}$$

together with whatever boundary conditions are appropriate. The electric and magnetic fields are

$$\mathbf{E}(\mathbf{r},t) = -\frac{\partial \mathbf{A}}{\partial t} = -[\dot{\alpha}(t)\mathbf{A}_0(\mathbf{r}) + \dot{\alpha}^*(t)\mathbf{A}_0^*(\mathbf{r})], \tag{3.2.9}$$

$$\mathbf{B}(\mathbf{r},t) = \nabla \times \mathbf{A} = \alpha(t)\nabla \times \mathbf{A}_0(\mathbf{r}) + \alpha^*(t)\nabla \times \mathbf{A}_0^*(\mathbf{r}). \tag{3.2.10}$$

The field energy is

$$E_{\text{field}} = \frac{1}{2}\int d^3r(\epsilon_0\mathbf{E}^2 + \frac{1}{\mu_0}\mathbf{B}^2) = \frac{1}{2}\epsilon_0\int d^3r(\mathbf{E}^2 + c^2\mathbf{B}^2) = 2\epsilon_0\omega^2|\alpha|^2, \tag{3.2.11}$$

where, to obtain the last equality, we have chosen the normalization $\int d^3r|\mathbf{A}_0(\mathbf{r})|^2 = 1$. For an effectively unbounded region of space, we will work with plane-wave fields with the periodic boundary conditions reviewed in Section 3.3.

Exercise 3.3: What other assumptions for the mode functions $\mathbf{A}_0(\mathbf{r})$ are needed to derive (3.2.11)?

The frequency ω and the function $\mathbf{A}_0(\mathbf{r})$ define a *mode* of the field. $\mathbf{A}(\mathbf{r},t)$, $\mathbf{E}(\mathbf{r},t)$, and $\mathbf{B}(\mathbf{r},t)$ are fully specified by ω, $\mathbf{A}_0(\mathbf{r})$, and $\alpha(0)$. Defining

$$p = \omega\sqrt{\epsilon_0}(\alpha + \alpha^*) \quad \text{and} \quad q = i\sqrt{\epsilon_0}(\alpha - \alpha^*), \tag{3.2.12}$$

we can write

$$E_{\text{field}} = \frac{1}{2}(p^2 + \omega^2 q^2) \tag{3.2.13}$$

and, from (3.2.7),

$$\dot{p} = -\omega^2 q, \quad \dot{q} = p. \tag{3.2.14}$$

These equations imply that a mode of the field is a harmonic oscillator with "momentum" p and "coordinate" q. In other words, a field mode is described by the Hamiltonian and the canonical equations of motion

$$H = \frac{1}{2}(p^2 + \omega^2 q^2), \tag{3.2.15}$$

$$\dot{q} = \frac{\partial H}{\partial p}, \quad \dot{p} = -\frac{\partial H}{\partial q}. \tag{3.2.16}$$

The frequency of the oscillator is the same as that of the field mode, and it has "unit mass" according to our definitions of p and q.

More generally, we expand the field in a complete set of modes, each having a frequency ω_j and an associated transverse mode function $\mathbf{A}_j(\mathbf{r})$ satisfying $\nabla^2\mathbf{A}_j + (\omega_j^2/c^2)\mathbf{A}_j = 0$:

$$\mathbf{A}(\mathbf{r},t) = \sum_j [\alpha_j(t)\mathbf{A}_j(\mathbf{r}) + \text{cc}], \tag{3.2.17}$$

with $\dot{\alpha}_j(t) = -i\omega_j\alpha_j(t)$, where cc stands for "complex conjugate." The eigenfunctions $\mathbf{A}_j(\mathbf{r})$ corresponding to different eigenfrequencies may be taken to be orthogonal (and normalized):

$$\int d^3 r \mathbf{A}_j^*(\mathbf{r}) \cdot \mathbf{A}_i(\mathbf{r}) = \delta_{ij}. \tag{3.2.18}$$

Then, (3.2.15) generalizes to

$$H = \sum_j \frac{1}{2}(p_j^2 + \omega_j^2 q_j^2), \tag{3.2.19}$$

$$p_j = \omega_j \sqrt{\epsilon_0}(\alpha_j + \alpha_j^*), \quad q_j = i\sqrt{\epsilon_0}(\alpha_j - \alpha_j^*), \tag{3.2.20}$$

$$\dot{q}_j = \frac{\partial H}{\partial p_j}, \quad \dot{p}_j = -\frac{\partial H}{\partial q_j}. \tag{3.2.21}$$

The Hamiltonian for the field is the Hamiltonian for a set of independent harmonic oscillators.

Of course, this Hamiltonian approach to the classical electromagnetic field amounts to nothing more than a reformulation of the Maxwell equations in the absence of any sources. It is not the approach taken in traditional expositions of classical electrodynamics, but it allows us to proceed straightforwardly to the quantization of the field.

In a model sometimes used to allow for a spatially varying permittivity, it is assumed that the medium is non-absorbing and non-dispersive and that $\mu \cong \mu_0$. In this model, $-\nabla \times (\nabla \times \mathbf{A}_j) + \omega_j^2 \epsilon_0 \mu_0 \mathbf{A}_j = -\nabla(\nabla \cdot \mathbf{A}_j) + \nabla^2 \mathbf{A}_j + \omega_j^2 \epsilon_0 \mu_0 \mathbf{A}_j = \nabla^2 \mathbf{A}_j + (\omega_j^2/c^2)\mathbf{A}_j = 0$ is replaced by

$$-\nabla \times (\nabla \times \mathbf{A}_j) + \omega_j^2 \mu_0 \epsilon(\mathbf{r})\mathbf{A}_j = 0, \tag{3.2.22}$$

and a gauge-fixing condition follows from the identity $\nabla \cdot [\nabla \times (\nabla \times \mathbf{A}_j)] = 0$:

$$\nabla \cdot [\epsilon(\mathbf{r})\mathbf{A}_j(\mathbf{r})] = 0. \tag{3.2.23}$$

The modes are normalized according to

$$\int d^3 r \epsilon(\mathbf{r})\mathbf{A}_i^*(\mathbf{r}) \cdot \mathbf{A}_j(\mathbf{r}) = \delta_{ij}, \tag{3.2.24}$$

and the field is quantized along the lines described in Section 3.3.[2]

3.3   Field Quantization: Energy and Momentum

In the quantum theory of the field, we take q_j and p_j to be operators satisfying the commutation relations $[q_i, p_j] = i\hbar\delta_{ij}$, $[q_i, q_j] = [p_i, p_j] = 0$. Because q_j and p_j for the field are, respectively, coordinates and momenta of harmonic oscillators, we *postulate* that, for a quantum description of the field, we need only regard them as operators like those for any other harmonic oscillator in quantum theory. Thus, we make the replacements $\alpha_j \to C_j a_j$, $\alpha_j^* \to C_j^* a_j^\dagger$ and require that the operators a_j and $a_j^\dagger$ satisfy

$$[a_i, a_j^\dagger] = \delta_{ij}, \tag{3.3.1}$$

[2] See, for instance, R. J. Glauber and M. Lewenstein, Phys. Rev. A **43**, 467 (1991).

so that q_j and p_j defined by (3.2.20) satisfy the canonical commutation relations for q's and p's. We can do this by choosing $C_j = \sqrt{\hbar/2\epsilon_0\omega_j}$. Then, the quantized-field versions of (3.2.17) and (3.2.19) are

$$\mathbf{A}(\mathbf{r},t) = \sum_j \sqrt{\frac{\hbar}{2\epsilon_0\omega_j}} \left[\mathbf{A}_j(\mathbf{r})a_j(t) + \mathbf{A}_j^*(\mathbf{r})a_j^\dagger(t) \right], \qquad (3.3.2)$$

$$H = \frac{1}{2} \sum_j \hbar\omega_j(a_j^\dagger a_j + a_j a_j^\dagger) = \sum_j \hbar\omega_j(a_j^\dagger a_j + \frac{1}{2}), \qquad (3.3.3)$$

and the electric field operator, for instance, is

$$\mathbf{E}(\mathbf{r},t) = -\frac{\partial \mathbf{A}}{\partial t} = i \sum_j \sqrt{\frac{\hbar\omega_j}{2\epsilon_0}} \left[\mathbf{A}_j(\mathbf{r})a_j(t) - \mathbf{A}_j^*(\mathbf{r})a_j^\dagger(t) \right]. \qquad (3.3.4)$$

As will be discussed in Section 3.5, the *quantized* electric field (3.3.4) comports nicely with the observed "wave–particle duality" of light. The mode functions $\mathbf{A}_j(\mathbf{r})$ account for the spatial variations of the field in exactly the same way as in classical electromagnetic theory—these mode functions are the same mode functions appearing in classical theory. But the temporal variations of the field in the Heisenberg picture follow those of the *operators* $a_j(t)$ and $a_j^\dagger(t)$. It is through these operators that the field at every point in space becomes a *quantum* field.

For fields in effectively unlimited space—that is, when we can ignore any effects of boundaries on the electric and magnetic fields—it is convenient to expand the field in plane-wave mode functions:

$$\mathbf{A}_{\mathbf{k}\lambda}(\mathbf{r}) = \frac{1}{\sqrt{V}}\mathbf{e}_{\mathbf{k}\lambda}e^{i\mathbf{k}\cdot\mathbf{r}}. \qquad (3.3.5)$$

Here, V is a "quantization volume" such that

$$\int_V d^3r |\mathbf{A}_{\mathbf{k}\lambda}(\mathbf{r})|^2 = 1. \qquad (3.3.6)$$

The (real) unit vector $\mathbf{e}_{\mathbf{k}\lambda}$ obviously characterizes the polarization of the plane wave. $\nabla \cdot \mathbf{A} = 0$ implies $\mathbf{k} \cdot \mathbf{e}_{\mathbf{k}\lambda} = 0$, so that, for each $\mathbf{k}$, we can define two polarization vectors, which we label with λ ($= 1, 2$) and take to be orthogonal:

$$\mathbf{e}_{\mathbf{k}\lambda} \cdot \mathbf{e}_{\mathbf{k}\lambda'} = \delta_{\lambda\lambda'}. \qquad (3.3.7)$$

Thus, we have the orthonormality condition

$$\int_V d^3r \mathbf{A}_{\mathbf{k}\lambda}^*(\mathbf{r}) \cdot \mathbf{A}_{\mathbf{k}'\lambda'}(\mathbf{r}) = \delta_{\mathbf{k}\mathbf{k}'}^3 \delta_{\lambda\lambda'}. \qquad (3.3.8)$$

In terms of these (linear polarization) modes,

$$\mathbf{A}(\mathbf{r}, t) = \sum_{\mathbf{k}\lambda} \sqrt{\frac{\hbar}{2\epsilon_0 \omega_k V}} \left[a_{\mathbf{k}\lambda}(t) e^{i\mathbf{k}\cdot\mathbf{r}} + a_{\mathbf{k}\lambda}^\dagger(t) e^{-i\mathbf{k}\cdot\mathbf{r}} \right] \mathbf{e}_{\mathbf{k}\lambda}, \qquad (3.3.9)$$

$$\dot{a}_{\mathbf{k}\lambda}(t) = -i\omega_k a_{\mathbf{k}\lambda}(t), \quad \omega_k = |\mathbf{k}|c = kc, \qquad (3.3.10)$$

$$\mathbf{E}(\mathbf{r}, t) = -\frac{\partial \mathbf{A}}{\partial t} = i \sum_{\mathbf{k}\lambda} \sqrt{\frac{\hbar\omega_k}{2\epsilon_0 V}} \left[a_{\mathbf{k}\lambda}(t) e^{i\mathbf{k}\cdot\mathbf{r}} - a_{\mathbf{k}\lambda}^\dagger(t) e^{-i\mathbf{k}\cdot\mathbf{r}} \right] \mathbf{e}_{\mathbf{k}\lambda}, \qquad (3.3.11)$$

$$\mathbf{B}(\mathbf{r}, t) = \nabla \times \mathbf{A} = i \sum_{\mathbf{k}\lambda} \sqrt{\frac{\hbar}{2\epsilon_0 \omega_k V}} \left[a_{\mathbf{k}\lambda}(t) e^{i\mathbf{k}\cdot\mathbf{r}} - a_{\mathbf{k}\lambda}^\dagger(t) e^{-i\mathbf{k}\cdot\mathbf{r}} \right] \mathbf{k} \times \mathbf{e}_{\mathbf{k}\lambda}, \qquad (3.3.12)$$

and the field Hamiltonian is

$$H = \frac{1}{2} \sum_{\mathbf{k}\lambda} (p_{\mathbf{k}\lambda}^2 + \omega_k^2 q_{\mathbf{k}\lambda}^2) = \sum_{\mathbf{k}\lambda} \hbar\omega_k (a_{\mathbf{k}\lambda}^\dagger a_{\mathbf{k}\lambda} + \frac{1}{2}), \qquad (3.3.13)$$

$$p_{\mathbf{k}\lambda} = \omega_k \sqrt{\epsilon_0} (a_{\mathbf{k}\lambda} + a_{\mathbf{k}\lambda}^\dagger), \quad q_{\mathbf{k}\lambda} = i\sqrt{\epsilon_0} (a_{\mathbf{k}\lambda} - a_{\mathbf{k}\lambda}^\dagger), \qquad (3.3.14)$$

$$[a_{\mathbf{k}\lambda}(t), a_{\mathbf{k}'\lambda'}^\dagger(t')] = [a_{\mathbf{k}\lambda}(0), a_{\mathbf{k}'\lambda'}^\dagger(0)] e^{-i\omega_k t} e^{i\omega_{k'} t'} = \delta_{\mathbf{k}\mathbf{k}'}^3 \delta_{\lambda\lambda'} e^{i\omega_k(t'-t)}. \qquad (3.3.15)$$

Exercise 3.4: Justify the assumption that the mode functions in (3.3.5) are a complete set, that is, that the transverse vector potential in free space can be expressed in the form given in (3.3.9).

The introduction of the volume V here is just an artifice that avoids having to deal with mode functions whose norm (squared modulus integrated over all space) is infinite. In this "box normalization," we pretend that space is divided into boxes of sides L_x, L_y, and L_z, and that the field satisfies the periodic boundary conditions

$$\mathbf{A}(x + L_x, y + L_y, z + L_z) = \mathbf{A}(x, y, z). \qquad (3.3.16)$$

This implies

$$k_x = 2\pi n_x/L_x, \quad k_y = 2\pi n_y/L_y, \quad k_z = 2\pi n_z/L_z \quad (n_x, n_y, n_z \text{ integers}). \qquad (3.3.17)$$

In calculating transition probabilities, interaction energies, and other effects of electromagnetic interactions in free space, we can then work with discrete sums of modes of finite norm, keeping $V = L_x L_y L_z$ finite, and if the box is sufficiently large, we account for the fact that there is practically a continuum of modes by replacing mode summations by integrations:

$$\sum_{\mathbf{k}} \to \sum_{n_x} \sum_{n_y} \sum_{n_z} \to \frac{L_x L_y L_z}{(2\pi)^3} \int \int \int dk_x dk_y dk_z = \frac{V}{(2\pi)^3} \int d^3 k. \qquad (3.3.18)$$

In the idealization of unbounded space, the box is obviously "sufficiently large"; more generally, we can use (3.3.18) whenever there is an approximately continuous distribution of modes of interest. If calculating the spontaneous emission rate on some transition of an atom inside a cavity, for example, we can use (3.3.18) if the transition wavelength is small compared to the distance of the atom from any wall of the cavity. Then, the emission rate does not depend on exactly where the atom is positioned inside the cavity and is the same as in unbounded space. Otherwise, we just work from beginning to end with a discrete set of mode functions $\mathbf{A}_j(\mathbf{r})$ satisfying the boundary conditions imposed by the cavity.

It should be obvious that the field *operators* satisfy Maxwell's equations; we have, in fact, *presumed* that Maxwell's equations are satisfied in quantum theory. The big difference from classical electromagnetic theory, of course, is that the quantized fields are operators, and, furthermore, that these operators generally do not commute. From the commutation relations for the a's and $a^\dagger$'s, we can easily derive commutation relations for the fields (see Section 3.10).

3.3.1   Linear Momentum

Because the field operators satisfy Maxwell's equations, we can follow familiar classical arguments to obtain expressions for other operators of physical interest. The operator corresponding to the linear momentum of the field in free space, for instance, is

$$\mathbf{P} = \epsilon_0 \mu_0 \int d^3 r (\mathbf{E} \times \mathbf{H}) = \frac{1}{2} \sum_{\mathbf{k}\lambda} \hbar \mathbf{k} [a_{\mathbf{k}\lambda} a_{\mathbf{k}\lambda}^\dagger + a_{\mathbf{k}\lambda}^\dagger a_{\mathbf{k}\lambda}] = \sum_{\mathbf{k}\lambda} \hbar \mathbf{k} a_{\mathbf{k}\lambda}^\dagger a_{\mathbf{k}\lambda}. \qquad (3.3.19)$$

We have used $\sum_{\mathbf{k}\lambda} \frac{1}{2}\hbar\mathbf{k} = 0$: the contribution to $\mathbf{P}$ from a mode with wave vector $\mathbf{k}$ is cancelled by the contribution from the mode with wave vector $-\mathbf{k}$.[3] Eigenstates of the field Hamiltonian H are obviously also eigenstates of the field linear momentum $\mathbf{P}$. For example, an energy eigenstate $|n\rangle$ of a plane-wave mode with wave vector $\mathbf{k}$ in free space has an energy $n\hbar\omega$ and a linear momentum $n\hbar\mathbf{k}$.

The fact that, in an eigenstate of the field Hamiltonian, the excitations (photons) of a field of frequency ω have energy $\hbar\omega$ is obviously consistent with Einstein's inference of "light quanta" from the form of the Planck spectrum; similarly, the fact that these excitations have linear momentum of magnitude $\hbar\omega/c$ is consistent with his deduction that atoms must recoil with this momentum when they emit or absorb a light quantum (see Section 2.8). Moreover, the commutator (3.3.1) confirms that photons are

---

[3] Because $\mathbf{E}$ and $\mathbf{H}$ do not commute (see Section 3.10), $\mathbf{P}$ is often defined as $\frac{1}{2}\epsilon_0\mu_0 \int d^3 r (\mathbf{E} \times \mathbf{H} - \mathbf{H} \times \mathbf{E})$ in order to make it "manifestly" Hermitian.

bosons; in particular, the expression (3.1.15) shows explicitly that an n-photon state is symmetric with respect to permutations of the n photons, as required for bosons.

Everyone agrees that the photon momentum is $\hbar\omega/c$ in vacuum. How it depends on the refractive index of a dielectric medium is an old question that still attracts interest (Section 1.7).

3.3.2   Angular Momentum (Spin)

According to classical electromagnetic theory, the angular momentum of the field in vacuum is[4]

$$\mathbf{L} = \epsilon_0 \int d^3r [\mathbf{r} \times (\mathbf{E} \times \mathbf{B})] = \epsilon_0 \int d^3r \, \mathbf{E} \times \mathbf{A} + \epsilon_0 \sum_{i=1}^{3} \int d^3r E_i \mathbf{r} \times \nabla A_i. \quad (3.3.20)$$

The first term in the second equality is independent of the origin of coordinates and is therefore termed the "intrinsic" angular momentum of the field. For the quantized field, the operators $\mathbf{E}$ and $\mathbf{A}$ do not, in general, commute (see Section 3.10), and so the conventional approach is to define the intrinsic angular momentum by the manifestly Hermitian expression

$$\mathbf{L}_s = \frac{1}{2}\epsilon_0 \int d^3r (\mathbf{E} \times \mathbf{A} - \mathbf{A} \times \mathbf{E}). \quad (3.3.21)$$

From (3.3.9), (3.3.11), and some algebra, we arrive at

$$\mathbf{L}_s = i\hbar \sum_{\mathbf{k}} \hat{\mathbf{k}}(a_{\mathbf{k}2}^\dagger a_{\mathbf{k}1} - a_{\mathbf{k}1}^\dagger a_{\mathbf{k}2}), \quad (3.3.22)$$

where the unit vector $\hat{\mathbf{k}} = \mathbf{e}_{\mathbf{k}1} \times \mathbf{e}_{\mathbf{k}2}$.

The operator (3.3.22) does not commute with the photon number operator $a_{\mathbf{k}\lambda}^\dagger a_{\mathbf{k}\lambda}$, and, consequently, an energy eigenstate $|n_{\mathbf{k}\lambda}\rangle$ of the field is not an eigenstate of $\mathbf{L}_s$. But we can construct an energy eigenstate state of the field that is also an eigenstate of $\mathbf{L}_s$ by introducing the complex polarization unit vectors

$$\mathbf{e}_{\mathbf{k},+1} = -\frac{1}{\sqrt{2}}(\mathbf{e}_{\mathbf{k}1} + i\mathbf{e}_{\mathbf{k}2}) \quad \text{and} \quad \mathbf{e}_{\mathbf{k},-1} = \frac{1}{\sqrt{2}}(\mathbf{e}_{\mathbf{k}1} - i\mathbf{e}_{\mathbf{k}2}), \quad (3.3.23)$$

which satisfy $\mathbf{e}_{\mathbf{k}\alpha}^* \cdot \mathbf{e}_{\mathbf{k}\alpha'} = \delta_{\alpha\alpha'}$, $\mathbf{e}_{\mathbf{k}\alpha}^* \times \mathbf{e}_{\mathbf{k}\alpha'} = i\alpha\hat{\mathbf{k}}\delta_{\alpha\alpha'}$, $\alpha = \pm1$.[5] Our original polarization vectors $\mathbf{e}_{\mathbf{k}\lambda}$ ($\lambda = 1, 2$) correspond to orthogonal *linear* polarizations, whereas

---

[4] Proof of the second equality is straightforward and may be found in textbooks such as *Mandel and Wolf (1995)*. These expressions for field angular momentum are gauge-invariant because $\mathbf{A}$ here is the transverse part of the vector potential, which is gauge-invariant. (A gauge transformation adds a longitudinal field ($\nabla\chi$) to $\mathbf{A}$, and thereby changes only its longitudinal part.)

[5] Recall the definitions in (2.4.2).

the complex polarization vectors $\mathbf{e}_{\mathbf{k}\alpha}$ ($\alpha = \pm 1$) correspond to opposite *circular* polarizations. We also define photon annihilation operators for the circular polarization modes $(\mathbf{k}, \alpha)$:

$$a_{\mathbf{k},+1} = -\frac{1}{\sqrt{2}}(a_{\mathbf{k}1} - i a_{\mathbf{k}2}) \quad \text{and} \quad a_{\mathbf{k},-1} = \frac{1}{\sqrt{2}}(a_{\mathbf{k}1} + i a_{\mathbf{k}2}). \tag{3.3.24}$$

In terms of these operators,

$$\mathbf{L}_s = \hbar \sum_{\mathbf{k}} \hat{\mathbf{k}}(a^\dagger_{\mathbf{k},+1} a_{\mathbf{k},+1} - a^\dagger_{\mathbf{k},-1} a_{\mathbf{k},-1}) = \sum_{\mathbf{k}\alpha} \alpha \hbar \hat{\mathbf{k}} a^\dagger_{\mathbf{k}\alpha} a_{\mathbf{k}\alpha}, \tag{3.3.25}$$

and similarly

$$H = \sum_{\mathbf{k}\alpha} \hbar \omega_k a^\dagger_{\mathbf{k}\alpha} a_{\mathbf{k}\alpha}, \tag{3.3.26}$$

$$\mathbf{P} = \sum_{\mathbf{k}\alpha} \hbar \mathbf{k} a^\dagger_{\mathbf{k}\alpha} a_{\mathbf{k}\alpha}, \tag{3.3.27}$$

for the field Hamiltonian and linear momentum operators, and

$$\mathbf{A}(\mathbf{r}, t) = \sum_{\mathbf{k}\alpha} \sqrt{\frac{\hbar}{2\epsilon_0 \omega_k V}} \left[a_{\mathbf{k}\alpha}(t) \mathbf{e}_{\mathbf{k}\alpha} e^{i\mathbf{k}\cdot\mathbf{r}} + a^\dagger_{\mathbf{k}\alpha}(t) \mathbf{e}^*_{\mathbf{k}\alpha} e^{-i\mathbf{k}\cdot\mathbf{r}} \right], \tag{3.3.28}$$

$$\mathbf{E}(\mathbf{r}, t) = i \sum_{\mathbf{k}\alpha} \sqrt{\frac{\hbar \omega_k}{2\epsilon_0 V}} \left[a_{\mathbf{k}\alpha}(t) \mathbf{e}_{\mathbf{k}\alpha} e^{i\mathbf{k}\cdot\mathbf{r}} - a^\dagger_{\mathbf{k}\alpha}(t) \mathbf{e}^*_{\mathbf{k}\alpha} e^{-i\mathbf{k}\cdot\mathbf{r}} \right] \tag{3.3.29}$$

for the vector potential and electric field operators, respectively.

For circularly polarized plane-wave mode functions, therefore, we have states that are eigenstates not only of field energy and linear momentum but also of intrinsic angular momentum or *spin*. According to (3.3.25), the component of the photon spin along the propagation direction, or, in other words, the photon *helicity*, is $\pm\hbar$; a photon is a boson with spin 1, and this spin is either parallel or anti-parallel to the propagation direction. For linearly polarized plane-wave modes, a photon may be described as being in a 50/50 superposition of right-hand (helicity -1) and left-hand (helicity $+1$) circular polarization states. Elliptical polarization involves superpositions with generally unequal portions of the two spin eigenstates.

The spin angular momentum of photons is directly related to the selection rules governing the emission and absorption of light (e.g., $\Delta m = 0, \pm 1$ for the change in the magnetic quantum number m). Experimental proof that photons have spin *and that this spin is 1* was provided by Raman and Bhagavantam in 1931. They cited a communication from S. N. Bose indicating that, in including a factor of 2 in the counting of states in his derivation of the blackbody spectum, "he envisioned the possibility of the [photon] possessing besides energy $h\nu$ and linear momentum $h\nu/c$ also an intrinsic spin or angular momentum $\pm h/2\pi$ round an axis parallel to the direction of its motion."[6] In other words, the factor of 2 accounted for two possible helicities of a photon.

[6] C. V. Raman and S. Bhagavantam, Indian J. Phys. **6**, 353 (1931), p. 354.

The experiments measured the depolarization of scattered light from molecular gases and liquids. Depolarization is possible because molecules do not have the spherical symmetry of atoms, and can undergo a change in their rotational state when they scatter light. Calculations based on the Kramers–Heisenberg dispersion formula for the polarizability tensor (Section 2.3) and spectroscopic selection rules for transition electric dipole moments, assuming that photons have spin 1, were in excellent agreement with the measured depolarization of Rayleigh-scattered light from gases of linear molecules such as N_2, O_2, and CO_2 and the much larger depolarizations measured in light scattering by the liquids CS_2 and C_6H_6.

The Raman–Bhagavantam experiment answered a question that arose following the discovery of electron spin: can a photon likewise have an intrinsic angular momentum? Einstein seemed doubtful that it could. The importance of the work did not escape the attention of *Time* magazine (July 6, 1931): "The practical value of this discovery lies in the fertile future. It may someday explain why the Sun turns a tomato red, a face tan; why an ape is not a man."

Exercise 3.5: Consider a circularly polarized plane wave propagating in the z direction. For such a wave, $\mathbf{E} \times \mathbf{B}$ points in the z direction, and therefore $\mathbf{r} \times (\mathbf{E} \times \mathbf{B})$ has no z component. How then can the definition $\mathbf{L} = \epsilon_0 \int d^3r [\mathbf{r} \times (\mathbf{E} \times \mathbf{B})]$ of the total angular momentum for the field lead to intrinsic angular momentum with a non-vanishing z component? (Note: in writing the second equality in (3.3.20), it has been assumed that surface terms vanish at infinity.)

3.4   Quantized Fields in Dielectric Media

The quantization procedure shown in Section 3.3 can be generalized to the case of dielectric media. We again start with a (classical) monochromatic field of frequency ω and consider a homogeneous, isotropic, non-absorbing medium characterized by a dielectric permittivity $\epsilon(\omega)$ and a magnetic permeability $\mu(\omega)$. (The generalization to absorbing media is discussed in Section 7.6.) Using $\mathbf{D} = \epsilon(\omega)\mathbf{E}$, and $\mathbf{B} = \mu(\omega)\mathbf{H}$, we find from Maxwell's equations that the Coulomb-gauge vector potential $\mathbf{A}$ ($\mathbf{B} = \nabla \times \mathbf{A}$) satisfies

$$\nabla^2 \mathbf{A} + \omega^2 \epsilon \mu \mathbf{A} = 0. \tag{3.4.1}$$

As in the case of a vacuum, we write

$$\mathbf{A}(\mathbf{r}, t) = \alpha(t)\mathbf{A}_0(\mathbf{r}) + \alpha^*(t)\mathbf{A}_0^*(\mathbf{r}) \tag{3.4.2}$$

with $\alpha(t) = \alpha(0)e^{-i\omega t}$ and

$$\mathbf{E}(\mathbf{r}, t) = -[\dot{\alpha}(t)\mathbf{A}_0(\mathbf{r}) + \dot{\alpha}^*(t)\mathbf{A}_0^*(\mathbf{r})), \tag{3.4.3}$$

$$\mathbf{H}(\mathbf{r}, t) = \frac{1}{\mu}[\alpha(t)\nabla \times \mathbf{A}_0(\mathbf{r}) + \alpha^*(t)\nabla \times \mathbf{A}_0^*(\mathbf{r})], \tag{3.4.4}$$

$$\nabla^2 \mathbf{A}_0 + \omega^2 \epsilon \mu \mathbf{A}_0 = 0. \tag{3.4.5}$$

We review below, again, the fact that the cycle-averaged energy density in the presence of such a field in a dielectric medium may be defined as

$$u_\omega = \frac{1}{4}\left[\frac{d}{d\omega}(\epsilon\omega)\mathbf{E}^2 + \frac{d}{d\omega}(\mu\omega)\mathbf{H}^2\right] \tag{3.4.6}$$

if the medium does not absorb radiation of frequency ω.[7] Using vector identities for the (normalized) mode functions similar to those used to obtain (3.2.11) for a field in a vacuum (Exercise 3.3), we obtain

$$E_{\text{field}} = \int d^3r\, u_\omega = |\alpha|^2\left[\frac{d}{d\omega}(\epsilon\omega) + \frac{\epsilon}{\mu}\frac{d}{d\omega}(\mu\omega)\right] = \frac{2nn_g}{\mu c^2}w^2|\alpha|^2, \tag{3.4.7}$$

where $n(\omega) = \sqrt{\epsilon\mu/\epsilon_0\mu_0} = c\sqrt{\epsilon\mu}$ is the (real) refractive index at frequency ω, and

$$n_g(\omega) = n(\omega) + \omega\frac{dn}{d\omega} = \frac{d}{d\omega}(n\omega) \tag{3.4.8}$$

is the group index: the group velocity at frequency ω is $c/n_g(\omega)$ (see Section 1.6).

We proceed as before, defining a canonical momentum p and a coordinate q ($\dot{p} = -\omega^2 q$, $\dot{q} = p$):

$$p = \omega\left(\frac{nn_g}{\mu c^2}\right)^{1/2}(\alpha + \alpha^*) \quad\text{and}\quad q = i\left(\frac{nn_g}{\mu c^2}\right)^{1/2}(\alpha - \alpha^*), \tag{3.4.9}$$

in terms of which E_{field} is the energy of a harmonic oscillator:

$$E_{\text{field}} = \frac{1}{2}(p^2 + \omega^2 q^2). \tag{3.4.10}$$

Then, we "quantize" by replacing p and q by *operators* satisfying the canonical commutation relation $[q, p] = i\hbar$. This is done by introducing operators a and $a^\dagger$ via the replacements

$$\alpha = \left(\frac{\hbar\mu c^2}{2nn_g}\right)^{1/2} a \quad\text{and}\quad \alpha^* = \left(\frac{\hbar\mu c^2}{2nn_g}\right)^{1/2} a^\dagger \tag{3.4.11}$$

and requiring $[a, a^\dagger] = 1$. The Hamiltonian for our monochromatic *quantum* field becomes

$$H = \frac{1}{2}(p^2 + \omega^2 q^2) = \frac{1}{2}\hbar\omega(aa^\dagger + a^\dagger a) = \hbar\omega(a^\dagger a + \frac{1}{2}), \tag{3.4.12}$$

and the electric and magnetic field operators are therefore

$$\mathbf{E}(\mathbf{r},t) = i\sqrt{\frac{\mu c^2\hbar\omega}{2nn_g}}\left[a(t)\mathbf{A}_0(\mathbf{r}) - a^\dagger(t)\mathbf{A}_0^*(\mathbf{r})\right], \tag{3.4.13}$$

---

[7] Recall also the derivation in Section 1.6.

$$\mathbf{H}(\mathbf{r}, t) = \sqrt{\frac{\hbar c^2}{2nn_g\mu\omega}} \left[a(t)\nabla \times \mathbf{A}_0(\mathbf{r}) + a^\dagger(t)\nabla \times \mathbf{A}_0^*(\mathbf{r}) \right]. \tag{3.4.14}$$

The multimode generalization for the electric field, for instance, is

$$\mathbf{E}(\mathbf{r}, t) = i\sum_{\mathbf{k}\lambda} \sqrt{\frac{\mu c^2\hbar\omega_k}{2nn_gV}} \left[a_{\mathbf{k}\lambda}(t)e^{i\mathbf{k}\cdot\mathbf{r}} - a_{\mathbf{k}\lambda}^\dagger(t)e^{-i\mathbf{k}\cdot\mathbf{r}} \right] \mathbf{e}_{\mathbf{k}\lambda} \tag{3.4.15}$$

when we use linearly polarized plane-wave modes. This reduces to the expression for the vacuum electric field operator when we set $\mu = \mu_0$, and $n = n_g = 1$, for every mode. The formula (3.4.15) may be regarded as an expression for the electric field in terms of the normal modes, or polaritons, of the coupled matter–field system.[8]

For later reference, we calculate the expectation value of $\mathbf{E}^2(\mathbf{r}, t)$ for the vacuum state of the field ($\langle a_{\mathbf{k}\lambda}^\dagger(t)a_{\mathbf{k}'\lambda'}(t)\rangle = \langle a_{\mathbf{k}\lambda}(t)a_{\mathbf{k}'\lambda'}(t)\rangle = 0$, $\langle a_{\mathbf{k}\lambda}(t)a_{\mathbf{k}'\lambda'}^\dagger(t)\rangle = \delta_{\mathbf{k},\mathbf{k}'}^3\delta_{\lambda\lambda'}$), assuming a non-magnetic medium:

$$\langle \mathbf{E}^2(\mathbf{r}, t)\rangle = \sum_{\mathbf{k}\lambda} \left(\frac{\mu_0 c^2\hbar\omega_k}{2nn_gV} \right) \to \frac{V}{8\pi^3} \sum_{\lambda=1,2} \int d^3k \left(\frac{\hbar\omega}{2\epsilon_0 nn_gV} \right). \tag{3.4.16}$$

Using

$$\int d^3k = 4\pi \int dk\, k^2 = 4\pi \int d\omega \frac{dk}{d\omega} k^2$$

$$= 4\pi \int d\omega \frac{d}{d\omega}\left(\frac{n\omega}{c} \right) \frac{n^2\omega^2}{c^2} = \frac{4\pi}{c^3} \int d\omega\, n_g n^2\omega^2 \tag{3.4.17}$$

$$\tag{3.4.18}$$

for the k-integration, we obtain, putting $n = n_R$ to remind us that the refractive index has been assumed to be real,

$$\langle \mathbf{E}^2(\mathbf{r}, t)\rangle = \frac{\hbar}{2\pi^2\epsilon_0 c^3} \int_0^\infty d\omega\, n_R(\omega)\omega^3. \tag{3.4.19}$$

Of course, this diverges if we allow all field modes to contribute, no matter how high their frequencies, since $n(\omega) = \sqrt{\epsilon(\omega)/\epsilon_0} \to 1$ as $\omega \to \infty$ (Section 6.6).

In the absence of absorption and any free currents, the (classical) electromagnetic energy density u and the Poynting vector $\mathbf{S} = \mathbf{E} \times \mathbf{H}$ satisfy Poynting's theorem: $\nabla \cdot \mathbf{S} + \partial u/\partial t = 0$, or

$$\frac{\partial u}{\partial t} = -\nabla \cdot (\mathbf{E} \times \mathbf{H}) = \mathbf{E} \cdot \frac{\partial \mathbf{D}}{\partial t} + \mathbf{H} \cdot \frac{\partial \mathbf{B}}{\partial t}. \tag{3.4.20}$$

---

[8] P. W. Milonni, J. Mod. Opt. **42**, 1991 (1995). For a more rigorous analysis of this field quantization procedure, and the extension of the theory to the case of *inhomogeneous and dissipative* media, see A. Drezet, Phys. Rev. A **95**, 023832 (2017), and also L. G. Suttorp and A. J. van Wonderen, Europhys. Lett. **67**, 766 (2004), and references therein.

Consider a narrow band of frequencies around ω, and write the electric field as (the real part of)

$$\mathbf{E}(\mathbf{r}, t) = \mathbf{E}_\omega(\mathbf{r}, t)e^{-i\omega t} = \int_{-\infty}^{\infty} d\Omega \mathbf{e}_\omega(\mathbf{r}, \Omega)e^{-i(\omega+\Omega)t} \qquad (3.4.21)$$

under the assumption that $\mathbf{E}_\omega$ varies slowly in time compared with $\exp(-i\omega t)$. Then,

$$\begin{aligned}
\mathbf{D}(\mathbf{r}, t) &= \int_{-\infty}^{\infty} d\Omega \epsilon(\omega + \Omega)\mathbf{e}_\omega(\mathbf{r}, \Omega)e^{-i(\omega+\Omega)t} \\
&\cong \int_{-\infty}^{\infty} d\Omega \left(\epsilon(\omega) + \Omega\frac{d\epsilon}{d\omega} \right) \mathbf{e}_\omega(\mathbf{r}, \Omega)e^{-i(\omega+\Omega)t} \\
&= \epsilon(\omega)\mathbf{E}(\mathbf{r}, t) + i\frac{d\epsilon}{d\omega}e^{-i\omega t}\frac{\partial \mathbf{E}_\omega}{\partial t} \qquad (3.4.22)
\end{aligned}$$

and

$$\begin{aligned}
\frac{\partial \mathbf{D}}{\partial t} &\cong \epsilon\frac{\partial \mathbf{E}}{\partial t} + \omega\frac{d\epsilon}{d\omega}\frac{\partial \mathbf{E}_\omega}{\partial t}e^{-i\omega t} = \left(\epsilon\frac{\partial \mathbf{E}_\omega}{\partial t} - i\omega\epsilon\mathbf{E}_\omega + \omega\frac{d\epsilon}{d\omega}\frac{\partial \mathbf{E}_\omega}{\partial t} \right)e^{-i\omega t} \\
&= \left[\frac{d}{d\omega}(\epsilon\omega)\frac{\partial \mathbf{E}_\omega}{\partial t} - i\omega\epsilon\mathbf{E}_\omega \right]e^{-i\omega t}. \qquad (3.4.23)
\end{aligned}$$

It then follows that

$$\mathbf{E} \cdot \frac{\partial \mathbf{D}}{\partial t} \cong \frac{1}{4}\frac{d}{d\omega}(\epsilon\omega)\frac{\partial}{\partial t}|\mathbf{E}_\omega|^2, \qquad (3.4.24)$$

and, together with the analogous expression for $\mathbf{H} \cdot \partial\mathbf{B}/\partial t$ and (3.4.20), that the cycle-averaged field energy density at a frequency ω where absorption is negligible is given by (3.4.6), as derived a bit differently and more generally in Section 1.6; recall from the discussion there that (3.4.6) is the total (field plus material) energy density. The factors $d(\epsilon\omega)/d\omega$ and $d(\mu\omega)/d\omega$ are positive for any passive medium, as are the quantities appearing in the square roots in expressions such as (3.4.15).

3.5 Photons and Interference

In Section 2.8.2, we reviewed how Einstein, well before the formal development of quantum theory, concentrated on momentum fluctuations of particles in thermal equilibrium with radiation to deduce that light has both wave and particle attributes. The quantization of the field in Section 3.3 follows essentially the original work of Born, Heisenberg, Jordan, Dirac, and others in 1925–7. As a quantum theory of the electromagnetic field per se, that work has required no substantial modifications, that is, there have been no experiments suggesting anything "wrong" with the theory. Indeed, some of its more (classically) peculiar aspects have, if anything, taken on greater significance over the years.

One thing that was quickly recognized at the advent of field quantization was how naturally it accounts for the "wave–particle duality" of light. Consider the operator for the energy density of the field in free space, which is proportional to the square of the electric field operator. In the case of a single mode of the field, we obtain, from (3.3.4),

$$\mathbf{E}^2(\mathbf{r}, t) = \frac{\hbar\omega}{2\epsilon_0}\{|\mathbf{A}_0(\mathbf{r})|^2[a^\dagger(t)a(t) + a(t)a^\dagger(t)] - \mathbf{A}_0^2(\mathbf{r})a(t)a(t)$$
$$+ \mathbf{A}_0^*(\mathbf{r})^2 a^\dagger(t)a^\dagger(t)\}. \tag{3.5.1}$$

The single-mode field Hamiltonian is $H = \hbar\omega(a^\dagger a + 1/2)$, and, in the absence of any sources, the Heisenberg equation of motion for the operator $a(t)$ is

$$i\hbar\dot{a} = [a, H] = \hbar\omega(aa^\dagger a - a^\dagger aa) = \hbar\omega(aa^\dagger - a^\dagger a)a = \hbar\omega a, \tag{3.5.2}$$

and, therefore, $\dot{a} = -i\omega a$, and

$$a(t) = a(0)\exp(-i\omega t) \equiv a\exp(-i\omega t), \tag{3.5.3}$$

consistent, of course, with the *classical* equation given in (3.2.7) for the time dependence of the field mode amplitude. Therefore,

$$\mathbf{E}^2(\mathbf{r}, t) = \frac{\hbar\omega}{2\epsilon_0}\{|\mathbf{A}_0(\mathbf{r})|^2[a^\dagger a + aa^\dagger] - \mathbf{A}_0^2(\mathbf{r})aae^{-2i\omega t} + \mathbf{A}_0^*(\mathbf{r})^2 a^\dagger a^\dagger e^{2i\omega t}\}. \tag{3.5.4}$$

The expectation value of this operator when the field (harmonic oscillator) is in an eigenstate $|n\rangle$, that is, a state with precisely n photons, follows from the expectation values

$$\langle n|a^\dagger a + aa^\dagger|n\rangle = \langle n|[2a^\dagger a + 1]|n\rangle = 2n + 1 \tag{3.5.5}$$

and $\langle n|aa|n\rangle = \langle n|a^\dagger a^\dagger|n\rangle = 0$, which is to say that the expectation value of the field energy density at $\mathbf{r}$ is

$$\langle u(\mathbf{r})\rangle = \epsilon_0\langle n|\mathbf{E}^2(\mathbf{r}, t)|n\rangle = \hbar\omega|\mathbf{A}_0(\mathbf{r})|^2(n + \frac{1}{2}), \tag{3.5.6}$$

and the expectation value of the field energy over all space is

$$\langle H\rangle = \int d^3r\langle u(\mathbf{r})\rangle = \hbar\omega(n + \frac{1}{2}), \tag{3.5.7}$$

since $\int d^3r|\mathbf{A}_0(\mathbf{r})|^2 = 1$.

The zero-point field energy $\hbar\omega/2$ does not appear to be directly measurable.[9] In a more careful analysis of what we actually measure in photon-counting experiments, for example, zero-point energy does not even appear in our equations (see Appendix C). Such an analysis requires consideration of the detection apparatus, which is not needed in the present discussion. We postpone further consideration of zero-point field energy and simply replace (3.5.6) and (3.5.7) by what we assert for now are the measurable

---

[9] But it *is* possible to measure effects attributable to *differences* in zero-point field energy—the Casimir effect described in Section 7.3 being a prime example.

energy density and total energy for a single mode of the field in an energy eigenstate state $|n\rangle$:

$$\langle u(\mathbf{r})\rangle = \hbar\omega|\mathbf{A}_0(\mathbf{r})|^2 n, \tag{3.5.8}$$

$$\langle H\rangle = \hbar\omega n. \tag{3.5.9}$$

Now, simplify even further and suppose that $n = 1$: $\langle u(\mathbf{r})\rangle = \hbar\omega|\mathbf{A}_0(\mathbf{r})|^2$ and $\langle H\rangle = \hbar\omega$. What this means according to the quantized-field theory is that there is a *probability* $|\mathbf{A}_0(\mathbf{r})|^2$ per unit volume that a photon will be found at point $\mathbf{r}$, while the probability that a photon will be found *somewhere* is unity. A similar conclusion applies when a single-photon field is incident on an opaque screen with (for example) two slits: the classical interference pattern of the transmitted light defines the *probability distribution* for finding the single photon. Where the intensity is high, there is a high probability of finding a photon, and, where it is low, there is a low probability of finding a photon. At a point of complete destructive interference, there is *zero* probability of detecting a photon. When we repeat this single-photon two-slit experiment over and over again, the accumulated photon counts are found to be distributed according to the classical wave interference pattern. Or, if we perform the experiment once with an n-photon field, $n \gg 1$, we observe the entire interference pattern given by classical optics. We might say, as Dirac famously did in his *Principles of Quantum Mechanics*, that each photon interferes only with itself—that photons do not interfere with each other. This certainly seems true in this particular type of interference experiment, whereas, in other types of experiments, the statement can be a source of confusion, as we noted in Section 2.9. But the point here is that the quantized-field theory neatly accounts for the wave–particle duality of light in *all* optical interference experiments that have ever been performed. The interference patterns of classical optics are predicted when the number of photons is large; the quantized-field theory correctly predicts, as far as we know, all interference effects, regardless of how many photons are involved. Moreover, as discussed below, the quantized-field theory allows for states of the field with properties having no counterparts in classical electromagnetic theory.

The buildup of a classical interference pattern from a probability distribution for individual photons has been demonstrated experimentally. The earliest experiments were performed by the twenty-four-year-old Geoffrey Taylor, a grandson of George Boole, in a room of his parents' house. The purpose of the experiments, which were suggested by J. J. Thomson, was to determine whether ordinary diffraction phenomena are modified when light is very dim and perhaps describable in terms of "indivisible units" of energy. In one experiment, the light was so dim that one could say, from today's perspective, that, at any given time, no more than one photon was incident on a photographic plate—the exposure time was about three months, during part of which Taylor was away on a yacht. What he found was an intensity pattern of the same form expected from classical optics. In other words, he (indirectly) observed the buildup of an interference pattern by successive photons ("indivisible units"). This came as a surprise to Thomson.

Taylor's short paper was his first in a long and illustrious career that saw his name attached to a variety of phenomena (e.g., the Rayleigh–Taylor instability).[10] It seems to have been the only work he reported in the field of optics. Among his other contributions was the introduction of correlation functions in the statistical theory of turbulence.[11] Taylor was described in a biographical note written after his death as "a happy man who spent a long life doing what he wanted most to do and doing it supremely well."[12]

In more recent times, it has become fairly straightforward to observe directly the buildup of an interference pattern as photons are counted one at a time. In one such experiment, light from a 1 mW He–Ne laser ($\lambda = 632.8$ nm) was directed through a small hole in a box whose inside walls were painted black. Neutral density filters reduced the power of the light by a factor of $\alpha = 5 \times 10^{-11}$. In photon language, one might say that the attenuated light consisted of

$$\dot{q} = \frac{\alpha \times (\text{laser power})}{hc/\lambda} = 1.5 \times 10^5 \text{ photons/s}, \qquad (3.5.10)$$

corresponding to an average separation $d = c/\dot{q} = 2$ km between photons. Then one could say that at most one photon at a time was incident on three slits housed in the box. After passing through the slits, the light was reflected from a mirror at the far end of the box and directed onto a photon-counting CCD camera. Figure 3.1 shows the observed results for three different acquisition times. The top frame, for instance, was obtained with an acquisition time of $1/30$ s, for which one estimates an expected number q of photons equal to $(1.5 \times 10^5 \text{ photons/s})(1/30 \text{ s})$ times the quantum efficiency of the detector, which was about 0.001 at the 632.8 nm wavelength; thus, $q \approx 5$ for an acquisition time of $1/30$ s.

3.6 Quantum States of the Field and Their Statistical Properties

In this section, we review a few important examples of quantum field states, beginning with the most "classical-like."

3.6.1 Coherent States

From (3.1.17) we see that, for the ground state of the harmonic oscillator, the uncertainty product $\Delta q \Delta p = \hbar/2$, which is the smallest value allowed by the Heisenberg uncertainty relation. States with minimal uncertainty products and having the same

---

[10] G. I. Taylor, Proc. Camb. Phil. Soc. **15**, 114 (1909).

[11] G. I. Taylor, Lond, Math. Soc. Proc. **20**, 196 (1920). This work strongly influenced Norbert Wiener's work on the application of harmonic analysis to Brownian motion; see N. Wiener, *I Am a Mathematician*, MIT Press, Cambridge, Massachusetts, 1970, pp. 36–37. Autocorrelation functions were used earlier by Einstein, in Arch. Sci. Phys. Natur. **37**, 254 (1914). In this paper, Einstein also introduced the concept of a power spectrum and the relation between it and a correlation function that later came to be known as the Wiener–Khintchine theorem (see Section 6.4.1). See, for instance, A. M. Yaglom, IEEE ASSP Mag. **4**, 7 (October, 1987) or the article entitled "The History of Noise" by L. Cohen, IEEE Signal Proc. Mag. **22**, 20 (2005).

[12] G. K. Batchelor, J. Fluid Mech. **173**, 1 (1986), p. 1.

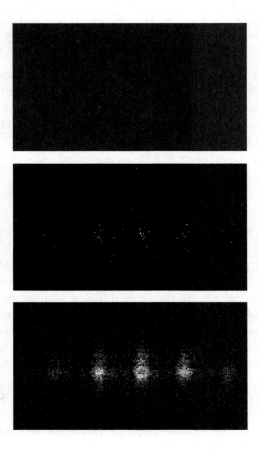

Fig. 3.1 Experimental results for the buildup of a three-slit interference pattern as photons are counted one at a time. The three frames correspond, from top to bottom, to acquisition (or "buildup") times of 1/30 s, 1 s, and 100 s, respectively. From http://phy-page-imac.princeton.edu/~page/single_ photon.html, accessed October 19, 2018.

"width" $\Delta q = \sqrt{\hbar/2\omega}$ as the ground state, are *coherent states* of the harmonic oscillator. We will show in Section 3.12 that all coherent states $|\alpha\rangle$ of the harmonic oscillator may be defined by

$$a|\alpha\rangle = \alpha|\alpha\rangle, \tag{3.6.1}$$

that is, the coherent states are eigenstates of the lowering operator a with (complex) eigenvalues α. There is no restriction on the possible values of α; the ground ("vacuum") state is the particular coherent state with $\alpha = 0$.

From $a = (p - i\omega q)/\sqrt{2\hbar\omega}$ and (3.6.1), it follows that the wave function $\psi_\alpha(q) = \langle q|\alpha\rangle$ describing a coherent state in the coordinate representation satisfies the differential equation

$$\left(\frac{\hbar}{i}\frac{d}{dq} - i\omega q\right)\psi_\alpha(q) = \sqrt{2\hbar\omega}\,\alpha\psi_\alpha(q), \quad \alpha = (\langle p\rangle - i\omega\langle q\rangle)/\sqrt{2\hbar\omega}. \tag{3.6.2}$$

The solution of this equation for $\psi_\alpha(q)$, normalized so that $\int_{-\infty}^{\infty} dq|\psi_\alpha(q)|^2 = 1$, is

$$\psi_\alpha(q) = \left(\frac{\omega}{\pi\hbar}\right)^{1/4} e^{-\omega(q-\langle q\rangle)^2/2\hbar} e^{i\langle p\rangle q/\hbar}. \tag{3.6.3}$$

Comparing this with the ground-state wave function (see (3.1.16)),

$$\psi_0(q) = \left(\frac{\omega}{\pi\hbar}\right)^{1/4} e^{-\omega q^2/2\hbar}, \tag{3.6.4}$$

we see that the coherent state $|\alpha\rangle$ is a "displaced" ground state, the value of q at which $|\psi(q)|^2$ has its maximum value being displaced by $\langle q\rangle$. We therefore have three equivalent ways of defining coherent states: (1) as minimum-uncertainty-product states having the same spatial width as the ground state, (2) as eigenstates of a, and (3) as displaced ground states.

If a harmonic oscillator at $t = 0$ is described by the coherent state $|\alpha\rangle$, then, at times $t \neq 0$,

$$\langle a(t)\rangle = \langle a\rangle e^{-i\omega t} = \alpha e^{-i\omega t} = |\alpha|e^{-i(\omega t+\theta)}, \tag{3.6.5}$$

and therefore

$$\langle q(t)\rangle = i\sqrt{\frac{\hbar}{2\omega}}\left(\langle a(t)\rangle - \langle a^\dagger(t)\rangle\right) = \sqrt{\frac{2\hbar}{\omega}}|\alpha|\sin(\omega t + \theta), \tag{3.6.6}$$

$$\langle p(t)\rangle = \sqrt{\frac{\hbar\omega}{\omega}}|\alpha|\left(\langle a(t)\rangle + \langle a^\dagger(t)\rangle\right) = \sqrt{2\hbar\omega}|\alpha|\cos(\omega t + \theta), \tag{3.6.7}$$

or

$$\langle q(t)\rangle = \langle q(0)\rangle\cos\omega t + \frac{1}{\omega}\langle p(0)\rangle\sin\omega t, \tag{3.6.8}$$

$$\langle p(t)\rangle = \langle p(0)\rangle\cos\omega t - \omega\langle q(0)\rangle\sin\omega t. \tag{3.6.9}$$

This is the same time dependence for the coordinate and the momentum as obtained in the classical theory of the harmonic oscillator. Coherent states of the harmonic oscillator were first introduced in 1926 by Schrödinger in response to remarks by Lorentz to the effect that Schrödinger's wave functions did not appear to predict motions of the type that might be expected from classical mechanics. Schrödinger obtained two-parameter $(\langle q\rangle, \langle p\rangle)$ solutions, of the form shown in (3.6.3), of the Schrödinger equation for the harmonic oscillator—Gaussians having the same width Δq as the ground state, and with expectation values $\langle q\rangle$ and $\langle p\rangle$ given by (3.6.8) and (3.6.9). The fact that the expectation values of $q(t)$ and $p(t)$ in a coherent state oscillate exactly according to the classical equations for the coordinate and momentum, and that $\Delta q \Delta p$ has, for all times, the smallest possible value allowed by quantum theory, suggest that coherent states are the most "classical-like" states of the harmonic oscillator.

If a single-mode electromagnetic field, for example, is described by the coherent state $|\alpha\rangle$, then (see (3.3.4))

$$\langle \mathbf{E}(\mathbf{r}, t) \rangle = i\sqrt{\frac{\hbar}{2\epsilon_0}} \left[\mathbf{A}(\mathbf{r})\alpha(t) - \mathbf{A}^*(\mathbf{r})\alpha^*(t) \right]$$

$$= i\sqrt{\frac{\hbar}{2\epsilon_0}} \left[\mathbf{A}(\mathbf{r})\alpha(0)e^{-i\omega t} - \mathbf{A}^*(\mathbf{r})\alpha^*(0)e^{i\omega t} \right], \qquad (3.6.10)$$

that is, the expectation value of the electric field operator oscillates exactly like the electric field of a classical monochromatic field (and likewise, of course, for the magnetic field). In the case of a state of definite photon number n, in contrast, $\langle n|a|n \rangle = \sqrt{n}\langle n|n-1 \rangle = 0$, and $\langle \mathbf{E}(\mathbf{r}, t) \rangle = 0$.

Schrödinger's coherent-state wave function, given in (3.6.3), shows that a coherent state is a displaced ground state. It is convenient to introduce the unitary *displacement operator*

$$D(\alpha) = e^{\alpha a^\dagger - \alpha^* a} = e^{-|\alpha|^2/2} e^{\alpha a^\dagger} e^{-\alpha^* a}, \qquad (3.6.11)$$

which has the properties

$$D(-\alpha) = D^\dagger(\alpha) = D^{-1}(\alpha) \qquad (3.6.12)$$

and

$$D^{-1}(\alpha) a D(\alpha) = a + \alpha, \quad D^{-1}(\alpha) a^\dagger D(\alpha) = a^\dagger + \alpha^*. \qquad (3.6.13)$$

The second equality of (3.6.11), for example, follows from the Baker–Hausdorff theorem for two operators A and B such that $[A, [A, B]] = [B, [A, B]] = 0$:

$$e^{A+B} = e^A e^B e^{-\frac{1}{2}[A,B]}.[13] \qquad (3.6.14)$$

Then,

$$[a, D(\alpha)] = [a, e^{-\frac{1}{2}|\alpha|^2} e^{\alpha a^\dagger} e^{-\alpha^* a}] = e^{-\frac{1}{2}|\alpha|^2} [a, e^{\alpha a^\dagger}] e^{-\alpha^* a}$$

$$= e^{-\frac{1}{2}|\alpha|^2} \alpha e^{\alpha a^\dagger} e^{-\alpha^* a} = \alpha D(\alpha), \qquad (3.6.15)$$

and, consequently,

$$D(\alpha) a |0\rangle = 0 = (a - \alpha) D(\alpha)|0\rangle, \qquad (3.6.16)$$

or $D(\alpha)|0\rangle = |\alpha\rangle$.[14] In other words, any coherent state $|\alpha\rangle$ is a "displaced" ground state.

---

[13] This can be verified by defining $f(x) = e^{xA} e^{xB}$ (x a c-number) and differentiating with respect to x, giving $df/dx = Ae^{xA}e^{xB} + e^{xA}e^{xB}B = (A + e^{xA}Be^{-xA})f(x)$. Then use (2.2.18), and solve the resulting differential equation with the initial condition $f(0) = 1$.

[14] We use here the commutation relation $[a, e^{\alpha a^\dagger}] = \alpha e^{\alpha a^\dagger}$, a special case of the general identity $[a, f(a, a^\dagger)] = \partial f/\partial a^\dagger$ satisfied by any function $f(a, a^\dagger)$ that can be expressed as a power series in a and $a^\dagger$.

For reasons stated earlier, states $|n\rangle$ of definite photon number—*Fock states*—form a complete set. In particular, we can express a coherent state in terms of Fock states:

$$|\alpha\rangle = \sum_{n=0}^{\infty} c_n |n\rangle, \tag{3.6.17}$$

with

$$
\begin{aligned}
c_n &= \langle n|\alpha\rangle = \langle n|D(\alpha)|0\rangle = e^{-|\alpha|^2/2} \langle n|e^{\alpha a^\dagger} e^{-\alpha^* a}|0\rangle \\
&= e^{-|\alpha|^2/2} \langle n|e^{\alpha a^\dagger}|0\rangle = e^{-|\alpha|^2/2} \sum_{m=0}^{\infty} \frac{\alpha^m}{m!} \langle n|(a^\dagger)^m|0\rangle \\
&= e^{-|\alpha|^2/2} \sum_{m=0}^{\infty} \frac{\alpha^m}{m!} \sqrt{m!} \langle n|m\rangle = \frac{\alpha^n}{\sqrt{n!}} e^{-|\alpha|^2/2},
\end{aligned}
\tag{3.6.18}
$$

that is,

$$|\alpha\rangle = e^{-|\alpha|^2/2} \sum_{n=0}^{\infty} \frac{\alpha^n}{\sqrt{n!}} |n\rangle, \tag{3.6.19}$$

since $(a^\dagger)^m|0\rangle = \sqrt{m!}|m\rangle$, and $\langle n|m\rangle = \delta_{nm}$. It follows that different coherent states are not orthogonal:

$$
\begin{aligned}
\langle \beta|\alpha\rangle &= e^{-(|\alpha|^2+|\beta|^2)/2} \sum_{m=0}^{\infty} \sum_{n=0}^{\infty} \frac{(\beta^*)^m}{\sqrt{m!}} \frac{\alpha^n}{\sqrt{n!}} \langle m|n\rangle \\
&= e^{-(|\alpha|^2+|\beta|^2)/2} \sum_{n=0}^{\infty} \frac{(\alpha\beta^*)^n}{n!} \\
&= e^{-(|\alpha|^2+|\beta|^2)/2} e^{\alpha\beta^*},
\end{aligned}
\tag{3.6.20}
$$

and

$$|\langle \beta|\alpha\rangle|^2 = e^{-|\alpha-\beta|^2}. \tag{3.6.21}$$

The coherent states are easily shown to form a complete set:

$$
\begin{aligned}
\int d^2\alpha |\alpha\rangle\langle\alpha| &= \int d^2\alpha \, e^{-|\alpha|^2} \sum_{m=0}^{\infty} \sum_{n=0}^{\infty} \frac{(\alpha^*)^m}{\sqrt{m!}} \frac{\alpha^n}{\sqrt{n!}} |m\rangle\langle n| \\
&= \sum_{m=0}^{\infty} \sum_{n=0}^{\infty} \int_0^\infty dr \, r \, e^{-r^2} \frac{r^{m+n}}{\sqrt{m!n!}} |m\rangle\langle n| \int_0^{2\pi} d\phi \, e^{i(n-m)\phi} \\
&= 2\pi \sum_{n=0}^{\infty} \frac{1}{n!} |n\rangle\langle n| \int_0^\infty dr \, r^{2n+1} e^{-r^2} = \pi \sum_{n=0}^{\infty} |n\rangle\langle n| = \pi,
\end{aligned}
\tag{3.6.22}
$$

or

$$\frac{1}{\pi} \int d^2\alpha |\alpha\rangle\langle\alpha| = 1. \tag{3.6.23}$$

An arbitrary state $|\psi\rangle$ can therefore be expressed as

$$|\psi\rangle = \frac{1}{\pi} \int d^2\alpha \langle \alpha|\psi\rangle |\alpha\rangle. \qquad (3.6.24)$$

In particular, a coherent state $|\beta\rangle$ can be expressed as

$$|\beta\rangle = \frac{1}{\pi} \int d^2\alpha \langle \alpha|\beta\rangle |\alpha\rangle = \frac{1}{\pi} \int d^2\alpha e^{-|\alpha|^2/2 - |\beta|^2/2} e^{\beta\alpha^*} |\alpha\rangle. \qquad (3.6.25)$$

Thus, a coherent state can be written as a superposition of other coherent states. In this sense, coherent states are said to be *overcomplete*.

From (3.6.19),

$$\langle \alpha|\psi\rangle = \langle \psi|\alpha\rangle^* = e^{-|\alpha|^2/2} \sum_{n=0}^{\infty} \frac{(\alpha^*)^n}{\sqrt{n!}} \langle n|\psi\rangle \qquad (3.6.26)$$

for any state vector $|\psi\rangle$. Writing $|\psi\rangle$ as an expansion in Fock states, we have

$$|\psi\rangle = \sum_{n=0}^{\infty} d_n |n\rangle = \sum_{n=0}^{\infty} \langle n|\psi\rangle |n\rangle, \qquad (3.6.27)$$

and

$$\langle \alpha|\psi\rangle = e^{-|\alpha|^2/2} \sum_{n=0}^{\infty} \frac{(\alpha^*)^n}{\sqrt{n!}} d_n = e^{-|\alpha|^2/2} f(\alpha^*), \qquad (3.6.28)$$

$$f(\alpha^*) = \sum_{n=0}^{\infty} d_n \frac{\alpha^{*n}}{\sqrt{n!}}. \qquad (3.6.29)$$

We can therefore express (3.6.24) as

$$|\psi\rangle = \frac{1}{\pi} \int d^2\alpha e^{-|\alpha|^2/2} f(\alpha^*) |\alpha\rangle. \qquad (3.6.30)$$

The infinite series defining $f(\alpha^*)$ converges for all finite values of $|\alpha|$. In other words, $f(\alpha^*)$ is analytic everywhere in the finite complex plane: it is an *entire function*. On the other hand, it is not difficult to show that if $f(\alpha^*)$ is an entire analytic function and can therefore be expressed as a power series that converges for all finite values of $|\alpha|$, (3.6.30) can be inverted to give

$$f(\alpha^*) = e^{|\alpha|^2/2} \langle \alpha|\psi\rangle. \qquad (3.6.31)$$

The expansion (3.6.30) of an arbitrary state vector $|\psi\rangle$ in terms of coherent states thus establishes a one-to-one correspondence $|\psi\rangle \leftrightarrow f$ between Hilbert-space vectors and entire functions.

According to (3.6.18), the probability $p_n = |c_n|^2$ that exactly n photons will be found in a measurement of the photon number when the field is in a single-mode coherent state $|\alpha\rangle$ is

$$p_n = \frac{|\alpha|^2}{n!} e^{-|\alpha|^2} = \frac{\langle n \rangle^n}{n!} e^{-\langle n \rangle}. \tag{3.6.32}$$

This is a Poisson distribution with mean photon number $\langle n \rangle = |\alpha|^2$ and variance

$$\langle \Delta n^2 \rangle = \langle n^2 \rangle - \langle n \rangle^2 = \langle n \rangle. \tag{3.6.33}$$

One reason why coherent states of the field are important in quantum optics is that laser radiation can often be regarded in effect as a coherent-state field (see Section 3.6.3).

We have considered for simplicity only a single mode of the field. The multimode generalization is straightforward and involves basically just an extension of the notation. For example, a state in which each mode k of the field is described by a coherent state $|\alpha_k\rangle$ is written as

$$|\{\alpha\}\rangle = \prod_k |\alpha_k\rangle. \tag{3.6.34}$$

3.6.2   Classically Prescribed Sources

Quantum properties of light follow from the quantum properties of its sources. Since coherent states are evidently the most classical-like quantum states of light, it might be expected that a source that is well described in a completely classical fashion will produce coherent-state light, and this can, in fact, be shown rather generally.[15] Consider, for example, a Hertzian electric dipole source in an otherwise empty cavity in which only a single mode of the field can oscillate. The Hamiltonian for the system, with the dipole moment $\mathbf{d}(t)$ regarded as a classically prescribed, deterministic function of time, is presumed here to be

$$H = \hbar\omega a^\dagger a - \mathbf{d}(t) \cdot \mathbf{E} = \hbar\omega a^\dagger a - i[F(t)a - F^*(t)a^\dagger] = H_0 - i[F(t)a - F^*(t)a^\dagger]. \tag{3.6.35}$$

Here, ω is the cavity mode frequency, and the c-number $F(t)$ is defined by

$$F(t) = \sqrt{\frac{\hbar\omega}{2\epsilon_0}} \mathbf{A}_0(\mathbf{r}_0) \cdot \mathbf{d}(t), \tag{3.6.36}$$

where $\mathbf{A}_0(\mathbf{r}_0)$ is the cavity mode function (see (3.3.4)) evaluated at the (fixed) position $\mathbf{r}_0$ of the dipole. The state vector $|\psi_I(t)\rangle$ in the interaction picture satisfies (2.1.14). Define

$$|\phi(t)\rangle = e^{(i/\hbar)\int_0^t dt' h_I(t')} |\psi_I(t)\rangle, \tag{3.6.37}$$

which satisfies

[15] R. J. Glauber, Phys. Rev. **84**, 395 (1951).

$$i\hbar\frac{\partial}{\partial t}|\phi(t)\rangle = \left[-h_I(t) + e^{(i/\hbar)\int_0^t dt' h_I(t')} h_I(t) e^{-(i/\hbar)\int_0^t dt' h_I(t')}\right]|\phi(t)\rangle. \qquad (3.6.38)$$

From the definition given in (2.1.12), the interaction Hamiltonian in the interaction picture is

$$\begin{aligned}
h_I(t) &= -i[F(t)U_0^\dagger(t)aU_0(t) - F^*(t)U_0^\dagger(t)a^\dagger U_0(t)] \\
&= -i[F(t)ae^{-i\omega t} - F^*(t)a^\dagger e^{i\omega t}],
\end{aligned} \qquad (3.6.39)$$

since $U_0^\dagger(t)aU_0(t)$ is just the Heisenberg operator $ae^{-i\omega t}$ whose time evolution is governed by the unperturbed Hamiltonian H_0. Now,

$$[h_I(t'), h_I(t'')] = -2i\text{Im}\left[F^*(t')F^{(}t'')e^{i\omega(t'-t'')}\right], \qquad (3.6.40)$$

and, from the general operator formula (2.2.18),

$$\begin{aligned}
e^{(i/\hbar)\int_0^t dt' h_I(t')} h_I(t) e^{-(i/\hbar)\int_0^t dt'' h_I(t'')} &= h_I(t) + (i/\hbar)\int_0^t dt'[h_I(t'), h_I(t)] \\
&= h_I(t) + f(t),
\end{aligned} \qquad (3.6.41)$$

where

$$f(t) = \frac{2}{\hbar}\text{Im}\int_0^t dt' F(t)F^*(t')e^{i\omega(t'-t)} \qquad (3.6.42)$$

is an ordinary (c-number) function of time. Then,

$$i\hbar\frac{\partial}{\partial t}|\phi(t)\rangle = f(t)|\phi(t)\rangle, \qquad (3.6.43)$$

$$|\phi(t)\rangle = e^{-(i/\hbar)\int_0^t dt' f(t')}|\phi(0)\rangle = e^{-i\theta(t)}|\phi(0)\rangle = e^{-i\theta(t)}|\psi_I(0)\rangle, \qquad (3.6.44)$$

$$\begin{aligned}
|\psi_I(t)\rangle &= e^{-i\theta(t)} e^{-(i/\hbar)\int_0^t dt' h_I(t')}|\psi_I(0)\rangle = e^{-i\theta(t)} e^{\beta(t)a^\dagger - \beta^*(t)a}|\psi_I(0)\rangle \\
&= e^{-i\theta(t)} D[\beta(t)]|\psi(0)\rangle,
\end{aligned} \qquad (3.6.45)$$

where D is the displacement operator defined by (3.6.11), and

$$\beta(t) = \frac{1}{\hbar}\int_0^t dt' F^*(t')e^{i\omega t'}. \qquad (3.6.46)$$

If the field is initially in the vacuum state $|0\rangle$ of no photons, therefore,

$$|\psi_I(t)\rangle = e^{-i\theta(t)} D[\beta(t)]|0\rangle = e^{-i\theta(t)}|\beta(t)\rangle, \qquad (3.6.47)$$

where $|\beta(t)\rangle$ is a displaced ground state, that is, a *coherent state*.

The Schrödinger-picture state vector is

$$|\psi(t)\rangle = U_0(t)|\psi_I(t)\rangle = e^{-i\Theta(t)} \sum_{n=0}^{\infty} \frac{\beta^n(t)}{\sqrt{n!}} e^{-|\beta(t)|^2/2} U_0(t)|n\rangle$$

$$= e^{-i\Theta(t)} e^{-|\beta(t)|^2/2} \sum_{n=0}^{\infty} \frac{\beta^n(t)}{\sqrt{n!}} e^{-in\omega t}|n\rangle = e^{-i\Theta(t)}|\beta(t)e^{-i\omega t}\rangle. \quad (3.6.48)$$

Thus, the field from a classically prescribed source is in a coherent state, $|\alpha\rangle = |\beta(t)e^{-i\omega t}\rangle$. $\beta(t)e^{-i\omega t}$ is just the expectation value of $a(t)$: from the Heisenberg equation of motion,

$$\dot{a}(t) = -i\omega a(t) + \frac{1}{\hbar}F^*(t), \quad (3.6.49)$$

which follows from the Hamiltonian (3.6.35),

$$\langle a(t)\rangle = \langle a(0)\rangle e^{-i\omega t} + \frac{1}{\hbar}\int_0^t dt' F^*(t')e^{i\omega(t'-t)}, \quad (3.6.50)$$

which is $\beta(t)e^{-i\omega t}$ for the initial state $|0\rangle$ of the field ($\langle a(0)\rangle = \langle 0|a(0)|0\rangle = 0$). The expectation value of the electric field produced by the classical electric dipole moment $\mathbf{d}(t)$ is

$$\langle \mathbf{E}(\mathbf{r},t)\rangle = i\sqrt{\frac{\hbar\omega}{2\epsilon_0}} \left[\mathbf{A}_0(\mathbf{r})\langle a(t)\rangle - \mathbf{A}_0^*(\mathbf{r})\langle a^\dagger(t)\rangle\right]$$

$$= \frac{i\omega}{2\epsilon_0}\mathbf{A}_0(\mathbf{r})e^{-i\omega t}\mathbf{A}_0^*(\mathbf{r}_0) \cdot \int_0^t dt'\mathbf{d}(t')e^{i\omega t'} + \text{complex conjugate}, \quad (3.6.51)$$

which has exactly the form of the electric field calculated using classical electromagnetic theory. The photon probability distribution associated with the field at time t is

$$p_n(t) = |\langle n|\psi(t)\rangle|^2 = \frac{1}{n!}|\beta(t)|^{2n}e^{-|\beta(t)|^2/2}. \quad (3.6.52)$$

This or any other model in which a source is treated classically but the field is treated quantum mechanically cannot be taken too far, since a prescribed time evolution for the source does not allow for the back reaction of the quantized field on the source, that is, for the radiation reaction field. The model just described, for example, allows photon emission to occur without the source losing energy, because the source experiences no radiation reaction.

Exercise 3.6: (a) Suppose the field in our model with a classical source is not in its vacuum state $|0\rangle$ at $t = 0$ but in a coherent state $|\alpha\rangle$. What is the field state at times $t > 0$? (b) Does the conclusion that a classically prescribed source produces a coherent state of the field depend on our point-dipole model, or does it apply more generally to an extended source? (c) Using the Hamiltonian approach of Section 3.2, show that the right side of (3.6.51) is indeed the electric field given by classical electromagnetic theory.

3.6.3 Laser Light

The field from a laser operating steadily on a single resonator mode has a very stable amplitude but a fluctuating phase. The nearly constant amplitude can be understood by assuming that the rate of amplification R_A of the field by stimulated emission is proportional to the difference $p_2 - p_1$ of the upper- and lower-state probabilities of the lasing transition and varies with the intensity I in the resonator as $R_A \propto (1 + I/I_{\text{sat}})^{-1}$.[16] The amplification rate balances the attenuation rate due to output coupling of the radiation from the resonator and other effects, implying that the intensity and the electric field amplitude are constant. In addition to amplification due to stimulated emission, however, there is spontaneous emission which, while typically adding very little to the field amplitude, causes the phase to fluctuate. The laser field can be described, in effect, as if it has a fixed amplitude $|\alpha|$ like a coherent state, but with a randomly varying phase $\theta(t)$: $\alpha = |\alpha| \exp[i\theta(t)]$. From (3.6.19), the photon number distribution for such a state is

$$p_n = |\langle n|\alpha\rangle|^2 = \frac{|\alpha|^2}{n!} e^{-|\alpha|^2}. \qquad (3.6.53)$$

This Poisson distribution with mean photon number $\langle n \rangle = |\alpha|^2$ is identical to that for a coherent state $|\alpha\rangle$, and has been well demonstrated in photon–counting experiments with laser radiation.

So because of spontaneous emission, a laser cannot produce a coherent state of the field. However, as discussed in Section 6.9, the phase diffusion can ideally be slow enough that laser radiation can be treated, in effect, as a coherent-state field with a constant α.[17]

The Poisson distribution for the photon number probability p_n of a single-mode laser in steady-state operation can be derived as follows.[18] Let $P_{1,n}$ denote the joint

---

[16] Recall equations (2.7.15) and (2.7.16).

[17] Perhaps the first suggestion that the intensity distribution of laser radiation could be very different from the Gaussian distribution of chaotic sources came from M. J. E. Golay, Proc. IRE **49**, 958 (1961), who also remarked that for laser radiation a "substantial random drift in phase may occur."

[18] See, for instance, *Sargent, Scully, and Lamb*; *Scully and Zubairy*; or *Loudon*. We follow closely the treatment in *Loudon*.

probability that (1) an atom is in the lower-energy Level 1 of the lasing transition and (2) there are n photons in the laser cavity, and let $P_{2,n}$ denote the joint probability that (1) an atom is in the higher-energy Level 2 and (2) there are n photons in the cavity. The rate of change of $P_{2,n}$ due to photon absorption and emission is

$$\dot{P}_{2,n} = -R(n+1)P_{2,n} + R(n+1)P_{1,n+1}. \tag{3.6.54}$$

R is the rate of spontaneous emission into the single field mode, and Rn is the rate of absorption and stimulated emission in the presence of n photons. The first term on the right is the rate of decrease of $P_{2,n}$ due to stimulated and spontaneous emission from atoms in Level 2. The second term is the rate of increase of $P_{2,n}$ due to absorption by atoms in Level 1 when there are $n+1$ photons in the field; this absorption reduces the photon number from $n+1$ to n as an atom makes a transition from Level 1 to Level 2, thereby increasing $P_{2,n}$. For $\dot{P}_{1,n+1}$, similarly,

$$\dot{P}_{1,n+1} = -R(n+1)P_{1,n+1} + R(n+1)P_{2,n}. \tag{3.6.55}$$

The first term is the rate of decrease of $P_{1,n+1}$ due to absorption when there are $n+1$ photons in the field and an atom is in Level 1, and the second term is the rate of increase of $P_{1,n+1}$ by stimulated and spontaneous emission from an atom in Level 2 when there are n photons.

Equations (3.6.54) and (3.6.55) are obviously not the whole story. The upper level, Level 2, must be continuously excited if there is to be steady-state lasing. We model this pumping by adding a term rp_n to the right side of (3.6.54). This assumes that Level 2 is populated at the expense of populations in other energy levels not explicitly included in our model, and therefore that the rate of increase of $P_{2,n}$ is simply proportional to the probability p_n that there n photons in the field, irrespective of the populations of Levels 1 and 2. We must also allow for the depletion of the populations of Levels 1 and 2 by processes other than absorption and emission in the laser field. We denote the rates at which Levels 1 and 2 are depleted by these other processes by γ_1 and γ_2, respectively. With these modifications, we replace (3.6.54) and (3.6.55) by

$$\dot{P}_{2,n} = rp_n - R(n+1)P_{2,n} + R(n+1)P_{1,n+1} - \gamma_2 P_{2,n} \tag{3.6.56}$$

and

$$\dot{P}_{1,n+1} = -R(n+1)P_{1,n+1} + R(n+1)P_{2,n} - \gamma_1 P_{1,n+1}, \tag{3.6.57}$$

respectively. The steady-state solutions are then

$$P_{2,n} = \frac{rp_n[R(n+1)+\gamma_1]}{R(n+1)(\gamma_1+\gamma_2)+\gamma_1\gamma_2}, \tag{3.6.58}$$

$$P_{1,n+1} = \frac{rp_n R(n+1)}{R(n+1)(\gamma_1+\gamma_2)+\gamma_1\gamma_2}. \tag{3.6.59}$$

We also write an equation for the rate of change of the photon number probability p_n due to emission and absorption by the $\mathcal{N}$ active atoms of the laser medium:

$$\dot{p}_n^{\text{rad}} = \mathcal{N}R[P_{1,n+1}(n+1) - P_{1,n}n] + \mathcal{N}R[P_{2,n-1}n - P_{2,n}(n+1)]. \tag{3.6.60}$$

The two terms give the rates of change of p_n caused by emission and absorption by atoms in Levels 1 and 2, respectively. p_n will also change as photons escape from the laser cavity: a photon loss rate $dn/dt = -Cn$, where C is a constant characteristic of the particular cavity, implies that

$$\dot{p}_n^{\text{loss}} = -Cnp_n + C(n+1)p_{n+1} \tag{3.6.61}$$

due to cavity loss. In steady-state lasing, $\dot{p}_n^{\text{rad}} + \dot{p}_n^{\text{loss}} = 0$. From this condition, together with the expressions (3.6.58) and (3.6.59) for $P_{2,n}$ and $P_{1,n+1}$,

$$(\mathcal{R}_n + Cn)p_n = \mathcal{R}_{n-1}p_{n-1} + C(n+1)p_{n+1}, \tag{3.6.62}$$

$$\mathcal{R}_n \equiv \frac{\mathcal{N}Rr\gamma_1(n+1)}{R(n+1)(\gamma_1 + \gamma_2) + \gamma_1\gamma_2}. \tag{3.6.63}$$

For a steady-state photon number n such that

$$n \gg \frac{1}{R}\frac{\gamma_1\gamma_2}{\gamma_1 + \gamma_2}, \tag{3.6.64}$$

$$\mathcal{R}_n \cong \mathcal{R}_{n-1} \equiv \mathcal{R}, \tag{3.6.65}$$

and we can replace (3.6.62) by

$$(\mathcal{R} + Cn)p_n = \mathcal{R}p_{n-1} + C(n+1)p_{n+1}. \tag{3.6.66}$$

Assuming a solution of the form

$$p_n = p_0\frac{x^n}{n!}, \tag{3.6.67}$$

we obtain from (3.6.66) a quadratic equation for x, from which we deduce $x = \mathcal{R}/C$. Imposing the requirement that $\sum_{n=0}^{\infty} p_n = 1$, we identify $p_0 = e^{-\mathcal{R}/C}$, and, finally,

$$p_n = e^{-\mathcal{R}/C}\frac{(\mathcal{R}/C)^n}{n!} = \frac{\langle n\rangle^n}{n!}e^{-\langle n\rangle}. \tag{3.6.68}$$

If, instead of (3.6.64), we use

$$n \ll \frac{1}{R}\frac{\gamma_1\gamma_2}{\gamma_1 + \gamma_2}, \tag{3.6.69}$$

then

$$\mathcal{R}_n \cong \frac{\mathcal{N}Rr}{\gamma_2}(n+1) = K(n+1), \quad \mathcal{R}_{n-1} \cong Kn, \tag{3.6.70}$$

and, from (3.6.62),

$$[K(n+1) + Cn]p_n = Knp_{n-1} + C(n+1)p_{n+1}. \tag{3.6.71}$$

Assume now a solution of the form

$$p_n = p_0 x^n. \tag{3.6.72}$$

Then, (3.6.71) leads to a quadratic equation for x, from which we deduce $x = K/C$ and, using $\sum_{n=0}^{\infty} p_n = 1$, we obtain $p_0 = 1 - K/C$ and

$$p_n = \left(1 - \frac{K}{C}\right)\left(\frac{K}{C}\right)^n. \tag{3.6.73}$$

Since

$$\langle n \rangle = \sum_{n=0}^{\infty} np_n = \frac{K}{C}\frac{1}{1 - K/C}, \tag{3.6.74}$$

p_n has the Bose–Einstein form

$$p_n = \frac{\langle n \rangle^n}{[\langle n \rangle + 1]^{n+1}}. \tag{3.6.75}$$

Note that (3.6.73) only makes sense if $K < C$, which means that the rate at which photons in the cavity are generated by stimulated emission is less than the rate at which they are lost. Since $K = C$ is the gain–loss threshold condition for laser oscillation, the Bose–Einstein distribution (3.6.75) applies when the laser is operating below threshold, whereas the Poisson distribution (3.6.68) applies when the laser is operating well above threshold.

3.6.4 Squeezed States

The coherent states of the harmonic oscillator have the smallest possible uncertainty product $\Delta q \Delta p$ and the same width Δq as the ground state. Of course, it is possible to satisfy the minimum uncertainty product condition $\Delta q \Delta p = \hbar/2$ with different values of Δq and Δp. For example, Δq can be "squeezed" below its ground-state value while Δp is correspondingly larger and such that $\Delta q \Delta p \geq \hbar/2$.

Consider a single mode of the field and choose the mode function $\mathbf{A}_0(\mathbf{r})$ such that the electric field operator is

$$\mathbf{E}(\mathbf{r}, t) = i\mathbf{E}_0(\mathbf{r})(ae^{-i\omega t} - a^\dagger e^{i\omega t}) = 2\mathbf{E}_0(X_1 \cos\omega t + X_2 \sin\omega t). \tag{3.6.76}$$

We defined the dimensionless "coordinate" and "momentum" operators

$$X_1 = \frac{i}{2}(a - a^\dagger), \quad X_2 = \frac{1}{2}(a + a^\dagger) \tag{3.6.77}$$

satisfying

$$[X_1, X_2] = \frac{i}{2}, \tag{3.6.78}$$

and therefore, from the general uncertainty relation (3.12.6),

$$\Delta X_1 \Delta X_2 \geq \frac{1}{4}. \tag{3.6.79}$$

The field is in a *squeezed state* if

$$\Delta X_i < \frac{1}{2} \quad (i = 1 \text{ or } 2). \tag{3.6.80}$$

For a coherent state $|\alpha\rangle$,

$$\Delta X_1^2 = \langle\alpha|X_1^2|\alpha\rangle - \langle\alpha|X_1|\alpha\rangle^2$$
$$= -\frac{1}{4}\langle\alpha|(aa - 2a^\dagger a - 1 + a^\dagger a^\dagger)|\alpha\rangle + \frac{1}{4}(\alpha - \alpha^*)^2 = 1/4, \tag{3.6.81}$$

and, similarly, $\Delta X_2^2 = 1/4$. So, a coherent state is not a squeezed state.

Based on the heuristic number–phase uncertainty relation $\Delta n \Delta \phi \geq 1/2$, a Fock state $|n\rangle$ might be said to exhibit squeezing in the sense that $\Delta n = 0$.[19] But, from (3.6.77),

$$\Delta X_1^2 = -\frac{1}{4}\langle n|(aa - 2a^\dagger a - 1 + a^\dagger a^\dagger)|n\rangle + \frac{1}{4}\langle n|(a - a^\dagger)|n\rangle^2 = \frac{1}{4}(2n + 1), \tag{3.6.82}$$

and the same result is obtained for ΔX_2^2. A Fock state is therefore not a "quadrature" squeezed state.

A squeezed state in which one quadrature component is squeezed at the expense of the other (see (3.6.80)) and the uncertainty product has its smallest possible value (1/4) allowed by the uncertainty relation is said to be a *squeezed vacuum* state. Figure 3.2 illustrates two other types of squeezed states, *phase squeezed* and *amplitude squeezed*. Phase-squeezed light allows more accurate interferometry than coherent-state light, whereas amplitude-squeezed light allows photodetection with a noise level below the Poissonian, "shot-noise" limit for coherent light.

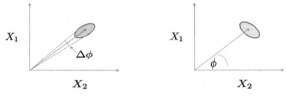

Fig. 3.2 Phase-squeezed (left) and amplitude-squeezed (right) states.

---

[19] For a careful analysis of the concept of phase of a quantized field, see D. T. Pegg and S. M. Barnett, Phys. Rev. A **39**, 1665 (1989); J. Mod. Opt. **44**, 225 (1997).

Squeezed states can be generated in nonlinear optical processes such as parametric down-conversion—a "pump" laser beam of frequency ω_3 is incident on a crystal and results in the generation of fields of frequency ω_1 and ω_2 such that $\omega_3 = \omega_1 + \omega_2$.[20] In the degenerate case we will consider here, $\omega_1 = \omega_2 = \omega$ and $\omega_3 = 2\omega$, and, in the simplest description of down-conversion, we take the interaction Hamiltonian to be

$$H_I = -\frac{1}{2} i\hbar (C a_3 a^{\dagger 2} - C^* a_3^\dagger a^2), \qquad (3.6.83)$$

that is, the annihilation (creation) of a pump photon is accompanied by the creation (annihilation) of two photons of frequency ω. The constant C is proportional to the nonlinear susceptibility characterizing the crystal. If the fraction of the pump beam's power that is down-converted is very small, as is most often the case, it is reasonable as a first approximation to ignore any depletion of the pump beam and to regard it as a classically prescribed field, which we will assume to be constant. In this approximation, we replace $C a_3(t)$ by a numerical constant Ω, which we take for simplicity to be real, times $\exp(-i\omega_3 t)$:

$$H_I = -\frac{1}{2} i\hbar\Omega (a^{\dagger 2} e^{-2i\omega t} - a^2 e^{2i\omega t}). \qquad (3.6.84)$$

The complete Hamiltonian describing the time evolution of the down-converted light is then

$$H = \hbar\omega a^\dagger a - \frac{1}{2} i\hbar\Omega (a^{\dagger 2} e^{-2i\omega t} - a^2 e^{2i\omega t}). \qquad (3.6.85)$$

The frequency Ω is proportional to the electric field strength of the pump as well as the nonlinear susceptibility. With this Hamiltonian and the commutator $[a, a^{\dagger 2}] = 2a^\dagger$, the Heisenberg equation of motion for $a(t)$ is

$$\dot{a} = \frac{1}{i\hbar}[a, H] = -i\omega a - \Omega a^\dagger e^{2i\omega t}. \qquad (3.6.86)$$

The solution of this equation is

$$a(t) = \big[a(0)\cosh\Omega t - a^\dagger(0)\sinh\Omega t\big] e^{-i\omega t}, \qquad (3.6.87)$$

and it is easily checked that $[a(t), a^\dagger(t)] = [a(0), a^\dagger(0)] = 1$.

The electric field operator for the down-converted light is

$$\mathbf{E}(t) = i\mathbf{E}_0[a(t) - a^\dagger(t)] = 2\mathbf{E}_0\big[X_1 \cos\omega t + X_2 \sin\omega t\big], \qquad (3.6.88)$$

where

$$X_1 = \frac{i}{2} e^{\Omega t}[a(0) - a^\dagger(0)], \qquad (3.6.89)$$

$$X_2 = \frac{1}{2} e^{-\Omega t}[a(0) + a^\dagger(0)], \qquad (3.6.90)$$

[20] D. Stoler, Phys. Rev. D **1**, 3217 (1970); Phys. Rev. Lett. **33**, 1397 (1974); L. A. Wu, H. J. Kimble, J. L. Hall, and H. Wu, Phys. Rev. Lett. **57**, 2520 (1986).

and

$$X_1^2 = -\frac{1}{4}e^{2\Omega t}\left[a(0)a(0) - a(0)a^\dagger(0) - a^\dagger(0)a(0) + a^\dagger(0)a^\dagger(0)\right], \tag{3.6.91}$$

$$X_2^2 = \frac{1}{4}e^{-2\Omega t}\left[a(0)a(0) + a(0)a^\dagger(0) + a^\dagger(0)a(0) + a^\dagger(0)a^\dagger(0)\right]. \tag{3.6.92}$$

Suppose that, at $t = 0$, there is no down-converted light, that is, that the initial state of the down-converted light is the vacuum state $|0\rangle$. Then,

$$\langle X_1 \rangle = \langle X_2 \rangle = 0 \tag{3.6.93}$$

and

$$\langle X_1^2 \rangle = \frac{1}{4}e^{2\Omega t}\langle 0|a(0)a^\dagger(0)|0\rangle = \frac{1}{4}e^{2\Omega t}, \tag{3.6.94}$$

$$\langle X_2^2 \rangle = \frac{1}{4}e^{-2\Omega t}\langle 0|a(0)a^\dagger(0)|0\rangle = \frac{1}{4}e^{-2\Omega t}, \tag{3.6.95}$$

$$\Delta X_1^2 = \frac{1}{4}e^{2\Omega t}, \qquad \Delta X_2^2 = \frac{1}{4}e^{-2\Omega t}, \qquad \Delta X_1 \Delta X_2 = 1/4. \tag{3.6.96}$$

In other words, the field generated in degenerate parametric down-conversion is in a squeezed vacuum state. Squeezing has been observed in various nonlinear optical processes and, to a smaller degree, in some other physical systems, such as, for example, in the displacements of atoms in crystal lattices irradiated by short pulses of light.

Exercise 3.7: How are (3.6.93)–(3.6.96) changed if the down-converted light is initially in a coherent state rather than the vacuum state?

3.6.5   Thermal and Chaotic Radiation

The average number of photons in a single mode of blackbody radiation is

$$\langle n(\omega) \rangle = \frac{1}{e^{\hbar\omega/k_B T} - 1}. \tag{3.6.97}$$

As discussed in Section 2.8, the ratio of the rates of spontaneous and stimulated emission at frequency ω is

$$A/[B\rho(\omega)) = 1/\langle n(\omega)\rangle = e^{\hbar\omega/k_B T} - 1 \tag{3.6.98}$$

in thermal equilibrium at temperature T. The spectrum of solar radiation, for example, is well approximated by the Planck distribution with $T = 5800$ K; for this temperature and a wavelength of 600 nm, the ratio (3.6.98) is about 62. We estimate therefore that about 98% of the light from the Sun is due to spontaneous radiative processes.

Because $\langle n(\omega) \rangle << 1$, the fluctuations in the photon number at these temperatures and wavelengths are associated predominantly with the particle term in the Einstein fluctuation formula given in (2.8.32).

For the field at frequency ω in thermal equilibrium at temperature T, the photon occupation probability p_n is easily deduced from the Boltzmann distribution and the fact that the energy of the state with n photons is $E_n = (n + 1/2)\hbar\omega$:

$$p_n = e^{-E_n/k_B T} / \sum_{n=0}^{\infty} e^{-E_n/k_B T} = e^{-(n+\frac{1}{2})\hbar\omega/k_B T} / \sum_{n=0}^{\infty} e^{-(n+\frac{1}{2})\hbar\omega/k_B T}$$

$$= e^{-n\hbar\omega/k_B T} / \sum_{n=0}^{\infty} e^{-n\hbar\omega/k_B T} = e^{-n\hbar\omega/k_B T}(1 - e^{-\hbar\omega/k_B T})$$

$$= \frac{\langle n \rangle^n}{[\langle n \rangle + 1]^{n+1}}, \tag{3.6.99}$$

where $\langle n \rangle = \sum_{n=0}^{\infty} np_n$ is defined by (3.6.97). Thus, for example,

$$\langle n^2 \rangle = \sum_{n=0}^{\infty} n^2 p_n = 2\langle n \rangle^2 + \langle n \rangle \tag{3.6.100}$$

and $\langle \Delta n^2 \rangle = \langle n \rangle^2 + \langle n \rangle$. These results, which we have already discussed in Section 2.8, are of course well-understood consequences of Bose–Einstein statistics for photons in thermal equilibrium.

From a classical perspective, we might think of radiation from a source in thermal equilibrium as being produced by a large collection of dipole oscillators that are continuously absorbing and emitting radiation. As a result of collisions in a gas, for example, the dipole oscillations keep getting interrupted and reoriented, such that the radiation from the gas is homogeneous, isotropic, unpolarized, and well described for any frequency and polarization as the sum of a large number of fields whose phases vary randomly and independently over the interval $[0, 2\pi]$. We therefore assume that the electric field for a single frequency, propagation direction, and polarization can be expressed as

$$E(t) = e^{-i\omega t} \sum_{j=1}^{N} A_j e^{i\phi_j} = (X + iY)e^{-i\omega t}, \tag{3.6.101}$$

where

$$X = \sum_{j=1}^{N} A_j \cos \phi_j, \quad Y = \sum_{j=1}^{N} A_j \sin \phi_j, \tag{3.6.102}$$

$\langle \cos \phi_j \rangle = \langle \sin \phi_j \rangle = \langle \cos \phi_j \sin \phi_j \rangle = 0$, and $\langle \cos \phi_i \cos \phi_j \rangle = \langle \sin \phi_i \sin \phi_j \rangle = \frac{1}{2}\delta_{ij}$. Then,

$$\langle X \rangle = \sum_{j=1}^{N} A_j \langle \cos \phi_j \rangle = 0, \tag{3.6.103}$$

$$\langle X^2 \rangle = \sum_{i=1}^{N} \sum_{j=1}^{N} A_i A_j \langle \cos \phi_i \cos \phi_j \rangle = \frac{1}{2} \sum_{j=1}^{N} A_j^2, \qquad (3.6.104)$$

and, likewise, $\langle Y \rangle = 0$ and $\langle Y^2 \rangle = \langle X^2 \rangle$. The central limit theorem for the sum of a large number of independent random variables implies a Gaussian probability distribution for X:

$$p_X(X) = \frac{1}{\sigma \sqrt{2\pi}} e^{-X^2/2\sigma^2}, \quad \sigma^2 = \langle X^2 \rangle, \qquad (3.6.105)$$

and likewise for the probability distribution $p_Y(Y)$ for Y. And since X and Y are independent random variables, their joint probability distribution $p_{XY}(X,Y)$ is

$$p_{XY}(X,Y) = p_X(X)p_Y(Y) = \frac{1}{2\pi\sigma^2} e^{-(X^2+Y^2)/2\sigma^2}. \qquad (3.6.106)$$

Writing $X = R\cos\theta$, $Y = R\sin\theta$, and defining the joint probability distribution $p_{R\theta}(R,\theta)$ such that

$$p_{R\theta}(R,\theta)dRd\theta = p_{XY}(X,Y)dXdY = p_{XY}(X,Y)RdRd\theta, \qquad (3.6.107)$$

we obtain the probability distribution for R:

$$p_R(R) = \int_0^{2\pi} p_{R\theta}(R,\theta)d\theta = \frac{R}{\sigma^2} e^{-R^2/2\sigma^2}. \qquad (3.6.108)$$

The (cycle-averaged) intensity $I = \frac{1}{2}(X^2 + Y^2) = \frac{1}{2}R^2$, and its probability distribution $p(I)$ satisfies $p(I)dI = p_R(R)dR$; therefore, $p(I) = (1/\sigma^2)e^{-I/\sigma^2}$, or, since $\sigma^2 = \frac{1}{2}\langle R^2 \rangle = \langle I \rangle$,

$$p(I) = \frac{1}{\langle I \rangle} e^{-I/\langle I \rangle} \qquad (3.6.109)$$

is the probability distribution of the total intensity of a large number of fields with random, uncorrelated phases. It follows that

$$\langle I^n \rangle = \int_0^{\infty} I^n p(I)dI = n!\langle I \rangle^n, \qquad (3.6.110)$$

which generalizes to the equation (2.9.1). This classical model thus reproduces the wave term in the Einstein fluctuation formula for thermal radiation.

In fact, this model, based on the central limit theorem, applies not only to radiation from a source in thermal equilibrium but to any radiation produced by a "chaotic" source consisting of a large number of radiators whose fields vary randomly and independently. A fluorescent lamp, for example, can be regarded as a chaotic source. Our derivation of (3.6.109) and (3.6.110) applies to a single mode of radiation from any chaotic source, and, in particular, to a quasi-monochromatic, polarized beam from such a source. For an arbitrary degree of polarization P, $0 \leq P \leq 1$, the intensity distribution is found by similar arguments to be

$$p(I) = \frac{1}{P\langle I \rangle} \left(e^{-(2I/[(1+P)\langle I \rangle])} - e^{-(2I/[(1-P)\langle I \rangle])} \right). \qquad (3.6.111)$$

P is defined as $|I_x - I_y|/(I_x + I_y)$, where I_x and I_y are the intensities of two statistically independent fields propagating in the z direction and polarized in the (orthogonal) x and y directions, respectively.

Exercise 3.8: (a) Show that this definition of the degree of polarization is independent of how the x and y axes are chosen. (b) Show that $p(I) = \int_0^I dI_x p(I_x) p(I - I_x)$ and derive (3.6.111).

We can recover the Bose–Einstein distribution that accounts for both "wave" and "particle" fluctuations by the following argument. We assume first that, if the intensity of the field is I, the average photon number in the field is κI, where κ is a constant. We assume next that the photon probability distribution follows the Poisson distribution characteristic of particle shot noise, that is, that the probability that there are n photons in the field is

$$p_n(I) = \frac{\langle n \rangle^n}{n!} e^{-\langle n \rangle} = \frac{(\kappa I)^n}{n!} e^{-\kappa I} \qquad (3.6.112)$$

if the intensity is I. Finally, we assume that the intensity I has a probability distribution given by (3.6.109), and therefore we define the photon probability distribution as

$$p_n = \int_0^\infty dI P(I) \frac{(\kappa I)^n}{n!} e^{-\kappa I} = \int_0^\infty dI \frac{1}{\langle I \rangle} e^{-I/\langle I \rangle} \frac{(\kappa I)^n}{n!} e^{-\kappa I}, \qquad (3.6.113)$$

which is

$$p_n = \frac{[\kappa \langle I \rangle]^n}{(\kappa \langle I \rangle + 1)^{n+1}} = \frac{\langle n \rangle^n}{(\langle n \rangle + 1)^{n+1}}. \qquad (3.6.114)$$

Thus, by invoking both wave and particle concepts, we obtain the Bose–Einstein distribution (3.6.99). While this derivation is only heuristic, it finds support in the more rigorous derivation of (3.6.114) given in Section 3.8 (see (3.8.22)).

The Bose–Einstein distribution for photons is the thermal equilibrium distribution for *indistinguishable* particles, with no restriction on the number of particles that can occupy a given energy level or on the total number of particles. Because of the absence of any restriction on the number of particles, there is no "chemical potential" characteristic of the more general form of the Bose–Einstein distribution. These properties of photons are, of course, implicit in the derivation leading to (3.6.99); in particular, each energy level E_n in that derivation is characterized only by the number n of photons, with no distinctions among the photons.

The quantum theory of the field accounts naturally for the fact that photons of each field mode are indistinguishable bosons. In the simplest approach to field quantization we have followed, this is simply a consequence of the fact that each mode is a harmonic oscillator: the n-particle (photon) state $|n\rangle$ as expressed by (3.1.15) is symmetric with respect to any permutations of the n particles. By considering the angular momentum of the field, we showed that these bosons have spin 1.

3.6.6　Mandel's Q Parameter

The variance in the single-mode photon number, corresponding in the quantized-field theory to the number operator $n = a^\dagger a$, can be expressed as

$$\langle \Delta n^2 \rangle = \langle a^\dagger a a^\dagger a \rangle - \langle a^\dagger a \rangle^2 = \langle a^\dagger (a^\dagger a + 1) a \rangle - \langle a^\dagger a \rangle^2$$
$$= \langle \Delta n^2 \rangle_{\text{particles}} + \langle \Delta n^2 \rangle_{\text{waves}}, \tag{3.6.115}$$

with

$$\langle \Delta n^2 \rangle_{\text{particles}} = \langle a^\dagger a \rangle = \langle n \rangle \tag{3.6.116}$$

and

$$\langle \Delta n^2 \rangle_{\text{waves}} = \langle a^\dagger a^\dagger a a \rangle - \langle a^\dagger a \rangle^2, \tag{3.6.117}$$

for any state of the field. This separation into "wave" and "particle" parts follows closely the distinction between wave and particle contributions to the Einstein fluctuation formula for thermal radiation (Section 2.8.2). For a thermal radiation field, $\langle \Delta n^2 \rangle_{\text{waves}} = \langle n \rangle^2$, or $\langle n^2 \rangle = 2\langle n \rangle^2 + \langle n \rangle$, the factor 2 implying photon bunching. If the field is described by a coherent state, $\langle \Delta n^2 \rangle_{\text{waves}} = 0$, and $\langle \Delta n^2 \rangle = \langle \Delta n^2 \rangle_{\text{particles}} = \langle n \rangle$. A coherent state therefore exhibits no "wave fluctuations," and so no photon bunching, whereas thermal radiation has both wave and particle fluctuation characteristics.

In the case of a (Fock) state of definite photon number, $\langle \Delta n^2 \rangle = 0$ and therefore, from (3.6.115), $\langle \Delta n^2 \rangle_{\text{waves}} = -\langle n \rangle$. A negative $\langle \Delta n^2 \rangle_{\text{waves}}$ is characteristic of a quantum state of the field that has no classical analog, as discussed in Section 3.8; thus, a Fock state can be said to have no classical analog. The *Mandel Q parameter*,

$$Q = \frac{\langle \Delta n^2 \rangle_{\text{waves}}}{\langle n \rangle} = \frac{\langle \Delta n^2 \rangle - \langle n \rangle}{\langle n \rangle}, \tag{3.6.118}$$

provides a simple measure of nonclassicality, a negative Q being associated with a field considered to be nonclassical (see Section 3.8).[21] Thus, a coherent state has a Q of 0, thermal radiation has a Q of $\langle n \rangle$, and, for a Fock state $Q = -1$, the most negative possible value of Q. Q is negative whenever the photon statistics are sub-Poissonian, that is, whenever $\langle \Delta n^2 \rangle < \langle n \rangle$.

---

[21] L. Mandel, Opt. Lett. **4**, 205 (1979).

3.6.7 Photon Counting

We now consider the probability $P_n(T)$ of counting n photons during some integration time T, given the field intensity incident on a photodetector. By "counting photons," what we really mean is that a photon triggers an event, such as the emission of a photoelectron, and the average number of such events ("clicks") in a time T is equal to, or at least proportional to, the average number of photons incident on the photodetector during the time T. Ideally, the probability distribution for the clicks would faithfully represent the photon probability distribution of the incident field.

We assume that the probability $p(t)\Delta t$ of "counting a photon" in a small time interval Δt is proportional to the cycle-averaged intensity $I(t)$ of the field incident on the detector:

$$p(t)\Delta t = \eta I(t)\Delta t, \qquad (3.6.119)$$

where η depends on the efficiency and other properties of the particular photodetector. We are, for now, taking a semiclassical approach and assuming, among other things, that we can ignore the possibility that the field intensity might vary significantly over the surface of the detector.

Suppose photons are counted in successive time intervals of duration T. The number recorded in each interval will vary, and, after perhaps thousands of such intervals, we will have recorded, in effect, a probability distribution for the number of photons counted in a time T. In our semiclassical description, this probability will depend on the integral of the intensity over the time T, as we now show.

Let $P_n(t)$ be the probability of counting n photons in a time interval $[0, t]$. If Δt in (3.6.119) is chosen to be sufficiently small that the probability of counting more than one photon during the time Δt is negligible,

$$P_n(t + \Delta t) = P_{n-1}(t)p(t)\Delta t + P_n(t)[1 - p(t)\Delta t]. \qquad (3.6.120)$$

That is, the probability of counting n photons in the time interval $[t, t+\Delta t]$ is equal to (1) the probability $P_{n-1}(t)$ of counting $n-1$ photons in the time t, times the probability $p(t)\Delta t$ of counting one more photon in the time Δt, plus (2) the probability $P_n(t)$ of counting n photons in the time t, times the probability $[1 - p(t)\Delta t]$ of not counting any photons in the time Δt. If we take $\Delta t \to 0$, (3.6.120) becomes

$$\frac{dP_n}{dt} = p(t)[P_{n-1}(t) - P_n(t)] = \eta I(t)[P_{n-1}(t) - P_n(t)], \qquad (3.6.121)$$

with the solution

$$P_n(T) = \frac{[X(T)]^n}{n!}e^{-X(T)}, \quad X(T) = \eta \int_0^T I(t')dt', \qquad (3.6.122)$$

for the probability $P_n(T)$ of counting n photons in the time interval from $t = 0$ to $t = T$.

In an actual photon–counting measurement, there will be a large number of time intervals of duration T, each with a different "starting time" t. To account for this, we replace (3.6.122) by

$$P_n(T) = \left\langle \frac{1}{n!} X^n e^{-X} \right\rangle, \tag{3.6.123}$$

where now

$$X = \eta \int_t^{t+T} I(t')dt', \tag{3.6.124}$$

and $\langle \ldots \rangle$ denotes an average over the starting times. The formula (3.6.123) for $P_n(T)$ was originally derived by Mandel.[22] If X has a probability distribution $P(X)$,

$$P_n(T) = \int_0^\infty \frac{1}{n!} X^n e^{-X} P(X)dX, \tag{3.6.125}$$

which generalizes (3.6.113) (or (3.8.22)).

If X is independent of t, (3.6.123) reduces to a Poisson distribution:

$$P_n(T) = \frac{\langle n \rangle^n}{n!} e^{-\langle n \rangle}, \tag{3.6.126}$$

with an average count

$$\langle n \rangle = X = \eta \int_t^{t+T} I(t')dt', \tag{3.6.127}$$

independent of t and proportional to the average number of photons incident on the detector during a time T. If $I(t)$ has the constant value I_0, for example, $\langle n \rangle = \eta I_0 T$. Even in this case of constant intensity, we can only specify a (Poisson) probability distribution for $P_n(T)$; this comes ultimately from the fact that transitions between quantum states of the detector material can only be described probabilistically.

More generally, a Poisson distribution follows, regardless of the actual photon probability distribution p_n, whenever the counting time interval T is sufficiently large, compared with a correlation time τ characterizing the intensity fluctuations, that $X \cong \eta \langle I \rangle T \equiv \langle n \rangle$, where $\langle I \rangle$ is the average intensity in the counting interval:

$$P_n(T) \cong \frac{\langle n \rangle^n}{n!} e^{-\langle n \rangle} = \frac{1}{n!} (\eta \langle I \rangle T)^n e^{-\eta \langle I \rangle T}. \tag{3.6.128}$$

If, instead, the counting interval T is very *short* compared to τ, with τ typically the inverse bandwidth of the incident field, $X \cong \eta I(t)T$, and the probability distribution for X relates directly to the probability distribution $P(I)$ for the intensity I. In this case, (3.6.125) can be replaced by

$$P_n(T) = \int_0^\infty \frac{1}{n!} [\eta I T]^n e^{-\eta I T} P(I)dI. \tag{3.6.129}$$

Such an expression was previously assumed (see (3.6.113)) on intuitive grounds for the case of chaotic radiation.

---

[22] L. Mandel, Proc. Phys. Soc. **72**, 1037 (1958); **74**, 233 (1959). See *Mandel and Wolf (1995)* or *Loudon* for more detailed treatments of photon counting.

In the derivation of (3.6.123), the field was treated classically. In the quantized-field approach we require the field intensity *operator*, which in the plane-wave approximation is

$$I = \frac{c\hbar\omega}{V}a^\dagger a, \qquad (3.6.130)$$

that is, the velocity of light c times the photon number density operator $a^\dagger a/V$ times the photon energy $\hbar\omega$. From the expression (3.3.11) for the free-space electric field operator, we identify the photon annihilation and creation parts of a single mode of the electric field,

$$E^{(+)}(\mathbf{r}, t) = i\sqrt{\frac{\hbar\omega_k}{2\epsilon_0 V}}ae^{-i\omega t}e^{i\mathbf{k}\cdot\mathbf{r}} \qquad (3.6.131)$$

and

$$E^{(-)}(\mathbf{r}, t) = -i\sqrt{\frac{\hbar\omega_k}{2\epsilon_0 V}}a^\dagger e^{i\omega t}e^{-i\mathbf{k}\cdot\mathbf{r}}, \qquad (3.6.132)$$

respectively. In terms of these operators, the intensity operator (3.6.130) is the normally ordered product

$$I(t) = 2\epsilon_0 cE^{(-)}(\mathbf{r}, t)E^{(+)}(\mathbf{r}, t), \qquad (3.6.133)$$

and the semiclassical result (3.6.123) is replaced by

$$P_n(T) = \left\langle : \frac{X^n}{n!}e^{-X} : \right\rangle, \qquad (3.6.134)$$

where the operator

$$X = 2\epsilon_0 c\eta \int_t^{t+T} I(t')dt'.^{23} \qquad (3.6.135)$$

The notation $: A :$ here means that field annihilation and creation operators are to be written in normal order, that is, with all a's appearing to the right of $a^\dagger$'s:

$$: a^\dagger a :\, \equiv a^\dagger a, \quad : aa^\dagger :\, \equiv a^\dagger a, \quad : (a^\dagger a)^2 :\, \equiv a^\dagger a^\dagger aa, \quad \text{and so on.} \qquad (3.6.136)$$

3.7 The Density Operator

In quantum theory, we associate with a measurable dynamical variable a Hermitian operator A.[24] In addition, it is asserted that any single measurement of A will yield one of the (real) eigenvalues a of A. If the eigenvalue a is measured, the system is described immediately afterwards by the eigenstate $|a\rangle$. Prior to the measurement, the state describing the system may be unknown, but if a measurement of A yields a, we can say with certainty that the system is then "in" the eigenstate $|a\rangle$ of A: the measurement has given us new information about the system, such that, after

---

[23] P. L. Kelley and W. H. Kleiner, Phys. Rev. **A136**, 316 (1964); see also *Mandel and Wolf (1995)*, Chapter 14.

[24] But not every Hermitian operator we can define represents an observable!

the measurement, we can say with certainty that the system is in an eigenstate of A. The eigenstates $|a\rangle$ of an *observable* associated with a Hermitian operator A form a complete set, so that any state $|\psi\rangle$ of the system is linearly dependent on these eigenstates:

$$|\psi\rangle = \sum_a c_a |a\rangle \quad (\sum_a |c_a|^2 = 1), \tag{3.7.1}$$

the sum being over all the eigenstates $|a\rangle$. The expectation value of A when the system is described by such a linear superposition is

$$\langle A \rangle_\psi = \langle \psi | A | \psi \rangle = \sum_a |c_a|^2 a. \tag{3.7.2}$$

$|c_a|^2 = |\langle a|\psi\rangle|^2$ is the probability that a measurement of A will result in the eigenvalue a.

When we say that a system is described by a state vector (or wave function) $|\psi\rangle$, it is implied that the only uncertainties in our knowledge of the system are those inherent in quantum theory. $|\psi\rangle$ contains the maximal knowledge about the system allowed by quantum theory. It is presumed that, in theory anyway, sufficiently many measurements have been made on the system to provide us with this maximal knowledge. In general, however, we will not have the maximal knowledge about the system allowed by quantum theory and so will only be able to say that there are certain probabilities p_ψ for the system to be "in" states described by the state vector $|\psi\rangle$. In that case, the expectation value of the observable A is

$$\langle A \rangle = \sum_\psi p_\psi \langle \psi | A | \psi \rangle = \sum_\psi p_\psi \langle A \rangle_\psi. \tag{3.7.3}$$

In general, therefore, we must deal with two types of uncertainty, one being the inescapable, quantum-mechanical uncertainty when the state vector is known precisely, and the other arising from our imprecise knowledge as to what exactly is the state vector. If, as a result of sufficiently many measurements, we can describe the system by a state vector $|\psi\rangle$, then the only uncertainties in our predictions about the system are purely quantum mechanical. A system described by a state vector is said to be in a *pure state*. If our knowledge is insufficient to describe the system by a state vector, and we can only assign probabilities p_ψ to different state vectors, then the uncertainties in our predictions arise from a mixture of the two types of uncertainty. Our description of the system in this case is said to be based on a *mixed state*.

To work with mixed states, we introduce the (Hermitian) *density operator*, a superposition of projection operators:

$$\rho = \sum_\psi p_\psi |\psi\rangle\langle\psi| \quad (|\psi\rangle\langle\psi| = 1), \tag{3.7.4}$$

where the p_ψ, being probabilities, are real numbers satisfying $p_\psi \geq 0$, and $\sum_\psi p_\psi = 1$. We can, of course, represent ρ by using any complete set of states $\{|i\rangle\}$ of the system:

$$\rho = 1\rho1 = \sum_i \sum_j |i\rangle\langle i| \Big(\sum_\psi p_\psi |\psi\rangle\langle\psi| \Big) |j\rangle\langle j| = \sum_\psi p_\psi \langle i|\psi\rangle\langle\psi|j\rangle |i\rangle\langle j|$$

$$= \sum_i \sum_j \rho_{ij} |i\rangle\langle j|, \tag{3.7.5}$$

where

$$\rho_{ij} = \sum_\psi p_\psi \langle i|\psi\rangle\langle\psi|j\rangle \tag{3.7.6}$$

are the elements of the *density matrix* for the particular representation (complete set $\{|i\rangle\}$ of basis states) chosen. Obviously, $\rho_{ii} \geq 0$ in any representation, since both p_ψ and $|\langle i|\psi\rangle|^2$ are positive-definite.

The expectation value of an observable A is expressed in terms of the density operator by

$$\langle A \rangle = \sum_\psi p_\psi \langle\psi|A|\psi\rangle = \sum_\psi p_\psi \sum_i \sum_j \langle\psi|i\rangle\langle i|A|j\rangle\langle j|\psi\rangle$$

$$= \sum_i \sum_j \rho_{ji} A_{ij} = \sum_j (\rho A)_{jj} = \text{Tr}(\rho A). \tag{3.7.7}$$

The trace of the density operator, in any representation, is unity:

$$\text{Tr}\rho = \text{Tr} \sum_\psi p_\psi |\psi\rangle\langle\psi| = \sum_\psi p_\psi \sum_i \langle i|\psi\rangle\langle\psi|i\rangle = \sum_\psi p_\psi \langle\psi|\psi\rangle = \sum_\psi p_\psi = 1. \tag{3.7.8}$$

Another basic property follows from the fact that ρ is Hermitian and therefore can always be diagonalized by a unitary transformation. That is, we can always choose a representation $\{|k\rangle\}$ in which the density matrix is diagonal. In such a representation, the "purity" is

$$\text{Tr}\rho^2 = \sum_k \rho_{kk}^2. \tag{3.7.9}$$

Since $\text{Tr}\rho = 1$, that is, $\sum_k \rho_{kk} = 1$, and $\sum_k \rho_{kk}^2 \leq \Big(\sum_k \rho_{kk} \Big)^2$,

$$\text{Tr}\rho^2 \leq 1. \tag{3.7.10}$$

The equality holds when we have a pure state: $\rho = |\psi\rangle\langle\psi|$ and

$$\text{Tr}\rho^2 = \text{Tr}\Big(|\psi\rangle\langle\psi|\psi\rangle|\psi\rangle \Big) = \text{Tr}\Big(|\psi\rangle\langle\psi| \Big) = \langle\psi|\psi\rangle = 1. \tag{3.7.11}$$

A useful measure of the extent to which a state characterized by a density operator ρ differs from a pure state is the *von Neumann entropy* defined by

$$S = -\text{Tr}(\rho \log \rho) = -\sum_i p_i \log p_i, \tag{3.7.12}$$

where the p_i are the eigenvalues of the matrix ρ. For a pure state, $\rho = |\psi\rangle\langle\psi|$, and $S = 0$, its smallest possible value.

For a single field mode (harmonic oscillator) in thermal equilibrium, for example, the density operator is

$$\rho = Z^{-1} e^{-H/k_B T}, \quad H = \hbar\omega(a^\dagger a + 1/2), \tag{3.7.13}$$

where the partition function is

$$Z = \mathrm{Tr}(e^{-H/k_B T}). \tag{3.7.14}$$

We can express ρ in terms of the complete set of Fock states:

$$\begin{aligned}
\rho = 1\rho 1 &= \sum_{m=0}^{\infty}\sum_{n=0}^{\infty} |m\rangle\langle m|\rho|n\rangle\langle n| \\
&= Z^{-1} \sum_{m=0}^{\infty}\sum_{n=0}^{\infty} |m\rangle\langle m|e^{-\hbar\omega(a^\dagger a + 1/2)/k_B T}|n\rangle\langle n| \\
&= Z^{-1} e^{-\hbar\omega/2k_B T} \sum_{n=0}^{\infty} e^{-n\hbar\omega/k_B T}|n\rangle\langle n|,
\end{aligned} \tag{3.7.15}$$

with

$$\begin{aligned}
Z &= \sum_{n=0}^{\infty} e^{-\hbar\omega(a^\dagger a + 1/2)/k_B T} = e^{-\hbar\omega/2k_B T} \sum_{n=0}^{\infty} e^{-n\hbar\omega/k_B T} \\
&= e^{-\hbar\omega/2k_B T}(1 - e^{-\hbar\omega/k_B T})^{-1}
\end{aligned} \tag{3.7.16}$$

and therefore (see (3.6.99)) the density matrix for a single mode of a thermal field has the Fock-state representation

$$\rho = \sum_{n=0}^{\infty} \frac{\langle n\rangle^n}{(\langle n\rangle + 1)^{n+1}} |n\rangle\langle n| = \sum_{n=0}^{\infty} p_n |n\rangle\langle n|, \tag{3.7.17}$$

$$\langle n\rangle = (e^{\hbar\omega/k_B T} - 1)^{-1}. \tag{3.7.18}$$

3.7.1   Characteristic Function

As in classical probability theory, we can introduce a *characteristic function* that may be used to derive a probability distribution. For an observable A, the characteristic function is defined as

$$C_A(\xi) = \langle e^{i\xi A}\rangle = \mathrm{Tr}\big[\rho(t)e^{i\xi A}\big] = \sum_{\psi} p_\psi \langle\psi(t)|e^{i\xi A}|\psi(t)\rangle. \tag{3.7.19}$$

It follows from this definition that the expectation values of powers of A are given by

$$\langle A^n\rangle = \frac{\partial^n}{\partial(i\xi)^n} C_A(\xi)\Big|_{\xi=0}, \tag{3.7.20}$$

assuming $C_A(\xi)$ is differentiable. Since $|\psi(t)\rangle$ can be expressed as $\sum_n a_n(t)|A_n\rangle$, where $\{|A_n\rangle\}$ is the complete (and orthornormal) set of eigenstates of A, with corresponding (real) eigenvalues A_n,

$$
\begin{aligned}
C_A(\xi) &= \sum_\psi p_\psi \sum_m \sum_n a_m^*(t) a_n(t) e^{i\xi A_n} \langle A_m | A_n \rangle \\
&= \sum_\psi p_\psi \sum_n |a_n(t)|^2 e^{i\xi A_n} = \sum_\psi p_\psi \sum_n |\langle A_n | \psi(t)\rangle|^2 e^{i\xi A_n} \\
&= \sum_n p(A_n) e^{i\xi A_n}.
\end{aligned}
\tag{3.7.21}
$$

$p(A_n)$ is the probability that a measurement of the observable A will yield the numerical value A_n. We have derived (3.7.21) as if A has a purely discrete spectrum, but, of course, it might have a continuous spectrum. Assuming that the probability distribution function $p(\mathcal{A})$ for obtaining the numerical value $\mathcal{A}$ is square-integrable, we can use (3.7.21) to express it as the Fourier transform of the characteristic function:

$$
p(\mathcal{A}) = \frac{1}{2\pi} \int_{-\infty}^{\infty} d\xi C_A(\xi) e^{-i\xi \mathcal{A}}.
\tag{3.7.22}
$$

In Exercise 3.9 this formula is used to obtain the probability distribution of the electric field of thermal radiation.

The moments shown in (3.7.20) may not always exist. For example, for a (square-integrable) Lorentzian distribution

$$
p(\mathcal{A}) = \frac{\beta/\pi}{(\mathcal{A} - \mathcal{A}_0)^2 + \beta^2}, \qquad -\infty < \mathcal{A} < \infty,
\tag{3.7.23}
$$

has infinite moments. In this example,

$$
C_A(\xi) = \int_{-\infty}^{\infty} d\mathcal{A} p(\mathcal{A}) e^{i\xi \mathcal{A}} = e^{-\beta|\xi|} e^{i\mathcal{A}_0 \xi}
\tag{3.7.24}
$$

is not differentiable. The Lorentzian distribution is very often a good *approximation* to spectral lineshapes, for example, but it should not be taken too seriously for frequencies far from line center.

3.7.2 Generating Function for Photon Number Probability

Another useful function is

$$
p(\xi) = \sum_{n=0}^{\infty} p_n \xi^n,
\tag{3.7.25}
$$

the moment generating function for the photon number probability p_n of a single-mode field, which is related to the characteristic function $C(\xi)$ for the photon number probability:

$$
C(\xi) = p(e^{i\xi}).
\tag{3.7.26}
$$

It is also related to the normally ordered expectation value

$$F_N(\xi) = \sum_{r=0}^{\infty} \frac{\xi^r}{r!} \langle a^{\dagger r} a^r \rangle \tag{3.7.27}$$

as follows. Write the state vector as a sum over the complete set of photon number states $|n\rangle$:

$$|\psi\rangle = \sum_{n=0}^{\infty} c_n |n\rangle, \quad p_n = |c_n|^2. \tag{3.7.28}$$

Since

$$\langle a^\dagger a \rangle = \sum_{m=0}^{\infty} c_m^* c_n \langle m | a^\dagger a | n \rangle = \sum_{n=0}^{\infty} |c_n|^2 n = \sum_{n=0}^{\infty} p_n n = \langle n \rangle, \tag{3.7.29}$$

$$\langle a^{\dagger 2} a^2 \rangle = \langle n(n-1) \rangle, \tag{3.7.30}$$

and so forth,

$$F_N(\xi) = 1 + \xi \langle n \rangle + \frac{\xi^2}{2!} \langle n(n-1) \rangle + \ldots = \langle (1 + e^\xi)^n \rangle$$

$$= \sum_{n=0}^{\infty} p_n (1+\xi)^n = p(\xi + 1). \tag{3.7.31}$$

For later use, we calculate the generating function $p(\xi)$ for the Poisson and the Bose–Einstein distributions. For the Poisson distribution,

$$p(\xi) = F_N(\xi - 1) = \sum_{n=0}^{\infty} \frac{\langle n \rangle^n}{n!} e^{-\langle n \rangle} \xi^n = e^{(\xi - 1)\langle n \rangle}, \tag{3.7.32}$$

whereas, for the Bose–Einstein distribution,

$$p(\xi) = \frac{1}{\langle n \rangle + 1} \sum_{n=0}^{\infty} \left(\frac{\xi \langle n \rangle}{\langle n \rangle + 1} \right)^n = \frac{1}{1 - (\xi - 1)\langle n \rangle}. \tag{3.7.33}$$

We also introduce the anti-normally ordered expectation value

$$F_A(\xi) = \sum_{r=0}^{\infty} \frac{\xi^r}{r!} \langle a^r a^{\dagger r} \rangle, \tag{3.7.34}$$

which is easily shown to be related to the generating function $p(\xi)$ by

$$F_A(\xi) = \frac{1}{1 - \xi} p\left(\frac{1}{1 - \xi} \right), \tag{3.7.35}$$

or, making the replacement $\xi \to 1 - 1/\xi$,

$$p(\xi) = \frac{1}{\xi} F_A\left(1 - \frac{1}{\xi} \right). \tag{3.7.36}$$

$F_N(\xi)$ and $F_A(\xi)$ will be found to be useful for calculating the photon statistics of linear attenuators and amplifiers, respectively (Section 6.10).

3.8   Coherent-State Representation of the Density Operator

Of course, any complete set of states can be used in defining the density operator. In terms of coherent states,

$$\rho = |\psi\rangle\langle\psi| = \frac{1}{\pi^2} \int\int d^2\alpha d^2\beta |\alpha\rangle\langle\alpha|\rho|\beta\rangle\langle\beta|, \qquad (3.8.1)$$

where we have used (3.6.24). Proceeding in a manner similar to that leading to (3.6.30), we express $\langle\alpha|\rho|\beta\rangle$ in terms of Fock states:

$$\begin{aligned} \langle\alpha|\rho|\beta\rangle &= \sum_{m=0}^{\infty}\sum_{n=0}^{\infty}\langle\alpha|m\rangle\langle m|\rho|n\rangle\langle n|\beta\rangle \\ &= \sum_{m=0}^{\infty}\sum_{n=0}^{\infty}\frac{\alpha^{*m}}{\sqrt{m!}}e^{-|\alpha|^2/2}\frac{\beta^n}{\sqrt{n!}}e^{-|\beta|^2/2}\langle m|\rho|n\rangle \\ &= e^{-(|\alpha|^2+|\beta|^2)/2}\sum_{m=0}^{\infty}\sum_{n=0}^{\infty}\rho_{mn}\frac{\alpha^{*m}\beta^n}{\sqrt{m!n!}} \\ &\equiv e^{-(|\alpha|^2+|\beta|^2)/2}R(\alpha^*,\beta), \end{aligned} \qquad (3.8.2)$$

and therefore

$$\rho = \frac{1}{\pi^2}\int\int d^2\alpha d^2\beta e^{-(|\alpha|^2+|\beta|^2)/2}R(\alpha^*,\beta)|\alpha\rangle\langle\beta|. \qquad (3.8.3)$$

We showed earlier that an arbitrary state vector has a unique expansion in terms of coherent states, the function $f(\alpha)$ in (3.6.30) being an entire analytic function. By similar arguments, we can assume that $R(\alpha^*,\beta)$ is an entire analytic function of the complex variables α^* and β. Then, similar to the fact that an analytic function like $f(\alpha)$ that is zero on any real interval must be zero everywhere, the function $R(\alpha^*,\beta)$ must be zero everywhere if it is zero for $\beta = \alpha$. So if

$$\rho_1 = \frac{1}{\pi^2}\int\int d^2\alpha d^2\beta e^{-(|\alpha|^2+|\beta|^2)/2}R_1(\alpha^*,\beta)|\alpha\rangle\langle\beta|, \qquad (3.8.4)$$

$$\rho_2 = \frac{1}{\pi^2}\int\int d^2\alpha d^2\beta e^{-(|\alpha|^2+|\beta|^2)/2}R_2(\alpha^*,\beta)|\alpha\rangle\langle\beta|, \qquad (3.8.5)$$

and $R_1(\alpha^*,\alpha) = R_2(\alpha^*,\alpha)$, $R_1(\alpha^*,\beta) - R_2(\alpha^*,\beta)$ must be identically zero, and therefore $\rho_1 = \rho_2$. It follows that the diagonal elements $\langle\alpha|\rho|\alpha\rangle = \exp(-2|\alpha|^2)R(\alpha^*,\alpha)$ of ρ in the coherent-state representation determine the operator ρ uniquely. In a way similar to the derivation of (3.6.31), one easily shows that (3.8.3) can be inverted to give

$$R(\alpha^*,\beta) = e^{(|\alpha|^2+|\beta|^2)/2}\langle\alpha|\rho|\beta\rangle. \qquad (3.8.6)$$

Equation (3.8.3) is a non-diagonal representation of the density operator. We very often require expectation values of *normally ordered products* in which creation operators $a^\dagger$ appear to the left of annihilation operators a; the simplest example of a normally ordered product is the photon number operator $a^\dagger a$. Evaluation of normally ordered expectation values is simplified if the coherent-state representation of the density operator is diagonal, that is, if

$$\rho = \int P(\alpha)|\alpha\rangle\langle\alpha|d^2\alpha. \tag{3.8.7}$$

Then, for example,

$$\langle a^\dagger a\rangle = \text{Tr}(\rho a^\dagger a) = \text{Tr}(a^\dagger a\rho) = \int P(\alpha)\langle\alpha|a^\dagger a|\alpha\rangle d^2\alpha = \int P(\alpha)|\alpha|^2 d^2\alpha, \tag{3.8.8}$$

since $a|\alpha\rangle = \alpha|\alpha\rangle$ and $\langle\alpha|a^\dagger = \langle\alpha|\alpha^*$. Similarly,

$$\langle(a^\dagger)^m a^n\rangle = \int P(\alpha)(\alpha^*)^m \alpha^n d^2\alpha. \tag{3.8.9}$$

The relevance of expectation values of normally ordered field products stems in part from their appearance in the theory of photodetection based—as all practical photodetectors are—on the absorption of light (see Appendix C).

It is remarkable that the density operator for *any* field state has a diagonal coherent-state representation. The diagonal coherent-state representation (3.8.7) in terms of coherent states is commonly called the *P representation* of the density operator. We can obtain an expression for $P(\alpha)$ as follows. We use (3.6.21) to write $\langle\alpha|\rho|\alpha\rangle$, which, as we have noted, uniquely determines ρ, to write

$$\langle\alpha|\rho|\alpha\rangle = \int P(\beta)\langle\alpha|\beta\rangle\langle\beta|\alpha\rangle d^2\beta = \int P(\beta)|\langle\alpha|\beta\rangle|^2 d^2\beta$$
$$= \int P(\beta)e^{-|\alpha-\beta|^2} d^2\beta. \tag{3.8.10}$$

Expressing the complex numbers α and β in terms of real numbers x, y, and x', y' ($\alpha = x + iy$ and $\beta = x' + iy'$), we re-write (3.8.10) as

$$\langle x, y|\rho|x, y\rangle = \int_{-\infty}^{\infty}\int_{-\infty}^{\infty} P(x', y')e^{-[(x-x')^2+(y-y')^2]}dx'dy'. \tag{3.8.11}$$

This takes a simpler form in terms of the Fourier transforms

$$\tilde{\rho}(X, Y) = \frac{1}{2\pi}\int_{-\infty}^{\infty}\int_{-\infty}^{\infty} dx dy e^{i(Xy-Yx)}\langle x, y|\rho|x, y\rangle \tag{3.8.12}$$

and

$$\tilde{P}(X, Y) = \frac{1}{2\pi}\int_{-\infty}^{\infty}\int_{-\infty}^{\infty} dx dy e^{i(Xy-Yx)} P(x, y). \tag{3.8.13}$$

Straightforward algebra and an integration yield

$$\tilde{P}(X, Y) = \frac{1}{\pi}e^{(X^2+Y^2)/4}\tilde{\rho}(X, Y) \tag{3.8.14}$$

and

$$P(x,y) = \frac{1}{2\pi^2} \int_{-\infty}^{\infty} \int_{-\infty}^{\infty} dX dY e^{-i(Xy-Yx)} e^{(X^2+Y^2)/4} \tilde{\rho}(X,Y). \tag{3.8.15}$$

Consider again the example of a single mode of thermal radiation, for which a diagonal representation of the density operator in terms of Fock states is given by (3.7.17). From (3.7.17) and (3.6.26),

$$\langle \alpha | \rho | \alpha \rangle = \sum_{n=0}^{\infty} p_n \langle \alpha | n \rangle \langle n | \alpha \rangle = \sum_{n=0}^{\infty} p_n |\langle n | \alpha \rangle|^2 = \sum_{n=0}^{\infty} p_n e^{-|\alpha|^2} \frac{|\alpha|^{2n}}{n!}$$

$$= \frac{e^{-|\alpha|^2}}{\langle n \rangle + 1} \sum_{n=0}^{\infty} \frac{1}{n!} \left(\frac{\langle n \rangle}{\langle n \rangle + 1} |\alpha|^2 \right)^n = (1-s)e^{-|\alpha|^2(1-s)}, \tag{3.8.16}$$

or, equivalently,

$$\langle x, y | \rho | x, y \rangle = (1-s)e^{-(x^2+y^2)(1-s)}, \tag{3.8.17}$$

where, again, $\langle n \rangle = (e^{\hbar\omega/k_B T} - 1)^{-1}$ and we have defined $s = e^{-\hbar\omega/k_B T}$. It now follows from (3.8.12)–(3.8.15) that

$$P(x,y) = \frac{1}{\pi \langle n \rangle} e^{-(x^2+y^2)/\langle n \rangle}, \tag{3.8.18}$$

or

$$P(\alpha) = \frac{1}{\pi \langle n \rangle} e^{-|\alpha|^2/\langle n \rangle}. \tag{3.8.19}$$

This is the P distribution function for a single mode of thermal radiation. From it, we recover familiar results such as

$$\langle a^\dagger a \rangle = \int P(\alpha) |\alpha|^2 d^2\alpha = \frac{1}{\pi \langle n \rangle} \int |\alpha|^2 e^{-|\alpha|^2/\langle n \rangle} d^2\alpha = \frac{1}{\pi \langle n \rangle} 2\pi \int_0^\infty dr r^3 e^{-r^2/\langle n \rangle}$$

$$= \langle n \rangle = (e^{\hbar\omega/k_B T} - 1)^{-1} \tag{3.8.20}$$

and

$$p_n = \text{Tr}(\rho | n \rangle \langle n |) = \int P(\alpha) |\langle n | \alpha \rangle|^2 d^2\alpha$$

$$= \int \frac{1}{\pi \langle n \rangle} e^{-|\alpha|^2/\langle n \rangle} \frac{|\alpha|^{2n}}{n!} e^{-|\alpha|^2} d^2\alpha = \frac{\langle n \rangle^n}{(\langle n \rangle + 1)^{n+1}}. \tag{3.8.21}$$

Writing $|\alpha|^2 = \kappa I$ and $\langle n \rangle = \langle \kappa I \rangle$, we can cast the integral in the first line in the form

$$p_n = \int_0^\infty dI \frac{1}{\langle I \rangle} e^{-\kappa I} \frac{(\kappa I)^n}{n!} e^{-I/\langle I \rangle}. \tag{3.8.22}$$

This is exactly the integral (3.6.113) arrived at in our heuristic derivation of p_n in Section 3.6. As noted there, the Gaussian distribution ((3.6.106) or (3.8.19)) is characteristic of any "chaotic" field, a thermal radiation field being but one example.

Thermal radiation is also an example of a *stationary* field, in the sense that the description of it in terms of a density operator ρ is independent of time, that is, ρ commutes with the Hamiltonian. In other words, ρ is a function only of $a^\dagger a$, not of a or $a^\dagger$ separately. As such, a stationary field is described by a density operator that is diagonal in the Fock-state basis, and in the P representation of the density operator for a stationary field, $P(\alpha)$ depends only on $|\alpha|^2$, as in (3.8.19) for thermal radiation.

Equation (3.8.7) and $\mathrm{Tr}\rho = 1$ imply that

$$\int P(\alpha)d^2\alpha = 1, \tag{3.8.23}$$

suggesting the interpretation of $P(\alpha)$ as a probability density for α. This is often a useful interpretation, but, in general, $P(\alpha)$ is not positive-definite for all α and so cannot be regarded as a probability density. Consider, for example, a state of the field for which the Mandel Q parameter is negative. If $P(\alpha)$ is the P distribution for such a state, then, from (3.6.117),

$$\langle a^\dagger a^\dagger aa\rangle - \langle a^\dagger a\rangle^2 = \langle a^\dagger a^\dagger aa\rangle - 2\langle a^\dagger a\rangle\langle a^\dagger a\rangle + \langle a^\dagger a\rangle^2$$

$$= \int P(\alpha)\big(|\alpha|^4 - 2|\alpha|^2\langle a^\dagger a\rangle + \langle a^\dagger a\rangle^2\big)\, d^2\alpha$$

$$= \int P(\alpha)\big(|\alpha|^2 - \langle a^\dagger a\rangle\big)^2 d^2\alpha \tag{3.8.24}$$

is negative and therefore $P(\alpha)$ cannot be positive-definite for all values of α. A P distribution that is not positive-definite cannot be a probability distribution in the classical sense and is therefore characteristic of a field having properties that cannot be accounted for in purely classical terms. As noted following (3.6.118), a Fock state is an example of such a nonclassical field. The P distribution for a Fock state $|n\rangle$ is easily deduced from (3.8.10) and $\rho = |n\rangle\langle n|$:

$$\langle\alpha|\rho|\alpha\rangle = |\langle n|\alpha\rangle|^2 = \frac{|\alpha|^{2n}}{n!}e^{-|\alpha|^2} = \int P(\xi)e^{-(|\alpha|^2+|\xi|^2)}e^{-(\alpha^*\xi+\alpha\xi^*)}d^2\xi, \tag{3.8.25}$$

implying that

$$P(\xi) = \frac{1}{n!}e^{|\xi|^2}\frac{\partial^{2n}}{\partial\xi^n\partial\xi^{*n}}\delta^2(\xi). \tag{3.8.26}$$

The P representation of the density operator for a multimode field has the form

$$\rho = \int P(\{\alpha_\beta\})|\{\alpha_\beta\}\rangle\langle\{\alpha_\beta\}|\prod_\beta d^2\alpha_\beta. \tag{3.8.27}$$

Here, $a_\beta|\alpha_\beta\rangle = \alpha_\beta|\alpha_\beta\rangle$, and $\{\alpha_\beta\}$ stands for $\alpha_1, \alpha_2, \dots$. For thermal radiation,

$$P(\{\alpha_\beta\}) = \prod_\beta \frac{1}{\pi\langle n_\beta\rangle}e^{-|\alpha_\beta|^2/\langle n_\beta\rangle}, \tag{3.8.28}$$

where $\langle n_\beta\rangle$ is the average photon number of mode β; this is the multimode generalization of (3.8.19).

Exercise 3.9: (a) Using the P representation of the density operator for a chaotic field, show that

$$\langle e^{\mu^* a^\dagger - \mu a} \rangle = e^{-|\mu|^2(\langle n \rangle + 1/2)}, \qquad \langle n \rangle = \langle a^\dagger a \rangle.$$

(Hint: use (3.6.14) to put the operator $e^{\mu^* a^\dagger - \mu a}$ in normal order.) (b) Using (3.7.22), show that the probability distribution for the single-mode electric field $E = iC(a - a^\dagger)$ for chaotic light is

$$P(\mathcal{E}) = \sqrt{\frac{1}{2\pi \langle E^2 \rangle}} e^{-\mathcal{E}^2/2\langle E^2 \rangle}.$$

(c) Consider a squeezed state for which $\Delta X_i < 1/2$, $i = 1$ or 2, with X_i defined by (3.6.77). Show that $P(\alpha)$ cannot be a positive-definite distribution function.

The diagonal representation of the density operator in the coherent-state basis was introduced independently by R. J. Glauber and E. C. G. Sudarshan.[25] The fact that the expectation value of a function $f^{(N)}(a, a^\dagger)$ in which the field annihilation and creation operators are normally ordered has the expectation value

$$\langle f^{(N)}(a, a^\dagger) \rangle = \int P(\alpha) f^{(N)}(\alpha, \alpha^*) d^2\alpha = \langle f^{(N)}(\alpha, \alpha^*) \rangle_P \qquad (3.8.29)$$

shows a formal equivalence between quantum-mechanical expectation values $\langle \ldots \rangle$ of normally ordered field operators and corresponding c-number averages $\langle \ldots \rangle_P$ with respect to $P(\alpha)$. Thus, if $P(\alpha)$ is regarded as a probability distribution, expectation values of normally ordered field operators reduce, in effect, to classical averages with respect to the distribution $P(\alpha)$ of values of α over the complex plane. This was first recognized by Sudarshan and was later referred to as the *optical equivalence theorem*. As we have noted, $P(\alpha)$ cannot always be regarded as a probability distribution.

The P representation facilitates the calculation of expectation values of normally ordered operator products which, for reasons given above and in Appendix C, appear frequently in quantum optics. One can introduce other representations of the density operator (e.g., the Q and Wigner representations) to facilitate the calculation of expectation values of differently ordered photon creation and annihilation operators (e.g., anti-normal or symmetric).[26]

3.9 Correlation Functions

Consider, for example, the normally ordered "first-order" correlation function

$$G^{(1)}(x_1; x_2) = \langle E^{(-)}(\mathbf{r}_1, t_1) E^{(+)}(\mathbf{r}_2, t_2) \rangle = \langle E^{(-)}(x_1) E^{(+)}(x_2) \rangle. \qquad (3.9.1)$$

x_j denotes $(\mathbf{r}_j, t_j)$, and $E^{(+)}(\mathbf{r}, t)$ $(= E^{(-)}(\mathbf{r}, t)^\dagger)$ is a single polarization component of the "positive-frequency" part of the electric field operator, which, for convenience, we

[25] R. J. Glauber, Phys. Rev. Lett. **10**, 84 (1963); E. C. G. Sudarshan, Phys. Rev. Lett. **10**, 277 (1963).
[26] See, for instance, *Mandel and Wolf (1995)* or *Scully and Zubairy*.

write here simply as

$$E^{(+)}(\mathbf{r}, t) = i \sum_\beta C_\beta(\mathbf{r}) a_\beta e^{-i\omega_\beta t}. \tag{3.9.2}$$

Then, for a chaotic field, which has the P distribution (3.8.28), which depends only on the moduli $|\alpha_\beta|^2$,

$$
\begin{aligned}
G^{(1)}(x_1; x_2) &= \sum_\beta \sum_\gamma C_\gamma^*(\mathbf{r}_1) C_\beta(\mathbf{r}_2) \langle a_\gamma^\dagger a_\beta \rangle e^{i\omega_\gamma t_1} e^{-i\omega_\beta t_2} \\
&= \sum_\beta C_\beta^*(\mathbf{r}_1) C_\beta(\mathbf{r}_2) \langle n_\beta \rangle e^{i\omega_\beta(t_1 - t_2)}.
\end{aligned}
\tag{3.9.3}
$$

$\langle n_\beta \rangle = \langle a_\beta^\dagger a_\beta \rangle$ is the average photon number at frequency ω_β. With more algebra, we also find, using the Gaussian function (3.8.28), that the "second-order" normally ordered correlation function is

$$
\begin{aligned}
G^{(2)}(x_1; x_2; x_3; x_4) &= \langle E^{(-)}(x_1) E^{(-)}(x_2) E^{(+)}(x_3) E^{(+)}(x_4) \rangle \\
&= G^{(1)}(x_1; x_3) G^{(1)}(x_2; x_4) + G^{(1)}(x_1; x_4) G^{(1)}(x_2; x_3)
\end{aligned}
\tag{3.9.4}
$$

for a chaotic field. This is a general result for any field described by a density operator with a P representation having the Gaussian form (3.8.28): all correlation functions $G^{(n)}(x_1; x_2; \ldots; x_{2n})$, which involve a normally ordered product of n $E^{(-)}$'s with n $E^{(+)}$'s, can be expressed as sums of products of first-order correlation functions. The first-order correlation function $G^{(1)}$ thus determines all higher-order correlation functions. In particular,

$$G^{(n)}(x; x; \ldots; x) = n! [G^{(1)}(x, x)]^n. \tag{3.9.5}$$

This normally ordered correlation function has the same form as the expression (3.6.110), which was derived in a classical model for chaotic light.

The density matrix for a multimode *coherent* field is simply $|\{\alpha_\beta\}\rangle\langle\{\alpha_\beta\}|$. For such a field,

$$\langle E^{(+)}(\mathbf{r}, t) \rangle = i \sum_\beta C_\beta(\mathbf{r}) \alpha_\beta e^{-i\omega_\beta t} = \mathcal{E}(\mathbf{r}, t), \tag{3.9.6}$$

$$G^{(1)}(x_1; x_2) = \mathcal{E}^*(x_1) \mathcal{E}(x_2), \tag{3.9.7}$$

$$G^{(2)}(x_1; x_2; x_3; x_4) = \mathcal{E}^*(x_1) \mathcal{E}^*(x_2) \mathcal{E}(x_3) \mathcal{E}(x_4). \tag{3.9.8}$$

All higher-order correlation functions $G^{(n)}$ are similarly factorized.

$G^{(1)}(x_1, x_2)$ is used to describe first-order interference effects observed, for example, in Michelson and Mach–Zehnder interferometers, whereas $G^{(2)}(x_1; x_2; x_3; x_4)$ relates to second-order interference effects such as photon bunching. Correlation functions beyond second order have thus far not been of much practical interest, as the kind of measurements they describe are very complicated.

Exercise 3.10: (a) Show that the expectation value of the electric field for chaotic radiation is 0. (b) Verify (3.9.4). (c) How are (3.9.4) and (3.9.5) related to the Einstein fluctuation formula and the Hanbury Brown–Twiss "photon bunching" effect?

We have simplified by ignoring polarization. More generally, one defines normally ordered correlation functions as follows:

$$G_{ij}^{(1)}(x_1; x_2) = \langle E_i^{(-)}(x_1) E_j^{(+)}(x_2) \rangle, \tag{3.9.9}$$

$$G_{ijk\ell}^{(2)}(x_1; x_2; x_3; x_4)) = \langle E_i^{(-)}(x_1) E_j^{(-)}(x_2) E_k^{(+)}(x_3) E_\ell^{(+)}(x_4) \rangle, \tag{3.9.10}$$

and so forth, where the subscripts denote components of the (operator) electric field vector.

As already noted, normally ordered field correlation functions are related to measurements involving the absorption of photons. Practical photon detectors work in one way or another by photon absorption. *Anti*-normally ordered correlation functions would be used to describe photon detectors based on photon *emission*.

Normally ordered field correlation functions for blackbody radiation have been evaluated analytically and numerically.[27] For example,

$$\langle \mathbf{E}^{(-)}(\mathbf{r}, t) \cdot \mathbf{E}^{(+)}(\mathbf{r}, t + \tau) \rangle = \sum_{\mathbf{k}\lambda} \frac{\hbar\omega_k}{2\epsilon_0 V} \frac{e^{-i\omega_k \tau}}{e^{\hbar\omega_k/k_B T} - 1}, \tag{3.9.11}$$

which follows from (3.9.3) and a plane-wave expansion of the field ($|C_\beta(\mathbf{r})|^2 \to |C_{\mathbf{k}\lambda}|^2 = \hbar\omega_k/2\epsilon_0 V$). In the mode-continuum limit ($V \to \infty$),

$$\langle \mathbf{E}^{(-)}(\mathbf{r}, t) \cdot \mathbf{E}^{(+)}(\mathbf{r}, t + \tau) \rangle = \frac{\hbar}{2\pi^2 \epsilon_0 c^3} \int_0^\infty \frac{d\omega \omega^3 e^{-i\omega\tau}}{e^{\hbar\omega/k_B T} - 1}. \tag{3.9.12}$$

This is very small for τ greater than a few multiples of $\hbar/k_B T$, which is $\approx 3 \times 10^{-14}$ s at room temperature. In terms of the wavelength $\lambda_{\max} \approx \hbar c/k_B T$, where the blackbody spectrum peaks, the correlation function (3.9.12) is nearly zero for times τ much greater than $\lambda_{\max}/c$. Likewise, $\langle \mathbf{E}^{(-)}(\mathbf{r}, t) \cdot \mathbf{E}^{(+)}(\mathbf{r} + \mathbf{R}, t) \rangle$ is very small for R much greater than $\lambda_{\max}$.

A symmetrically ordered correlation function, similarly, is

$$\begin{aligned}
\langle \mathbf{E}(\mathbf{r}, t) \cdot \mathbf{E}(\mathbf{r}, t + \tau) \rangle &= \langle \mathbf{E}^{(-)}(\mathbf{r}, t) \cdot \mathbf{E}^{(+)}(\mathbf{r}, t + \tau) \rangle + \langle \mathbf{E}^{(+)}(\mathbf{r}, t) \cdot \mathbf{E}^{(-)}(\mathbf{r}, t + \tau) \rangle \\
&= \frac{\hbar}{2\pi^2 \epsilon_0 c^3} \int_0^\infty d\omega \omega^3 \left(\frac{e^{-i\omega\tau}}{e^{\hbar\omega/k_B T} - 1} + \frac{e^{i\omega\tau}}{e^{\hbar\omega/k_B T} - 1} + e^{i\omega\tau} \right) \\
&= \frac{\hbar}{\pi^2 \epsilon_0 c^3} \int_0^\infty \frac{d\omega \omega^3 \cos\omega\tau}{e^{\hbar\omega/k_B T} - 1} + \frac{\hbar}{2\pi^2 \epsilon_0 c^3} \int_0^\infty d\omega \omega^3 e^{i\omega\tau} \tag{3.9.13}
\end{aligned}$$

for blackbody radiation; the last, "zero-point" term, which does not appear in the normally ordered correlation function (3.9.12), follows from the fact that $\langle a_\beta a_\beta^\dagger \rangle = \langle a_\beta^\dagger a_\beta \rangle + 1 =$

---

[27] See, for instance, *Mandel and Wolf (1995)*, Sections 13.1.6 and 13.1.7.

$\langle n_\beta \rangle + 1 = (e^{\hbar \omega_\beta / k_B T} - 1)^{-1} + 1$ for each field mode. In particular, the energy density expectation value

$$
\begin{aligned}
\epsilon_0 \langle \mathbf{E}^2(\mathbf{r}, t) \rangle &= \frac{\hbar}{\pi^2 c^3} \int_0^\infty \frac{d\omega \omega^3}{e^{\hbar \omega / k_B T} - 1} + \frac{\hbar}{2\pi^2 c^3} \int_0^\infty d\omega \omega^3 \\
&= \frac{\hbar}{\pi^2 c^3} \int_0^\infty d\omega \omega^3 \left(\frac{1}{e^{\hbar \omega / k_B T} - 1} + \frac{1}{2} \right) \\
&= \int_0^\infty d\omega \rho(\omega) + \int_0^\infty d\omega \rho_0(\omega) = \frac{\pi^2 k_B^4}{15 \hbar^3 c^3} T^4 + \int_0^\infty d\omega \rho_0(\omega), \quad (3.9.14)
\end{aligned}
$$

where $\rho(\omega)$ is the Planck blackbody spectrum, given in (2.8.3), and $\rho_0(\omega) = \hbar \omega^3 / 2\pi^2 c^3$ is the spectrum of the vacuum, zero-point ($T = 0$) field.[28]

The temperature-dependent term in (3.9.13) is

$$
\frac{\hbar}{\pi^2 \epsilon_0 c^3} \int_0^\infty \frac{d\omega \omega^3 \cos \omega \tau}{e^{\hbar \omega / k_B T} - 1} = \frac{\hbar}{\pi^2 \epsilon_0 c^3} \left(b^4 \mathrm{csch}^2(b\tau) \left[3 \mathrm{csch}^2(b\tau) + 2 \right] - \frac{3}{\tau^4} \right), \qquad (3.9.15)
$$

$b = \pi k_B T / \hbar$. We evaluate the temperature-independent (zero-point) term as follows:

$$
\frac{\hbar}{2\pi^2 \epsilon_0 c^3} \int_0^\infty d\omega \omega^3 e^{i\omega \tau} \equiv \frac{\hbar}{2\pi^2 \epsilon_0 c^3} \lim_{a \to 0} \int_0^\infty d\omega \omega^3 e^{i\omega(\tau + ia)} = \frac{\hbar}{\pi^2 \epsilon_0 c^3} \times \frac{3}{\tau^4}. \qquad (3.9.16)
$$

Then

$$
\langle \mathbf{E}(\mathbf{r}, t) \cdot \mathbf{E}(\mathbf{r}, t + \tau) \rangle = \frac{\hbar}{\pi^2 \epsilon_0 c^3} b^4 \mathrm{csch}^2(b\tau) \left[3 \mathrm{csch}^2(b\tau) + 2 \right]. \qquad (3.9.17)
$$

For $T \to 0$,

$$
\langle \mathbf{E}(\mathbf{r}, t) \cdot \mathbf{E}(\mathbf{r}, t + \tau) \rangle \to \frac{3\hbar}{\pi^2 \epsilon_0 c^3 \tau^4} = \frac{0.13}{\tau(\mathrm{ps})^4} \ \mathrm{V}^2 / \mathrm{m}^2. \qquad (3.9.18)
$$

3.10   Field Commutators and Uncertainty Relations

The commutation relation (3.3.15) for the photon annihilation and creation operators implies commutation relations for the electric and magnetic field operators. Thus, from (3.3.9),

$$
\begin{aligned}
[A_i(\mathbf{r}_1, t_1), A_j(\mathbf{r}_2, t_2)] &= \frac{\hbar}{2\epsilon_0 V} \sum_{\mathbf{k}\lambda} \frac{1}{\omega_k} [e^{i\mathbf{k} \cdot (\mathbf{r}_1 - \mathbf{r}_2)} e^{-i\omega_k c(t_1 - t_2)} \\
&\quad - e^{-i\mathbf{k} \cdot (\mathbf{r}_1 - \mathbf{r}_2)} e^{i\omega_k c(t_1 - t_2)}] e_{\mathbf{k}\lambda i} e_{\mathbf{k}\lambda j} \\
&= -\frac{i\hbar}{c\epsilon_0 V} \sum_{\mathbf{k}} e^{i\mathbf{k} \cdot (\mathbf{r}_1 - \mathbf{r}_2)} \frac{1}{k} \sin kc(t_1 - t_2) \\
&\quad \times \sum_\lambda e_{\mathbf{k}\lambda i} e_{\mathbf{k}\lambda j}, \qquad (3.10.1)
\end{aligned}
$$

where we have used the fact that $\mathbf{k}$ and $-\mathbf{k}$ contribute equally to the summation over plane-wave modes. Since the electric field operator in free space is $\mathbf{E} = -\partial \mathbf{A} / \partial t$,

---

[28] See, for example, equation (4.5.21), and Section 7.4.

$$[E_i(\mathbf{r}_1, t_1), E_j(\mathbf{r}_2, t_2)] = -\frac{i\hbar}{c\epsilon_0 V}\frac{\partial^2}{\partial t_1 \partial t_2}\sum_{\mathbf{k}} e^{i\mathbf{k}\cdot(\mathbf{r}_1-\mathbf{r}_2)}\frac{1}{k}\sin kc(t_1-t_2)$$

$$\times \sum_{\lambda} e_{\mathbf{k}\lambda i}e_{\mathbf{k}\lambda j}. \tag{3.10.2}$$

It is shown below that this is equivalent to

$$[E_i(\mathbf{r}_1, t_1), E_j(\mathbf{r}_2, t_2)] = \frac{i\hbar c}{\epsilon_0}\left(\frac{\delta_{ij}}{c^2}\frac{\partial^2}{\partial t_1\partial t_2} - \frac{\partial^2}{\partial r_{1i}\partial r_{2j}}\right)D(|\mathbf{r}_1-\mathbf{r}_2|, t_1-t_2), \tag{3.10.3}$$

$$D(r, t) = \frac{1}{4\pi r}\big[\delta(r+ct) - \delta(r-ct)\big]. \tag{3.10.4}$$

To obtain (3.10.3), we first show that

$$\sum_{\lambda} e_{\mathbf{k}\lambda i}e_{\mathbf{k}\lambda j} = \delta_{ij} - \frac{k_i k_j}{k^2} \tag{3.10.5}$$

as a consequence of the fact that, since $\hat{\mathbf{k}}$, $\mathbf{e}_{\mathbf{k}1}$, and $\mathbf{e}_{\mathbf{k}2}$ are three orthogonal unit vectors,

$$\mathbf{a} = (\hat{\mathbf{k}}\cdot\mathbf{a})\hat{\mathbf{k}} + \sum_{\lambda=1}^{2}(\mathbf{a}\cdot\mathbf{e}_{\mathbf{k}\lambda})\mathbf{e}_{\mathbf{k}\lambda} \quad (\hat{\mathbf{k}} = \mathbf{k}/k) \tag{3.10.6}$$

for any vector $\mathbf{a}$. Then, for two arbitrary vectors $\mathbf{a}$ and $\mathbf{b}$, the dot product

$$\sum_{i} a_i b_i = \sum_{i}\sum_{j} a_i b_j \hat{k}_i \hat{k}_j + \sum_{i}\sum_{j} a_i b_j \sum_{\lambda} e_{\mathbf{k}\lambda i}e_{\mathbf{k}\lambda j}, \tag{3.10.7}$$

and (3.10.5) follows.

Next, we use (3.3.18) to replace (3.10.2) by

$$[E_i(\mathbf{r}_1, t_1), E_j(\mathbf{r}_2, t_2)] = -\frac{i\hbar}{8\pi^3 c\epsilon_0}\frac{\partial^2}{\partial t_1\partial t_2}\int d^3k\left(\delta_{ij} - \frac{k_i k_j}{k^2}\right)e^{i\mathbf{k}\cdot(\mathbf{r}_1-\mathbf{r}_2)}$$

$$\times \frac{1}{k}\sin kc(t_1-t_2)$$

$$= -\frac{i\hbar}{8\pi^3 c\epsilon_0}\int d^3k\left(\delta_{ij}\frac{\partial^2}{\partial t_1\partial t_2} + c^2 k_i k_j\right)e^{i\mathbf{k}\cdot(\mathbf{r}_1-\mathbf{r}_2)}$$

$$\times \frac{1}{k}\sin kc(t_1-t_2)$$

$$= -\frac{i\hbar}{8\pi^3 c\epsilon_0}\int d^3k\left(\delta_{ij}\frac{\partial^2}{\partial t_1\partial t_2} - c^2\frac{\partial^2}{\partial r_{1i}\partial r_{2j}}\right)e^{i\mathbf{k}\cdot(\mathbf{r}_1-\mathbf{r}_2)}$$

$$\times \frac{1}{k}\sin kc(t_1-t_2). \tag{3.10.8}$$

One conventionally defines

$$D(r,t) = -\left(\frac{1}{2\pi}\right)^3 \int d^3k\, e^{i\mathbf{k}\cdot\mathbf{r}} \frac{1}{k} \sin kct$$

$$= -\frac{1}{8\pi^3} 2\pi \int_0^\infty dk\, k \int_0^\pi d\theta \sin\theta\, e^{ikr\cos\theta} \sin kct$$

$$= -\frac{1}{4\pi^2} \int_0^\infty dk\, k \int_{-1}^1 du\, e^{ikru} \sin kct = -\frac{1}{2\pi^2 r} \int_0^\infty dk \sin kr \sin kct$$

$$= \frac{1}{4\pi^2 r} \int_0^\infty dk [\cos k(r+ct) - \cos k(r-ct)]$$

$$= \frac{1}{4\pi r} [\delta(r+ct) - \delta(r-ct)], \tag{3.10.9}$$

in terms of which (3.10.8) takes the form (3.10.3).

One similarly obtains other "Pauli–Jordan" commutation relations such as

$$[B_i(\mathbf{r}_1,t_1), B_j(\mathbf{r}_2,t_2)] = [E_i(\mathbf{r}_1,t_1), E_j(\mathbf{r}_2,t_2)], \tag{3.10.10}$$

$$[E_i(\mathbf{r}_1,t_1), B_j(\mathbf{r}_2,t_2)] = -\frac{i\hbar}{\epsilon_0 c} \epsilon_{ijk} \frac{\partial^2}{\partial t_1 \partial r_{2k}} D(|\mathbf{r}_1 - \mathbf{r}_2|, t_1 - t_2), \tag{3.10.11}$$

where ϵ_{ijk} is the (three-dimensional) Levi–Civita symbol.[29] Thus,

$$[E_i(\mathbf{r}_1,t_1), B_i(\mathbf{r}_2,t_2)] = 0. \tag{3.10.12}$$

Another important commutation relation follows by differentiation of (3.10.1) with respect to t_2 and then setting $t_2 = t_1 = t$. Since $\partial \mathbf{A}(\mathbf{r},t)/\partial t = -\mathbf{E}^\perp(\mathbf{r},t)$, we obtain the equal-time commutation relation

$$[A_i(\mathbf{r}_1,t), E_j^\perp(\mathbf{r}_2,t)] = -\frac{i\hbar}{\epsilon_0 V} \frac{V}{(2\pi)^3} \int d^3k \left(\delta_{ij} - \frac{k_i k_j}{k^2}\right) e^{i\mathbf{k}\cdot(\mathbf{r}_1 - \mathbf{r}_2)}$$

$$= -\frac{i\hbar}{\epsilon_0} \delta_{ij}^\perp(\mathbf{r}_1 - \mathbf{r}_2). \tag{3.10.13}$$

(The transverse and longitudinal delta functions, $\delta_{ij}^\perp(\mathbf{r})$ and $\delta_{ij}^\parallel(\mathbf{r})$, respectively, are reviewed in Appendix B.)

These field commutation relations imply constraints on the simultaneous measurability of fields at the two space-time points $(\mathbf{r}_1,t_1)$ and $(\mathbf{r}_2,t_2)$. Thus, (3.10.3) implies that the electric field strength can be measured simultaneously at $(\mathbf{r}_1,t_1)$ and $(\mathbf{r}_2,t_2)$ whenever these points cannot be "connected" by a signal propagating with the velocity c, that is, whenever $|\mathbf{r}_1 - \mathbf{r}_2| \neq \pm c(t_1 - t_2)$. If $(\mathbf{r}_1,t_1)$ and $(\mathbf{r}_2,t_2)$ are so connected, the electric field operators at these two space-time points do not commute and therefore cannot be simultaneously measured. The equal-time commutators $[E_i(\mathbf{r}_1,t), E_j(\mathbf{r}_2,t)]$ and $[B_i(\mathbf{r}_1,t), B_j(\mathbf{r}_2,t)]$ vanish (see (3.10.2)) and so it is possible to measure with arbitrary accuracy the electric (or magnetic) field strength at two points in space at any

---

[29] $\epsilon_{ijk} = 1$ if ijk is an even permutation of 123 (e.g., $\epsilon_{123} = \epsilon_{312} = 1$), and $\epsilon_{ijk} = -1$ if ijk is an odd permutation of 123 (e.g., $\epsilon_{213} = -1$); $\epsilon_{ijk} = 0$ if any two of the indices are equal.

one instant of time. The commutator (3.10.11), however, does not vanish for $t_1 = t_2$, and therefore it is not possible to simultaneously measure *different* components of the electric and magnetic fields at different points in space at a fixed instant of time; this is discussed further below. These matters were rigorously analyzed by Bohr and Rosenfeld.[30]

The notion that $E_x(\mathbf{r}_1, t)$ and $E_x(\mathbf{r}_2, t)$, for instance, can be measured simultaneously, to any degree of accuracy, was disputed by Landau and Peierls, who argued, in particular, that the field of a moving "test body" would interfere with the very field to be measured (inferred) from its effect on that test body. According to Schweber, "Bohr disagreed sharply with the view expressed by Landau and Peierls, and in a famous paper with Rosenfeld demonstrated that their claims were wrong . . . . Bohr and Rosenfeld pointed out that measurements at a point are not possible because extended test bodies have to be used to make field measurements. The only observables are space averages of the field operators. They showed that one could in principle construct devices that would measure one component of the field averaged over a finite volume (or over a finite time) to any degree of accuracy. Bohr and Rosenfeld further demonstrated the consistency of the uncertainty relations implied by the commutation rules of the electromagnetic field components."[31]

The fact that field operators at different points in space and at different times do not, in general, commute, and therefore that there are limitations to the accuracy with which fields at different points and times can be "simultaneously" measured, is required for the consistency of quantum theory: the Heisenberg position–momentum uncertainty relation for an electron, for example, must imply an uncertainty relation for any electromagnetic field from the electron. In other words, we cannot consistently treat matter quantum mechanically and radiation classically—or vice versa. Semiclassical radiation theory, for instance, is fundamentally inconsistent in this sense, and it is therefore not surprising that it fails to describe every aspect of the interaction of light with matter.

For particles with finite mass, the commutation relation $[x, p_x] = i\hbar$ implies the minimal uncertainty product $\Delta x \Delta p_x = \hbar/2$. There are famous thought experiments, such as the Heisenberg microscope, that nicely illustrate this; they invoke in one way or another the de Broglie relation $\lambda = h/|\mathbf{p}|$. Now, it is all well and good to say that the electromagnetic field commutation relations imply minimal uncertainty products for field measurements, but to support this conclusion by analyses of thought experiments is no easy matter. Pais wrote that, "From decades of involvement with quantum field theory I can testify that [the Bohr–Rosenfeld paper] has been read by very very few of the *aficionados*. The main reason is, I think, that even by Bohr's standards this paper

---

[30] N. Bohr and L. Rosenfeld, Det. Kgl. dansk. Vid. Selskab **12**, No. 8 (1933). An English translation appears in *Quantum Theory and Measurement*, J. A. Wheeler and W. H. Zurek, eds., Princeton University Press, Princeton, 1983, pp. 479–534. See also *Heitler*, Section 9, and N. Bohr and L. Rosenfeld, Phys. Rev. **78**, 794 (1950).

[31] *Schweber*, p. 112.

is very difficult to penetrate. It takes inordinate care and patience to follow Bohr's often quite complex gyrations with test bodies. In addition the main message of the article: in quantum electrodynamics measurability is compatible with the commutation relations, just as in quantum mechanics, is cheered with enthusiasm—after which the pros continue with whatever calculations they are engaged in. As a friend of Bohr's and mine once said to me: 'It is a very good paper that one does not have to read. You just have to know it exists.' Nevertheless men like Pauli and Heitler did read it with great care."[32]

Bohr and Rosenfeld demonstrated how "in quantum electrodynamics measurability is compatible with the commutation relations, just as in quantum mechanics."[33] That is, they showed that the accuracies with which electric and magnetic field strengths can, in principle, be measured are compatible with the uncertainty relations implied by the field commutators. We will take a decidedly pedestrian approach to get at the main points of their analysis.

Consider first what the field commutators tell us about the accuracy with which the fields can be measured. Any measurement of electric field strength, for instance, requires an average over a finite region of space as well as a finite interval of time—a point first emphasized, it seems, by Heisenberg. We therefore define the operator

$$A = \frac{1}{\Omega_1} \int_{\Omega_1} d^3r_1 dt_1 E_x(\mathbf{r}_1, t_1). \tag{3.10.14}$$

Here, $\Omega_1 = V_1 T_1$, where V_1 and T_1 are, respectively, a spatial volume element and a time interval involved in the measurement of the x component of the electric field. Likewise,

$$B = \frac{1}{\Omega_2} \int_{\Omega_2} d^3r_2 dt_2 E_x(\mathbf{r}_2, t_2) \tag{3.10.15}$$

represents the operator corresponding to the space-time average of the x component of the electric field in a volume $\Omega_2 = V_2 T_2$ of space-time. The general uncertainty relation (see (3.12.6)) $\Delta A \Delta B \geq \frac{1}{2} |\langle A, B \rangle|$ then implies that the minimal uncertainty product for the space-time-averaged electric fields in Ω_1 and Ω_2 is

---

[32] *Pais (1991)*, p. 362.
[33] Ibid.

$$\Delta \overline{E}_x(\Omega_1) \Delta \overline{E}_x(\Omega_2) = \frac{1}{2} \frac{1}{\Omega_1 \Omega_2} \left| \int_{\Omega_1} d^3 r_1 dt_1 \int_{\Omega_2} d^3 r_2 dt_2 \langle E_x(\mathbf{r}_1, t_1), E_x(\mathbf{r}_2, t_2) \rangle \right|$$

$$= \frac{1}{2} \frac{1}{\Omega_1 \Omega_2} \left| \int_{\Omega_1} d\Omega_1 \int_{\Omega_2} d\Omega_2 \frac{i\hbar c}{\epsilon_0} \left(\frac{1}{c^2} \frac{\partial^2}{\partial t_1 \partial t_2} - \frac{\partial^2}{\partial x_1 \partial x_2} \right) \right.$$

$$\left. \times D(|\mathbf{r}_1 - \mathbf{r}_2|, t_1 - t_2) \right|$$

$$= \frac{\hbar c}{8\pi\epsilon_0} \frac{1}{\Omega_1 \Omega_2} \left| \int_{\Omega_1} d\Omega_1 \int_{\Omega_2} d\Omega_2 \left(\frac{\partial^2}{\partial x_1 \partial x_2} - \frac{1}{c^2} \frac{\partial^2}{\partial t_1 \partial t_2} \right) \right.$$

$$\left. \times \frac{\delta(|\mathbf{r}_2 - \mathbf{r}_1| - c(t_2 - t_1))}{|\mathbf{r}_2 - \mathbf{r}_1|} \right| \tag{3.10.16}$$

if, in using (3.10.3) for $[E_x(\mathbf{r}_1, t_1), E_x(\mathbf{r}_2, t_2)]$, we assume that $t_2 > t_1$.

Consider now an electric field measurement employing a test body of charge Q uniformly distributed over a volume V_1. The electric field $\overline{E}_x(\Omega_1)$, an average over the volume V_1 and a time interval T_1, is inferred from the measured change in the linear momentum of the test body. This momentum change is just the average force $Q\overline{E}_x(\Omega_1)$ on the test body times the time T_1: $QT_1\overline{E}_x(\Omega_1)$. But there must be some uncertainty $\Delta\overline{E}_x(\Omega_1)$ in the average field found in this way because, according to the Heisenberg uncertainty relation, there must be an uncertainty of at least $\hbar/2\Delta x$ in the measured momentum, Δx being the uncertainty in the initial position of the test body. In other words, there is an uncertainty of at least

$$\Delta \overline{E}_x(\Omega_1) = \frac{\hbar}{2QT_1\Delta x} \tag{3.10.17}$$

in the x component of the average electric field inferred from the force this electric field exerts on the test body. The position uncertainty Δx of the test body also implies an uncertainty of at least $\Delta\phi(\mathbf{r}_2, t_2) \cong [\partial\phi(\mathbf{r}_2, t_2)/\partial x_1]\Delta x$ in the scalar potential at $(\mathbf{r}_2, t_2)$ $(t_2 > t_1)$, due to the test body:

$$\Delta\phi(\mathbf{r}_2, t_2) = \frac{\partial}{\partial x_1} \left[\frac{1}{4\pi\epsilon_0} \int_{V_1} d^3 r_1 \int_{T_1} dt_1 \frac{Q}{V_1} \frac{\delta(t_2 - t_1 - |\mathbf{r}_2 - \mathbf{r}_1|/c)}{|\mathbf{r}_2 - \mathbf{r}_1|} \right] \Delta x$$

$$= \frac{c}{4\pi\epsilon_0} \frac{Q}{V_1} \Delta x \int d\Omega_1 \frac{\partial}{\partial x_1} \frac{\delta(|\mathbf{r}_2 - \mathbf{r}_1| - c(t_2 - t_1))}{|\mathbf{r}_2 - \mathbf{r}_1|}, \tag{3.10.18}$$

and this implies an uncertainty in the average electric field over a volume element V_2 and a time interval T_2. The magnitude of this uncertainty is

$$\Delta\overline{E}_x(\Omega_2) = \frac{1}{V_2 T_2} \left| \int_{V_2} d^3 r_2 \int_{T_2} dt_2 \frac{\partial}{\partial x_2} \Delta\phi(\mathbf{r}_2, t_2) \right| = \frac{c}{4\pi\epsilon_0} \frac{Q}{V_1 V_2 T_2} \Delta x$$

$$\times \left| \int_{\Omega_1} d\Omega_1 \int_{\Omega_2} d\Omega_2 \frac{\partial^2}{\partial x_1 \partial x_2} \frac{\delta(|\mathbf{r}_2 - \mathbf{r}_1| - c(t_2 - t_1))}{|\mathbf{r}_2 - \mathbf{r}_1|} \right|. \tag{3.10.19}$$

Multiplication of this expression by (3.10.17) yields

$$\Delta \overline{E}_x(\Omega_1)\Delta \overline{E}_x(\Omega_2) = \frac{\hbar c}{8\pi\epsilon_0} \frac{1}{\Omega_1\Omega_2}$$
$$\times \left| \int_{\Omega_1} d\Omega_1 \int_{\Omega_2} d\Omega_2 \frac{\partial^2}{\partial x_1 \partial x_2} \frac{\delta(|\mathbf{r}_2 - \mathbf{r}_1| - c(t_2 - t_1))}{|\mathbf{r}_2 - \mathbf{r}_1|} \right|, \quad (3.10.20)$$

which is independent of the charge Q of our test body. This result, which we obtained by considering the scalar potential associated with the test body, gives exactly the part of (3.10.16) with $\partial^2/\partial x_1 \partial x_2$ in the integrand. The remaining part of (3.10.16), the part involving $\partial^2/\partial t_1 \partial t_2$, may be shown in a similar fashion to come from the *vector* potential. Thus, we draw the desired conclusion that the minimal uncertainty product $\Delta \overline{E}_x(\Omega_1)\Delta \overline{E}_x(\Omega_2)$ deduced from the position–momentum uncertainty relation for the *test body* is indeed just that expected from the uncertainty relation for the *electric field* implied by the commutator (3.10.3).

The argument just presented makes a number of simplifications, including the assumptions that the average external field $\overline{E}_x$ at the test body can be described in an essentially classical way, that the radiation reaction field acting on the test body can be ignored, and that the test body itself can be treated as a rigid body. It has also been assumed that the commutation relation (3.10.3), which is derived for the ("free") field in vacuum, holds also—at least to a sufficient degree of approximation—in the presence of the test body. These and other assumptions are justified in the Bohr–Rosenfeld paper.[34]

It is also instructive to indicate the consistency of the commutation relation (3.10.11) between different orthogonal components of the electric and magnetic field operators and the uncertainties in measurements of these components implied by uncertainty relations for material particles. From (3.10.11) it follows that

$$[E_x(\mathbf{r}_1, t), B_y(\mathbf{r}_2, t)] = -\frac{i\hbar}{\epsilon_0} \frac{\partial}{\partial z_2} \delta^3(\mathbf{r}_1 - \mathbf{r}_2); \quad (3.10.21)$$

this equal-time free-field commutator is most easily deduced directly from the expressions (3.3.11) and (3.3.12) for the electric and magnetic field operators, respectively. We now define field operators averaged over a small region of space of volume $V \sim (\delta\ell)^3$ and derive, following the same approach leading to (3.10.16), the uncertainty relation

$$\Delta \overline{E}_x \Delta \overline{B}_y > \frac{\hbar}{2\epsilon_0} \left(\frac{1}{\delta\ell}\right)^4 \quad (3.10.22)$$

---

[34] An essential feature of their argument is that interference of the field from the test body with the field to be measured by its effect on the test body—a concern raised by Landau and Peierls—can be eliminated in principle by including an oppositely charged body constrained by mechanical forces to follow exactly the motion of the test body; the additional body cancels the field of the test body. Freeman Dyson (see Int. J. Mod. Phys. **28**, 1330041 (2013)) has argued that the non-existence of negative *masses* precludes such a cancellation of a *gravitational* field, and, therefore, that we cannot conclude, based on the Bohr–Rosenfeld analysis, that the gravitational field must be quantized.

for the fields E_x and B_y averaged over V. This relation was obtained in 1929 by Heisenberg.[35]

Heisenberg gave another example showing that the uncertainty relation (3.10.22) is compatible with the position–momentum uncertainty relation for a charged particle. He considered two charged-particle beams with velocities $\pm v_y$ and subject to an electric field E_x and a magnetic field B_z (see Figure 3.3) in a small volume. The width of each beam is $\delta\ell$ along z, and $d < \delta\ell$ along x, and the two beams are separated by $\sim \delta\ell$. $\delta\ell$ is sufficiently small that the electric and magnetic fields in the volume $\delta\ell^3$ are approximately homogeneous. Each particle has a charge e and a mass m, and the

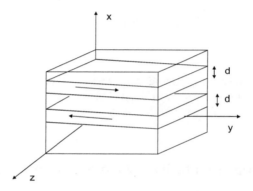

Fig. 3.3 Two oppositely traveling charged-particle beams, each of extent $\sim d$ along the x direction, are deflected by an electric field along the x direction and a magnetic field along the z direction. The volume of the box shown is $\delta\ell^3$. From *Heisenberg*, Figure 13, with permission.

force exerted on it is $F = ma = e(E_x \pm v_y B_z)$, so that its displacement along x is

$$x = \frac{1}{2}at^2 = \frac{e}{2m}(E_x \pm v_y B_z)\left(\frac{\delta\ell}{v_y}\right)^2 = \frac{e}{2m}\left(E_x \pm \frac{p_y}{m}B_z\right)\left(\frac{m\delta\ell}{p_y}\right)^2 \qquad (3.10.23)$$

after traversing the distance $\delta\ell$ along y. The fields E_x and B_z can be inferred from the observation of the angular deflections

$$\alpha_\pm \cong \frac{x}{\delta\ell} = \frac{e}{2p_y}\left(E_x \pm \frac{p_y}{m}B_z\right)\frac{m\delta\ell}{p_y}. \qquad (3.10.24)$$

For instance,

$$B_z = \frac{p_y}{e}\frac{1}{\delta\ell}(\alpha_+ - \alpha_-). \qquad (3.10.25)$$

Since the "diffraction" angle of each particle is given by its de Broglie wavelength $h/|\mathbf{p}| \cong h/p_y$ divided by the width d, the angular divergence $\alpha_+ - \alpha_- \sim 2\pi\hbar/(p_y d)$, and this implies an uncertainty in the inferred B_z:

---

[35] See *Heisenberg*, equation (38). In the Gaussian units used by Heisenberg, the factor $\hbar/2\epsilon_0$ in (3.10.22) is replaced by $2\pi\hbar c$.

$$\Delta B_z \approx \frac{2\pi\hbar}{ed}\frac{1}{\delta\ell}. \tag{3.10.26}$$

Each particle experiences the electric field of the charges in the other beam. Since the beam separation $\approx \delta\ell$, the uncertainty in this field is

$$\Delta E_x > \frac{1}{4\pi\epsilon_0}\frac{e}{\delta\ell^2}\frac{d}{\delta\ell}, \tag{3.10.27}$$

and therefore

$$\Delta E_x \Delta B_z > \frac{\hbar}{2\epsilon_0}\left(\frac{1}{\delta\ell}\right)^4, \tag{3.10.28}$$

consistent with (3.10.21) and (3.10.22).

Exercise 3.11: (a) Using the formal identity $\delta'(r)/r = -2\pi\delta^3(\mathbf{r})$, show that (3.10.21) follows from (3.10.11). (b) Using again the uncertainty relation $\Delta A \Delta B \geq \frac{1}{2}|\langle A, B\rangle|$, verify (3.10.22).

3.11 Complementarity: Wave and Particle Descriptions of Light

In contrast to the quantitative character of the uncertainty relation, Bohr's complementarity principle was essentially qualitative: certain concepts—the wave and particle descriptions of electrons or light, for example—are mutually exclusive ("complementary"). Whether we are led to regard light in terms of particle or wave concepts depends on the type of measurement we make. In selecting a particular experimental arrangement, we are, in effect, *choosing* which of these complementary concepts is best suited to interpret the results.

Consider again the recoil of an atom when it radiates spontaneously (Section 2.3), and imagine an experiment that allows us to determine the direction of the atom's recoil with a high degree of accuracy. Then, by conservation of linear momentum, we also know with a high degree of accuracy the direction of emission of radiation; in such an experiment, the radiation is described as highly directional. If, however, the experiment is such as to confine the atom to a region of space that is small compared with the radiation wavelength, the observed recoil direction of the atom does not sharply define the direction of emission of radiation. "The particular solution of the Maxwell equation which represents the emitted radiation depends on the accuracy with which the coordinates of the center of mass of the atom are known."[36] It is pointless as a practical matter to regard the radiation as directional or non-directional, or in terms of particles or waves, without specifying the experimental arrangement.

[36] *Heisenberg*, p. 92.

A famous example illustrating the complementary nature of wave and particle attributes is the diffraction of electrons in a double-slit experiment. We observe the buildup of an interference pattern in an experiment that cannot determine which of the slits an electron passes through. But any measurement that *can* fully determine which path an electron takes results in a washing-out of the interference pattern. For instance, if we infer the path of an electron by measuring the recoil momentum of the plate containing the two slits, we find that the consequent uncertainty in the position of the plate is enough to wash out the interference pattern. The same considerations apply to the diffraction of light.[37]

Intimations of complementary appear in classical wave theory. Consider the classical theory of the double-slit experiment based on the far-field (Fraunhofer) diffraction integral for a monochromatic field. Assume a placement of slits such that the amplitude transmission function for the field incident on the input plane $z = 0$ is

$$t(x', y') = \text{rect}\left(\frac{x'}{X}\right)\left[\text{rect}\left(\frac{y' - \Delta/2}{Y}\right) + \text{rect}\left(\frac{y' + \Delta/2}{Y}\right)\right], \tag{3.11.1}$$

where the rectangle function $\text{rect}(x) = 1$ if $|x| \leq 1/2$, and 0 otherwise. Then, for uniform illumination of each slit, the field at the point (x, y, z) in the observation plane is

$$\mathcal{E}(x, y, z) \cong -\frac{ie^{ikz}}{\lambda z}e^{ik(x^2 + y^2)/2z}XY\text{sinc}\left(\frac{xX}{\lambda z}\right)\text{sinc}\left(\frac{yY}{\lambda z}\right)\left(\mathcal{E}_1 e^{-i\pi y\Delta/\lambda z} + \mathcal{E}_2 e^{i\pi y\Delta/\lambda z}\right)$$

$$\equiv F(x, y, z)\left(\mathcal{E}_1 e^{-i\pi y\Delta/\lambda z} + \mathcal{E}_2 e^{i\pi y\Delta/\lambda z}\right), \tag{3.11.2}$$

where $\lambda = 2\pi/k$ is the wavelength and $\mathcal{E}_1$ and $\mathcal{E}_2$ are the constant field amplitudes at the slits.[38] In terms of the intensities I_1 and I_2 *at the slits*, the intensity at the observation plane is

$$I(x, y, z) = |F(x, y, z)|^2\left(I_1 + I_2 + \epsilon_0 c\text{Re}\left(\mathcal{E}_1^* \mathcal{E}_2 e^{2\pi iy\Delta/\lambda z}\right)\right), \tag{3.11.3}$$

or

$$I(x, y, z) = |F(x, y, z)|^2\left(I_1 + I_2 + 2\sqrt{I_1 I_2}\cos\Phi\right), \tag{3.11.4}$$

where we have written $\mathcal{E}_1 = |\mathcal{E}_1|e^{i\Phi_1}$, $\mathcal{E}_2 = |\mathcal{E}_2|e^{i\Phi_2}$, and defined $\Phi = 2\pi y\Delta/\lambda z+$. The fringe *visibility* of the interference pattern at the observation plane, a measure of wave interference, is

---

[37] N. Bohr, in *Albert Einstein, Philosopher–Scientist*, ed. P. A. Schilpp, Harper, New York, 1949, Vol. I, pp. 199–241. This example is discussed at some length in *Feynman, Leighton, and Sands*, Vol. III, Chapters 1 and 2. Along these lines, see also W. K. Wooters and W. H. Zurek, Phys. Rev. D **19**, 473 (1979), who show that a considerable degree of both wave and particle (and position and momentum) information is possible without violating the uncertainty relation in the double-slit experiment.

[38] See, for instance, J. F. Goodman, *Introduction to Fourier Optics*, fourth edition, W. H. Freeman and Company, New York, 2017.

$$V = \frac{I_{\max} - I_{\min}}{I_{\max} + I_{\min}} = \frac{2\sqrt{I_1 I_2}}{I_1 + I_2}. \tag{3.11.5}$$

We also define the *contrast*

$$K = \frac{|I_1 - I_2|}{I_1 + I_2}, \tag{3.11.6}$$

and, from these definitions, it follows that

$$K^2 + V^2 = 1. \tag{3.11.7}$$

This shows the complementary relation between the difference in field intensities at the slits and the degree of interference at the observation plane: the greater the contrast or "distinguishability" of the fields at the two slits, the less the wave interference.

More generally, the field undergoes fluctuations, and what is measured is an average $\langle \ldots \rangle$ over the fluctuations. Then, we replace (3.11.4) by

$$\langle I(x,y,z) \rangle = |F(x,y,z)|^2 \Big[\langle I_1 \rangle + \langle I_2 \rangle + 2\sqrt{\langle I_1 \rangle \langle I_2 \rangle} \mathrm{Re}\big(\gamma e^{2\pi i y \Delta / \lambda z}\big)\Big], \tag{3.11.8}$$

where the "complex degree of coherence"[39]

$$\gamma \equiv \frac{c\epsilon_0 \langle \mathcal{E}_1^* \mathcal{E}_2 \rangle}{\sqrt{\langle I_1 \rangle \langle I_2 \rangle}}. \tag{3.11.9}$$

Defining the visibility V similarly in terms of average intensities, we obtain the generalization of (3.11.7):

$$K^2 + V^2 = 1 - \frac{4\langle I_1 \rangle \langle I_2 \rangle}{(\langle I_1 \rangle + \langle I_2 \rangle)^2}\big(1 - |\gamma|^2\big). \tag{3.11.10}$$

Since $0 \le |\gamma| \le 1$, we have, in general,

$$K^2 + V^2 \le 1. \tag{3.11.11}$$

A general relation $K^2 + V^2 \le 1$ holds in quantum theory when there are two possible paths to a final state.[40] K and V represent "which-path" distinguishability and interference, respectively, and $K^2 + V^2 \le 1$ expresses a complementary relation between path distinguishability and quantum-mechanical interference. It follows from the summation of probability amplitudes, and the fact that the probability for the two-channel process is the squared modulus of the sum, analogous in the classical double-slit example to the squaring of the sum of two fields to obtain an intensity.

---

[39] See, for instance, *Born and Wolf*. The relation $0 \le |\gamma| \le 1$ follows from the Schwarz inequality.

[40] Such a relation appears, for instance, in R. J. Glauber, in *New Techniques and Ideas in Quantum Measurement Theory*, D. M. Greenberger, ed., Ann. N. Y. Acad. Sci. **480**, pp. 336–372 (1986); D. M. Greenberger and A. Yasin, Phys. Lett. A **128**, 391 (1988); L. Mandel, Opt. Lett. **16**, 1882 (1991); G. Jaeger, A. Shimony, and L. Vaidman, Phys. Rev. A **51**, 54 (1995); and B. G. Englert, Phys. Rev. Lett. **77**, 2154 (1996).

The formula $E = h\nu$ for the energy of a photon already presents us with a sort of wave–particle duality: E is the energy of a photon, and ν is the frequency of a wave. It was suggested by Bohr, Kramers, and Slater in 1924 that, when an atom makes a spontaneous transition from a state with energy E_2 to one with energy $E_1 = E_2 - h\nu$, the radiation might take the form of a continuous, classical wave of finite duration rather than an instantaneously emitted quantum of light; energy in the Bohr–Kramers–Slater theory is conserved only statistically, not in individual transitions. This was rejected by Einstein, who suggested an experiment to test whether the emission of radiation is instantaneous or temporally continuous. An idealized conception of the experiment is shown in Figure 3.4. An atom with velocity v passes a slit of width d, and emits radiation into a spectroscope behind the slit. According to Heisenberg, this experiment presented a paradox: from the point of view in which the radiation is in the form of a continuous wave, the radiation reaching the spectroscope has the frequency bandwidth

$$\Delta\nu \sim \frac{v}{d}, \tag{3.11.12}$$

since the wave packet reaching the spectroscope has a finite duration d/v. But, from the point of view in which the radiation is emitted instantaneously in the form of a light quantum of frequency ν, it might seem that there should be no such spectral width: "The atom emits monochromatic radiation, the energy of each particle of which is $h\nu$, and the diaphragm (because of its great mass) will not be able to change the energy of the particles."[41] What is missing from that argument is the diffraction of light by the slit and the Doppler effect. If θ is the angular spread of the light due to the slit, the Doppler shift at the spectroscope is

$$\Delta\nu \approx \frac{\nu v}{c}\sin\theta \approx \frac{\nu v}{c}\frac{\lambda}{d} = \frac{v}{d}, \tag{3.11.13}$$

in agreement with (3.11.12). In other words, the same frequency spread is predicted by the "continuous wave" and "instantaneous" pictures of the radiation from the atom.

The experiment originally suggested by Einstein involved canal rays passing a wire grid. (A "canal ray," in the terminology of the time, consists of a stream of atoms and ions passing through a hole in the cathode of an electric discharge and into the vacuum behind the cathode.) Einstein soon realized, in contrast to his initial expectation that such an experiment would confirm the "instantaneous" nature of the radiation, that it should yield the same result regardless of whether the radiation is instantaneous or continuous. In 1926 E. Rupp reported experiments in which he used a Michelson interferometer to measure the coherence length $c/\Delta\nu$ of radiation from canal rays passing a slit. Rupp's experiments were said to support Einstein's revised expectation that the coherence length should be as expected from the continuous-wave theory. Heisenberg in his book cites Rupp's experiment in a section titled "The Experiment of Einstein and Rupp," but there is reason to believe that this and other experiments reported by Rupp, who at the time was regarded as one of the leading experimental physicists in the world, were fraudulent.[42]

---

[41] *Heisenberg*, p. 79.

[42] J. van Dongen, Hist. Stud. Phys. Biol Sci. **37**, Supplement, 73 (2007); A. P. French, Phys. Perspect. **1**, 3 (1999).

Fig. 3.4 Schematic of the "Einstein–Rupp experiment" in which an atom passes a slit of width d. A spectroscope placed behind the slit is used to analyze radiation from the atom.

The Bohr–Kramers–Slater theory theory had already been ruled out by the Bothe–Geiger experiment in 1924. According to the Bohr–Kramers–Slater theory, there is a (random) difference between the time of recoil of an electron and the time at which the scattered photon appears. Bothe and Geiger performed coincidence-counting experiments showing that the two events occur within a time interval no greater than about a millisecond.[43] In more recent times, this time interval has been found to be on the order of a few picoseconds or less.

We have reviewed a few examples of how uncertainty relations ensure the internal consistency of quantum theory. We will recall here one more example—again involving light—that convinced Einstein, after Bohr's analysis, that quantum mechanics is, at least, a consistent theory.[44] This thought experiment was presented at the sixth Solvay Conference in 1930 by Einstein, as an argument *against* the internal consistency of quantum theory. He considered a box filled with radiation and containing a clock that opens a small hole in the box at a certain time and for a very short time interval, during which a photon escapes from the box. By weighing the box before and after the photon escapes, the energy E of the photon can be determined using $E = mc^2$. Since the time interval can be precisely controlled by the clock, and the change Δm in the mass of the box can apparently be precisely measured, there appears to be a violation of the energy–time uncertainty relation: both the photon energy and the time at which it leaves the box can be measured with arbitrarily high precision, and therefore its energy and arrival time at a plane outside the box can be predicted with arbitrarily small uncertainties.[45] Einstein's argument came as "quite a shock to Bohr" and made him "extremely unhappy," but, during a sleepless night, he constructed a rebuttal.[46]

Bohr reasoned as follows. The box can be supposed to be suspended by a spring balance and has attached to it a pointer, so that its position can be read off a ruler fixed on the structure supporting the spring balance (see Figure 3.5). The weight of the

[43] W. Bothe and H. Geiger, Z. Phys. **26**, 44 (1924).

[44] It did not change Einstein's fundamental objections to quantum theory, which we review in Section 5.8.

[45] See *Pais (1982)*, pp. 446–448.

[46] Recollections of L. Rosenfeld, as quoted in *Pais (1991)*, p. 427.

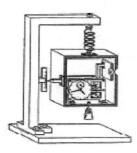

Fig. 3.5 Drawing by Niel's Bohr of Einstein's "clock-in-a-box" thought experiment. From *Pais (1982)*, p. 447, with permission.

box after the hole is closed can then be determined by adding or removing smaller and smaller weights that serve to bring the balance back to the "zero" displacement position before the hole is opened. There will be some imprecision in this zero displacement, and this uncertainty, Δx, implies an uncertainty $\Delta p \geq \hbar/2\Delta x$ in the vertical momentum of the box. Now, it takes some time τ to add or remove the last weight $(\Delta m)g$ in the balancing procedure, during which an impulse on the order of $(\Delta m)g\tau$ or less is imparted to the box; hence

$$\Delta p < (\Delta m)gT, \tag{3.11.14}$$

where $T(>\tau)$ is the time taken to carry out the entire balancing process, so that

$$\Delta x \geq \frac{\hbar}{2\Delta p} \geq \frac{\hbar}{2(\Delta m)gT}. \tag{3.11.15}$$

Bohr next invokes a consequence of general relativity. If a clock is displaced by Δx along a gravitational field, its rate changes. Over the course of a time interval T, the change in the time it records is approximately

$$\Delta t = \frac{1}{c^2}g\Delta x T. \tag{3.11.16}$$

From (3.11.15), therefore,

$$\Delta t \geq \frac{\hbar}{2\Delta mc^2} = \frac{\hbar}{2\Delta E}, \tag{3.11.17}$$

or $\Delta E \Delta t \geq \hbar/2$. ΔE is the uncertainty in the energy of the photon, and Δt is the uncertainty in the time of passage of the photon due to the uncertainty in the clock rate associated with the uncertainty Δx. The energy–time uncertainty relation is therefore satisfied.

Bohr's argument is not entirely compelling to everyone. Why should it be necessary to account for a consequence of general relativity—the change in the clock rate in a gravitational field—in order to support a consequence of quantum theory? This and other aspects of Bohr's analysis have been addressed in the literature. It has been argued, for example, that the change in the time recorded by a clock that is displaced in a gravitational field follows from the assumption that energy has weight, and, as such, does not require general relativity. We will summarize the argument.[47]

Denote by H_c the Hamiltonian for the clock's internal variables, which do not depend on its center of mass. If a mass E/c^2 is associated with an energy E, then the appropriate Hamiltonian when the clock is at a height x in a uniform gravitational field characterized by a local gravitational acceleration g is

$$H = H_c + \left(M + \frac{H_c}{c^2}\right)gx + H_e, \qquad (3.11.18)$$

where M is the mass of the clock, and H_e includes all contributions to the total clock Hamiltonian that do not depend on the *internal* variables of the clock. Now, an operator A corresponding to any internal clock variable, or a function thereof, commutes with H_e and x, and therefore its Heisenberg equation of motion is

$$\frac{dA}{dt} = \frac{1}{i\hbar}[A, H] = \frac{1}{i\hbar}\left(1 + \frac{gx}{c^2}\right)[A, H_c]. \qquad (3.11.19)$$

If there were no gravitational field, the equation of motion would, of course, be

$$\frac{dA}{dt} = \frac{1}{i\hbar}[A, H_c]. \qquad (3.11.20)$$

Therefore, if $A = f(t)$ is a solution for A in the absence of any gravitational field, the solution for A when the clock is at rest in a gravitational field will, under identical initial conditions, be

$$A = f[t(1 + \frac{gx}{c^2})]. \qquad (3.11.21)$$

In other words, the change in the time recorded by the clock when it is displaced by Δx in the gravitational field is

$$\Delta t = T\frac{g\Delta x}{c^2} \qquad (3.11.22)$$

for a given time interval T; this is formula (3.11.16), which is valid to order $1/c^2$, according to general relativity.

Bohr thought about the photon-in-a-box experiment for many years. He published his response to Einstein nearly two decades after his first (oral) presentation of it, and is said

---

[47] W. G. Unruh and G. I. Opat, Am. J. Phys. **47**, 743 (1979). The analysis in this paper is more general than our adaptation in that only the proportionality of energy and weight is invoked. The authors remark that the argument goes through in the same way whether one uses quantum mechanics (commutators) or classical Hamiltonian mechanics (Poisson brackets).

to have sketched the experiment on his blackboard on the last day of his life (November 18, 1962).[48]

Exercise 3.12: Is it necessary for Einstein's argument, or for Bohr's rebuttal, that the escaping radiation consists of a single photon? What if the box is filled with a gas—do Einstein's argument and Bohr's rebuttal carry over if the clockwork allows *atoms* to escape from the box? What if the experiment is arranged so that the clock is held fixed *outside* the box and opens and closes the hole by remote control?

3.12 More on Uncertainty Relations

In the examples in Sections 3.10 and 3.11, "uncertainties" such as Δx and Δp were not precisely defined. Uncertainty relations for non-commuting observables were first shown to follow from the formalism of quantum theory by Kennard, who *defined* the uncertainties as standard deviations.[49] We begin by reviewing a derivation of the uncertainty relation for two observables A and B, taking the standard deviation (square root of the variance) as the measure of "uncertainty." Define the variances $\Delta A^2 = \langle\psi|\mathcal{A}^2|\psi\rangle$, and $\Delta B^2 = \langle\psi|\mathcal{B}^2|\psi\rangle$, for a state $|\psi\rangle$, where $\mathcal{A} = A - \langle A\rangle$, and $\mathcal{B} = B - \langle B\rangle$. Define also the states $|\psi_\mathcal{A}\rangle = \mathcal{A}|\psi\rangle$, and $|\psi_\mathcal{B}\rangle = \mathcal{B}|\psi\rangle$, in terms of which

$$\Delta A^2 \Delta B^2 = \langle\psi|\mathcal{A}^2|\psi\rangle\langle\psi|\mathcal{B}^2|\psi\rangle = \langle\psi_\mathcal{A}|\psi_\mathcal{A}\rangle\langle\psi_\mathcal{B}|\psi_\mathcal{B}\rangle \geq |\langle\psi_\mathcal{A}|\psi_\mathcal{B}\rangle|^2, \qquad (3.12.1)$$

where, in the last step, we use the Schwarz inequality.[50] Next, we use the identity

$$\langle\psi_\mathcal{A}|\psi_\mathcal{B}\rangle = \langle\psi|\mathcal{A}\mathcal{B}|\psi\rangle = \frac{1}{2}\langle\psi|(\mathcal{A}\mathcal{B}+\mathcal{B}\mathcal{A})|\psi\rangle + \frac{1}{2}\langle\psi|(\mathcal{A}\mathcal{B}-\mathcal{B}\mathcal{A})|\psi\rangle. \qquad (3.12.2)$$

The first term on the right in the second equality is real, since it is the expectation value of a Hermitian operator; the second term is the expectation value of an *anti*-Hermitian operator and is therefore a purely imaginary number. Therefore,

$$|\langle\psi_\mathcal{A}|\psi_\mathcal{B}\rangle|^2 = \frac{1}{4}|\langle\psi|(\mathcal{A}\mathcal{B}+\mathcal{B}\mathcal{A})|\psi\rangle|^2 + \frac{1}{4}|\langle\psi|[\mathcal{A},\mathcal{B}]|\psi\rangle|^2, \qquad (3.12.3)$$

and, from (3.12.1) and $[\mathcal{A},\mathcal{B}] = [A,B]$,

$$\Delta A^2 \Delta B^2 \geq \frac{1}{4}|\langle\psi|[A,B]|\psi\rangle|^2 + \frac{1}{4}|\langle\psi|(\mathcal{A}\mathcal{B}+\mathcal{B}\mathcal{A})|\psi\rangle|^2, \qquad (3.12.4)$$

as first derived by Schrödinger in 1930.[51] If, for example, $[A,B] = [x,p] = i\hbar$,

---

[48] J. Kalckar, ed., *Foundations of Quantum Physics II (1933–1958)*, Elsevier, Amsterdam, 1996, p. 286.

[49] E. H. Kennard, Z. Phys. **44**, 326 (1927). What are now called squeezed states also appeared in this paper; see M. M. Nieto, Phys. Lett. A **229**, 135 (1997).

[50] $|\langle\phi|\chi\rangle|^2 \leq \langle\phi|\phi\rangle\langle\chi|\chi\rangle$ for any two ket vectors $|\phi\rangle$ and $|\chi\rangle$.

[51] See, for instance, I. Kim and G. Mahler, Phys. Lett. A **269**, 287 (2000), and references therein.

$$\Delta A^2 \Delta B^2 \geq \frac{\hbar^2}{4} + \frac{1}{4}|\langle\psi|\delta x \delta p + \delta p \delta x|\psi\rangle|^2, \tag{3.12.5}$$

where $\delta x = x - \langle x \rangle$, and $\delta p = p - \langle p \rangle$. The second (state-dependent) term on the right, the covariance of δx and δp, is usually ignored; then, the usual Heisenberg uncertainty relation expressing a lower bound on $\Delta x \Delta p$ follows. Similarly, the second term in (3.12.4) is usually ignored, and a lower bound on $\Delta A^2 \Delta B^2$ is expressed as

$$\Delta A^2 \Delta B^2 \geq \frac{1}{4}|\langle\psi|[A,B]|\psi\rangle|^2. \tag{3.12.6}$$

This is the "generalized" uncertainty relation, as first derived by Robertson.[52]

For a minimal value of the uncertainty product $\Delta A \Delta B$, we require

$$\Delta A^2 \Delta B^2 = \frac{1}{4}|\langle\psi|[A,B]|\psi\rangle|^2, \tag{3.12.7}$$

and

$$\langle\psi|(\mathcal{A}\mathcal{B} + \mathcal{B}\mathcal{A})|\psi\rangle = 0. \tag{3.12.8}$$

We also require, from (3.12.1), that $\langle\psi_\mathcal{A}|\psi_\mathcal{A}\rangle\langle\psi_\mathcal{B}|\psi_\mathcal{B}\rangle = |\langle\psi_\mathcal{A}|\psi_\mathcal{B}\rangle|^2$, which, in turn, requires that the ket vectors $|\psi_\mathcal{A}\rangle$ and $|\psi_\mathcal{B}\rangle$ be "parallel":

$$\mathcal{A}|\psi\rangle = \lambda\mathcal{B}|\psi\rangle, \tag{3.12.9}$$

where λ is a complex number. Now, from (3.12.8),

$$\langle\psi|(\mathcal{A}\mathcal{B} + \mathcal{B}\mathcal{A})|\psi\rangle = (\lambda^* + \lambda)\langle\psi_\mathcal{B}|\psi_\mathcal{B}\rangle = 0, \tag{3.12.10}$$

that is, $\lambda = i\kappa$, κ real. Then, (3.12.9) takes the form

$$(A - \langle A\rangle)|\psi\rangle = i\kappa(B - \langle B\rangle)|\psi\rangle, \tag{3.12.11}$$

or

$$(A - i\kappa B)|\psi\rangle = (\langle A\rangle - i\kappa\langle B\rangle)|\psi\rangle. \tag{3.12.12}$$

The conclusion is that, for $|\psi\rangle$ to be a state for which the product $\Delta A \Delta B$ is minimized, it must satisfy the eigenvalue equation given in (3.12.12).

In the case $A = p$, $B = q$, and $[q, p] = i\hbar$, (3.12.12) becomes

$$(p - i\kappa q)|\psi\rangle = (\langle p\rangle - i\kappa\langle q\rangle)|\psi\rangle, \tag{3.12.13}$$

and (3.12.7) reduces to $\Delta q^2 \Delta p^2 = \hbar^2/4$. For the harmonic oscillator, comparison with (3.6.2) shows that, in order for $\langle q|\psi\rangle$ to have the same spatial width Δq as the ground-state wave function $\psi_0(q)$, we must put $\kappa = \omega$. Then, (3.12.13) is equivalent to $a|\psi\rangle = \alpha|\psi\rangle$, $\alpha = (\langle p\rangle - i\omega\langle q\rangle)/\sqrt{2\hbar\omega}$, that is, our minimal uncertainty (coherent) state must be an eigenstate of the annihilation operator a.

[52] H. P. Robertson, Phys. Rev. **34**, 573 (1929).

Heisenberg originally inferred uncertainty relations from consideration of the disturbance of a system caused by an observation or measurement, as in the "Heisenberg microscope." But the derivation of the uncertainty relation (3.12.6) invokes no physics; the inequality (3.12.6) is just a mathematical requirement for two Hermitian operators A and B with variances ΔA^2 and ΔB^2 as defined with respect to some $|\psi\rangle$. The physical application is made when we identify $\langle A \rangle$ as an expectation value of an observable A of a system described by the state vector $|\psi\rangle$. An operational interpretation of the uncertainty relation then follows: if we measure A for a large number of identically prepared systems, we will generally obtain different numerical values for each system and will record a distribution of values and a standard deviation ΔA. If, instead, we measure B following preparation of the systems in the same way we will likewise record a distribution of values and a standard deviation ΔB. If $[A, B] \neq 0$, the product $\Delta A \Delta B$ must have the lower bound (3.12.6). This lower bound does not imply that a measurement of A must "disturb" a measurement of B, or vice versa, since the measurements of A and B are not made simultaneously on the same individual systems.

This operational interpretation of uncertainty relations leaves much room for further discussion, bearing on different interpretations of quantum theory itself. Do uncertainty relations apply to single measurements on individual systems or only to ensembles of identically prepared systems? The latter viewpoint is emphatically expressed by Pagels:

An important warning must be stated regarding Heisenberg's uncertainty relation: it does not apply to a single measurement on a single particle, although people often think of it that way. Heisenberg's relation is a statement about a statistical average over lots of measurements of position and momenta . . . the uncertainty Δp or Δq has meaning only if you repeat measurements. Some people imagine that quantum objects like the electron are 'fuzzy' because we cannot measure their position and momentum simultaneously and they therefore lack objectivity, but that way of thinking is inaccurate.[53]

There is the related question of whether a pure state $|\psi\rangle$ (and the very notion of probability in quantum theory!) characterizes an individual system or an ensemble; essentially the same issue has long been debated in classical probability theory.[54] Because the question does not bear directly on analyses of actual experiments, it is not always explicitly addressed in textbooks.[55]

[53] H. Pagels, *The Cosmic Code. Quantum Physics as the Language of Nature*, Bantam Books, New York, 1984, p. 71.

[54] E. T. Jaynes, *Probability Theory. The Logic of Science*, Cambridge University Press, Cambridge, 2005, especially Chapter 10.

[55] Exceptions include *Schiff* (p. 15: "ψ will be regarded as describing the behavior of a single particle or photon, not the statistical distribution of a number of such quanta") and *Gottfried and Yan* (p. 40: "we will assume that the quantum state is associated with individual systems"). In contrast, the view is taken in *Ballentine* (p. 175), for instance, that "a pure state describes the statistical properties of an *ensemble* of similarly prepared systems."

3.12.1   Simultaneous Measurement of Non-Commuting Observables

Dispersions Δx and Δp of canonically conjugate variables like position x and momentum p are *inherent* to any quantum state, prior to any measurement, and must satisfy $\Delta x \Delta p \geq \hbar/2$. Any measurement of x and p is limited by this uncertainty relation. In a *simultaneous* measurement of x and p in the sense described below, there is a further limitation—a doubling of the minimal product of the dispersions allowed by the uncertainty relation.[56]

   A simultaneous measurement of two observables of a system requires, as with any measurement, a coupling to some measurement apparatus. Imagine repeated simultaneous measurements of

$$X = \alpha x + A_x \tag{3.12.14}$$

and

$$P = \beta p + A_p, \tag{3.12.15}$$

where α and β are real numbers and A_x and A_p are Hermitian operators associated with intrinsic noise involved in an actual measurement. We assume that values of X and P can be read out separately and simultaneously by meters attached to the measurement apparatus, and that the values registered by the meters are normalized such that $\alpha = \beta = 1$. For simplicity, we assume also that the measurement does not introduce any bias; then, $\langle X \rangle = \langle x \rangle$, $\langle P \rangle = \langle p \rangle$ and, therefore, $\langle A_x \rangle = \langle A_p \rangle = 0$. Since X and P can be measured simultaneously,

$$[X, P] = 0. \tag{3.12.16}$$

With another reasonable assumption that x and A_p, and p and A_x, are uncorrelated, so that $[x, A_p] = [p, A_x] = 0$, it follows from (3.12.16) that

$$[A_x, A_p] = -i\hbar, \tag{3.12.17}$$

and therefore, from (3.12.6),

$$\Delta A_x^2 \Delta A_p^2 = \langle A_x^2 \rangle \langle A_p^2 \rangle \geq \hbar^2/4. \tag{3.12.18}$$

From (3.12.14) and (3.12.15), $\langle X^2 \rangle = \langle x^2 \rangle + \langle A_x^2 \rangle = \langle x^2 \rangle + \Delta A_x^2$, $\langle P^2 \rangle = \langle p^2 \rangle + \langle A_p^2 \rangle = \langle p^2 \rangle + \Delta A_p^2$, and

$$\Delta X^2 \Delta P^2 = \Delta x^2 \Delta p^2 + \Delta A_x^2 \Delta A_p^2 + \Delta x^2 \Delta A_p^2 + \Delta p^2 \Delta A_x^2$$
$$= \left(\Delta x^2 + \Delta A_x^2 \right) \left(\Delta p^2 + \Delta A_p^2 \right). \tag{3.12.19}$$

Thus, $\Delta X^2 = \Delta x^2 + \Delta A_x^2$ has a contribution Δx^2 expressing a variance in the initial state of the system *prior to any measurement*, and a contribution ΔA_x^2 resulting from

---

[56] E. Arthurs and J. L. Kelly, Jr., Bell Systems Tech. J. **44**, 725 (1965). See also E. Arthurs and M. S. Goodman, Phys. Rev. Lett. **60**, 2447 (1988).

the coupling to the measurement apparatus; similarly, $\Delta P^2 = \Delta p^2 + \Delta A_p^2$ has a contribution Δp^2 from a variance in the initial state prior to a measurement, and a contribution ΔA_p^2 from coupling to the measurement apparatus.[57] The uncertainty relation imposed by these "inherent" and "measurement" contributions is

$$\Delta X^2 \Delta P^2 = \Delta x^2 \Delta p^2 + \Delta A_x^2 \Delta A_p^2 + \Delta x^2 \Delta A_p^2 + \Delta p^2 \Delta A_x^2$$
$$\geq \Delta x^2 \Delta p^2 + \Delta A_x^2 \Delta A_p^2 + 2\left[\Delta x^2 \Delta p^2 \Delta A_x^2 \Delta A_p^2\right]^{1/2}, \qquad (3.12.20)$$

or, since $\Delta x^2 \Delta p^2 \geq \hbar^2/4$, and $\Delta A_x^2 \Delta A_p^2 \geq \hbar^2/4$,

$$\Delta X^2 \Delta P^2 \geq \left(\frac{\hbar^2}{4} + \frac{\hbar^2}{4} + 2\frac{\hbar^2}{4}\right) = \hbar^2, \qquad (3.12.21)$$

and

$$\Delta X \Delta P \geq \hbar. \qquad (3.12.22)$$

The lower bound on this uncertainty product is twice that imposed by the Heisenberg uncertainty relation $\Delta x \Delta p \geq \hbar/2$. This "penalty" incurred in a simultaneous measurement of two non-commuting observables is a consequence of the inevitable noise involved in a measurement. The noise is inevitable because, without the noise operators A_x and A_p, the commutation relation (3.12.16) would be impossible.

3.12.2 Energy–Time Uncertainty Relations

The energy–time uncertainty relation

$$\Delta E \Delta t \geq \hbar/2 \qquad (3.12.23)$$

is frequently invoked to interpret such time-dependent effects as the radiative decay of an excited atomic state: a lifetime τ_R, identified as Δt, implies a spectral width $\Delta \omega = \Delta E/\hbar \approx 1/\tau_R$ of the spontaneously emitted radiation, consistent with observation. Similarly, the identification of Δt in (3.12.23) with a mean collision time τ_c in a gas implies a collision-broadened spectral width $\Delta \omega \approx 1/\tau_c$. In the context of "virtual particles," one encounters statements to the effect that "the Heisenberg uncertainty principle $\Delta E \Delta t \approx h$ permits an amount of energy, ΔE, to be 'borrowed' for the duration Δt."[58]

Such interpretations are often criticized because, unlike the uncertainty relation $\Delta x \Delta p \geq \hbar/2$, (3.12.23) cannot be derived from a commutation relation between two observables; there is no Hermitian "time operator" T canonically conjugate to the Hamiltonian H. In fact, as shown by Aharonov and Bohm, "there is no reason inherent in the principles of quantum theory why the energy of a system cannot be measured in as short a time as we please."[59]

---

[57] For discussion of this point and an insightful analysis of joint measurements of position and momentum in single-slit diffraction, see M. G. Raymer, Am. J. Phys. **62**, 986 (1994). See also C. Y. She and H. Heffner, Phys. Rev. **152**, 1103 (1966).

[58] P. C. W. Davies, *The Accidental Universe*, Cambridge University Press, Cambridge, 1982, p. 106.

[59] Y. Aharonov and D. Bohm, Phys. Rev. **122**, 1649 (1961), p. 1651; Phys. Rev. **134**, B1417 (1964).

Suppose there were a Hermitian "time operator" T satisfying $[T, H] = i\hbar$. Then, an uncertainty relation $\Delta H \Delta T \geq \hbar/2$ would follow in consequence of (3.12.6). Now, since

$$He^{i\alpha T/\hbar} = [H, e^{i\alpha T/\hbar}] + e^{i\alpha T/\hbar} H = -i\hbar \frac{\partial}{\partial T} e^{i\alpha T/\hbar} + e^{i\alpha T/\hbar} H$$

$$= e^{i\alpha T/\hbar}(H + \alpha), \tag{3.12.24}$$

it would also follow that, if $H|E\rangle = E|E\rangle$, then

$$He^{i\alpha T/\hbar}|E\rangle = (E + \alpha)e^{i\alpha T/\hbar}|E\rangle. \tag{3.12.25}$$

In other words, $E + \alpha$ would be an allowed eigenvalue of H. But since α can be any real number, that would be inconsistent with the requirement that the Hamiltonian must be bounded from below (see Exercise 3.1). Therefore, there can be no Hermitian operator T such that $[T, H] = i\hbar$.[60]

Application of (3.12.23) to radiation seems to make better sense when we combine the relation $E = \hbar\omega$ for photons with the idea well known from classical theory that a pulse of duration Δt has a frequency bandwidth $\Delta\omega \approx 1/\Delta t$. Suppose, for example, that a single-photon pulse of duration Δt is incident on a medium that is so strongly absorbing that there is a high probability that the pulse (photon) will be absorbed. Of course, we can only calculate the rate of absorption, not the time at which the absorption occurs, for which there is an "uncertainty" $\approx \Delta t \approx 1/\Delta\omega$. And the energy absorbed is also imprecisely defined: $\Delta E = \hbar\Delta\omega$, and therefore $\Delta E \Delta t \approx \hbar$. In the case of radiative decay, similarly, we can know the lifetime τ_R but cannot predict the exact time for the occurrence of a "quantum jump" and the emission of a photon, and so there is an uncertainty $\Delta t \approx \tau_R$ in the time of emission. Assuming the applicability of (3.12.23), this Δt, together with $E = \hbar\omega$, implies the familiar relation $\Delta\omega \approx 1/\tau_R$.

With regard to the "borrowing" of energy for a time Δt, consider the following example. A two-state atom at time $t = 0$ is presumed to be in its ground state and to interact with a single-mode *vacuum* field of frequency ω. The probability at time $t > 0$ that the atom is in its excited state is

$$P(t) \propto \frac{\sin^2\left[\frac{1}{2}(\omega + \omega_0)t\right]}{\left[\frac{1}{2}(\omega + \omega_0)\right]^2}, \tag{3.12.26}$$

where ω_0 is the atom's transition frequency.[61] For sufficiently short times, there appears to be a significant excitation probability, but, of course, calling it an "energy-nonconserving" process ignores the simple fact that the Hamiltonian of the atom-plus-field system does not change in time. In a Feynman diagram, the process corresponds

---

[60] This argument follows *Ballentine*, p. 239, where it is attributed to Pauli.

[61] See, for instance, (2.3.3). The resonant term makes no contribution when the calculation is done with field quantization, whereas there is a non-rotating-wave-approximation term involving $\omega_{fi} + \omega = \omega_0 + \omega$ that results from $\langle aa^\dagger \rangle = 1$ for the vacuum-field state.

to an energy-level (Lamb) shift that appears pictorially as a "virtual" transition in which the atom emits a photon and then absorbs a photon (from the vacuum!). It is not clear what this really has to do with "uncertainties" ΔE and Δt.

An energy–time uncertainty relation is sometimes inferred from the uncertainty relation between position and momentum. Suppose we know the velocity v of an electron and measure the time $t = x/v$ it takes for it to move some distance x. An uncertainty Δx in the measured distance implies an uncertainty $\Delta t = \Delta x/v \geq \hbar/2v\Delta p$ in the elapsed time, or, since $\Delta E = \Delta(\frac{1}{2}mv^2) = p\Delta v$ when x is measured, $\Delta E \Delta t \geq \hbar/2$.

Heisenberg emphasized that "the uncertainty relation does not refer to the past; if the velocity of the electron is at first known and the position then accurately measured, the position for times previous to the measurement may be calculated. Then for these past times $\Delta p \Delta q$ is smaller than the usual limiting value, but this knowledge of the past is of a purely speculative character, since it can never (because of the unknown change in momentum caused by the position measurement) be used as an initial condition in any calculation of the future progress of the electron and thus cannot be subjected to experimental verification. It is a matter of personal belief whether such a calculation concerning the past history of the electron can be ascribed any physical reality or not."[62] In the example just given, both the velocity at some earlier time and the initial position relative to the displacement we measure can be assumed to be precisely known.

In all these examples, Δt is regarded in the usual way as an "external" parameter, independent of any observables of the system. A relation of the form (3.12.23) can be derived by defining a Δt_A for each observable A of a system.[63] Consider an observable A that is not a constant of the motion, so that $[A, H] \neq 0$. From (3.12.6),

$$\Delta E \Delta A \geq \frac{1}{2}|\langle [A, H] \rangle| = \frac{1}{2}|\langle \frac{dA}{dt} \rangle|. \qquad (3.12.27)$$

The second equality follows from the Heisenberg equation of motion, $i\hbar dA/dt = [A, H]$. We now *define* Δt_A as a time interval over which the observable A undergoes a significant variation:

$$\Delta t_A = \langle A \rangle / |\langle dA/dt \rangle|. \qquad (3.12.28)$$

We then obtain the *Mandelstam–Tamm relation*

$$\Delta E \Delta t_A \geq \hbar/2. \qquad (3.12.29)$$

Aside from the fact that this relation applies to a particular observable A, it is not really an uncertainty relation in the usual sense, because Δt_A characterizes a rate of variation of A and not an uncertainty inherent in a time measurement. Unlike, (3.12.23), however, it expresses a general connection between a dispersion ΔE in energy and a particular time interval Δt_A, and it follows rigorously from basic principles of quantum theory.

---

[62] *Heisenberg*, p. 20.
[63] L. Mandelstam and I. Tamm, J. Phys. USSR **9**, 249 (1945).

References and Suggested Additional Reading

Some of the many works on the quantized electromagnetic field in optical physics are *Klauder and Sudarshan*; *Louisell*; *Loudon*; *Mandel and Wolf (1995)*; *Scully and Zubairy*; *Vogel, Welsch, and Wallentowitz*; and *Garrison and Chiao*.

The simplified approach to field quantization in dielectric media in Section 3.4 follows P. W. Milonni, J. Mod. Opt. **42**, 1991 (1995). See also P. D. Drummond, Phys. Rev. A **42**, 6845 (1990); R. J. Glauber and M. Lewenstein, Phys. Rev. A **43**, 467 (1991); *Ginzburg*; and *Garrison and Chiao*.

The discussion of coherent states of the field and the diagonal representation of the density matrix is based mainly on papers by R. J. Glauber (Phys. Rev. **130**, 2529 (1963) and Phys. Rev. **131**, 2766 (1963)) and on *Klauder and Sudarshan*.

Reviews of squeezed states include D. F. Walls, Nature **306**, 141 (1983); R. Loudon and P. L. Knight, J. Mod. Opt. **34**, 709 (1987); and U. L. Andersen, G. Tobias, C. Marquardt, and G. Leuchs, Phys. Scr. **91**, 053001 (2016).

For history and different interpretations of the uncertainty principle, focusing especially on the views of Heisenberg and Bohr, see, for instance, J. Hilgevoord and J. Uffink, "The Uncertainty Principle," The Stanford Encyclopedia of Philosophy (Winter 2016 Edition), ed. E. N. Zalta, http://plato.stanford.edu/entries/qt-uncertainty/, accessed September 12, 2018, and references therein. Simultaneous measurements of non-commuting observables are reviewed by Y. Yamamoto and H. A. Haus, Rev. Mod. Phys. **58**, 1001 (1986), and S. Stenholm, Ann. Phys. **218**, 233 (1992). For a review of energy–time uncertainty relations, with references to hundreds of papers, see V. V. Dodonov and A. V. Dodonov, Phys. Scr. **90**, 074049 (2015).

Many influential papers on photon statistics, field fluctuations, and classical and quantum theories of optical coherence are reprinted in *Mandel and Wolf (1970)*.

4

Interaction Hamiltonian and Spontaneous Emission

In Chapter 2, the atom–field interaction was formulated in the electric dipole approximation. The atom was treated quantum mechanically, the field classically. This semiclassical approach is, of course, fundamentally inconsistent: observables associated with charged particles satisfy uncertainty relations, while field observables do not. In this chapter, atoms and fields are both described consistently according to quantum theory. We focus on spontaneous emission, the origin of most of the light from the Sun, the source of some of the most significant effects of noise in photonics, and one of the phenomena most often cited in connection with the inadequacies of semiclassical radiation theories.

It is usually more convenient in quantum optics, and often more generally in non-relativistic quantum electrodynamics, to work with the electric dipole form of the Hamiltonian rather than the more fundamental minimal coupling form. We begin by considering again these Hamiltonians and their equivalence, this time with the quantized field.

4.1 Atom–Field Hamiltonian: Why Minimal Coupling?

The interaction Hamiltonian for a point particle of charge e and mass m in an electromagnetic field is obtained in classical electrodynamics by replacing the linear momentum $\mathbf{p}$ in the free-particle Hamiltonian by $\mathbf{p} - e\mathbf{A}$, where $\mathbf{A}$ is the vector potential. In other words, the kinetic energy term $\mathbf{p}^2/2m$ is replaced by $(\mathbf{p} - e\mathbf{A})^2/2m$, implying the interaction Hamiltonian (Section 2.2)

$$H_I = -\frac{e}{m}\mathbf{A} \cdot \mathbf{p} + \frac{e^2}{2m}\mathbf{A}^2. \qquad (4.1.1)$$

For reasons given below, this procedure of replacing $\mathbf{p}$ by $\mathbf{p} - e\mathbf{A}$ to obtain the interaction Hamiltonian is called *minimal coupling*, although it was used long before it came to be called minimal coupling. The momentum $\mathbf{p}$ $(= m\mathbf{v} + e\mathbf{A})$ in the minimal coupling Hamiltonian is the momentum conjugate to the coordinate $\mathbf{x}$. It is the momentum that is conserved if the Lagrangian is independent of $\mathbf{x}$. We take (4.1.1) to be the interaction Hamiltonian in classical electrodynamics because we obtain from it the correct Lorentz force law, $\mathbf{F} = e(\mathbf{E} + \mathbf{v} \times \mathbf{B})$. In (non-relativistic) quantum

An Introduction to Quantum Optics and Quantum Fluctuations. Peter W. Milonni.
© Peter W. Milonni 2019. Published in 2019 by Oxford University Press.
DOI:10.1093/oso/9780199215614.001.0001

electrodynamics, we simply carry over the same Hamiltonian but take $\mathbf{x}$, $\mathbf{p}$, and $\mathbf{A}$ to be (Hermitian) operators.

A deeper perspective on minimal coupling comes from the *gauge principle*. The Schrödinger equation for a free particle of mass m is

$$i\hbar\frac{\partial}{\partial t}\psi(\mathbf{x},t) = -\frac{\hbar^2}{2m}\nabla^2\psi(\mathbf{x},t) \tag{4.1.2}$$

in the coordinate representation, and the same equation is satisfied by the wave function $\psi'(\mathbf{x},t) = \psi(\mathbf{x},t)e^{i\theta}$, with θ a constant. This form invariance of the Schrödinger equation is a *global* phase invariance. The gauge principle requires invariance under a *local* phase transformation: $\psi'(\mathbf{x},t) = \psi(\mathbf{x},t)e^{ie\chi(\mathbf{x},t)}$, with e the charge of the particle and $\chi(\mathbf{x},t)$ depending on both time and the position of the particle. Consider, for example, a local phase transformation such that χ is independent of time. $\psi'(\mathbf{x},t) = \psi(\mathbf{x},t)e^{ie\chi(\mathbf{x})}$ is not a solution of the free-particle Schrödinger equation, so, to respect the gauge principle, we seek a Schrödinger equation that *is* invariant under such a local phase transformation. Such an equation is

$$i\hbar\frac{\partial}{\partial t}\psi(\mathbf{x},t) = -\frac{\hbar^2}{2m}\mathcal{D}^2\psi(\mathbf{x},t), \tag{4.1.3}$$

where

$$\mathcal{D}(\mathbf{x},t) = \nabla - \frac{ie}{\hbar}\mathbf{A}(\mathbf{x},t) \tag{4.1.4}$$

and the vector field $\mathbf{A}(\mathbf{x},t)$ transforms as

$$\mathbf{A} \to \mathbf{A}' = \mathbf{A} + \nabla\chi \tag{4.1.5}$$

when the local phase transformation

$$\psi(\mathbf{x},t) \to \psi'(\mathbf{x},t) = \psi(\mathbf{x},t)e^{ie\chi(\mathbf{x})} \tag{4.1.6}$$

is applied to ψ. Under these transformations, (4.1.3) becomes a Schrödinger equation of the same form:

$$i\hbar\frac{\partial}{\partial t}\psi'(\mathbf{x},t) = -\frac{\hbar^2}{2m}\mathcal{D}'^2\psi(\mathbf{x},t)$$

$$\mathcal{D}' = \nabla - \frac{ie}{\hbar}\mathbf{A}'. \tag{4.1.7}$$

In other words, the replacement

$$\nabla \to \mathcal{D} = \nabla - \frac{ie}{\hbar}\mathbf{A} \tag{4.1.8}$$

in the free-particle Schrödinger equation results in a Schrödinger equation that is invariant under the local phase transformation (4.1.6), provided that $\mathbf{A}$ is transformed according to (4.1.5). Since $\mathbf{p} = (\hbar/i)\nabla$, (4.1.8) is equivalent to $\mathbf{p} \to \mathbf{p} - e\mathbf{A}$. Thus, the gauge principle leads us from the free-particle Schrödinger equation to a Schrödinger equation describing an interaction of the particle with a vector field that must transform in a particular way (see (4.1.5)) under a local phase transformation. Of course, this transformation is just the familiar gauge transformation of the vector potential (Section 1.3). So, by requiring invariance under a local phase transformation, we deduce that a charge e interacts with a "gauge field" $\mathbf{A}$ that happens to be just the familiar vector potential. The gauge principle in this sense suggests the *existence* of such a "force field."

The term "minimal" coupling in electromagnetic interactions was evidently coined by Murray Gell-Mann as "a principle that is given wide, though usually tacit acceptance."[1] Minimal coupling involves charges but not, directly, higher multipole moments, which appear as a *consequence* of the minimal coupling. In the gauge theories of modern particle physics, the "charges," or coupling constants, are, of course, not necessarily electric, but could be the color charges of quantum chromodynamics, for example. The "covariant derivative" $\mathcal{D}_\mu$, which reduces to $\mathcal{D}$ in our simple example, defines a *minimal* coupling for *point* particles because, more generally, we could add to $\mathcal{D}_\mu$ terms that are consistent with the gauge principle but would imply additional interactions that are not included (or needed) in electrodynamics.

The complete Hamiltonian for a charged particle and the electromagnetic field is obtained by adding the field energy to $(\mathbf{p} - e\mathbf{A})^2/2m$:

$$H = \frac{1}{2m}(\mathbf{p} - e\mathbf{A})^2 + \frac{1}{2}\int d^3r(\epsilon_0\mathbf{E}^2 + \mu_0\mathbf{H}^2). \qquad (4.1.9)$$

We work in the Coulomb gauge: $\mathbf{A}$ is transverse ($\nabla \cdot \mathbf{A} = 0, \mathbf{A} = \mathbf{A}^\perp$) and therefore gauge-invariant (Section 3.3.2). In terms of the transverse and longitudinal parts of the electric field ($\mathbf{E} = \mathbf{E}^\perp + \mathbf{E}^\parallel$, $\mathbf{E}^\perp = -\partial\mathbf{A}/\partial t$, $\mathbf{E}^\parallel = -\nabla\phi$), the integral over $\mathbf{E}^2$ in (4.1.9) is

$$\int d^3r\mathbf{E}^2 = \int d^3r(\mathbf{E}^\perp + \mathbf{E}^\parallel)^2$$

$$= \int d^3r(\mathbf{E}^{\perp 2} + \mathbf{E}^{\parallel 2}) + 2\int d^3r(\mathbf{E}^\perp \cdot \mathbf{E}^\parallel). \qquad (4.1.10)$$

The last integral vanishes (see also Appendix B):

$$\int d^3r(\mathbf{E}^\perp \cdot \mathbf{E}^\parallel) = -\int d^3r(\mathbf{E}^\perp \cdot \nabla\phi) = \int d^3r\phi(\nabla \cdot \mathbf{E}^\perp) = 0, \qquad (4.1.11)$$

where we have integrated by parts and used the fact that $\nabla \cdot \mathbf{E}^\perp$ is identically zero. Furthermore,

$$\int d^3r\mathbf{E}^{\parallel 2} = \int d^3r(\nabla\phi \cdot \nabla\phi) = \int d^3r[\nabla \cdot (\phi\nabla\phi) - \phi\nabla^2\phi]$$

$$= -\int d^3r\phi\nabla^2\phi = \frac{1}{\epsilon_0}\int d^3r\rho\phi, \qquad (4.1.12)$$

---

[1] M. Gell-Mann, Suppl. Nuovo Cim. **4**, 848 (1956), p. 853.

where we have used the divergence theorem, the assumption that ϕ (like $\mathbf{E}^\perp$) vanishes at infinity, and the Poisson equation ($\nabla^2 \phi = -\rho/\epsilon_0$) for the Coulomb-gauge scalar potential. Therefore,

$$\int d^3 r \mathbf{E}^2 = \int d^3 r (\mathbf{E}^{\perp 2} + \mathbf{E}^{\|2}) = \int d^3 r \mathbf{E}^{\perp 2} + \frac{1}{\epsilon_0} \int d^3 r \rho \phi \tag{4.1.13}$$

and

$$H = \frac{1}{2m}(\mathbf{p} - e\mathbf{A})^2 + \frac{1}{2} \int d^3 r (\epsilon_0 \mathbf{E}^{\perp 2} + \mu_0 \mathbf{H}^2) + \frac{1}{2}\epsilon_0 \int d^3 r \mathbf{E}^{\|2}$$

$$= \frac{1}{2m}(\mathbf{p} - e\mathbf{A})^2 + \frac{1}{2} \int d^3 r (\epsilon_0 \mathbf{E}^{\perp 2} + \mu_0 \mathbf{H}^2) + \frac{1}{2} \int d^3 r \rho \phi. \tag{4.1.14}$$

The last contribution to H is just the Coulomb energy of a charge density ρ in a scalar potential ϕ. For a charge distribution in free space, this energy can be expressed in terms of ρ alone using the solution we obtained in Chapter 1 for the Poisson equation (see 1.3.39):

$$\frac{1}{2} \int d^3 r \rho \phi = \frac{1}{8\pi\epsilon_0} \int d^3 r \int d^3 r' \frac{\rho(\mathbf{r}, t)\rho(\mathbf{r}', t)}{|\mathbf{r} - \mathbf{r}'|}. \tag{4.1.15}$$

For a collection of $\mathcal{N}$ point charges q_i at positions $\mathbf{r}_i(t)$, for example,

$$\rho(\mathbf{r}, t) = \sum_{i=1}^{\mathcal{N}} q_i \delta^3(\mathbf{r} - \mathbf{r}_i), \tag{4.1.16}$$

and

$$\frac{1}{2} \int d^3 r \rho \phi = \frac{1}{4\pi\epsilon_0} \sum_{i}^{\mathcal{N}} \sum_{j>i}^{\mathcal{N}} \frac{q_i q_j}{|\mathbf{r}_i - \mathbf{r}_j|}, \tag{4.1.17}$$

if we ignore (infinite) Coulomb self-energies. The Hamiltonian for the entire system of charges and the electromagnetic field is then

$$H = \sum_{i=1}^{\mathcal{N}} \frac{1}{2m_i} \left[\mathbf{p}_i - q_i \mathbf{A}(\mathbf{r}_i)\right]^2$$

$$+ \frac{1}{2} \int d^3 r (\epsilon_0 \mathbf{E}^{\perp 2} + \mu_0 \mathbf{H}^2) + \frac{1}{4\pi\epsilon_0} \sum_{i}^{\mathcal{N}} \sum_{j>i}^{\mathcal{N}} \frac{q_i q_j}{|\mathbf{r}_i - \mathbf{r}_j|}. \tag{4.1.18}$$

This minimal coupling Hamiltonian is applicable in both classical and quantum electrodynamics. As remarked earlier, we make the transition from classical to quantum theory by taking the particles' coordinates and momenta, as well as the vector potential $\mathbf{A}$ and therefore the electric and magnetic fields, to be Hermitian operators in a Hilbert space. In the Coulomb gauge used here, it is only necessary to quantize the (transverse) vector potential, from which the transverse fields $\mathbf{E}^\perp$ and $\mathbf{H}$ follow.

The effect of the scalar potential, and therefore the longitudinal part of the electric field, is contained entirely in *instantaneous Coulomb interactions between the charges*, the last term in (4.1.18). As discussed in Section 1.3.2 and Appendix A, the electromagnetic interactions between particles are, nevertheless, properly retarded in the Coulomb gauge.

Note also that, in the Coulomb gauge, the momentum operators $\mathbf{p}_i$ commute with $\mathbf{A}$, even though $\mathbf{A}$ for each particle in the Hamiltonian is evaluated at the coordinate $\mathbf{r}_i$ of that particle. In the coordinate representation, for example,

$$\mathbf{p}_i \cdot \mathbf{A}\psi \to \frac{\hbar}{i}\nabla \cdot (\mathbf{A}\psi) = \frac{\hbar}{i}(\nabla \cdot \mathbf{A})\psi + \frac{\hbar}{i}\mathbf{A} \cdot \nabla\psi = \frac{\hbar}{i}\mathbf{A} \cdot \nabla_i\psi = \mathbf{A} \cdot \mathbf{p}_i\psi. \quad (4.1.19)$$

4.1.1 Maxwell's Equations for Field Operators

We are interested primarily in electrons bound to nuclei, in the approximation that charge-neutral atoms act as electric dipoles in their interaction with the field. Consider to begin with the case of a single, stationary atom at the fixed position $\mathbf{r}_0$, which, in the electric dipole approximation we now make, is the same as the position of the nucleus. The Heisenberg equation of motion for the transverse electric field operator is

$$i\hbar\frac{\partial}{\partial t}E_i^\perp(\mathbf{r}, t) = [E_i^\perp(\mathbf{r}), H]. \quad (4.1.20)$$

Since H is now given by (4.1.18) with $\mathcal{N} = 1$ and $\mathbf{A} = \mathbf{A}(\mathbf{r}_0)$,

$$i\hbar\frac{\partial}{\partial t}E_i^\perp(\mathbf{r}) = \frac{1}{2m}[E_i^\perp(\mathbf{r}), [\mathbf{p} - e\mathbf{A}(\mathbf{r}_0)]^2] + \frac{\mu_0}{2}\int d^3r'[E_i^\perp(\mathbf{r}), \mathbf{H}^2(\mathbf{r}')]. \quad (4.1.21)$$

For notational simplicity here, we do not indicate time dependence, as all the operators refer to the same time. Now, since the commutator $[E_i^\perp(\mathbf{r}), A_j(\mathbf{r}_0)]$ is just a c-number (see (3.10.13)), and the field operator $E_i^\perp(\mathbf{r})$ commutes with the atom operator $\mathbf{p}$,

$$[E_i^\perp(\mathbf{r}), [\mathbf{p} - e\mathbf{A}(\mathbf{r}_0)]^2] = 2[E_i^\perp(\mathbf{r}), p_j - eA_j(\mathbf{r}_0)][p_j - eA_j(\mathbf{r}_0)]$$
$$= -2e[E_i^\perp(\mathbf{r}), A_j(\mathbf{r}_0)]m\dot{x}_j, \quad (4.1.22)$$

where m and e $(= -|e|)$ are, respectively, the mass and charge of the electron, and $m\dot{\mathbf{x}} = m\mathbf{v} = \mathbf{p} - e\mathbf{A}(\mathbf{r}_0)$, the kinetic (as opposed to canonical) momentum; $\mathbf{x}$ may be regarded as the displacement of the electron from the nucleus, the two opposite charges forming an electric dipole moment $e\mathbf{x}$. Similarly,

$$[E_i^\perp(\mathbf{r}), \mathbf{H}^2(\mathbf{r}')] = 2[E_i^\perp(\mathbf{r}), H_j(\mathbf{r}')]H_j(\mathbf{r}'), \quad (4.1.23)$$

with implied summation over repeated indices (j). Next, we use the equal-time commutation relations (3.10.13) and (3.10.21),

$$[E_i^\perp(\mathbf{r}), A_j(\mathbf{r}_0)] = \frac{i\hbar}{\epsilon_0}\delta_{ij}^\perp(\mathbf{r} - \mathbf{r}_0) \quad (4.1.24)$$

and

$$[E_i^{\perp}(\mathbf{r}), H_j(\mathbf{r}')] = -i\hbar c^2 \epsilon_{ijk} \frac{\partial}{\partial r_k'} \delta^3(\mathbf{r} - \mathbf{r}'), \qquad (4.1.25)$$

whence

$$\frac{\partial}{\partial t} E_i^{\perp}(\mathbf{r}) = -\frac{1}{\epsilon_0} e\dot{x}_j \delta_{ij}^{\perp}(\mathbf{r} - \mathbf{r}_0) + \frac{1}{\epsilon_0} \epsilon_{ijk} \frac{\partial}{\partial r_j} H_k(\mathbf{r}), \qquad (4.1.26)$$

or since

$$\epsilon_{ijk} \frac{\partial}{\partial r_j} H_k(\mathbf{r}) = (\nabla \times \mathbf{H})_i, \qquad (4.1.27)$$

$$\nabla \times \mathbf{H} = \epsilon_0 \frac{\partial \mathbf{E}^{\perp}}{\partial t} + \frac{\partial \mathbf{P}^{\perp}}{\partial t}. \qquad (4.1.28)$$

We have defined the single-atom electric dipole moment density (which is the "polarization" in the electric dipole approximation)

$$\mathbf{P}(\mathbf{r}) = e\mathbf{x}\delta^3(\mathbf{r} - \mathbf{r}_0), \qquad (4.1.29)$$

the transverse part of which is

$$P_j^{\perp}(\mathbf{r}) = ex_i \delta_{ij}^{\perp}(\mathbf{r} - \mathbf{r}_0). \qquad (4.1.30)$$

In similar fashion, we obtain from the Heisenberg equation for $\mathbf{H}$ the Maxwell equation

$$\nabla \times \mathbf{E}^{\perp} = -\mu_0 \frac{\partial \mathbf{H}}{\partial t}. \qquad (4.1.31)$$

These results are easily generalized to the case of $\mathcal{N}$ atoms: (4.1.29) and (4.1.30) are replaced by

$$\mathbf{P}(\mathbf{r}) = \sum_{\ell=1}^{\mathcal{N}} e\mathbf{x}_\ell \delta^3(\mathbf{r} - \mathbf{r}_\ell), \qquad (4.1.32)$$

$$P_j^{\perp}(\mathbf{r}) = \sum_{\ell=1}^{\mathcal{N}} ex_{\ell i} \delta_{ij}^{\perp}(\mathbf{r} - \mathbf{r}_\ell). \qquad (4.1.33)$$

$e\mathbf{x}_\ell$ is the dipole moment operator for the ℓth atom at the fixed position $\mathbf{r}_\ell$. Now, since, by assumption, there are no free, unbound charges, $\nabla \cdot \mathbf{D} = \nabla \cdot (\epsilon_0 \mathbf{E}^{\parallel} + \mathbf{P}^{\parallel}) = 0$ everywhere. Therefore,

$$\mathbf{E}^{\parallel} = -\frac{1}{\epsilon_0} \mathbf{P}^{\parallel}. \qquad (4.1.34)$$

Noting furthermore that $\nabla \times \mathbf{E}^{\perp} = \nabla \times \mathbf{E}$, we can write (4.1.28) and (4.1.31) equivalently as

$$\nabla \times \mathbf{H} = \epsilon_0 \frac{\partial \mathbf{E}}{\partial t} + \frac{\partial \mathbf{P}}{\partial t} = \frac{\partial \mathbf{D}}{\partial t} \qquad (4.1.35)$$

and

$$\nabla \times \mathbf{E} = -\mu_0 \frac{\partial \mathbf{H}}{\partial t} = -\frac{\partial \mathbf{B}}{\partial t}, \qquad (4.1.36)$$

respectively. $\mathbf{B} = \nabla \times \mathbf{A}$ implies $\nabla \cdot \mathbf{B} = 0$, and so we have demonstrated for our purposes here that the four Maxwell equations for the electromagnetic field *operators* are satisfied when the time evolution is governed by the minimal coupling Hamiltonian.

4.2 Electric Dipole Hamiltonian

The transformation from the minimal coupling Hamiltonian to the electric dipole form was carried out in Section 2.2, in classical theory as well as in semiclassical radiation theory. We now derive the electric dipole form of the Hamiltonian when the charges and the field are both treated quantum mechanically. We first consider the simplest case of interest, a single one-electron atom coupled to the electromagnetic field. The minimal coupling Hamiltonian (4.1.18) reduces to

$$H = \frac{1}{2m}(\mathbf{p} - e\mathbf{A})^2 + V(\mathbf{x}) + \frac{1}{2}\int d^3r(\epsilon_0\mathbf{E}^{\perp 2} + \mu_0\mathbf{H}^2), \qquad (4.2.1)$$

where

$$V(\mathbf{x}) = -\frac{1}{4\pi\epsilon_0}\frac{e^2}{\mathbf{x}}. \qquad (4.2.2)$$

It is assumed that there is no longitudinal field other than the Coulomb field of the nucleus. The nucleus is assumed to be stationary and therefore its kinetic energy is not included in H. In the electric dipole approximation, $\mathbf{A}$ is evaluated at the fixed location $\mathbf{r}_0$ of the nucleus in the first term on the right side of (4.2.1). In the semiclassical approach of Section 2.2, the vector potential $\mathbf{A}(\mathbf{r}_0)$ was a (classical) c-number, but the electron coordinate $\mathbf{x}$ was a Hermitian operator. Now both are Hermitian operators, as are $\mathbf{E}^{\perp}$ and $\mathbf{H}$.

The transformation to the electric dipole form of the Hamiltonian can be approached in different ways.[2] We choose to follow closely the semiclassical approach of Section 2.2, employing now the unitary operator

$$\mathcal{U} = e^{-iS/\hbar} = e^{ie\mathbf{A}(\mathbf{r}_0)\cdot\mathbf{x}/\hbar}.[3] \qquad (4.2.3)$$

In this approach, we replace the (atom-and-field) state vector $|\psi\rangle$, whose time evolution is governed by the minimal coupling Hamiltonian H, by the state vector $|\psi'\rangle$ defined by

$$|\psi\rangle = \mathcal{U}|\psi'\rangle. \qquad (4.2.4)$$

As in Section 2.2, we use the time-dependent Schrödinger equation

$$i\hbar\frac{\partial}{\partial t}|\psi\rangle = H|\psi\rangle \qquad (4.2.5)$$

to determine the time evolution of the state vector $|\psi'\rangle$. Now, however, the transformation from $|\psi\rangle$ to $|\psi'\rangle$ involves the *operator* (quantized) Coulomb-gauge vector potential

---

[2] See, for instance, J. R. Ackerhalt and P. W. Milonni, J. Opt. Soc. Am. B **1**, 116 (1984), and references therein.

[3] E. A. Power and S. Zienau, Phil. Trans. R. Soc. **A251**, 427 (1959). See also *Power*.

$\mathbf{A}(\mathbf{r}_0)$. Since $\mathbf{A}(\mathbf{r}_0)$, $\mathbf{x}$, and therefore $\mathcal{U}$ are independent of time in the Schrödinger picture, it follows from (4.2.4), (4.2.5), and the unitarity of $\mathcal{U}$ that

$$i\hbar \frac{\partial}{\partial t}|\psi'\rangle = \mathcal{U}^\dagger H \mathcal{U}|\psi'\rangle = H'|\psi'\rangle. \tag{4.2.6}$$

From the general operator identity (2.2.18), the commutation relation (3.10.13), and the fact that the vector potential commutes with the magnetic field, it follows straightforwardly that

$$\mathcal{U}^\dagger \mathbf{p} \mathcal{U} = \mathbf{p} - \frac{ie}{\hbar}[\mathbf{A}(\mathbf{r}_0) \cdot \mathbf{x}, \mathbf{p}] = \mathbf{p} + e\mathbf{A}(\mathbf{r}_0),$$

$$\mathcal{U}^\dagger \mathbf{x} \mathcal{U} = \mathbf{x},$$

$$\mathcal{U}^\dagger \mathbf{E}^\perp(\mathbf{r})\mathcal{U} = \mathbf{E}^\perp(\mathbf{r}) - \frac{1}{\epsilon_0}\mathbf{P}^\perp(\mathbf{r}),$$

$$\mathcal{U}^\dagger \mathbf{B}(\mathbf{r})\mathcal{U} = \mathbf{B}(\mathbf{r}) = \mu_0 \mathbf{H}(\mathbf{r}), \tag{4.2.7}$$

and therefore

$$H' = \frac{1}{2m}[\mathbf{p} + e\mathbf{A}(\mathbf{r}_0) - e\mathbf{A}(\mathbf{r}_0)]^2 + V(\mathbf{x}) + \frac{1}{2}\int d^3r\left[\epsilon_0(\mathbf{E}^\perp - \frac{1}{\epsilon_0}\mathbf{P}^\perp)^2 + \mu_0 \mathbf{H}^2)\right]$$

$$= \frac{1}{2m}\mathbf{p}^2 + V(\mathbf{x}) - \int d^3r(\mathbf{E}^\perp \cdot \mathbf{P})$$

$$+ \frac{1}{2}\int d^3r(\epsilon_0 \mathbf{E}^{\perp 2} + \mu_0 \mathbf{H}^2) + \frac{1}{2\epsilon_0}\int d^3r \mathbf{P}^{\perp 2}. \tag{4.2.8}$$

$\mathbf{P}$ and $\mathbf{P}^\perp$ are given by (4.1.32) and (4.1.33), with $\mathcal{N} = 1$, and $\mathbf{r}_1 = \mathbf{r}_0$. We have derived H' using the Schrödinger picture, but it goes without saying that it can just as well be used in the Heisenberg picture, for instance.

The transformed Hamiltonian (4.2.8), like (4.2.1), does not include the Coulomb energy

$$\frac{1}{2}\epsilon_0 \int d^3r \mathbf{E}^{\|2} = \frac{1}{2}\epsilon_0 \int d^3r\left(-\frac{1}{\epsilon_0}\mathbf{P}^\|\right)^2 = \frac{1}{2\epsilon_0}\int d^3r \mathbf{P}^{\|2}. \tag{4.2.9}$$

Adding this term to (4.2.8), and generalizing to $\mathcal{N}$ atoms, we obtain the general form of the electric dipole Hamiltonian:

$$H' = \sum_{i=1}^{\mathcal{N}}\left[\frac{\mathbf{p}_i^2}{2m} + V(\mathbf{x}_i)\right] - \int d^3r(\mathbf{E}^\perp \cdot \mathbf{P})$$

$$+ \int d^3r(\epsilon_0 \mathbf{E}^{\perp 2} + \mu_0 \mathbf{H}^2) + \frac{1}{2\epsilon_0}\int d^3r \mathbf{P}^2$$

$$= H_A - \int d^3r(\mathbf{E}^\perp \cdot \mathbf{P}) + \int d^3r(\epsilon_0 \mathbf{E}^{\perp 2} + \mu_0 \mathbf{H}^2) + \frac{1}{2\epsilon_0}\int d^3r \mathbf{P}^2, \tag{4.2.10}$$

with H_A the Hamiltonian for $\mathcal{N}$ atoms, and $\mathbf{P}$ defined by (4.1.32).

The electric dipole form of the Hamiltonian for the interaction of an atom with the electromagnetic field was apparently first employed by Maria Göppert-Mayer, who used it to treat two-photon absorption some 30 years before it was observed in experiments.[4] She derived the electric dipole form using the classical Lagrangian approach of Section 2.2. A difference between the classical or semiclassical derivations of the electric dipole form of the Hamiltonian from the minimal coupling form, compared to the approach here where the field is quantized, is worth noting: the term $(1/2\epsilon_0) \int d^3r \mathbf{P}^{\perp 2}$ does not appear in classical or semiclassical derivations. It does not appear for the simple reason that the commutation relation (3.10.13) between the vector potential and the transverse electric field is responsible for the appearance of $\mathbf{P}^{\perp}$ in $\mathcal{U}^{\dagger} \mathbf{E}^{\perp} \mathcal{U} = \mathbf{E}^{\perp} - (1/\epsilon_0) \mathbf{P}^{\perp}$, which in turn results in the term $(1/2\epsilon_0) \int d^3r \mathbf{P}^{\perp 2}$ in H'. If the field is not quantized, there is no such (non-vanishing) field commutator because the field variables are then c-numbers, not operators.

The formal equivalence of the minimal coupling and electric dipole Hamiltonians means, obviously, that physical predictions cannot depend on which form is used. For example, the expectation value in state $|\psi\rangle$ of an observable $\mathcal{O}(t)$, whose time evolution is governed by the Hamiltonian H, is identical to the expectation value in the transformed state $|\psi'\rangle = \mathcal{U}^{\dagger}|\psi\rangle$ of $\mathcal{O}'(t) = \mathcal{U}^{\dagger}\mathcal{O}(t)\mathcal{U}$, whose time evolution is determined by the Hamiltonian H':

$$\langle\psi|\mathcal{O}(t)|\psi\rangle = \langle\psi'|\mathcal{U}^{\dagger}\mathcal{O}(t)\mathcal{U}|\psi'\rangle = \langle\psi'|\mathcal{O}'(t)|\psi'\rangle. \tag{4.2.11}$$

Just as in the semiclassical theory of Section 2.2, the approach here involves a transformation of states from $|\psi\rangle$ to $|\psi'\rangle$ along with the transformation of the Hamiltonian from H to H'. The same considerations about whether a transformation of states is needed to ensure physically meaningful results of calculations performed with the minimal coupling and electric dipole Hamiltonians apply in both the semiclassical and the quantized-field formulations. But, as also discussed in that section, calculations based on the minimal coupling and electric dipole Hamiltonians will not necessarily yield exactly the same results when approximations, such as a restriction to an incomplete set of states, are made.

So, why bother to introduce the electric dipole Hamiltonian, when the more fundamental minimal coupling Hamiltonian is perfectly adequate? Apart from the fact that most workers in quantum optics use it, the electric dipole Hamiltonian has significant advantages. For one thing, the vector potential does not appear, and therefore neither do any questions about gauge invariance. For another, as emphasized by Power and Zienau, it is generally more convenient to use the electric dipole Hamiltonian for the calculation of interatomic electric dipole interactions.[5] This is clear from the fact that the last term on the right side of (4.2.10) is just a sum of *single-atom* (divergent!) self-energies:

[4] M. Göppert-Mayer, Ann. Physik **401**, 273 (1931).

[5] E. A. Power and S. Zienau, Phil. Trans. R. Soc. **A251**, 427 (1959). See also *Power*.

$$\frac{1}{2\epsilon_0} \int d^3r' \mathbf{P}^2(\mathbf{r}') = \frac{1}{2\epsilon_0} \sum_{\ell=1}^{\mathcal{N}} e^2 \mathbf{x}_\ell^2 \delta^3(0). \tag{4.2.12}$$

Interatomic interactions therefore come from the field $\mathbf{E}^\perp$ *in the Hamiltonian* H', whereas, when the minimal coupling Hamiltonian is used, there are contributions to these interactions from both the vector potential $\mathbf{A}$ and the instantaneous Coulomb interactions described by the last term on the right side of (4.1.18), the two contributions together ensuring properly retarded interactions (see Appendix A).

Recall that $\mathbf{E}^\perp$ in the minimal coupling Hamiltonian H satisfies Maxwell's equations. In particular, (4.1.28) and (4.1.31) imply that

$$\nabla^2 \mathbf{E}^\perp - \frac{1}{c^2} \frac{\partial^2 \mathbf{E}^\perp}{\partial t^2} = \frac{1}{\epsilon_0 c^2} \frac{\partial^2 \mathbf{P}^\perp}{\partial t^2} \quad \text{(from the minimal coupling Hamiltonian)}. \tag{4.2.13}$$

But a different wave equation is found for the operator $\mathbf{E}^\perp$ that appears in the electric dipole Hamiltonian H':

$$\nabla^2 \mathbf{E}^\perp - \frac{1}{c^2} \frac{\partial^2 \mathbf{E}^\perp}{\partial t^2} = \frac{1}{\epsilon_0} \nabla^2 \mathbf{P}^\perp \quad \text{(from the electric dipole Hamiltonian)}, \tag{4.2.14}$$

as shown below. Now, recall that $\epsilon_0 \mathbf{E}^\parallel + \mathbf{P}^\parallel = 0$ (see (4.1.34)) and that $\nabla \cdot \mathbf{D} = 0$ when the free-charge density $\rho = 0$. Then, (4.2.14) is the wave equation for $1/\epsilon_0$ times the electric displacement $\mathbf{D} = \epsilon_0 \mathbf{E} + \mathbf{P} = \epsilon_0 \mathbf{E}^\perp + \mathbf{P}^\perp$, so we might write $\mathbf{D}/\epsilon_0$ instead of $\mathbf{E}^\perp$ in H'. This distinction, however, is inconsequential for purposes like calculating the effects on an atom of an applied field or the field from other atoms. Write $\mathbf{D}/\epsilon_0$ instead of $\mathbf{E}^\perp$ in the Hamiltonian (4.2.10) and use (4.1.29) and (4.2.12) to derive the Heisenberg equations of motion for the electron coordinate and momentum operators for the atom at $\mathbf{r} = 0$: $\dot{x}_i = p_i/m$ and

$$\begin{aligned}
\dot{p}_i &= -\nabla_i V(\mathbf{x}) + \frac{e}{\epsilon_0} D_i + \frac{e^2}{2\epsilon_0}[p_i, \mathbf{x}^2]\delta^3(0) \\
&= -\nabla_i V(\mathbf{x}) + \frac{e}{\epsilon_0} D_i - \frac{e^2}{\epsilon_0} x_i \delta^3(0) \\
&= -\nabla_i V(\mathbf{x}) + \frac{e}{\epsilon_0}[D_i - P_i] = \nabla_i V(\mathbf{x}) + e\mathcal{E}_i, \tag{4.2.15}
\end{aligned}$$

where the operator $\mathcal{E}_i$ is the ith component of the *electric field* at the position of the atom. It is therefore the electric field that acts on the atom, even though the electric displacement appears in the electric dipole Hamiltonian.[6] Along this line, we will show in Section 4.3 that, at a point away from an atom in free space, where $\mathbf{P} = 0$, the field $\mathbf{E}^\perp/\epsilon_0$ calculated from the Hamiltonian (4.2.10) has exactly the familiar form of the electric field from an electric dipole. This is the electric field that can act on other atoms.

---

[6] More generally, the electric field will have a longitudinal component, in which case it will satisfy (6.6.74).

The transformation

$$\mathcal{U}^\dagger \mathbf{E}^\perp \mathcal{U} = \mathbf{E}^\perp - \mathbf{P}^\perp/\epsilon_0 \tag{4.2.16}$$

follows from

$$\mathcal{U}^\dagger a_{\mathbf{k}\lambda}\mathcal{U} = a_{\mathbf{k}\lambda} + ie\Big(\frac{1}{2\epsilon_0 \hbar \omega_k V}\Big)^{1/2} \mathbf{x} \cdot \mathbf{e}_{\mathbf{k}\lambda}, \tag{4.2.17}$$

$$[\mathcal{U}^\dagger a_{\mathbf{k}\lambda}\mathcal{U}, \mathcal{U}^\dagger a_{\mathbf{k}'\lambda'}^\dagger \mathcal{U}] = [a_{\mathbf{k}\lambda}, a_{\mathbf{k}'\lambda'}^\dagger] = \delta_{\mathbf{k}\mathbf{k}'}^3 \delta_{\lambda\lambda'}. \tag{4.2.18}$$

The second term on the right side of (4.2.17) leads to the second term on the right side of (4.2.16). Thus, the first term on the right side of (4.2.16), the operator $\mathbf{E}^\perp$ appearing in the electric dipole Hamiltonian, can be expressed in the same form as the transverse electric field operator in the minimal coupling Hamiltonian (recall (3.3.11)):

$$\mathbf{E}^\perp(\mathbf{r}) = i\sum_{\mathbf{k}\lambda} \sqrt{\frac{\hbar\omega_k}{2\epsilon_0 V}} \Big[a_{\mathbf{k}\lambda}e^{i\mathbf{k}\cdot\mathbf{r}} - a_{\mathbf{k}\lambda}^\dagger e^{-i\mathbf{k}\cdot\mathbf{r}}\Big]\mathbf{e}_{\mathbf{k}\lambda}, \tag{4.2.19}$$

and similarly for $\mathbf{H}(\mathbf{r})$. The operator $a_{\mathbf{k}\lambda}$ here is actually the transformed operator (4.2.17), but the distinction between these operators—and between the photons associated with the two equivalent Hamiltonians—is unimportant for our purposes: all that matters as we work with either the minimal coupling Hamiltonian or the electric dipole Hamiltonian is the commutation relation (4.2.18) used to derive Heisenberg equations of motion.

Using (4.1.33), (4.2.12), and (4.2.19), we write the electric dipole Hamiltonian (4.2.10) as

$$H' = H_A + H_F - ie\sum_{\ell=1}^{N}\sum_{\mathbf{k}\lambda}\Big(\frac{\hbar\omega_k}{2\epsilon_0 V}\Big)^{1/2}\big[a_{\mathbf{k}\lambda}e^{i\mathbf{k}\cdot\mathbf{r}_\ell} - a_{\mathbf{k}\lambda}^\dagger e^{-i\mathbf{k}\cdot\mathbf{r}_\ell}\big]\mathbf{x}_\ell \cdot \mathbf{e}_{\mathbf{k}\lambda}$$

$$+ \frac{1}{2\epsilon_0}\sum_{\ell=1}^{N}e^2\mathbf{x}_\ell^2\delta^3(\mathbf{r}=\mathbf{r}_\ell),$$

$$H_A = \sum_{\ell=1}^{N}\Big(\frac{\mathbf{p}_\ell^2}{2m} + V(\mathbf{x}_\ell)\Big),$$

$$H_F = \int d^3r\big(\epsilon_0\mathbf{E}^{\perp 2} + \mu_0\mathbf{H}^2\big) = \sum_{\mathbf{k}\lambda}\hbar\omega_k(a_{\mathbf{k}\lambda}^\dagger a_{\mathbf{k}\lambda} + 1/2). \tag{4.2.20}$$

We will now show directly from this Hamiltonian that $\mathbf{E}^\perp$ satisfies (4.2.14). The Heisenberg equation of motion for $a_{\mathbf{k}\lambda}$ is

$$i\hbar\dot{a}_{\mathbf{k}\lambda} = [a_{\mathbf{k}\lambda}, H'] = \hbar\omega_k a_{\mathbf{k}\lambda} + ie\sum_{\ell=1}^{N}\Big(\frac{\hbar\omega_k}{2\epsilon_0 V}\Big)^{1/2}e^{-i\mathbf{k}\cdot\mathbf{r}_\ell}\mathbf{x}_\ell(t)\cdot\mathbf{e}_{\mathbf{k}\lambda}; \tag{4.2.21}$$

therefore,

$$\ddot{a}_{\mathbf{k}\lambda} = -\omega_k^2 a_{\mathbf{k}\lambda} + e \sum_{\ell=1}^{\mathcal{N}} \sum_{i=1}^{3} \left(\frac{\omega_k}{2\epsilon_0 \hbar V}\right)^{1/2} e^{-i\mathbf{k}\cdot\mathbf{r}_\ell}(-i\omega_k x_{\ell i} + \dot{x}_{\ell i}) e_{\mathbf{k}\lambda i}, \qquad (4.2.22)$$

and, from (4.2.19),

$$
\begin{aligned}
\left(\nabla^2 - \frac{1}{c^2}\frac{\partial^2}{\partial t^2}\right) E_j^\perp(\mathbf{r}, t) &= -\frac{ie}{c^2} \sum_{\ell=1}^{\mathcal{N}} \sum_{i=1}^{3} \sum_{\mathbf{k}\lambda} \left(\frac{\hbar\omega_k}{2\epsilon_0 V}\right)^{1/2} \left(\frac{\omega_k}{2\epsilon_0 \hbar V}\right)^{1/2} e^{i\mathbf{k}\cdot(\mathbf{r}-\mathbf{r}_\ell)} \\
&\quad \times \left[-i\omega_k x_{\ell i} + \dot{x}_{\ell i}\right] e_{\mathbf{k}\lambda i} e_{\mathbf{k}\lambda j} + \mathrm{hc} \\
&= -\frac{e}{\epsilon_0} \sum_{\ell=1}^{\mathcal{N}} \sum_{i=1}^{3} \frac{1}{V} \sum_{\mathbf{k}} k^2 e^{i\mathbf{k}\cdot(\mathbf{r}-\mathbf{r}_\ell)} x_{\ell i}\left(\delta_{ij} - \frac{k_i k_j}{k^2}\right) \\
&= \frac{1}{\epsilon_0}\nabla^2 \sum_{\ell=1}^{\mathcal{N}} \sum_{i=1}^{3} e x_{\ell i}\left[\frac{1}{V}\sum_{\mathbf{k}} e^{i\mathbf{k}\cdot(\mathbf{r}-\mathbf{r}_\ell)}\left(\delta_{ij} - \frac{k_i k_j}{k^2}\right)\right], \qquad (4.2.23)
\end{aligned}
$$

where "hc" stands for Hermitian conjugate. We have used the identity (3.10.5) and the fact that $k^2 = \omega_k^2/c^2$ for the field in free space. Finally, we let the quantization volume $V \to \infty$ and use (3.3.18) to identify the term in square brackets in (4.2.23) as the transverse delta function, $\delta_{ij}^\perp(\mathbf{r} - \mathbf{r}_\ell)$, thus concluding that $\mathbf{E}^\perp$ satisfies (4.2.14).

A few points about the electric dipole Hamiltonian (4.2.10) are worth emphasizing. The vector potential $\mathbf{A}$ does not appear, there is no question as to the gauge invariance of H' or of calculations based on it, and H_A is the Hamiltonian for the unperturbed, non-interacting atoms. For a single one-electron atom, its eigenstates in the coordinate representation are the familiar hydrogenic Schrödinger wave functions $\phi_{n\ell m}(r, \theta, \phi)$. As discussed in the semiclassical framework in Section 2.2, these are the initial, time-independent states $|\psi'\rangle$ to use in Heisenberg-picture calculations; to obtain the same results with calculations based on the minimal coupling Hamiltonian, we use the transformed states $|\psi\rangle$ given by (4.2.4). Electric dipole interactions among the atoms are fully described by the interaction $-\mathbf{E}^\perp \cdot \mathbf{P}$; there is no need to add non-retarded, electrostatic dipole–dipole interactions, as must be done when the minimal coupling Hamiltonian is used. And although the field $\mathbf{E}^\perp$ appearing in the electric dipole Hamiltonian is actually $\mathbf{D}/\epsilon_0$, the field acting on atoms is, as expected, the electric field.

Finally we observe that the Hamiltonian (4.2.20) has two glaring infinities, one from the term involving the delta function with zero argument, and the other from the "zero-point energy"

$$\sum_{\mathbf{k}\lambda} \frac{1}{2}\hbar\omega_k \qquad (4.2.24)$$

in H_F. We will address the first infinity in Section 4.8. The zero-point energy is just a (c-)number and therefore commutes with everything else in the Hamiltonian and does not affect Heisenberg equations of motion; it can be omitted from the Hamiltonian by simply redefining the zero of energy. We will say more about it in Chapter 7.

4.3   The Field of an Atom

The formal solution of the Heisenberg equation of motion (4.2.21) for a single one-electron atom at $\mathbf{r} = 0$ is

$$a_{\mathbf{k}\lambda}(t) = a_{\mathbf{k}\lambda}(0)e^{-i\omega_k t} + \frac{e}{\hbar}\left(\frac{\hbar\omega_k}{2\epsilon_0 V}\right)^{1/2}\mathbf{e}_{\mathbf{k}\lambda}\cdot\int_0^t dt'\mathbf{x}(t')e^{i\omega_k(t'-t)}. \tag{4.3.1}$$

We now use this formal solution in (4.2.19). Since the field given by (4.2.19) is obviously a transverse field, we drop the redundant superscript ($\perp$) and write

$$\mathbf{E}(\mathbf{r},t) = \mathbf{E}_0(\mathbf{r},t) + \mathbf{E}_S(\mathbf{r},t), \tag{4.3.2}$$

where we have defined

$$\mathbf{E}_0(\mathbf{r},t) = i\sum_{\mathbf{k}\lambda}\left(\frac{\hbar\omega_k}{2\epsilon_0 V}\right)^{1/2}\left[a_{\mathbf{k}\lambda}(0)e^{-i(\omega_k t - \mathbf{k}\cdot\mathbf{r})} - a_{\mathbf{k}\lambda}^\dagger(0)e^{i(\omega_k t - \mathbf{k}\cdot\mathbf{r})}\right]\mathbf{e}_{\mathbf{k}\lambda} \tag{4.3.3}$$

and

$$\begin{aligned}
E_{Sj}(\mathbf{r},t) &= -2e\sum_i\sum_{\mathbf{k}\lambda}\left(\frac{\omega_k}{2\epsilon_0 V}\right)e^{i\mathbf{k}\cdot\mathbf{r}}e_{\mathbf{k}\lambda i}e_{\mathbf{k}\lambda j}\int_0^t dt'x_i(t')\sin\omega_k(t'-t) \\
&= -2e\sum_i\sum_{\mathbf{k}}\left(\frac{\omega_k}{2\epsilon_0 V}\right)\left(\delta_{ij} - \frac{k_i k_j}{k^2}\right)e^{i\mathbf{k}\cdot\mathbf{r}}\int_0^t dt'x_i(t')\sin\omega_k(t'-t) \\
&\to -\frac{e}{\epsilon_0}\left(\frac{1}{2\pi}\right)^3\sum_i\int d^3k\,\omega_k\left(\delta_{ij} - \frac{k_i k_j}{k^2}\right)e^{i\mathbf{k}\cdot\mathbf{r}} \\
&\quad\times\int_0^t dt'x_i(t')\sin\omega_k(t'-t).
\end{aligned} \tag{4.3.4}$$

We have used (3.10.5) and, in the last step, replaced the summation over wave vectors $\mathbf{k}$ by an integration in k space, as in (3.3.18):

$$\sum_{\mathbf{k}} \to \frac{V}{(2\pi)^3}\int d^3k. \tag{4.3.5}$$

$t = 0$ is a time when the atom–field interaction is artificially turned on. For present purposes, this just means that $\mathbf{E}_0(\mathbf{r},t)$ is a "free field" that exists independently of the atom, whereas $\mathbf{E}_S(\mathbf{r},t)$ is the "source field" due to the atom.

For now, we are only interested in the source part of $\mathbf{E}(\mathbf{r},t)$. Using (4.3.16) below, we can express (4.3.4) as

$$\begin{aligned}
\mathbf{E}_S(\mathbf{r},t) = -\frac{ec}{2\pi^2\epsilon_0}\int_0^t dt'x(t')\int_0^\infty dk\,k^3\Big\{&\left[\hat{\mathbf{d}} - (\hat{\mathbf{d}}\cdot\hat{\mathbf{r}})\hat{\mathbf{r}}\right]\frac{\sin kr}{kr} \\
&+ \left[\hat{\mathbf{d}} - 3(\hat{\mathbf{d}}\cdot\hat{\mathbf{r}})\hat{\mathbf{r}}\right]\left(\frac{\cos kr}{k^2 r^2} - \frac{\sin kr}{k^3 r^3}\right)\Big\}\sin kc(t'-t),
\end{aligned} \tag{4.3.6}$$

where we have used $\omega_k = kc$ and introduced $\hat{\mathbf{d}}$, the unit vector in the (fixed) direction of $\mathbf{x}(t)$ ($\mathbf{x}(t) = x(t)\hat{\mathbf{d}}$). The k integration brings in delta functions and derivatives of delta functions. For example,

$$\int_0^\infty dk k^3 \frac{\sin kr}{k^3 r^3} \sin kc(t' - t) = \frac{1}{2r^3} \int_0^\infty dk \big[\cos kc(t' - t - r/c) - \cos kc(t' - t + r/c)\big]$$

$$= -\frac{\pi}{2cr^3}\big[\delta(t' - t + r/c) - \delta(t' - t - r/c)\big]. \tag{4.3.7}$$

For $r > 0$, only the first delta function will contribute to the integral over t' in (4.3.6), so we retain only this delta function and take

$$\int_0^\infty dk k^3 \frac{\sin kr}{k^3 r^3} \sin kc(t' - t) \to -\frac{\pi}{2cr^3}\delta(t' - t + r/c) \tag{4.3.8}$$

in (4.3.6). Likewise,

$$\int_0^\infty dk k^3 \frac{\cos kr}{k^2 r^2} \sin kc(t' - t) = \frac{1}{cr^2} \frac{\partial}{\partial t'} \int_0^\infty dk \cos kr \cos kc(t' - t)$$

$$\to \frac{\pi}{2c^2 r^2} \frac{\partial}{\partial t'}\delta(t' - t + r/c), \tag{4.3.9}$$

and

$$\int_0^\infty dk k^3 \frac{\sin kr}{kr} \sin kc(t' - t) = -\frac{1}{c^2 r} \frac{\partial^2}{\partial t'^2} \int_0^\infty dk \sin kr \cos kc(t' - t)$$

$$\to \frac{\pi}{2c^3 r} \frac{\partial^2}{\partial t'^2}\delta(t' - t + r/c) \tag{4.3.10}$$

in (4.3.6). Thus, $\mathbf{E}(\mathbf{r}, t)$ has the familiar form of the retarded electric field of an electric dipole:

$$\mathbf{E}(\mathbf{r}, t) = \mathbf{E}_S(\mathbf{r}, t) = -\frac{1}{4\pi\epsilon_0}\Big\{\frac{1}{c^2 r}\Big[\hat{\mathbf{d}} - (\hat{\mathbf{d}} \cdot \hat{\mathbf{r}})\hat{\mathbf{r}}\Big]e\ddot{x}(t - r/c)$$

$$+ \Big[\hat{\mathbf{d}} - 3(\hat{\mathbf{d}} \cdot \hat{\mathbf{r}})\hat{\mathbf{r}}\Big]\Big[\frac{1}{cr^2}e\dot{x}(t - r/c) + \frac{1}{r^3}ex(t - r/c)\Big]\Big\} \tag{4.3.11}$$

for $t > r/c$.

The integration over $\mathbf{k}$ space in (4.3.4) is over all $k = |\mathbf{k}|$ and all solid angles about $\mathbf{k}$:

$$\sum_i \int d^3 k \omega_k \Big(\delta_{ij} - \frac{k_i k_j}{k^2}\Big)d_i e^{i\mathbf{k}\cdot\mathbf{r}} = \sum_i \int dk k^2 \omega_k \int d\Omega_{\mathbf{k}} \Big(\delta_{ij} - \frac{k_i k_j}{k^2}\Big)d_i e^{i\mathbf{k}\cdot\mathbf{r}}, \tag{4.3.12}$$

$d_i = ex_i$. The term with δ_{ij} is easily evaluated by taking the "z" axis of k space to be along $\mathbf{r} = r\hat{\mathbf{r}}$:

$$\int dk k^2 \omega_k \mathbf{d} \int d\Omega_{\mathbf{k}} e^{i\mathbf{k}\cdot\mathbf{r}} = \int dk k^2 \omega_k \mathbf{d}(2\pi)\int_0^\pi d\theta \sin\theta e^{ikr\cos\theta}$$

$$= 4\pi \int dk k^2 \omega_k \Big(\mathbf{d}\frac{\sin kr}{kr}\Big). \tag{4.3.13}$$

The remaining term is the jth component of the vector

$$
\begin{aligned}
\mathbf{F} &= -\int dk k^2 \omega_k \frac{1}{k^2} \int d\Omega_{\mathbf{k}} (\mathbf{d} \cdot \mathbf{k}) \mathbf{k} e^{i\mathbf{k} \cdot \mathbf{r}} \\
&= -\int dk k^2 \omega_k \frac{1}{k^2} \int_0^{2\pi} d\phi \int_0^{\pi} d\theta \sin\theta (d_x k_x + d_y k_y + d_z k_z)(k_x \hat{\mathbf{x}} + k_y \hat{\mathbf{y}} + k_z \hat{\mathbf{z}}) e^{ikr\cos\theta} \\
&= -\int dk k^2 \omega_k \int_0^{2\pi} d\phi \int_0^{\pi} d\theta \sin\theta (d_x \cos^2\phi \sin^2\theta \hat{\mathbf{x}} + d_y \sin^2\phi \sin^2\theta \hat{\mathbf{y}} \\
&\quad + d_z \cos^2\theta \hat{\mathbf{z}}) e^{ikr\cos\theta} \\
&= -4\pi \Big\{ (\mathbf{d} \cdot \hat{\mathbf{r}})\hat{\mathbf{r}} \frac{\sin kr}{kr} + [3(\mathbf{d} \cdot \hat{\mathbf{r}})\hat{\mathbf{r}} - \mathbf{d}]\Big(\frac{\cos kr}{k^2 r^2} - \frac{\sin kr}{k^3 r^3}\Big)\Big\},
\end{aligned}
\tag{4.3.14}
$$

and therefore (4.3.12) is the jth component of

$$
4\pi \int d^3 k \omega_k \Big\{ [\mathbf{d} - (\mathbf{d} \cdot \hat{\mathbf{r}})\hat{\mathbf{r}}] \frac{\sin kr}{kr} + [\mathbf{d} - 3(\mathbf{d} \cdot \hat{\mathbf{r}})\hat{\mathbf{r}}]\Big(\frac{\cos kr}{k^2 r^2} - \frac{\sin kr}{k^3 r^3}\Big)\Big\}.
\tag{4.3.15}
$$

We note also that these expressions imply

$$
\frac{1}{4\pi} \int d\Omega_{\mathbf{k}} \Big(\delta_{ij} - \frac{k_i k_j}{k^2}\Big) e^{i\mathbf{k} \cdot \mathbf{r}} = (\delta_{ij} - \hat{r}_i \hat{r}_j) \frac{\sin kr}{kr} + (\delta_{ij} - 3\hat{r}_i \hat{r}_j)\Big(\frac{\cos kr}{k^2 r^2} - \frac{\sin kr}{k^3 r^3}\Big).
\tag{4.3.16}
$$

Exercise 4.1: Show that part of the transverse electric field in the minimal coupling Hamiltonian (4.2.1) for a single atom in the dipole approximation is a static, nonretarded field that varies with the distance r from the atom as $1/r^3$.

The field (4.3.11) has exactly the form familiar from classical electromagnetism as the retarded electric field at any point away from a Hertzian electric dipole $e\mathbf{x}(t)$ (see (1.4.20)). This is consistent with the remarks in Section 4.2 that the field $\mathbf{E}^{\perp}(\mathbf{r}, t)$ in the electric dipole Hamiltonian acts as the electric field operator—as long as we are not concerned with self-interactions associated with the *radiation reaction* field $\mathbf{E}(0, t)$, which we will denote by $\mathbf{E}_{\mathrm{RR}}(t)$. This field follows from (4.3.6) with $r \to 0$, using

$$
\lim_{r \to 0} \Big\{ [\hat{\mathbf{d}} - (\hat{\mathbf{d}} \cdot \hat{\mathbf{r}})\hat{\mathbf{r}}] \frac{\sin kr}{kr} + [\hat{\mathbf{d}} - 3(\hat{\mathbf{d}} \cdot \hat{\mathbf{r}})\hat{\mathbf{r}}]\Big(\frac{\cos kr}{k^2 r^2} - \frac{\sin kr}{k^3 r^3}\Big)\Big\} = \frac{2}{3}\hat{\mathbf{d}}.
\tag{4.3.17}
$$

Thus,

$$
\begin{aligned}
\mathbf{E}_{\mathrm{RR}}(t) &= -\frac{e}{2\pi^2 \epsilon_0 c^3}\Big(\frac{2}{3}\hat{\mathbf{d}}\Big) \int_0^t dt' x(t') \frac{\partial^3}{\partial t'^3} \int_0^{\infty} d\omega \cos\omega(t' - t) \\
&= -\frac{e}{3\pi \epsilon_0 c^3} \int_0^t dt' \mathbf{x}(t') \frac{\partial^3}{\partial t'^3} \delta(t' - t).
\end{aligned}
\tag{4.3.18}
$$

Integrating by parts, and discarding terms that arise from the artificial sudden turn-on of the electromagnetic interaction at time $t = 0$, we obtain

$$\mathbf{E}_{\mathrm{RR}}(t) = \frac{e}{3\pi\epsilon_0 c^3}\left[\frac{1}{2}\,\dddot{\mathbf{x}}\,(t) - \ddot{\mathbf{x}}(t)\delta(t=0) - \mathbf{x}(t)\ddot{\delta}(t=0)\right]$$

$$= \frac{e}{6\pi\epsilon_0 c^3}\,\dddot{\mathbf{x}}\,(t) - \frac{e\Lambda}{3\pi^2\epsilon_0 c^3}\ddot{\mathbf{x}}(t) + \frac{e\Lambda^3}{9\pi^2\epsilon_0 c^3}\mathbf{x}(t), \qquad (4.3.19)$$

where we have used

$$\int_0^t dt'\,\dddot{\mathbf{x}}\,(t')\delta(t'-t) = \frac{1}{2}\,\dddot{\mathbf{x}}\,(t) \qquad (4.3.20)$$

and *defined*

$$\delta(t=0) = \lim_{T\to 0}\frac{1}{\pi}\int_0^\Lambda d\omega\cos\omega T = \frac{1}{\pi}\Lambda, \qquad (4.3.21)$$

$$\ddot{\delta}(t=0) = -\lim_{T\to 0}\frac{1}{\pi}\int_0^\Lambda d\omega\omega^2\cos\omega T = -\frac{1}{3\pi}\Lambda^3, \qquad (4.3.22)$$

and $\dot{\delta}(t=0) = 0$, since $\delta(t)$ is formally an even function of t. Λ is a high-frequency cutoff introduced to avert an ultraviolet catastrophe. The idea here is that the divergences we incur without such a cutoff can be "renormalized" or otherwise eliminated in a more complete analysis. In particular, the non-relativistic theory leading to (4.3.19) breaks down for frequencies ω greater than $mc^2/\hbar$. Moreover, the interaction of our dipole with arbitrarily high frequencies contradicts the dipole approximation in which the atom is assumed to interact with fields of wavelength large compared with its dimensions. We must introduce a high-frequency cutoff in $\mathbf{E}_{\mathrm{RR}}(t)$ to stay consistent with the assumptions made in deriving it.

To interpret the different parts of the radiation reaction field (4.3.19), consider the Heisenberg equation of motion, for a single atom at $\mathbf{r}_\ell = 0$, that follows from the electric dipole Hamiltonian (4.2.20):

$$m\ddot{x}_i = -\nabla_i V(\mathbf{x}) + eE_i(0,t) - \frac{e^2}{\epsilon_0}x_j\delta_{ij}^\perp(0)$$

$$= -\nabla_i V(\mathbf{x}) + eE_{0i}(0,t) + eE_{\mathrm{RR}i}(0,t) - \frac{e^2}{\epsilon_0}x_j\delta_{ij}^\perp(0). \qquad (4.3.23)$$

The last contribution arises from $(e^2/2\epsilon_0)\mathbf{x}^2\delta^3(0)$ in (4.2.20), after we drop a (divergent) Coulomb self-interaction energy $(e^2/2\epsilon_0)x_j\delta_{ij}^\parallel(0)$. From (4.3.19),

$$m\ddot{x}_i = -\nabla_i V(\mathbf{x}) + eE_{0i}(0,t) + \frac{e^2}{6\pi\epsilon_0 c^3}\,\dddot{x}_i$$

$$- \frac{e^2\Lambda}{3\pi^2\epsilon_0 c^3}\ddot{x}_i + \frac{e^2\Lambda^3}{9\pi^2\epsilon_0 c^3}x_i - \frac{e^2}{\epsilon_0}\delta_{ij}^\perp(0)x_j. \qquad (4.3.24)$$

From the definition (B.9) of the transverse delta function, and (4.3.15) and (4.3.17),

$$\frac{e^2}{\epsilon_0}\delta_{ij}^\perp(0)x_j = \frac{e^2}{\epsilon_0}\left(\frac{1}{2\pi}\right)^3\int d^3k[\mathbf{x} - (\mathbf{x}\cdot\mathbf{k})\mathbf{k}/k^2]_i$$

$$= \frac{e^2}{\epsilon_0}\left(\frac{1}{2\pi}\right)^3 4\pi\left(\frac{2}{3}x_i\right)\frac{1}{c^3}\int_0^\Lambda d\omega\omega^2 = \frac{2e^2\Lambda^3}{9\pi^2\epsilon_0 c^3}x_i, \qquad (4.3.25)$$

and so we see that there is a cancellation of the last two terms on the right side of (4.3.24), leaving

$$m_{\text{obs}}\ddot{\mathbf{x}} = -\nabla V + e\mathbf{E}_0(0, t) + \frac{e^2}{6\pi\epsilon_0 c^3}\,\dddot{\mathbf{x}} \tag{4.3.26}$$

and showing that the radiation reaction field acting on the dipole is

$$\mathbf{E}_{\text{RR}}(t) = -\frac{e\Lambda}{3\pi^2\epsilon_0 c^3}\ddot{\mathbf{x}} + \frac{e}{6\pi\epsilon_0 c^3}\,\dddot{\mathbf{x}}\,. \tag{4.3.27}$$

We have defined the *observed mass* of the electron:

$$m_{\text{obs}} = m + \frac{e^2\Lambda}{3\pi^2\epsilon_0 c^3} = m_{\text{bare}} + \frac{e^2\Lambda}{3\pi^2\epsilon_0 c^3} = m_{\text{bare}} + m_{\text{em}}. \tag{4.3.28}$$

The point here is that the electron mass m that we have heretofore written in our equations does not include the effect of the interaction of the electron with its own radiation reaction field, which evidently increases the inertial mass of the electron. The observed electron mass $m_{\text{obs}} = 9.18 \times 10^{-31}$ kg therefore includes an *electromagnetic mass* m_{em} in addition to a "bare mass" m_{bare} of presumably non-electromagnetic provenance. m_{em} diverges if we let the high-frequency cutoff $\Lambda \to \infty$, but that would violate our non-relativistic approximation. The fully relativistic quantum-electrodynamical theory still gives a divergent electromagnetic mass, although the divergence is "only" logarithmic:

$$m_{\text{em}} = \frac{3m_{\text{obs}}\alpha}{2\pi}\log\left(\frac{\Lambda}{m_{\text{obs}}c^2}\right), \tag{4.3.29}$$

where $\alpha = (4\pi\epsilon_0)^{-1}e^2/\hbar c$ is the fine-structure constant. In any event, we know that, for whatever reason, m_{obs} is finite! Thus, we "renormalize" the mass and rewrite (4.3.26) as

$$m\ddot{\mathbf{x}} = -\nabla V + e\mathbf{E}_0(0, t) + \frac{e^2}{6\pi\epsilon_0 c^3}\,\dddot{\mathbf{x}} \quad (m = m_{\text{obs}}). \tag{4.3.30}$$

Not surprisingly, the radiation reaction field $(e/6\pi\epsilon_0 c^3)\,\dddot{\mathbf{x}}$ has exactly the same form obtained classically in (1.11.14)) based on a power-balancing argument. The same equation of motion for $\mathbf{x}$ follows from the minimal coupling form of the Hamiltonian.

Exercise 4.2: Consider an electric dipole moment $\mathbf{p}(t) = e\mathbf{x}(t)$. Evaluate the electric field given in (1.4.20) for $r \to 0$ by making Taylor series expansions of $\mathbf{p}(t - r/c)$ and its derivatives, and compare the result with the radiation reaction field given in (4.3.19).

4.4 Spontaneous Emission

You might wonder what [my father] got out of it all. I went to MIT. I went to Princeton. I came home, and he said, "Now you've got a science education. I always wanted to know something that I have never understood; and so, my son, I want you to explain it to me." I said, "Yes."

He said, "I understand that they say that light is emitted from an atom when it goes from one state to another, from an excited state to a state of lower energy."

I said, "That's right."

"And light is a kind of particle, a photon, I think they call it."

"Yes."

"So if the photon comes out of the atom when it goes from the excited to the lower state, the photon must have been in the atom in the excited state."

I said, "Well, no."

He said, "Well, how do you look at it so you can think that a particle photon comes out without it having been there in the excited state?"

I thought a few minutes, and I said, "I'm sorry; I don't know. I can't explain it to you."

— Richard Feynman[7]

Spontaneous emission is the process responsible for most of the light around us. We previously noted, in connection with (3.6.98), that about 98% of the light from the Sun, for instance, is the result of spontaneous emission. It has not been found possible, despite notable attempts, to formulate a viable theory of spontaneous emission—one that is consistent with experiments and observations—without field quantization. The first successful treatment of spontaneous emission, by Dirac in 1927, was pivotal not only as an early application of quantum field theory, but also as the first example of how quantum theory could account for the creation of particles, in this case photons.[8] "Dirac's successful treatment of the spontaneous emission of radiation confirmed the universal character of quantum mechanics."[9]

A complete theory of spontaneous emission must yield not only the rate and the spectrum of radiation from an atom, but also the changes in atomic energy levels and the corresponding shift in the radiation frequency due to the atom–field coupling. The

---

[7] R. P. Feynman, The Physics Teacher, September 1969, p. 319.

[8] P. A. M. Dirac, Proc. R. Soc. Lond. A **114**, 243 (1927).

[9] S. Weinberg, Daedalus **106**, 17 (1977), p. 22.

observed shifts of the energy levels are generally small—roughly a part in a million in the case of the famous Lamb shift in hydrogen—whereas, for reasons closely related to the divergence in the radiation reaction field found in Section 4.3, the calculated shifts are finite (and in spectacular agreement with experiment) only after renormalization. In the following, we will address these different aspects of spontaneous emission, first in the Schrödinger picture and then in the Heisenberg picture.

4.4.1 The Schrödinger Picture

Suppose we have an atom in free space, at a fixed position $\mathbf{r} = 0$, in an excited state, $|2\rangle$, and that the atom can make an electric dipole transition to a lower, stable state, $|1\rangle$. Using the notation introduced in Sections 2.3 and 2.5, we write the atomic Hamiltonian for the two-state-atom) model as

$$H_A = E_1\sigma_{11} + E_2\sigma_{22} \tag{4.4.1}$$

and the interaction Hamiltonian as

$$
\begin{aligned}
H_I &= -e\mathbf{x} \cdot \mathbf{E}(0) = -ie \sum_{\mathbf{k}\lambda} \left(\frac{\hbar\omega_k}{2\epsilon_0 V}\right)^{1/2} (a_{\mathbf{k}\lambda} - a_{\mathbf{k}\lambda}^\dagger)(\mathbf{x} \cdot \mathbf{e}_{\mathbf{k}\lambda}) \\
&= -ie \sum_{\mathbf{k}\lambda} \left(\frac{\hbar\omega_k}{2\epsilon_0 V}\right)^{1/2} (a_{\mathbf{k}\lambda} - a_{\mathbf{k}\lambda}^\dagger)(\mathbf{d}_{12}\sigma_{12} + \mathbf{d}_{21}\sigma_{21}) \cdot \mathbf{e}_{\mathbf{k}\lambda} \\
&= -i\hbar \sum_{\mathbf{k}\lambda} (C_{\mathbf{k}\lambda}\sigma_{12} + C_{\mathbf{k}\lambda}^*\sigma_{21})(a_{\mathbf{k}\lambda} - a_{\mathbf{k}\lambda}^\dagger),
\end{aligned}
\tag{4.4.2}
$$

$$C_{\mathbf{k}\lambda} = \frac{1}{\hbar}\left(\frac{\hbar\omega_k}{2\epsilon_0 V}\right)^{1/2} \mathbf{d}_{12} \cdot \mathbf{e}_{\mathbf{k}\lambda}. \tag{4.4.3}$$

We will henceforth work exclusively with the electric dipole form of the Hamiltonian and will no longer distinguish it from the minimal coupling Hamiltonian by labeling it with a "prime." We omit for now the pesky divergent term $(e^2/2\epsilon_0)\mathbf{x}^2\delta^3(0)$ from the Hamiltonian (4.2.20) but will account for it later on. Thus, we will work for now with the Hamiltonian

$$H = E_1\sigma_{11} + E_2\sigma_{22} + \sum_\beta \hbar\omega_\beta a_\beta^\dagger a_\beta - i\hbar \sum_\beta (C_\beta\sigma_{12} + C_\beta^*\sigma_{21})(a_\beta - a_\beta^\dagger). \tag{4.4.4}$$

The zero-point field energy has been thrown away by an appropriate choice of the zero of energy, and, to simplify the notation a bit, we have let the subscript β stand for $\mathbf{k}\lambda$: $C_\beta = C_{\mathbf{k}\lambda}$, and so on. We denote the initial state of interest by $|2\rangle|\{0\}\rangle$—the atom is in state $|2\rangle$, and the field is in the vacuum state $|\{0\}\rangle$ of no photons, that is, the field state for which $a_{\mathbf{k}\lambda}|\{0\}\rangle = 0$ for all modes $\mathbf{k}, \lambda$; the photon annihilation operators $a_{\mathbf{k}\lambda}$ are those appearing in the electric field operator (4.2.19). From its excited state $|2\rangle$, the atom can emit a photon into some mode β and, in so doing, drop to its lower state $|1\rangle$. We denote the state of the atom–field system after this transition by $|1\rangle|1_\beta\rangle$.

For the initial state assumed, there are no other one-photon states generated directly (that is, to first order in the atom–field coupling) by the Hamiltonian (4.4.4). The state vector at time t therefore has the form

$$|\psi(t)\rangle = b_2(t)|2\rangle|\{0\}\rangle + \sum_\beta b_\beta(t)|1\rangle|1_\beta\rangle. \tag{4.4.5}$$

Now,

$$H|\psi(t)\rangle = \sum_\beta (E_1 + \hbar\omega_\beta)b_\beta(t)|1\rangle|1_\beta\rangle + E_2 b_2(t)|2\rangle|\{0\}\rangle + i\hbar \sum_\beta C_\beta b_2(t)|1\rangle|1_\beta\rangle$$
$$- i\hbar \sum_\beta C_\beta^* b_\beta(t)|2\rangle|\{0\}\rangle, \tag{4.4.6}$$

and so the Schrödinger equation reduces to the following coupled equations for the state probability amplitudes:

$$\dot{b}_2(t) = -\frac{i}{\hbar}E_2 b_2(t) - \sum_\beta C_\beta^* b_\beta(t), \tag{4.4.7}$$

$$\dot{b}_\beta(t) = -\frac{i}{\hbar}(E_1 + \hbar\omega_\beta)b_\beta(t) + C_\beta b_2(t). \tag{4.4.8}$$

In one more simplification of notation, we reset the zero of energy such that $E_2 \to 0$, and $E_1 \to E_1 - E_2 = -\hbar\omega_0$, ω_0 being the transition (angular) frequency, and replace (4.4.7) and (4.4.8) by

$$\dot{b}_2(t) = -\sum_\beta C_\beta^* b_\beta(t), \tag{4.4.9}$$

$$\dot{b}_\beta(t) = -i(\omega_\beta - \omega_0)b_\beta(t) + C_\beta b_2(t), \tag{4.4.10}$$

respectively.

Consider first the very special case in which there is somehow only one possible field mode. In this model (Section 5.3), the excitation, initially in the atom, oscillates between the atom and the field: the solutions for the occupation probabilities of the states $|2\rangle|\{0\}\rangle$ and $|1\rangle|1_\beta\rangle$ with $b_2(t = t_0) = 1$ and $b_\beta(t = t_0) = 0$ are, respectively,

$$|b_2(t)|^2 = \cos^2 \frac{1}{2}\Omega t + \frac{\Delta^2}{\Omega^2} \sin^2 \frac{1}{2}\Omega t, \tag{4.4.11}$$

$$|b_\beta(t)|^2 = \frac{4|C_\beta|^2}{\Omega^2} \sin^2 \frac{1}{2}\Omega t, \tag{4.4.12}$$

where, for convenience, we set the "initial condition time" $t_0 = 0$ here and below and define $\Delta = \omega_0 - \omega_\beta$, $\Omega = (4|C_\beta|^2 + \Delta^2)^{1/2}$. The physics here is not that of *irreversible* emission into a continuum of field modes.

It is the far more common situation of irreversible spontaneous emission that is of interest here. We seek the probability $|b_2(t)|^2$ at the "long" times of practical interest—certainly times much greater than $1/\omega_0$. For the initial state we have chosen, it follows from (4.4.9) and (4.4.10) that $b_2(t)$ satisfies the integro-differential equation

$$\dot{b}_2(t) = -\sum_\beta |C_\beta|^2 \int_0^t dt' b_2(t') e^{i(\omega_\beta - \omega_0)(t'-t)}, \tag{4.4.13}$$

or

$$\dot{b}_2(t) = -\frac{V}{8\pi^3} \int d^3k \left(\frac{\omega_k}{2\hbar\epsilon_0 V}\right) \sum_{\lambda=1}^2 |\mathbf{d}_{12} \cdot \mathbf{e}_{\mathbf{k}\lambda}|^2 \int_0^t dt' b_2(t') e^{i(\omega_k - \omega_0)(t'-t)}$$

$$= -\frac{1}{16\pi^3\hbar\epsilon_0 c^3} \int_0^\infty d\omega \omega^3 \left[\int d\Omega_{\mathbf{k}} \left(|\mathbf{d}_{12}|^2 - |\mathbf{d}_{12} \cdot \mathbf{k}|^2/k^2\right)\right]$$

$$\times \int_0^t dt' b_2(t') e^{i(\omega - \omega_0)(t'-t)}$$

$$= -\frac{|\mathbf{d}_{12}|^2}{16\pi^3\hbar\epsilon_0 c^3} \frac{8\pi}{3} \int_0^\infty d\omega \omega^3 \int_0^t dt' b_2(t') e^{i(\omega - \omega_0)(t'-t)}. \tag{4.4.14}$$

Here, we used the fact that

$$\mathbf{d}_{12} = (\mathbf{d}_{12} \cdot \mathbf{k})\mathbf{k}/k^2 + \sum_{\lambda=1}^2 (\mathbf{d}_{12} \cdot \mathbf{e}_{\mathbf{k}\lambda})\mathbf{e}_{\mathbf{k}\lambda}, \tag{4.4.15}$$

since $\mathbf{k}/k$, $\mathbf{e}_{\mathbf{k}1}$, and $\mathbf{e}_{\mathbf{k}2}$ are orthogonal unit vectors in three-dimensional space. Therefore,

$$|\mathbf{d}_{12}|^2 = |\mathbf{d}_{12} \cdot \mathbf{k}|^2/k^2 + \sum_\lambda |\mathbf{d}_{12} \cdot \mathbf{e}_{\mathbf{k}\lambda}|^2, \tag{4.4.16}$$

which we used in writing (4.4.14).

In a first, simplest approximation, we replace $b_2(t')$ in the integrand by its initial value, 1:

$$\dot{b}_2(t) \cong -\frac{|\mathbf{d}_{12}|^2}{6\pi^2\hbar\epsilon_0 c^3} \int_0^\infty d\omega \omega^3 \int_0^t dt' e^{i(\omega - \omega_0)(t'-t)}, \tag{4.4.17}$$

implying

$$\frac{d}{dt}|b_2(t)|^2 = b_2^* \dot{b}_2 + \dot{b}_2^* b_2 \cong \dot{b}_2 + \dot{b}_2^* \cong -\frac{|\mathbf{d}_{12}|^2}{3\pi^2\hbar\epsilon_0 c^3} \int_0^\infty d\omega \omega^3 \frac{\sin(\omega - \omega_0)t}{\omega - \omega_0}. \tag{4.4.18}$$

Introduce the new integration variable $x = (\omega - \omega_0)t$, and define $T = \omega_0 t$, the time in units of $1/\omega_0$:

$$\int_0^\infty d\omega \omega^3 \frac{\sin(\omega - \omega_0)t}{\omega - \omega_0} = \omega_0^3 \int_{-T}^\infty dx \left(\frac{x}{T} + 1\right)^3 \frac{\sin x}{x}. \tag{4.4.19}$$

For the "long" times of interest, for example, $T = 1000$, this is well approximated by $\pi\omega_0^3$. For a transition wavelength of 600 nm, or $\omega_0 = 2\pi c/\lambda = 3.14 \times 10^{15}$ rad/s,

$T = 1000$ corresponds to $t = 3.2 \times 10^{-13}$ s, much longer than $1/\omega_0$ but still much shorter than typical atomic radiative lifetimes ($\sim 10^{-8}$ s). For such timescales, we can approximate (4.4.18) with

$$\frac{d}{dt}|b_2(t)|^2 = -A_{21},\tag{4.4.20}$$

where

$$A_{21} = \frac{|\mathbf{d}_{12}|^2 \omega_0^3}{3\pi\hbar\epsilon_0 c^3} = \frac{1}{4\pi\epsilon_0}\frac{4|\mathbf{d}_{12}|^2 \omega_0^3}{3\hbar c^3}\tag{4.4.21}$$

is the well-known formula for the rate of spontaneous emission (that is, the Einstein A coefficient). Of course, the approximation in (4.4.20) only makes sense for times short compared with the radiative lifetime $1/A_{21}$ of the transition but long compared with the inverse of the transition frequency ω_0.

The fact that the rate of change of $b_2(t)$ over the times of interest is indicated by (4.4.20) to be relatively slow suggests a better approximation of (4.4.13):

$$\dot{b}_2(t) \cong -\left(\sum_\beta |C_\beta|^2 \int_0^t dt' e^{i(\omega_\beta - \omega_0)(t'-t)}\right) b_2(t).\tag{4.4.22}$$

In this approximation, unlike the case in (4.4.14), the field is allowed to act back on the atom (radiation reaction), but only in an effectively "Markovian" way in which the back action at time t exhibits no "memory" of earlier times t'. From the expression appearing in (4.4.14) for the factor in curly brackets,

$$\begin{aligned}
\frac{1}{b_2(t)}\frac{db_2}{dt} &\cong -\frac{|\mathbf{d}_{12}|^2}{6\pi^2\hbar\epsilon_0 c^3}\int_0^\infty d\omega\,\omega^3 \int_0^t dt' e^{i(\omega-\omega_0)(t'-t)} \\
&= -\frac{|\mathbf{d}_{12}|^2}{6\pi^2\hbar\epsilon_0 c^3}\left[\int_0^\infty d\omega\,\omega^3 \frac{\sin(\omega-\omega_0)t}{\omega-\omega_0} - i\int_0^\infty d\omega\,\omega^3\frac{1-\cos(\omega-\omega_0)t}{\omega-\omega_0}\right] \\
&\cong -\frac{1}{2}A_{21} + i\frac{|\mathbf{d}_{12}|^2}{6\pi^2\hbar\epsilon_0 c^3}\int_0^\infty d\omega\,\omega^3\frac{1-\cos(\omega-\omega_0)t}{\omega-\omega_0},
\end{aligned}\tag{4.4.23}$$

where we again approximate the integral (4.4.19) by $\pi\omega_0^3$. Thus,

$$|b_2(t)|^2 = e^{-A_{21}t}.\tag{4.4.24}$$

For times $t \ll 1/A_{21}$, this is consistent with (4.4.20). The approximation that the excited-state probability decays exponentially according to (4.4.24) is usually called the *Weisskopf–Wigner approximation*.

The integrand in the second term on the right side of (4.4.23) vanishes for $\omega = \omega_0$, and so we can replace the integral by its Cauchy principal part (f). For the "long" times of interest ($t \gg 1/\omega_0$), the cosine term varies very rapidly and makes no contribution to the integral; therefore

$$\fint_0^\infty d\omega\,\omega^3\frac{1-\cos(\omega-\omega_0)t}{\omega-\omega_0} \cong \fint_0^\infty \frac{\omega^3 d\omega}{\omega-\omega_0},\tag{4.4.25}$$

and

$$\frac{1}{b_2(t)}\frac{db_2}{dt} \cong -\frac{1}{2}A_{21} - i\delta E_2/\hbar, \tag{4.4.26}$$

$$b_2(t) \cong e^{-A_{21}t/2}e^{-i\delta E_2 t/\hbar}, \tag{4.4.27}$$

where

$$\delta E_2 = -\frac{|\mathbf{d}_{12}|^2}{6\pi^2\epsilon_0 c^3}\int_0^\infty \frac{\omega^3 d\omega}{\omega - \omega_0} \tag{4.4.28}$$

is a shift in the energy E_2 of the initial excited state. It is wildly divergent, in part because in deriving it using the electric dipole Hamiltonian, (4.2.8), we have in our two-state-atom model ignored the self-energy $(1/2\epsilon_0)\int d^3r \mathbf{P}^{\perp 2}$. A more complete, multistate derivation (see Section 4.8) results in the replacement of (4.4.28) by

$$\delta E_2 = -\frac{|\mathbf{d}_{12}|^2\omega_0^2}{6\pi^2\epsilon_0 c^3}\int_0^\infty \frac{\omega d\omega}{\omega - \omega_0} \tag{4.4.29}$$

in the two-state model; this is exactly the shift derived using the minimal coupling Hamiltonian. It is still divergent, but the divergence can be tamed (mass-renormalized) to obtain a finite level shift, as shown in Section 4.8. For now, we simply assert that, to remain consistent with the non-relativistic approximation used here, we should cut off the upper limit of integration at some high frequency Λ and replace (4.4.29) by the finite expression

$$\delta E_2 = -\frac{|\mathbf{d}_{12}|^2\omega_0^2}{6\pi^2\epsilon_0 c^3}\int_0^\Lambda \frac{\omega d\omega}{\omega - \omega_0} \tag{4.4.30}$$

The integral over t' that appears when we make a Markovian approximation as in (4.4.22) is encountered frequently, and it will be convenient for later purposes to record its long-time ($t \gg 1/\omega_0$) approximation as derived above:

$$\int_0^t dt' e^{i(\omega-\omega_0)(t'-t)} \cong \pi\delta(\omega - \omega_0) - i\mathrm{P}\Big(\frac{1}{\omega - \omega_0}\Big) \tag{4.4.31}$$

and, likewise,

$$\int_0^t dt' e^{i(\omega+\omega_0)(t'-t)} \cong -i\mathrm{P}\Big(\frac{1}{\omega + \omega_0}\Big) \tag{4.4.32}$$

for $\omega, \omega_0 > 0$. The symbol P indicates that the Cauchy principal value is to be taken in integrations over ω.

Another way to write these expressions is also very useful. Recall that, whenever we have an integral along the real axis such that the integrand has a simple pole at a point $x_0 \pm i\epsilon$ just above (below) the real axis, we can bypass the pole using an integration path that includes a small semicircle of radius ϵ, centered at x_0, lying below (above) the real axis. Then,

$$\int \frac{f(x)dx}{x - x_0 \mp i\epsilon} = \int \frac{f(x)dx}{x - x_0} \pm i\pi f(x_0), \qquad (4.4.33)$$

or, formally,

$$\frac{1}{x - x_0 \mp i\epsilon} = P \frac{1}{x - x_0} \pm i\pi\delta(x - x_0). \qquad (4.4.34)$$

It has been assumed in the foregoing that the spontaneous emission occurs in free space. More generally, the field depends on the position $\mathbf{r}$ of the atom; recall the expression (3.3.4), for instance, for the electric field operator. Thus, for an atom at $\mathbf{r}$, the coupling constant (4.4.3) is replaced by

$$C_\beta(\mathbf{r}) = \frac{1}{\hbar}\left(\frac{\hbar\omega_\beta}{2\epsilon_0}\right)^{1/2}\mathbf{d}_{12} \cdot \mathbf{A}_\beta(\mathbf{r}), \qquad (4.4.35)$$

and (4.4.22) by

$$\dot{b}_2(t) \cong -\left(\sum_\beta |C_\beta(\mathbf{r})|^2 \int_0^t dt' e^{i(\omega_\beta - \omega_0)(t' - t)}\right)b_2(t)$$

$$= -\left(\sum_\beta \frac{\omega_\beta}{2\hbar\epsilon_0}|\mathbf{d}_{12} \cdot \mathbf{A}_\beta(\mathbf{r})|^2 \int_0^t dt' e^{i(\omega_\beta - \omega_0)(t' - t)}\right)b_2(t), \qquad (4.4.36)$$

and the spontaneous emission rate is

$$A_{21}(\mathbf{r}) = \sum_\beta \frac{\pi\omega_\beta}{\hbar\epsilon_0}|\mathbf{d}_{12} \cdot \mathbf{A}_\beta(\mathbf{r})|^2 \delta(\omega_\beta - \omega_0), \qquad (4.4.37)$$

which, of course, reduces to the $\mathbf{r}$-independent A coefficient when (recall (3.3.5))

$$A_\beta(\mathbf{r}) \to A_{\mathbf{k}\lambda}(\mathbf{r}) = \frac{1}{\sqrt{V}}\mathbf{e}_{\mathbf{k}\lambda}e^{i\mathbf{k}\cdot\mathbf{r}}. \qquad (4.4.38)$$

This position dependence of the spontaneous emission rate is generally negligible if the distance of the atom from any material surface is large compared to the transition wavelength (Section 7.11). Then the spontaneous emission rate is effectively the same as if the atom were in unbounded free space.

The energy levels of an atom are similarly dependent in general on the location of the atom. For a two-state atom, a formula for the position-dependent level shift follows simply from (4.4.36) and (4.4.31). The two-state model does not provide a generally realistic model for level shifts, but the formula obtained in the model is easily generalized. The shift δE_g of the ground-state energy, for example, is given, before any renormalization, by

$$\delta E_g(\mathbf{r}) = -\sum_\beta \sum_n \frac{\omega_\beta}{2\epsilon_0}\frac{|\mathbf{d}_{ng} \cdot \mathbf{A}_\beta(\mathbf{r})|^2}{\omega_\beta + \omega_{ng}} = -\frac{\hbar}{2\pi}\sum_n \int_0^\infty \frac{d\omega A_{ng}(\omega, \mathbf{r})}{\omega + \omega_{ng}}. \qquad (4.4.39)$$

$A_{ng}(\omega, \mathbf{r})$ is given by the formula for the spontaneous emission rate for the transition $n \to g$ with the transition dipole moment $\mathbf{d}_{ng}$ but with the transition frequency ω_{ng}

in that formula replaced by ω. The summation is over all states n. An example of such a position-dependent shift is the Casimir–Polder energy of an atom near a conducting plate (Section 7.3.3).

We now use the (approximate) solution $b_2(t) = \exp(-A_{21}t/2)\exp(-iE_2t/\hbar)$ in (4.4.7) to obtain an expression for the probability at time t that a photon has been emitted. For times much longer than the radiative lifetime, $t \gg 1/A_{21}$, this is

$$\sum_\beta |b_\beta(\infty)|^2 = \sum_\beta \frac{|C_\beta|^2}{(\omega_\beta - \omega_0)^2 + A_{21}^2/4}$$

$$\rightarrow \frac{|\mathbf{d}_{12}|^2}{6\pi^2\hbar\epsilon_0 c^3}\int_0^\infty \frac{\omega^3 d\omega}{(\omega - \omega_0)^2 + A_{21}^2/4}$$

$$= \int_0^\infty S(\omega)d\omega, \tag{4.4.40}$$

$$S(\omega) = \frac{A_{21}}{2\pi}\frac{\omega^3/\omega_0^3}{(\omega - \omega_0)^2 + A_{21}^2/4}. \tag{4.4.41}$$

For $A_{21}/\omega_0 \ll 1$, the case of physical interest, $S(\omega)$ is sharply peaked near $\omega = \omega_0$ and describes quite well the observed Lorentzian-like lineshapes for frequencies ω near the peak transition frequency ω_0, that is, for $|\omega - \omega_0|$ not much greater than the "natural linewidth" $A_{21}/2$. But $S(\omega)$ obviously cannot be valid for all frequencies, as its integral over all frequencies is infinite.

What went wrong? The divergence of the integrated lineshape is sometimes attributed to a breakdown of the dipole approximation, which takes account only of field wavelengths large compared with atomic dimensions. While this approximation indeed breaks down for sufficiently small wavelengths $\lambda = 2\pi c/\omega$, it cannot be held responsible for the divergence of the integrated lineshape: (4.4.9) and (4.4.10) imply that

$$|b_2(t)|^2 + \sum_\beta |b_\beta(t)|^2 = |b_2(0)|^2 + \sum_\beta |b_\beta(0)|^2 = 1, \tag{4.4.42}$$

for all times, whereas (4.4.41) obviously contradicts this. Clearly, then, the fault lies not with the dipole approximation or the two-state-atom model as such but in the approximations we have made in solving (4.4.9) and (4.4.10). In particular, exponential decay of the excited-state probability violates (4.4.42) (and therefore the unitary time evolution required by the Schrödinger equation) and leads inescapably to a Lorentzian-type frequency distribution $S(\omega)$ whose integral over all frequencies is infinite.

In fact, exponential decay of the excited-state probability is strictly forbidden by quantum theory, as can be seen from the following argument.[10] Let $U(t) = \exp(-iHt/\hbar)$ be the time evolution operator for a system that, at time $t = 0$, is in some nonstationary state $|\psi(0)\rangle$. The probability amplitude for this state at any later time t is

$$b(t) = \langle\psi(0)|\psi(t)\rangle = \langle\psi(0)|U(t)|\psi(0)\rangle. \tag{4.4.43}$$

[10] L. A. Khalfin, Sov. Phys. JETP **6**, 1053 (1958).

Denote the complete set of (orthonormal) eigenstates of the Hamiltonian H of the system by $|E\rangle$. The energy eigenvalues E of any physical system must be bounded from below; otherwise, there would be no stable ground state. Denote the lowest-energy eigenvalue of H by $E_{\min}$. From the completeness relation $\int dE|E\rangle\langle E| = 1$,

$$
\begin{aligned}
b(t) &= \int_{E_{\min}}^{\infty} dE' \int_{E_{\min}}^{\infty} dE \langle\psi(0)|E'\rangle\langle E'|U(t)|E\rangle\langle E|\psi(0)\rangle \\
&= \int_{E_{\min}}^{\infty} dE\, e^{-iEt/\hbar} |\langle E|\psi(0)\rangle|^2 \\
&= \int_{-\infty}^{\infty} dE\, w(E) e^{-iEt/\hbar},
\end{aligned}
\tag{4.4.44}
$$

where $w(E) = |\langle E|\psi(0)\rangle|^2$ for $E \geq E_{\min}$, $w(E) = 0$ for $E < E_{\min}$. Now we invoke the Paley–Wiener theorem, which, for our purposes, can be stated as follows: if $b(t)$ is square-integrable, a necessary and sufficient condition that its Fourier transform $w(E)$ exists and is bounded from below is

$$
\int_{-\infty}^{\infty} \frac{|\log|b(t)||}{1 + t^2} dt < \infty.
\tag{4.4.45}
$$

Since this condition would be violated if $|b(t)|$ were to fall off exponentially at large t, purely exponential decay cannot occur.

A different form of the lineshape function can be derived as follows. From (4.4.10), the initial condition $b_\beta(0) = 0$, and the assumption that $b_2(t) = 0$ for $t < 0$,

$$
|b_\beta(\infty)|^2 = |C_\beta|^2 \left| \int_{-\infty}^{\infty} dt\, b_2(t) e^{i(\omega_\beta - \omega_0)t} \right|^2.
\tag{4.4.46}
$$

Now write $b_2(t)$ in terms of its Fourier transform, $\hat{b}_2(\omega)$:

$$
\hat{b}_2(\omega) = \int_0^{\infty} dt\, b_2(t) e^{i\omega t},
\tag{4.4.47}
$$

$$
b_2(t) = \frac{1}{2\pi} \int_{-\infty}^{\infty} d\omega\, \hat{b}_2(\omega) e^{-i\omega t},
\tag{4.4.48}
$$

whence

$$
\begin{aligned}
|b_\beta(\infty)|^2 &= \frac{1}{4\pi^2} |C_\beta|^2 \left| \int_{-\infty}^{\infty} dt \int_{-\infty}^{\infty} d\omega\, \hat{b}_2(\omega) e^{i(\omega_\beta - \omega_0 - \omega)t} \right|^2 \\
&= \frac{1}{4\pi^2} |C_\beta|^2 \left| \int_{-\infty}^{\infty} d\omega\, \hat{b}_2(\omega)(2\pi)\delta(\omega - \omega_0 + \omega_\beta) \right|^2 \\
&= |C_\beta|^2 |\hat{b}_2(\omega_\beta - \omega_0)|^2.
\end{aligned}
\tag{4.4.49}
$$

Next return to (4.4.9) and (4.4.10) and take Laplace transforms. The Laplace transform of $b_2(t)$ is found to be

$$\tilde{b}_2(s) = \frac{1}{s + \sum_{\beta'} |C_{\beta'}|^2 [s + i(\omega_{\beta'} - \omega_0)]^{-1}}, \tag{4.4.50}$$

and then from the formula

$$\hat{b}_2(\omega) = \lim_{\epsilon \to 0^+} \tilde{b}_2(-i\omega + \epsilon) \tag{4.4.51}$$

relating the Fourier transform to the Laplace transform,

$$\hat{b}_2(\omega_\beta - \omega_0) = \frac{i}{\omega_\beta - \omega_0 + i\epsilon + \sum_{\beta'} |C_{\beta'}|^2 [\omega_{\beta'} - \omega_\beta - i\epsilon]^{-1}}. \tag{4.4.52}$$

Using once again

$$\sum_\beta |C_\beta|^2 (\ldots) \to \frac{|\mathbf{d}_{12}|^2}{6\pi^2 \hbar \epsilon_0 c^3} \int_0^\infty d\omega \omega^3 (\ldots), \tag{4.4.53}$$

and (4.4.34),

$$\lim_{\epsilon \to 0^+} \frac{1}{\omega_{\beta'} - \omega_\beta - i\epsilon} = \mathrm{P}\Big(\frac{1}{\omega_{\beta'} - \omega_\beta}\Big) + i\pi\delta(\omega_{\beta'} - \omega_\beta), \tag{4.4.54}$$

we can replace the last term in the denominator of (4.4.52) by

$$\tilde{\Delta}(\omega_\beta) + i\gamma(\omega_\beta) = \frac{|\mathbf{d}_{12}|^2 \omega_\beta^3}{6\pi^2 \hbar \epsilon_0 c^3} \Big(\fint \frac{\omega^3 d\omega}{\omega - \omega_\beta} + i\pi\omega_\beta^3\Big) = \tilde{\Delta}(\omega_\beta) + \frac{i}{2} A_{21} \Big(\frac{\omega_\beta}{\omega_0}\Big)^3$$
$$= \tilde{\Delta}(\omega_\beta) + i\gamma(\omega_\beta). \tag{4.4.55}$$

Finally, then, using (4.4.49) and (4.4.53), we obtain the probability at long times that there is a photon:

$$\sum_\beta |b_\beta(\infty)|^2 = \sum_\beta \Big|\frac{|C_\beta|^2}{[\omega_\beta + \tilde{\Delta}(\omega_\beta) - \omega_0]^2 + \gamma^2(\omega_\beta)}$$
$$\to \frac{|\mathbf{d}_{12}|^2}{6\pi^2 \hbar \epsilon_0 c^3} \int_0^\infty \frac{\omega^3 d\omega}{[\omega + \tilde{\Delta}(\omega) - \omega_0]^2 + \gamma^2(\omega)}$$
$$= \frac{A_{21}}{2\pi} \int_0^\infty \frac{(\omega^3/\omega_0^3) d\omega}{[\omega + \tilde{\Delta}(\omega) - \omega_0]^2 + A_{21}^2 \omega^6/4\omega_0^6}. \tag{4.4.56}$$

If we ignore the unphysical, divergent frequency shift $\tilde{\Delta}(\omega)$, we are left with the lineshape function

$$S(\omega) = \frac{A_{21}}{2\pi} \frac{\omega^3/\omega_0^3}{(\omega - \omega_0)^2 + A_{21}^2 \omega^6/4\omega_0^6}, \tag{4.4.57}$$

whose integral over all frequencies is finite and approximately 1 for $A_{21}/\omega_0 \ll 1$. In other words, the probability that a photon is eventually emitted is 1 in this limit.

Exercise 4.3: What assumptions and approximations were made in obtaining (4.4.56), compared to (4.4.41)?

Frequency-dependent functions, such as $\tilde{\Delta}(\omega)$ and $\gamma(\omega)$ in (4.4.56), can be expected to appear in any formulation of spectral lineshape theory meant to improve on the simple Lorentzian form in which frequency shifts and linewidths are independent of ω. Relating to this, we recall that a linewidth factor $\propto \omega^3$ appears as a consequence of radiative reaction in (1.11.16), which is the classical expression for the response of a dipole oscillator to an *applied* field of frequency ω, and that such frequency dependence is required for consistency with the optical theorem.

What is most relevant as a practical matter is the radiative lineshape for frequencies $\omega \approx \omega_0$. As already remarked, the basic Lorentzian form

$$S(\omega) \propto [(\omega - \omega_0)^2 + \gamma^2]^{-1} \tag{4.4.58}$$

with $\gamma \ll \omega_0$ generally describes very well the observed spectra around such frequencies. If the lineshape is calculated using the minimal coupling Hamiltonian under the same assumptions (no photons initially, approximate exponential decay of the initially excited atomic state) used to obtain (4.4.41), one obtains

$$S(\omega) = \frac{A_{21}}{2\pi} \frac{\omega/\omega_0}{(\omega - \omega_0)^2 + A_{21}^2/4}. \tag{4.4.59}$$

This differs (for $\omega \neq \omega_0$) from the lineshape (4.4.41) calculated with the electric dipole Hamiltonian in having a factor ω in the numerator instead of ω^3. This is easily understood from the fact that $\mathbf{E} = -\partial \mathbf{A}/\partial t$: the component of $\mathbf{A}$ at any field frequency ω is proportional to $1/\omega$ times the corresponding component of $\mathbf{E}$, and therefore, with the minimal coupling interaction $-(e/mc)\mathbf{A} \cdot \mathbf{p}$, the atom–field coupling constant is proportional to $\omega^{-1/2}$, as opposed to the electric dipole coupling constant (see (4.4.3)), which is proportional to $\omega^{1/2}$.[11]

Which lineshape is (more) correct? The question cannot be easily answered because the lineshape obtained from either Hamiltonian will generally depend on the way the atom is excited.

In one of his papers on the energy difference between the $2s_{1/2}$ and $2p_{1/2}$ levels of the hydrogen atom (the Lamb shift), Lamb wrote that, if the lineshape function is calculated using the minimal coupling Hamiltonian instead of the electric dipole form, an additional factor $[(\omega_0/\omega)^2]$ appears . . . This would give rise to a significant distortion of the resonance curve, and it is therefore important to choose the correct form for the analysis

---

[11] In the dipole approximation, the $(e^2/2mc^2)\mathbf{A}^2$ part of the minimal coupling Hamiltonian does not include any atom operators and therefore does not contribute to probability amplitudes for transitions between different atomic states.

of the data. Of course, the difference between the perturbations $\mathbf{E} \cdot \mathbf{r}$ and $-\mathbf{A} \cdot \mathbf{p}/m$ just corresponds to a gauge transformation under which the theory is known to be invariant, so that both perturbations must lead to the same physical predictions. Nevertheless, a closer examination shows that the usual interpretation of probability amplitudes is valid only in the former gauge, and no additional factor $[(\omega_0/\omega)^2]$ actually occurs.[12]

Radiative lineshapes cannot be meaningfully calculated without accounting for how the atoms are excited. The lineshapes (4.4.41) and (4.4.59), for example, were obtained in the idealization that a two-state atom is known with certainty to be in its excited state at the time $t = 0$. In fact this does not even come close to describing the Lamb shift experiments. In the original experiments the difference in the hydrogen $2s_{1/2}$ and $2p_{1/2}$ energy levels—which is predicted by the Dirac equation without coupling of the atom to the electromagnetic field to be zero—was measured by inducing a transition between the $2s_{1/2}$ state and the $2p_{1/2}$ state with a microwave field in the presence of a variable DC magnetic field that Zeeman-shifts the transition frequency into (or out of) resonance with the microwave frequency. For such an experiment Fried showed that, when all intermediate p states of the atom are included in calculations with the minimal coupling Hamiltonian, the lineshape obtained is in close agreement with that obtained with the electric dipole Hamiltonian.[13] This illustrates the important point made earlier: the two forms of the Hamiltonian in principle lead to equivalent physical predictions, but this equivalence does not in general hold in approximate calculations.

We have assumed that our two-state atom is in the upper-energy state $|2\rangle$ at $t = 0$. Suppose instead that the atom is initially in the lower-energy state $|1\rangle$ and that again the field is in the vacuum state $|\{0\}\rangle$. In this case, to first order in the atom–field coupling, the Hamiltonian (4.4.4) couples the initial state $|1\rangle|\{0\}\rangle$ to states $|2\rangle|1_\beta\rangle$ in which the atom is in the upper-energy state and there is one photon in a mode β of the field, and the state vector at any time t has the form

$$|\psi(t)\rangle = b_1(t)|1\rangle|\{0\}\rangle + \sum_\beta b_\beta(t)|2\rangle|1_\beta\rangle, \qquad (4.4.60)$$

with the time-dependent Schrödinger equation now taking the form

$$\dot{b}_1(t) = -C_\beta b_\beta(t),$$
$$\dot{b}_\beta(t) = -i(\omega_\beta + \omega_0)b_\beta(t) + C_\beta^* b_1(t), \qquad (4.4.61)$$

where we set the zero of energy such that the energy of the lower state of the (unperturbed) atom is zero. Then,

[12] W. E. Lamb, Jr., Phys. Rev. **85**, 259 (1952), p. 268.

[13] Z. Fried, Phys. Rev. A **8**, 2835 (1973). See also W. E. Lamb, Jr., R. R. Schlicher, and M. O. Scully, Phys. Rev. A **36**, 2763 (1987).

$$\dot{b}_1(t) = -\sum_{\beta} |C_\beta|^2 \int_0^t dt' b_1(t') e^{i(\omega_\beta + \omega_0)(t'-t)}$$

$$\cong -\left(\sum_{\beta} |C_\beta|^2 \int_0^t dt' e^{i(\omega_\beta + \omega_0)(t'-t)}\right) b_1(t)$$

$$\rightarrow -\left(\frac{|\mathbf{d}_{12}|^2}{6\pi^2 \hbar \epsilon_0 c^3} \int_0^\infty d\omega \omega^3 \int_0^t dt' e^{i(\omega_\beta + \omega_0)(t'-t)}\right) b_1(t)$$

$$= -i(\delta E_1/\hbar) b_1(t). \tag{4.4.62}$$

We have again made the Markovian approximation and the "long-time" approximation (4.4.32), and have identified a shift in the ground-state energy due to the interaction of the atom with the electromagnetic field:

$$\delta E_1 = -\frac{|\mathbf{d}_{12}|^2}{6\pi^2 \epsilon_0 c^3} \int_0^\infty \frac{\omega^3 d\omega}{\omega + \omega_0}. \tag{4.4.63}$$

The atom–field interaction giving this level shift can be interpreted as an emission of a virtual ("energy nonconserving") photon followed by its reabsorption, since

$$|C_\beta|^2 \propto |\mathbf{d}_{12} \cdot \mathbf{e}_\beta|^2$$
$$= \langle 1|\langle\{0\}|(\mathbf{d} \cdot \mathbf{e}_\beta) a_\beta |2\rangle |1_\beta\rangle \times \langle 2|\langle 1_\beta|(\mathbf{d} \cdot \mathbf{e}_\beta) a_\beta^\dagger |1\rangle |\{0\}\rangle \tag{4.4.64}$$

is the probability amplitude for the transition $|1\rangle|\{0\}\rangle \rightarrow |2\rangle|1_\beta\rangle$ in which a virtual photon is created $(a_\beta^\dagger)$ and the atom jumps to the upper state, times the probability amplitude for the transition $|2\rangle|1_\beta\rangle \rightarrow |1\rangle|\{0\}\rangle$ in which the same virtual photon is annihilated (a_β) and the atom jumps back down to the lower state.

As in the case of the upper-state shift δE_2 (see (4.4.30)), we replace (4.4.63) by

$$\delta E_1 = -\frac{|\mathbf{d}_{12}|^2 \omega_0^2}{6\pi^2 \epsilon_0 c^3} \int_0^\Lambda \frac{\omega d\omega}{\omega + \omega_0}. \tag{4.4.65}$$

This lower-state shift δE_1 in our two-state-atom model differs from the upper-state shift δE_2 given by (4.4.30) in having an "antiresonant" denominator $\omega + \omega_0$ instead of a "resonant" denominator $\omega - \omega_0$. More meaningful radiative energy-level shifts are found only when we allow for all the energy levels of an atom *and* renormalize, as discussed below.

It is straightforward in the Schrödinger picture to go beyond the two-state-atom model as long as only one or a few photon states have significant occupation probabilities. In the radiative cascade indicated in Figure 4.1, for example, the state vector of the atom–field system is well approximated by the superposition of the states $|2\rangle|\{0\}\rangle$ and $|1\rangle|1_\beta\rangle$ *and* the states $|0\rangle|1_\beta, 1_{\beta'}\rangle$ in which the atom is in its lowest-energy state $|0\rangle$ and there is a photon in each of the field modes β and β'. This results in a coupling between the probability amplitudes for the states $|1\rangle|1_\beta\rangle$ and $|0\rangle|1_\beta, 1_{\beta'}\rangle$, and, from the (approximate) solution for the probability amplitudes $b_\beta(t)$ of the $|1\rangle|1_\beta\rangle$ states,

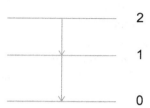

Fig. 4.1 A radiative cascade: an atom in State 2 can drop to State 1, which can then drop to a stable state, State 0. The linewidth γ of the $2 \to 1$ transition is equal to half the sum of the spontaneous emission rates for the $2 \to 1$ and $1 \to 0$ transitions.

it follows that the lineshape of the $2 \to 1$ transition is approximately a Lorentzian function with linewidth $\gamma = (A_{21} + A_{10})/2$, where A_{21} and A_{10} are the spontaneous emission rates (Einstein A coefficients) for the $2 \to 1$ and $1 \to 0$ transitions. This generalizes the lineshape (4.4.41) to the case where the lower state of the transition can itself undergo a radiative decay. We will derive this result in the Heisenberg picture below, where we allow for an arbitrary number of atomic states and for initial field states that are not devoid of photons.

4.4.2 Heisenberg Picture: Two-State Atom

As in Section 2.5, we write the two-state-atom Hamiltonian as $(\hbar\omega_0/2)\sigma_z$ and define the lowering and raising operators $\sigma = \sigma_{12}$ and $\sigma^\dagger = \sigma_{21}$ satisfying the commutation relations (2.5.5). In terms of these operators, the Hamiltonian (4.4.4) for a single two-state atom and the transverse electromagnetic field is

$$H = \frac{1}{2}\hbar\omega_0\sigma_z + \sum_\beta \hbar\omega_\beta a_\beta^\dagger a_\beta - i\hbar \sum_\beta C_\beta(\sigma + \sigma^\dagger)(a_\beta - a_\beta^\dagger), \qquad (4.4.66)$$

and the Heisenberg equations of motion that follow from it are

$$\dot{\sigma} = -i\omega_0\sigma + \sum_\beta C_\beta(\sigma_z a_\beta - a_\beta^\dagger \sigma_z), \qquad (4.4.67)$$

$$\dot{\sigma}_z = 2\sum_\beta C_\beta(\sigma - \sigma^\dagger)a_\beta - 2\sum_\beta C_\beta a_\beta^\dagger(\sigma - \sigma^\dagger), \qquad (4.4.68)$$

$$\dot{a}_\beta = -i\omega_\beta a_\beta + C_\beta(\sigma + \sigma^\dagger). \qquad (4.4.69)$$

We have now taken C_β (where $\beta = (\mathbf{k}, \lambda)$; see (4.4.3)) and $\mathbf{d}_{12}$ to be real. This will not be true in general, but it slightly simplifies the calculations below and does not affect the physical interpretations drawn from these calculations.

The equal-time atom–field commutators $[a_\beta, \sigma] = [a_\beta^\dagger, \sigma] = [a_\beta, \sigma_z] = 0$, and, therefore, we can put products of atom operators with equal-time field operators in any order we please: $\sigma(t)a_\beta(t) = a_\beta(t)\sigma(t)$, $a_\beta^\dagger(t)\sigma_z(t) = \sigma_z(t)a_\beta^\dagger(t)$, and so on. In writing (4.4.67)–(4.4.69), we have used normal ordering (Section 3.8): photon annihilation operators a_β are put to the right of all other operators, and photon creation operators $a_\beta^\dagger$ are put to the left in these equations.[14]

Approximate solutions of these operator equations are obtained in the same fashion as the approximate solutions for probability amplitudes in the Schrödinger picture. The formal solution of (4.4.69) is

$$a_\beta(t) = a_\beta(0)e^{-i\omega_\beta t} + C_\beta \int_0^t dt'\, \sigma(t')e^{i\omega_\beta(t'-t)} + C_\beta \int_0^t dt'\, \sigma^\dagger(t')e^{i\omega_\beta(t'-t)}, \quad (4.4.70)$$

and we use it in (4.4.67) to obtain an integro-differential equation for $\sigma(t)$:

$$\dot{\sigma}(t) = -i\omega_0\sigma(t) + \sum_\beta C_\beta[\sigma_z(t)a_\beta(0)e^{-i\omega_\beta t} - a_\beta^\dagger(0)\sigma_z(t)e^{i\omega_\beta t}]$$

$$+ \sum_\beta C_\beta^2 \int_0^t dt'\, \sigma_z(t)\sigma(t')e^{i\omega_\beta(t'-t)}$$

$$+ \sum_\beta C_\beta^2 \int_0^t dt'\, \sigma_z(t)\sigma^\dagger(t')e^{i\omega_\beta(t'-t)}$$

$$- \sum_\beta C_\beta^2 \int_0^t dt'\, \sigma^\dagger(t')\sigma_z(t)e^{-i\omega_\beta(t'-t)}$$

$$- \sum_\beta C_\beta^2 \int_0^t dt'\, \sigma(t')\sigma_z(t)e^{-i\omega_\beta(t'-t)}. \quad (4.4.71)$$

We now make the approximations

$$\sigma_z(t)\sigma(t') \cong \sigma_z(t)\sigma(t)e^{-i\omega_0(t'-t)} = -\sigma(t)e^{-i\omega_0(t'-t)} \quad (4.4.72)$$

and

$$\sigma_z(t)\sigma^\dagger(t') \cong \sigma_z(t)\sigma^\dagger(t)e^{i\omega_0(t'-t)} = \sigma^\dagger(t)e^{i\omega_0(t'-t)}, \quad (4.4.73)$$

and likewise for the Hermitian conjugates, in the integrands. That is, we assume that the operators $\sigma(t')$ and $\sigma^\dagger(t')$ evolve in time *approximately* according to their free, unperturbed evolution. We have used the operator identities $\sigma_z(t)\sigma(t) = -\sigma(t)$ and $\sigma_z(t)\sigma^\dagger(t) = \sigma^\dagger(t)$. Thus,

---

[14] The commutation of equal-time atom and field operators is required in order to obtain Maxwell's equations for the electric and magnetic field operators (see Section 4.1.1). Of course, it doesn't matter how we order equal-time products of a's (or $a^\dagger$'s) with respect to each other, because $[a_\beta(t), a_\gamma(t)] = 0$.

$$\dot{\sigma}(t) \cong -i\omega_0\sigma(t) + \sum_\beta C_\beta[\sigma_z(t)a_\beta(0)e^{-i\omega_\beta t} - a_\beta^\dagger(0)\sigma_z(t)e^{i\omega_\beta t}]$$

$$- \sigma(t) \sum_\beta C_\beta^2 \int_0^t dt' \left(e^{i(\omega_\beta - \omega_0)(t'-t)} + e^{-i(\omega_\beta + \omega_0)(t'-t)}\right)$$

$$+ \sigma^\dagger(t) \sum_\beta C_\beta^2 \int_0^t dt' \left(e^{-i(\omega_\beta - \omega_0)(t'-t)} + e^{i(\omega_\beta + \omega_0)(t'-t)}\right). \qquad (4.4.74)$$

We replace $\sum_\beta$ by an integration over all (positive) frequencies ω as in (4.4.14), and use (4.4.31) and (4.4.32) to obtain

$$\dot{\sigma}(t) \cong -i(\omega_0 + \Delta_{21} - iA_{21}/2)\sigma(t) - i(\Delta_{21} + iA_{21}/2)\sigma^\dagger(t)$$
$$+ \sum_\beta C_\beta[\sigma_z(t)a_\beta(0)e^{-i\omega_\beta t} - a_\beta^\dagger(0)\sigma_z(t)e^{i\omega_\beta t}], \qquad (4.4.75)$$

$$\Delta_{21} = (\delta E_2 - \delta E_1)/\hbar. \qquad (4.4.76)$$

δE_2 and δE_1 are the upper- and lower-state energy shifts defined by (4.4.30) and (4.4.65), respectively, when we modify them as discussed in Section 4.4.1, and therefore Δ_{21} is the radiative shift in the transition *frequency*. For now, we will again assume that this frequency shift is included in the definition of ω_0.

Since the time evolution of $\sigma(t)$ is approximately an under-damped oscillation at frequency $\omega_0 \gg A_{21}$, we make a rotating-wave approximation (see Section 2.5) and ignore the "counter-rotating" terms involving $\sigma^\dagger(t)$ and $a_\beta^\dagger(0)\sigma_z(t)e^{+i\omega_\beta t}$ in (4.4.75). It is presumed in making this approximation that the rapidly oscillating terms $a_\beta^\dagger(t)\sigma^\dagger(t)$ and $\sigma(t)a_\beta(t)$ in (4.4.68) have a negligible effect on the time evolution of $\sigma_z(t)$, which consequently is slowly varying compared with $\exp(i\omega_0 t)$. This is reasonable on physical grounds, as $\sigma(t)a_\beta(t)$, for instance, is associated with a "non-energy-conserving" transition of the atom from the upper state to the lower state with the simultaneous annihilation of a photon. This approximation was made in effect in our Schrödinger-picture treatment of upper-state radiative decay when we truncated our Hilbert space by assuming the state vector has the form (4.4.5). Thus,

$$\dot{\sigma}(t) \cong -i(\omega_0 - iA_{21}/2)\sigma(t) + \sum_\beta C_\beta\sigma_z(t)a_\beta(0)e^{-i\omega_\beta t}, \qquad (4.4.77)$$

$$\sigma(t) \cong \sigma(0)e^{-i\omega_0 t}e^{-A_{21}t/2}$$
$$- \sum_\beta C_\beta e^{-i\omega_\beta t} \int_0^t dt'\sigma_z(t')a_\beta(0)e^{i(\omega_\beta - \omega_0 - iA_{21}/2)(t'-t)}. \qquad (4.4.78)$$

Recall that the expectation values of $\sigma^\dagger\sigma$ and $\sigma\sigma^\dagger$ are, respectively, upper- and lower-state occupation probabilities, and that $\sigma_z = \sigma^\dagger\sigma - \sigma\sigma^\dagger$ (see Section 2.5). If we use (4.4.78) to obtain an expression for $\sigma^\dagger\sigma$, and take an expectation value for an initial state $|\psi\rangle$ of the atom–field system in which $a_\beta(0)|\psi\rangle = 0$, that is, for an initial state with no photons, we get simply

$$\langle\sigma^\dagger(t)\sigma(t)\rangle = \langle\sigma^\dagger(0)\sigma(0)\rangle e^{-A_{21}t}. \tag{4.4.79}$$

In other words, the upper-state occupation probability decays exponentially, at the rate A_{21}, from its value at time $t = 0$. This is just what we calculated in the Schrödinger picture.

Exercise 4.4: Show that (4.4.79) follows directly from (4.4.75), without the rotating-wave approximation.

In the Schrödinger picture, we calculated the emission lineshape by considering the probability $\sum_\beta |b_\beta(\infty)|^2$ that, at long times $(t \gg A_{21}^{-1})$, there is a photon in the field (see (4.4.41)). In the Heisenberg picture, we can similarly define a lineshape by considering the expectation value $\sum_\beta \langle a_\beta^\dagger(t)a_\beta(t)\rangle$ for $t \gg A_{21}^{-1}$. In the rotating-wave approximation, we drop the counter-rotating term in (4.4.70) and take

$$a_\beta(t) \cong a_\beta(0)e^{-i\omega_\beta t} + C_\beta \int_0^t dt'\, \sigma(t')e^{i\omega_\beta(t'-t)}. \tag{4.4.80}$$

Then, we use the rotating-wave approximation in (4.4.77) to obtain

$$\sum_\beta \langle a_\beta^\dagger(t)a_\beta(t)\rangle \cong \langle\sigma^\dagger(0)\sigma(0)\rangle \sum_\beta C_\beta^2 \int_0^t dt' \int_0^t dt''\, e^{-i(\omega_\beta-\omega_0)(t''-t')} e^{-A_{21}(t'+t'')/2}$$

$$= \langle\sigma^\dagger(0)\sigma(0)\rangle \sum_\beta C_\beta^2 \left| \int_0^t dt'\, e^{i(\omega_\beta-\omega_0+iA_{21}/2)t'} \right|^2, \tag{4.4.81}$$

and, for $t \gg A_{21}^{-1}$,

$$\sum_\beta \langle a_\beta^\dagger(t)a_\beta(t)\rangle \cong \langle\sigma^\dagger(0)\sigma(0)\rangle \sum_\beta \frac{C_\beta^2}{(\omega_\beta-\omega_0)^2 + A_{21}^2/4}$$

$$\to \langle\sigma^\dagger(0)\sigma(0)\rangle \int_0^\infty S(\omega)d\omega, \tag{4.4.82}$$

where the lineshape function $S(\omega)$ is defined by (4.4.41) and, as discussed earlier, is only meaningful for frequencies ω near the transition frequency ω_0.

The main approximation in treating spontaneous emission in the Heisenberg picture was made in (4.4.72) and (4.4.73); in the Schrödinger-picture treatment, the equivalent approximation was made in (4.4.22). As a result of this approximation, the upper-state occupation probability at time t decreases at a rate that depends only on the occupation probability at the same time t:

$$\frac{d}{dt}\langle\sigma^\dagger(t)\sigma(t)\rangle = -A_{21}\langle\sigma^\dagger(t)\sigma(t)\rangle. \tag{4.4.83}$$

In this sense, the approximation is Markovian: the predicted behavior of the atom at time t has no "memory" of past behavior. Referring back to (4.4.74), we note that this approximation as used here involves the interaction of the atom with the part of the electromagnetic field that is independent of field operators $a_\beta(0)$ and $a_\beta^\dagger(0)$, that is, it relates only to the "source" part of the field corresponding to the second and third terms on the right side of (4.4.70). In other words, our treatment of spontaneous emission suggests that this ubiquitous phenomenon is attributable to the interaction of the atom with its radiation reaction field. We discuss this further in Section 4.5.

4.4.3 Non-Exponential Decay

We arrived at the approximation of exponential decay in both the Schrödinger and the Heisenberg pictures by assuming the atom is only mildly perturbed by its radiation reaction field; recall (4.4.22), (4.4.72), and (4.4.73). Another way of arriving at this approximation is to cast the calculations in terms of Fourier or Laplace transforms. The inverse of the Laplace transform (4.4.50) of the upper-state probability amplitude, for instance, is

$$b_2(t) = \frac{1}{2\pi i}\int_{-\infty}^{\infty}\frac{e^{izt}dz}{z-\sum_\beta|C_\beta|^2(z+\omega_\beta-\omega_0-i\epsilon)^{-1}} \quad (\epsilon\to 0^+), \tag{4.4.84}$$

and the approximation of exponential decay follows when we ignore z in the sum $\sum_\beta$ over the field modes, use (4.4.54), and make the familiar replacement of the summation over β by an integration over a continuum of field modes in free space, that is,

$$\sum_\beta\frac{|C_\beta|^2}{\omega_\beta-\omega_0-i\epsilon}\to\frac{|\mathbf{d}_{12}|^2}{6\pi^2\hbar\epsilon_0 c^3}\int_0^\Lambda d\omega\frac{\omega^3}{\omega-\omega_0-i\epsilon}\to\frac{i}{2}A_{21}-\delta E_2/\hbar. \tag{4.4.85}$$

The exponential decay of $b_2(t)$ results in this calculation from a simple pole at $z = iA_{21}/2$ (and Cauchy's theorem). The neglect of z in the sum over field modes in (4.4.84) is sometimes called the *pole approximation*.

The pole approximation assumes there is a continuum of field modes that can take up the energy initially stored in the excited atom. Then, the transfer of energy from the atom to the field is effectively irreversible. In the opposite extreme, where the atom interacts with only a single field mode, the pole approximation fails miserably. In that case, from (4.4.84), we recover the expression (4.4.11) for the upper-state probability:

$$|b_2(t)|^2 = \left|\frac{1}{2\pi i}\int_{-\infty}^{\infty}\frac{e^{izt}dz}{z-(|C_\beta|^2[z-\Delta-i\epsilon]^{-1})}\right|^2 = \cos^2\frac{1}{2}\Omega t+\frac{\Delta^2}{\Omega^2}\sin^2\frac{1}{2}\Omega t. \tag{4.4.86}$$

In intermediate cases, where the atom is assumed to interact with some finite number of modes, the initial excitation energy can return quasiperiodically to the atom, generally less frequently as the number of modes increases.

Unless an atom is within a cavity that supports only a small number of modes near the transition wavelength, the assumption that it "sees" a continuum of modes is an excellent one, as is the approximation of exponential decay in spontaneous emission. But, even in the mode-continuum approximation, there must be corrections to perfectly exponential decay (see Section 4.4.1). Such corrections follow from (4.4.84) when we do not make the pole approximation. They are found to take the form of long-time $(t \gg A_{21}^{-1})$, power-law corrections to exponential decay. A typical, approximate result for the deviation from purely exponential decay is[15]

$$b_2(t) \cong e^{-A_{21}t/2} + \Big(\frac{A_{21}}{2\pi\omega_0}\Big)\Big(\frac{1}{\omega_0 t}\Big)^2 e^{i\omega_0 t}. \tag{4.4.87}$$

Because the ratio A_{21}/ω_0 is generally very small for atoms, such a power-law correction to exponential decay is exceedingly difficult to observe: by the time the correction has any significance, there is precious little excited-state probability or radiation left to measure.[16]

4.5   Radiation Reaction and Vacuum-Field Fluctuations

As remarked at the end of Section 4.4.2, the spontaneous emission rate and the radiative frequency shift appearing in (4.4.75) are the result of the source part of the field acting on the atom. The remaining part of the field acting on the atom is the source-free field involving the operators $a_\beta(0)$ and $a_\beta^\dagger(0)$; in the absence of any applied field from sources other than the atom, the source-free field is the fluctuating quantum "vacuum" field. In our calculation, in other words, spontaneous emission and the radiative frequency shift are attributable to radiation reaction, just as the radiative damping of an oscillating dipole is attributed to radiation reaction in classical electrodynamics, where there is no quantum vacuum field.

4.5.1   Normal Ordering

To elaborate on these points, let us write $\sigma_x(t')$ for $\sigma(t') + \sigma^\dagger(t')$ (see (2.5.8)) and recast (4.4.71) as

---

[15] See, for instance, P. L. Knight and P. W. Milonni, Phys. Lett. **56A**, 275 (1976); P. L. Knight, Phys. Lett. **61A**, 25 (1977); and references therein. See also D. S. Onley and A. Kumar, Am. J. Phys. **60**, 432 (1992) for an analytical and numerical study of deviations from exponential decay for a two-state system coupled to a continuum.

[16] Predicted power-law corrections to exponential decay after over 20 lifetimes following pulsed-laser excitation of dissolved organic materials have been observed by C. Rothe, S. I. Hintschich, and A. P. Monkman, Phys. Rev. Lett. **96**, 163601 (2006). As the authors were careful to point out, however, these corrections to exponential decay were made possible by additional, *nonradiative* line broadenings.

$$\dot{\sigma}(t) = -i\omega_0\sigma(t) - i\frac{d_{12}}{\hbar}\left[\sigma_z(t)E_0^{(+)}(t) + E_0^{(-)}(t)\sigma_z(t)\right]$$

$$- i\frac{d_{12}}{\hbar}\left[\sigma_z(t)E_{RR}^{(+)}(t) + E_{RR}^{(-)}(t)\sigma_z(t)\right], \qquad (4.5.1)$$

where the operators (see (4.3.3))

$$\frac{d_{12}}{\hbar}E_0^{(+)}(t) = i\sum_\beta C_\beta a_\beta(0)e^{-i\omega_\beta t} \qquad (4.5.2)$$

and

$$\frac{d_{12}}{\hbar}E_0^{(-)}(t) = -i\sum_\beta C_\beta a_\beta^\dagger(0)e^{i\omega_\beta t} \qquad (4.5.3)$$

are the positive- and negative-frequency parts of the source-free field, and we have defined the source-field operators[17]

$$\frac{d_{12}}{\hbar}E_{RR}^{(+)}(t) = i\int_0^t dt'\sigma_x(t')\sum_\beta C_\beta^2 e^{i\omega_\beta(t'-t)}, \qquad (4.5.4)$$

$$\frac{d_{12}}{\hbar}E_{RR}^{(-)}(t) = -i\int_0^t dt'\sigma_x(t')\sum_\beta C_\beta^2 e^{-i\omega_\beta(t'-t)}. \qquad (4.5.5)$$

Now,

$$\frac{d_{12}}{\hbar}E_{RR}(t) = \frac{1}{\hbar}\mathbf{d}_{12}\cdot\mathbf{E}_{RR}(t) = \frac{d_{12}}{\hbar}\left(E_{RR}^{(+)}(t) + E_{RR}^{(-)}(t)\right)$$

$$= -2\int_0^t dt'\sigma_x(t')\sum_\beta C_\beta^2\sin\omega_\beta(t'-t)$$

$$\rightarrow -2\frac{d_{12}^2}{6\pi^2\hbar\epsilon_0 c^3}\int_0^t dt'\sigma_x(t')\int_0^\infty d\omega\,\omega^3\sin\omega(t'-t)$$

$$= \frac{d_{12}}{\hbar}\times\left(\frac{-1}{3\pi\epsilon_0 c^3}\right)\int_0^t dt'(d_{12}\sigma_x(t'))\frac{\partial^3}{\partial t'^3}\delta(t'-t). \qquad (4.5.6)$$

Comparison with (4.3.18) shows that this is just $(d_{12}/\hbar)$ times the classical expression for the radiation reaction field of an electric dipole moment $d_{12}\sigma_x(t)$, that is, it has the same form as the classical radiation reaction field of a Hertzian dipole, except, of course, that now the dipole moment is an operator—in the present case, a Hermitian operator in the Hilbert space of our two-state atom. This supports the claim that spontaneous emission and the radiative frequency shift are attributable to radiation reaction.

---

[17] Note that $E_{RR}^{(+)}(t)$ and $E_{RR}^{(-)}(t)$ are only *approximately* the positive- and negative-frequency parts, respectively, of $E_{RR}(t)$. $E_{RR}^{(+)}(t)$, for instance, has (negative-frequency) Fourier components $\exp(i\omega t)$, $\omega > 0$.

But these considerations also reveal an important distinction between the classically familiar formula for the radiation reaction field and the source field acting on an atom in quantum electrodynamics with normal ordering: the source fields $E_{\mathrm{RR}}^{(+)}(t)$ and $E_{\mathrm{RR}}^{(-)}(t)$ in (4.5.1) do not appear symmetrically as $E_{\mathrm{RR}}(t) = E_{\mathrm{RR}}^{(+)}(t) + E_{\mathrm{RR}}^{(-)}(t)$. In particular, the operators $\sigma_z(t)$ and $E_{\mathrm{RR}}^{(\pm)}(t)$ do not commute. In the same approximations made in (4.4.72) and (4.4.73),

$$\frac{d_{12}}{\hbar} E_{\mathrm{RR}}^{(+)}(t) = i \int_0^t dt' [\sigma(t') + \sigma^\dagger(t')] \sum_\beta C_\beta^2 e^{i\omega_\beta (t'-t)}$$

$$\cong i\sigma(t) \int_0^t dt' \sum_\beta C_\beta^2 e^{i(\omega_\beta - \omega_0)(t'-t)} + i\sigma^\dagger(t) \int_0^t dt' \sum_\beta C_\beta^2 e^{i(\omega_\beta + \omega_0)(t'-t)}$$

$$\cong \frac{i}{2} A_{21}\sigma(t) - \frac{1}{\hbar}\delta E_2 \sigma(t) - \frac{1}{\hbar}\delta E_1 \sigma^\dagger(t) \tag{4.5.7}$$

when we go to the mode-continuum limit and use (4.4.31) and (4.4.32). Similarly,[18]

$$\frac{d_{12}}{\hbar} E_{\mathrm{RR}}^{(-)}(t) \cong -\frac{i}{2} A_{21}\sigma^\dagger(t) - \frac{1}{\hbar}\delta E_2 \sigma^\dagger(t) - \frac{1}{\hbar}\delta E_1 \sigma(t). \tag{4.5.8}$$

When these expressions are used in (4.5.1), they, of course, reproduce the (approximate) equation (4.4.75) for $\sigma(t)$ when we use the operator identities $\sigma_z(t)\sigma(t) = -\sigma(t)$, and so on. Because of this non-commutative algebra, the atom, according to our calculation, does not "see" directly the classical-like radiation reaction field $E_{\mathrm{RR}}(t) = E_{\mathrm{RR}}^{(+)}(t) + E_{\mathrm{RR}}^{(-)}(t)$, and, of course, responds quite differently to the separate parts $E_{\mathrm{RR}}^{(+)}(t)$ and $E_{\mathrm{RR}}^{(-)}(t)$ of $E_{\mathrm{RR}}(t)$.

The approximate expressions (4.5.7) and (4.5.8) help to shed some light on the nature of the approximations (4.4.72) and (4.4.73). Addition of (4.5.7) and (4.5.8) gives

$$\frac{1}{\hbar} d_{12} E_{\mathrm{RR}}(t) = \frac{1}{\hbar} d_{12} E_{\mathrm{RR}}^{(+)}(t) + \frac{1}{\hbar} d_{12} E_{\mathrm{RR}}^{(-)}(t)$$

$$\cong \frac{i}{2} A_{21}[\sigma(t) - \sigma^\dagger(t)] - \frac{1}{\hbar}(\delta E_2 + \delta E_1)[\sigma(t) + \sigma^\dagger(t)]. \tag{4.5.9}$$

Return to (4.3.27) and use in that equation $\mathbf{d}_{12}\sigma_x(t) = \mathbf{d}_{12}[\sigma(t) + \sigma^\dagger(t)]$ for the electric dipole moment $e\mathbf{x}(t)$:

$$\frac{1}{\hbar} \mathbf{d}_{12} \cdot \mathbf{E}_{\mathrm{RR}}(t) = \frac{d_{12}^2}{6\pi\hbar\epsilon_0 c^3} \dddot{\sigma}_x(t) - \frac{d_{12}^2 \Lambda}{3\pi^2 \hbar\epsilon_0 c^3} \ddot{\sigma}_x(t). \tag{4.5.10}$$

Since $\sigma(t)$ varies approximately as $e^{-i\omega_0 t}$, and $\Lambda \gg \omega_0$,

---

[18] δE_1 and δE_2 should be understood here to be modified as in Section 4.4.1.

$$\frac{1}{\hbar}\mathbf{d}_{12} \cdot \mathbf{E}_{RR}(t) \cong \frac{id_{12}^2\omega_0^3}{6\pi\hbar\epsilon_0 c^3}[\sigma(t) - \sigma^\dagger(t)] + \frac{\omega_0^2 d_{12}^2 \Lambda}{3\pi^2\hbar\epsilon_0 c^3}[\sigma(t) + \sigma^\dagger(t)]$$

$$\cong \frac{i}{2}A_{21}[\sigma(t) - \sigma^\dagger(t)]$$

$$+ \frac{d_{12}^2\omega_0^2}{6\pi^2\hbar\epsilon_0 c^3}\left[\int_0^\Lambda d\omega\omega\left(\frac{1}{\omega - \omega_0} + \frac{1}{\omega + \omega_0}\right)\right][\sigma(t) + \sigma^\dagger(t)]. \quad (4.5.11)$$

This is equivalent to (4.5.9). In particular, the approximation that $\sigma(t)$ varies as $e^{-i\omega_0 t}$ leads us to approximate $\dddot{\sigma}_x(t)$ by $-\omega_0^2\dot{\sigma}_x(t) \cong i\omega_0^3[\sigma(t) - \sigma^\dagger(t)]$. This is analogous in the classical theory of radiation reaction to the replacement of a third derivative of the electron coordinate $\mathbf{x}$ by a term involving only a first derivative of $\mathbf{x}$, and is closely related to the "Landau–Lifshitz equation" in the classical theory of radiation reaction in the case of a linear oscillator (see Section 6.8).

It is also interesting to consider the harmonic-oscillator model that follows from (4.5.1) by replacing $\sigma_z(t)$ by -1 (Section 2.5):

$$\dot{\sigma}(t) = -i\omega_0\sigma(t) + i\frac{d_{12}}{\hbar}[E_0(t) + E_{RR}(t)], \quad (4.5.12)$$

$$\dot{\sigma}^\dagger(t) = i\omega_0\sigma(t) - i\frac{d_{12}}{\hbar}[E_0(t) + E_{RR}(t)], \quad (4.5.13)$$

and, in terms of $\sigma_x = \sigma + \sigma^\dagger$,

$$\ddot{\sigma}_x(t) + \omega_0^2\sigma_x(t) = \frac{2d_{12}\omega_0}{\hbar}[E_0(t) + E_{RR}(t)]. \quad (4.5.14)$$

Here, $E_0(t) = E_0^{(+)}(t) + E_0^{(-)}(t)$, and $E_{RR}(t)$ is given by (4.5.9). If there is no externally applied field, that is, if the field is initially in the vacuum state, then $\langle E_0(t)\rangle = 0$, and[19]

$$\langle\ddot{\sigma}_x(t)\rangle + \omega_0^2\langle\sigma_x(t)\rangle = \frac{2d_{12}\omega_0}{\hbar}\langle E_{RR}(t)\rangle$$

$$\cong iA_{21}\omega_0[\langle\sigma(t)\rangle - \langle\sigma^\dagger(t)\rangle] - \frac{2\omega_0}{\hbar}(\delta E_2 + \delta E_1)\langle\sigma_x(t)\rangle$$

$$\cong -A_{21}\langle\dot{\sigma}_x(t)\rangle - \frac{2\omega_0}{\hbar}(\delta E_2 + \delta E_1)\langle\sigma_x(t)\rangle. \quad (4.5.15)$$

Assuming A_{21} and the (renormalized) level shifts $\delta E_{1,2}$ are small compared with ω_0, we conclude that oscillations of $\langle\sigma_x(t)\rangle$ in the harmonic-oscillator model are damped at the rate $A_{21}/2$ and that the oscillation frequencies are

$$\omega_0' \cong \omega_0 + (\delta E_2 + \delta E_1)/\hbar. \quad (4.5.16)$$

Unlike the two-state atom, the radiative frequency shift in the harmonic-oscillator model of the atom is $1/\hbar$ times the *sum* of the two energy-level shifts rather than the difference.

Aside from the aforementioned details relating to the non-commuting operators $\mathbf{E}_{RR}^{(+)}(t)$ and $\mathbf{E}_{RR}^{(-)}(t)$, we can conclude that spontaneous emission and the radiative

---

[19] Ibid.

frequency shift are caused by the back action on the atom of its own source field, or, in other words, by radiation reaction. This is consistent with the earliest interpretations of spontaneous emission. In his 1927 paper Dirac, for instance, remarked, "The present theory, since it gives a proper account of spontaneous emission, must presumably give the effect of radiation reaction on the emitting system."[20] Such an interpretation is evident as well in a paper published the same year by Landau (see Section 4.8), and, in an earlier paper, van Vleck stated that "the correspondence principle for emission correlates quantum theory spontaneous emission with classical theory emission due to the radiation force $2e^2\ddot{v}/3c^3$" $(v = \dot{x})$.[21]

But later, and especially after developments relating to the Lamb shift and its interpretation, spontaneous emission was said to be a consequence of vacuum-field fluctuations. Welton, in an influential paper, stated that spontaneous emission "can be thought of as forced emission taking place under the action of the fluctuating field," and, later, Weisskopf remarked similarly that "spontaneous emission appears as a forced emission caused by the zero-point oscillations of the electromagnetic field, which are always present, even in a space without any photons."[22]

Suppose spontaneous emission is indeed a "forced emission" caused by the zero-point electromagnetic field. "Forced emission" presumably means stimulated emission, so consider the rate $R_{21} = B_{21}\rho_0(\omega_0)$ of stimulated emission due to the zero-point field, where B_{21} is the Einstein coefficient for stimulated emission (Section 2.8), and $\rho_0(\omega_0)$ is the spectral energy density of the zero-point field at the transition frequency ω_0. Since, in free space, the number of electromagnetic modes per unit volume and in the infinitesimal frequency interval $[\omega, \omega + d\omega]$ is ω^2/π^2c^3, as reviewed below, and each mode has a zero-point energy $(1/2)\hbar\omega$,

$$\rho_0(\omega_0) = \left(\frac{1}{2}\hbar\omega_0\right)\left(\frac{\omega_0^2}{\pi^2c^3}\right). \tag{4.5.17}$$

Therefore,

$$R_{21} = B_{21}\left(\frac{\hbar\omega_0^3}{2\pi^2c^3}\right). \tag{4.5.18}$$

But we recall from Einstein's theory of thermal radiation (see Section 2.8) that $B_{21} = A_{21}\pi^2c^3/\hbar\omega_0^3$, and so we conclude that the rate of "forced emission caused by the zero-point oscillations of the electromagnetic field" is actually only half the rate of spontaneous emission:

$$R_{21} = \frac{1}{2}A_{21}. \tag{4.5.19}$$

This suggests that we cannot treat the zero-point oscillations of the electromagnetic field in the same way as we would oscillatory *applied* fields. As Schiff remarked in

[20] P. A. M. Dirac, Proc. R. Soc. Lond. A **114**, 243 (1927), p. 265.

[21] J. H. van Vleck, Phys. Rev. **24**, 330 (1924), p. 338.

[22] T. A. Welton, Phys. Rev. **74**, 1157 (1948), p. 1157; V. F. Weisskopf, Physics Today **34**, 69 (November, 1981), p. 70.

the last sentence of his textbook, the zero-point oscillations of the field "are twice as effective in producing emissive transitions as are real photons and are of course incapable of producing absorptive transitions."[23]

The number of field modes in a volume V and in the interval $[k, k + dk]$ of field wave numbers in (isotropic) free space is $2 \times [V/(2\pi)^3]d^3k = 2[V/(2\pi)^3](4\pi k^2 dk) = (\pi^2\omega^2 c^3)V d\omega$, the factor of 2 coming from the fact that there are two independent polarizations for each $\mathbf{k}$; $k = |\mathbf{k}| = \omega/c$. In other words, the sum over field modes is

$$\sum_{\beta} = 2 \times \frac{V}{(2\pi)^3} \int d^3k = V \int d\omega \left(\frac{\omega^2}{\pi^2 c^3}\right) \tag{4.5.20}$$

and, therefore, the (infinite) zero-point energy of the field is

$$\sum_{\beta} \frac{1}{2}\hbar\omega_\beta = V \int d\omega \frac{1}{2}\hbar\omega \left(\frac{\omega^2}{\pi^2 c^3}\right) = V \int d\omega \rho_0(\omega) \tag{4.5.21}$$

in the mode-continuum limit appropriate to free space.

So, although spontaneous emission can be attributed to the action on the atom of its own source field, it seems that the zero-point oscillations of the source-free, vacuum field might also have a role in "stimulating" the emission. But this possibility raises more questions. Why does the source field alone seem sufficient to give the spontaneous emission rate A_{21}? And if we could resolve the issue of the 1/2 in (4.5.19) and take the stimulating effect of the zero-point oscillations of the vacuum field seriously, how would we explain Schiff's observation that these oscillations "are of course incapable of producing absorptive transitions," that is, why is there no spontaneous *absorption*?

These questions could be said to have been answered, to a considerable extent, when it was shown that the physical interpretation of spontaneous emission, and, in particular, the contributions we associate with radiation reaction and the zero-point oscillations of the vacuum field, depend on how we choose to order *commuting* atom and field operators. We now review these considerations.

4.5.2 Symmetric Ordering

As already noted, we can put products of atom operators with equal-time field operators in any order we like. For example, we can order them symmetrically and write (4.5.1) equivalently as

---

[23] *Schiff*, p. 533.

$$\dot{\sigma}(t) = -i\omega_0\sigma(t) - i\frac{d_{12}}{2\hbar}\Big\{\sigma_z(t)\big[E_0^{(+)}(t) + E_0^{(-)}(t)\big]$$

$$+ \big[E_0^{(+)}(t) + E_0^{(-)}(t)\big]\sigma_z(t)\Big\} - i\frac{d_{12}}{2\hbar}\Big\{\sigma_z(t)\big[E_{\mathrm{RR}}^{(+)}(t) + E_{\mathrm{RR}}^{(-)}(t)\big]$$

$$+ \big[E_{\mathrm{RR}}^{(+)}(t) + E_{\mathrm{RR}}^{(-)}(t)\big]\sigma_z(t)\Big\}$$

$$= -i\omega_0\sigma(t) - i\frac{d_{12}}{2\hbar}\big[\sigma_z(t)E_0(t) + E_0(t)\sigma_z(t)\big]$$

$$- i\frac{d_{12}}{2\hbar}\big[\sigma_z(t)E_{\mathrm{RR}}(t) + E_{\mathrm{RR}}(t)\sigma_z(t)\big]. \tag{4.5.22}$$

The Pauli two-state operators $\sigma(t)$ and $\sigma^\dagger(t)$ in the radiation reaction field (4.5.9) satisfy

$$\sigma(t)\sigma_z(t) + \sigma_z(t)\sigma(t) = \sigma^\dagger(t)\sigma_z(t) + \sigma_z(t)\sigma^\dagger(t) = 0. \tag{4.5.23}$$

Therefore, in the approximation given in (4.5.9), the radiation reaction field makes no contribution to (4.5.22):

$$\dot{\sigma}(t) \cong -i\omega_0\sigma(t) - i\frac{d_{12}}{2\hbar}\big[\sigma_z(t)E_0(t) + E_0(t)\sigma_z(t)\big]. \tag{4.5.24}$$

In other words, the effects on $\sigma(t)$ of the coupling of the atom to the electromagnetic field can apparently be attributed now not to radiation reaction but to the source-free vacuum field $E_0(t)$.

It is not difficult to calculate these effects of the vacuum field with symmetric ordering. The calculation is simplified if we work with expectation values over an initial state $|\psi\rangle$ with no photons. Then, $E_0^{(+)}(t)|\psi\rangle = \langle\psi|E_0^{(-)}(t) = 0$ and, from (4.5.24),

$$\langle\dot{\sigma}(t)\rangle \cong -i\omega_0\langle\sigma(t)\rangle - \frac{id_{12}}{2\hbar}\big[\langle\sigma_z(t)E_0^{(-)}(t)\rangle + \langle E_0^{(+)}(t)\sigma_z(t)\rangle\big]$$

$$= -\frac{id_{12}}{2\hbar}\big[\langle\sigma_z(t)E_0^{(-)}(t)\rangle + \mathrm{cc}\big]. \tag{4.5.25}$$

The symmetrically ordered equation for $\sigma_z(t)$ that follows from (4.4.68) is

$$\dot{\sigma}_z(t) = -\frac{id_{12}}{\hbar}\Big\{\big[\sigma(t) - \sigma^\dagger(t)\big]E_0(t) + E_0(t)\big[\sigma(t) - \sigma^\dagger(t)\big]\Big\}$$

$$- \frac{id_{12}}{\hbar}\Big\{\big[\sigma(t) - \sigma^\dagger(t)\big]E_{\mathrm{RR}}(t) + E_{\mathrm{RR}}(t)\big[\sigma(t) - \sigma^\dagger(t)\big]\Big\}$$

$$\cong -\frac{id_{12}}{\hbar}\Big\{\big[\sigma(t) - \sigma^\dagger(t)\big]E_0(t) + E_0(t)\big[\sigma(t) - \sigma^\dagger(t)\big]\Big\} - A_{21}. \tag{4.5.26}$$

This approximation is based on (4.5.7), (4.5.8), and the two-state-atom-operator algebra, including the identity $\sigma^\dagger(t)\sigma(t) + \sigma(t)\sigma^\dagger(t) = 1$, from which the term $-A_{21}$ in (4.5.26) derives. Formal integration of (4.5.26) then implies

$$\langle\sigma_z(t)E_0^{(-)}(t)\rangle = -\frac{2id_{12}}{\hbar}\int_0^t dt'\Big\langle\big[\sigma(t') - \sigma^\dagger(t')\big]E_0^{(+)}(t')E_0^{(-)}(t)\Big\rangle. \tag{4.5.27}$$

This follows from $\langle E_0^{(-)}(t)\rangle = 0$ and the fact that equal-time atom and field operators commute. From (4.5.2) and (4.5.3),

$$E_0^{(+)}(t')E_0^{(-)}(t)|\psi\rangle = \frac{\hbar^2}{d_{12}^2}\sum_\beta\sum_\gamma C_\beta C_\gamma e^{i\omega_\gamma t}e^{-i\omega_\beta t'}a_\beta(0)a_\gamma^\dagger(0)|\psi\rangle$$

$$= \frac{\hbar^2}{d_{12}^2}\sum_\beta C_\beta^2 e^{-i\omega_\beta(t'-t)}|\psi\rangle \tag{4.5.28}$$

for the initial state $|\psi\rangle$ in which the field is in the vacuum state. Therefore,

$$\frac{d_{12}}{2\hbar}\langle\sigma_z(t)E_0^{(-)}(t)\rangle = -i\sum_\beta C_\beta^2\int_0^t dt'\langle\sigma(t')-\sigma^\dagger(t')\rangle e^{-i\omega_\beta(t'-t)}$$

$$\cong -i\langle\sigma(t)\rangle\sum_\beta C_\beta^2\int_0^t dt' e^{-i(\omega_\beta+\omega_0)(t'-t)}$$

$$+ i\langle\sigma^\dagger(t)\rangle\sum_\beta C_\beta^2\int_0^t dt' e^{-i(\omega_\beta-\omega_0)(t'-t)} \tag{4.5.29}$$

in the Markovian approximation, and (4.5.25) takes the form

$$\langle\dot\sigma(t)\rangle \cong -i\omega_0\langle\sigma(t)\rangle - \langle\sigma(t)\rangle\sum_\beta C_\beta^2\int_0^t dt'\left(e^{-i(\omega_\beta+\omega_0)(t'-t)} + e^{i(\omega_\beta-\omega_0)(t'-t)}\right)$$

$$+ \langle\sigma^\dagger(t)\rangle\sum_\beta C_\beta^2\int_0^t dt'\left(e^{-i(\omega_\beta-\omega_0)(t'-t)} + e^{i(\omega_\beta+\omega_0)(t'-t)}\right). \tag{4.5.30}$$

But this is exactly what we get from (4.4.74) and (4.4.75) when we take expectation values for an initial state with no photons: whereas normal ordering leads us to attribute the radiative damping and frequency shift in spontaneous emission to radiation reaction, symmetric ordering leads us to associate these same "radiative corrections" with the effect on the atom of the vacuum quantum field. The latter interpretation rests simply on the fact that the expectation value of the *anti*-normal-ordered product $E_0^{(+)}(t')E_0^{(-)}(t)$ of source-free electric field operators acting on the vacuum state of the field in (4.5.28) does not vanish.

4.5.3 Dissipation of Excited-State Energy

In the Heisenberg-picture approach, we focused on the time evolution of the two-state-atom lowering operator $\sigma(t)$. Consider now the operator $\sigma_z(t) = \sigma^\dagger(t)\sigma(t)-\sigma(t)\sigma^\dagger(t) = \sigma_{22}(t) - \sigma_{11}(t)$, corresponding to the two-state-atom population difference. Its time evolution is given by (4.4.68) as

$$\dot{\sigma}_z(t) = -\frac{2id_{12}}{\hbar} \left\{ E^{(-)}(t) \left[\sigma(t) - \sigma^\dagger(t) \right] + \left[\sigma(t) - \sigma^\dagger(t) \right] E^{(+)}(t) \right\}$$

$$= -\frac{2id_{12}}{\hbar} \left\{ E_0^{(-)}(t) \left[\sigma(t) - \sigma^\dagger(t) \right] + \left[\sigma(t) - \sigma^\dagger(t) \right] E_0^{(+)}(t) \right\}$$

$$- \frac{2id_{12}}{\hbar} \left\{ E_{\mathrm{RR}}^{(-)}(t) \left[\sigma(t) - \sigma^\dagger(t) \right] + \left[\sigma(t) - \sigma^\dagger(t) \right] E_{\mathrm{RR}}^{(+)}(t) \right\}. \quad (4.5.31)$$

We have chosen here to put the field operators in normal order. From (4.5.7),

$$\frac{d_{12}}{\hbar} \left[\sigma(t) - \sigma^\dagger(t) \right] E_{\mathrm{RR}}^{(+)}(t) \cong -\frac{i}{2} A_{21} \sigma^\dagger(t) \sigma(t) + \frac{1}{\hbar} \delta E_2 \sigma^\dagger(t) \sigma(t)$$

$$- \frac{1}{\hbar} \delta E_1 \sigma(t) \sigma^\dagger(t), \quad (4.5.32)$$

and, likewise, from (4.5.8),

$$\frac{d_{12}}{\hbar} E_{\mathrm{RR}}^{(-)}(t) \left[\sigma(t) - \sigma^\dagger(t) \right] \cong -\frac{i}{2} A_{21} \sigma^\dagger(t) \sigma(t) - \frac{1}{\hbar} \delta E_2 \sigma^\dagger(t) \sigma(t)$$

$$+ \frac{1}{\hbar} \delta E_1 \sigma(t) \sigma^\dagger(t). \quad (4.5.33)$$

In writing (4.5.32) and (4.5.33), we have used the two-state-atom-operator property $\sigma(t)\sigma(t) = \sigma^\dagger(t)\sigma^\dagger(t) = 0$. Thus,

$$\dot{\sigma}_z(t) \cong -\frac{2id_{12}}{\hbar} \left\{ E_0^{(-)}(t) \left[\sigma(t) - \sigma^\dagger(t) \right] + \left[\sigma(t) - \sigma^\dagger(t) \right] E_0^{(+)}(t) \right\}$$

$$- 2A_{21} \sigma^\dagger(t)\sigma(t), \quad (4.5.34)$$

or, for the expectation value over an initial atom–field state in which the field is in its vacuum state,

$$\langle \dot{\sigma}_z(t) \rangle \cong -2A_{21} \langle \sigma^\dagger(t)\sigma(t) \rangle = -A_{21} \left[\langle \sigma_z(t) \rangle + 1 \right], \quad (4.5.35)$$

$$\langle \sigma_z(t) \rangle \cong -1 + \left[\langle \sigma_z(0) \rangle + 1 \right] e^{-A_{21}t}, \quad (4.5.36)$$

which follows from the identity $\sigma^\dagger(t)\sigma(t) = (1/2)[\sigma_z(t)+1]$. We have merely reproduced (4.4.83): the upper-state probability $(1/2)[\langle \sigma_z(t) \rangle + 1]$ decreases exponentially at the rate A_{21}. In this calculation employing normally ordered field operators, the dissipation of the excited-state energy of the atom is attributable to radiation reaction.

But we can just as well calculate $\langle \sigma_z(t) \rangle$ using the symmetrically ordered equation (4.5.26) for $\sigma_z(t)$. Since this is by now a familiar refrain, we simply state, without showing the details that can be found in papers cited at the end of the chapter, that one obtains

$$\langle \dot{\sigma}_z(t) \rangle \cong -A_{21} - \frac{2d_{12}^2}{\hbar^2} \langle \sigma_z(t) \rangle \int_0^t dt' \left[\langle E_0^{(+)}(t') E_0^{(-)}(t) \rangle \right.$$

$$\left. + \langle E_0^{(+)}(t) E_0^{(-)}(t') \rangle \right] \cos \omega_0 (t' - t). \quad (4.5.37)$$

The term $-A_{21}$ comes from (4.5.26), and was derived from the radiation reaction field. The integral, however, depends on the vacuum-field expectation values $\langle E_0^{(+)}(t')E_0^{(-)}(t)\rangle$ and $\langle E_0^{(+)}(t)E_0^{(-)}(t')\rangle$; from (4.5.2) and (4.5.3), respectively,

$$\frac{d_{12}^2}{\hbar^2}\langle E_0^{(+)}(t')E_0^{(-)}(t)\rangle = \sum_\beta C_\beta^2 e^{-i\omega_\beta(t'-t)}, \tag{4.5.38}$$

$$\frac{d_{12}^2}{\hbar^2}\langle E_0^{(+)}(t)E_0^{(-)}(t')\rangle = \sum_\beta C_\beta^2 e^{i\omega_\beta(t'-t)}. \tag{4.5.39}$$

Exercise 4.5: Show that

$$\frac{2d_{12}^2}{\hbar^2}\int_0^t dt'\left[\langle E_0^{(+)}(t')E_0^{(-)}(t)\rangle + \langle E_0^{(+)}(t)E_0^{(-)}(t')\rangle\right]\cos\omega_0(t'-t) \cong \frac{d_{12}^2\omega_0^3}{3\pi\hbar\epsilon_0 c^3} = A_{21}$$

for $t \gg 1/\omega_0$.

From (4.5.37) and Exercise 4.5,

$$\langle\dot\sigma_z(t)\rangle \cong -A_{21} - A_{21}\langle\sigma_z(t)\rangle = \langle\dot\sigma_z(t)\rangle_{\rm RR} + \langle\dot\sigma_z(t)\rangle_{\rm VF}$$
$$= -A_{21}\left[\langle\sigma_z(t)\rangle + 1\right], \tag{4.5.40}$$

where we identify contributions from radiation reaction and the vacuum field as

$$\langle\dot\sigma_z(t)\rangle_{\rm RR} = -A_{21}, \tag{4.5.41}$$

and

$$\langle\dot\sigma_z(t)\rangle_{\rm VF} = -A_{21}\langle\sigma_z(t)\rangle, \tag{4.5.42}$$

respectively. If the atom is initially in its excited state ($\langle\sigma_z(0)\rangle = +1$), the contributions of radiation reaction and the vacuum field to $\langle\dot\sigma_z(0)\rangle$ add to give the total decay rate

$$\langle\dot\sigma_z(0)\rangle = \langle\dot\sigma_z(0)\rangle_{\rm RR} + \langle\dot\sigma_z(0)\rangle_{\rm VF} = -A_{21} - A_{21}\langle\sigma_z(0)\rangle = -2A_{21}. \tag{4.5.43}$$

Since $\langle\sigma_z(0)\rangle = p_2(0) - p_1(0) = 2p_2(0) - 1$, where p_2 and p_1 are, respectively, the two-state-atom upper- and lower-state occupation probabilities, (4.5.43) implies that the upper-state occupation probability decays at the rate A_{21}. If the atom is initially in the lower (ground) state ($\langle\sigma_z(0)\rangle = -1$), then the contributions of radiation reaction and the vacuum field cancel:

$$\langle\dot\sigma_z(0)\rangle = \langle\dot\sigma_z(0)\rangle_{\rm RR} + \langle\dot\sigma_z(0)\rangle_{\rm VF} = -A_{21} - A_{21}(-1) = 0. \tag{4.5.44}$$

It is because of this cancellation that there is no "spontaneous absorption" from the vacuum field by an atom in its ground state.

4.6   Fluctuations, Dissipation, and Commutators

We have shown that the radiative linewidth and a frequency shift in spontaneous emission can be attributed entirely to the atom's radiation reaction when we use normal ordering of field operators, or entirely to the vacuum field acting on the atom when we use symmetric ordering. We showed similarly that we can attribute the loss of the atom's energy in spontaneous emission entirely to radiation reaction when we use normal ordering, and that *both* radiation reaction and the vacuum field appear to "stimulate" this energy loss when we use symmetric ordering. "Spontaneous absorption" from the vacuum field is precluded by the damping effect of radiation reaction.

Of course, normal ordering and symmetric ordering are but two of an infinite number of possible orderings of the atom and field operators, and different orderings lead to different but equally acceptable viewpoints as to the "cause" of spontaneous emission and the associated linewidth and frequency shift. We can even choose operator orderings that lead us to an interpretation of the Einstein A coefficient in which, say, iA_{21} ($i = \sqrt{-1}$) comes from radiation reaction, and $(1-i)A_{21}$ comes from the vacuum field. It turns out, however, that there is no ordering that allows us to attribute the loss of the atom's energy in spontaneous emission solely to the vacuum field acting on the atom.

It appears, then, that the original association of spontaneous emission with radiation reaction alone was too restrictive, as was the later interpretation of spontaneous emission as "a forced emission caused by the zero-point oscillations of the electromagnetic field." In fact, both radiation reaction and the zero-point (vacuum) electromagnetic field are required for the consistency of the quantum theory of atom–field interactions. This can be seen by considering a simpler problem—a model for a *free* electron interacting with the field. The (non-relativistic) equation for the x-component of the electron coordinate is (see (4.3.30))

$$\ddot{x} - \tau_e \dddot{x}\,(t) = \frac{e}{m}E_0(t) = \frac{ie}{m}\sum_\beta \left(\frac{\hbar\omega_\beta}{2\epsilon_0 V}\right)^{1/2}[a_\beta(0)e^{-i\omega_\beta t} - a_\beta^\dagger(0)e^{i\omega_\beta t}]e_\beta; \quad (4.6.1)$$

$\tau_e = e^2/6\pi\epsilon_0 mc^3$, m is the observed electron mass, and e_β is the x-component of the linear polarization unit vector for mode β. Using the solution

$$x(t) = -\frac{ie}{m}\sum_\beta \left(\frac{\hbar\omega_\beta}{2\epsilon_0 V}\right)^{1/2}\left(\frac{a_\beta(0)e^{-i\omega_\beta t}}{\omega_\beta^2 + i\tau_e\omega_\beta^3} - \frac{a_\beta^\dagger(0)e^{i\omega_\beta t}}{\omega_\beta^2 - i\tau_e\omega_\beta^3}\right)e_\beta; \quad (4.6.2)$$

and $p(t) = m\dot{x}(t)$, together with the commutation relations for the photon annihilation and creation operators, we obtain, after taking $V \to \infty$,

$$[x(t), p(t)] = 4i\tau_e\pi^2 c^3 \int_0^\infty \frac{d\omega\,\rho_0(\omega)}{\omega^3(1 + \tau_e^2\omega^2)} = 2i\hbar\tau_e \int_0^\infty \frac{d\omega}{1 + \tau_e^2\omega^2} = i\hbar, \quad (4.6.3)$$

as required for the consistency of the theory.[24]

---

[24] Inconsistencies relating to (4.6.1) and (4.6.2), ignored here, are reviewed in Section 6.8.

The fact that this commutation relation is preserved for all times depends on the relation between radiation reaction and the vacuum field: the former goes as the *third* derivative of the electron coordinate, while the spectral energy density $\rho_0(\omega)$ goes as the *third* power of ω. If the damping force were of the ohmic type, proportional to $\dot{x}$, the spectrum of the corresponding fluctuating, source-free force would have to vary as the *first* derivative of ω in order for the commutation relation to be preserved. That is, if, instead of using (4.6.1), we consider a particle described by

$$m\ddot{x}(t) + \gamma\dot{x}(t) = F(t), \tag{4.6.4}$$

with

$$F(t) = \int_0^\infty d\omega A(\omega)[a(\omega)e^{-i\omega t} + a^\dagger(\omega)e^{i\omega t}], \tag{4.6.5}$$

$$[a(\omega), a^\dagger(\omega')] = \delta(\omega - \omega'), \quad [a(\omega), a(\omega')] = [a^\dagger(\omega), a^\dagger(\omega')] = 0, \tag{4.6.6}$$

then, for $\gamma t/m \gg 1$, for example, it follows easily from (4.6.4) that[25]

$$[x(t), m\dot{x}(t)] = \frac{2i}{m}\int_0^\infty \frac{d\omega A^2(\omega)}{\omega(\omega^2 + \gamma^2/m^2)}. \tag{4.6.7}$$

According to the (zero-temperature) fluctuation–dissipation theorem for this example,[26]

$$A^2(\omega) = \gamma\hbar\omega/\pi, \tag{4.6.8}$$

and (4.6.7) then gives $[x(t), m\dot{x}(t)] = [x(t), p(t)] = i\hbar$. Consistency here is tied to the fact that the spectrum $A^2(\omega)$ of the fluctuating force goes as the first power of ω, while the dissipative force is proportional to the first derivative of $x(t)$. Note that (4.6.6) and (4.6.8) imply that

$$[F(t), F(t')] = \int_0^\infty d\omega A^2(\omega)(e^{-i\omega(t-t')} - e^{i\omega(t-t')}) = -\frac{2i\gamma\hbar}{\pi}\int_0^\infty d\omega\omega\sin\omega(t-t')$$

$$= 2i\gamma\hbar\frac{\partial}{\partial t}\delta(t-t'). \tag{4.6.9}$$

The commutation relation $[x(t), p(t)] = i\hbar$ in the model described by (4.6.4) and (4.6.5) can be proven for all times.[27] Using

$$\dot{x} = p/m, \tag{4.6.10}$$

$$\dot{p} = -(\gamma/m)p + F(t), \tag{4.6.11}$$

and writing $Z(t) = [x(t), p(t)]$, we obtain from (4.6.4)

$$\frac{dZ}{dt} = -\frac{\gamma}{m}Z(t) + [x(t), F(t)], \tag{4.6.12}$$

[25] P. W. Milonni, Phys. Lett. **82A**, 225 (1981).

[26] H. B. Callen and T. A. Welton, Phys. Rev. **83**, 34 (1951). See also H. B. Callen and R. F. Greene, Phys. Rev. **86**, 702 (1952), and Section 6.6.

[27] H. Dekker, Phys. Lett. A **119**, 201 (1986). See also P. T. Leung and K. Young, Phys. Lett. A **121**, 4 (1987).

$$Z(t) = Z(0)e^{-\gamma t/m} + \int_0^t dt'[x(t'), F(t')]e^{\gamma(t'-t)/m}. \tag{4.6.13}$$

Assuming the particle operators at time $t = 0$—when the interaction with the fluctuating force is presumed to be switched on—commute with $F(t)$, we obtain straightforwardly from (4.6.4) and (4.6.5) that

$$[x(t), F(t)] = -\frac{1}{m}\int_0^\infty d\omega A^2(\omega)\left(\frac{1}{\omega^2 + i\gamma\omega/m} - \frac{1}{\omega^2 - i\gamma\omega/m}\right)$$

$$= \frac{2i\gamma}{m^2}\frac{\gamma\hbar}{\pi}\int_0^\infty \frac{d\omega\omega^2}{\omega^4 + \gamma^2\omega^2/m^2} = i\hbar\gamma/m, \tag{4.6.14}$$

and therefore, from (4.6.13), with $Z(0) = [x(0), p(0)] = i\hbar$,

$$[x(t), p(t)] = i\hbar e^{-\gamma t/m} + i\hbar(1 - e^{-\gamma t/m}) = i\hbar. \tag{4.6.15}$$

The preservation of this canonical commutation relation follows also for the more general model in which the particle experiences a force $-dV(x)/dx$ in addition to the dissipative force $-\gamma\dot{x}$ and the "noise" force $F(t)$:[28]

$$\dot{x} = p/m, \tag{4.6.16}$$

$$\dot{p} = -\frac{dV}{dx} - (\gamma/m)p + F(t). \tag{4.6.17}$$

If the fluctuating force $F(t)$ is given by (4.6.5), (4.6.6), and (4.6.8), and the system (particle) experiences the fluctuating force after some time $t = 0$, then $[x(t), p(t)] = i\hbar$ for all times $t > 0$.[29]

The fluctuation–dissipation theorem is the subject of Section 6.6. For our purposes here, its main import is that there is necessarily a relation between radiation reaction and the fluctuating vacuum field, and, in particular, between the source and source-free fields acting on an atom. This relation allows us to describe spontaneous emission in different ways, depending on how we choose to order equal-time atom and field operators.

Quite a few people have been uncomfortable with the notion of zero-point fluctuations of quantum fields. Pauli, for example, stated that

. . . the zero-point energy of the vacuum derived from the quantized field becomes infinite, a result which is directly connected with the fact that the system considered has an infinite number of degrees of freedom. It is clear that this zero-point energy has no physical reality, for instance it is not the source of a gravitational field. Formally it is easy to subtract constant infinite terms which are independent of the state considered and never change; nevertheless it seems to me that already this result is an indication that a fundamental change in the concepts underlying the present theory of quantized fields will be necessary.[30]

Jaynes invoked the different interpretations of spontaneous emission that follow from different operator orderings to suggest that vacuum-field fluctuations are not necessarily "real":

---

[28] C. W. Gardiner, IBM J. Res. Develop. **32**, 127 (1988).

[29] See also Section 6.6.1.

[30] W. Pauli, "Exclusion Principle and Quantum Mechanics," http://www.nobelprize.org/prizes/physics/1945/pauli/lecture, accessed September 24, 2018.

This independence of the initial ordering is, then, just a very simple, general, and elegant fluctuation–dissipation theorem; but let me suggest a different physical interpretation from the usual one. This complete interchangeability of source-field effects and vacuum-fluctuation effects does not show that vacuum fluctuations are "real." It shows that source-field effects are the same as if vacuum fluctuations were present. For many years, starting with Einstein's relation between diffusion coefficient and mobility, theoreticians have been discovering a steady stream of close mathematical connections between stochastic problems and dynamical problems. It has taken us a long time to recognize that [quantum electrodynamics] was just another example of this.[31]

Neither of these opinions comports with our current understanding of the vacuum in quantum field theory.

4.7 Spontaneous Emission and Semiclassical Theory

In semiclassical radiation theory, as usually formulated, there is no nontrivial vacuum field, and the field acting on an atom in a vacuum is assumed either to be non-existent or to be the atom's own field acting back on the atom (radiation reaction). Under the latter assumption, the field acting on the atom, like all electromagnetic fields in semiclassical radiation theory, is taken to be a c-number, not an operator. To formulate a semiclassical theory of spontaneous emission, consider first the vacuum expectation value of $\sigma(t)$, which, from (4.5.1), satisfies

$$\langle \dot{\sigma}(t) \rangle = -i\omega_0 \langle \sigma(t) \rangle - i\frac{d_{12}}{\hbar}\left[\langle \sigma_z(t) E_{\mathrm{RR}}^{(+)}(t) \rangle + \langle E_{\mathrm{RR}}^{(-)}(t) \sigma_z(t) \rangle\right]. \qquad (4.7.1)$$

Since, in semiclassical theory, $(d_{12}/\hbar)E_{\mathrm{RR}}^{(+)}(t)$ and $(d_{12}/\hbar)E_{\mathrm{RR}}^{(-)}(t)$ have to be replaced by c-numbers, let us use (4.5.7) and (4.5.8) to replace them by

$$\frac{d_{12}}{\hbar}\langle E_{\mathrm{RR}}^{(+)}(t)\rangle = \frac{i}{2}A_{21}\langle \sigma(t)\rangle - \frac{1}{\hbar}\delta E_2 \langle \sigma(t)\rangle - \frac{1}{\hbar}\delta E_1 \langle \sigma^{\dagger}(t)\rangle \qquad (4.7.2)$$

and

$$\frac{d_{12}}{\hbar}\langle E_{\mathrm{RR}}^{(-)}(t)\rangle = -\frac{i}{2}A_{21}\langle \sigma^{\dagger}(t)\rangle - \frac{1}{\hbar}\delta E_2 \langle \sigma^{\dagger}(t)\rangle - \frac{1}{\hbar}\delta E_1 \langle \sigma(t)\rangle, \qquad (4.7.3)$$

respectively, in the Markovian approximation. Then, in the rotating-wave approximation in which we drop terms proportional to $\langle \sigma^{\dagger}(t) \rangle$ on the right side of (4.7.1),

$$\langle \dot{\sigma}(t) \rangle = -i\omega_0 \langle \sigma(t) \rangle + \frac{1}{2}A_{21}\langle \sigma_z(t)\rangle\langle \sigma(t)\rangle + \frac{i}{\hbar}\left[\delta E_2 \langle \sigma_z(t)\rangle\langle \sigma(t)\rangle + \delta E_1 \langle \sigma(t)\rangle\langle \sigma_z(t)\rangle\right]$$

$$= -i\omega_0 \langle \sigma(t) \rangle + \left(\frac{1}{2}A_{21} + i\overline{\Delta}\right)\langle \sigma_z(t)\rangle\langle \sigma(t)\rangle \qquad (4.7.4)$$

where we have defined

$$\overline{\Delta} = (\delta E_2 + \delta E_1)/\hbar. \qquad (4.7.5)$$

δE_2 and δE_1 are the (non-renormalized) two-state-atom energy-level shifts, defined by (4.4.28) and (4.4.63), obtained in quantum electrodynamics. Next, we write $\langle \sigma(t) \rangle =$

[31] E. T. Jaynes, "Electrodynamics Today," http://bayes.wustl.edu/etj/articles/electrodynamics.today.pdf, p. 12, accessed September 24, 2018.

$(1/2)[x(t)+iy(t)]\exp(-i\omega_0 t)$ and assume that the real quantities $x(t)$, $y(t)$, and $z(t) = \langle\sigma_z(t)\rangle$ are slowly varying in time compared with $\exp(-i\omega_0 t)$, whence

$$\dot{x}(t) = \frac{1}{2}x(t)z(t) - \overline{\Delta}y(t)z(t),$$

$$\dot{y}(t) = \frac{1}{2}y(t)z(t) + \overline{\Delta}x(t)z(t). \qquad (4.7.6)$$

To derive a semiclassical equation for $z(t)$, we proceed similarly: in (4.5.31), with the substitutions (4.7.2) and (4.7.3), we replace $\langle\sigma^\dagger(t)\sigma(t)\rangle$ by $\langle\sigma^\dagger(t)\rangle\langle\sigma(t)\rangle = (1/4)[x^2(t) + y^2(t)]$, which, from the "conservation of probability" condition $x^2(t)+y^2(t)+z^2(t) = 1$, is $(1/4)[1 - z^2(t)]$. We thus obtain

$$\dot{z}(t) = -\frac{1}{2}A_{21}[1 - z^2(t)]. \qquad (4.7.7)$$

Equations (4.7.6)–(4.7.7) describe spontaneous emission according to the *neoclassical* version of semiclassical radiation theory, which includes the radiation reaction field of an atom.[32]

The neoclassical theory gives the correct Einstein coefficient A_{21} for spontaneous emission. But, according to (4.7.7), a two-state atom in a vacuum and in its upper-energy state ($z = 1$) will remain in this excited state forever ($\dot{z} = 0$), in contrast to the quantum-electrodynamical prediction that the probability of the atom remaining excited decreases with time. The reason for this is clear: if the atom at $t = 0$ is in its excited state (or, for that matter, in its ground state) the radiation reaction field acting on it, according to neoclassical theory, vanishes, because $\langle\sigma(0)\rangle = \langle\sigma^\dagger(0)\rangle = 0$ and therefore $\langle E_{RR}^{(+)}(0)\rangle = \langle E_{RR}^{(-)}(0)\rangle = 0$. In more physical terms, the radiation reaction field in neoclassical theory vanishes for an atom in one of its energy eigenstates because the average value of the electric dipole moment er is zero.

The nonlinear equations (4.7.6)–(4.7.7) also predict that the radiative frequency shift of the atom, once it is acted upon by some perturbation that makes $z(0) < 1$ and therefore allows for the atom to lose energy to the electromagnetic field $[\frac{1}{2}\hbar\omega_0\dot{z}(0) < 0]$, will be time dependent. Moreover, the frequency $\overline{\Delta}$ that appears in this shift in neo-classical theory is the *sum* of the energy-level shifts δE_2 and δE_1, whereas, in quantum electrodynamics, the frequency shift Δ_{21} is the difference in the energy-level shifts (see (4.4.76)), and does not depend on time. From the perspective of quantum electrodynamics, these predictions of neoclassical theory stem from the decorrelation of expectation values of two-state-atom operators in the neoclassical formulation, that is, from the fact that it replaces the radiation reaction field *operator* by its expectation value. In the first line of (4.7.4), for example, the neoclassical theory has replaced $\langle\sigma(t)\sigma_z(t)\rangle$ and $\langle\sigma_z(t)\sigma(t)\rangle$ with $\langle\sigma(t)\rangle\langle\sigma_z(t)\rangle$, whereas, in the quantum-electrodynamical calculation,

---

[32] C. R. Stroud, Jr. and E. T. Jaynes, Phys. Rev. A **1**, 106 (1970). Solutions of the neoclassical equations and their implications are discussed in detail in this paper.

these expectation values are, respectively, $\langle \sigma(t) \rangle$ and $-\langle \sigma(t) \rangle$, since $\sigma(t)\sigma_z(t) = \sigma(t)$, and $\sigma_z(t)\sigma(t) = -\sigma(t)$.

The most compelling evidence against neoclassical theory comes from experiments on the radiation *emitted* by atoms. In Section 2.3, we mentioned examples of the failure of any theory that treats the electromagnetic field classically, and, in Sections 5.7 and 5.8, we discuss examples of the failure of any theory in which expectation values of products of two atom operators are "de-correlated" and replaced by the product of two expectation values.[33]

4.8   Multistate Atoms

We have, for the most part in this chapter, taken atoms to be two-state systems. This can be a realistic model, as the atoms could be prepared as two-state systems by optical pumping (see Figure 2.4). More generally, we could sensibly assume that two energy levels, with a transition frequency close to resonance with an applied field, determine, to a large extent, the response of the atom to the field. In spontaneous emission from an excited state to a single ground state, the two-state model is again sensible, provided we are not concerned with radiative level shifts. And, in our discussion of the roles of radiation reaction and vacuum-field fluctuations in spontaneous emission, the two-state model was useful as a simplified description of a real atom: including all possible states of the atom would complicate our equations without affecting our conclusions about the "cause" of spontaneous emission.

We will now go beyond the two-state model to describe an atom interacting with the quantized field in the electric dipole approximation. The Hamiltonian (4.2.20) in the case of a single atom at $\mathbf{r} = 0$ is

$$H = H_A + H_F - \mathbf{d} \cdot \mathbf{E}(0) + \frac{1}{2\epsilon_0} \mathbf{d}^2 \delta^3(\mathbf{r} = 0), \qquad (4.8.1)$$

where $\mathbf{d} = e\mathbf{x}$ is the electric dipole moment operator. As in our treatment of the two-state atom we have dropped the "prime" on H, it being understood that we are working with the electric dipole form of the atom–field Hamiltonian. The last term in (4.8.1) is

$$\frac{1}{2\epsilon_0} \int d^3r \mathbf{P}^2(\mathbf{r}) = \frac{1}{2\epsilon_0} \int d^3r \mathbf{P}^\perp(\mathbf{r})^2 + \frac{1}{2\epsilon_0} \int d^3r \mathbf{P}^\parallel(\mathbf{r})^2, \qquad (4.8.2)$$

and the last term here is a divergent Coulomb self-energy of the type that can be properly dealt with only in the fully relativistic theory of radiative corrections to atomic energy levels; for present purposes, we will just assume that whatever effect it

---

[33] Neoclassical theory can be related to the nonlinear wave equations considered by E. Fermi, Rend. Lincei **5**, 795 (1927), and S. Weinberg, Phys. Rev. Lett. **62**, 485 (1989). See also K. Wódiewicz and M. O. Scully, Phys. Rev. A **42**, 5111 (1990), and references therein.

has on the energy levels has already been included in the numerical values we assign to them. Now,

$$
\begin{aligned}
\frac{1}{2\epsilon_0} \int d^3r \mathbf{P}^\perp(\mathbf{r})^2 &= \frac{1}{2\epsilon_0} \int d^3r \mathbf{P}(\mathbf{r}) \cdot \mathbf{P}^\perp(\mathbf{r}) = \frac{1}{2\epsilon_0} \int d^3r \left[\mathbf{d}\delta^3(\mathbf{r}) \right] \cdot \mathbf{P}^\perp(\mathbf{r}) \\
&= \frac{1}{2\epsilon_0} \mathbf{d} \cdot \mathbf{P}^\perp(0) = \frac{1}{2\epsilon_0} d_i d_j \delta_{ij}^\perp(\mathbf{r} = 0) \\
&= \frac{1}{2\epsilon_0} \left(\frac{1}{2\pi} \right)^3 \int d^3k \left[\mathbf{d}^2 - (\mathbf{k} \cdot \mathbf{d})^2 / k^2 \right] \\
&= \frac{1}{6\pi^2 \epsilon_0 c^3} \mathbf{d}^2 \int d\omega \omega^2
\end{aligned}
\tag{4.8.3}
$$

is also divergent, since the integration over ω runs from 0 to ∞. We will, however, retain this term. The divergent energy-level shifts it produces will be found to combine with another divergent shift, coming from the $-\mathbf{d} \cdot \mathbf{E}$ interaction, to give a weaker divergence. Now, since

$$
\mathbf{d}^2 = \sum_i \sum_j |i\rangle\langle i|\mathbf{d}^2|j\rangle\langle j| = \sum_i \sum_j (\mathbf{d}^2)_{ij} \sigma_{ij},
\tag{4.8.4}
$$

where, as in Section 2.1, we have introduced sums over complete sets of states, we can write (4.8.3) as

$$
\frac{1}{2\epsilon_0} \int d^3r \mathbf{P}^\perp(\mathbf{r})^2 = \frac{1}{6\pi^2 \epsilon_0 c^3} \sum_i \sum_j (\mathbf{d}^2)_{ij} \sigma_{ij} \int d\omega \omega^2.
\tag{4.8.5}
$$

Again, as in Section 2.1,

$$
H_A = \sum_i E_i \sigma_{ii},
\tag{4.8.6}
$$

and

$$
\mathbf{d} = \sum_i \sum_j \mathbf{d}_{ij} \sigma_{ij};
\tag{4.8.7}
$$

therefore,

$$
\begin{aligned}
H = \sum_i E_i \sigma_{ii} + H_F &- \sum_i \sum_j \mathbf{d}_{ij} \sigma_{ij} \cdot \mathbf{E}^\perp(0) \\
&+ \frac{1}{6\pi^2 \epsilon_0 c^3} \sum_i \sum_j (\mathbf{d}^2)_{ij} \sigma_{ij} \int d\omega \omega^2,
\end{aligned}
\tag{4.8.8}
$$

or, from (4.2.19) and (4.2.20),

$$
\begin{aligned}
H = \sum_i E_i \sigma_{ii} &+ \sum_{\mathbf{k}\lambda} \omega_k (a_{\mathbf{k}\lambda}^\dagger a_{\mathbf{k}\lambda} + 1/2) \\
&- i \sum_{\mathbf{k}\lambda} \sum_i \sum_j \sqrt{\frac{\hbar\omega_k}{2\epsilon_0 V}} (\mathbf{d}_{ij} \cdot \mathbf{e}_{\mathbf{k}\lambda})(\sigma_{ij} a_{\mathbf{k}\lambda} - a_{\mathbf{k}\lambda}^\dagger \sigma_{ij}) \\
&+ \frac{1}{6\pi^2 \epsilon_0 c^3} \sum_i \sum_j (\mathbf{d}^2)_{ij} \sigma_{ij} \int d\omega \omega^2.
\end{aligned}
\tag{4.8.9}
$$

This is the electric dipole Hamiltonian for a single, neutral atom and the transverse electromagnetic field in free space. We have put the field operators in normal order— the $a_{\mathbf{k}\lambda}$'s to the right and the $a^\dagger_{\mathbf{k}\lambda}$'s to the left, in products of (commuting, equal-time) atom and field operators.

4.8.1 Heisenberg Equations of Motion, and Vacuum Expectations

Heisenberg equations of motion for the atom and field operators follow from the commutation relations $[\sigma_{ij}, \sigma_{k\ell}] = \delta_{jk}\sigma_{i\ell} - \delta_{i\ell}\sigma_{kj}$ ((2.3.13)), $[a_{\mathbf{k}\lambda}, a^\dagger_{\mathbf{k}'\lambda'}] = \delta^3_{\mathbf{k}\mathbf{k}'}\delta_{\lambda\lambda'}$, and the vanishing of (equal-time) commutators of atom operators with field operators:

$$\dot{\sigma}_{ij} = \frac{1}{i\hbar}[\sigma_{ij}, H]$$

$$= -i\omega_{ji}\sigma_{ij} - \sum_{\mathbf{k}\lambda}\sqrt{\frac{\omega_k}{2\epsilon_0\hbar V}}\sum_\ell(\mathbf{d}_{j\ell}\cdot\mathbf{e}_{\mathbf{k}\lambda})(\sigma_{i\ell}a_{\mathbf{k}\lambda} - a^\dagger_{\mathbf{k}\lambda}\sigma_{i\ell})$$

$$+ \sum_{\mathbf{k}\lambda}\sqrt{\frac{\omega_k}{2\epsilon_0\hbar V}}\sum_\ell(\mathbf{d}_{\ell i}\cdot\mathbf{e}_{\mathbf{k}\lambda})(\sigma_{\ell j}a_{\mathbf{k}\lambda} - a^\dagger_{\mathbf{k}\lambda}\sigma_{\ell j})$$

$$- \frac{i}{6\pi^2\epsilon_0\hbar c^3}\sum_\ell\left[(\mathbf{d}^2)_{j\ell}\sigma_{i\ell} - (\mathbf{d}^2)_{\ell i}\sigma_{\ell j}\right]\int d\omega\omega^2, \tag{4.8.10}$$

$$\dot{a}_{\mathbf{k}\lambda} = \frac{1}{i\hbar}[a_{\mathbf{k}\lambda}, H] = -i\omega_k a_{\mathbf{k}\lambda} + \sqrt{\frac{\omega_k}{2\epsilon_0\hbar V}}\sum_i\sum_j(\mathbf{d}_{ij}\cdot\mathbf{e}_{\mathbf{k}\lambda})\sigma_{ij}. \tag{4.8.11}$$

Except for the term involving $\int d\omega\omega^2$, (4.8.10) is seen to have the same form as the Heisenberg equation of motion obtained in semiclassical radiation theory when we recall the expression

$$\mathbf{E}(t) = i\sum_{\mathbf{k}\lambda}\sqrt{\frac{\hbar\omega_k}{2\epsilon_0 V}}[a_{\mathbf{k}\lambda} - a^\dagger_{\mathbf{k}\lambda}]\mathbf{e}_{\mathbf{k}\lambda} \tag{4.8.12}$$

for the quantized electric field acting on the atom in free space. The difference between (4.8.10) and (2.3.16), apart from the obvious one that the field in (4.8.10) is quantized, comes from the term $(1/2\epsilon_0)\int d^3r\mathbf{P}^{\perp 2}$ that distinguishes the electric dipole Hamiltonian of the quantized-field theory from that of semiclassical radiation theory, as discussed following (4.2.10).

We have used normal ordering in (4.8.10). For an atom in vacuum, we now proceed exactly as we did in the two-state model when we chose normal ordering: using the formal solution

$$a_{\mathbf{k}\lambda}(t) = a_{\mathbf{k}\lambda}(0)e^{-i\omega_k t} + \sum_i\sum_j\sqrt{\frac{\omega_k}{2\epsilon_0\hbar V}}(\mathbf{d}_{ij}\cdot\mathbf{e}_{\mathbf{k}\lambda})\int_0^t dt'\sigma_{ij}(t')e^{i\omega_k(t'-t)} \tag{4.8.13}$$

of (4.8.11) in (4.8.10), and taking expectation values over an initial (bare) atom–field state in which there are no photons, we obtain

$$\langle \dot{\sigma}_{ij} \rangle = -i\omega_{ji}\langle \sigma_{ij}(t) \rangle - \sum_{\mathbf{k}\lambda} \sum_{\ell mn} C_{\mathbf{k}\lambda j\ell} C_{\mathbf{k}\lambda mn} \int_0^t dt' \langle \sigma_{i\ell}(t)\sigma_{mn}(t') \rangle e^{i\omega_k(t'-t)}$$

$$+ \sum_{\mathbf{k}\lambda} \sum_{\ell mn} C_{\mathbf{k}\lambda j\ell} C^*_{\mathbf{k}\lambda mn} \int_0^t dt' \langle \sigma_{nm}(t')\sigma_{i\ell}(t) \rangle e^{-i\omega_k(t'-t)}$$

$$+ \sum_{\mathbf{k}\lambda} \sum_{\ell mn} C_{\mathbf{k}\lambda \ell i} C_{\mathbf{k}\lambda mn} \int_0^t dt' \langle \sigma_{\ell j}(t)\sigma_{mn}(t') \rangle e^{i\omega_k(t'-t)}$$

$$- \sum_{\mathbf{k}\lambda} \sum_{\ell mn} C_{\mathbf{k}\lambda \ell i} C^*_{\mathbf{k}\lambda mn} \int_0^t dt' \langle \sigma_{nm}(t')\sigma_{\ell j}(t) \rangle e^{-i\omega_k(t'-t)}$$

$$- \frac{i}{6\pi^2 \epsilon_0 \hbar c^3} \sum_\ell \left[(\mathbf{d}^2)_{j\ell}\langle \sigma_{i\ell}(t) \rangle - (\mathbf{d}^2)_{\ell i}\langle \sigma_{\ell j}(t) \rangle \right] \int d\omega\, \omega^2, \qquad (4.8.14)$$

$$C_{\mathbf{k}\lambda mn} = \sqrt{\frac{\omega_k}{2\epsilon_0 \hbar V}} (\mathbf{d}_{mn} \cdot \mathbf{e}_{\mathbf{k}\lambda}), \qquad (4.8.15)$$

and again $\omega_{ji} = (E_j - E_i)/\hbar$, the (angular) transition frequency between the atomic states with energy eigenvalues E_j and E_i. Aside from the dipole approximation, (4.8.14) is (non-relativistically) "exact."

4.8.2 The Markovian Approximation

Again following the route taken in the two-state model, we make the Markovian approximation,

$$\sigma_{mn}(t') \cong \sigma_{mn}(t)e^{-i\omega_{nm}(t'-t)}, \qquad (4.8.16)$$

in (4.8.14). Then, for example,

$$\int_0^t dt' \langle \sigma_{i\ell}(t)\sigma_{mn}(t') \rangle e^{i\omega_k(t'-t)} \cong \langle \sigma_{i\ell}(t)\sigma_{mn}(t) \rangle \int_0^t dt' e^{i(\omega_k-\omega_{nm})(t'-t)}$$

$$= \delta_{\ell m}\langle \sigma_{in}(t) \rangle \int_0^t dt' e^{i(\omega_k-\omega_{nm})(t'-t)}, \quad (4.8.17)$$

where we use the relation $\sigma_{i\ell}(t)\sigma_{mn}(t) = \delta_{\ell m}\sigma_{in}(t)$. This is obviously true for the unperturbed atom $[\sigma_{i\ell}(0)\sigma_{mn}(0) = |i\rangle\langle \ell|m\rangle\langle n| = \delta_{\ell m}|i\rangle\langle n| = \delta_{\ell m}\sigma_{in}(0)]$, and is therefore taken as true for all times as a consequence of the way Heisenberg-picture operators A_H evolve in time $[A_H(t) = U^\dagger(t)A_H(0)U(t)$, where $U(t)$ is the (unitary) time evolution operator]. In fact, it is true within the Markovian approximation:

$$\sigma_{i\ell}(t)\sigma_{mn}(t) \cong \sigma_{i\ell}(0)\sigma_{mn}(0)e^{-i(\omega_{nm}+\omega_{\ell i})t} = \delta_{\ell m}\sigma_{in}(0)e^{-i(\omega_{nm}+\omega_{mi})t}$$

$$= \delta_{\ell m}\sigma_{in}(0)e^{-i\omega_{ni}t} \cong \delta_{\ell m}\sigma_{in}(t). \qquad (4.8.18)$$

In this approximation, (4.8.14) becomes

$$\langle \dot{\sigma}_{ij} \rangle = -i\omega_{ji}\langle \sigma_{ij}(t) \rangle - \sum_m \sum_n \Big[\Gamma_{jmmn}(t)\langle \sigma_{in}(t) \rangle - \Gamma^*_{mjin}(t)\langle \sigma_{nm}(t) \rangle$$

$$- \Gamma_{mijn}(t)\langle \sigma_{mn}(t) \rangle + \Gamma^*_{immn}(t)\langle \sigma_{nj}(t) \rangle \Big]$$

$$- \frac{i}{6\pi^2 \epsilon_0 \hbar c^3} \sum_\ell \big[(\mathbf{d}^2)_{j\ell}\langle \sigma_{i\ell}(t) \rangle - (\mathbf{d}^2)_{\ell i}\langle \sigma_{\ell j}(t) \rangle \big] \int d\omega\omega^2, \qquad (4.8.19)$$

$$\Gamma_{\ell mnp}(t) = \sum_{\mathbf{k}\lambda} C_{\mathbf{k}\lambda\ell m} C_{\mathbf{k}\lambda np} \int_0^t dt' e^{i(\omega_k - \omega_{pn})(t'-t)}$$

$$= \sum_{\mathbf{k}\lambda} \frac{\omega_k}{2\epsilon_0 \hbar V} (\mathbf{d}_{\ell m} \cdot \mathbf{e}_{\mathbf{k}\lambda})(\mathbf{d}_{np} \cdot \mathbf{e}_{\mathbf{k}\lambda}) \int_0^t dt' e^{i(\omega_k - \omega_{pn})(t'-t)}$$

$$= \sum_{\mathbf{k}} \frac{\omega_k}{2\epsilon_0 \hbar V} \big[\mathbf{d}_{\ell m} \cdot \mathbf{d}_{np}$$

$$- (\mathbf{d}_{\ell m} \cdot \mathbf{k})(\mathbf{d}_{np} \cdot \mathbf{k})/k^2 \big] \int_0^t dt' e^{i(\omega_k - \omega_{pn})(t'-t)}. \qquad (4.8.20)$$

$\Gamma_{\ell mnp}(t)$ is evaluated in the mode-continuum limit ($V \to \infty$) in the familiar fashion (see Section 4.3). From (4.4.31),

$$\Gamma_{\ell mnp}(t) \cong \frac{\mathbf{d}_{\ell m} \cdot \mathbf{d}_{np}}{6\pi^2 \epsilon_0 \hbar c^3} \int_0^\infty d\omega\omega^3 \Big[\pi\delta(\omega - \omega_{pn}) - i\mathrm{P}\frac{1}{\omega - \omega_{pn}} \Big]$$

$$= \frac{\omega_{pn}^3 \mathbf{d}_{\ell m} \cdot \mathbf{d}_{np}}{6\pi \epsilon_0 \hbar c^3} \theta(\omega_{pn}) - \frac{i\mathbf{d}_{\ell m} \cdot \mathbf{d}_{np}}{6\pi^2 \epsilon_0 \hbar c^3} \int_0^\infty \frac{d\omega\omega^3}{\omega - \omega_{pn}}$$

$$= \beta_{\ell mnp} - i\gamma_{\ell mnp} \qquad (4.8.21)$$

for times $t \gg |\omega_{pn}|^{-1}$; θ is the (Heaviside) unit step function. Note that $\beta_{\ell mnp} = \gamma_{\ell mnp} = 0$ for $\ell = m$ or $n = p$ (because $\mathbf{d}_{nn} = 0$), and that $\beta_{m\ell np} = \beta^*_{\ell mnp}$, $\gamma_{m\ell np} = \gamma^*_{\ell mnp}$ (because $\mathbf{d}_{\ell m} = \mathbf{d}^*_{m\ell}$).

Collecting and arranging terms, we write (4.8.19) as

$$\langle \dot{\sigma}_{ij}(t) \rangle = -i\Big\{ \omega_{ji} - \sum_m (\gamma_{jmmj} - \gamma_{immi})$$

$$+ \frac{1}{6\pi^2 \epsilon_0 \hbar c^3} \big[(\mathbf{d}^2)_{jj} - (\mathbf{d}^2)_{ii} \big] \int d\omega\omega^2 \Big\} \langle \sigma_{ij}(t) \rangle$$

$$- \sum_m \big(\beta_{jmmj} + \beta_{immi} \big) \langle \sigma_{ij}(t) \rangle - \sum_m \sum_{n \neq j} \big(\beta_{jmmn} - i\gamma_{jmmn} \big) \langle \sigma_{in}(t) \rangle$$

$$- \sum_m \sum_{n \neq i} \big(\beta^*_{immn} + i\gamma^*_{immn} \big) \langle \sigma_{nj}(t) \rangle + \sum_m \sum_n \big(\beta^*_{mjin} + i\gamma^*_{mjin} \big) \langle \sigma_{nm}(t) \rangle$$

$$+ \sum_m \sum_n \big(\beta_{mijn} - i\gamma_{mijn} \big) \langle \sigma_{mn}(t) \rangle - \frac{i}{6\pi^2 \epsilon_0 \hbar c^3} \Big[\sum_{n \neq j} (\mathbf{d}^2)_{jn} \langle \sigma_{in}(t) \rangle$$

$$- \sum_{n \neq i} (\mathbf{d}^2)_{ni} \langle \sigma_{nj}(t) \rangle \Big] \int d\omega\omega^2. \qquad (4.8.22)$$

4.8.3   Level Shifts, and Renormalization

Consider first

$$\sum_m \gamma_{jmmj} - \frac{1}{6\pi^2\epsilon_0\hbar c^3}(\mathbf{d}^2)_{jj}\int d\omega\omega^2 = \frac{1}{6\pi^2\epsilon_0\hbar c^3} \times$$

$$\left[\sum_m \mathbf{d}_{jm}\cdot\mathbf{d}_{mj}\fint\frac{d\omega\omega^3}{\omega - \omega_{jm}} - (\mathbf{d}^2)_{jj}\int d\omega\omega^2\right]. \qquad (4.8.23)$$

From $\mathbf{d}_{jm}\cdot\mathbf{d}_{mj} = |\mathbf{d}_{jm}|^2$ and

$$(\mathbf{d}^2)_{jj} = \langle j|\mathbf{d}\cdot\mathbf{d}|j\rangle = \langle j|\left(\sum_{mn}\mathbf{d}_{nm}|n\rangle\langle m|\right)\cdot\left(\sum_{\ell p}\mathbf{d}_{\ell p}|\ell\rangle\langle p|\right)|j\rangle$$

$$= \sum_{mn}\sum_{\ell p}\delta_{jn}\delta_{\ell m}\delta_{pj}\mathbf{d}_{nm}\cdot\mathbf{d}_{\ell p} = \sum_m |\mathbf{d}_{jm}|^2, \qquad (4.8.24)$$

$$\sum_m \gamma_{jmmj} - \frac{1}{6\pi^2\epsilon_0\hbar c^3}(\mathbf{d}^2)_{jj}\int d\omega\omega^2 = \frac{1}{6\pi^2\epsilon_0\hbar c^3}\sum_m |\mathbf{d}_{jm}|^2\left\{\fint\frac{d\omega\omega^3}{\omega - \omega_{jm}} - \int d\omega\omega^2\right\}$$

$$= \frac{1}{6\pi^2\epsilon_0\hbar c^3}\sum_m |\mathbf{d}_{jm}|^2\omega_{jm}\fint\frac{d\omega\omega^2}{\omega - \omega_{jm}}$$

$$= -\frac{1}{\hbar}\delta E_j, \qquad (4.8.25)$$

and, likewise,

$$\sum_m \gamma_{immi} - \frac{1}{6\pi^2\epsilon_0\hbar c^3}(\mathbf{d}^2)_{ii}\int d\omega\omega^2 = \frac{1}{6\pi^2\epsilon_0\hbar c^3}\sum_m |\mathbf{d}_{im}|^2\omega_{im}\fint\frac{d\omega\omega^2}{\omega - \omega_{im}}$$

$$= -\frac{1}{\hbar}\delta E_i. \qquad (4.8.26)$$

Then,

$$\omega_{ji} - \sum_m(\gamma_{jmmj} - \gamma_{immi}) + \frac{1}{6\pi^2\epsilon_0\hbar c^3}\left[(\mathbf{d}^2)_{jj} - (\mathbf{d}^2)_{ii}\right]\int d\omega\omega^2$$

$$= \frac{1}{\hbar}\left[(E_j + \delta E_j) - (E_i + \delta E_i)\right], \qquad (4.8.27)$$

and we identify δE_j (δE_i) as a radiative shift in the atom's energy level E_j (E_i).

We now argue that, to obtain observable energy-level shifts, we should first subtract off the (unobservable) transverse self-energy E_{free} of a *free* electron. We can derive E_{free} in different ways; the easiest is to assert that a free electron has (approximately) a continuum of possible energies, so that E_{free} follows from δE_j (for any j) when we ignore the transition frequencies ω_{jm} compared with all field frequencies ω:

$$\delta E_{\text{free}} = -\frac{1}{6\pi^2\epsilon_0 c^3} \sum_m |\mathbf{d}_{jm}|^2 \omega_{jm} \int d\omega\omega. \tag{4.8.28}$$

(See Exercise 4.6 for another way to derive E_{free}.) Then, we replace δE_j by

$$\delta E_j' = \delta E_j - E_{\text{free}} = -\frac{1}{6\pi^2\epsilon_0 c^3} \sum_m |\mathbf{d}_{jm}|^2 \omega_{jm} \left(\fint \frac{d\omega\omega^2}{\omega - \omega_{jm}} - \int d\omega\omega \right)$$

$$= -\frac{1}{6\pi^2\epsilon_0 c^3} \sum_m \omega_{jm}^2 |\mathbf{d}_{jm}|^2 \fint \frac{d\omega\omega}{\omega - \omega_{jm}}. \tag{4.8.29}$$

We can further reduce the degree of divergence by *mass renormalization*. In our original Hamiltonian (4.8.1), $H_A = \mathbf{p}^2/2m_{\text{obs}} + V(\mathbf{x})$, with m_{obs} the observed electron mass. But part of this observed mass comes from the coupling of the electron to the electromagnetic field, as discussed in Section 4.3. To avoid double counting of the contribution of this electromagnetic mass (m_{em}) to the electron's kinetic energy when we "turn on" the coupling of the atom to the field in our calculations, we should subtract from δE_j the energy

$$\left\langle j \left| \left(\frac{\mathbf{p}^2}{2m_{\text{obs}}} - \frac{\mathbf{p}^2}{2m_{\text{bare}}} \right) \right| j \right\rangle = \left\langle j \left| \left(\frac{\mathbf{p}^2}{2(m_{\text{bare}} + m_{\text{em}})} - \frac{\mathbf{p}^2}{2m_{\text{bare}}} \right) \right| j \right\rangle$$

$$\cong -\frac{m_{\text{em}}}{2m_{\text{obs}}^2} \langle j | \mathbf{p}^2 | j \rangle = -\frac{m_{\text{em}}}{2m_{\text{obs}}^2} \sum_m |\mathbf{p}_{jm}|^2. \tag{4.8.30}$$

It is assumed that m_{em}, whatever its (finite) value, is small compared with the observed mass m_{obs} and the bare (non-electromagnetic) mass m_{bare}.

With mass renormalization, therefore, obtain the "observable" level shift

$$\delta E_j^{\text{obs}} = \delta E_j' + \frac{m_{\text{em}}}{2m^2} \sum_n |\mathbf{p}_{jn}|^2$$

$$= -\frac{e^2}{6\pi^2\epsilon_0 m^2 c^3} \sum_n |\mathbf{p}_{jn}|^2 \left(\fint_0^\Lambda \frac{d\omega\omega}{\omega - \omega_{jn}} - \int_0^\Lambda d\omega \right)$$

$$= \left(\frac{2\alpha}{3\pi} \right) \left(\frac{1}{mc} \right)^2 \sum_n (E_n - E_j) |\mathbf{p}_{jn}|^2 \log \left| \frac{\Lambda}{\omega_{jn}} \right| \tag{4.8.31}$$

for $\Lambda \gg |\omega_{jn}|$, where again α is the fine-structure constant. We have used the identity (recall (2.3.39))

$$\omega_{jn}^2 |\mathbf{d}_{jn}|^2 = e^2 |\mathbf{p}_{jn}|^2 / m^2, \tag{4.8.32}$$

and now, as usual, let m denote the observed mass. Importantly, because of its logarithmic dependence on the cutoff frequency, δE_j^{obs} is not very sensitive to the precise value of Λ. Taking it to be $\sim mc^2/\hbar$ presumes that δE_j^{obs} is primarily a non-relativistic effect.

Exercise 4.6: Using the sum rule (2.3.37), show that δE_{free} as defined by (4.8.28) is equal to the vacuum-field expectation value $\langle e^2 \mathbf{A}^2/2m \rangle$, regardless of the state of the atom. (Recall that the energy $e^2 \mathbf{A}^2/2m$ appears explicitly in the minimal coupling form of the Hamiltonian but not in the electric dipole form we are using.)

In what was once called "the most important discovery in the history of quantum electrodynamics,"[34] Bethe evaluated (4.8.31), with $\hbar \Lambda = mc^2$, for the 2s level of the hydrogen atom; he obtained $\delta E_{2s}^{\text{obs}} \cong 1040$ MHz, in excellent agreement with the "Lamb shift" measured by Lamb and Retherford.[35] In his numerical evaluation of the Lamb shift, Bethe replaced the logarithm in (4.8.31) by an average value:

$$\delta E_j^{\text{obs}} \cong \left(\frac{2\alpha}{3\pi} \right) \left(\frac{1}{mc} \right)^2 \log \frac{mc^2}{|E_n - E_j|_{\text{avg}}} \sum_n |\mathbf{p}_{jn}|^2 (E_n - E_j). \tag{4.8.33}$$

The sum over the complete set of states can be simplified. Note first that

$$\sum_n |\mathbf{p}_{jn}|^2 (E_n - E_j) = \sum_n \langle j | \mathbf{p} H_A - H_A \mathbf{p} | n \rangle \cdot \langle n | \mathbf{p} | j \rangle, \tag{4.8.34}$$

where $H_A = p^2/2m + V(r)$, and that $[\mathbf{p}, H_A] = -i\hbar \nabla H_A = -i\hbar \nabla V$. Thus

$$\sum_n |\mathbf{p}_{jn}|^2 (E_n - E_j) = -i\hbar \sum_n \langle j | \nabla V | n \rangle \cdot \langle n | \mathbf{p} | j \rangle = -\frac{i\hbar}{2} \langle j | [\nabla V, \mathbf{p}] | j \rangle$$

$$= -\frac{i\hbar}{2} (i\hbar) \langle j | \nabla^2 V | j \rangle = \frac{\hbar^2}{2} \int d^3 r |\psi_j(\mathbf{r})|^2 \nabla^2 V$$

$$= \frac{\hbar^2 e^2}{2\epsilon_0} |\psi_j(0)|^2, \tag{4.8.35}$$

since $V(r) = -e^2/4\pi\epsilon_0 r$, $\nabla^2 V = (e^2/\epsilon_0)\delta^3(\mathbf{r})$, and therefore

$$\delta E_j^{\text{obs}} \cong \frac{\alpha}{3\pi\epsilon_0} \left(\frac{e\hbar}{mc} \right)^2 |\psi_j(0)|^2 \log \frac{mc^2}{|E_n - E_j|_{\text{avg}}}. \tag{4.8.36}$$

The factor $|\psi_j(0)|^2$ implies, correctly, that the s level $(2s_{1/2}, \psi(0) \neq 0)$ is Lamb-shifted in hydrogen, but not the p level $(2p_{1/2}, \psi(0) = 0)$.

For purposes here, it would be too great a digression to discuss the actual Lamb shift in any detail, and, besides, a proper treatment requires fully relativistic quantum

[34] R. P. Feynman, "The Development of the Space-Time View of Quantum Electrodynamics," http://www.nobelprize.org/prizes/physics/1965/feynman/lecture, accessed May 11, 2018.

[35] H. A. Bethe, Phys. Rev. **72**, 241 (1947). The essential feature of Bethe's calculation—mass renormalization—was based on ideas of H. A. Kramers (*Schweber*, 2012).

electrodynamics.[36] In what follows, we will just assume that Lamb shifts have been included in the definitions of the transition frequencies ω_{ji}.

What about the terms involving "generalized frequency shift" coefficients γ in (4.8.22) that do not couple $\langle \sigma_{ij}(t) \rangle$ to itself, and the terms with $\int d\omega \omega^2$ in the last line? These terms couple different $\langle \sigma_{ij}(t) \rangle$'s in a way analogous to a linear coupling among classical harmonic oscillators. In the classical case, the coupling results in normal modes with oscillation frequencies different from those of the uncoupled oscillators. The coupling among the $\langle \sigma_{ij}(t) \rangle$'s in (4.8.22), analogously, leads to consideration of states of the atom that are "dressed" by its interaction with the vacuum field. Equations can be derived, based on these dressed states, that are similar in form to (4.8.22) but do not involve any couplings by generalized frequency shifts.[37] The net effect of the generalized frequency shifts is very small unless there are eigenstates of the atom that are nearly degenerate and have the same angular momentum quantum numbers. Since radiative frequency shifts are only of secondary interest here, we will ignore the generalized frequency shift terms altogether and replace (4.8.22) by[38]

$$\langle \dot{\sigma}_{ij}(t) \rangle = -i\omega_{ji}\langle \sigma_{ij}(t) \rangle - \sum_m \left(\beta_{jmmj} + \beta_{immi} \right) \langle \sigma_{ij}(t) \rangle$$

$$- \sum_m \sum_{n \neq j} \beta_{jmmn} \langle \sigma_{in}(t) \rangle - \sum_m \sum_{n \neq i} \beta_{immn}^* \langle \sigma_{nj}(t) \rangle$$

$$+ \sum_m \sum_n \beta_{mjin}^* \langle \sigma_{nm}(t) \rangle + \sum_m \sum_n \beta_{mijn} \langle \sigma_{mn}(t) \rangle. \qquad (4.8.37)$$

Exercise 4.7: We derived δE_j^{obs} as given by (4.8.31) in the electric dipole approximation, which breaks down at high frequencies. Does the dipole approximation fail at frequencies larger or smaller than Bethe's cutoff frequency? (Calculations that go beyond the dipole approximation by including retardation indicate that Bethe's estimate of the Lamb shift is so accurate because (1) the most significant contributions come from frequencies not affected by his choice for the cutoff, and (2) at much higher frequencies, retardation itself introduces an effective cutoff. See C. K. Au and G. Feinberg, Phys. Rev. A **9**, 1794 (1974).)

4.8.4 Radiative Decay Rates and Landau Terms

Setting $i = j$ and using the fact that $\beta_{miim}^* = \beta_{miim}$, we find from (4.8.37) that the rate of change of the occupation probability $\langle \sigma_{ii}(t) \rangle$ of state i is

---

[36] The original (late 1940s), fully relativistic calculations of the Lamb shift of the 2s level of hydrogen are reviewed in *Milonni (1994)*. We return to Bethe's expression (4.8.36) in relation to vacuum-field fluctuations and zero-point energy in Section 7.4.

[37] D. A. Cardimona and C. R. Stroud, Jr., Phys. Rev. A **27**, 2456 (1983).

[38] If the generalized frequency shifts are significant, we obtain equations of the same form as (4.8.37) but with the β coefficients having slightly different numerical values.

$$\langle \dot{\sigma}_{ii}(t) \rangle = -2 \sum_m \beta_{immi} \langle \sigma_{ii}(t) \rangle + 2 \sum_m \beta_{miim} \langle \sigma_{mm}(t) \rangle$$

$$- \sum_m \sum_{n \neq i} \left(\beta_{immn} \langle \sigma_{in}(t) \rangle + \beta^*_{immn} \langle \sigma_{ni}(t) \rangle \right)$$

$$+ \sum_m \sum_{n \neq m} \left(\beta^*_{miin} + \beta_{niim} \right) \langle \sigma_{nm}(t) \rangle. \tag{4.8.38}$$

The rate

$$2\beta_{immi} = \frac{\omega_{im}^3 |\mathbf{d}_{im}|^2}{3\pi \epsilon_0 \hbar c^3} \quad (E_i > E_m)$$

$$= 0 \quad (\text{when } E_i < E_m) \tag{4.8.39}$$

is the Einstein A coefficient A_{im} for spontaneous emission from state i to a lower-lying state m (see (4.4.21)). If there are no near degeneracies, the off-diagonal expectation values $\langle \sigma_{in}(t) \rangle$, $\langle \sigma_{ni}(t) \rangle$, and $\langle \sigma_{nm}(t) \rangle$ in (4.8.39) are approximately oscillatory at frequencies much larger than the spontaneous emission rates and can be ignored in a rotating-wave approximation, since they oscillate rapidly about zero during times on the order of radiative lifetimes. Then, (4.8.38) can be approximated by the simple rate equation

$$\dot{P}_i(t) = -A_i P_i(t) + \sum_{m, E_m > E_i} A_{mi} P_m(t), \tag{4.8.40}$$

where $P_n(t) = \langle \sigma_{nn}(t) \rangle$, and

$$A_i = \sum_{m, E_m < E_i} A_{im} \tag{4.8.41}$$

is the radiative decay rate of state i. This is just what one would expect: the population of each state decreases by radiative transitions to lower-lying states and increases by radiative transitions into it from higher-lying states.

According to (4.8.38), however, things are not that simple: the time evolution of $\langle \sigma_{ii}(t) \rangle$ depends not only on occupation probabilities but also on the off-diagonal terms in the last two lines of (4.8.38). Their contribution to the time evolution of $\langle \sigma_{ii}(t) \rangle$ is negligible in general only if there are no near degeneracies; otherwise, the simple rate equation (4.8.41) does not correctly describe the evolution of $\langle \sigma_{ii}(t) \rangle$. Similarly, it is seen from (4.8.37) that there are contributions to the time evolution of $\langle \sigma_{ij}(t) \rangle$, $i \neq j$, that can be ignored in a rotating-wave approximation but not, in general, if there are near degeneracies. Contributions to (4.8.37) involving $\langle \sigma_{mn}(t) \rangle$ with $m \neq i$ or $n \neq j$, and to (4.8.38) involving $\sigma_{mn}(t) \rangle$ with $m \neq n$, may be called *Landau terms*, since they were apparently first derived by Landau in a paper especially noteworthy for its introduction of the density-matrix formalism to treat a quantum system that cannot be adequately described by a state vector.[39] In our derivation of (4.8.37) and

---

[39] L. D. Landau, Z. Phys. **45**, 430 (1927). An English translation appears in *Collected Papers of L. D. Landau*, ed. D. ter Haar, Gordon and Breach, New York, 1965, pp. 8–18. The density matrix was also introduced in 1927, in a more general context, by von Neumann, Nachr. Ges. Wiss. Göttingen **1**, 245 (1927).

(4.8.38), based on normal ordering and the Markovian approximation, the appearance of the Landau terms has an easy interpretation: associated with each transition is a total radiation reaction field that depends on the transition dipole moment, the frequency of that transition, *and* the dipole moments and frequencies of all other allowed transitions, resulting in a coupling of different transitions via the coefficients

$$\beta_{\ell mnp} = \frac{\omega_{pn}^3}{3\pi\epsilon_0\hbar c^3}\mathbf{d}_{\ell m}\cdot\mathbf{d}_{np}\theta(\omega_{pn}), \tag{4.8.42}$$

where $\theta(\omega)$ is the unit step function. Different operator orderings allow for interpretations in which the Landau terms are related to the vacuum-field fluctuations, just as in the two-state model of the atom (see Section 4.5).

An example with just three states, as in Figure 4.2, is instructive. Consider the power

$$P(t) = \sum_{\mathbf{k}\lambda}\hbar\omega_k\frac{d}{dt}\langle a_{\mathbf{k}\lambda}^\dagger(t)a_{\mathbf{k}\lambda}(t)\rangle \tag{4.8.43}$$

radiated by such a three-state atom. Again assuming that all radiative frequency shifts have been accounted for, we obtain from this expression, (4.8.11), and (4.8.13) the formula

$$P(t)/\hbar\omega_{31} \cong 2\beta_{22}\langle\sigma_{22}(t)\rangle + 2\beta_{33}\langle\sigma_{33}(t)\rangle + 4\beta_{23}\mathrm{Re}\big[\langle\sigma_{23}(t)\rangle\big], \tag{4.8.44}$$

for the rate of photon emission, where we have made the Markovian approximation and defined

$$\beta_{22} = \beta_{2112} = \frac{\omega_{31}^2}{6\pi\epsilon_0\hbar c^3}|\mathbf{d}_{12}|^2, \quad \beta_{33} = \beta_{3113} = \frac{\omega_{31}^2}{6\pi\epsilon_0\hbar c^3}|\mathbf{d}_{13}|^2, \tag{4.8.45}$$

and

$$\beta_{23} = \beta_{2113} = \frac{\omega_{31}^2}{6\pi\epsilon_0\hbar c^3}\mathbf{d}_{13}\cdot\mathbf{d}_{21}, \tag{4.8.46}$$

and assumed for simplicity that β_{23} is real. In defining β_{22}, β_{33}, and β_{23}, we have also assumed that $\omega_{21} \approx \omega_{31}$. Equation (4.8.44) is obtained under the assumption of an initial state devoid of any photons, so that $\langle a_{\mathbf{k}\lambda}^\dagger(0)a_{\mathbf{k}\lambda}(0)\rangle = 0$ for all field modes $\mathbf{k}, \lambda$. The equations needed to evaluate the photon emission rate (4.8.44) follow from (4.8.37) and (4.8.38):

$$\langle\dot\sigma_{22}(t)\rangle = -2\beta_{22}\langle\sigma_{22}(t)\rangle - 2\beta_{23}\mathrm{Re}\big[\langle\sigma_{23}(t)\rangle\big],$$
$$\langle\dot\sigma_{33}(t)\rangle = -2\beta_{33}\langle\sigma_{33}(t)\rangle - 2\beta_{23}\mathrm{Re}\big[\langle\sigma_{23}(t)\rangle\big],$$
$$\langle\dot\sigma_{23}(t)\rangle = -i\big[\omega_{32} - i(\beta_{22} + \beta_{33})\big]\langle\sigma_{23}(t)\rangle - \beta_{23}\big[\langle\sigma_{22}(t)\rangle + \langle\sigma_{33}(t)\rangle\big]. \tag{4.8.47}$$

Figure 4.3 shows the calculated emission rate (4.8.44) when $\beta_{22} = 4\beta_{33}$, $\beta_{23} = 2\beta_{33}$, $\omega_{32} = 25\beta_{33}$, and $\langle\sigma_{11}(0)\rangle = 0$, $\langle\sigma_{22}(0)\rangle = \langle\sigma_{33}(0)\rangle = \langle\sigma_{23}(0)\rangle = 1/2$. The emission rate exhibits beats rather than a simple monotonic decay with time. It is obvious from

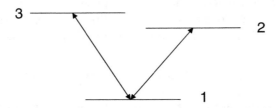

Fig. 4.2 A three-state atom with allowed electric dipole transitions between States 1 and 2 and between States 1 and 3.

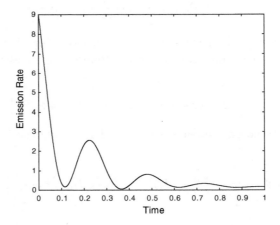

Fig. 4.3 The calculated spontaneous photon emission rate (4.8.44) for the three-state atom depicted in Figure 4.2, with the initial conditions $\langle \sigma_{11}(0) \rangle = 0$, $\langle \sigma_{22}(0) \rangle = \langle \sigma_{33}(0) \rangle = 1/2$, and $\langle \sigma_{23}(0) \rangle = 1/2$. The parameters in (4.8.47) are defined in the text, and time is in units of $1/\beta_{33}$.

(4.8.47) that such beats would not occur without the Landau terms involving β_{23}: if the transition dipole moments $\mathbf{d}_{13}$ and $\mathbf{d}_{21}$ are orthogonal, $\beta_{23} = 0$, and no beats are predicted. What is also interesting is that beats are predicted even if only one of the two upper-energy states in Figure 4.2 is initially populated, that is, if $\langle \sigma_{22}(0) \rangle = 1$ or $\langle \sigma_{33}(0) \rangle = 1$, provided that $\beta_{23} \neq 0$.[40]

---

[40] The parameters were chosen to correspond to those assumed in Figure 2 of G. C. Hegerfeldt and M. B. Plenio, Quantum Opt. **6**, 15 (1994). The calculation we have described leads to the same results obtained in the Monte Carlo (or "quantum trajectory") approach taken by these authors.

Exercise 4.8: (a) Derive (4.8.44) and (4.8.47). (b) Derive the corresponding equations that describe the emission rate for the transition scheme shown in Figure 4.4. Are there beats in that case?

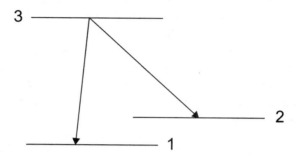

Fig. 4.4 A three-state atom with allowed electric dipole transitions between States 1 and 3 and between States 2 and 3.

The rate (4.4.21) at which the excited-state probability of an atom decreases, and the probability of emitting a photon increases, is one of the simplest but most important things we can calculate in quantum electrodynamics. Dirac's calculation of this rate "confirmed the universal character of quantum mechanics" because it showed that quantum theory could account for the creation of excitations—or "particles"—of a quantum field.[41] Although the formula for this rate is derived in neoclassical theory, for example, it seems impossible to explain *all* experimental observations relating to spontaneously emission within the framework of classical electromagnetic theory. The correct description of the process ultimately responsible for phenomena ranging from the light-up of fireflies to the start-up of a laser, not to mention most of the light from the Sun, evidently requires quantum electrodynamics.

We began this chapter by reviewing the transformation from the minimal coupling Hamiltonian to the (usually) more convenient electric dipole form when the field is quantized. Then we discussed, among other things, the approximations leading to the ("Weisskopf–Wigner)" prediction of exponential decay of excited-state probability, physical interpretations of spontaneous emission involving the vacuum electromagnetic field and radiation reaction, and a fairly general "multistate" description of spontaneous emission. We have considered only "the basics" and have ignored important effects that have been, and continue to be, subjects of pure and applied research. Some of these, including spontaneous emission "noise" and the modification of single-atom spontaneous emission by applied fields, other atoms, reflecting surfaces, or a host medium, are broached in the following chapters.

---

[41] S. Weinberg, Daedalus **106**, 17 (1977), p. 22.

References and Suggested Additional Reading

The transformation from the minimal coupling Hamiltonian to the electric dipole form when the field is quantized has been the subject of many analyses since the paper by E. A. Power and S. Zienau, Phil. Trans. R. Soc. **A251**, 427 (1959). Among these are L. Davidovich and H. M. Nussenzveig, in *Foundations of Radiation Theory and Quantum Electrodynamics*, ed. A. O. Barut, Plenum Press, New York, 1980, pp. 83–108, which focuses on some subtle aspects of natural lineshape theory, including those relating to the different forms of the Hamiltonian; R. G. Woolley, Mol. Phys. **22**, 1013 (1971); J. Savolainen and S. Stenholm, Am. J. Phys. 40, 667 (1972); and J. R. Ackerhalt and P. W. Milonni, J. Opt. Soc. Am. B **1**, 116 (1984), which reviews different approaches to the transformation.

In his paper entitled "The Damping Problem in Wave Mechanics," which was cited in Section 4.8 in connection with the "Landau terms," Landau made approximations that implied exponential decay of the initially excited state in spontaneous emission theory. A different approach to the approximation of exponential decay was taken later by V. Weisskopf and E. Wigner (Z. Phys. **63**, 54 (1930)). For an analysis of this approximation, see, for instance, P. R. Berman and G. W. Ford, Adv. Atom. Molec. Opt. Phys. **59**, 175 (2010), and the papers cited in Section 4.4.3.

Our discussion of different operator orderings in the Heisenberg-picture treatment of spontaneous emission and the radiative frequency shift follows I. R. Senitzky, Phys. Rev. Lett. **31**, 955 (1973); P. W. Milonni, J. R. Ackerhalt, and W. A. Smith, Phys. Rev. Lett. **31**, 958 (1973); and P. W. Milonni and W. A. Smith, Phys. Rev. A **11**, 814 (1975). Symmetric ordering is appealing for the purpose of interpreting these effects along classical lines, since the source and vacuum fields both appear as Hermitian operators; see J. Dalibard, J. Dupont-Roc, and C. Cohen-Tannoudji, J. Physique **43**, 1617 (1982). The paper by Milonni and Smith shows how symmetric ordering can be used to relate the quantum theory of the electromagnetic field to classical "stochastic electrodynamics." See also T. H. Boyer, Phys. Rev. D **11**, 809 (1975).

5

Atoms and Light: Quantum Theory

In this chapter, we apply the quantum theory of atom–field interactions in examples where "external" fields induce transitions in atoms. Absorption and stimulated emission are shown to result from the interference of an incident field with the field scattered by an atom. The interaction of a two-state atom and a single-mode field is formulated in the rotating-wave approximation. Originally developed "to clarify the relationship between the quantum theory of radiation . . . and the semiclassical theory," this *Jaynes–Cummings model* reveals distinctly quantum effects due to the "granular," photonic character of the field.[1]

The "dressed states" obtained by diagonalizing the Jaynes–Cummings Hamiltonian have been useful in treatments of resonant atom–field interactions, including the calculation of the spectrum of radiation from a two-state atom driven by a coherent, monochromatic field. We use them to deduce some basic features of this resonance fluorescence spectrum, and then calculate the spectrum as well as the coincidence rate for counting photons of the radiation scattered by the atom. The fact that two photons cannot be counted simultaneously in resonance fluorescence—*photon anti-bunching*—cannot be explained within classical electromagnetic theory.

While semiclassical radiation theory, Bloch equations, and rate equations as discussed in Chapter 2 provide the theoretical framework for most practical applications of laser–matter interactions and spectroscopy, many current and anticipated applications require consideration of the quantum nature of light, and some of these applications would be inconceivable without field quantization. Cavity quantum electrodynamics, resonance fluorescence, photon anti-bunching, and related quantum effects considered in this chapter provide conceptually instructive examples of some distinctly quantum-mechanical effects in the interaction of atoms and fields.

The last part of the chapter focuses on one of the hallmarks of quantum theory: entanglement. We calculate the polarization correlations of photons emitted in an atomic cascade transition, an example of entanglement that has been among the most influential in experiments testing the very foundations of quantum theory.

---

[1] E. T. Jaynes and F. W. Cummings, Proc. IEEE **51**, 89 (1963), p. 89.

An Introduction to Quantum Optics and Quantum Fluctuations. Peter W. Milonni.
© Peter W. Milonni 2019. Published in 2019 by Oxford University Press.
DOI:10.1093/oso/9780199215614.001.0001

5.1 Optical Bloch Equations for Expectation Values

We return now to the Heisenberg equation of motion, (4.5.1), and consider expectation values over an initial state $|\psi\rangle = |\psi_{\text{atom}}\rangle \otimes |\psi_{\text{field}}\rangle$ in which the field is *not* in its zero-point, vacuum state:

$$\langle \dot{\sigma}(t)\rangle = -i\omega_0\langle \sigma(t)\rangle - i\frac{d_{12}}{\hbar}\langle \sigma_z(t)E_0^{(+)}(t) + E_0^{(-)}(t)\sigma_z(t)\rangle$$

$$- i\frac{d_{12}}{\hbar}\langle \sigma_z(t)E_{\text{RR}}^{(+)}(t) + E_{\text{RR}}^{(-)}(t)\sigma_z(t)\rangle$$

$$\cong -i(\omega_0 - i\beta)\langle \sigma(t)\rangle - i\frac{d_{12}}{\hbar}\langle \sigma_z(t)E_0^{(+)}(t) + E_0^{(-)}(t)\sigma_z(t)\rangle \quad (5.1.1)$$

in the Markovian approximation for the effect (spontaneous emission) of the radiation reaction field. We have defined $\beta = A_{21}/2$ and included any (renormalized) radiative frequency shifts in the definition of ω_0.

The field state that is easiest to work with in many systems of interest in quantum optics and spectroscopy—and the one which often best describes the field of a laser—is the coherent state. We will assume for simplicity a single-mode coherent state of an applied field with annihilation operator $a(t)$ and frequency ω:

$$|\psi_{\text{field}}\rangle = |\alpha\rangle, \quad a(0)|\alpha\rangle = \alpha|\alpha\rangle. \quad (5.1.2)$$

Then,

$$\frac{d_{12}}{\hbar}E_0^{(+)}(t)|\psi_{\text{field}}\rangle = iCa(0)e^{-i\omega t}|\alpha\rangle = iC\alpha e^{-i\omega t}|\psi_{\text{field}}\rangle$$

$$= \Omega^{(+)}(t)|\psi_{\text{field}}\rangle, \quad (5.1.3)$$

$$\langle \psi_{\text{field}}|\frac{d_{12}}{\hbar}E_0^{(-)}(t) = \Omega^{(-)}(t)\langle \psi_{\text{field}}|, \quad (5.1.4)$$

where the (c-number) functions $\Omega^{(\pm)}(t)$ are defined by

$$\Omega^{(+)}(t) = iC\alpha e^{-i\omega t}, \quad (5.1.5)$$

$$\Omega^{(-)}(t) = -iC\alpha^* e^{i\omega t}. \quad (5.1.6)$$

$\langle \sigma(t)\rangle$ then evolves according to the equation

$$\langle \dot{\sigma}(t)\rangle = -i(\omega_0 - i\beta)\langle \sigma(t)\rangle - i\big[\Omega^{(+)}(t) + \Omega^{(-)}(t)\big]\langle \sigma_z(t)\rangle. \quad (5.1.7)$$

Now,

$$\Omega^{(+)}(t) + \Omega^{(-)}(t) = \frac{d_{12}}{\hbar}\langle \psi_{\text{field}}|E_0^{(+)}(t) + E_0^{(-)}(t)|\psi_{\text{field}}\rangle = \frac{d_{12}}{\hbar}\langle E_0(t)\rangle = \Omega(t), \quad (5.1.8)$$

and therefore

$$\langle \dot{\sigma}(t)\rangle = -i(\omega_0 - i\beta)\langle \sigma(t)\rangle - i\Omega(t)\langle \sigma_z(t)\rangle. \quad (5.1.9)$$

From (4.5.34), we obtain in the same way

$$\langle \dot{\sigma}_z(t)\rangle = -2\beta\big[1 + \langle \sigma_z(t)\rangle\big] - 2i\Omega(t)\big[\langle \sigma(t)\rangle - \langle \sigma^\dagger(t)\rangle\big] \quad (5.1.10)$$

in the Markovian approximation.

Equations (5.1.9) and (5.1.10) are identical to the equations for $\langle\sigma(t)\rangle$ and $\langle\sigma_z(t)\rangle$ derived in semiclassical radiation theory in Chapter 2, and in the rotating-wave approximation, we obtain from (5.1.9) and (5.1.10) the optical Bloch equations (2.7.1)–(2.7.3). We obtained (5.1.9) and (5.1.10) for the case where spontaneous emission is the only damping mechanism, but, of course, we can modify these equations, as in Section 2.7, to allow for collisions and other damping processes. The point is that if the applied field is well described by a coherent state, the equations for the expectation values of $\sigma(t)$ and $\sigma_z(t)$ are exactly those derived semiclassically. We have shown this for a single-mode field, but the conclusion holds as well for a multimode coherent state of the field. The optical Bloch equations, and their multistate generalizations, describe absorption and stimulated emission processes in atoms and molecules irradiated by any field that is well characterized by a coherent state or that can be treated approximately as a prescribed classical field.

We noted in Section 3.6.3 that the field of a single-mode laser can be considered to have a stable amplitude but a randomly varying phase and acts in effect as a coherent-state field if the phase diffusion rate is sufficiently slow. More generally, laser radiation can be described as a partially coherent field with a phase and Rabi frequency in the optical Bloch equations treated as stochastic variables.[2]

5.2 Absorption and Stimulated Emission as Interference Effects

Absorption and stimulated emission are correctly and satisfactorily treated by (5.1.9) and (5.1.10) and their straightforward generalizations to include more than two atomic states, inhomogeneous broadening, and so on. Given the obvious importance and ubiquity of absorption and stimulated emission, however, it seems worthwhile to describe these processes in a less formal, more physical way. That is the purpose of this section. To simplify matters a bit, we will, for this purpose, temporarily depart from the fully quantum theory and take a semiclassical approach, treating the applied field as a classically prescribed c-number function. To simplify further, we will take this field to be a monochromatic plane wave. Neither of these simplifications affect this section's conclusion that absorption and stimulated emission can be regarded as destructive and constructive interference, respectively, of the applied field and the field of the dipole moment induced in the atom by the applied field; as Lamb said,

When stimulated emission by an atom is treated, either quantum mechanically or by a suitable classical model, one finds that the numbers of photons in those modes of the radiation field that were initially excited are increased by the interaction. On the other hand, the electromagnetic field radiated by such an atom is found to have the appropriate multipole character and shows no trace of the above augmentation of the incident wave. In order to get amplification of the incident wave it is necessary to consider the interference of the incident and radiated waves.[3]

---

[2] See, for instance, J. H. Eberly, Phys. Rev. Lett. **37**, 1387 (1976).

[3] W. E. Lamb, Jr., in *Advances in Quantum Electronics*, ed. J. R. Singer, Columbia University, New York, 1961, p. 370.

The electromagnetic energy in free space at any time t is

$$W(t) = \int d^3 r \left[\frac{1}{2} \epsilon_0 \mathbf{E}^2(\mathbf{r}, t) + \frac{1}{2} \mu_0 \mathbf{B}^2(\mathbf{r}, t) \right] = \epsilon_0 \int d^3 r \mathbf{E}^2(\mathbf{r}, t). \tag{5.2.1}$$

Suppose the electric field $\mathbf{E}(\mathbf{r}, t) = \mathbf{E}_d(\mathbf{r}, t) + \mathbf{E}_i(\mathbf{r}, t)$, with $\mathbf{E}_d(\mathbf{r}, t)$ the electric field from an electric dipole at $\mathbf{r} = 0$, and $\mathbf{E}_i(\mathbf{r}, t)$ the "incident" (applied) electric field due to other sources of radiation (e.g., a laser).[4] Then,

$$W(t) = \epsilon_0 \int d^3 r \mathbf{E}_i^2(\mathbf{r}, t) + \epsilon_0 \int d^3 r \mathbf{E}_d^2(\mathbf{r}, t) + 2\epsilon_0 \int d^3 r \mathbf{E}_i(\mathbf{r}, t) \cdot \mathbf{E}_d(\mathbf{r}, t). \tag{5.2.2}$$

The first two terms are just the energies associated with the incident field and the field from the dipole, respectively, whereas the energy

$$W_{\text{int}}(t) = 2\epsilon_0 \int d^3 r \mathbf{E}_i(\mathbf{r}, t) \cdot \mathbf{E}_d(\mathbf{r}, t) \tag{5.2.3}$$

results from the interference of the incident and dipole fields. We assume

$$\mathbf{E}_i(\mathbf{r}, t) = \hat{\mathbf{x}} E_0 \cos(\omega_i t - \mathbf{k}_i \cdot \mathbf{r}), \quad \mathbf{k}_i = k_i \hat{\mathbf{z}} = \omega_i \hat{\mathbf{z}}/c, \tag{5.2.4}$$

where $\hat{\mathbf{x}}$ and $\hat{\mathbf{z}}$ are orthogonal unit vectors. Then,

$$W_{\text{int}}(t) = 2\epsilon_0 E_0 \cos \omega_i t \int d^3 r \hat{\mathbf{x}} \cdot \mathbf{E}_d(\mathbf{r}, t) \cos \mathbf{k}_i \cdot \mathbf{r}$$
$$+ 2\epsilon_0 E_0 \sin \omega_i t \int d^3 r \hat{\mathbf{x}} \cdot \mathbf{E}_d(\mathbf{r}, t) \sin \mathbf{k}_i \cdot \mathbf{r}. \tag{5.2.5}$$

To evaluate (5.2.5), we use the formula (4.3.4) for the field of an electric dipole moment $\mathbf{p} = e\mathbf{x} = \hat{\mathbf{x}} p(t)$ along the assumed direction $\hat{\mathbf{x}}$ of polarization of the incident field:

$$\hat{\mathbf{x}} \cdot \mathbf{E}_d(\mathbf{r}, t) = -\frac{1}{\epsilon_0} \left(\frac{1}{2\pi} \right)^3 \int d^3 k \omega_k [1 - (\mathbf{k} \cdot \hat{\mathbf{x}})^2 / k^2] e^{i\mathbf{k} \cdot \mathbf{r}} \int_0^t dt' p(t') \sin \omega_k (t' - t). \tag{5.2.6}$$

Then,

$$W_{\text{int}}(t) = -2E_0 \int d^3 k \omega_k [1 - (\mathbf{k} \cdot \hat{\mathbf{x}})^2 / k^2] \delta^3(\mathbf{k} - \mathbf{k}_i)$$
$$\times \int_0^t dt' p(t') \sin \omega_k (t' - t) \cos \omega_i t \tag{5.2.7}$$

after integration over all space. The delta function $\delta^3(\mathbf{k} - \mathbf{k}_i)$ requires $|\mathbf{k}| = \omega_k/c = |\mathbf{k}_i| = \omega_i/c$ and, consequently,

---

[4] The analysis here follows M. Cray, M.-L. Shih, and P. W. Milonni, Am. J. Phys. **50**, 1016 (1982).

$$\sin \omega_k(t' - t) \cos \omega_i t \rightarrow \frac{1}{2} \sin \omega_i t' \tag{5.2.8}$$

when we average over times t long compared with $1/\omega_i$. Then,

$$W_{\text{int}}(t) = -\omega_i E_0 \int_0^t dt' p(t') \sin \omega_i t', \tag{5.2.9}$$

and the rate at which field energy changes because of the interference of the incident field and the field radiated by the electric dipole is

$$\dot{W}_{\text{int}}(t) = -\omega_i E_0 p(t) \sin \omega_i t. \tag{5.2.10}$$

In this semiclassical approach, we write

$$p(t) = d \big[u(t) \cos \omega_i t - v(t) \sin \omega_i t \big], \tag{5.2.11}$$

as in (2.5.32), and

$$\dot{W}_{\text{int}}(t) = -\omega_i d E_0 \big[u(t) \cos \omega_i t \sin \omega_i t - v(t) \sin^2 \omega_i t \big] \rightarrow \frac{1}{2} \omega_i d E_0 v(t) \tag{5.2.12}$$

upon time averaging, assuming $u(t)$ and $v(t)$ vary slowly compared with $\exp(i\omega_i t)$. This gives the rate of change of energy in the electromagnetic field as a result of the interference of the incident field and the dipole field of the atom. The rate of change of the atom's energy as a result of this interference is therefore

$$\frac{1}{2}\hbar\omega_0 \langle \dot{\sigma}_z(t) \rangle \cong \frac{1}{2}\hbar\omega_i \langle \dot{\sigma}_z(t) \rangle = \frac{1}{2}\hbar\omega_i \langle \dot{w}(t) \rangle = -\dot{W}_{\text{int}}(t), \tag{5.2.13}$$

or

$$\dot{w}(t) \cong -\frac{d E_0}{\hbar} v(t) = -\Omega v(t) \tag{5.2.14}$$

in the notation of Chapter 2. This is just the rate of change of the population difference $w(t)$ in the optical Bloch equations (see (2.7.3)), which account for absorption and stimulated emission, as described in Chapter 2. We have therefore shown that the atom's energy changes as a consequence of the interference of the incident field with the field of the dipole moment it induces in the atom. Absorption and stimulated emission result from an induced dipole moment not in phase with the incident field, but *in quadrature* with it, as is evident from (5.2.11) and (5.2.12).

Stimulated emission is sometimes presented as a distinctly quantum effect, but it also occurs in classical mechanical systems.[5]

As the operating principle of lasers, stimulated emission is of great scientific, technological, and economic importance. For nearly four decades after it was postulated by Einstein in his derivation of the Planck spectrum, it attracted surprisingly little interest. In 1924, J. H. Van Vleck referred to "spontaneous emission" and "induced emission" in connection with a

---

[5] See, for instance, B. Fain and P. W. Milonni, J. Opt. Soc. Am. B **4**, 78 (1987).

correspondence principle for absorption, and characterized the induced emission as "negative absorption."[6] In the same year, Tolman wrote in a paper on the lifetimes of excited states that this negative absorption should result in an amplification of radiation.[7] In 1937, Menzel included effects of "stimulated emissions" and recognized the possibility of amplification of radiation, but remarked that this would "probably never occur in practice."[8] And, in 1940, V. A. Fabrikant, in his doctoral dissertation, noted that there could be amplification of radiation if there were more atoms in an excited level than a lower level.[9]

We have already noted (see Exercise 2.3) that the refractive index is changed when excited states are populated. The "negative dispersion" that can result in this case, and its connection with stimulated emission, was first recognized by Kramers.[10]

Radar research during World War II resulted in higher-power microwave sources that were employed in post-war microwave spectroscopy. In connection with their work on the energy difference (Lamb shift) between the $2S_{1/2}$ and $2P_{1/2}$ levels of the hydrogen atom (see Section 4.8), Lamb and Retherford analyzed microwave-induced transitions between the $2S_{1/2}$ and $2P_{3/2}$ levels and remarked that, under the conditions assumed, "negative absorption" would be expected at the 2.7 cm transition wavelength.[11]

Why did it take so long before serious consideration was given to the practical application of stimulated emission, that is, to the possibility of making masers and—much more importantly, as it turned out—lasers?[12] One important reason, it seems, was that scientists found it difficult to imagine going sufficiently far from thermal equilibrium to make amplification by stimulated emission practical; according to Schawlow, "people did not continue Ladenburg and Kopfermann's studies on anomalous dispersion because they believed so firmly in equilibrium that they thought it was impossible to go so far away from it as to have negative absorption."[13]

5.3   The Jaynes–Cummings Model

In the Jaynes–Cummings model, a two-state atom is coupled to just one field mode, and the interaction is treated within the rotating-wave approximation. We have already introduced this model in the context of spontaneous emission in Section 4.4; there, the rotating-wave approximation was implicit in our restriction of the Hilbert space to states in which there is at most one photon. Apart from its conceptual simplicity, the Jaynes–Cummings model is of particular interest as an exactly solvable model for a nontrivial, experimentally accessible system.

The Jaynes–Cummings Hamiltonian follows from (4.4.66) when we make the rotating-wave approximation by dropping the terms $\sigma^\dagger a$ and $\sigma^\dagger a^\dagger$:

$$H = \frac{1}{2}\hbar\omega_0\sigma_z + \hbar\omega a^\dagger a + i\hbar C(\sigma a^\dagger - \sigma^\dagger a). \tag{5.3.1}$$

[6] J. H. van Vleck, Phys. Rev. **24**, 330 (1924).

[7] R. C. Tolman, Phys. Rev. **23**, 693 (1924).

[8] D. H. Menzel, Astrophys. J. **85**, 330 (1937), p. 335.

[9] See M. Bertolotti, *Masers and Lasers. An Historical Approach*, second edition, CRC Press, Boca Raton, Florida, 2015, pp. 156-159.

[10] H. A. Kramers, Z. Phys. **31**, 681 (1925).

[11] W. E. Lamb, Jr. and R. C. Retherford, Phys. Rev. **76**, 549 (1950), p. 570.

[12] The first maser was demonstrated in 1953, and the first laser was demonstrated in 1960.

[13] Quoted in M. Bertolotti, *Masers and Lasers. An Historical Approach*, second edition, CRC Press, Boca Raton, Florida, 2015, p. 28.

The rotating-wave approximation eliminates counter-rotating terms in the Heisenberg equations of motion. It also results in conservation of the total atom–field "excitation number" $\sigma^\dagger \sigma + a^\dagger a$:

$$\frac{d}{dt}(\sigma^\dagger \sigma + a^\dagger a) = 0. \tag{5.3.2}$$

This follows from the (rotating-wave-approximation) Heisenberg equations of motion,

$$\dot{\sigma}_z = -2C(\sigma a^\dagger + \sigma^\dagger a), \tag{5.3.3}$$

$$\dot{\sigma} = -i\omega_0 \sigma + C\sigma_z a, \tag{5.3.4}$$

$$\dot{a} = -i\omega a + C\sigma. \tag{5.3.5}$$

In the case of exact resonance between the two-state atom and the field, $\omega = \omega_0$, it follows from (5.3.3)–(5.3.5), the operator algebra of the two-state operators, and the fact that equal-time atom and field operators commute, that

$$\ddot{\sigma}_z + 2C^2 \sigma_z(t) + 4C^2 \sigma_z(t) N(t) = 0, \tag{5.3.6}$$

where the operator

$$N(t) = \frac{1}{2}\sigma_z(t) + a^\dagger(t)a(t) = \sigma^\dagger(t)\sigma(t) + a^\dagger(t)a(t) - \frac{1}{2}, \tag{5.3.7}$$

which commutes with $\sigma_z(t)$, is seen from (5.3.2) to be a constant of the motion. Suppose, for example, that the two-state atom at the initial time $t = 0$ is in the upper state $|2\rangle$, while the field is in the Fock state $|n\rangle$ of photon number n. Then, since $\sigma_z(0)|\psi\rangle = 1$, and $a^\dagger(0)a(0)|\psi\rangle = n$, for this initial state $|\psi\rangle = |2\rangle|n\rangle$, the expectation value $\langle\psi|\sigma_z(t)N(t)|\psi\rangle$ is

$$\langle\sigma_z(t)N(t)\rangle = \langle\psi|\frac{1}{2}\sigma_z(t)\sigma_z(0)|\psi\rangle + \langle\psi|\sigma_z(t)a^\dagger(0)a(0)|\psi\rangle = (n + \frac{1}{2})\langle\sigma_z(t)\rangle, \tag{5.3.8}$$

and therefore, from (5.3.6),

$$\langle\ddot{\sigma}_z(t)\rangle + 4C^2(n+1)\langle\sigma_z(t)\rangle = 0, \tag{5.3.9}$$

with the solution

$$\langle\sigma_z(t)\rangle = \cos(2C\sqrt{n+1}t). \tag{5.3.10}$$

Recalling that $\langle\sigma_z(t)\rangle = \langle\sigma_{22}(t)\rangle - \langle\sigma_{11}(t)\rangle$ is the difference in the upper- and lower-state probabilities $P_2(t)$ and $P_1(t)$, respectively, we deduce from (5.3.10) that

$$P_2(t) = \cos^2(C\sqrt{n+1})t), \quad P_1(t) = \sin^2(C\sqrt{n+1})t), \tag{5.3.11}$$

and

$$\langle a^\dagger(t)a(t)\rangle = \langle N(0)\rangle - \frac{1}{2}\langle\sigma_z(t)\rangle = n + \sin^2(C\sqrt{n+1}t). \tag{5.3.12}$$

If $n = 0$, that is, if there are no photons initially, (5.3.11) and (5.3.12) just reproduce the results (4.4.11) and (4.4.12), obtained in the Schrödinger picture. For $n \gg 1$, $\langle a^\dagger(t)a(t)\rangle \cong n$, and

$$P_2(t) = \cos^2(C\sqrt{n}t), \qquad P_1(t) = \sin^2(C\sqrt{n}t). \tag{5.3.13}$$

If we identify $2C\sqrt{n}$ with the Rabi frequency $\mathbf{d} \cdot \mathbf{E}_0/\hbar$, as discussed below, these are just the Rabi oscillations predicted by semiclassical radiation theory (see (2.6.3) and (2.6.4)): the fully quantized Jaynes–Cummings model in the limit of high photon number makes the same predictions for $P_2(t)$ and $P_1(t)$ that one makes without quantizing the field. In this limit, stimulated emission is much stronger than spontaneous emission. The 1 in $n + 1$ in (5.3.11) and (5.3.12) corresponds to spontaneous emission, which is not correctly accounted for by semiclassical radiation theory.

The interaction Hamiltonian (5.3.1) for a two-state atom coupled to a single-mode field,

$$H_{\text{int}} = -\mathbf{d}\sigma_x \cdot \mathbf{E} = -\mathbf{d}(\sigma + \sigma^\dagger) \cdot \mathbf{E}, \tag{5.3.14}$$

is replaced by

$$H_{\text{int}}^{\text{RWA}} = -\mathbf{d}\sigma \cdot \mathbf{E}^{(-)} - \mathbf{d}\sigma^\dagger \cdot \mathbf{E}^{(+)} \tag{5.3.15}$$

in the Jaynes–Cummings model. The electric field operator for a single field mode described by a (presumed real) mode function equal to $\mathbf{A}(\mathbf{r}_0)$ at the position $\mathbf{r}_0$ of the atom is, from (3.3.4),

$$\mathbf{E}(\mathbf{r}_0, t) = i\left(\frac{\hbar\omega}{2\epsilon_0}\right)^{1/2}[a(t) - a^\dagger(t)]\mathbf{A}(\mathbf{r}_0), \tag{5.3.16}$$

from which we identify

$$\mathbf{d} \cdot \mathbf{E}^{(+)}(t) = i\left(\frac{\hbar\omega}{2\epsilon_0}\right)^{1/2}a(t)\mathbf{d} \cdot \mathbf{A}(\mathbf{r}_0) = i\hbar C a(t), \tag{5.3.17}$$

$$\mathbf{d} \cdot \mathbf{E}^{(-)}(t) = -i\left(\frac{\hbar\omega}{2\epsilon_0}\right)^{1/2}a^\dagger(t)\mathbf{d} \cdot \mathbf{A}(\mathbf{r}_0) = -i\hbar C a^\dagger(t), \tag{5.3.18}$$

$$C = \frac{1}{\hbar}\left(\frac{\hbar\omega}{2\epsilon_0}\right)^{1/2}\mathbf{d} \cdot \mathbf{A}(\mathbf{r}_0). \tag{5.3.19}$$

To facilitate comparison with the semiclassical theory, replace $a(t)$ by $-i\sqrt{n}e^{-i\omega t}$ in (5.3.16), which is then replaced by a "classical" (c-number) field $\mathbf{E}_{\text{cl}}(\mathbf{r}_0, t)$ such that

$$\mathbf{d} \cdot \mathbf{E}_{\text{cl}}(\mathbf{r}_0, t) = 2\left(\frac{\hbar\omega}{2\epsilon_0}\right)^{1/2}\sqrt{n}\mathbf{d} \cdot \mathbf{A}(\mathbf{r}_0) = \mathbf{d} \cdot \mathbf{E}_0 \cos\omega t, \tag{5.3.20}$$

so that we can write

$$C\sqrt{n} = \frac{1}{2\hbar}\mathbf{d} \cdot \mathbf{E}_0 = \Omega_n/2. \tag{5.3.21}$$

Ω_n is the Rabi frequency defined in semiclassical radiation theory in Chapter 2 (see (2.5.3)). The oscillations of the upper- and lower-state occupation probabilities in

(5.3.13) are then the same as those given in semiclassical theory by (2.6.3) and (2.6.4) with $\omega = \omega_0$.

If we assume the initial state in which the field is in a Fock state of n photons and the two-state atom is in its *lower*-energy eigenstate, we obtain, instead of (5.3.11),

$$P_2(t) = \sin^2(C\sqrt{n})t, \quad P_1(t) = \cos^2(C\sqrt{n})t. \tag{5.3.22}$$

The identification of $2C\sqrt{n}$ with the Rabi frequency puts the Jaynes–Cummings solutions for the two-state-atom occupation probabilities in the same form as the semiclassical solutions. In particular, for an initial vacuum field, $n = 0$, the Jaynes–Cummings model and semiclassical theory both predict that the two-state atom remains forever in its lower-energy state.

The solutions (5.3.11) and (5.3.12) show that an initially excited atom in a vacuum field will undergo Rabi oscillations as it exchanges excitation with the field. C (or $2C$) is said to be the *vacuum Rabi frequency*. It depends on the mode function of the field evaluated at the position $\mathbf{r}_0$ of the atom. If the atom happens to be positioned at a node of the field ($\mathbf{A}(\mathbf{r}_0, t) = 0$), $C = 0$ and the atom does not interact at all with the single-mode field; an initially excited atom will remain forever excited in the absence of any other modes or interactions.

More general solutions for $\langle \sigma_z(t) \rangle$ when the field is initially in a Fock state $|n\rangle$ are easily obtained. If the detuning $\Delta = \omega_0 - \omega \neq 0$, for example,

$$\langle \sigma_z(t) \rangle = \frac{\Delta^2 + \Omega_n^2(1 + 1/n)\cos\sqrt{\Omega_n^2(1 + 1/n) + \Delta^2}\,t}{\Omega_n^2(1 + 1/n) + \Delta^2} \tag{5.3.23}$$

if the atom is initially in the upper state ($\langle \sigma_z(0) \rangle = +1$), and

$$\langle \sigma_z(t) \rangle = -\frac{\Delta^2 + \Omega_n^2\cos\sqrt{\Omega_n^2 + \Delta^2}\,t}{\Omega_n^2 + \Delta^2} \tag{5.3.24}$$

if the atom is initially in the lower state ($\langle \sigma_z(0) \rangle = -1$).

Exercise 5.1: Show that

$$\ddot{\sigma}_z(t) + 2C^2\sigma_z(t) + 4C^2\sigma_z(t)N(t) = -2iC\Delta X(t),$$

and

$$\dot{X}(t) = -\frac{i\Delta}{2C}\dot{\sigma}_z(t),$$

with $X(t) = \sigma^\dagger(t)a(t) - a^\dagger(t)\sigma(t)$, and derive from these equations the solutions (5.3.23) and (5.3.24).

It is not hard to derive solutions for $\langle \sigma_z(t) \rangle$ when the initial field state is not simply a vacuum or a Fock state $|n\rangle$. Suppose the initial field state can be expressed as a superposition of Fock states:

$$|\psi_F\rangle = \sum_{n=0}^{\infty} c_n |n\rangle. \tag{5.3.25}$$

We will focus again on $\langle \sigma_z(t) \rangle = \langle \psi | \sigma_z(t) | \psi \rangle$, $|\psi\rangle = |\psi_A\rangle |\psi_F\rangle$. $|\psi_A\rangle$, the initial state of the atom, will be taken to be the lower state $|1\rangle$, and we will simplify further by assuming the atom and the field are in perfect resonance ($\Delta = 0$). We write (5.3.6) as

$$\ddot{\sigma}_z(t) + M^2 \sigma_z(t) = 0, \tag{5.3.26}$$

where the *operator* $M^2 = 4C^2[\sigma^\dagger(t)\sigma(t) + a^\dagger(t)a(t)] = 4C^2[\sigma^\dagger(0)\sigma(0) + a^\dagger(0)a(0)]$ is a constant of the motion and commutes with $\sigma_z(t)$ for all times $t \geq 0$. We are interested in the expectation value of $\sigma_z(t)$ over the initial state $|\psi\rangle = |1\rangle|\psi_F\rangle$:

$$\langle \sigma_z(t) \rangle = \langle \sigma_z(0) \cos Mt \rangle. \tag{5.3.27}$$

This applies whenever the atom is initially in one of its two eigenstates, in which case $\langle \dot{\sigma}_z(0) \rangle = -2C\langle [\sigma(0)a^\dagger(0) + \sigma^\dagger(0)a(0)] \rangle = 0$. Now, since $\sigma^\dagger(0)\sigma(0)|1\rangle = 0$, and $a^\dagger(0)a(0)|n\rangle = n|n\rangle$,

$$M^2|\psi\rangle = \sum_{n=0}^{\infty} c_n M^2 |n\rangle |1\rangle = 4C^2 \sum_{n=0}^{\infty} n c_n |n\rangle |1\rangle, \tag{5.3.28}$$

$$M^4|\psi\rangle = (4C^2)^2 \sum_{n=0}^{\infty} n^2 c_n |n\rangle |1\rangle, \tag{5.3.29}$$

and so on for higher powers of M^2, and so we deduce that

$$\sigma_z(0) \cos Mt |\psi\rangle = \sigma_z(0) \sum_{n=0}^{\infty} c_n \left[1 - (4nC^2t^2)/2! + (4nC^2t^2)^2/4! - \ldots \right] |n\rangle |1\rangle$$

$$= -\sum_{n=0}^{\infty} c_n \cos(2C\sqrt{n}t)|n\rangle|1\rangle \tag{5.3.30}$$

and, since $\langle m|n \rangle = \delta_{mn}$, and $\langle 1|1 \rangle = 1$,

$$\langle \sigma_z(t) \rangle = \langle \psi | \sigma_z(0) \cos Mt | \psi \rangle = -\sum_{n=0}^{\infty} |c_n|^2 \cos(2C\sqrt{n}t). \tag{5.3.31}$$

Of particular interest are initial states $|\psi\rangle$ in which the field is described by a coherent state $|\alpha\rangle$; then (recall (3.6.18))

$$|c_n|^2 = \frac{|\alpha|^{2n}}{n!} e^{-|\alpha|^2} = \frac{\langle n \rangle^n}{n!} e^{-\langle n \rangle}, \tag{5.3.32}$$

$$\langle \sigma_z(t) \rangle = -\sum_{n=0}^{\infty} \frac{\langle n \rangle^n}{n!} e^{-\langle n \rangle} \cos\left(2C\sqrt{n}\,t\right), \tag{5.3.33}$$

when the two-state atom is initially in its lower-energy eigenstate. If the two-state atom is initially in its higher-energy eigenstate, and, again, $\Delta = 0$, we obtain, similarly,

$$\langle \sigma_z(t) \rangle = \sum_{n=0}^{\infty} \frac{\langle n \rangle^n}{n!} e^{-\langle n \rangle} \cos\left(2C\sqrt{n+1}\,t\right). \tag{5.3.34}$$

Although a coherent field state is, in most respects, the most "classical-like," these solutions do not look much like the perfectly sinusoidal oscillations of $\langle \sigma_z(t) \rangle$ predicted by semiclassical radiation theory when the atom is irradiated by a monochromatic field. In contrast, the solution (5.3.24) in the case of a Fock state of the field reduces for the case under consideration ($\Delta = 0$) to $\langle \sigma_z(t) \rangle = -\cos(\Omega_n t)$, just the sort of sinusoidal oscillation predicted by semiclassical theory. We do recover more semiclassical-like behavior with an initial coherent state of the field, however, when the average photon number $\langle n \rangle = |\alpha|^2$ is large: since the variance $\langle (\Delta n)^2 \rangle = \langle n^2 \rangle - \langle n \rangle^2 = \langle n \rangle$ for the photon number Poisson distribution (5.3.32), $\langle (\Delta n)^2 \rangle^{1/2}/\langle n \rangle = 1/\langle n \rangle^{1/2} \ll 1$. For $\langle n \rangle \gg 1$, therefore, (5.3.33) and (5.3.34) can be expected to exhibit at least some similarity to the Rabi oscillations of semiclassical radiation theory. Figure 5.1 shows $\langle \sigma_z(t) \rangle$ as given by (5.3.34) for $\langle n \rangle = 4, 9$, and 99. There are Rabi-like oscillations for small times, and these oscillations are seen to persist for longer times as the mean photon number $\langle n \rangle$ increases. But the oscillations are found to eventually "collapse" no matter how large the mean photon number.

An initial field produced by a "chaotic" source (see Section 3.6) is also of special interest. One obtains

$$\langle \sigma_z(t) \rangle = \frac{1}{\langle n \rangle + 1} \sum_{n=0}^{\infty} \left(\frac{\langle n \rangle}{\langle n \rangle + 1}\right)^n \cos(2C\sqrt{n+1}\,t) \tag{5.3.35}$$

when it is assumed that the atom is initially in the upper state at $t = 0$, $\Delta = 0$, and the average initial photon number is $\langle n \rangle$; for a thermal field, $\langle n \rangle = 1/[\exp(\hbar\omega/k_B T - 1]$ (see Section 3.6). For such an initial field state, the variance $\langle (\Delta n)^2 \rangle = \langle n \rangle^2 + \langle n \rangle$ (recall (3.6.100)) and $\langle (\Delta n)^2 \rangle^{1/2}/\langle n \rangle = (1 + 1/\langle n \rangle)^{1/2}$. The photon number distribution in this case, unlike the coherent-state case for large $\langle n \rangle$, is broad, and, consequently, the time evolution of the system exhibits little evidence of Rabi oscillations (see Figure 5.2).

One of the aims of the Jaynes–Cummings paper was "to clarify the relationship between the quantum theory of radiation, where the electromagnetic field-expansion coefficients satisfy commutation relations, and the semiclassical theory, where the electromagnetic field is

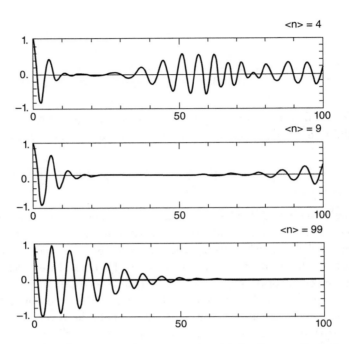

Fig. 5.1 An atom's population inversion versus time in the Jaynes–Cummings model with $\Delta = 0$. At $t = 0$, the two-state atom is in its upper state, and the field is in a coherent state with $|\alpha|^2 = \langle n \rangle = 4, 9$, and 99. Time is in units of $(2C\langle n \rangle^{1/2}/\hbar)^{-1}$. From B. W. Shore and P. L. Knight, J. Mod. Opt. **40**, 1195 (1993), with permission.

considered as a definite function of time rather than as an operator."[14] We discussed in Section 4.7 their conclusion that "semiclassical theory [leads] to a prediction of spontaneous emission, with the same decay rate as given by quantum electrodynamics, described by the Einstein A coefficients."[15]

While semiclassical radiation theory sometimes falls short vis-à-vis experiment (see Sections 2.3, 4.7, and 5.8), it remains the primary—indeed, one that is practically indispensable—approach for analyzing the light–matter interactions that are most directly relevant to spectroscopy and other applications. The work of Jaynes, his students, and many others shows how well it serves as a conceptual as well as a calculational framework; their work has also stimulated further theoretical work and experiments that have sharpened our understanding of the quantum nature of light.

The Jaynes–Cummings model has itself stimulated much theoretical work and some remarkable experiments. In one such experiment, a velocity-selected beam of rubidium atoms entered a superconducting microwave cavity after excitation by linearly polarized ultraviolet radiation to a high ("Rydberg") energy level ($63\mathrm{p}_{3/2}$), such that the 21.6 GHz ($63\mathrm{p}_{3/2}$–$61\mathrm{d}_{5/2}$) transition frequency was resonant with a cavity mode, and the atoms were approximately two-state systems. Atoms exiting the cavity in the $63\mathrm{p}_{3/2}$ level were detected by field ionization, and, by selection of different atom velocities and therefore atom–field interaction times t, the upper-state probability $P_2(t)$ of the approximately two-state atom could be determined and compared with the predictions of the Jaynes–Cummings model with $\langle n \rangle \cong 2$ for $\omega = 21.6/2\pi$

---

[14] E. T. Jaynes and F. W. Cummings, Proc. IEEE **51**, 89 (1963), p. 89.

[15] Ibid.

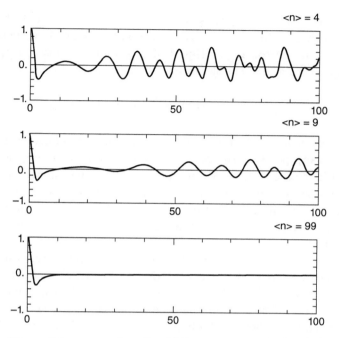

Fig. 5.2 As in Figure 5.1, except that the field is initially in a thermal state with mean photon number $\langle n \rangle$. From B. W. Shore and P. L. Knight, J. Mod. Opt. **40**, 1195 (1993), with permission.

GHz and the cavity temperature $T = 2.5$ K. Excellent agreement was found with the theory for the accessible range of interaction times (~ 30–140 µs).[16] Section 5.4 focuses on another prediction of the Jaynes–Cummings model that has been experimentally verified.

5.4   Collapses and Revivals

Expressions such as (5.3.33) and (5.3.35), or, more generally, "quasiperiodic" functions

$$x(t) = \sum_{n=0}^{N} a_n \cos(\omega_n t + \phi_n), \tag{5.4.1}$$

where the ω_n (> 0) are linearly independent ("incommensurate"), have the property that, given $x(t)$ at *any* time t, there are an infinite number of times t' such that $x(t')$ comes as close to $x(t)$ as we wish.[17] This property of functions with discrete Fourier spectra implies the *quantum recurrence theorem*: "Let us consider a system with *discrete* energy eigenvalues E_n: if $\psi(t_0)$ is its state vector at the time t_0 and ϵ is any positive number, at least one T will exist such that the norm $||\psi(T) - \psi(t_0)||$

---

[16] G. Rempe, H. Walthe, and N. Klein, Phys. Rev. Lett. **58**, 353 (1987).

[17] N need not be finite, and the a_n may be taken to be positive. Proofs of this recurrence theorem, and more precise statements of it, may be found in the mathematical literature on quasiperiodic functions.

of the vector $\psi(T) - \psi(t_0)$ is smaller than ϵ."[18] The proof of this theorem follows straightforwardly from the recurrence property of quasiperiodic functions and the fact that the state vector at any time t of a system with discrete energy eigenvalues E_n and eigenvectors $|\phi_n\rangle$ has the form

$$|\psi(t)\rangle = \sum_{n=0}^{\infty} c_n e^{-iE_n t/\hbar} |\phi_n\rangle. \qquad (5.4.2)$$

This is analogous to the Poincaré recurrence theorem of classical mechanics, according to which any point in the phase space of a system occupying a *finite volume* will be revisited as closely as we wish if we wait long enough. In the classical realm, the Poincaré recurrence times can be prohibitively long—longer, for instance, than the age of the universe—and of no practical interest. In quantum systems, similarly, the recurrence times can be extremely large when the energy eigenvalues are densely distributed.

It was found in numerical studies that recurrent "collapse" and "revival" phenomena can occur in the time evolution of quantities such as $\langle \sigma_z(t) \rangle$ in the Jaynes–Cummings model.[19] The former phenomenon is evident in the last two plots of Figure 5.1: after some Rabi oscillations, the population inversion in the case of an initial coherent state of the field "collapses." Eberly et al. generalized some analytical approximations of Cummings for this collapse effect.[20] The envelope of the oscillations for large $\langle n \rangle$ has a Gaussian form, and, for $\Delta = 0$, the collapse time of the population inversion for large $\langle n \rangle$ is given approximately by $t_c = 1/C$, that is, by the inverse of the vacuum Rabi frequency. The number of oscillations within the envelope increases as $\langle n \rangle$ increases, but the collapse time remains the same. It was noted that "collapse occurs only if the upper state is occupied (that is, only if spontaneous emission can occur)," indicating that collapse can be associated with quantum fluctuations of the field.[21] It can, however, be described by a semiclassical model with a fluctuating field.

The "revival" effect, a distinctly quantum effect stemming from the quantum nature of the field, was discovered in computations of $\langle \sigma_z(t) \rangle$ for much longer times than in earlier studies. The effect is clearly seen in the middle plot of Figure 5.3, which shows $\langle \sigma_z(t) \rangle$ for an initial state in which the two-state atom is excited, the field is in a coherent state with $\langle n \rangle = 16$, and $\Delta = 0$. The revival frequency decreases with increasing $\langle n \rangle$; an approximate formula for the revival period for $\Delta = 0$ and initial coherent states of the field is[22]

$$T_r = \frac{2\pi}{C} \langle n \rangle^{1/2} = 2\pi \langle n \rangle^{1/2} t_c. \qquad (5.4.3)$$

[18] P. Bocchieri and A. Loinger, Phys. Rev. **107**, 337 (1957), p. 337.

[19] J. H. Eberly, N. B. Narozhny, and J. J. Sanchez-Mondragon, Phys. Rev. Lett. **44**, 1323 (1980).

[20] F. W. Cummings, Phys. Rev. **140**, A1051 (1965).

[21] J. H. Eberly, N. B. Narozhny, and J. J. Sanchez-Mondragon, Phys. Rev. Lett. **44**, 1323 (1980), p. 1325.

[22] N. B. Narozhny, J. J. Sanchez-Mondragon, and J. H. Eberly, Phys. Rev. A **23**, 236 (1981).

It might seem from the middle plot of Figure 5.3 that the atom–field dynamics can somehow be "static" after a collapse and before a revival. What is actually happening is that other things besides $\sigma_z(t)$—in particular the atom's dipole expectation value—vary significantly during these time intervals.[23]

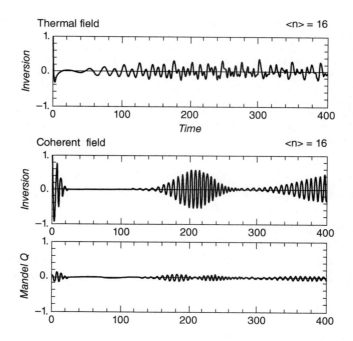

Fig. 5.3 Population inversion versus time in the Jaynes–Cummings model with $\Delta = 0$, $\langle n \rangle = 16$, and an initial state in which the atom is excited. In the top plot the field is initially in a chaotic (e.g., thermal) state, whereas, in the middle plot, it is in a coherent state. The bottom plot shows the Mandel Q parameter for the coherent-state case. Time is in units of $(2C\langle n \rangle^{1/2}/\hbar)^{-1}$. From B. W. Shore and P. L. Knight, J. Mod. Opt. **40**, 1195 (1993), with permission.

The bottom plot of Figure 5.3 shows the Mandel Q parameter (3.6.118) calculated for the coherent-state example. The fact that it dips below 0 at various times reflects sub-Poissonian, nonclassical behavior of the field, which also exhibits squeezing in certain time intervals during its evolution.

Collapse and revival effects predicted by the Jaynes–Cummings model have been observed in "micromaser" experiments of the type mentioned in Section 5.3. Based on a relation between the Mandel Q parameters (normalized variances) for the number of photons and the number

---

[23] Ibid.

of ground-state atoms in the cavity, these experiments have also demonstrated nonclassical, sub-Poissonian counting statistics for both photons and atoms.[24]

Collapse and revival effects have been observed in other experiments besides those inspired by work on the Jaynes–Cummings model. In one experiment potassium atoms were excited by a 20 ps ultraviolet pump pulse to create a coherent superposition of Rydberg states with principal quantum numbers ≈ 65, forming a radially localized wave packet that can be described as a "shell" that oscillates between the nucleus and an outer turning point; the high principal quantum numbers involved imply that the states forming the wave packet are nearly equally spaced in energy and that the oscillation period is approximately the period expected classically. If a second pulse with electric field $\mathbf{E}(t)$ is applied, the rate at which the atom absorbs energy is proportional to $\mathbf{v}(t) \cdot \mathbf{E}(t)$, where $\mathbf{v}(t)$ is the electron velocity. Near the outer turning point, $\mathbf{v}$ is small, but, near the nucleus, it is large, and the atom can absorb strongly and become ionized. The oscillations of the wave packet can therefore be probed by counting the ions generated as a function of the time delay between the two pulses. Collapses and revivals in the time evolution of the wave packet were predicted from solutions of the time-dependent Schrödinger equation and also expected intuitively because the excited Rydberg states are nearly but not exactly equally spaced. Such behavior was observed.[25]

For initial chaotic states of the field, numerical simulations do not show any discernible collapses and revivials. For initial coherent states of the field, revivals appear to overlap in time intervals much longer than shown in Figures 5.1–5.3. For such times, there is no clear pattern of collapses and revivals, and any recurrent behavior is of the general "quantum recurrence" nature characteristic of functions of the type (5.4.1). Such functions appear, of course, in many different areas and have been studied by mathematicians for many years. Among the interesting properties of $x(t)$ is the average frequency $L(q)$ with which it takes on some particular value q; this is defined as

$$L(q) = \lim_{T \to \infty} \frac{1}{T} N_T(q), \tag{5.4.4}$$

where $N_T(q)$ is the number of zeros of the function $x(t) - q$. As shown by Kac,

$$L(q) = \frac{1}{2\pi^2} \int_{-\infty}^{\infty} \int_{-\infty}^{\infty} d\alpha d\eta \eta^{-2} \cos q\alpha \left[\prod_{n=0}^{N} J_0(|a_n|\alpha) \right.$$
$$\left. - \prod_{n=0}^{N} J_0(|a_n|\sqrt{\alpha^2 + \eta^2 \omega_n^2}) \right], \tag{5.4.5}$$

with J_0 the zeroth-order Bessel function.[26] The probability distribution of $x(t)$ can also be expressed in terms of zeroth-order Bessel functions:

$$P(x) = \frac{1}{2\pi} \int_{-\infty}^{\infty} d\alpha e^{-i\alpha x} \prod_{n=0}^{N} J_0(|a_n|\alpha). \tag{5.4.6}$$

These formulas have been applied to the Jaynes–Cummings population inversions (5.3.34) and (5.3.35) to derive approximate analytical expressions for their long-time

[24] G. Rempe, F. Schmidt-Kaler, and H. Walther, Phys. Rev. Lett. **64**, 2483 (1990).

[25] J. A. Yeazell, M. Mallalieu, and C. R. Stroud, Jr., Phys. Rev. Lett. **64**, 2007 (1990).

[26] For a "shortened proof for physicists," see P. Mazur and E. W. Montroll, J. Math. Phys. **1**, 70 (1960).

probability distributions and for the average frequencies with which they have particular values q.[27]

5.5 Dressed States

The eigenstates $|1\rangle|n\rangle \equiv |1; n\rangle$, and $|2\rangle|n\rangle \equiv |2; n\rangle$, of the *uncoupled* atom–field system in the Jaynes–Cummings model are a complete set and can be used to represent the rotating-wave-approximation Hamiltonian (5.3.1) as an infinite-dimensional, block-diagonal Hermitian matrix with 2×2 elements

$$\langle 2; n-1|H|2; n-1\rangle = \frac{1}{2}\hbar\omega_0 + (n-1)\hbar\omega = (n-\frac{1}{2})\hbar\omega + \frac{1}{2}\hbar\Delta,$$

$$\langle 2; n-1|H|1; n\rangle = -i\hbar C\sqrt{n},$$

$$\langle 1; n|H|2; n-1\rangle = i\hbar C\sqrt{n},$$

$$\langle 1; n|H|1; n\rangle = -\frac{1}{2}\hbar\omega_0 + n\hbar\omega = (n-\frac{1}{2})\hbar\omega - \frac{1}{2}\hbar\Delta. \tag{5.5.1}$$

For each n, we can diagonalize the corresponding 2×2 matrix and express its two orthogonal (and normalized) eigenvectors in terms of the (orthonormal) eigenstates of the uncoupled system. We denote these eigenvectors by $|n+\rangle$ and $|n-\rangle$, and the corresponding eigenvalues by $E(n\pm)$:

$$|n+\rangle = \cos\theta_n|2; n-1\rangle + \sin\theta_n|1; n\rangle,$$

$$|n-\rangle = -\sin\theta_n|2; n-1\rangle + \cos\theta_n|1; n\rangle,$$

$$H|n\pm\rangle = E(n\pm)|n\pm\rangle, \tag{5.5.2}$$

with

$$E(n\pm) = (n-\frac{1}{2})\hbar\omega \pm \hbar\sqrt{C^2 n + \Delta^2/4} \tag{5.5.3}$$

being the energy eigenvalues of the *coupled* atom–field system. The "mixing angle" θ_n satisfies

$$\tan 2\theta_n = \frac{2C\sqrt{n}}{\Delta}. \tag{5.5.4}$$

If there are initially n photons and the atom is in the lower state $|1\rangle$, the rotating-wave-approximation Hamiltonian allows transitions to the state in which the atom is in the upper state $|2\rangle$ and there are $n-1$ photons. Neither of these states are stationary states of the coupled atom–field system; in the Heisenberg picture, we obtained exact solutions for their time-dependent occupation probabilities. We have now obtained the stationary states of the coupled system, the so-called *dressed states*, whose energy eigenvalues are $E(n\pm)$.

---

[27] F. T. Hioe, J. Stat. Phys. **30**, 467 (1983). This article includes historical remarks relating to (5.4.1) and its appearance in some other fields.

Of course, we can turn things around and express the "bare" states as superpositions of dressed states:

$$|2; n - 1\rangle = \cos\theta_n|n+\rangle - \sin\theta_n|n-\rangle,$$
$$|1; n\rangle = \sin\theta_n|n+\rangle + \cos\theta_n|n-\rangle, \qquad (5.5.5)$$

and we can use either the bare states or the dressed states in calculations. Suppose, for example, that we have an initial state $|\psi(0)\rangle = |2; n - 1\rangle$. We have already used the bare-state basis to calculate the probability that the atom–field system is in such a state at any later time t. The calculation is just as easy using the dressed-state basis,

$$|\psi(0)\rangle = |2; n - 1\rangle = \cos\theta_n|n+\rangle - \sin\theta_n|n-\rangle, \qquad (5.5.6)$$

and the state vector at time t is

$$|\psi(t)\rangle = e^{-iHt/\hbar}|\psi(0)\rangle = \cos\theta_n e^{-iE(n+)t/\hbar}|n+\rangle - \sin\theta_n e^{-iE(n-)t/\hbar}|n-\rangle. \qquad (5.5.7)$$

The probability that the system is in the initial state (5.5.6) at time t is

$$P_2(t) = |\langle\psi(0)|\psi(t)\rangle|^2 = \left| \cos^2\theta_n e^{-iE(n+)t/\hbar} + \sin^2\theta_n e^{-iE(n-)t/\hbar} \right|^2. \qquad (5.5.8)$$

This is equivalent to (5.3.23), which is the result that we derived for $\langle\sigma_z(t)\rangle$ ($= P_2(t) - P_1(t) = 2P_2(t) - 1$) using the bare-state basis.

If $\Delta^2/4 \gg C^2 n$,

$$E(n+) \cong (n - 1)\hbar\omega + \frac{1}{2}\hbar\omega_0 + \frac{\hbar C^2 n}{\Delta}, \qquad (5.5.9)$$

$$E(n-) \cong n\hbar\omega - \frac{1}{2}\hbar\omega_0 - \frac{\hbar C^2 n}{\Delta}. \qquad (5.5.10)$$

In this limit, $\theta_n \cong 0$ if $\Delta > 0$, and $|n+\rangle$ is approximately $|2; n - 1\rangle$ with a small admixture of $|1; n\rangle$. $E(n+)$ is therefore approximately the energy $(n - 1)\hbar\omega + \frac{1}{2}\hbar\omega_0$ of the "bare" state $|2; n - 1\rangle$, but with a small additional energy $\hbar C^2 n/\Delta$. Likewise $|n-\rangle$ is approximately $|1; n\rangle$ plus a small admixture of $|2; n-1\rangle$, and $E(n-)$ is approximately the energy $n\hbar\omega - \frac{1}{2}\hbar\omega_0$ of the bare state $|1; n\rangle$ plus a small additional energy $-\hbar C^2 n/\Delta$. If $\Delta < 0$, similarly, $\theta_n \cong \pi/2$ and $|n+\rangle \cong |1; n\rangle$, $|n-\rangle \cong -|2; n - 1\rangle$.

A simple interpretation of the energies $\pm\hbar C^2 n/\Delta$ follows from the AC Stark shifts δE_2 and δE_1 of the upper and lower states, respectively, of a two-state atom, which we derived in semiclassical radiation theory, and the rotating-wave approximation (see (2.5.40) and (2.5.41)):

$$\delta E_2 = \frac{d^2 E_0^2}{4\hbar\Delta}, \qquad \delta E_1 = -\frac{d^2 E_0^2}{4\hbar\Delta}. \qquad (5.5.11)$$

If we now identify dE_0 with $2\hbar C\sqrt{n}$ (see (5.3.21)), we can write these two-state-atom AC Stark shifts as

$$\delta E_2 = \frac{\hbar C^2 n}{\Delta}, \quad \delta E_1 = -\frac{\hbar C^2 n}{\Delta}. \tag{5.5.12}$$

From the remarks following (5.5.9) and (5.5.10), we identify the energy shift in the state in which the atom is in the upper state and there are n photons,

$$\delta E(2; n) \cong \frac{\hbar C^2}{\Delta}(n + 1), \tag{5.5.13}$$

and the shift in the state in which the atom is in the *lower* state and there are n photons,

$$\delta E(1; n) \cong -\frac{\hbar C^2}{\Delta} n. \tag{5.5.14}$$

Thus, our semiclassical argument, supplemented by (5.3.21), does not account for the 1 in the factor $n + 1$ appearing in $\delta E(2; n)$, that is, it does not predict the level shift C^2/Δ of the upper state of the atom when there are no photons in the field. That shift is a Lamb shift (or, equivalently, an AC Stark shift due to *the vacuum* electric field) in the two-state model of an atom. The semiclassical model does predict, in agreement with (5.5.14), that there is no shift of the lower state of the atom when there are no photons.[28]

The atom in this "dispersive limit" ($\Delta^2/4 \gg C^2 n$) acts in effect as a non-absorbing dielectric "medium," and, as such, can affect the frequency of the field mode. Suppose, for example, that the field frequency is that of a single-mode cavity of volume V and that there is a small density N of atoms in the cavity, so that the refractive index is (recall (1.5.22))

$$n(\omega) \cong 1 + N\alpha(\omega)/2\epsilon_0. \tag{5.5.15}$$

The phase velocity of light in the cavity is $c/n(\omega) = \lambda\omega/2\pi n(\omega)$, and since the wavelength λ is fixed by the size of the cavity, the cavity mode frequency is changed from its vacuum value ω to $\omega/n(\omega)$, that is, there is a frequency shift

$$\delta\omega = \frac{\omega}{n(\omega)} - \omega = -\omega\left[\frac{n(\omega) - 1}{n(\omega)}\right] \cong -\omega[n(\omega) - 1] \cong -\frac{N\omega}{2\epsilon_0}\alpha(\omega)$$
$$= -\frac{\omega}{2\epsilon_0 V}\alpha(\omega) \tag{5.5.16}$$

due to the presence of a single atom in the cavity ($N = 1/V$). Equation (2.5.35) with $\mathbf{d}_{12}^2 = 3d^2$ then implies that the mode frequency is shifted to

$$\delta\omega_2 = \frac{\omega d^2}{2\epsilon_0 \hbar V \Delta} = C^2/\Delta \tag{5.5.17}$$

---

[28] Actually, there is a ground-state (Lamb) shift, but it involves a *non-resonant* denominator $\omega_0 + \omega$ rather than Δ, and such a small (Bloch–Siegert) shift does not appear in the rotating-wave approximation, within which $|1; 0\rangle$ is an eigenstate of the Hamiltonian.

due to the presence in the cavity of an atom in state $|2\rangle$.[29] (We have used the definition of C given by (5.3.19) with $\mathbf{A}^2(\mathbf{r}_0) \approx 1/V$.) Similarly, when the atom in the cavity is in state $|1\rangle$, the polarizability changes sign, and

$$\delta\omega_1 \cong -C^2/\Delta. \tag{5.5.18}$$

These frequency shifts also follow from (5.5.13) and (5.5.14):

$$\frac{1}{\hbar}[\delta E(2; n+1) - \delta E(2; n)] = C^2/\Delta = \delta\omega_2, \tag{5.5.19}$$

and, likewise,

$$\frac{1}{\hbar}[\delta E(1; n+1) - \delta E(1; n)] = -C^2/\Delta = \delta\omega_1. \tag{5.5.20}$$

The opposite, "strong coupling" limit $\Delta^2/4 \ll C^2 n$, is realized in particular when $\Delta = 0$:

$$|n\pm\rangle = \frac{1}{\sqrt{2}}\Big(\pm |2; n-1\rangle + |1; n\rangle\Big) \tag{5.5.21}$$

and

$$E(n\pm) = (n - \frac{1}{2})\hbar\omega \pm \hbar C\sqrt{n}. \tag{5.5.22}$$

The states $|2; n-1\rangle$ and $|1; n\rangle$, which, in the absence of atom–field coupling, have the same energy $n\hbar\omega$ when $\Delta = 0$, are superpositions of dressed states with energies (5.5.22) having *dynamic Stark shifts* $\pm\hbar C\sqrt{n}$. The state

$$|2; 0\rangle = \frac{1}{\sqrt{2}}\big[|1+\rangle - |1-\rangle\big], \tag{5.5.23}$$

for example, becomes a doublet with a *vacuum Rabi splitting* $E(1+) - E(1-) = 2\hbar C$ when $\Delta = 0$.

Thanks to years of remarkable experimental efforts, effects of the type just described—and many more—have actually been observed. In the first observation of vacuum Rabi splitting, for example, a beam of ground-state cesium atoms was injected into a high-Q optical resonator (3.2 mm long) and the transmission intensity of a weak, tunable laser beam injected through one of the mirrors was measured for different values of the average number $\mathcal{N}$ of atoms in the cavity. With no atoms in the resonator, the transmitted intensity had a Lorentzian spectral shape centered at a single resonator mode frequency, near the frequency of a two-state transition of the atoms; fields not resonant with a resonator frequency are simply reflected. With atoms in the resonator, however, the spectrum of transmitted intensity exhibited two peaks, consistent with vacuum Rabi splitting. Success in the experiment depended on a number of achievements, including a resonator of sufficiently high quality to minimize photon loss, and a vacuum Rabi frequency C sufficiently large compared with the photon loss rate and the rate of spontaneous emission of the atoms into other modes of the open resonator. The comparison with the simple theory predicting the vacuum Rabi splitting was complicated by

---

[29] See also Section 7.3.5.

the fact that the atoms experienced a spatially varying field in the resonator and that there were fluctuations in the number of atoms in the resonator. Numerical simulations taking such complications into account were found to be in very good agreement with the measured transmission data for different average numbers of atoms, including $\mathcal{N} \cong 1$.[30]

5.6 Resonance Fluorescence

In resonance fluorescence, an atom is excited by light and emits light of the same wavelength. It's an old subject, but it remains of interest and importance in a variety of applications. Before lasers, studies of resonance fluorescence involved relatively weak interactions of atoms and molecules with light. With lasers came the possibility of observing resonance fluorescence in fields that interact strongly with atoms and molecules and affect them, and the emitted light, in ways that could not previously be realized. Here, we will focus on the conceptually simplest scenario—a two-state atom driven by a strong, quasi-monochromatic field tuned near the atomic transition frequency.

Consider first the system from the perspective of dressed states in the strong-coupling regime, assuming $\Delta = 0$. The state $|2; n\rangle$ has a total (atom-plus-photon) excitation number $n+1$, and, in free space, it can decay by spontaneous emission, into any field mode, to a state with excitation number n, which can decay to a state with excitation number $n - 1$, and so on. For each n, the decay can proceed from one of the dressed states $|n\pm\rangle$, which are separated in energy by $\Omega_n = \hbar C \sqrt{n}$, to one of the dressed states $|n - 1, \pm\rangle$, which are separated in energy by $\Omega_{n-1} = \hbar C \sqrt{n - 1}$. For $n \gg 1$, $\Omega_{n-1} \cong \Omega_n = \Omega$, and there are, therefore, two transitions with a frequency ω: one transition with frequency $\omega + \Omega$, and another transition with frequency $\omega - \Omega$ (see Figure 5.4). The spectrum of radiation from a two-state atom driven by a sufficiently strong monochromatic field of frequency $\omega = \omega_0$ will therefore have peaks at ω and $\omega \pm \Omega$, where Ω is the (on-resonance) Rabi frequency.

Of course, these three peaks will have finite widths. To calculate the actual spectrum of the resonance fluorescence, we can proceed using either the dressed-state or the bare-state basis. We will do a Heisenberg-picture calculation using bare states.

The expectation value of the rate of change of energy in the electromagnetic field is

$$P(t) = \frac{\partial}{\partial t} \sum_\beta \hbar \omega_\beta \langle a_\beta^\dagger(t) a_\beta(t) \rangle, \tag{5.6.1}$$

and, with (4.4.69) and $\sigma_x(t) = \sigma(t) + \sigma^\dagger(t)$, this is

$$P(t) = 2\mathrm{Re} \sum_\beta C_\beta \hbar \omega_\beta \langle a_\beta^\dagger(0) \sigma_x(t) \rangle e^{i\omega_\beta t}$$

$$+ 2\mathrm{Re} \sum_\beta C_\beta^2 \hbar \omega_\beta \int_0^t dt' \langle \sigma_x(t') \sigma_x(t) \rangle e^{-i\omega_\beta(t'-t)} \tag{5.6.2}$$

---

[30] R. J. Thompson, G. Rempe, and H. J. Kimble, Phys. Rev. Lett. **68**, 1132 (1992).

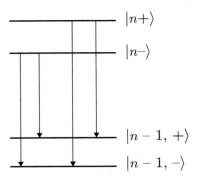

Fig. 5.4 Spontaneous decay transitions from a dressed doublet $|n, \pm\rangle$ to a dressed doublet $|n-1, \pm\rangle$ having one less total excitation number, when the photon number $n \gg 1$. In this limit, the two doublets have the same energy separation, equal to $\hbar$ times the Rabi frequency Ω, and there are two allowed transitions at frequency ω and one at each of the frequencies $\omega - \Omega$ and $\omega + \Omega$.

for a two-state atom. The first term is the rate of change of field energy resulting from stimulated emission and absorption induced by an applied field. Consider, for example, a single initially occupied field mode described by a coherent state with an expectation value $\langle E_0^{(-)}(t) + E_0^{(+)}(t) \rangle = E_0 \cos \omega t$ for the electric field. From (4.5.3) for $E_0^{(-)}(t)$, this implies

$$2\text{Re} \sum_\beta C_\beta \hbar \omega_\beta \langle a_\beta^\dagger(0)\sigma_x(t)\rangle e^{i\omega_\beta t} = 2\hbar\omega\text{Re}\left[\frac{id}{2\hbar}E_0 e^{i\omega t}\langle\sigma_x(t)\rangle\right]$$

$$= -\omega E_0 p(t) \sin \omega t, \qquad (5.6.3)$$

where $p(t) = d\langle \sigma_x(t)\rangle$ is the dipole expectation value. Comparison with (5.2.10) shows that this is indeed the rate of change of field energy resulting from stimulated emission and absorption.

The second contribution to $P(t)$, $P_s(t)$, is the power radiated, or scattered, out of the applied field. In the mode-continuum limit, and with the introduction of the slowly varying [compared with $\sigma(t)$] operator $S(t) = \sigma(t) \exp(i\omega t)$, where $\omega \cong \omega_0$ is the frequency of the applied field,

$$P_s(t) = \frac{2d^2}{6\pi^2\epsilon_0 c^3}\text{Re}\int_0^\infty d\nu \nu^4 \int_0^t dt' \langle S^\dagger(t')S(t)\rangle e^{i(\nu-\omega)(t'-t)} \qquad (5.6.4)$$

in the rotating-wave approximation. Anticipating that the spectrum of the radiated field will be concentrated near the frequency ω, we approximate this by

$$P_s(t) = \frac{d^2\omega^4}{3\pi^2\epsilon_0 c^3}\text{Re}\int_0^\infty d\nu \int_0^t dt' \langle S^\dagger(t')S(t)\rangle e^{i(\nu-\omega)(t'-t)}. \qquad (5.6.5)$$

We can simply further: for sufficiently long atom–field interaction times—times much longer than a radiative lifetime—the atom is in a stationary regime where $\langle S^\dagger(t')S(t)\rangle$

is a function only of $\tau = t - t'$. In this stationary regime, the radiated power is independent of time:

$$P_s = \frac{d^2\omega^4}{3\pi^2\epsilon_0 c^3}\mathrm{Re}\int_0^\infty d\nu \int_0^\infty d\tau \langle S^\dagger(t_0)S(t_0+\tau)\rangle e^{i(\nu-\omega)\tau}, \qquad (5.6.6)$$

where t_0 is some time at and after which the atom is in the stationary regime and therefore

$$\langle \dot{S}(t_0)\rangle = \langle \dot{S}^\dagger(t_0)\rangle = \langle \dot{\sigma}_z(t_0)\rangle = 0. \qquad (5.6.7)$$

Note that, within the approximation made in going from (5.6.4) to (5.6.5),

$$P_s = \frac{d^2\omega^4}{3\pi^2\epsilon_0 c^3}\times \pi\langle S^\dagger(t_0)S(t_0)\rangle = A\hbar\omega\langle S^\dagger(t_0)S(t_0)\rangle = A\hbar\omega P_2(t_0), \qquad (5.6.8)$$

that is, the scattered power in the stationary regime is the rate A of spontaneous emission times the steady-state upper-state occupation probability $P_2(t_0)$ times the energy $\hbar\omega$.

The equations of motion for the two-state-atom operators $S(t)$ and $\sigma_z(t)$, within the Markovian approximation for spontaneous emission, follow from (4.4.77) and (4.5.34):

$$\dot{S}(t) = -i(\Delta - iA/2)S(t) - i\frac{d}{\hbar}\sigma_z(t)E_0^{(+)}(t)e^{i\omega t}, \qquad (5.6.9)$$

and

$$\dot{\sigma}_z(t) = -A[1+\sigma_z(t)] - \frac{2id}{\hbar}\Big[E_0^{(-)}(t)S(t)e^{-i\omega t} - S^\dagger(t)E_0^{(+)}(t)e^{i\omega t}\Big]. \qquad (5.6.10)$$

We have used the expression (4.5.2) for $E_0^{(+)}(t)$ and made the rotating-wave approximation. We now use these equations to calculate the power spectrum of the resonance fluorescence. Based on (5.6.6), we define this, somewhat arbitrarily, as

$$p(\nu) = 2\mathrm{Re}\int_0^\infty d\tau\langle S^\dagger(t_0)S(t_0+\tau)\rangle e^{i(\nu-\omega)\tau} = 2\mathrm{Re}\int_0^\infty d\tau g(\tau)e^{i(\nu-\omega)\tau}. \qquad (5.6.11)$$

This requires calculation of the correlation function

$$g(\tau) = \langle S^\dagger(t_0)S(t_0+\tau)\rangle, \qquad (5.6.12)$$

which, from (5.6.9), satisfies

$$\frac{\partial}{\partial\tau}g(\tau) = -i(\Delta - iA/2)g(\tau) - i\frac{d}{\hbar}\langle S^\dagger(t_0)\sigma_z(t_0+\tau)E_0^{(+)}(t_0+\tau)\rangle e^{i\omega(t_0+\tau)}. \qquad (5.6.13)$$

Let the initial state $|\psi\rangle$ be such that the field is in a coherent state, that is, an eigenstate of $E_0^{(+)}(t)$, such that

$$E_0^{(+)}(t)|\psi\rangle = \frac{i}{2}E_0 e^{-i\omega t}|\psi\rangle \qquad (5.6.14)$$

and therefore that the initially excited field mode has an expectation value $\langle E_0^{(+)}(t) + E_0^{(-)}(t)\rangle = E_0 \sin \omega t$. Then,

$$\frac{\partial}{\partial \tau} g(\tau) = -i(\Delta - iA/2)g(\tau) + \frac{1}{2}\Omega h(\tau). \qquad (5.6.15)$$

We have defined the Rabi frequency $\Omega = dE_0/\hbar$ and introduced a second two-state-atom correlation function,

$$h(\tau) = \langle S^\dagger(t_0)\sigma_z(t_0 + \tau)\rangle. \qquad (5.6.16)$$

From (5.6.10) we obtain an equation for $h(\tau)$:

$$\frac{\partial}{\partial \tau} h(\tau) = -A\langle S^\dagger(t_0)\rangle - \Omega\langle S^\dagger(t_0)S^\dagger(t_0 + \tau)\rangle$$
$$- \frac{2id}{\hbar}\langle S^\dagger(t_0)E_0^{(-)}(t_0 + \tau)S(t_0 + \tau)\rangle e^{-i\omega(t_0+\tau)}. \qquad (5.6.17)$$

Equation (5.6.9) implies that $S^\dagger(t)$ depends only on what $E_0^{(-)}(t')$ is at times $t' \le t$. And since (see Exercise 5.2),

$$\left[\frac{d}{\hbar}E_0^{(+)}(t), \frac{d}{\hbar}E_0^{(-)}(t')\right]e^{i\omega_0(t-t')} \cong A\delta(t - t'), \qquad (5.6.18)$$

the commutator

$$[S^\dagger(t_0), E_0^{(-)}(t_0 + \tau)] \cong 0. \qquad (5.6.19)$$

Furthermore, $\langle\psi|E_0^{(-)}(t_0 + \tau) = -(i/2)E_0 e^{i\omega(t_0+\tau)}\langle\psi|$, and therefore

$$\frac{\partial}{\partial \tau} h(\tau) \cong -A\langle S^\dagger(t_0)\rangle - \Omega\langle S^\dagger(t_0)S^\dagger(t_0 + \tau)\rangle$$
$$- \frac{2id}{\hbar}\langle E_0^{(-)}(t_0 + \tau)S^\dagger(t_0)S^\dagger(t_0 + \tau)\rangle e^{-i\omega(t_0+\tau)}$$
$$= -A\langle S^\dagger(t_0)\rangle - \Omega f(\tau) - \Omega g(\tau). \qquad (5.6.20)$$

We have introduced one more correlation function:

$$f(\tau) = \langle S^\dagger(t_0)S^\dagger(t_0 + \tau)\rangle. \qquad (5.6.21)$$

Using again (5.6.9) and (5.6.19), we find that $f(\tau)$ satisfies

$$\frac{\partial}{\partial \tau} f(\tau) \cong i(\Delta + iA/2)f(\tau) + \frac{1}{2}\Omega h(\tau). \qquad (5.6.22)$$

Exercise 5.2: Show that (5.6.18) is valid up to corrections involving derivatives of $\delta(t-t')$ divided by powers of ω_0. (Such corrections are negligible for test functions whose variations during a time interval $\sim \omega_0^{-1}$ are negligible.)

The (approximate) equations (5.6.15), (5.6.20), and (5.6.22) comprise a closed set of equations that can be solved for $g(\tau)$ and therefore $p(\nu)$:[31]

$$\frac{\partial}{\partial \tau} g(\tau) = -i(\Delta - iA/2)g(\tau) + \frac{1}{2}\Omega h(\tau), \tag{5.6.23}$$

$$\frac{\partial}{\partial \tau} h(\tau) = -A\langle S^\dagger(t_0)\rangle - \Omega f(\tau) - \Omega g(\tau), \tag{5.6.24}$$

$$\frac{\partial}{\partial \tau} f(\tau) = i(\Delta + iA/2)f(\tau) + \frac{1}{2}\Omega h(\tau). \tag{5.6.25}$$

The initial conditions are

$$g(0) = \langle S^\dagger(t_0)S(t_0)\rangle = \frac{1}{2}[1 + \langle \sigma_z(t_0)\rangle], \tag{5.6.26}$$

$$h(0) = \langle S^\dagger(t_0)\sigma_z(t_0)\rangle = -\langle S^\dagger(t_0)\rangle, \tag{5.6.27}$$

$$f(0) = \langle S^\dagger(t_0)S^\dagger(t_0)\rangle = 0, \tag{5.6.28}$$

and these expectation values are easily evaluated from the stationary (steady-state) solutions of the optical Bloch equations,

$$\langle \dot{S}(t)\rangle = -i(\Delta - iA/2)\langle S(t)\rangle + \frac{1}{2}\langle \sigma_z(t)\rangle, \tag{5.6.29}$$

$$\langle \dot{\sigma}_z\rangle = -A[1 + \langle \sigma_z(t)\rangle] - \Omega\big[\langle S(t)\rangle + \langle S^\dagger(t)\rangle\big], \tag{5.6.30}$$

which follow when expectation values over the initial state $|\psi\rangle$ are taken on both sides of (5.6.9) and (5.6.10):

$$g(0) = \frac{\Omega^2/4}{\Delta^2 + A^2/4 + \Omega^2/2}, \tag{5.6.31}$$

and

$$h(0) = \frac{\Omega(A/2 + i\Delta)/2}{\Delta^2 + A^2/4 + \Omega^2/2}. \tag{5.6.32}$$

If, for example, $\Delta = 0$, and $\Omega \gg A$,

$$g(\tau) \cong (A/2\Omega)^2 + \frac{1}{2}\left(e^{-A\tau/2} + e^{-3A\tau/4}\cos\Omega\tau\right) \tag{5.6.33}$$

---

[31] P. L. Knight and P. W. Milonni, Phys. Rep. **66**, 21 (1980). See also *Mandel and Wolf (1995)*, Section 15.6.

and

$$p(\nu) = 2\pi(A/2\Omega)^2\delta(\nu - \omega) + \frac{A/4}{(\nu - \omega)^2 + A^2/4}$$
$$+ \frac{3A/16}{(\nu - \omega - \Omega)^2 + 9A^2/16} + \frac{3A/16}{(\nu - \omega + \Omega)^2 + 9A^2/16}. \qquad (5.6.34)$$

To interpret the part of $p(\nu)$ that goes as $\delta(\nu - \omega)$, write $g(\tau) = g_e(\tau) + g_i(\tau)$,

$$g_e(\tau) = \langle S^\dagger(t)0)\rangle\langle S(t_0 + \tau)\rangle, \qquad (5.6.35)$$

$$g_i(\tau) = \langle S^\dagger(t_0)S(t_0 + \tau)\rangle - \langle S^\dagger(t)0)\rangle\langle S(t_0 + \tau)\rangle, \qquad (5.6.36)$$

and define the "elastic" part of the resonance fluorescence spectrum as

$$p_e(\nu) = 2\mathrm{Re}\int_0^\infty d\tau g_e(\tau)e^{i(\nu-\omega)\tau}. \qquad (5.6.37)$$

Now, $\langle S(t_0 + \tau)\rangle = \langle S(t_0)\rangle = -h(0)$, and so

$$p_e(\nu) = 2|h(0)|^2\mathrm{Re}\int_0^\infty d\tau e^{i(\nu-\omega)\tau} = 2\pi|h(0)|^2\delta(\nu - \omega)$$
$$= 2\pi\frac{\Omega^2(A^2/4 + \Delta^2)/4}{(\Omega^2/2 + \Delta^2 + A^2/4)^2}\delta(\nu - \omega)$$
$$\cong 2\pi(A/2\Omega)^2\delta(\nu - \omega) \qquad (\Delta = 0, \Omega \gg A). \qquad (5.6.38)$$

Thus, the part of the spectrum having a delta-function peak at the frequency ω of the applied field is obtained if we "decorrelate" the correlation function $g(\tau)$ by replacing the expectation value of the product of the operators $S(t_0)$ and $S(t_0 + \tau)$ with the product of the expectation values $\langle S(t_0)\rangle$ and $\langle S(t_0+\tau)\rangle$. It is the (complete) spectrum we would obtain if we defined the spectrum based on the Fourier transform of the steady-state dipole moment calculated from the optical Bloch equations, as we would do in semiclassical radiation theory.

The "inelastic" part of the spectrum,

$$p_i(\nu) = 2\mathrm{Re}\int_0^\infty d\tau g_i(\tau)e^{i(\nu-\omega)\tau}, \qquad (5.6.39)$$

has the triplet structure expected from the dressed-state perspective. The peaks in the spectrum are at the applied field frequency ω and at the "Rabi sidebands" $\omega \pm \Omega$. For $\Delta = 0$, and $\Omega \gg A$, $p_i(\nu)$ is given by the last three terms in (5.6.34):

$$p_i(\nu) = \frac{A/4}{(\nu - \omega)^2 + A^2/4} + \frac{3A/16}{(\nu - \omega - \Omega)^2 + 9A^2/16}$$
$$+ \frac{3A/16}{(\nu - \omega + \Omega)^2 + 9A^2/16}. \qquad (5.6.40)$$

The integral over all frequencies of the first term is twice the integrals of the second and third terms, consistent with the dressed-state picture, whereas the peak value of the first term is three times the peak values of the second and third terms.

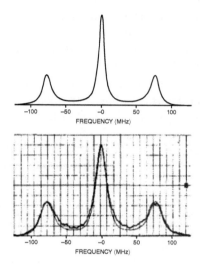

Fig. 5.5 Theory (top) and experimental results (bottom) for the spectrum of resonance fluorescence by a two-state atom driven by a laser field with frequency equal to the atom's transition frequency. The solid curve in the bottom figure is a convolution of the theoretical spectrum with the instrumental lineshape of width 9.5 MHz. For the field strength in the experiment, the elastic part of the spectrum is negligible and not included in the plot of the theoretical spectrum. From R. E. Grove, F. Y. Wu, and S. Ezekiel, Phys. Rev. A **15**, 227 (1977), with permission.

The theory of the triplet structure of the resonance fluorescence spectrum of a two-state atom in a strong, coherent monochromatic field was developed in the 1970s.[32] In the first experiments, the (sodium) atoms were not prepared as two-state systems, but the triplet structure was clearly demonstrated.[33] These were soon followed by experiments in which optical pumping prepared the atoms as two-state systems, and the measured spectra were in very close agreement with theory (see Figure 5.5).[34] More recent observations of the triplet structure of strong-field resonance fluorescence have been made in quantum dots and artificial atoms.[35]

In the older (pre-laser) literature, observed resonance fluorescence spectra had no Rabi sidebands because the applied fields were too weak. The weak-field theory is worked out in Heitler's treatise.[36] There it is shown that, for narrowband applied fields,

[32] A. I. Burshtein, Sov. Phys. JETP **21**, 567 (1965); A. I. Burshtein, Sov. Phys. JETP **22**, 939 (1966); M. Newstein, Phys. Rev. **167**, 89 (1968); B. R. Mollow, Phys. Rev. **188**, 1969 (1969).

[33] F. Schuda, C. R. Stroud, Jr., and M. Hercher, J. Phys. B: Atom. Molec. Phys. **7**, L198 (1974).

[34] R. E. Grove, F. Y. Wu, and S. Ezekiel, Phys. Rev. A **15**, 227 (1977).

[35] See, for instance, O. Astafiev, A. M. Zagoskin, A. A. Abdumalikov,Jr., Y. A. Pashkin, T. Yamamoto, K. Inomata, Y. Nakamura, and J. S. Tsai, Science **327**, 840 (2010).

[36] W. Heitler, *The Quantum Theory of Radiation*, third edition, Oxford University Press, London, 1960, pp. 196–204.

the scattered radiation is similarly narrowband, whereas for broadband excitation the scattered radiation has the natural lineshape for spontaneous emission.[37] These results are easily derived in our Heisenberg-picture approach. If the upper-state occupation probability when the atom is irradiated by a low-intensity field is small, the atom is well approximated by the harmonic-oscillator model, and we can replace $\sigma_z(t)$ by -1 (Section 2.5.2). Using this approximation in (4.4.74), making the rotating-wave approximation, and ignoring radiative frequency shifts (or assuming they are included in the definition of ω_0), we obtain straightforwardly, for $t_0 \to \infty$,

$$\langle \sigma^\dagger(t_0)\sigma(t_0 + \tau) \rangle = \sum_\beta C_\beta^2 \frac{\langle a_\beta^\dagger(0)a_\beta(0) \rangle}{(\omega_\beta - \omega_0)^2 + A^2/4} e^{-i\omega_\beta \tau} \tag{5.6.41}$$

and

$$p(\nu) = 2\mathrm{Re} \int_0^\infty d\tau \langle \sigma^\dagger(t_0)\sigma(t_0 + \tau) \rangle e^{i(\nu - \omega)\tau}$$

$$= 2\pi \sum_\beta C_\beta^2 \frac{\langle a_\beta^\dagger(0)a_\beta(0) \rangle}{(\omega_\beta - \omega_0)^2 + A^2/4} \delta(\omega_\beta - \nu). \tag{5.6.42}$$

In the extreme narrowband limit where the applied field is taken to be monochromatic (and instrumental effects on the lineshape are ignored),

$$p(\nu) = 2\pi \frac{C^2 \langle n \rangle}{(\omega - \omega_0)^2 + A^2/4} \delta(\nu - \omega) = 2\pi \frac{\Omega^2/4}{\Delta^2 + A^2/4} \delta(\nu - \omega), \tag{5.6.43}$$

where $\langle n \rangle = \langle a^\dagger(0)a(0) \rangle$ is the average photon number of the driving field of frequency ω and $\Omega = 2C\sqrt{\langle n \rangle}$ (see (5.3.21)). This reproduces the elastic part of the resonance fluorescence spectrum (see (5.6.38)) for $\Omega \ll A$. For a broadband driving field, however, $\langle a_\beta^\dagger(0)a_\beta(0) \rangle$ can be assumed to be a relatively flat function of frequency, and, in this case, (5.6.42) implies that the spectrum of resonance fluorescence is just the natural lineshape:

$$p(\nu) \propto \frac{1}{(\nu - \omega_0)^2 + A^2/4}. \tag{5.6.44}$$

Exercise 5.3: (a) Derive (5.6.41) and (5.6.42). (b) The peak laser field intensity used in obtaining the spectrum shown in Figure 5.5 was 640 mW/cm^2. Given that the spontaneous emission lifetime for the upper state of the sodium atom in the experiments is 16 ns, calculate the on-resonance Rabi frequency and compare it with the separation of the sidebands from the center frequency (corresponding to the wavelength 589 nm) in the figure.

---

[37] The former is termed the "Heitler effect"; in particular, the existence of a fluorescence linewidth smaller than the natural linewidth was demonstrated in a "pedagogical experiment" by H. M. Gibbs and T. N. C. Venkatesan, Opt. Commun. **17**, 87 (1976).

5.7 Photon Anti-Bunching in Resonance Fluorescence

The photon bunching in the intensity correlations of thermal radiation was discussed in Section 2.9. We related the bunching to the "wave" term in the Einstein fluctuation formula for thermal radiation. In Section 3.6, we identified "wave" and "particle" parts of the variance in the photon number of a single-mode field and deduced that radiation described by a coherent state has no "wave" fluctuations and exhibits no photon bunching. We introduced the Mandel Q parameter such that a field with a negative Q has properties that cannot be explained by classical radiation theory (see Section 3.8) and noted that thermal and coherent states of the field both have non-negative Q's, whereas a squeezed state, for example, can have a negative Q. The most obvious example of a "nonclassical" field state is a Fock state, that is, a field having a definite photon number and therefore no "particle" fluctuations. We turn now to another example of field nonclassicality: photon *anti*-bunching and, in particular, the photon anti-bunching in the resonance fluorescence of an atom driven by a coherent-state monochromatic field.

In Section 2.9 we surmised that the photon (or actually photoelectron) coincidence rate measured in the experiment shown in Figure 2.6 (with $I_1 = I_2$, and $\tau = 0$), corresponds to

$$\langle n^2 \rangle - \langle n \rangle = \langle a^\dagger a a^\dagger a \rangle - \langle a^\dagger a \rangle = \langle a^\dagger a^\dagger a a \rangle, \tag{5.7.1}$$

the expectation value of a *normally* ordered product of photon annihilation and creation operators; this assumes, among other things, a single-mode field, and photon counters that respond instantaneously to an incident field. If a time difference τ is allowed for, and the single-mode assumption is dropped, one finds that the coincidence rate is proportional to (see Appendix C)

$$g^{(2)}(\tau) = \langle E^{(-)}(t_0) E^{(-)}(t_0 + \tau) E^{(+)}(t_0 + \tau) E^{(+)}(t_0) \rangle \tag{5.7.2}$$

within the approximation of an ideal detector. Since each polarization component of the (transverse) electric field operator associated with any point $\mathbf{r}$ in space has the form (recall (4.2.19))

$$E(t) = i \sum_\beta [A_\beta a_\beta(t) - A_\beta^* a_\beta^\dagger(t)], \tag{5.7.3}$$

and $a_\beta(t)$ and $a_\beta^\dagger(t)$ vary *approximately* as $e^{-i\omega_\beta t}$ and $e^{i\omega_\beta t}$, respectively,

$$E^{(+)}(t) = i \sum_\beta A_\beta a_\beta(t) \tag{5.7.4}$$

and $E^{(-)}(t) = E^{(+)}(t)^\dagger$ are, approximately, the positive- and negative-frequency parts, respectively, of $E(t)$. (The form of the $\mathbf{r}$-dependent A_β's is of no importance for present purposes.)

We imagine an experiment like that shown in Figure 2.6, but with the "incident light" being the fluorescent light of a two-state atom driven by a coherent monochromatic field of frequency ω, as in Section 5.6. The electric fields at t_0 and $t_0 + \tau$ in

$g^{(2)}(\tau)$ are the fields at the photomultipliers PM1 and PM2, taken for simplicity to be essentially point-like detectors, practically equidistant from the atom. The electric field of the atom in the radiation zone follows from (4.3.11):

$$E(t) \propto e\ddot{x}(t - r/c) = d\ddot{\sigma}_x(t - r/c) = d[\ddot{\sigma}(t - r/c) + \ddot{\sigma}^\dagger(t - r/c)]$$
$$\cong -\omega^2 d[\sigma(t - r/c) + \sigma^\dagger(t - r/c)], \tag{5.7.5}$$

and therefore, in a rotating-wave approximation,

$$E^{(+)}(t) \propto \sigma(t - r/c) \cong \sigma(t)e^{i\omega r/c}, \quad E^{(-)}(t) \propto \sigma^\dagger(t - r/c) \cong \sigma^\dagger(t)e^{-i\omega r/c} \tag{5.7.6}$$

and

$$g^{(2)}(\tau) \propto \langle \sigma^\dagger(t_0)\sigma^\dagger(t_0 + \tau)\sigma(t_0 + \tau)\sigma(t_0) \rangle$$
$$= \langle S^\dagger(t_0)S^\dagger(t_0 + \tau)S(t_0 + \tau)S(t_0) \rangle \tag{5.7.7}$$

in steady state, with t_0 and $S(t)$ having been defined in Section 5.6. It follows immediately from the fermionic operator identity $\sigma^2(t) = S^2(t) = 0$ that

$$g^{(2)}(0) = 0. \tag{5.7.8}$$

The probability of counting two photons simultaneously is therefore zero. The physical interpretation of this anti-bunching is clear: the atom can only emit a photon from its excited state; having done so, it cannot emit another photon until it has been pumped back up to the excited state by the driving field, and this repopulation of the excited state takes a finite amount of time.

To calculate the correlation function $g^{(2)}(\tau)$, we first note that, since $S^\dagger(t)S(t) = \sigma^\dagger(t)\sigma(t) = (1/2)[1 + \sigma_z(t)]$,

$$\langle S^\dagger(t_0)S^\dagger(t_0 + \tau)S(t_0 + \tau)S(t_0) \rangle = \frac{1}{2}\langle S^\dagger(t_0)S(t_0) \rangle + \frac{1}{2}\langle S^\dagger(t_0)\sigma_z(t_0 + \tau)S(t_0) \rangle$$
$$= \frac{1}{2}g(0) + \frac{1}{2}\langle S^\dagger(t_0)\sigma_z(t_0 + \tau)S(t_0) \rangle, \tag{5.7.9}$$

where $g(\tau)$ is the correlation function (5.6.12). The calculation of the correlation function $G(\tau) = \langle S^\dagger(t_0)\sigma_z(t_0 + \tau)S(t_0) \rangle$ and, therefore, $g^{(2)}(\tau)$ now proceeds in the same fashion as the calculation of $g(\tau)$. We define two more correlation functions, $F(\tau) = \langle S^\dagger(t_0)S^\dagger(t_0 + \tau)S(t_0) \rangle$, and $H(\tau) = \langle S^\dagger(t_0)S(t_0 + \tau)S(t_0) \rangle$, and find that $F(\tau), G(\tau)$, and $H(\tau)$ satisfy the closed set of equations

$$\frac{\partial}{\partial \tau}F(\tau) = i(\Delta + iA/2)F(\tau) + \frac{1}{2}\Omega G(\tau),$$

$$\frac{\partial}{\partial \tau}G(\tau) = -Ag(0) - AG(\tau) - \Omega F(\tau) - \Omega H(\tau),$$

$$\frac{\partial}{\partial \tau}H(\tau) = -i(\Delta - iA/2)H(\tau) + \frac{1}{2}\Omega G(\tau), \tag{5.7.10}$$

with initial conditions

$$F(0) = \langle S^\dagger(t_0)S^\dagger(t_0)S(t_0) = 0,$$
$$G(0) = \langle S^\dagger(t_0)\sigma_z(t_0)S(t_0)\rangle = -\langle S^\dagger(t_0)S(t_0)\rangle = -g(0),$$
$$H(0) = \langle S^\dagger(t_0)S(t_0)S(t_0)\rangle = 0. \tag{5.7.11}$$

For $\Delta = 0$, for example,

$$g^{(2)}(\tau) = g(0)^2\left[1 - e^{-3A\tau/4}\left(\cos\Omega'\tau + \frac{3A}{4\Omega'}\sin\Omega'\tau\right)\right], \tag{5.7.12}$$

where we have defined $\Omega' = \sqrt{\Omega^2 - A^2}$ and used the steady-state solution

$$g(0) = \langle S^\dagger(t_0)S(t_0)\rangle = \frac{\frac{1}{2}\Omega^2}{\Omega^2 + A^2/2} \tag{5.7.13}$$

of the optical Bloch equations. Equation (5.7.12) can be expressed as

$$g^{(2)}(\tau) = \left[\frac{1}{2} + \frac{1}{2}\langle\sigma_z(t_0)\rangle\right]\left[\frac{1}{2} + \frac{1}{2}\langle\sigma_z(\tau)\rangle_G\right], \tag{5.7.14}$$

where $\langle\sigma_z(\tau)\rangle_G$ is defined as the expectation value of the population difference $\sigma_z(\tau)$ when the atom is in its ground state at $\tau = 0$. In this form, $g^{(2)}(\tau)$ is seen to be the steady-state probability that the atom is in its upper state, $[\frac{1}{2} + \frac{1}{2}\langle\sigma_z(t_0)\rangle]$, times the probability $[\frac{1}{2} + \frac{1}{2}\langle\sigma_z(\tau)\rangle_G]$ that, at time τ, the *initially unexcited* atom is in its upper state. It thus expresses in a more quantitative way the physical interpretation of the photon anti-bunching given following (5.7.8).

Photon anti-bunching cannot be explained by semiclassical radiation theory, or any theory that treats the field classically: for a single-mode field $g^{(2)}(0) \propto \langle a^\dagger a^\dagger aa\rangle$, $a = a(0)$, and $g^{(2)}(0) = 0$ implies that the Mandel Q parameter $(= \langle a^\dagger a^\dagger aa\rangle - \langle a^\dagger a\rangle^2)$ is negative, a signature of nonclassicality as discussed in Section 3.8.

As emphasized by Glauber, "The variety of fields encountered in the quantum theory is simply much larger than that allowed by classical theory," and whereas classical theory can account for the Hanbury Brown–Twiss correlations, or photon bunching, it does not allow for the possibility of "negative Hanbury Brown–Twiss correlations."[38] Stoler discussed further the negative correlation effect, which he called "anti-bunching," and suggested that it could be observed in the light from a degenerate parametric oscillator.[39] Soon after several independent predictions of photon anti-bunching in resonance fluorescence, the effect was confirmed by Kimble et al. in an experiment quite similar in concept to the idealized version we have described.[40]

---

[38] R. J. Glauber, in *Quantum Optics and Electronics*, C. DeWitt, A. Blandin, and C. Cohen Tannoudji, eds., Gordon and Breach, New York, 1965, p. 155.

[39] D. Stoler, Phys. Rev. Lett. **33**, 1397 (1974).

[40] H. J. Carmichael and D. F. Walls, J. Phys. B: Atom. Molec. Phys. **9**, 1199 (1976); H. J. Kimble and L. Mandel, Phys. Rev. A **13**, 2123 (1976); C. Cohen-Tannoudji, in *Frontiers in Laser Spectroscopy*, R. Balian, S. Haroche, and S. Liberman, eds., North-Holland, Amsterdam, 1977, Volume I, pp. 3–102; H. J. Kimble, M. Dagenais, and L. Mandel, Phys. Rev. Lett. **39**, 691 (1977).

Observations of distinctly quantum effects in the electromagnetic field generally require consideration of "second-order" correlation functions such as $g^{(2)}(\tau)$, as opposed to "first-order" correlation functions such as $g^{(1)}(\tau) = \langle E^{(-)}(t)E^{(+)}(t+\tau)\rangle$. The Mandel Q parameter, for example, is defined in terms of a second-order correlation function, and we will see in Section 5.8 that the photon polarization correlations studied in the context of the Einstein–Podolsky–Rosen (EPR) "paradox" likewise involve $g^{(2)}$. For another example, consider again the example given in Figure 2.1: a single excited atom makes a single, spontaneous transition and radiates light in the presence of two ideal point-like photodetectors, D_1 at position $\mathbf{r}_1$, and D_2 at position $\mathbf{r}_2$. The probability of counting a photon at D_1 at time t_1 *and* of counting a photon at D_2 at time t_2 is proportional to

$$g^{(2)}(\mathbf{r}_1,t_1;\mathbf{r}_2,t_2) = \langle E^{(-)}(\mathbf{r}_1,t_1)E^{(-)}(\mathbf{r}_2,t_2)E^{(+)}(\mathbf{r}_2,t_2)E^{(+)}(\mathbf{r}_1,t_1)\rangle. \qquad (5.7.15)$$

(For simplicity we are considering only a single polarization component of the field.) Using again (5.7.6), and assuming exponential decay of the excited state, we find that

$$E^{(+)}(\mathbf{r}_1,t_1) \propto \sigma(t_1 - r_1/c) \cong \sigma(t_1)e^{i\omega_0 r_1/c} \cong \sigma(0)e^{-At_1/2}e^{i\omega_0 r_1/c} \qquad (5.7.16)$$

and

$$E^{(-)}(\mathbf{r}_1,t_1) \propto \sigma^\dagger(t_1 - r_1/c) \cong \sigma^\dagger(0)e^{-At_1/2}e^{-i\omega_0 r_1/c}; \qquad (5.7.17)$$

and that therefore

$$g^{(2)}(\mathbf{r}_1,t_1;\mathbf{r}_2,t_2) \propto \langle \sigma^\dagger(0)\sigma^\dagger(0)\sigma(0)\sigma(0)\rangle = 0, \qquad (5.7.18)$$

meaning that a single atom emitting a single photon cannot produce photon counts at two detectors. As in the case of photon anti-bunching (for $\tau = 0$), this can be viewed formally as just a consequence of the operator identity $\sigma^2(0) = 0$. We again modeled the atom as a two-state system, made the rotating-wave approximation, assumed purely exponential decay, ignored different polarization components of the field, and so on, but the conclusion that a single photon cannot be registered at two detectors is unaffected by these simplifications. In classical wave theory, in contrast, the field from the atom is a real, measurable wave spread out over space and perfectly capable of being observed at two (or more) detectors. Experiments similar in concept to that given in Figure 2.1, have been performed; in these experiments, a dim beam of light— so dim that its energy corresponds to that of a single photon—is incident on a beam splitter, and each resulting beam is directed to a photodetector. No "extra" photon coincidences, which would be expected according to classical theory, were found.[41]

[41] J. F. Clauser, Phys. Rev. D **9**, 853 (1974). See also P. Grangier, G. Roger, and A. Aspect, Europhys. Lett. **1**, 173 (1986).

5.8   Polarization Correlations of Photons from an Atomic Cascade

Quantum-optical experiments have had a large role in studies in the conceptual foundations of quantum theory. In especially influential experiments, polarization correlations of photons from atomic cascade transitions have been measured. The basic idea is shown in Figure 5.6. An atom emits radiation at frequencies ω_A and ω_B when it undergoes spontaneous emission from a level with angular momentum quantum number $J = 0$ to a level with $J = 1$ and then to a level with $J = 0$. The polarizers are oriented to transmit light polarized at angles ϕ_A and ϕ_B with respect to an axis perpendicular to the z (quantization) axis; the photons are ideally detected following propagation in the $\pm z$ directions, although the light is not restricted to propagation in these directions. We label the initial state of the cascade transition, for which $J = M = 0$, as $|4\rangle$. The intermediate states, with $J = 1, M = 1$ and $J = 1, M = -1$, are denoted by $|2\rangle$ and $|3\rangle$, respectively, and the final state with $J = M = 0$ is denoted by $|1\rangle$. (The state with $M = 0$ in the $J = 1$ level can be ignored because transitions to and from it do not occur with emission of transversely polarized light propagating in the $\pm z$ directions.) The electric dipole moment operator for the three-level (four-state) atom of Figure 5.6 is given by (4.8.7):

$$\mathbf{d} = \mathbf{d}_{12}\sigma_{12} + \mathbf{d}_{12}^*\sigma_{12}^\dagger + \mathbf{d}_{13}\sigma_{13} + \mathbf{d}_{13}^*\sigma_{13}^\dagger + \mathbf{d}_{24}\sigma_{24} + \mathbf{d}_{24}^*\sigma_{24}^\dagger + \mathbf{d}_{34}\sigma_{34} + \mathbf{d}_{34}^*\sigma_{34}^\dagger. \quad (5.8.1)$$

The dipole matrix element $\mathbf{d}_{12}$ is between a state with $M = 0$ and a state with $M' = 1$. In the notation of Section 2.4, $q = -1$, and so

$$\mathbf{d}_{12} = |\mathbf{d}_{12}| \frac{1}{\sqrt{2}} (\hat{\mathbf{x}} - i\hat{\mathbf{y}}) \quad (5.8.2)$$

and, likewise,

$$\mathbf{d}_{13} = |\mathbf{d}_{13}| \frac{1}{\sqrt{2}} (\hat{\mathbf{x}} + i\hat{\mathbf{y}}), \quad \mathbf{d}_{24} = |\mathbf{d}_{24}| \frac{1}{\sqrt{2}} (\hat{\mathbf{x}} + i\hat{\mathbf{y}}), \quad \mathbf{d}_{34} = |\mathbf{d}_{34}| \frac{1}{\sqrt{2}} (\hat{\mathbf{x}} - i\hat{\mathbf{y}}). \quad (5.8.3)$$

$\hat{\mathbf{x}}$ and $\hat{\mathbf{y}}$ are unit vectors orthogonal to the z axis and to each other.

For propagation in the $\pm z$ directions, and for points whose distance from the atom is large compared with the two transition wavelengths, the electric field operator for the field from the atom is (see (4.3.11))

$$\mathbf{E}(z, t) = -\frac{1}{4\pi\epsilon_0} \frac{1}{c^2 r} \ddot{\mathbf{d}}(t - z/c), \quad (5.8.4)$$

and the positive-frequency part of this field operator is

$$\begin{aligned}
\mathbf{E}^{(+)}(z, t) &\cong -\frac{1}{4\pi\epsilon_0} \frac{1}{c^2 r} \Big[\mathbf{d}_{12}\ddot{\sigma}_{12}(t - z/c) + \mathbf{d}_{13}\ddot{\sigma}_{13}(t - z/c) \\
&\qquad + \mathbf{d}_{24}\ddot{\sigma}_{24}(t - z/c) + \mathbf{d}_{34}\ddot{\sigma}_{34}(t - z/c) \Big] \\
&\cong \frac{1}{4\pi\epsilon_0} \frac{1}{c^2 r} \Big\{ \omega_B^2 \big[\mathbf{d}_{12}\sigma_{12}(t) + \mathbf{d}_{13}\sigma_{13}(t) \big] e^{-i\omega_B(t - z/c)} \\
&\qquad + \omega_A^2 \big[\mathbf{d}_{24}\sigma_{24}(t) + \mathbf{d}_{34}\sigma_{34}(t) \big] e^{-i\omega_A(t - z/c)} \Big\}
\end{aligned} \quad (5.8.5)$$

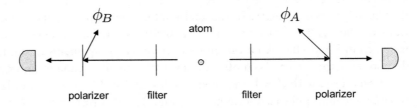

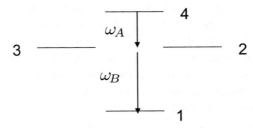

Fig. 5.6 Setup (top) for measuring polarization correlations of photons emitted in an atomic cascade transition (bottom).

in the rotating-wave approximation. Retention of other terms does not affect the polarization correlations we now calculate.

The probability of counting photons of frequencies ω_A and ω_B is (ideally) proportional to

$$R_{AB} = \langle E_A^{(-)}(t) E_B^{(-)}(t) E_B^{(+)}(t) E_A^{(+)}(t) \rangle \tag{5.8.6}$$

for the setup of Figure 5.6. $E_A^{(+)}(t)$ is the part of $\mathbf{E}^{(+)}$ at the detector following the ω_A filter and polarizer:

$$
\begin{aligned}
E_A^{(+)}(t) &= \mathbf{E}^{(+)}(z_A, t) \cdot (\hat{\mathbf{x}} \cos \phi_A + \hat{\mathbf{y}} \sin \phi_A) \\
&\cong \frac{1}{4\pi\epsilon_0} \frac{\omega_A^2}{c^2 r} \Big[\sigma_{24}(t) \mathbf{d}_{24} \cdot (\hat{\mathbf{x}} \cos \phi_A \\
&\quad + \hat{\mathbf{y}} \sin \phi_A) + \sigma_{34}(t) \mathbf{d}_{34} \cdot (\hat{\mathbf{x}} \cos \phi_A + \hat{\mathbf{y}} \sin \phi_A) \Big] e^{-i\omega_A(t - z_A/c)} \\
&= \frac{1}{4\pi\epsilon_0} \frac{\omega_A^2}{c^2 r} \frac{1}{\sqrt{2}} |\mathbf{d}_{24}| \Big[\sigma_{24}(t)(\cos \phi_A + i \sin \phi_A) \\
&\quad + \sigma_{34}(t)(\cos \phi_A - i \sin \phi_A) \Big] e^{-i\omega_A(t - z_A/c)} \\
&= \frac{1}{4\pi\epsilon_0} \frac{\omega_A^2}{c^2 r} \frac{1}{\sqrt{2}} |\mathbf{d}_{24}| \Big[\sigma_{24}(t) e^{i\phi_A} + \sigma_{34}(t) e^{-i\phi_A} \Big] e^{-i\omega_A(t - z_A/c)}, \tag{5.8.7}
\end{aligned}
$$

since $|\mathbf{d}_{34}| = |\mathbf{d}_{24}|$ (see Exercise 5.4). z_A is the distance from the atom to the detector at the right in Figure 5.6. In the same way, we obtain the part of $\mathbf{E}^{(+)}$ at the detector following the ω_B filter and polarizer:

$$E_B^{(+)}(t) = \mathbf{E}^{(+)}(z_B, t) \cdot (\hat{\mathbf{x}} \cos \phi_B + \hat{\mathbf{y}} \sin \phi_B)$$

$$\cong \frac{1}{4\pi\epsilon_0} \frac{\omega_B^2}{c^2 r} \frac{1}{\sqrt{2}} |\mathbf{d}_{12}| \left[\sigma_{12}(t) e^{-i\phi_B} + \sigma_{13}(t) e^{i\phi_B} \right] e^{-i\omega_B(t - z_B/c)}, \quad (5.8.8)$$

since $|\mathbf{d}_{13}| = |\mathbf{d}_{12}|$. Thus, since $\sigma_{ij}\sigma_{mn} = |i\rangle\langle j|m\rangle\langle n| = \delta_{jm}\sigma_{in}$,

$$E_B^{(+)}(t)E_A^{(+)}(t) \propto \left[\sigma_{12}(t) e^{-i\phi_B} + \sigma_{13}(t) e^{i\phi_B} \right] \left[\sigma_{24}(t) e^{i\phi_A} + \sigma_{34}(t) e^{-i\phi_A} \right]$$

$$\times e^{-i(\omega_A + \omega_B)t} e^{i\omega_A z_A/c} e^{i\omega_B z_B/c}$$

$$= \left[\sigma_{14}(t) e^{i(\phi_A - \phi_B)} \right.$$

$$\left. + \sigma_{14}(t) e^{-i(\phi_A - \phi_B)} \right] e^{-i(\omega_A + \omega_B)t} e^{i\omega_A z_A/c} e^{i\omega_B z_B/c}, \quad (5.8.9)$$

$$E_A^{(-)}(t)E_B^{(-)}(t) = \left[E_B^{(+)}(t)E_A^{(+)}(t) \right]^\dagger \propto \left[\sigma_{41}(t) e^{-i(\phi_A - \phi_B)} + \sigma_{41}(t) e^{i(\phi_A - \phi_B)} \right]$$

$$\times e^{i(\omega_A + \omega_B)t} e^{-i\omega_A z_A/c} e^{-i\omega_B z_B/c}, \quad (5.8.10)$$

and, finally,

$$R_{AB} \propto \langle \sigma_{41}(t)\sigma_{14}(t) \rangle \left(e^{i(\phi_A - \phi_B)} + e^{-i(\phi_A - \phi_B)} \right)^2$$

$$= \langle \sigma_{44}(t) \rangle \times 4 \cos^2(\phi_A - \phi_B). \quad (5.8.11)$$

This is just the upper-state occupation probability $\langle \sigma_{44}(t) \rangle$ times the factor $\cos^2(\phi_A - \phi_B)$ that specifies the polarization correlations of the ω_A and ω_B photons:

$$R_{AB} \propto \cos^2(\phi_A - \phi_B). \quad (5.8.12)$$

If, for instance, the two polarizers are oriented in the same direction ($\phi_A - \phi_B = 0$), the probability of detecting both photons is maximized, whereas if the polarizers are oriented orthogonally ($\phi_A - \phi_B = \pi/2$), the probability is zero.

The polarization correlation follows in our Heisenberg-picture calculation from the operator identity $\sigma_{ij}\sigma_{mn} = |i\rangle\langle j|m\rangle\langle n| = \delta_{jm}\sigma_{in}$. The non-commutative algebra of the σ_{ij} operators is associated with nonclassical properties of the field from the atom. Not surprisingly, the prediction (5.8.12) of quantum theory, which has been confirmed experimentally, cannot be derived from semiclassical radiation theory.[42] It can be interpreted as follows. Consider

$$\langle 2|\mathbf{d}|4 \rangle \cdot (\hat{\mathbf{x}} \cos \phi_A + \hat{\mathbf{y}} \sin \phi_A) = \frac{|\mathbf{d}_{24}|}{\sqrt{2}} (\hat{\mathbf{x}} + i\hat{\mathbf{y}}) \cdot (\hat{\mathbf{x}} \cos \phi_A + \hat{\mathbf{y}} \sin \phi_A)$$

$$\propto |\mathbf{d}_{24}| e^{\phi_A}. \quad (5.8.13)$$

This is (proportional to) the probability amplitude for the atomic transition $4 \to 2$ with the detection of a photon of frequency ω_A that passes through the A polarizer.

---

[42] J. F. Clauser, Phys. Rev. A **6**, 49 (1972).

Similarly, $|\mathbf{d}_{12}|e^{-i\Phi_B}$ is the probability amplitude for the atomic transition $2 \to 1$ with the detection of a photon of frequency ω_B that passes through the B polarizer. The product of these two amplitudes is the amplitude for the transition $4 \to 2 \to 1$ with the detection of photons of frequencies ω_A and ω_B in the setup of Figure 5.6. Likewise, $|\mathbf{d}_{34}|e^{-i\Phi_A}$ is the amplitude for the transition $4 \to 3$ with the detection of a photon of frequency ω_A, $|\mathbf{d}_{13}|e^{i\Phi_B}$ is the amplitude for the transition $3 \to 1$ with the detection of a photon of frequency ω_B, and the product of these two amplitudes is the amplitude for the process $4 \to 3 \to 1$ with the detection of photons of frequencies ω_A and ω_B. The sum $|\mathbf{d}_{24}||\mathbf{d}_{12}|e^{i(\Phi_A-\Phi_B)} + |\mathbf{d}_{34}||\mathbf{d}_{13}|e^{-i(\Phi_A-\Phi_B)}$ is then (proportional to) the probability amplitude for the process $4 \to 1$ with the detection of both ω_A and ω_B photons *when we cannot distinguish whether the cascade emission occurred via the path* $4 \to 2 \to 1$ *or* $4 \to 3 \to 1$. This is an example of two fundamental rules of quantum theory: (1) the probability amplitude for two successive processes is the product of the amplitudes for the individual processes, and (2) the amplitude for a process that can occur in different *indistinguishable* ways is the sum of the amplitudes for the individual ways.[43]

Exercise 5.4: (a) What is the justification for taking $|\mathbf{d}_{13}| = |\mathbf{d}_{12}|$ and $|\mathbf{d}_{34}| = |\mathbf{d}_{24}|$? (b) In writing (5.8.1), we have ignored constant phase factors that could be included in the state vectors $|j\rangle$, $j = 1, 2, 3, 4$, without affecting the norms $\langle j | j \rangle$. In other words, we ignored the fact that we could add a constant phase factor $e^{i\Phi_j}$ to each wave function $\psi_j(\mathbf{r})$ used to compute the dipole matrix elements $\mathbf{d}_{ij} = \int d^3r\, \psi_i^*(\mathbf{r}) e\mathbf{r} \psi_j(\mathbf{r})$. Show that the result (5.8.11) is independent of any such phase factors.

The polarization correlations can be understood as a consequence of the algebra of the non-commuting operators σ_{ij} or of the rules for combining probability amplitudes. Of course either perspective brings into clear focus the differences between classical and quantum theories. Is one perspective "better" or "more fundamental" than the other? Dirac felt so: "The question arises whether the non-commutation is really the main new idea of quantum mechanics. Previously I always thought it was but recently I have begun to doubt it and to think that maybe from the physical point of view, the non-commutation is not the only important idea and there is perhaps some deeper idea, some deeper change in our ordinary concepts which is brought about by quantum mechanics . . . I believe [the] concept of the probability amplitude is perhaps the most fundamental concept of quantum theory . . . The immediate effect of the existence of these probability amplitudes is to give rise to interference phenomena. If some process can take place in various ways, by various channels, as people say, what we must do is to calculate the probability amplitude for each of these channels. Then add all the probability amplitudes, and only after we have done this addition do we form the square of the modulus and get the total result for the probability taking place. You see that that result is quite different from what we should have if we had taken the square of the modulus of the individual terms referring to the various channels. It is this difference which

---

[43] R. P. Feynman, *Theory of Fundamental Processes*, W. A. Benjamin, New York, 1961.

gives rise to the phenomenon of interference, which is all pervading in the atomic world."[44] Feynman, similarly, felt that "far more fundamental" than non-commuting operators for the difference between classical and quantum theories was "the discovery that in nature the laws of combining probabilities were *not* those of the classical probability theory of Laplace."[45]

As emphasized in Section 2.9, in connection with Dirac's remark about "photon interference," interference occurs not between photons per se, but between probability amplitudes. Shorthand statements such as "a single photon cannot be split by a beam splitter," or "the two photons emitted in a $J = 0 \rightarrow 1 \rightarrow 0$ atomic cascade transition tend to be emitted in the same direction or in opposite directions," must always be understood to refer to probability amplitudes and not to hypothetical "buckshot" particles of light having a physical significance independent of observation.

The two photons in the $J = 0 \rightarrow 1 \rightarrow 0$ cascade tend to be emitted in the same direction or in opposite directions. To see why, consider the probability amplitude for the transition $4 \rightarrow 1$ with the emission of a photon polarized along $\mathbf{e}_1$ followed by the emission of a photon polarized along $\mathbf{e}_2$:

$$A_{14} \propto \sum_M \langle 1 | \mathbf{d} \cdot \mathbf{e}_2 | J = 1, M \rangle \langle J = 1, M | \mathbf{d} \cdot \mathbf{e}_1 | 4 \rangle. \tag{5.8.14}$$

Only $J = 1$ intermediate states contribute, so the intermediate states effectively make a complete set; therefore,

$$A_{14} \propto \langle 1 | \mathbf{d} \cdot \mathbf{e}_2 \mathbf{d} \cdot \mathbf{e}_1 | 4 \rangle \propto \langle 1 | \mathbf{d}^2 | 4 \rangle \mathbf{e}_1 \cdot \mathbf{e}_2, \tag{5.8.15}$$

$$|A_{14}|^2 \propto |\mathbf{e}_1 \cdot \mathbf{e}_2|^2 = \sum_{i,j} e_{1i} e_{1j} e_{2i} e_{2j}, \tag{5.8.16}$$

since the initial and final states are both spherically symmetric ($J = 0$). Next, we sum over the different possible polarizations λ of the ω_A photons and the different possible polarizations λ' of the ω_B photons:

$$\sum_{\lambda,\lambda'} |A_{14}|^2 = \sum_{i,j} \sum_\lambda e_{1\lambda i} e_{1\lambda j} \sum_{\lambda'} e_{2\lambda' i} e_{2\lambda' j}, \tag{5.8.17}$$

and recall (3.10.5):

$$\sum_\lambda e_{\mathbf{k}\lambda i} e_{\mathbf{k}\lambda j} = \delta_{ij} - k_i k_j / k^2. \tag{5.8.18}$$

The probability that the two photons are detected with an angle θ between their $\mathbf{k}$ vectors is therefore

$$\sum_{\lambda,\lambda'} |A_{14}|^2 \propto \sum_{i,j} \left(\delta_{ij} - \frac{k_{1i} k_{1j}}{k_1^2} \right) \left(\delta_{ij} - \frac{k_{2i} k_{2j}}{k_2^2} \right) = 1 + \frac{(\mathbf{k}_1 \cdot \mathbf{k}_2)^2}{k_1^2 k_2^2} = 1 + \cos^2 \theta. \tag{5.8.19}$$

[44] P. A. M. Dirac, *Fields Quanta* **3**, 139 (1972), p. 154.

[45] R. P. Feynman, in *Proceedings of the Second Berkeley Symposium on Mathematical Statistics and Probability*, ed. J. Neyman, University of California Press, Berkeley, 1951, p. 533.

Exercise 5.5: (a) Discuss the implications of (5.8.19) for the conservation of angular momentum. (b) Does the result (5.8.19) depend on the assumption that recoil of the atom is negligible?

We calculated the polarization correlations using the Heisenberg picture, but they can be deduced more directly as follows. Since $J = 0$ for both the initial and the final atomic states, the total angular momentum of the two photons must be zero. If we restrict consideration to propagation in opposite ($\pm z$) directions, there is no orbital angular momentum involved, so the total spin angular momentum of the two photons must be zero. And since there is a change in parity in the $J = 0 \rightarrow J = 1$ and the $J = 1 \rightarrow J = 0$ electric dipole transitions, the resulting two-photon state must have even parity, since parity is conserved (in electromagnetism). It must therefore have the form

$$|\psi\rangle = \frac{1}{\sqrt{2}}\big(|R_A\rangle|R_B\rangle + |L_A\rangle|L_B\rangle\big), \tag{5.8.20}$$

where R and L denote right- and left-hand circular polarization, respectively.[46] We can also express the two-photon state in a linear polarization basis: since

$$|R_A\rangle = \frac{1}{\sqrt{2}}\big(|x_A\rangle - i|y_A\rangle\big),$$

$$|R_B\rangle = \frac{1}{\sqrt{2}}\big(|x_B\rangle + i|y_B\rangle\big),$$

$$|L_A\rangle = \frac{1}{\sqrt{2}}\big(|x_A\rangle + i|y_A\rangle\big),$$

$$|L_B\rangle = \frac{1}{\sqrt{2}}\big(|x_B\rangle - i|y_B\rangle\big), \tag{5.8.21}$$

we can express the two-photon state equivalently as[47]

$$|\psi\rangle = \frac{1}{\sqrt{2}}\big(|x_A\rangle|x_B\rangle + |y_A\rangle|y_B\rangle\big). \tag{5.8.22}$$

Then, the probability of detecting an A photon when the polarizer at the right in Figure 5.6 is oriented to pass photons polarized along $\hat{\mathbf{x}}$ ($\hat{\mathbf{y}}$) with probability $\cos^2\phi_A$ ($\sin^2\phi_A$) *and* detecting a B photon when the polarizer at the left in Figure 5.6 is

---

[46] Helicity is reversed under spatial inversion, so that a state in which both photons have positive (negative) helicity is transformed under spatial inversion into a state in which both photons have negative (positive) helicity. The state (5.8.20) therefore has even parity.

[47] Recall the definitions given in (3.3.23) and that, in right- and left-circular polarization (in the convention we follow), the electric field vector rotates, respectively, clockwise and counter-clockwise, *as seen by a receiver*.

oriented to pass photons polarized along $\hat{\mathbf{x}}$ ($\hat{\mathbf{y}}$) with probability $\cos^2 \phi_B$ ($\sin^2 \phi_B$) is proportional to $(\cos \phi_A \cos \phi_B + \sin \phi_A \sin \phi_B)^2 = \cos^2(\phi_A - \phi_B)$, as was obtained in the Heisenberg picture.

This $\cos^2(\phi_A - \phi_B)$ polarization correlation of photons from a $J = 0 \rightarrow 1 \rightarrow 0$ atomic cascade was observed by Kocher and Commins, who noted that the experiment "is of interest as an example of a well-known problem in the quantum theory of measurement, first described by Einstein, Podolsky, and Rosen."[48]

Polarization correlations have been of interest in other physical systems. For example, Wheeler proposed a polarization correlation experiment to test the prediction that the γ-ray photons from the annihilation of singlet positronium have orthogonal linear polarizations, and Yang suggested that a measurement of polarization correlations could determine the symmetry properties of mesons that decay with the emission of two photons.[49] Wu and Shaknov were able to conclude from polarization correlations of 0.5-MeV γ photons, which were inferred from Compton scatterings, that the parity of the positronium ground state is negative.[50] For what follows, it is noteworthy that Bohm and Aharonov concluded from the data of Wu and Shaknov that the experiment "is explained adequately by the current quantum theory which implies distant correlations, of the type leading to the paradox of [Einstein, Podolsky, and Rosen], but not by any reasonable hypotheses implying a breakdown of the quantum theory that could avoid the paradox . . ."[51] Expressed in more contemporary parlance, what they concluded was that the two photons were created in an entangled state of polarization.

5.9 Entanglement

Quantum theory describes physical systems probabilistically, but its rules for calculating these probabilities are "*not* those of the classical probability theory of Laplace."[52] In the Heisenberg picture, this difference from classical theory is reflected in part in operator non-commutativity. In the Schrödinger or interaction pictures, it is reflected in the central importance of probability amplitudes. Superposition states like (5.8.22), which cannot be expressed as direct products of state vectors, capture the essence of this difference between classical and quantum theories. For such *entangled* states, different parts of the system are correlated in ways that seem to conflict with commonly held notions of objective "reality," as first discussed by Einstein, Podolsky, and Rosen.[53] Schrödinger introduced the term "entanglement" at the beginning of a paper entitled "Discussion of Probability Relations between Separated Systems," which seems to have been strongly influenced by the EPR paper:

When two systems, of which we know the states by their respective representatives, enter into temporary physical interaction due to known forces between them, and when after a

[48] C. A. Kocher and E. D. Commins, Phys. Rev. Lett. **18**, 575 (1967), p. 575.

[49] J. A. Wheeler, Ann. N. Y. Acad. Sci. **48**, 219 (1946); C. N. Yang, Phys. Rev. **77**, 242 (1950).

[50] C. S. Wu and I. Shaknov, Phys. Rev. **77**, 136 (1950).

[51] D. Bohm and Y. Aharonov, Phys. Rev. **108**, 1070 (1957), p. 1075. See also the remarks in J. F. Clauser and A. Shimony, Rep. Prog. Phys. **41**, 1881 (1978), and references therein.

[52] R. P. Feynman, in *Proceedings of the Second Berkeley Symposium on Mathematical Statistics and Probability*, ed. J. Neyman, University of California Press, Berkeley, 1951, p. 533.

[53] A. Einstein, B. Podolsky, and N. Rosen, Phys. Rev. **47**, 777 (1935).

time of mutual influence the systems separate again, then they can no longer be described in the same way as before, viz. by endowing each of them with a representative of its own. I would not call that *one* but rather *the* characteristic trait of quantum mechanics, the one that enforces its entire departure from classical lines of thought. By the interaction the two representatives (or ψ-functions) have become entangled.[54]

Einstein, Podolsky, and Rosen carefully defined what they meant by "reality" in the context of a physical theory: "If, without in any way disturbing a system, we can predict with certainty (that is, with probability equal to unity) the value of a physical quantity, then there exists an element of physical reality corresponding to this physical quantity."[55] They considered specifically a two-particle system prepared somehow in a state $|\psi\rangle$ that is an eigenstate of *both* $x_1 - x_2$ and $p_1 + p_2$, where x_j and p_j denote, respectively, the coordinate and momentum of Particle j along one spatial dimension (x); such an eigenstate of both $x_1 - x_2$ and $p_1 + p_2$ is certainly allowed, since $[x_1 - x_2, p_1 + p_2] = 0$. Now, if the momentum of Particle 1 is measured and found to have the numerical value P_1, the momentum of Particle 2 must be $-P_1$. But we could then measure the position of Particle 1, and if we find it to be X_1, the position of Particle 2 must be $X_1 + X_0$, where X_0 is the eigenvalue of $x_1 - x_2$ in the state $|\psi\rangle$. Thus, "without in any way disturbing" Particle 2, we can predict "with certainty" both its position and its momentum. The position and momentum of Particle 2 are therefore both "elements of physical reality." But, according to quantum theory, the position and momentum of Particle 2 cannot both have definite values because $[x_2, p_2] \neq 0$. Einstein, Podolsky, and Rosen concluded that quantum mechanics cannot be a *complete* theory because, according to them, a necessary condition for the completeness of a physical theory is that "every element of the physical reality must have a counterpart in the physical theory."[56]

Although quantum theory is not "complete" in the sense meant by Einstein, Podolsky, and Rosen, the EPR "paradox" does not imply any inconsistency within quantum theory. A measurement of the momentum of Particle 1 allows us to infer a definite value for the momentum of Particle 2, and a measurement of the position of Particle 1 then allows us to infer a definite value for the position of Particle 2. But that second measurement must result in an uncertainty in the momentum of Particle 1 *and therefore an uncertainty in the momentum of Particle 2*, because the position and momentum states of the two particles are entangled. This is not to say that this entanglement is not odd. The two particles in the EPR example could be arbitrarily far apart and feeling no interaction, and yet a measurement on one particle immediately affects the state of the other particle—or at least our state of knowledge about that state—by what Einstein called "spooky actions at a distance."[57]

[54] E. Schrödinger, Proc. Camb. Phil. Soc. **31**, 555 (1935); see also E. Schrödinger, Proc. Camb. Phil. Soc.**32**, 446 (1936).

[55] A. Einstein, B. Podolsky, and N. Rosen, Phys. Rev. **47**, 777 (1935), p. 777.

[56] Ibid.

[57] A. Einstein, letter to Max Born, 3 March 1947, translated and reprinted in *The Born–Einstein Letters*, Walker and Company, New York, 1971, p. 158.

However spooky, this consequence of entanglement cannot be used to transmit information. Suppose the two-photon entangled state (5.8.22) has been created by an atomic cascade transition and that observers Alice and Bob can make measurements of the fields of frequencies ω_A and ω_B, respectively, using the filters, polarizers, and photodetectors shown in Figure 5.6. Alice and Bob could agree beforehand that Alice will send a "0" or "1" message to Bob, depending on whether she measures x or y polarization. But if Bob detects an x-polarized photon, he can't be sure that detection of a y-polarized photon had not been possible, so he can't be sure of what Alice measured. This impossibility of (instantaneous) transmission of information can be shown more formally. Consider the density operator for the state (5.8.22):

$$\rho = |\psi\rangle\langle\psi| = \frac{1}{2}\big(|x_A\rangle\langle x_B| \otimes |x_A\rangle\langle x_B| + |x_A\rangle\langle x_B| \otimes |y_A\rangle\langle y_B|$$
$$+ |y_A\rangle\langle y_B| \otimes |x_A\rangle\langle x_B| + |y_A\rangle\langle y_B| \otimes |y_A\rangle\langle y_B|\big). \tag{5.9.1}$$

The reduced density matrix obtained by tracing over Alice's states is

$$\rho_B = \text{Tr}_A\rho = \frac{1}{2}\big(|x_B\rangle\langle x_B| + |y_B\rangle\langle y_B|\big) \tag{5.9.2}$$

in the linear polarization basis, and

$$\rho_B = \text{Tr}_A\rho = \frac{1}{2}\big(|R_B\rangle\langle R_B| + |L_B\rangle\langle L_B|\big) \tag{5.9.3}$$

in the circular polarization basis. These, of course, are just two different ways of expressing the same reduced density matrix: if Bob measures linear polarization, he is equally likely to find x or y polarization, and if he measures circular polarization he is equally likely to find left or right circular polarization, regardless of what Alice does. So, Alice's choice does not grant her the freedom to transmit information to Bob. And, given a single photon, Bob cannot distinguish between linear and circular polarization—there is no device that can determine the polarization parameters of a *single photon*. If that were possible, instantaneous communication would be too.

But if Bob has a field with $N \gg 1$ photons to work with, he can pass it through polarization-dependent beam splitters and thereby accurately determine whether a measurement of linear or circular polarization was made by Alice. If Bob performs his measurement by first passing his photon through an amplifier to generate a field with $N \gg 1$ photons in the same polarization state, therefore, he could decide with high confidence whether Alice measured linear or circular polarization; Alice could send him a message "superluminally."[58] This scheme fails, however, because the amplification process will inevitably result in spontaneously emitted photons with polarizations different from the incident photon; we cannot have stimulated emission without spontaneous emission and therefore cannot "clone" a single photon of *arbitrary* (a priori

[58] N. Herbert, Found. Phys. **12**, 1171 (1982).

unknown) polarization.[59] In fact, it is impossible to clone *any* arbitrary quantum state.[60]

The "no-cloning theorem" can be proved as follows.[61] Imagine a device that clones a polarization state $|s\rangle$ of a single photon by amplifying it to a state $|ss\rangle$ with two photons in exactly the same polarization state. Thus, $|A_0\rangle|s\rangle \to |A_s\rangle|ss\rangle$, where $|A_0\rangle$ and $|A_s\rangle$ are, respectively, the initial and final states of the amplifier. In particular, for x- and y-polarized photons,

$$|A_0\rangle|x\rangle \to |A_x\rangle|xx\rangle, \quad \text{and} \quad |A_0\rangle|y\rangle \to |A_y\rangle|yy\rangle, \tag{5.9.4}$$

and, since transformations in quantum theory are linear,

$$|A_0\rangle\big(\alpha|x\rangle + \beta|y\rangle\big) \to \alpha|A_x\rangle|xx\rangle + \beta|A_y\rangle|yy\rangle \tag{5.9.5}$$

for an arbitrary superposition state $\alpha|x\rangle + \beta|y\rangle$ of the original photon. But our imaginary cloning device should, by definition, produce the state in which both photons are described by the state $\alpha|x\rangle + \beta|y\rangle$, which contradicts (5.9.5), regardless of the possible final states $|A_x\rangle$ and $|A_y\rangle$ of the device. The linearity of quantum theory precludes the replication of an arbitrary quantum state. A single quantum cannot be cloned.

The no-cloning theorem must be dealt with in quantum computation, since it implies that information cannot be copied from one memory register to another with perfect fidelity. Imagine a single-photon field incident on an amplifier that ideally amplifies only single-frequency fields. Stimulated emission and spontaneous emission are equally likely. (Recall, for instance, (3.6.54).) Stimulated emission will result in two photons having the same polarization as the incident photon, whereas spontaneous emission is equally likely to result in a photon found to have the same polarization or an orthogonal polarization. There is, therefore, a probability 2/3 that the additional photon will be found to have the same probability as the incident photon, and a probability of 1/3 that it will be found to have the orthogonal polarization. The cloning fidelity, the probability that the outgoing photon is found to be in the same state as the incident photon, is, therefore, $(2/3)(1)+(1/3)(1/2) = 5/6$. This turns out to be the maximum possible cloning fidelity.[62] In fact, it is the maximum cloning fidelity that can be obtained without violating the Einstein causality condition that information cannot be transmitted faster than the speed of light in vacuum.[63]

[59] P. W. Milonni and M. L. Hardies, Phys. Lett. A **92**, 321 (1982); L. Mandel, Nature **304**, 188 (1983). The amplification could occur in a nonlinear process that does not involve "ordinary" stimulated emission by excited atoms.

[60] W. K. Wooters and W. H. Zurek, Nature **299**, 802 (1982); D. Dieks, Phys. Lett. A **92**, 371 (1982). A proof of the impossibility of cloning quantum states also appears in an earlier paper by J. L. Park, Found. Phys. **1**, 23 (1970). See also J. Ortigoso, Am. J. Phys. **86**, 201 (2018).

[61] Ibid.

[62] N. Gisin and S. Massar, Phys. Rev. Lett. **79**, 2153 (1997).

[63] N. Gisin, Phys. Lett. A **242**, 1 (1998).

Some states are, of course, more entangled than others. For example, the state

$$|\psi\rangle = \cos\alpha|x_A\rangle|x_B\rangle + \sin\alpha|y_A\rangle|y_B\rangle \tag{5.9.6}$$

with $|\alpha|^2 \ll 1$ is obviously less entangled than the state (5.8.22). One measure of entanglement is *entanglement entropy*, which is the von Neumann entropy (see (3.7.12)) of a *reduced* density matrix. Thus, the entanglement entropy of subsystem B with respect to subsystem A for the system with density matrix (5.9.1) is

$$S_B \equiv -\text{Tr}\big(\rho_B \log \rho_B\big) = -\sum_i p_i(B) \log p_i(B), \tag{5.9.7}$$

where ρ_B is the reduced density operator (5.9.2) and the $p_i(B)$ are the eigenvalues of ρ_B, expressed as a 2×2 matrix. These eigenvalues are both $1/2$, and so

$$S_B = -\frac{1}{2}\log\left(\frac{1}{2}\right) - \frac{1}{2}\log\left(\frac{1}{2}\right) = \log 2. \tag{5.9.8}$$

Such a non-zero value is the signature of entanglement. For the state (5.9.6) with $\alpha = 0$, a non-entangled state, the eigenvalues of ρ_B are 0 and 1, and, therefore, $S_B = 0$. In contrast, the value $\log 2$ for the state with $\alpha = \pi/4$, a *Bell state*, is maximally entangled (see Exercise 5.6). More generally, if the subsystems A and B each had $\mathcal{N}$ qubits ("quantum bits") instead of 2, that is, $2^{\mathcal{N}}$ states instead of 2, S_A would be $\mathcal{N}\log 2$. 2^{S_A} can therefore be interpreted as the number of entangled states.

Any pure state in a Hilbert space $\mathcal{H}_A \otimes \mathcal{H}_B$ of two systems A and B can be expressed as

$$|\psi\rangle = \sum_i \sum_j c_{ij}|\phi_i^A\rangle|\phi_j^B\rangle, \tag{5.9.9}$$

where the $|\phi_i^A\rangle$ and $|\phi_i^B\rangle$ are basis states in the Hilbert spaces $\mathcal{H}_A$ and $\mathcal{H}_B$ of the subsystems A and B, respectively. The *Schmidt decomposition theorem* says that, for any such pure state, there are orthonormal states $|i_A\rangle$ and $|i_B\rangle$ of the A and B subsystems such that

$$|\psi\rangle = \sum_i \sqrt{\lambda_i}|i_A\rangle|i_B\rangle, \quad 0 \le \lambda_i \le 1, \tag{5.9.10}$$

where the (real) Schmidt coefficients λ_i are unique to the state $|\psi\rangle$ and

$$\sum_i \lambda_i = 1, \tag{5.9.11}$$

so that $\langle\psi|\psi\rangle| = 1$. The number of terms in the summation is limited by the dimension of the smaller of the Hilbert spaces $\mathcal{H}_A$ and $\mathcal{H}_B$. The Schmidt decomposition provides a necessary and sufficient condition for entanglement: if two or more of the λ_i are non-vanishing, $|\psi\rangle$ is an entangled state. In the Schmidt basis, the reduced density matrices are

$$\rho_A = \sum_i \lambda_i |i_A\rangle\langle i_A|, \quad \text{and} \quad \rho_B = \sum_i \lambda_i |i_B\rangle\langle i_B|, \tag{5.9.12}$$

and so the λ_i are just the eigenvalues of the reduced density matrix ρ_A (or ρ_B). Similarly, the Schmidt basis vectors are the eigenvectors of the reduced density matrices. From (5.9.12) it is seen that $S_A = S_B$, which is a general property of pure states.

Another concept relating to entanglement is *concurrence*.[64] For pure states, the concurrence is

$$C \equiv \sqrt{2\left[1 - \text{Tr}(\rho_A^2)\right]}. \tag{5.9.13}$$

Equation (5.9.12) implies the relation

$$C = \sqrt{2\lambda_1\lambda_2} \tag{5.9.14}$$

between C and the Schmidt coefficients. A concurrence greater than 0 signifies entanglement. For the (maximally entangled) Bell state, for example, $C = 1$.

Following these perfunctory introductions to a few measures of entanglement, we turn now to aspects of entanglement that are closely related to the "paradox" of Einstein, Podolsky, and Rosen.

Exercise 5.6: (a) Compute the entanglement entropy, the Schmidt coefficients λ_1 and λ_2, and the concurrence C for the state (5.9.6). (b) Verify (5.9.14).

5.9.1 Hidden Variables and Von Neumann's Proof

As discussed earlier, Einstein, Podolsky, and Rosen raised a philosophical objection to quantum theory: there are two-particle quantum states such that the momentum and position of Particle 2 can be predicted with certainty, without measurements on Particle 2, by measurements made on Particle 1. They recognized, of course, that this is not inconsistent with quantum theory, since the position and momentum of Particle 2 are not both together being predicted with certainty. But since the position or momentum of Particle 2 *could* be predicted with certainty, Einstein, Podolsky, and Rosen regarded them as coexisting, objective realities. According to quantum theory, they cannot both have definite values, and therefore Einstein, Podolsky, and Rosen assert that the theory is "incomplete." Others have suggested modifications to quantum theory in which unknown (and perhaps unknowable) "hidden variables" determine precise, dispersionless values of all measurable physical quantities.

[64] W. K. Wooters, Phys. Rev. Lett. **80**, 2245 (1998).

In his book on the mathematical foundations of quantum theory in the form of "a theory of Hilbert space and the so-called Hermitian operators," von Neumann addresses the question of whether a hidden-variable theory could be consistent with the representation of "physical quantities" (position, momentum, energy, etc.) by operators in a Hilbert space.[65] He does not prove that hidden-variable theories are impossible; he proves that they must be "incompatible with certain qualitative fundamental postulates of quantum mechanics," which is "the only formal theory existing at the present time which orders and summarizes our experiences . . . in a half-way satisfactory manner."[66]

Von Neumann arrives at this conclusion by first deducing the existence of a density operator from a relatively small number of assumptions, independently of any consideration of hidden variables. It is assumed that there is a one-to-one correspondence between each physical quantity $\mathcal{R}$ and a Hermitian Hilbert-space operator R and that, (1) if $\mathcal{R} \geq 0$, its expectation value $\langle \mathcal{R} \rangle \geq 0$ and (2) if $\mathcal{R}, \mathcal{S}, \ldots$ are arbitrary physical quantities and $a, b, \ldots$ are real numbers, then

$$\langle a\mathcal{R} + b\mathcal{S} + \ldots \rangle = a\langle \mathcal{R} \rangle + b\langle \mathcal{S} + \ldots \rangle; \tag{5.9.15}$$

in addition, (3) if $\mathcal{R}$ "has the operator" R, the quantity $f(\mathcal{R})$ has the operator $f(R)$ and (4) if the physical quantities $\mathcal{R}, \mathcal{S}, \ldots$ have the operators $R, S, \ldots$, then the quantity $\mathcal{R} + \mathcal{S} + \ldots$ has the operator $R + S + \ldots$ Von Neumann does not *assume* that the expectation value of a physical quantity $\mathcal{R}$ in a quantum state $|\phi\rangle$ is $\langle \phi | R | \phi \rangle$, or, more generally, the trace of UR, where U is a density operator; this follows naturally from his assumptions, as we now show.[67]

The Hermitian operator R corresponding to the physical quantity $\mathcal{R}$ is expressed in terms of a complete set of states $|n\rangle$, as in (2.3.11):

$$R = \sum_{m,n} R_{nm} |n\rangle\langle m| = \sum_{m,n} R_{nm} \sigma_{nm}$$

$$= \sum_{n} R_{nn} \sigma_{nn} + \sum_{m,n;m<n} \mathrm{Re}[R_{mn}] u_{nm} + \sum_{m,n;m<n} \mathrm{Im}[R_{mn}] v_{nm}, \tag{5.9.16}$$

$$u_{nm} \equiv \sigma_{nm} + \sigma_{mn}, \quad v_{nm} \equiv -i[\sigma_{nm} - \sigma_{mn}]. \tag{5.9.17}$$

We have used the assumption that R is Hermitian: $R_{mn} = R_{nm}^{*}$. From assumptions (2) and (4),

[65] J. von Neumann, *Mathematical Foundations of Quantum Mechanics*, Princeton University Press, Princeton, 1955, English translation of the 1932 German edition by R. T. Beyer, p. viii.

[66] Ibid., p. 327.

[67] Our summary of von Neumann's analysis ignores technical details but, aside from notation, it adheres fairly closely to his Section 4.2 entitled "Proof of the Statistical Formulas." We follow von Neumann in writing $\mathcal{R}$ for a measurable physical quantity, R for the corresponding Hermitian operator, and U for the density operator.

$$\langle \mathcal{R} \rangle = \sum_n R_{nn} \langle \sigma_{nn} \rangle + \sum_{m,n;m<n} \mathrm{Re}[R_{mn}] \langle u_{nm} \rangle + \sum_{m,n;m<n} \mathrm{Im}[R_{mn}] \langle v_{nm} \rangle$$

$$= \sum_{m,n} U_{nm} R_{mn} = \sum_n (UR)_{nn} = \mathrm{Tr}(UR), \tag{5.9.18}$$

where the matrix elements $U_{mn} = \langle m|U|n \rangle$ are

$$U_{nn} = \langle \sigma_{nn} \rangle, \tag{5.9.19}$$

$$U_{mn} = \frac{1}{2}\big(\langle u_{mn} \rangle + i \langle v_{mn} \rangle\big), \quad m < n, \tag{5.9.20}$$

$$U_{nm} = \frac{1}{2}\big(\langle u_{mn} \rangle - i \langle v_{mn} \rangle\big), \quad m < n. \tag{5.9.21}$$

Thus, the expectation value of any physical quantity $\mathcal{R}$ is $\mathrm{Tr}(UR)$, where U is a Hermitian operator, independent of R, that characterizes the ensemble over which expectation values are taken. Assumption (1) implies that U is a "definite" operator, that is, $\langle \phi|U|\phi \rangle \geq 0$ for any (normalized) state vector $|\phi\rangle$. To prove this, take R to be the projection operator $P_\phi = |\phi\rangle\langle\phi|$; since $P_\phi = P_\phi^2 \geq 0$, $\mathrm{Tr}(UP_\phi) = \langle \phi|U|\phi \rangle \geq 0$.

Aside from the assumptions stated above and other considerations, von Neumann asks the reader to "forget the whole of quantum mechanics," and yet, as just shown, from these assumptions, he is able to deduce the existence of a density operator U and the fundamental *statistical* postulate $\langle \mathcal{R} \rangle = \mathrm{Tr}(UR)$ of quantum theory.[68] He comments several times on the assumption (5.9.15) in the case that $\mathcal{R}$ and $\mathcal{S}$ are not "simultaneously measurable," which he shows to be the case when R and S do not commute. If R and S commute, we can measure $\mathcal{R}$ and $\mathcal{S}$ simultaneously and obtain $\mathcal{R} + \mathcal{S}$ by simple addition. But if $[R, S] \neq 0$, $\mathcal{R} + \mathcal{S}$ can only be obtained by a separate measurement; as he explains on pages 309–310 of his book,

For example, the energy operator . . . of an electron moving in a potential field $V(x, y, z)$,

$$H_0 = \frac{(p^x)^2 + (p^y)^2 + (p^z)^2}{2m} + V(Q^x, Q^y, Q^z),$$

. . . . is a sum of two non-commuting operators

$$R = \frac{(p^x)^2 + (p^y)^2 + (p^z)^2}{2m}, \quad S = V(Q^x, Q^y, Q^z).$$

While the measurement of the quantity $\mathcal{R}$ belonging to R is a momentum measurement, and that of the quantity $\mathcal{S}$ belonging to S is a coordinate measurement, we measure the quantity $\mathcal{R} + \mathcal{S}$ belonging to $H_0 = R + S$ in an entirely different way: for example, by the measurement of the frequency of the spectral lines emitted by this (bound) electron, since these lines determine (by reason of the Bohr frequency relation) the energy levels, that is, the $\mathcal{R} + \mathcal{S}$ values. Nevertheless, under all circumstances,

$$\langle \mathcal{R} + \mathcal{S} \rangle = \langle \mathcal{R} \rangle + \langle \mathcal{S} \rangle.$$

[68] J. von Neumann, *Mathematical Foundations of Quantum Mechanics*, Princeton University Press, Princeton, 1955, English translation of the 1932 German edition by R. T. Beyer, p. 297.

Von Neumann regards the sum $\mathcal{R} + \mathcal{S}$ in the case that $[R, S] \neq 0$ as defined "only in an implicit way" by the fact that $\langle \mathcal{R} + \mathcal{S} \rangle = \langle \mathcal{R} \rangle + \langle \mathcal{S} \rangle$. $\langle \mathcal{R} \rangle$ and $\langle \mathcal{S} \rangle$ could be measured separately on ensembles of identically prepared systems and the results added to give $\langle \mathcal{R} + \mathcal{S} \rangle$. This may be taken as an operational definition of $\langle \mathcal{R} + \mathcal{S} \rangle$ when $[R, S] \neq 0$. *It is not required or assumed that $\mathcal{R} + \mathcal{S}$ has a value equal to the sum of the values of $\mathcal{R}$ and $\mathcal{S}$ when $[R, S] \neq 0$.* We return to this point below.

Based on the relation $\langle \mathcal{R} \rangle = \mathrm{Tr}(UR)$ for an ensemble characterized by the density operator U, von Neumann addresses the question of whether there are *dispersion-free* ensembles, that is, ensembles for which the variance

$$\langle \mathcal{R}^2 \rangle - \langle \mathcal{R} \rangle^2 = 0, \tag{5.9.22}$$

or

$$\mathrm{Tr}(UR^2) = \left[\mathrm{Tr}(UR) \right]^2, \tag{5.9.23}$$

for all Hermitian operators R. Consider the Hermitian operator $R = |\phi\rangle\langle\phi|$, $\langle\phi|\phi\rangle = 1$. From (5.9.23), $\langle\phi|U|\phi\rangle = \langle\phi|U|\phi\rangle^2$, or $\langle\phi|U|\phi\rangle = 0$ or 1. This must hold for all $|\phi\rangle$, and therefore U itself must be 0 or 1. $U = 0$ can be disregarded, since it would imply $\langle \mathcal{R} \rangle = 0$ for all $\mathcal{R}$. If $U = 1$, (5.9.23) gives

$$\langle (\mathcal{R} - \langle \mathcal{R} \rangle)^2 \rangle = \langle \mathcal{R}^2 \rangle - 2\langle \mathcal{R} \rangle^2 + \langle \mathcal{R} \rangle^2 \mathrm{Tr}(U) \neq 0, \tag{5.9.24}$$

and so, no matter how an ensemble characterized by U is chosen, not all physical quantities can have zero variance (dispersion). *There are no dispersion-free ensembles.*

What does this mean for hidden variables? If there were hidden variables, expectation values of a quantum system would actually be averages over ensembles of systems with different values of the hidden (unknown) variables. Each "sub-ensemble" would be characterized by fixed numerical values of all the hidden variables, that is, it would be dispersion-free. But von Neumann proved that there can be no dispersion-free ensembles, and therefore there can be no hidden variables consistent with his assumptions, starting from the most basic one that there is a one-to-one correspondence between physical quantities and Hermitian operators.

In summary, von Neumann considers a theoretical framework in which there is a one-to-one correspondence between each physical quantity $\mathcal{R}$ of a system and a Hermitian operator R. From assumptions (1)–(4), he proves that the expectation value $\langle \mathcal{R} \rangle$ in any such theory can be expressed (as in quantum theory) as $\mathrm{Tr}(UR)$, where U is a Hermitian density operator characterizing the statistical ensemble. He then proves that there can be no dispersion-free ensembles and therefore that it is not possible to construct a hidden-variable theory that is compatible with quantum theory, "the only formal theory existing at the present time which orders and summarizes our experiences . . . in a half-way satisfactory manner."[69]

[69] Ibid., p. 327.

Von Neumann's proof was sharply criticized by Bell, and many others have followed him in judging von Neumann's proof to be invalid or even, as Bell said, "foolish."[70] The criticism centers on von Neumann's assumption that $\langle \mathcal{R} + \mathcal{S} \rangle = \langle \mathcal{R} \rangle + \langle \mathcal{S} \rangle$.[71] Suppose R and S do not commute. In a dispersion-free state, a measurement of $\mathcal{R}$ results in a value r that must, according to quantum theory, be an eigenvalue of R, and a separate measurement of $\mathcal{S}$ results in a value s that must be an eigenvalue of S. But a measurement of $\mathcal{R} + \mathcal{S}$ will result in an eigenvalue of $R + S$ that differs from $r + s$.[72] It is true that the ensemble averages add up as $\langle R + S \rangle = \langle R \rangle + \langle S \rangle$, but this will not be the case in a single measurement. Now, if there were a dispersion-free state, every physical quantity $\mathcal{R}$ would have a fixed value equal to one of the eigenvalues of R, and this fixed value *would* be the same as the expectation value $\langle R \rangle$. Then von Neumann's assumption that $\langle \mathcal{R} + \mathcal{S} \rangle = \langle \mathcal{R} \rangle + \langle \mathcal{S} \rangle$ would have the erroneous implication that the eigenvalue of $R + S$ is equal to the sum of the eigenvalues of R and S. Thus, according to Bell and others, "it was the arbitrary assumption of a particular (and impossible) relation" that renders von Neumann's "no-hidden-variables" proof invalid.[73]

There are persuasive arguments that von Neumann's detractors have misunderstood or misconstrued what he actually proved.[74] As the quotation above from his book makes clear, von Neumann defines "only in an implicit way" the sum of two physical quantities whose operators do not commute, and there is no implication that the eigenvalue of $R + S$ is equal to the sum of the eigenvalues of R and S, as the critics assert.[75] $\mathcal{R} + \mathcal{S}$ must be measured "in an entirely different way" from $\mathcal{R}$ and $\mathcal{S}$ when $[R, S] \neq 0$. By defining $R + S$ in this implicit way, von Neumann *excludes* hidden-variable theories such that the numerical value of $\mathcal{R} + \mathcal{S}$ equals the sum $r + s$ of the values of $\mathcal{R}$ and $\mathcal{S}$, making the objection of Bell and others irrelevant.[76] But whatever one thinks of von Neumann's proof, one thing is sure: Bell's conclusion that the proof was erroneous led to Bell's theorem, and renewed interest in the conceptual foundations of quantum theory.

---

[70] J. S. Bell, Rev. Mod. Phys. **38**, 447 (1966); see also J. S. Bell, Found. Phys. **12**, 989 (1982).

[71] This equality of the *expectation value* of a sum and a sum of expectation values is, of course, valid in quantum mechanics.

[72] For example, for a spin-1/2 particle, the eigenvalues of σ_x and σ_y are ± 1, whereas the eigenvalues of $\sigma_x + \sigma_y$ are $\pm\sqrt{2}$.

[73] J. S. Bell, Rev. Mod. Phys. **38**, 447 (1966), p. 449.

[74] J. Bub, Found. Phys. **40**, 1333 (2010); D. Dieks, Stud. Hist. Phil. Mod. Phys. **60** 136 (2017).

[75] From our discussion above, it is seen that, in the derivation of $\langle \mathcal{R} \rangle = \text{Tr}(UR)$, von Neumann did not assume that hidden variables are eigenvalues of operators, nor, for that matter, did he refer to eigenvalues at all. But whether or not hidden variables are eigenvalues is not relevant to the question of the validity of von Neumann's proof.

[76] D. Dieks, Stud. Hist. Phil. Mod. Phys. **60** 136 (2017).

5.9.2   Bell's Theorem

Bohm put the gedankenexperiment of Einstein, Podolsky, and Rosen in sharper focus with a more practicable example of an entangled state: two spin-1/2 particles in the singlet spin state[77]

$$|\psi\rangle = \frac{1}{\sqrt{2}}\left(|+\rangle_1|-\rangle_2 - |-\rangle_1|+\rangle_2\right). \tag{5.9.25}$$

$|\pm\rangle_i$ is the state in which particle i has spin up $(+)$ or down $(-)$ along, say, the z direction. The entangled state (5.9.25) could result from the decay of a system of zero angular momentum into two spin-1/2 particles. Conservation of (zero) angular momentum implies by spherical symmetry that the direction z is arbitrary. The particles are assumed to be far apart and non-interacting.[78]

   If we measure the spin of Particle 1 in the z direction and find spin up, we will certainly measure spin down for Particle 2, and if we find spin down for Particle 1, we will certainly measure spin up for Particle 2. Obviously, we have the same situation here as in the original EPR example: since the particles are far apart and do not interact, measurements made on Particle 1 should not affect measurements on Particle 2. But a measurement of the spin of Particle 1 in the z direction allows us to predict with certainty the spin of Particle 2 in the z direction, which is therefore an "element of reality" as defined by Einstein, Podolsky, and Rosen. Similarly, by making a measurement of the spin of Particle 1 in the x direction, we can predict with certainty the spin of Particle 2 in the x direction, and so it, too, is an element of physical reality. Since both the x and the z components of Particle 2 can be predicted with probability 1, Einstein, Podolsky, and Rosen regard them as objective elements of reality existing independently of whether we observe them. But quantum theory does not allow both the x and the z components of the spin of Particle 2 to have definite values, since $[\sigma_{x2}, \sigma_{z2}] \neq 0$. According to Einstein, Podolsky, and Rosen, quantum theory is therefore not "complete."

   Bell posed the question of whether there could be hidden variables that determine the values of such physical quantities while agreeing with the statistical predictions of quantum theory. Imagine a theory in which the spin of Particle 1 in the direction of any unit vector $\mathbf{a}$ is a function $\frac{1}{2}A(\mathbf{a}, \lambda)$ $(\times \hbar)$ of $\mathbf{a}$ and a hidden variable or set of hidden variables λ. To bring such a theory into agreement with quantum mechanics for any direction $\mathbf{a}$, assume that $A(\mathbf{a}, \lambda)$ has only two possible values:

$$A(\mathbf{a}, \lambda) = \pm 1. \tag{5.9.26}$$

---

[77] D. Bohm, *Quantum Theory*, Prentice-Hall, New York, 1951, Sections 22.15–22.17.

[78] The minus sign between the two product states in (5.9.25) ensures that the eigenvalues of both $(\mathbf{s}_1 + \mathbf{s}_2)^2$ and $s_{1z} + s_{2z}$ are zero, where $\mathbf{s}_i$ is the angular momentum operator with components $(\hbar/2)(\sigma_{xi}, \sigma_{yi}, \sigma_{zi})$. Triplet states, which are symmetric under particle interchange, have $(\mathbf{s}_1 + \mathbf{s}_2)^2 = 2\hbar^2$ and therefore cannot result from the decay of a system of zero angular momentum.

We assume likewise that the spin of Particle 2 in any direction $\mathbf{b}$ is a function $\frac{1}{2}B(\mathbf{b}, \lambda)$ of $\mathbf{b}$ and λ and that this function can have only two possible values for any direction $\mathbf{b}$:

$$B(\mathbf{b}, \lambda) = \pm 1. \tag{5.9.27}$$

A crucial assumption of *locality* has been made here: $A(\mathbf{a}, \lambda)$ is independent of $\mathbf{b}$, and $B(\mathbf{b}, \lambda)$ is independent of $\mathbf{a}$. In other words, a measurement of the spin component of Particle 1 in any direction $\mathbf{a}$ is independent of the direction $\mathbf{b}$ in which the spin of Particle 2 is measured. This is perfectly reasonable, since we have specified that 1 and 2 are arbitrarily far apart and not interacting. The assumptions (5.9.26) and (5.9.27) put the hypothetical hidden-variable theory in agreement with quantum theory to the extent that a measurement of the spin component of either particle in any direction will give $\pm 1/2$ ($\times \hbar$). To ensure further agreement with quantum-mechanical predictions for the entangled state (5.9.25), assume also that

$$A(\mathbf{a}, \lambda) = -B(\mathbf{a}, \lambda), \tag{5.9.28}$$

that is, when the spins of the two particles are measured along the same axis, they will be found to be antiparallel. In order to mimic the probabilistic character of quantum mechanics, it is assumed that λ has a probability distribution $\rho(\lambda)$. Then, the average of the product of the measurements of the spins of Particles 1 and 2 in the $\mathbf{a}$ and $\mathbf{b}$ directions, respectively, is

$$E(\mathbf{a}, \mathbf{b}) \equiv \frac{1}{4} \int d\lambda \rho(\lambda) A(\mathbf{a}, \lambda) B(\mathbf{b}, \lambda) = -\frac{1}{4} \int d\lambda \rho(\lambda) A(\mathbf{a}, \lambda) A(\mathbf{b}, \lambda), \tag{5.9.29}$$

compared to the quantum-mechanical prediction (see Exercise 5.7)

$$E_{QM}(\mathbf{a}, \mathbf{b}) = \langle (\mathbf{s}_1 \cdot \mathbf{a})(\mathbf{s}_2 \cdot \mathbf{b}) \rangle = -\frac{1}{4}\mathbf{a} \cdot \mathbf{b}. \tag{5.9.30}$$

The average (5.9.29) predicted by a hidden-variable theory might conceivably equal the expectation value (5.9.30) predicted by quantum mechanics. But Bell, in what has become known as Bell's theorem, showed that no (local) hidden-variable theory satisfying (5.9.26)–(5.9.28) can be consistent with *every* statistical prediction of quantum mechanics.[79] Consider

$$E(\mathbf{a}, \mathbf{b}) - E(\mathbf{a}, \mathbf{c}) = -\frac{1}{4} \int d\lambda \rho(\lambda) \big[A(\mathbf{a}, \lambda)A(\mathbf{b}, \lambda) - A(\mathbf{a}, \lambda)A(\mathbf{c}, \lambda) \big]$$

$$= -\frac{1}{4} \int d\lambda \rho(\lambda) A(\mathbf{a}, \lambda) A(\mathbf{b}, \lambda) \big[1 - A(\mathbf{a}, \lambda)A(\mathbf{c}, \lambda) \big], \tag{5.9.31}$$

[79] J. S. Bell, Physics **1**, 195 (1964), reprinted in *Quantum Theory and Measurement*, J. A. Wheeler and W. H. Zurek, eds., Princeton University Press, Princeton, 1983, pp. 403–408. Other proofs of Bell's theorem, which do not refer to hidden variables but are always based on classical (positive-definite) probabilities, have appeared since Bell's work. But Bell's original proof arguably remains the most straightforward, and we follow it closely here.

where the second equality follows from $A^2(\mathbf{b}, \lambda) = 1$. Since $|A(\mathbf{a}, \lambda)A(\mathbf{b}, \lambda)| = 1$, and $1 - A(\mathbf{b}, \lambda)A(\mathbf{c}, \lambda)| \geq 0$,

$$|E(\mathbf{a}, \mathbf{b}) - E(\mathbf{a}, \mathbf{c})| = |\frac{1}{4} \int d\lambda \rho(\lambda)A(\mathbf{a}, \lambda)A(\mathbf{b}, \lambda)[1 - A(\mathbf{a}, \lambda)A(\mathbf{c}, \lambda)]|$$

$$\leq \frac{1}{4} \int d\lambda \rho(\lambda)[1 - A(\mathbf{b}, \lambda)A(\mathbf{c}, \lambda)], \tag{5.9.32}$$

or

$$|E(\mathbf{a}, \mathbf{b}) - E(\mathbf{a}, \mathbf{c})| \leq \frac{1}{4} + E(\mathbf{b}, \mathbf{c}). \tag{5.9.33}$$

Suppose, for example, that $\mathbf{a} \cdot \mathbf{b} = \mathbf{b} \cdot \mathbf{c} = \cos \pi/4$, and $\mathbf{a} \cdot \mathbf{c} = 0$. Then,

$$|E_{QM}(\mathbf{a}, \mathbf{b}) - E_{QM}(\mathbf{a}, \mathbf{c})| = |\langle (\mathbf{s}_1 \cdot \mathbf{a})(\mathbf{s}_2 \cdot \mathbf{b}) \rangle - \langle (\mathbf{s}_1 \cdot \mathbf{a})(\mathbf{s}_2 \cdot \mathbf{c}) \rangle|$$

$$= |-\frac{1}{4} \cos \pi/4 - 0| = \frac{1}{4} \frac{1}{\sqrt{2}}, \tag{5.9.34}$$

and

$$\frac{1}{4} + E_{QM}(\mathbf{b}, \mathbf{c}) = \frac{1}{4} + \langle (\mathbf{s}_1 \cdot \mathbf{b})(\mathbf{s}_2 \cdot \mathbf{c}) \rangle = \frac{1}{4} - \frac{1}{4} \frac{1}{\sqrt{2}}. \tag{5.9.35}$$

Since (5.9.34) is *greater than* (5.9.35), the quantum-mechanical expectation values do not satisfy the Bell inequality (5.9.33). In other words, a local hidden-variable theory cannot reproduce all the statistical predictions of quantum theory.

Exercise 5.7: Assuming $\mathbf{s}_1 = -\mathbf{s}_2$, and using the relation

$$\sigma_i \sigma_j = \delta_{ij} + i \sum \epsilon_{ijk} \sigma_k$$

satisfied by the Pauli operators, derive the expectation value given in (5.9.30). (ϵ_{ijk} is the Levi–Civita symbol defined in Section 3.10.)

Bell's theorem opened the possibility that the viability of local hidden-variable theories could be decided experimentally. But the derivation of the inequality (5.9.33) assumes the perfect anti-correlation (5.9.28) when $\mathbf{a} = \mathbf{b}$, which will not be realized in experiments with actual (imperfect) detection systems. A different Bell inequality can be derived without that assumption.[80] Consider two directions $\mathbf{a}, \mathbf{a}'$ for A, and two directions $\mathbf{b}, \mathbf{b}'$ for B. From the identity

$$E(\mathbf{a}, \mathbf{b}) - E(\mathbf{a}, \mathbf{b}') = \int d\lambda \rho(\lambda)A(\mathbf{a}, \lambda)B(\mathbf{b}, \lambda) - A(\mathbf{a}, \lambda)B(\mathbf{b}', \lambda)]$$

$$= \int d\lambda \rho(\lambda)A(\mathbf{a}, \lambda)B(\mathbf{b}, \lambda)[1 \pm A(\mathbf{a}', \lambda)B(\mathbf{b}', \lambda)]$$

$$- \int d\lambda \rho(\lambda)A(\mathbf{a}, \lambda)B(\mathbf{b}', \lambda)[1 \pm A(\mathbf{a}', \lambda)B(\mathbf{b}, \lambda)] \tag{5.9.36}$$

[80] J. F. Clauser, M. A. Horne, A. Shimony, and R. A. Holt, Phys. Rev. Lett. **23**, 880 (1969); J. S. Bell, in *Foundations of Quantum Mechanics*, ed. B. d'Espagnat, Academic Press, New York, 1971.

and

$$|A(\mathbf{a}, \lambda)| \leq 1, \quad |B(\mathbf{b}, \lambda)| \leq 1, \tag{5.9.37}$$

$$|E(\mathbf{a}, \mathbf{b}) - E(\mathbf{a}, \mathbf{b}')| \leq \int d\lambda \rho(\lambda) \big[1 \pm A(\mathbf{a}', \lambda) B(\mathbf{b}', \lambda) \big]$$
$$+ \int d\lambda \rho(\lambda) \big[1 \pm A(\mathbf{a}', \lambda) B(\mathbf{b}, \lambda) \big]. \tag{5.9.38}$$

Now, since $\int d\lambda \rho(\lambda) = 1$,

$$|E(\mathbf{a}, \mathbf{b}) - E(\mathbf{a}, \mathbf{b}')| \leq 2 \pm \big[E(\mathbf{a}', \mathbf{b}') + E(\mathbf{a}', \mathbf{b}) \big], \tag{5.9.39}$$

implying the inequality

$$|E(\mathbf{a}, \mathbf{b}) - E(\mathbf{a}, \mathbf{b}') + E(\mathbf{a}', \mathbf{b}) + E(\mathbf{a}', \mathbf{b}')| \leq 2. \tag{5.9.40}$$

Clauser et al. first suggested that, subject to very mild and natural assumptions, Bell inequalities could be tested by experiments measuring photon linear polarization correlations resulting from an atomic $J = 0 \rightarrow 1 \rightarrow 0$ radiative cascade transition (see Section 5.8).[81] Such experiments by Freedman and Clauser provided the first experimental evidence against local hidden-variable theories.[82] Experiments by Fry and Thomson gave further evidence against local hidden-variable theories and excellent agreement with the predictions of quantum theory.[83]

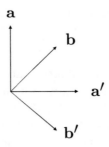

Fig. 5.7 Directions $\mathbf{a}, \mathbf{a}', \mathbf{b}, \mathbf{b}'$ for experiment testing the inequality (5.9.40). All four unit vectors lie in a plane perpendicular to the propagation directions $\pm\mathbf{z}$.

In experiments by Aspect et al., the inequality (5.9.40) was tested using two-channel polarizers to separate two linear polarizations, along orthogonal directions $\mathbf{a}$, $\mathbf{a}'$ for photons with propagation direction $\mathbf{z}$ and along orthogonal directions $\mathbf{b}$, $\mathbf{b}'$

[81] J. F. Clauser, M. A. Horne, A. Shimony, and R. A. Holt, Phys. Rev. Lett. **23**, 880 (1969).
[82] S. J. Freedman and J. F. Clauser, Phys. Rev. Lett. **28**, 938 (1972).
[83] E. S. Fry and R. C. Thompson, Phys. Rev. Lett. **37**, 465 (1976).

for photons with propagation direction $-\mathbf{z}$ (see Figure 5.7). The quantum-mechanical expectation value is

$$E_{QM}(\mathbf{a}, \mathbf{b}) = \cos 2\theta_{ab} \qquad (5.9.41)$$

in this case, where θ_{ab} is the angle between $\mathbf{a}$ and $\mathbf{b}$ (see Exercise 5.8). Therefore,

$$\Big[E(\mathbf{a}, \mathbf{b}) - E(\mathbf{a}, \mathbf{b}') + E(\mathbf{a}', \mathbf{b}) + E(\mathbf{a}', \mathbf{b}') \Big]_{QM}$$
$$= \cos 2\theta_{ab} - \cos 2\theta_{ab'} + \cos 2\theta_{a'b} + \cos 2\theta_{a'b'}. \qquad (5.9.42)$$

For $\theta_{ab} = \theta_{a'b} = \theta_{a'b'} = \pi/8$, and $\theta_{ab'} = 3\pi/8$, for example,

$$\Big[E(\mathbf{a}, \mathbf{b}) - E(\mathbf{a}, \mathbf{b}') + E(\mathbf{a}', \mathbf{b}) + E(\mathbf{a}', \mathbf{b}') \Big]_{QM} = 2\sqrt{2} = 2.828, \qquad (5.9.43)$$

which obviously violates the prediction (5.9.40) of a local hidden-variable theory.[84] When imperfect polarizer transmission coefficients and the finite solid angles subtended by the photomultipliers in the experiments are accounted for, the quantum-mechanical prediction becomes

$$\Big[E(\mathbf{a}, \mathbf{b}) - E(\mathbf{a}, \mathbf{b}') + E(\mathbf{a}', \mathbf{b}) + E(\mathbf{a}', \mathbf{b}') \Big]_{QM} = 2\sqrt{2} = 2.70 \pm 0.05, \qquad (5.9.44)$$

in excellent agreement with the experimental result, 2.697 ± 0.0515, but in strong violation of the inequality (5.9.40) predicted by a local hidden-variable theory.[85]

Exercise 5.8: Derive the expectation value (5.9.41).

Aspect et al. also performed experiments in which the polarizer angles at the A and B ends were switched independently between $\pi/8$ and $3\pi/8$ during time intervals short compared to the distance separating the A and B detectors. The results of these "delayed choice" experiments were again in strong violation of (5.9.40) and in excellent agreement with quantum theory.[86]

In light of the experiments cited and other, more recent experiments with different physical systems, there can be little doubt that local hidden-variable theories are untenable. This conclusion holds even if the presumed hidden variables are stochastic variables. It is the "realistic" feature of such theories, and not determinism, that puts them at odds with quantum theory.

Einstein may have disliked the fundamental indeterminism of quantum theory ("God playing dice"), but his principal "point of departure" from quantum theory

---

[84] It is easily shown that, with these angles, (5.9.42) has its maximum value.

[85] A. Aspect, P. Grangier, and G. Roger, Phys. Rev. Lett. **47**, 460 (1981), and Phys. Rev. Lett. **49**, 91 (1982); A. Aspect, J. Dalibard, and G. Roger, Phys. Rev. Lett. **49**, 1804 (1982).

[86] Ibid.

was "realistic" rather than "deterministic."[87] Even so, there is no indication that he was an advocate of hidden-variable theories; he wrote to Born in 1952 that such theories seemed "too cheap."[88] Nor is there any evidence that he doubted that the "spooky action-at-distance" correlations described in the EPR paper could be observed in actual experiments, as they later were.

The EPR paper has been studied and cited many times, and it was certainly influential to Bell. But its importance is not universally acknowledged. Pais, for example, writes that "this paper contains neither a paradox nor any flaw in logic. It simply concludes that objective reality is incompatible with the assumption that quantum mechanics is complete. This conclusion has not affected subsequent developments in physics, and it is doubtful that it ever will."[89]

There have been wildly divergent opinions about the significance of Bell's theorem. Feynman, for example, "had a somewhat complicated relationship with Bell's Theorem," and in 1983 said that "it is not a theorem that anybody thinks is of any particular importance. We who use quantum mechanics have been using it all the time. It is not an important theorem. It is simply a statement of something we know is true—a mathematical proof of it."[90] Stapp, in contrast, regarded Bell's theorem as "the most profound discovery of science."[91] Let the reader decide.

References and Suggested Additional Reading

Among the many books on the quantum theory of atom–field interactions are *Loudon*; *Mandel and Wolf*; *Scully and Zubairy* and *Garrison and Chiao*. *Shore* and *Berman and Malinovsky* focus primarily on laser–atom interactions and spectroscopy. *Budker, Kimball, and DeMille* is a nice collection of examples pertaining to atoms in electric and magnetic fields. *Haroche and Raimond* review studies of atom–field interactions in cavity quantum electrodynamics.

Coherent effects in optical spectra, with many references to some of the original work on resonance fluorescence and photon anti-bunching, are reviewed by P. L. Knight and P. W. Milonni, Phys. Rep. **66**, 21 (1980).

For a comprehensive review of theory relating to Bell's theorem and early experiments testing local hidden-variable theories vis-à-vis quantum theory, see J. F. Clauser and A. Shimony, Rep. Prog. Phys. **41**, 1881 (1978).

---

[87] W. Pauli, in *The Born–Einstein Letters*, Walker and Company, New York, 1971, p. 221.

[88] A. Einstein, *The Born–Einstein Letters*, p. 192.

[89] *Pais (1982)*, p. 456.

[90] A. Whitaker, Am. J. Phys. **84**, 493 (2016), p. 493.

[91] H. Stapp, Nuovo Cimento **40B**, 191 (1977), p. 271.

Quantum cloning and implications are reviewed by V. Scarani, S. Iblisdir, N. Gisin, and A. Acín, Rev. Mod. Phys. **77**, 1225 (2005).

Two clearly written books on quantum information science and applications are S. Stenholm and K.-A. Suominen, *Quantum Approach to Informatics*, Wiley–Interscience, Hoboken, New Jersey, 2005, and N. D. Mermin, *Quantum Computer Science. An Introduction*, Cambridge University Press, Cambridge, 2007.

6
Fluctuations, Dissipation, and Noise

Dissipation is inevitably accompanied by fluctuations, as we have seen in connection with blackbody radiation and spontaneous emission (see Sections 2.8 and 4.5). The dissipation and the fluctuations are related because they have a common origin. In spontaneous emission, for example, the dissipative radiation reaction in the coupling of the atom to the electromagnetic field is related to the quantum fluctuations of the field. In the Brownian motion of a particle suspended in a fluid, there is a drag force on the particle as it moves in the fluid, and this dissipative force is related to the fluctuations the particle experiences as a result of its collisions with the molecules of the fluid. The relation between the dissipation and the fluctuations is formalized in the fluctuation–dissipation theorem.

As noted in Section 2.8, the first examples of the connection of dissipation to fluctuations, and the precursors of a general fluctuation–dissipation theorem, appeared in Einstein's work on Brownian motion and blackbody radiation. We begin this chapter with a tutorial review of Einstein's theory of Brownian motion. This leads us to an elementary introduction to the Fokker–Planck and Langevin approaches to general types of "Brownian motion," with these involving fluctuations and dissipation. We discuss the equivalence of the two approaches and then derive the (Ford–Kac–Mazur) *quantum* Langevin equation. We next turn our attention to the seminal work of Nyquist on Johnson noise and its generalization in the form of the fluctuation–dissipation theorem of Callen and Welton. We discuss different expressions of the theorem and its application to oscillators interacting with a heat reservoir, and then a few examples of the interplay of fluctuations and dissipation, including the (Ford–O'Connell) theory of radiation reaction, the fundamental lower limit to the laser linewidth, and the quantum theory of linear (phase-insensitive) amplifiers, attenuators, and beam splitters.

6.1 Brownian Motion and Einstein's Relations

Robert Brown was not the first to observe the erratic movements of tiny particles (~ 1 μm) suspended in fluids. His 1827 discovery was that particles of inorganic as well as organic matter undergo this motion, dispelling the notion that it had something to do with living matter. Einstein, in 1905, and, later, von Smoluchowski, in 1906, developed a theory to explain Brownian motion, viewing it as a consequence of collisions of the suspended particles with the molecules of the surrounding fluid. The theory is, of

An Introduction to Quantum Optics and Quantum Fluctuations. Peter W. Milonni.
© Peter W. Milonni 2019. Published in 2019 by Oxford University Press.
DOI:10.1093/oso/9780199215614.001.0001

course, statistical in nature, as it involves huge numbers of bombarding molecules in random thermal motion.

Einstein was not the first to suggest that Brownian motion might be caused by molecular impacts, but he was the first to treat these impacts in a satisfactory, statistical way, in contrast to some earlier arguments against the molecular impact explanation.[1] One such argument was that a suspended particle of mass m in thermal equilibrium with the fluid at temperature T has, according to statistical mechanics, a mean-square velocity $3k_BT/m$, which would typically imply velocities much larger than what is observed in Brownian motion; this argument is addressed in Section 6.4.1. Here, we will briefly summarize a few of the main ideas and implications of Einstein's theory.

Einstein introduces a time interval τ that is small compared with any observation time interval but sufficiently large that the displacements of a suspended particle in two successive time intervals τ are uncorrelated. He defines a distribution $\phi(\Delta)$ such that the fraction of suspended particles that experience a displacement between Δ and $\Delta + d\Delta$ in a time interval τ is $\phi(\Delta)d\Delta$, with $\phi(\Delta)$ sharply peaked at $\Delta = 0$, and

$$\int_{-\infty}^{\infty} d\Delta\,\phi(\Delta) = 1, \quad \text{where } \phi(-\Delta) = \phi(\Delta). \tag{6.1.1}$$

From the definition of $\phi(\Delta)$, it follows that the probability $P(x, t+\tau)dx$ that a particle has an x coordinate between x and $x + dx$ at time $t + \tau$ satisfies

$$P(x, t + \tau)dx = dx \int_{-\infty}^{\infty} d\Delta\,\phi(\Delta)P(x + \Delta, t). \tag{6.1.2}$$

Expanding in powers of the small time τ and small displacements Δ,

$$P(x,t)+\tau\frac{\partial}{\partial t}P(x,t) \cong \int_{-\infty}^{\infty} d\Delta\,\phi(\Delta)\left[P(x,t)+\Delta\frac{\partial}{\partial x}P(x,t)+\frac{1}{2}\Delta^2\frac{\partial^2}{\partial x^2}P(x,t)\right]. \tag{6.1.3}$$

From (6.1.1) and the assumption that $\phi(\Delta)$ is sharply peaked at $\Delta = 0$,

$$\frac{\partial P}{\partial t} = D\frac{\partial^2 P}{\partial x^2}, \tag{6.1.4}$$

$$D = \frac{1}{\tau}\int_{-\infty}^{\infty} d\Delta\frac{\Delta^2}{2}\phi(\Delta). \tag{6.1.5}$$

---

[1] In the introduction to his 1905 paper Einstein wrote that "it is possible that the movements to be discussed here are identical with the so-called 'Brownian molecular motion'; however, the information available to me regarding the latter is so lacking in precision, that I can form no judgment in the matter." (*Fürth*, p. 1) In a second paper, from 1906, he remarked that, following the publication of the 1905 paper, he learned that, in 1888, M. Gouy and others "had been convinced by direct observation that the so-called Brownian motion is caused by the irregular movements of the molecules of the liquid" (*Fürth*, p. 19).

Einstein remarks that (6.1.4) "is the well-known differential equation for diffusion, and we recognize that D is the coefficient of diffusion," and writes the solution of this equation "for the problem of the diffusion outwards from a point":[2]

$$P(x,t) = \frac{1}{\sqrt{4\pi Dt}} e^{-x^2/4Dt}, \qquad \int_{-\infty}^{\infty} dx P(x,t) = 1. \tag{6.1.6}$$

In terms of the (Dirac) delta function—which was not used by physicists in 1905—this is the solution of the diffusion equation (6.1.4) satisfying[3]

$$P(x,0) = \lim_{t \to 0} \frac{1}{\sqrt{4\pi Dt}} e^{-x^2/4Dt} = \delta(x). \tag{6.1.7}$$

Note that $\langle x \rangle = 0$, and

$$\langle x^2 \rangle = \int_{-\infty}^{\infty} dx x^2 P(x,t) = 2Dt. \tag{6.1.8}$$

Einstein observes that "the mean displacement $[\langle x^2 \rangle^{1/2}]$ is therefore proportional to the square root of the time."[4]

He assumes that the particles can be treated as an ideal gas in thermal equilibrium at temperature T, and obtains the expression

$$D = \frac{k_B T}{6\pi\mu a} = \frac{k_B T}{\gamma} \tag{6.1.9}$$

for the diffusion coefficient, where a is the radius of the suspended particles, and μ is the coefficient of viscosity of the fluid. The denominator comes from the Stokes law for the drag force $F = -(6\pi\mu a)v = -\gamma v$ acting on a spherical particle of radius a moving with velocity v relative to a fluid with viscosity μ.

Equation (6.1.8) relates an effect observable on a macroscopic scale—diffusion—to the random walks of single microscopic particles, and specifically to the mean-square displacement associated with this random motion. Equation (6.1.9) relates this *fluctuation* effect to an average *dissipation* force. It is widely appreciated that the significance of the "Einstein relations" (6.1.8) and (6.1.9) extends far beyond the original context of the erratic motion of particles suspended in fluids. In particular, the relation between fluctuations and dissipation discovered by Einstein in his analysis of Brownian motion was eventually extended to a wide range of phenomena and formalized as the "fluctuation–dissipation theorem."

---

[2] *Fürth*, pp. 15–16.

[3] More generally, $P(x,0) = \delta(x - x_0)$, $P(x,t) = (1/\sqrt{4\pi Dt}) \exp[-(x - x_0)^2/4Dt]$, $\langle x \rangle = x_0$, and $\langle (x - x_0)^2 \rangle = 2Dt$.

[4] *Fürth*, p. 17.

The relation (6.1.9) may be derived from the balance between diffusion and drift of suspended particles of mass m and number density n. If the particles experience a force $F = -dV/dx$ (e.g., gravity) in addition to a drag force $-\gamma v$, there will be a drift velocity $v_d = F/\gamma$ along the x direction. The particles also undergo diffusion due to the random impacts of the molecules of the fluid; according to *Fick's law*, the particle flux along x due to diffusion is $J = -Ddn/dx$, where D is the diffusion coefficient. In equilibrium there is no net particle flux:

$$-D\frac{dn}{dx} + nv_d = -D\frac{dn}{dx} - n\frac{1}{\gamma}\frac{dV}{dx} = 0. \qquad (6.1.10)$$

In thermal equilibrium $n(x) \propto \exp[-V(x)/k_BT]$, whence $D = k_BT/\gamma$. This is often written as $D = Bk_BT$, where $B = 1/\gamma$ is the *mobility*.

The term "random walk," was introduced by the mathematician Karl Pearson in the same year as Einstein's (first) paper on Brownian motion, 1905. In a letter to *Nature*, "The Problem of the Random Walk," Pearson posed the question of finding the probability distribution for the (two-dimensional) random walk problem.[5] Rayleigh responded with the observation that the problem "is the same as that of the composition of N iso-periodic vibrations of unit amplitude and of phases distributed at random," which he had solved for large N in papers published in 1880 and 1899.[6] His solution for the (two-dimensional) random walk problem was similar in form to (6.1.6), prompting Pearson to famously remark that "the lesson of Lord Rayleigh's solution is that in open country the most probable place to find a drunken man who is at all capable of keeping on his feet is somewhere near his starting point!"[7]

Rayleigh also derived what would later be called the Fokker–Planck equation, derived by Fokker and Planck in their work on Brownian motion.[8]

Einstein's predictions, that the mean-square displacements in Brownian motion are inversely proportional to viscosity and proportional to time, were soon confirmed: Jean Perrin and his students "confirmed the distribution law for the probability of different displacements $P(x,t)$ in a quite unexceptionable manner."[9] Einstein had also suggested that the relation (6.1.9) and a measurement of the diffusion constant D could be used to determine k_B and therefore Avogadro's number N_A. Perrin et al. in their measurements inferred values of N_A in the range $5.6 - 8.8 \times 10^{23}$.[10]

6.2   The Fokker–Planck Equation

The diffusion equation (6.1.4) derived by Einstein is a special case of a Fokker–Planck equation. We will briefly review a standard derivation of the simplest example of a Fokker–Planck equation.

We define the probability $P(x,t)dx$ that a particle lies between x and $x + dx$ at time t, and a displacement probability $W(x,\xi)d\xi\delta t$ that a particle at x will undergo a displacement along the x direction by an amount between ξ and $\xi + d\xi$ in a time interval δt. We assume that $P(x,t)dx$ satisfies the intuitively plausible *master equation*

$$\frac{\partial P(x,t)}{\partial t} = -\left[\int_{-\infty}^{\infty} d\xi W(x,\xi)\right]P(x,t) + \int_{-\infty}^{\infty} d\xi W(x+\xi,-\xi)P(x+\xi,t). \quad (6.2.1)$$

---

[5] K. Pearson, Nature **72**, 294 (1905).

[6] Lord Rayleigh, Nature **72**, 318 (1905), p. 318.

[7] K. Pearson, Nature **72**, 342 (1905), p. 342.

[8] Lord Rayleigh, Phil. Mag. **32**, 424 (1891); A. D. Fokker, Ann. Physik **43**, 810 (1914); M. Planck, Abh. Preuss. Akad. Wiss. Berl. **24**, 324 (1917); see S. Chandrasekhar, Rev. Mod. Phys. **15**, 1 (1943).

[9] *Fürth*, p. 103.

[10] Ibid.

The first term on the right is the "loss" rate due to displacements from x by all possible amounts ξ, and the second term is the "gain" rate due to all possible displacements (by $-\xi$) that bring the particle back to x from $x + \xi$. It is assumed that $W(x, \xi)$ is sharply peaked at $\xi = 0$ and that $P(x, t)$ varies slowly enough with x that we can make a second-order Taylor-series expansion:

$$\frac{\partial P(x, t)}{\partial t} = \int_{-\infty}^{\infty} d\xi \Big\{ - W(x, \xi)P(x, t) + W(x - \xi, \xi)P(x - \xi, t) \Big\}$$

$$\cong \int_{-\infty}^{\infty} d\xi \Big\{ - W(x, \xi)P(x, t) + W(x, \xi)P(x, t) - \frac{\partial}{\partial x}\Big[\xi W(x, \xi)P(x, t)\Big]$$

$$+ \frac{1}{2}\frac{\partial^2}{\partial x^2}\Big[\xi^2 W(x, \xi)P(x, t)\Big]\Big\}, \tag{6.2.2}$$

and we obtain the Fokker–Planck equation as an approximation to the master equation:

$$\frac{\partial P(x, t)}{\partial t} = -\frac{\partial}{\partial x}\Big[A_1(x)P(x, t)\Big] + \frac{1}{2}\frac{\partial^2}{\partial x^2}\Big[A_2(x)P(x, t)\Big],$$

$$A_1(x) = \int_{-\infty}^{\infty} d\xi\, \xi W(x, \xi),$$

$$A_2(x) = \int_{-\infty}^{\infty} d\xi\, \xi^2 W(x, \xi). \tag{6.2.3}$$

This derivation of a Fokker–Planck equation is similar to Einstein's derivation of the diffusion equation. In particular, the displacements are presumed to be much more likely to be very small than large, and therefore $W(x, \xi)$ is taken to be strongly peaked at $\xi = 0$. Implicit in the "master equation" (6.2.1) is the assumption that $W(x, \xi)$ does not depend on the past history of a particle's trajectory, but only on where it is "now," that is, only on its present position x. In other words, as in the theory of spontaneous emission in Chapter 4, a Markovian approximation has been introduced: the rate of change of P in the master equation (6.2.1) does not depend on what P is at earlier times. In accordance with this approximation, the time interval δt should be understood to be small enough that a particle does not undergo an appreciable displacement, but large enough that the Markovian approximation can be made; like Einstein's τ, δt should be small but not too small.

The terms with $A_1(x)$ and $A_2(x)$ in the Fokker–Planck equation are called the "drift" and "diffusion" terms, respectively. The reason for associating $A_2(x)$ with diffusion is clear by comparison with the diffusion equation (6.1.4). The drift term vanishes if $W(x, \xi) = W(x, -\xi)$, that is, if positive and negative displacements along x are equally likely; then there is no net "drift" of the probability distribution along x. In Einstein's derivation of the diffusion equation, similarly, there is no drift because of his assumption that $\phi(\Delta) = \phi(-\Delta)$. In addition, his $\phi(\Delta)$, unlike $W(x, \xi)$, does not depend on x, and so he has $D\partial^2 P/\partial x^2$ in (6.1.4) rather than the diffusion term of the Fokker–Planck equation.

We now briefly review an example of the application of the Fokker–Planck equation. Consider particles undergoing (one-dimensional) Brownian motion in a fluid such that the drag force is $-\gamma v$, and suppose the particles are also subject to a restoring force $-Kx$. Assuming negligible particle mass, the equilibrium balancing of the average drag force and the restoring force implies that $Kx + \gamma\langle v(x)\rangle = 0$, or $\langle v(x)\rangle = -Kx/\gamma$. Taking this $\langle v(x)\rangle$ to be $A_1(x)$ in the Fokker–Planck equation, and $\langle x^2\rangle/\delta t = 2D = 2k_BT/\gamma$ (see (6.1.8) and (6.1.9)) to be $A_2(x)$, we have

$$\frac{\partial P}{\partial t} = \frac{K}{\gamma}\frac{\partial}{\partial x}(xP) + \frac{k_BT}{\gamma}\frac{\partial^2 P}{\partial x^2}. \tag{6.2.4}$$

The equilibrium distribution ($\partial P/\partial t = 0$) therefore satisfies

$$\frac{\partial}{\partial x}(xP) = -\frac{k_BT}{K}\frac{\partial^2 P}{\partial x^2}, \tag{6.2.5}$$

with the solution (normalized so that $\int_{-\infty}^{\infty}P(x)dx = 1$)

$$P(x) = \sqrt{\frac{K}{2\pi k_BT}}e^{-Kx^2/2k_BT}, \tag{6.2.6}$$

as expected for particles with potential energy $Kx^2/2$ in thermal equilibrium.[11]

From the time-dependent Fokker–Planck equation (6.2.4),

$$\langle x(t)\rangle = \int_{-\infty}^{\infty}dx\,xP(x,t) = x_0e^{-Kt/\gamma} \tag{6.2.7}$$

and

$$\langle x^2(t)\rangle = \int_{-\infty}^{\infty}dx\,x^2P(x,t) = x_0^2e^{-2Kt/\gamma} + \frac{k_BT}{K}\left(1 - e^{-2Kt/\gamma}\right). \tag{6.2.8}$$

Equation (6.2.7), for example, follows from

$$\frac{\partial}{\partial t}\langle x(t)\rangle = \int_{-\infty}^{\infty}dx\,x\frac{\partial P}{\partial t} = \int_{-\infty}^{\infty}dx\,x\left(\frac{K}{\gamma}P + \frac{K}{\gamma}x\frac{\partial P}{\partial x} + \frac{k_BT}{\gamma}\frac{\partial^2 P}{\partial x^2}\right)$$
$$= -\frac{K}{\gamma}\langle x(t)\rangle \tag{6.2.9}$$

after partial integrations, assuming $x^2P(x,t)$ and $x\partial^2 P(x,t)/\partial x^2 \to 0$ for $x \to \pm\infty$. For $K \to 0$,

$$\langle x^2(t)\rangle \to x_0^2 + (2k_BT/\gamma)t = x_0^2 + 2Dt, \tag{6.2.10}$$

consistent with (6.1.8) for the case where there is no restoring force. For $Kt/\gamma \to \infty$, $\frac{1}{2}K\langle x^2(t)\rangle \to \frac{1}{2}k_BT$, consistent with the equipartition theorem.

---

[11] Recall the representation $\delta(x) = \lim_{\alpha\to 0}\left[(\alpha\sqrt{2\pi})^{-1}e^{-x^2/2\alpha^2}\right]$ of the delta function. The solution (6.2.6) therefore approaches $\delta(x)$ as $T \to 0$.

The solution of the time-dependent Fokker–Planck equation describes the approach of the distribution to its equilibrium form. For example, for particles having the displacement $x = x_0$ at $t = 0$,

$$P(x,t) = \left[\frac{K}{2\pi k_B T(1 - e^{-2Kt/\gamma})} \right]^{1/2} \exp\left[-\frac{K(x - x_0 e^{-Kt/\gamma})^2}{2k_B T(1 - e^{-2Kt/\gamma})} \right], \qquad (6.2.11)$$

which approaches (6.2.6) when Kt/γ is large. Equations (6.2.7) and (6.2.8) can of course also be derived from (6.2.11).[12]

The master equation and the Fokker–Planck approximation to it are easily generalized to treat more than one spatial dimension. The simplistic derivation of the Fokker–Planck equation reviewed in this section presumes that higher-order terms in the Taylor expansion of the master equation are negligible. More precisely, it is assumed that $W(x, \xi)$ is sharply peaked at $\xi = 0$, that is, that large jumps in displacement are highly unlikely, and that *both* $W(x, \xi)$ and $P(x, t)$ vary sufficiently slowly with x that derivatives beyond the second are negligible.

The master equation itself is an approximation. In the quantum-mechanical context, it was introduced by Pauli under the assumption that the coefficients $c_i(t)$ in (2.1.9) have randomly varying phases. Then there are no "off-diagonal" contributions $c_i^*(t)c_j(t)$, $i \neq j$, to the time evolution. In particular, the time evolution described by the master equation is irreversible, as it does not have the time-reversal symmetry of the time evolution of a system described by the Schrödinger equation. In this sense, the rate equation (4.8.40), which was derived under the assumption that off-diagonal terms are negligible, can be regarded as a master equation.

In writing the Fokker–Planck equation (6.2.4), we assumed that the average velocity is "equilibrated" such that the restoring force equals the drag force. In the absence of a restoring force, the average velocity is zero. Suppose, instead, that we allow in that case for a variation of the velocity with time and write a Fokker–Planck equation for the *velocity* distribution $P(v, t)$ by analogy to (6.2.3). Assuming a drift term $A_1(v) = -(\gamma/m)$ (mean rate of velocity "displacement" according to $m\dot{v} = -\gamma v$), and a diffusion coefficient D_v, this equation is

$$\frac{\partial P}{\partial t} = \frac{\gamma}{m}\frac{\partial}{\partial v}(vP) + D_v \frac{\partial^2 P}{\partial v^2}. \qquad (6.2.12)$$

We choose D_v such that the equilibrium solution has the Maxwell–Boltzmann form:

$$P(v, \infty) = \sqrt{\frac{m}{2\pi k_B T}} e^{-mv^2/2k_B T}. \qquad (6.2.13)$$

Using this in (6.2.12) with $\partial P/\partial t = 0$ requires $D_v = \gamma k_B T/m^2$, so that (6.2.12) becomes

$$\frac{\partial P}{\partial t} = \frac{\gamma}{m}\left[\frac{\partial}{\partial v}(vP) + \frac{k_B T}{m}\frac{\partial^2 P}{\partial v^2} \right]. \qquad (6.2.14)$$

---

[12] For more rigorous treatments of some of the results reviewed in this section and Section 6.3, as well as references to original work, see the classic papers by G. E. Uhlenbeck and L. S. Ornstein, Phys. Rev. **36**, 823 (1930); S. Chandrasekhar, Rev. Mod. Phys. **15**, 1 (1943); and M. C. Wang and G. E. Uhlenbeck, Rev. Mod. Phys. **17**, 323 (1945).

The solution of this equation with the initial condition $P(v,0) = \delta(v - v_0)$ is

$$P(v,t) = \left[\frac{2\pi k_B T}{m}(1 - e^{-2t/\tau_c})\right]^{-1/2} \exp\left[-\frac{m}{2k_B T}\frac{(v - v_0 e^{-t/\tau_c})^2}{1 - e^{-2t/\tau_c}}\right], \qquad (6.2.15)$$

$$\tau_c = m/\gamma. \qquad (6.2.16)$$

From this solution, or just proceeding directly from (6.2.14) and partial integrations as in the derivation of $\langle x(t)\rangle$ and $\langle x^2(t)\rangle$ above, we obtain, for particles with initial velocity $v(t = 0) = v_0$,

$$\langle v(t)\rangle = v_0 e^{-t/\tau_c}, \qquad (6.2.17)$$

and

$$\langle v^2(t)\rangle = v_0^2 e^{-2t/\tau_c} + \frac{k_B T}{m}(1 - e^{-2t/\tau_c}). \qquad (6.2.18)$$

Multi-time averages $\langle v(t_1)v(t_2)\ldots v(t_n)\rangle$ can be calculated in a somewhat similar fashion. Consider the simplest such correlation function, $\langle v(t_1)v(t_2)\rangle$, for $t_2 \geq t_1$ and for particles with initial velocity $v(0) = v_0$:

$$\langle v(t_1)v(t_2)\rangle = \int_{-\infty}^{\infty} dv_1 \int_{-\infty}^{\infty} dv_2 v_1 v_2 P(v_2, t_2; v_1, t_1). \qquad (6.2.19)$$

Here, $P(v_2, t_2; v_1, t_1)$ is the joint probability distribution for the velocity to be v_2 at time t_2 and to be v_1 at time t_1, and has the factorized form

$$P(v_2, t_2; v_1, t_1) = P(v_2, t_2|v_1, t_1)P(v_1, t_1), \qquad (6.2.20)$$

where $P(v_2, t_2|v_1, t_1)$ is the probability that a particle has velocity v_2 at time t_2, given that it had the velocity v_1 at time t_1. The factorization (6.2.20) follows from the Markovian approximation: $P(v_2, t_2; v_1, t_1)$ depends only on the "initial" velocity v_1 at time t_1, and not at all on velocities at earlier times. Now, from (6.2.15),

$$P(v_1, t_1) = \frac{1}{\sigma(t_1)\sqrt{2\pi}} \exp\left[-(v_1 - v_0 e^{-t_1/\tau_c})^2/2\sigma^2(t_1)\right], \qquad (6.2.21)$$

$$\sigma^2(t) = \frac{k_B T}{m}(1 - e^{-2t/\tau_c}), \qquad (6.2.22)$$

and

$$P(v_2, t_2; v_1, t_1) = \frac{1}{\sigma(t_2)\sqrt{2\pi}} \exp\left[-(v_2 - v_1 e^{-(t_2-t_1)/\tau_c})^2/2\sigma^2(t_2)\right]. \qquad (6.2.23)$$

Then,

$$\begin{aligned}
\langle v(t_1)v(t_2)\rangle &= \frac{1}{2\pi\sigma(t_1)\sigma(t_2)} \int_{-\infty}^{\infty} dv_1 v_1 \exp\left[-(v_1 - v_0 e^{-t_1/\tau_c})^2/2\sigma^2(t_1)\right] \\
&\quad \times \int_{-\infty}^{\infty} dv_2 v_2 \exp\left[-(v_2 - v_1 e^{-(t_2-t_1)/\tau_c})^2/2\sigma^2(t_2)\right] \\
&= \left(v_0^2 - \frac{k_B T}{m}\right)e^{-(t_1+t_2)/\tau_c} + \frac{k_B T}{m}e^{-(t_2-t_1)/\tau_c}.
\end{aligned} \qquad (6.2.24)$$

Moments of $x(t) = x_0 + \int_0^t dt' v(t')$ can be calculated with a bit more effort. In particular, if we calculate $\langle [x(t) - x_0]^2 \rangle$ and then take another average $\langle \langle \ldots \rangle \rangle$ over an ensemble of particles having kinetic energy $\frac{1}{2} m v_0^2$ at $t = 0$ equal to the thermal equilibrium value $\frac{1}{2} k_B T$, we obtain

$$\langle \langle [x(t) - x_0]^2 \rangle \rangle = \frac{2 k_B T}{\gamma} \tau_c \Big(\frac{t}{\tau_c} + e^{-t/\tau_c} - 1 \Big). \tag{6.2.25}$$

Unlike the approximation (6.2.10), which does not account for a finite velocity relaxation time, this exhibits a crossover from a "ballistic" regime ($t \ll \tau_c$) in which $\langle \langle [x(t) - x_0]^2 \rangle \rangle$ increases quadratically with time, to a diffusive regime ($t \gg \tau_c$) in which $\langle \langle [x(t) - x_0]^2 \rangle \rangle$ increases only linearly with time. This is discussed in Section 6.3.

6.3   The Langevin Approach

Langevin, soon after Einstein and von Smoluchowski, developed his own phenomenological approach to Brownian motion, assuming that a particle of mass m, suspended in a fluid, experiences a drag force $-\gamma v$ as well as a *random* force $F_L(t)$ due to random collisions of the particle with the molecules of the fluid.[13] If the particle is free of any other forces, the *Langevin equation* describing the motion of the particle along a single direction is

$$m \frac{dv}{dt} = -\gamma v + F_L(t). \tag{6.3.1}$$

We assume that

$$\langle F_L(t) \rangle = 0 \tag{6.3.2}$$

and, with C to be determined,

$$\langle F_L(t_1) F_L(t_2) \rangle = C \delta(t_1 - t_2). \tag{6.3.3}$$

The effectively random nature of the collisions of the particle with the molecules of the fluid means there is no net force, on average, acting on the particle (see (6.3.2)); more precisely, the average of $F_L(t)$ over an ensemble of particles having the same velocity v_0 at $t = 0$ is assumed to vanish at all times.[14] Equation (6.3.3) says there is no correlation between the random forces $F_L(t_1)$ and $F_L(t_2)$, $t_1 \neq t_2$. This, too, is a reasonable assumption, since successive collisions of the particle with the molecules of the fluid can be expected to be correlated, if at all, only over time intervals very short compared with those of interest for observations of Brownian motion.

---

[13] P. Langevin, Compt. Rendus **146**, 530 (1908).

[14] We regard $\langle \ldots \rangle$ here as an ensemble average at some instant of time, but will assume that this ensemble average is equal to the average over a sufficiently long time for any single member, or "realization," of the ensemble.

For particles with velocity v_0 at $t = 0$,

$$v(t) = \frac{dx}{dt} = v_0 e^{-t/\tau_c} + \frac{1}{m} \int_0^t dt_1 F_L(t_1) e^{(t_1 - t)/\tau_c}, \quad \tau_c = m/\gamma, \tag{6.3.4}$$

and, from (6.3.2), $\langle v(t) \rangle = v_0 e^{-t/\tau_c}$. Likewise, from (6.3.3),

$$\langle v^2(t) \rangle = v_0^2 e^{-2t/\tau_c} + \frac{1}{m^2} \int_0^t dt_1 \int_0^t dt_2 \langle F_L(t_1) F_L(t_2) \rangle e^{(t_1 + t_2 - 2t)/\tau_c}$$

$$= v_0^2 e^{-2t/\tau_c} + \frac{C\tau_c}{2m^2} (1 - e^{-2t/\tau_c}). \tag{6.3.5}$$

We can now determine C by requiring that the equilibrium kinetic energy $\frac{1}{2} m \langle v^2 \rangle = C\tau_c/4m$ has the equipartition value $\frac{1}{2} k_B T$:

$$C = \frac{2k_B T m}{\tau_c} = 2\gamma k_B T, \tag{6.3.6}$$

so that

$$\langle v^2(t) \rangle = v_0^2 e^{-2t/\tau_c} + \frac{k_B T}{m} (1 - e^{-2t/\tau_c}), \tag{6.3.7}$$

in agreement with (6.2.18). Likewise, from (6.3.4), we obtain the velocity correlation function

$$\langle v(t_1) v(t_2) \rangle = v_0^2 e^{-(t_1 + t_2)/\tau_c} + \frac{C}{m^2} e^{-(t_1 + t_2)/\tau_c} \int_0^{t_1} dt' \int_0^{t_2} dt'' \delta(t' - t'') e^{(t' + t'')/\tau_c}$$

$$= \left(v_0^2 - \frac{k_B T}{m} \right) e^{-(t_1 + t_2)/\tau_c} + \frac{k_B T}{m} e^{-(t_2 - t_1)/\tau_c} \tag{6.3.8}$$

for $t_2 \geq t_1$, in agreement with (6.2.24).

From (6.3.4), similarly,

$$x(t) = x_0 + v_0 \tau_c (1 - e^{-t/\tau_c}) - \frac{\tau_c}{m} e^{-t/\tau_c} \int_0^t dt_1 F_L(t_1) e^{t_1/\tau_c} + \frac{\tau_c}{m} \int_0^t dt_1 F_L(t_1) \tag{6.3.9}$$

after an integration by parts. Therefore, $\langle x(t) \rangle = x_0 + v_0 \tau_c (1 - e^{-t/\tau_c})$ and, after an easy calculation using (6.3.3) and (6.3.6),

$$\langle [x(t) - x_0]^2 \rangle = v_0^2 \tau_c^2 (1 - e^{-t/\tau_c})^2 + \frac{k_B T m}{\gamma^2} (-3 - e^{-2t/\tau_c} + 4e^{-t/\tau_c}) + \frac{2k_B T}{\gamma} t. \tag{6.3.10}$$

For $t \ll \tau$,

$$\langle [x(t) - x_0] \rangle \cong v_0 t \quad \text{and} \quad \langle [x(t) - x_0]^2 \rangle \cong v_0^2 t^2, \tag{6.3.11}$$

whereas, for $t/\tau_c \to \infty$, $\langle [x(t) - x_0] \rangle \to v_0 \tau_c$, and

$$\langle [x(t) - x_0]^2 \rangle \to \frac{2k_B T}{\gamma} t = 2Dt, \tag{6.3.12}$$

in agreement with (6.1.8). If we take another average as in Section 6.2, replacing v_0 by 0 and v_0^2 by the equipartition value $k_B T/m$, we obtain $\langle\langle[x(t) - x_0]\rangle\rangle = 0$ and recover exactly the result (6.2.25):

$$\langle\langle[x(t) - x_0]^2\rangle\rangle = \frac{2k_B T}{\gamma}\tau_c\left(\frac{t}{\tau_c} + e^{-t/\tau_c} - 1\right). \tag{6.3.13}$$

For $t \ll \tau_c$,

$$\langle\langle[x(t) - x_0]^2\rangle\rangle \cong \frac{k_B T}{\gamma\tau_c}t^2 = \frac{k_B T}{m}t^2, \tag{6.3.14}$$

whereas, for $t/\tau_c \to \infty$, $\langle\langle[x(t) - x_0]^2\rangle\rangle \to 2Dt$.

It is only since around 2005 that the "ballistic" regime of Brownian motion (mean-square displacements proportional to t^2) has been experimentally accessible. In one remarkable experiment, a glass bead (diameter 3 μm) was suspended in air by laser beams (an "optical tweezer"). The interference pattern created by the interference between the light scattered by the bead and the non-scattered laser light changes with the particle displacement, resulting in an intensity difference at a detector that was used to monitor the position of the bead. The low viscosity of air makes τ_c much larger than for a particle suspended in a liquid and allows, in particular, the realization of the ballistic regime ($t \ll \tau_c$). Mean-square displacements independent of air pressure (in contrast to the diffusive regime), and well described by (6.3.14), were observed, and the measured distribution of instantaneous velocities was in very good agreement with a Maxwell–Boltzmann distribution with the root-mean-square velocity $\sqrt{k_B T/m} = 0.42$ mm/s.[15]

Exercise 6.1: Using the Langevin equation (6.3.1), show that

$$\langle v(t)F_L(t)\rangle = \frac{\gamma k_B T}{m},$$

$$\langle x(t)F_L(t)\rangle = 0,$$

$$\langle\langle[x(t_1) - x_0][x(t_2) - x_0]\rangle\rangle = D\left[2t_1 + \tau_c\left(e^{-2t_1/\tau_c} + e^{-2t_2/\tau_c} - e^{-(t_2-t_1)/\tau_c} - 1\right)\right],$$

assuming $t_2 > t_1 > 0$.

6.3.1 The Gaussian Model and Its Relation to the Fokker–Planck Equation

We have obtained the same mean-square velocity (6.3.7), velocity correlation function (6.3.8), and mean-square displacement (6.3.13), using the Fokker–Planck equation

---

[15] T. Li, S. Kheifets, D. Medellin, and M. G. Raizen, Science **328**, 1673 (2010). See also the references in this paper to earlier experiments.

(6.2.12) and the Langevin equation (6.3.1). The question arises whether solutions of these equations yield the same results for more complicated averages and, in particular, for higher-order moments. They do, provided we make an additional assumption about the "Langevin force" $F_L(t)$.[16] The additional assumption is that $F_L(t)$ is *Gaussian*, meaning that

$$\langle F_L(t_1)F_L(t_2) \ldots F_L(t_{2n+1})\rangle = 0, \tag{6.3.15}$$

which generalizes (6.3.2), and

$$\langle F_L(t_1)F_L(t_2) \ldots F_L(t_{2n})\rangle = \sum_{\substack{\text{all possible pairs}}} \langle F_L(t_i)F_L(t_j)\rangle\langle F_L(t_k)F_L(t_\ell)\rangle. \tag{6.3.16}$$

The connection of these conditions to elementary Gaussian statistics is not obvious, but is made plausible by recalling that a Gaussian probability distribution $p(x)$ with $\langle x\rangle = 0$ and $\langle x^2\rangle = \sigma^2$,

$$p(x) = \frac{1}{\sigma\sqrt{2\pi}}e^{-x^2/2\sigma^2}, \tag{6.3.17}$$

has vanishing odd moments, and even moments all expressed in terms of the second moment, σ^2:

$$\langle x^{2n-1}\rangle = 0, \quad n = 1, 2, 3, \ldots, \tag{6.3.18}$$

$$\langle x^{2n}\rangle = 1 \cdot 2 \cdot 3 \ldots (2n-1)(\sigma^2)^n. \tag{6.3.19}$$

The correlation functions (6.3.15) and (6.3.16), similarly, vanish when there are an odd number of F_L's, while, for an even number of F_L's, they depend only on the second-order correlation function, $\langle F_L(t_i)F_L(t_j)\rangle$. With the Gaussian assumption for $F_L(t)$, the "random process" $v(t)$ satisfying the Langevin equation (6.3.1) is also Gaussian, and therefore all its moments are determined by its first and second moments. The general proof of this statement follows along the same lines as the example in Exercise 6.2. With the identification (6.3.6), statistical properties deduced from the Langevin equation (6.3.1) will be equivalent to those deduced from the Fokker–Planck equation (6.2.12) and its (Gaussian) solution (6.2.15).

---

[16] See, for instance, *Van Kampen*.

Exercise 6.2: Using (6.3.16) with $n = 2$, that is,

$$\langle F_L(t_1)F_L(t_2)F_L(t_3)F_L(t_4)\rangle = \langle F_L(t_1)F_L(t_2)\rangle\langle F_L(t_3)F_L(t_4)\rangle$$

$$+\langle F_L(t_1)F_L(t_3)\rangle\langle F_L(t_2)F_L(t_4)$$

$$+ \langle F_L(t_1)F_L(t_4)\rangle\langle F_L(t_2)F_L(t_3)\rangle$$

$$= C^2\delta(t_1 - t_2)\delta(t_3 - t_4) + C^2\delta(t_1 - t_3)\delta(t_2 - t_4) + C^2\delta(t_1 - t_4)\delta(t_2 - t_3),$$

and the Langevin equation (6.3.1), show that

$$\langle [v(t) - v_0 e^{-t/\tau_c}]^4\rangle = 3\sigma^4(t),$$

where $\sigma(t)$ is defined by (6.2.22). Show that the same result follows from the Fokker–Planck equation.

6.4 Fourier Representation, Stationarity, and Power Spectrum

A periodic function $y(t)$ with a fundamental period T has a Fourier series representation:

$$y(t) = \sum_{k=1}^{\infty}(a_k \cos\omega_k t + b_k \sin\omega_k t), \quad \text{where } \omega_k = 2\pi k/T. \tag{6.4.1}$$

We are assuming there is no constant, "DC" component because the average value of $y(t)$ is zero or because (6.4.1) actually represents the deviation of $y(t)$ from its average value. The coefficients a_k and b_k are given by

$$a_k = \frac{2}{T}\int_0^T dt\, y(t)\cos\omega_k t, \quad \text{and} \quad b_k = \frac{2}{T}\int_0^T dt\, y(t)\sin\omega_k t. \tag{6.4.2}$$

If $y(t)$ is a randomly varying quantity, such as the displacement or velocity of a particle in Brownian motion, then, obviously, it cannot be periodic. But there are many situations where the basic physical process responsible for the random variations does not change in time. The random fluctuations of the velocity of a particle suspended in a fluid, for example, are caused by collisions with molecules having a constant mean-square velocity in thermal equilibrium. The statistical properties of the particle's velocity can then be assumed to be *stationary*—they do not depend on the time "origin" but only on time differences. According to (6.3.8), for example, the particle's velocity correlation function $\langle v(t_1)v(t_2) \rangle$ depends only on the time difference $t_2 - t_1$, not on t_1 or t_2 separately, when t_1 or t_2 is much greater than τ. The electric field correlation function $\langle \mathbf{E}^{(-)}(\mathbf{r}, t) \cdot \mathbf{E}^{(+)}(\mathbf{r}, t + \tau) \rangle$ for blackbody radiation, similarly, depends on τ but not t (recall (3.9.12)). In contrast, the random variations of the particle displacement in Brownian motion are not stationary, as is evident from Exercise 6.1.

If its random variations are statistically stationary, we can still express $y(t)$ as a Fourier series, provided we take the time T to be arbitrarily large *and* take the coefficients a_k and b_k themselves to be random variables. $\langle y(t) \rangle = 0$ implies that the averages $\langle a_k \rangle = \langle b_k \rangle = 0$. If we take $y(t)$ to be a Langevin random force $F_L(t)$, for example,

$$\begin{aligned}
\langle a_k^2 \rangle &= \frac{4}{T^2} \int_0^T dt_1 \int_0^T dt_2 \langle F_L(t_1) F_L(t_2) \rangle \cos \omega_k t_1 \cos \omega_k t_2 \\
&= \frac{4C}{T^2} \int_0^T dt_1 \int_0^T dt_2 \delta(t_1 - t_2) \cos \omega_k t_1 \cos \omega_k t_2 \\
&= \frac{4C}{T^2} \int_0^T dt_1 \cos^2 \omega_k t_1 \\
&= 2C/T.
\end{aligned} \tag{6.4.3}$$

More generally, from (6.3.16) and the fact that there are $1 \cdot 3 \cdot 5 \ldots (2n - 1)$ terms on the right side of that equation (see, e.g., Exercise 6.2),

$$\langle a_k^{2n} \rangle = 1 \cdot 3 \cdot 5 \ldots (2n - 1) \langle a_k^2 \rangle^n, \tag{6.4.4}$$

whereas

$$\langle a_k^{2n-1} \rangle = 0, \quad n = 1, 2, 3, \ldots. \tag{6.4.5}$$

Such moments characterize a Gaussian probability distribution: since the characteristic function

$$\begin{aligned}
\langle e^{i\xi a_k} \rangle &= 1 - \frac{\xi^2}{2!} \langle a_k^2 \rangle + \frac{\xi^4}{4!} \langle a_k^4 \rangle + \ldots = 1 - \frac{\xi^2}{2!} \langle a_k^2 \rangle + \frac{\xi^4}{4!} 3 \langle a_k^2 \rangle^2 + \ldots \\
&= \exp(-\xi^2 \langle a_k^2 \rangle / 2),
\end{aligned} \tag{6.4.6}$$

the probability distribution for a_k is (recall (3.7.22))

$$P(a_k) = \frac{1}{2\pi} \int_{-\infty}^{\infty} d\xi \langle e^{i\xi a_k} \rangle e^{-i\xi a_k} = \frac{1}{\sigma\sqrt{2\pi}} e^{-x^2/2\sigma^2}, \quad \sigma^2 = \langle a_k^2 \rangle. \tag{6.4.7}$$

The probability distribution for the b_k's in our example with $y(t) = F_L(t)$ obviously has the same Gaussian form and the same σ^2 ($= 2C/T$). The Gaussian probability distribution applies to *any* random variable a_k whose moments satisfy (6.4.4) and (6.4.5).

6.4.1 The Power Spectrum (Spectral Density)

The two-time (auto)correlation function provides one of the simplest and most important ways of characterizing a random process $y(t)$ in classical theory. Analogous correlation functions appear frequently in quantum statistical mechanics and quantum electrodynamics. In our quantum-electrodynamical treatment of resonance fluorescence in Section 5.6, for example, we found that the two-time correlation function of the atom's electric dipole moment determines the power radiated, and that its Fourier transform determines the power spectrum.[17] In that example, the atom experiences an externally applied field as well as quantum fluctuations attributable in part to the "vacuum" field, which plays a role analogous to the Langevin force F_L in the classical theory of Brownian motion.[18] The steady-state dipole correlation function depended on the two times only through their difference, and, likewise, we found that the velocity correlation function $\langle v(t_1)v(t_2)\rangle$ in the case of Brownian motion depends only on $t_2 - t_1$ when t_1 or t_2 is sufficiently large. We will now derive the *Wiener–Khintchine theorem* relating the "power spectrum" and correlation function $\langle y(t)y(t+\tau)\rangle$ of a *stationary* random process $y(t)$.[19] The analogous relation in quantum theory, where *operators* $p(t)$ and $p(t+\tau)$ do not, in general, commute, can be derived in a similar fashion.

We will continue to regard $\langle \ldots \rangle$ as an ensemble average and to assume that the random process $y(t)$ is such that this ensemble average is equivalent to an average over time for a single realization of the ensemble. We can proceed from the Fourier series (6.4.1), but, for a somewhat "cleaner" (albeit not mathematically rigorous) derivation, we will now express $y(t)$ as a Fourier *integral*:

$$y(t) = \int_{-\infty}^{\infty} Y(\omega)e^{-i\omega t}d\omega, \tag{6.4.8}$$

where

$$Y(\omega) = \frac{1}{2\pi}\int_{-\infty}^{\infty} y(t)e^{i\omega t}dt, \tag{6.4.9}$$

like $y(t)$, is a random "process." We will allow $y(t)$ to be complex, and define

$$g(\tau) = \langle y^*(t)y(t+\tau)\rangle, \tag{6.4.10}$$

which, under our assumption of stationarity, is independent of t. Now, since $\langle y^*(t')y(t)\rangle$ is a function only of $t - t' = \tau$, it follows that different Fourier components of the random process $y(t)$ are uncorrelated:

---

[17] See (5.6.2), (5.6.3), and (5.6.11).

[18] A *quantum* Langevin equation is considered in Section 6.5.

[19] As noted in Section 3.5, Einstein had deduced the theorem in 1914, well before Wiener or Khintchine.

$$\langle Y^*(\omega')Y(\omega)\rangle = \left(\frac{1}{2\pi}\right)^2 \int_{-\infty}^{\infty} dt \int_{-\infty}^{\infty} dt' \langle y^*(t')y(t)\rangle e^{i\omega t} e^{-i\omega' t'}$$

$$= \left(\frac{1}{2\pi}\right)^2 \int_{-\infty}^{\infty} dt' \int_{-\infty}^{\infty} d\tau \langle y^*(t')y(t'+\tau)\rangle e^{i\omega(t'+\tau)} e^{-i\omega' t'}$$

$$= \left(\frac{1}{2\pi}\right)^2 \int_{-\infty}^{\infty} d\tau g(\tau) e^{i\omega\tau} \int_{-\infty}^{\infty} dt' e^{i(\omega-\omega')t'}$$

$$= G(\omega)\delta(\omega - \omega'). \tag{6.4.11}$$

where the power spectrum, or spectral density, of $y(t)$ is defined as

$$G(\omega) = \frac{1}{2\pi} \int_{-\infty}^{\infty} g(\tau) e^{i\omega\tau} d\tau = \frac{1}{2\pi} \int_{-\infty}^{\infty} \langle y^*(t)y(t+\tau)\rangle e^{i\omega\tau} d\tau. \tag{6.4.12}$$

This is the Wiener–Khintchine theorem: the power spectrum $G(\omega)$ of a stationary random process $y(t)$ is the Fourier transform of the correlation function $\langle y^*(t)y(t+\tau)\rangle$.

If $y(t)$ is a real stationary process, $g(-\tau) = g(\tau)$, and therefore

$$G(\omega) = \frac{1}{\pi} \int_0^{\infty} \langle y(t)y(t+\tau)\rangle \cos\omega\tau d\tau. \tag{6.4.13}$$

An implicit assumption in our derivation is that $\int_{-\infty}^{\infty} |g(\tau)|d\tau < \infty$, so that the Fourier transform $G(\omega)$ exists. This condition is physically reasonable, since for large $|\tau|$, the correlation function $g(\tau)$ can be expected to decrease sufficiently rapidly to ensure absolute integrability. However, $|y(t)|$ itself is not integrable under our assumption of stationarity, which implies that $y(t)$ does not go to zero as $t \to \pm\infty$; our assumption that the Fourier transform $Y(\omega)$ (see (6.4.9)) exists is therefore unjustified. Of course, the Wiener–Khintchine theorem, which defines the power spectrum in terms of the Fourier transform of the correlation function of $y(t)$ rather than of $y(t)$ itself, has a more secure provenance than our derivation might suggest.[20]

If $y(t) = F_L(t)$, $\langle F_L(t)F_L(t+\tau)\rangle = C\delta(\tau)$ (see (6.3.3)), the spectral density $G(\omega)$ is independent of ω and is said to describe *white noise*:

$$G(\omega) = \frac{1}{2\pi} \int_{-\infty}^{\infty} C\delta(\tau) e^{i\omega\tau} d\tau = C/2\pi. \tag{6.4.14}$$

If $y(t) = v(t)$, where $\langle v(t_1)v(t_2)\rangle$ is given by (6.3.8) and at times much greater than τ_c depends on time only through $t_2 - t_1 = \tau > 0$, it follows from (6.4.13) that the power spectrum is a Lorentzian function of ω:

$$G(\omega) = \frac{1}{\pi} \int_0^{\infty} \left(\frac{k_B T}{m}\right) e^{-\tau/\tau_c} \cos\omega\tau d\tau = \frac{k_B T}{\pi m} \frac{1/\tau_c}{\omega^2 + 1/\tau_c^2}. \tag{6.4.15}$$

For such times, it follows from (6.4.13) that

---

[20] A more sophisticated approach to the Wiener–Khintchine theorem, along with some of its history, may be found in *Mandel and Wolf (1995)*.

$$\langle v^2 \rangle = 2 \int_0^\infty G(\omega)d\omega = \frac{2k_BT}{\pi m}\frac{1}{\tau_c}\int_0^\infty \frac{d\omega}{\omega^2 + 1/\tau_c^2} = \frac{k_BT}{m}, \tag{6.4.16}$$

which, of course, reproduces (6.3.8) for $t = t_1 = t_2 \gg \tau_c$. The low-frequency ($\omega \ll 1/\tau_c$) part of the power spectrum contributes a mean-square velocity fluctuation

$$\langle v^2 \rangle = \frac{2k_BT}{\pi m}\tau_c d\omega = \frac{2k_BT}{\pi m}\frac{m}{\gamma}(2\pi d\nu) = 4k_BTBd\nu, \tag{6.4.17}$$

where $B = 1/\gamma$ is the previously defined mobility, and $\nu = \omega/2\pi$ denotes (circular) frequency (as opposed to angular frequency ω).

As noted in Section 6.1, an argument against Einstein's explanation that Brownian motion, based on collisions of the suspended particles with molecules of the fluid, was that (6.4.16) predicts that the velocities of the suspended particles should be much larger than observed by eye. For a spherical particle of diameter 1 μm and density $\rho = 2$ g/cm^3, for example, $m \approx 10^{-12}$ g, and (6.4.16) gives $\sqrt{\langle v^2 \rangle} \approx 0.2$ cm/s, compared to the much smaller *observed* velocities $\sim 3 \times 10^{-4}$ cm/s. The difference can be explained by taking into account the finite response time τ_r involved in the visual observation of Brownian motion.[21] A response time τ_r implies an upper limit $\omega_r = 2\pi/\tau_r$ to the frequency response, and therefore that (6.4.16) should be replaced by

$$\langle v^2 \rangle = \frac{2k_BT}{\pi m}\frac{1}{\tau_c}\int_0^{\omega_r} \frac{d\omega}{\omega^2 + 1/\tau_c^2} = \frac{2k_BT}{\pi m}\tan^{-1}\left(\frac{2\pi\tau_c}{\tau_r}\right). \tag{6.4.18}$$

Assuming the fluid is water, which has a viscosity $\mu \cong 10^{-2}$ g cm^{-1} s^{-1} at 300 K, Stokes law (see Section 6.1) gives $\gamma \cong 3\pi \times 10^{-6}$ g/s, and $\tau_c = m/\gamma \approx 10^{-7}$ s, in our example. The human eye is generally assumed to have a response time of $\tau_r \approx 100$ ms, implying $2\pi\tau_c/\tau_r \ll 1$, and

$$\sqrt{\langle v^2 \rangle} \cong \sqrt{\frac{4k_BT}{m}\frac{\tau_c}{\tau_r}} \approx 4 \times 10^{-4} \text{ cm/s}, \tag{6.4.19}$$

which comports nicely with visual observations of Brownian motion.

6.5   The Quantum Langevin Equation

In Section 4.6, we considered the quantum-mechanical equation (4.6.4), which has the same form as the classical Langevin equation (6.3.1). The fluctuating force in (4.6.4) is represented by a Hermitian operator $F(t)$ having the explicit form (4.6.5) and satisfying the commutation relation (4.6.9). In writing the classical Langevin equation (6.3.1), in contrast, we simply *presumed* at the outset that the fluctuating force $F_L(t)$ satisfies (6.3.3), and we determined C by requiring that the equilibrium kinetic energy should satisfy the equipartition theorem.

[21] D. K. C. MacDonald, Phil. Mag. **41**, 814 (1950).

The quantum Langevin equation, (4.6.4), can be dealt with in much the same way as its classical counterpart (6.3.1), but with the crucial difference that $x(t)$ and $F(t)$ in (4.6.4) are operators. In fact, the formulas that we posited for $F(t)$ in (4.6.5), (4.6.6), and (4.6.8), lead to results that are similar in many respects to those derived from the classical Langevin equation. Consider, to begin with, the thermal-field expectation value

$$\frac{1}{2}\langle F(t)F(t') + F(t')F(t)\rangle = \frac{\gamma\hbar}{\pi}\int_0^\infty d\omega\,\omega \coth\left(\hbar\omega/2k_BT\right)\cos\omega(t-t'). \qquad (6.5.1)$$

This follows from (4.6.5) and (4.6.6), the assumption that the different reservoir oscillators responsible for the Langevin force (4.6.5) are uncorrelated, and that

$$\langle a^\dagger(\omega)a(\omega')\rangle = \frac{1}{e^{\hbar\omega/k_BT} - 1}\delta(\omega - \omega'), \qquad (6.5.2)$$

and

$$\langle a(\omega)a^\dagger(\omega')\rangle = \left(\frac{1}{e^{\hbar\omega/k_BT} - 1} + 1\right)\delta(\omega - \omega'). \qquad (6.5.3)$$

In the classical limit $\hbar\omega/k_BT \to 0$, $\coth\left(\hbar\omega/2k_BT\right) \to 2k_BT/\hbar\omega$ and

$$\frac{1}{2}\langle F(t)F(t') + F(t')F(t)\rangle = \frac{2\gamma\hbar}{\pi}\int_0^\infty d\omega\,\omega\left(\frac{k_BT}{\hbar\omega}\right)\cos\omega(t-t')$$
$$= 2\gamma k_BT\delta(t-t'). \qquad (6.5.4)$$

This quantum-mechanical expectation value for thermal equilibrium is equal to the classical average (6.3.3) when C in the latter is chosen to ensure that the average kinetic energy is $\frac{1}{2}k_BT$. The operator $F(t)$, furthermore, has Gaussian statistical properties analogous to (6.3.15) and (6.3.16), which we assumed for the classical Langevin force $F_L(t)$, with the overall time ordering of the non-commuting F factors maintained in each pair $\langle F(t_i)F(t_j)\rangle$. And since $[F(t), F(t')] \neq 0$ (see (4.6.9)), it is the *symmetrized* expectation value (6.5.4) that in the classical limit reduces to the classical average $\langle F_L(t)F_L(t')\rangle$.

Consider next $\langle v^2(t)\rangle = \langle \dot{x}^2(t)\rangle$. From (4.6.4) and the same assumptions that led to (6.5.1),

$$\langle v^2(t)\rangle = \langle v^2(0)\rangle e^{-2\gamma t/m}$$
$$+ \frac{\gamma\hbar}{\pi m^2}\int_0^\infty d\omega\,\frac{\omega \coth\left(\hbar\omega/2k_BT\right)}{\omega^2 + \gamma^2/m^2}\left(1 - 2e^{-\gamma t/m}\cos\omega t + e^{-2\gamma t/m}\right). \qquad (6.5.5)$$

In the classical limit $\hbar\omega/k_BT \to 0$, this has the same form as (6.3.7):

$$\langle v^2(t)\rangle = \langle v^2(0)\rangle e^{-2\gamma t/m} + \frac{2\gamma k_BT}{\pi m^2}\int_0^\infty \frac{d\omega}{\omega^2 + \gamma^2/m^2}\left(1 - 2e^{-\gamma t/m}\cos\omega t + e^{-2\gamma t/m}\right)$$
$$= \langle v^2(0)\rangle e^{-2t/\tau_c} + \frac{k_BT}{m}(1 - e^{-2t/\tau_c}), \quad \text{where } \tau_c = \gamma/m. \qquad (6.5.6)$$

The equilibrium kinetic energy has the value prescribed by the equipartition theorem at the temperature of the reservoir oscillators.

Since atoms can often be sensibly approximated by harmonic oscillators (recall the discussion in Section 2.5), it is also of interest to consider the quantum Langevin equation for a harmonic oscillator:

$$m\ddot{x}(t) + \gamma\dot{x}(t) + m\omega_0^2 x(t) = F(t) = \int_0^\infty d\omega \sqrt{\frac{\gamma\hbar\omega}{\pi}} \left[a(\omega)e^{-i\omega t} + a^\dagger(\omega)e^{i\omega t} \right]. \quad (6.5.7)$$

The expression for the Langevin force operator $F(t)$ here is the same as for the free particle; as discussed in Section 4.6, this ensures that the canonical commutation relation $[x(t), m\dot{x}(t)] = i\hbar$ holds at all times, as can be checked following the same approach used to derive equation (4.6.15):

$$\begin{aligned} [x(t), m\dot{x}(t)] &= [x(0), p(0)]e^{-\gamma t/m} \\ &\quad + \frac{2i\gamma\hbar}{\pi m}(1 - e^{-\gamma t/m}) \int_0^\infty \frac{d\omega\,\omega^2}{(\omega^2 - \omega_0^2)^2 + \gamma^2\omega^2/m^2} \\ &= i\hbar e^{-\gamma t/m} + i\hbar(1 - e^{-\gamma t/m}), \end{aligned} \quad (6.5.8)$$

and is, of course, independent of temperature. The preservation of a canonical commutation relation here and in the models below is not surprising, since essentially no approximations have been made and commutation relations do not change under the unitary time evolution: if $[A(0), B(0)] = C(0)$,

$$\begin{aligned} [A(t), B(t)] &= [U^\dagger(t)A(0)U(t), U^\dagger(t)B(0)U(t)] = U^\dagger(t)[A(0), B(0)]U(t) \\ &= U^\dagger(t)C(0)U(t) = C(t). \end{aligned} \quad (6.5.9)$$

The equilibrium expectation value of the oscillator Hamiltonian is found similarly. It has a rather complicated form:

$$\left\langle \frac{1}{2}m\dot{x}^2 + \frac{1}{2}m\omega_0^2 x^2 \right\rangle = \frac{\hbar\gamma}{2\pi m} \int_0^\infty d\omega \frac{\omega(\omega^2 + \omega_0^2)}{(\omega^2 - \omega_0^2)^2 + \gamma^2\omega^2/m^2} \coth \hbar\omega/2k_B T. \quad (6.5.10)$$

The temperature-dependent part in the classical limit is

$$\frac{\gamma k_B T}{\pi m} \int_0^\infty d\omega \frac{\omega^2 + \omega_0^2}{(\omega^2 - \omega_0^2)^2 + \gamma^2\omega^2/m^2} = k_B T, \quad (6.5.11)$$

as expected. In the weak-dissipation limit in which $\gamma/m \ll \omega_0$, similarly, the part of (6.5.10) that depends on T has the form $\hbar\omega_0/(e^{\hbar\omega_0/k_B T} - 1)$ for an oscillator of frequency ω_0 in thermal equilibrium at temperature T.

For arbitrary temperatures and dissipation strengths, expressions of physical interest for the oscillator—and the free particle—are generally quite complicated. In the zero-temperature limit, in fact, divergent results are obtained, simply because, as in our treatment of radiative level shifts of atoms in Section 4.8, we have in effect here made the dipole approximation but allowed the oscillators responsible for the Langevin force $F(t)$ to have infinitely high frequencies. If a high-frequency cutoff ω_c

is introduced for the reservoir oscillators, the energy (6.5.10) for $T = 0$, for example, becomes

$$\frac{\hbar\gamma}{2\pi m}\int_0^{\omega_c} d\omega \frac{\omega(\omega^2 + \omega_0^2)}{(\omega^2 - \omega_0^2)^2 + \gamma^2\omega^2/m^2} = \frac{\hbar\omega'}{\pi}\cos^{-1}\left(\frac{\gamma}{2m\omega_0}\right) + \frac{\hbar\gamma}{2\pi m}\log\left(\frac{\omega_c}{\omega_0}\right), \quad (6.5.12)$$

for $\omega' = (\omega_0^2 - \gamma^2/4m^2)^{1/2} > 0$, and it reduces to the zero-point energy $\frac{1}{2}\hbar\omega_0$ in the weak-dissipation limit.

6.5.1 Derivation of the Quantum Langevin Equation

We will now derive the quantum Langevin equation

$$m\ddot{x}(t) + \gamma\dot{x}(t) = -V'(x) + F(t), \quad V'(x) = \frac{d}{dx}V(x), \quad (6.5.13)$$

from a Hamiltonian for a particle ("system"), a reservoir of harmonic oscillators, and a particle-reservoir interaction:

$$H = H_{\text{sys}} + H_{\text{res}} + H_{\text{int}}, \quad (6.5.14)$$

$$H_{\text{sys}} = p^2/2m + V(x), \quad \text{where } [x, p] = i\hbar, \quad (6.5.15)$$

$$H_{\text{res}} = \sum_j \frac{1}{2}(p_j^2 + \omega_j^2 q_j^2), \quad \text{where } [q_i, p_j] = i\hbar\delta_{ij}, \quad (6.5.16)$$

and H_{int} is the particle–reservoir interaction. To simplify notation, we take all the reservoir oscillator masses m_j to be unity. We will assume, as is often done, a particle-reservoir interaction of the form

$$H_{\text{int}} = x\sum_j C_j q_j, \quad (6.5.17)$$

although, as discussed below, this is not strictly satisfactory. We specify later the coupling constant C_j between the particle and the reservoir oscillator j.

As usual, we define raising and lowering operators a_j and $a_j^\dagger$ for the reservoir oscillators:

$$a_j = \frac{i}{\sqrt{2\hbar\omega_j}}(p_j - iq_j), \quad \text{and} \quad a_j^\dagger = \frac{-i}{\sqrt{2\hbar\omega_j}}(p_j + iq_j), \quad \text{where } [a_i, a_j^\dagger] = \delta_{ij}, \quad (6.5.18)$$

$$q_j = \sqrt{\frac{\hbar}{2\omega_j}}(a_j + a_j^\dagger), \quad p_j = -i\sqrt{\frac{\hbar\omega_j}{2}}(a_j - a_j^\dagger), \quad (6.5.19)$$

in terms of which

$$H = p^2/2m + V(x) + \sum_j \frac{1}{2}(p_j^2 + \omega_j^2 q_j^2) + x\sum_j C_j q_j$$

$$= p^2/2m + V(x) + \sum_j \hbar\omega_j\left(a_j^\dagger a_j + \frac{1}{2}\right) + x\sum_j C_j\sqrt{\frac{\hbar}{2\omega_j}}(a_j + a_j^\dagger), \quad (6.5.20)$$

and we obtain the Heisenberg equations of motion

$$\dot{x} = p/m, \tag{6.5.21}$$

$$\dot{p} = -V'(x) - \sum_j C_j \sqrt{\frac{\hbar}{2\omega_j}}(a_j + a_j^\dagger), \tag{6.5.22}$$

$$\dot{a}_j = -i\omega_j a_j - iC_j \sqrt{\frac{1}{2\hbar\omega_j}}x.^{22} \tag{6.5.23}$$

The formal solution of (6.5.23) then implies the following equation for the operator $x(t)$:

$$m\ddot{x}(t) = -V'(x) + \sum_j C_j \sqrt{\frac{\hbar}{2\omega_j}}[a_j(0)e^{-i\omega_j t} + a_j^\dagger(0)e^{i\omega_j t}]$$

$$- \sum_j C_j^2 \frac{1}{\omega_j} \int_0^t dt' x(t')\sin\omega_j(t'-t). \tag{6.5.24}$$

We now take the reservoir-oscillator frequencies to be continuously and uniformly distributed and make the replacements $\omega_j \to \omega$, $C_j \to C(\omega)$, and $a_j(0) \to a(\omega)$; then, (6.5.24) becomes

$$m\ddot{x}(t) = -V'(x) + \int_0^\infty d\omega C(\omega)\sqrt{\frac{\hbar}{2\omega}}[a(\omega)e^{-i\omega t} + a^\dagger(\omega)e^{i\omega t}]$$

$$- \int_0^\infty d\omega \frac{C^2(\omega)}{\omega} \int_0^t dt' x(t')\sin\omega(t'-t). \tag{6.5.25}$$

The second term on the right side has the same form as the Langevin force operator (6.5.7) if we choose $C^2(\omega) = (2\gamma/\pi)\omega^2$, consistent with the fluctuation–dissipation theorem in the case of ohmic dissipation (see Section 6.6). Then,

$$m\ddot{x}(t) = -V'(x) + \int_0^\infty d\omega \sqrt{\frac{\gamma\hbar\omega}{\pi}}[a(\omega)e^{-i\omega t} + a^\dagger(\omega)e^{i\omega t}]$$

$$- \frac{2\gamma}{\pi} \int_0^\infty d\omega\omega \int_0^t dt' x(t')\sin\omega(t'-t)$$

$$= -V'(x) + F(t) + 2\gamma \int_0^t dt' x(t')\frac{\partial}{\partial t'}\delta(t'-t). \tag{6.5.26}$$

The last term has a divergent part:

---

[22] a_j differs from the definition of the lowering operator a used in (3.1.6) by a factor i. This difference is of no physical consequence. The definitions (6.5.18) are used here simply in order to obtain a Langevin force $F(t)$ (see (6.5.26)) of the same form as (4.6.5); otherwise, the expression for the Langevin force would have a minus sign between $a(\omega)\exp(-i\omega t)$ and $a^\dagger(\omega)\exp(i\omega t)$, and an overall factor i.

$$\int_0^t dt' x(t') \frac{\partial}{\partial t'} \delta(t' - t) = [x(t)\delta(0) - x(0)\delta(t) - \int_0^t dt' \dot{x}(t')\delta(t' - t)]$$

$$= [x(t)\delta(0) - x(0)\delta(t)] - \frac{1}{2}\dot{x}(t). \qquad (6.5.27)$$

To avoid the divergence we introduce a cutoff frequency ω_c and replace $\delta(0)$ by $(1/\pi) \int_0^{\omega_c} d\omega \cos(0) = \omega_c/\pi$. Then, for times of interest, sufficiently long after the time $(t = 0)$ when the particle is brought into contact with the reservoir,

$$m\ddot{x}(t) + \gamma \dot{x}(t) + V'(x) - \frac{2\gamma\omega_c}{\pi}x(t) = F(t). \qquad (6.5.28)$$

The last term on the left can be avoided by starting from a different Hamiltonian, H', which is obtained by adding $x^2 \sum_j (C_j^2/2\omega_j^2)$ to the original Hamiltonian H:

$$H' = H + x^2 \sum_j (C_j^2/2\omega_j^2)$$

$$= p^2/2m + V(x) + \sum_j \frac{1}{2}(p_j^2 + \omega_j^2 q_j^2) + x\sum_j C_j q_j + x^2 \sum_j (C_j^2/2\omega_j^2)$$

$$= p^2/2m + V(x) + \sum_j \frac{1}{2}[p_j^2 + \omega_j^2 (q_j + \frac{C_j x}{\omega_j^2})^2]. \qquad (6.5.29)$$

With this Hamiltonian, there is a contribution

$$(i\hbar)^{-1}[p, x^2]\sum_j C_j^2/2\omega_j^2 = -x\sum_j C_j^2/\omega_j^2 \qquad (6.5.30)$$

to $\dot{p}$ in addition to the expression for $m\ddot{x}$ that follows from the Hamiltonian H.[23] In the continuum model as above for the reservoir oscillators with a high-frequency cutoff ω_c, this contribution is

$$-x\int_0^{\omega_c} d\omega C^2(\omega)/\omega^2 = -\frac{2\gamma\omega_c}{\pi}x, \qquad (6.5.31)$$

and results in the replacement of (6.5.28) by the quantum Langevin equation (6.5.13):

$$m\ddot{x}(t) + \gamma\dot{x}(t) + V'(x) = F(t) + \frac{2\gamma\omega_c}{\pi}x(t) - \frac{2\gamma\omega_c}{\pi}x(t) = F(t), \qquad (6.5.32)$$

where, from the expression for $F(t)$ in (6.5.26),

$$[F(t), F(t')] = 2i\gamma\hbar\frac{\partial}{\partial t}\delta(t - t'), \qquad (6.5.33)$$

and, for thermal equilibrium at temperature T,

$$\frac{1}{2}\langle F(t)F(t') + F(t')F(t)\rangle = \frac{\gamma\hbar}{\pi}\int_0^\infty d\omega\omega \coth\left(\hbar\omega/2k_BT\right)\cos\omega(t - t'). \qquad (6.5.34)$$

[23] G. W. Ford, J. T. Lewis, and R. F. O'Connell, Phys. Rev. A **37**, 4419 (1988).

Exercise 6.3: Verify (6.5.33) and (6.5.34).

The Hamiltonian H' can be derived by a unitary transformation of a Hamiltonian

$$H'' = \frac{1}{2m}\left(p + \sum_j C_j p_j/\omega_j^2\right)^2 + V(x) + \sum_j \frac{1}{2}(p_j^2 + \omega_j^2 q_j^2) \tag{6.5.35}$$

as follows.[24] Define

$$\mathcal{U} = \exp(-ix\sum_j C_j p_j/\hbar\omega_j^2) \tag{6.5.36}$$

and use the operator identity (2.2.18) to obtain

$$\mathcal{U}^\dagger x \mathcal{U} = x, \quad \mathcal{U}^\dagger p \mathcal{U} = p - \sum_j C_j p_j/\omega_j^2,$$

$$\mathcal{U}^\dagger q_j \mathcal{U} = q_j + xC_j/\omega_j^2, \quad \mathcal{U}^\dagger p_j \mathcal{U} = p_j, \tag{6.5.37}$$

and

$$\mathcal{U}^\dagger H'' \mathcal{U} = p^2/2m + V(x) + \sum_j \frac{1}{2}\left[p_j^2 + \omega_j^2(q_j + xC_j/\omega_j^2)^2\right] = H'. \tag{6.5.38}$$

In the context of electromagnetic coupling, this is essentially just the transformation between the minimal coupling and electric dipole forms of the Hamiltonian (Sections 2.2 and 4.2), albeit we are working here with a model that allows only one space dimension. Thus H'' has the form of the minimal coupling Hamiltonian with "vector potential" $\sum_j A_j = \sum_j p_j/\omega_j$, and C_j/ω_j playing the role of a "charge" $-e$; $\omega_j q_j$ in this analogy acts as $E_j = -\dot{A}_j = -\dot{p}_j/\omega_j \ (= \omega_j q_j)$. H', similarly, has the form of the electric dipole Hamiltonian with an interaction $x\sum_j C_j q_j$ analogous to $-ex\sum_j E_j$, and the term $x^2\sum_j C_j^2/2\omega_j^2$ analogous to the (divergent) $\mathbf{P}^2$ term.[25]

Ford et al. observe that, because of the term proportional to x, the Hamiltonian H "has a grave defect: for a free particle, $V(x) = 0$, there is no lower bound on the energy. This means that there is no thermal equilibrium state; the bath [reservoir] is not passive."[26] The Hamiltonians H' and H'' obviously do not have this defect; as long as $p^2/2m + V(x)$ has a ground state they will be bounded from below.

6.5.2 Beyond Ohmic Dissipation

We have derived the quantum Langevin equation (6.5.13) for the special case of ohmic dissipation, that is, for a drag force $-\gamma\dot{x}$, by assuming a continuous distribution of

---

[24] G. W. Ford, J. T. Lewis, and R. F. O'Connell, Phys. Rev. A **37**, 4419 (1988).

[25] Recall (4.2.10) and (4.2.12).

[26] G. W. Ford, J. T. Lewis, and R. F. O'Connell, Phys. Rev. A **37**, 4419 (1988), p. 4425.

reservoir-oscillator frequencies and choosing $C^2(\omega) = (2\gamma/\pi)\omega^2$. More generally, we obtain from the Hamiltonian H' (or H''),

$$m\ddot{x}(t) + \int_{-\infty}^{t} dt'\, \dot{x}(t')\mu(t - t') + V'(x) = F(t), \tag{6.5.39}$$

$$\mu(t) = \sum_j \frac{C_j^2}{\omega_j^2} \cos\omega_j t, \tag{6.5.40}$$

$$F(t) = \sum_j C_j \sqrt{\frac{\hbar}{2\omega_j}} \left(a_j e^{-i\omega_j t} + a_j^\dagger e^{i\omega_j t}\right), \tag{6.5.41}$$

$$[a_i, a_j] = 0, \quad [a_i, a_j^\dagger] = \delta_{ij}. \tag{6.5.42}$$

From (6.5.41) and (6.5.42),

$$[F(t), F(t')] = -2i\sum_j \frac{\hbar C_j^2}{\omega_j} \sin\omega_j(t - t'), \tag{6.5.43}$$

and, assuming there are no correlations between reservoir oscillators,

$$\frac{1}{2}\langle F(t)F(t') + F(t')F(t)\rangle = \sum_j \frac{\hbar C_j^2}{\omega_j}\left(\langle a_j^\dagger a_j\rangle + \frac{1}{2}\right)\cos\omega_j(t - t')$$

$$= \sum_j \frac{\hbar C_j^2}{2\omega_j}\coth\frac{\hbar\omega_j}{2k_B T}\cos\omega_j(t - t') \tag{6.5.44}$$

for thermal equilibrium.

The model leading to the quantum Langevin equation (6.5.13) for ohmic dissipation is a relatively simple and very special one, and the most useful one for many systems of interest.[27] It follows from the more general quantum Langevin equation (6.5.39) when $\mu(t) = 2\gamma\delta(t), \gamma > 0$. In another, "quasi-continuum model," it is assumed that the reservoir-oscillator frequencies are discrete but evenly spaced: $\omega_j \to n/\rho$, $n = 0, 1, 2, \ldots$. To obtain a Langevin equation for $x(t)$ that reduces to (6.5.13) when the frequency spacing approaches zero, we also let

$$\mu(t - t') = \sum_j \frac{C_j^2}{\omega_j^2}\cos\omega_j(t - t') \to \beta^2\sum_{n=1}^{\infty}\cos[n\rho^{-1}(t' - t)] + \beta^2/2, \tag{6.5.45}$$

where β is another model parameter in addition to the "density of states" ρ. Then, from the Poisson summation formula,

---

[27] Some applications of this form of the quantum Langevin equation, and many references, may be found in R. F. O'Connell in *Fluctuations and Noise in Photonics and Quantum Optics III*, Proc. SPIE **5842**, P. R. Hemmer et al., eds. SPIE, Bellingham, WA, 2005, pp. 206–219.

$$\sum_{n=-\infty}^{\infty} e^{inx} = \sum_{n=-\infty}^{\infty} \delta(x/2\pi - n), \tag{6.5.46}$$

$$\mu(t - t') \to \pi\beta^2\rho \sum_{n=-\infty}^{\infty} \delta(t' - t + 2\pi n\rho), \tag{6.5.47}$$

and (6.5.39) in this model becomes the delay-differential equation

$$m\ddot{x}(t) + \gamma\dot{x}(t) + 2\gamma\sum_{n=1}^{\infty} \dot{x}(t - n\tau_\rho)\theta(t - n\tau_\rho) + V'(x) = F(t), \tag{6.5.48}$$

where $\theta(t)$ is the unit step function, $\tau_\rho = 2\pi\rho$, and $\gamma = \frac{1}{2}\pi\beta^2\rho$ has the form of a Fermi golden-rule rate. When the spacing of the reservoir-oscillator frequencies is very small, τ_ρ is very large, and, if we take $\tau_\rho \to \infty$ while keeping γ finite, (6.5.48) reduces to the form (6.5.13).[28]

The first careful and correct derivation of a quantum Langevin equation focused on a model with a distribution of reservoir oscillator frequencies resulting in the quantum Langevin equation for ohmic dissipation, but implicit in the formulation is the more general quantum Langevin equation (6.5.39).[29] Define the Fourier transform

$$\tilde{\mu}(z) = \int_0^\infty dt\mu(t)e^{izt}, \quad \text{where } \text{Im}(z) > 0, \tag{6.5.49}$$

in terms of which

$$\mu(t) = \frac{1}{2\pi}\int_{-\infty}^{\infty} d\omega\tilde{\mu}(\omega)e^{-i\omega t}, \tag{6.5.50}$$

$$\dot{\mu}(t) = -\frac{-i}{2\pi}\int_{-\infty}^{\infty} d\omega\omega\tilde{\mu}(\omega)e^{-i\omega t}.[30] \tag{6.5.51}$$

For the memory function given by (6.5.49), it follows from (6.5.43) that

$$[F(t), F(t')] = 2\hbar\dot{\mu}(t) = \frac{-i}{\pi}\int_{-\infty}^{\infty} d\omega\hbar\omega\tilde{\mu}(\omega)e^{-i\omega(t-t')}. \tag{6.5.52}$$

The fact that $\mu(t)$ is real implies that

$$\tilde{\mu}(-\omega) = \tilde{\mu}(\omega)^*, \tag{6.5.53}$$

and so the real part $\tilde{\mu}_R(\omega)$ of $\tilde{\mu}$ must be an even function of ω, and the imaginary part $\tilde{\mu}_I(\omega)$ must be an odd function of ω. Therefore,

---

[28] Quasi-continuum models have been considered in various contexts. See, for instance, P. W. Milonni. J. R. Ackerhalt, H. W. Galbraith, and M.-L. Shih, Phys. Rev. A **28**, 32 (1983), and references therein.

[29] G. W. Ford, M. Kac, and P. Mazur, J. Math. Phys. **6**, 504 (1965). See also G. W. Ford and M. Kac, J. Stat. Mech. **46**, 803 (1987).

[30] Since the definition of $\tilde{\mu}(z)$ only makes sense for $\text{Im}(z) > 0$ (otherwise, the integral diverges), $\tilde{\mu}(\omega)$ in these formulas and below should be understood to be $\tilde{\mu}(\omega + i0^+)$.

$$[F(t), F(t')] = -\frac{2i}{\pi} \int_0^\infty d\omega\, \hbar\omega\, \tilde{\mu}_R(\omega) \sin\omega(t - t'). \tag{6.5.54}$$

From (6.5.40),

$$\tilde{\mu}(\omega) = \sum_j \frac{C_j^2}{2\omega_j^2} \int_0^\infty dt \left(e^{i(\omega+\omega_j+i\epsilon)t} + e^{i(\omega-\omega_j+i\epsilon)t}\right) \quad (\epsilon \to 0^+)$$

$$= i \sum_j \frac{C_j^2}{2\omega_j^2} \left[\mathrm{P}\frac{1}{\omega + \omega_j} - i\pi\delta(\omega + \omega_j) + \mathrm{P}\frac{1}{\omega - \omega_j} - i\pi\delta(\omega - \omega_j)\right], \tag{6.5.55}$$

where we have used (4.4.34), and so

$$\tilde{\mu}_R(\omega) = \frac{\pi}{2} \sum_j \frac{C_j^2}{\omega_j^2} [\delta(\omega + \omega_j) + \delta(\omega - \omega_j)], \tag{6.5.56}$$

allowing (6.5.44) to be expressed as

$$\frac{1}{2}\langle F(t)F(t') + F(t')F(t)\rangle = \frac{1}{\pi} \int_0^\infty d\omega\, \hbar\omega\, \tilde{\mu}_R(\omega) \coth\left(\hbar\omega/2k_BT\right) \cos\omega(t-t'). \tag{6.5.57}$$

The function $\mu(t)$ must satisfy certain conditions, independent of any specific model Hamiltonian from which we can derive a quantum Langevin equation of the form (6.5.39).[31] Since $x(t)$ should not depend on $x(t')$ for $t' > t$, $\tilde{\mu}(z)$ must, from (6.5.49), be analytic in the upper half of the complex frequency plane (Im$(z) > 0$). In addition, (6.5.53) must hold, simply because $x(t)$ is a Hermitian operator. It has also been shown, from fairly general thermodynamic considerations, that $\tilde{\mu}_R(\omega) \geq 0$. These three properties follow without reference to any specific model for the coupling of the particle to the reservoir. Ford et al. call the specific model above, which satisfies these properties, the "independent oscillator model."[32] They "stress that this does not mean that in every physical situation in which the quantum Langevin equation (6.5.39) arises the actual bath is an [independent oscillator] bath, but rather that from a study of the equation and its solutions . . . one cannot tell the difference. It is remarkable that such a naive and simple model has such generality."[33]

It is obvious, and important, that the way the particle is coupled to the bath determines both the "memory" function $\mu(t)$ and the properties of the operator $F(t)$ representing the fluctuating Langevin force; those properties are determined by $\tilde{\mu}_R(\omega)$. $F(t)$ and $\mu(t)$ have the same physical origin—the coupling of the particle to the bath. The "particle" is more generally some system of interest that interacts with an "environment" represented by a reservoir ("bath") of harmonic oscillators. The assumption that the environment is not significantly affected by the system supports the modeling

[31] G. W. Ford, J. T. Lewis, and R. F. O'Connell, Phys. Rev. A **37**, 4419 (1988).

[32] Ibid., p. 4425.

[33] Ibid., p. 4423.

of the environment as a collection of harmonic oscillators: any physical system weakly perturbed from a stable equilibrium acts in effect as a harmonic oscillator.[34]

6.6   The Fluctuation–Dissipation Theorem

The Einstein relations (6.1.8) and (6.1.9) showed a connection between fluctuations and dissipation; this connection was to be found later in other contexts, including Einstein's work on blackbody radiation. In this work, he related the momentum fluctuations of bodies interacting with a thermal field to the drag force resulting from movement through the field (Section 2.8). In work on telephony and vacuum tube amplifiers, it was found that the voltage between terminals of conductors undergoes fluctuations that increase with the resistance (dissipation) R. A systematic experimental study entitled "Thermal agitation of electricity in conductors" was undertaken by Johnson, who expressed his results for mean-square current fluctuations in a formula that was soon after derived by Nyquist.[35] Nyquist's formula for *Johnson noise* in a frequency band $d\nu$ is

$$\langle (\Delta V)^2 \rangle = \langle V^2 \rangle - \langle V \rangle^2 = \langle V^2 \rangle = 4Rk_BT d\nu. \tag{6.6.1}$$

This has the same form as (6.4.17); as discussed below, this is no mere coincidence.

Nyquist's ingenious derivation of (6.6.1) may be summarized as follows. For two conductors I and II connected as in Figure 6.1, the emf V due to the thermal motion of charges in I results in a current $I = V/2R$ in the circuit, and an absorption of power I^2R in II. Power flows from II to I, and since the two conductors have the same equilibrium temperature T, the power flowing from I to II equals the power flowing from II to I. Nyquist argued that this balance must hold not only for the total power exchanged, but also for the power exchanged within any frequency band $d\nu$ of the fluctuating voltage, and that this fluctuating voltage due to "thermal agitation" in conductors must be a universal function of ν, R, and T.

To derive this function, Nyquist imagined the conductors to be connected by a lossless transmission line (see Figure 6.2) with an impedance equal to the resistance R at each terminal. With such impedance matching, the power flowing in each direction along the line will be absorbed without reflection by the resistance at each end.[36] Once thermal equilibrium is reached, energy is transported by traveling voltage waves from one conductor to the other and absorbed. Now let the line be disconnected from I and II. Then, the voltage is in the form of standing waves that vanish at the two ends, the

---

[34] A trivial argument from classical physics: a system characterized by a potential energy $V(X)$ and a stable equilibrium position X_0 will have a potential energy $V(X_0 + x) \cong V(X_0) + (1/2)kx^2$ for small displacements x from equilibrium, a stable equilibrium at X_0 implying $V'(X_0) = 0$ and $k = V''(X_0) > 0$.

[35] J. B. Johnson, Phys. Rev. **32**, 97 (1928), in which Johnson inferred a value for k_B that was within about 8% of the actual Boltzmann constant; H. Nyquist, Phys. Rev. **32**, 110 (1928).

[36] Nyquist remarked that power loss due to radiation can be avoided by arranging for one conductor to be internal to the other, presumably as in a coaxial cable.

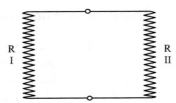

Fig. 6.1 Connected conductors I and II, each having a resistance R. From H. Nyquist, Phys. Rev. **32**, 110 (1928), with permission.

allowed frequencies being $nv/2\ell$, $n = 1, 2, 3, \ldots$, where v is the propagation velocity of voltage waves on the line of length ℓ. The number of possible frequencies in a band $d\nu$ is therefore $2\ell d\nu/v$; this is just the one-dimensional of (3.3.18), with $k = 2\pi\nu/v$, and the assumption that $2\ell d\nu/v \gg 1$. According to the equipartition theorem, each "oscillator" of frequency ν has an average energy $k_B T$ in thermal equilibrium, so the average energy in a frequency band $d\nu$ of voltage fluctuations must be $2\ell k_B T d\nu/v$ on the disconnected transmission line. The power P absorbed at each conductor before the transmission line was disconnected must therefore be $\ell k_B T d\nu/v$ divided by the wave transit time ℓ/v:

$$P = k_B T d\nu. \tag{6.6.2}$$

But since the current in each conductor is $V/2R$, the average power absorbed by each conductor can also be expressed as

$$\langle I^2 \rangle R = \frac{\langle V^2 \rangle}{(2R)^2} R, \tag{6.6.3}$$

and equating this to (6.6.2) gives Nyquist's equation, (6.6.1).

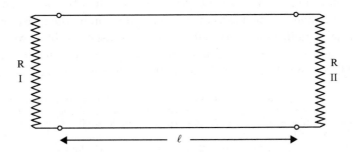

Fig. 6.2 Two conductors connected by a lossless transmission line. From H. Nyquist, Phys. Rev. **32**, 110 (1928), with permission.

Nyquist went on to show that (6.6.1) applies in the more general case in which the impedance is not purely resistive and the resistance may be a function $R(\nu)$ of frequency. He also noted that, if instead of the classical equipartition law, an average energy $h\nu(e^{h\nu/k_BT} - 1)^{-1}$ per oscillation degree of freedom is assumed, then (6.6.1) generalizes to

$$\langle V^2 \rangle = \frac{4R(\nu)h\nu d\nu}{e^{h\nu/k_BT} - 1},$$ (6.6.4)

which, of course, reduces to (6.6.1) when $k_BT \gg h\nu$. He concluded his paper with the remark that "within the ranges of frequency and temperature where experimental information is available this expression [(6.6.4)] is indistinguishable from that obtained from the equipartition law."[37]

Callen and Welton remarked that "the Nyquist relation is . . . of a form unique in physics, correlating a property of a system in equilibrium (that is, the voltage fluctuations) with a parameter which characterizes an irreversible process (that is, the electrical resistance). The equation, furthermore, gives not only the mean square fluctuating voltage, but provides, in addition, the frequency spectrum of the fluctuations."[38] They remarked further that it had "frequently been conjectured that the Nyquist relation can be extended to a general class of dissipative systems other than merely electrical systems."[39] They then generalized the Nyquist relation to derive what has long since been widely known as the *fluctuation–dissipation theorem*.[40]

The fluctuation–dissipation theorem can be derived in different ways. We will at first follow closely the original approach and notation of Callen and Welton. Consider a system subjected to a perturbation V. Let its unperturbed Hamiltonian be H_0, and let the effect of the perturbation be described by an interaction VQ, where Q will, in general, depend on coordinate and momentum variables characterizing the system of interest. Since V will be treated to begin with as a prescribed function of time, we can, in effect, take the total Hamiltonian to be

$$H = H_0 + VQ$$ (6.6.5)

without having to include a contribution from the energy of the system responsible for the force V. As the notation suggests, Q would be related to charge in a resistor, and V to voltage, in the example of Johnson noise.

It is assumed that the system can absorb energy when the perturbing "force" V is applied. The rate of energy absorption can be obtained in second-order perturbation theory by assuming a sinusoidal variation $V(t) = V_0 \cos \omega t$. Then, according to the

---

[37] H. Nyquist, Phys. Rev. **32**, 110 (1928), p. 113.

[38] H. B. Callen and T. A. Welton, Phys. Rev. **83**, 34 (1951), p. 34.

[39] Ibid.

[40] Nyquist's relation was not entirely "unique in physics"—recall the remarks in Sections 2.8.3 and 6.1 regarding Einstein's fluctuation–dissipation relations in blackbody radiation and Brownian motion.

golden rule (see equation (2.3.9)), the rate for the system to go from an initial energy state $|E_n\rangle$ to a final energy state $|E_n + \hbar\omega\rangle$ is

$$w_G(E_n) = \frac{\pi}{2\hbar} V_0^2 \big|\langle E_n + \hbar\omega|Q|E_n\rangle\big|^2 \rho(E_n + \hbar\omega), \qquad (6.6.6)$$

where $\rho(E)$ is the usual density (in energy) of final states. This is the rate for the gain of energy $\hbar\omega$ by the system. The transition rate for a loss of energy $\hbar\omega$, similarly, is

$$w_L(E_n) = \frac{\pi}{2\hbar} V_0^2 \big|\langle E_n - \hbar\omega|Q|E_n\rangle\big|^2 \rho(E_n - \hbar\omega), \qquad (6.6.7)$$

again assuming the system starts in an initial state of energy E_n. The net power to the system in equilibrium at temperature T is the difference between the power gain and loss, with each energy state $|E_n\rangle$ weighted by the Boltzmann factor $f(E_n)$:

$$\begin{aligned}
P(\omega) &= \sum_n \hbar\omega[w_G(E_n) - w_L(E_n)]f(E_n) \\
&= \frac{\pi\omega}{2} V_0^2 \sum_n \Big[\big|\langle E_n + \hbar\omega|Q|E_n\rangle\big|^2 \rho(E_n + \hbar\omega) \\
&\quad - \big|\langle E_n - \hbar\omega|Q|E_n\rangle\big|^2 \rho(E_n - \hbar\omega)\Big]f(E_n).
\end{aligned} \qquad (6.6.8)$$

We now use the fact that Q is Hermitian ($\langle E_n - \hbar\omega|Q|E_n\rangle = \langle E_n|Q|E_n - \hbar\omega\rangle$) and the assumption of a continuous distribution of energies. Replacing $\sum_n$ by $\int dE\rho(E)$ and using $\rho(E) = 0$ for $E < 0$ and $f(E + \hbar\omega) = f(E)e^{-\hbar\omega/k_B T}$, we obtain

$$P(\omega) = \frac{\pi\omega}{2} V_0^2 (1 - e^{-\hbar\omega/k_B T}) \int_0^\infty dE |\langle E + \hbar\omega|Q|E\rangle|^2 \rho(E + \hbar\omega)\rho(E)f(E) \quad (6.6.9)$$

for the power absorbed by the system when it experiences a periodic perturbing force $V_0 \cos\omega t$.

This is a general result, simply derived, for any system that is (a) *dissipative*, that is, that can absorb energy when a time-dependent perturbation is applied, and (b) *linear*, in the sense that the power absorbed is quadratic in the perturbation strength.[41] The power absorbed by such a system in response to an applied force $V_0 e^{-i\omega t}$ can also be expressed as

$$P(\omega) = \mathrm{Re}\langle V\dot{Q}\rangle, \qquad (6.6.10)$$

with $\dot{Q}$ linearly proportional to the force:

$$V = Z(\omega)\dot{Q}. \qquad (6.6.11)$$

The time-averaged power absorbed is therefore

$$P(\omega) = \frac{1}{2} V_0^2 \mathrm{Re}\Big(\frac{1}{Z(\omega)}\Big) = \frac{1}{2} V_0^2 \frac{R(\omega)}{|Z(\omega)|^2}, \qquad (6.6.12)$$

---

[41] Ibid.

where $R(\omega)$ is the real part of $Z(\omega)$. Equating this expression for the absorbed power to (6.6.9) gives the relation

$$R(\omega)/|Z(\omega)|^2 = \pi\omega(1 - e^{-\hbar\omega/k_B T})$$
$$\times \int_0^\infty dE |\langle E + \hbar\omega|Q|E\rangle|^2 \rho(E + \hbar\omega)\rho(E)f(E). \quad (6.6.13)$$

This has been derived under the assumption of a sinusoidal force $V(t)$, but it can easily be derived in the more general case in which the Fourier transforms of $V(t)$ and $\dot{Q}(t)$ are related by

$$V(\omega) = Z(\omega)\dot{Q}(\omega), \quad (6.6.14)$$

which holds for any linear dissipative system.[42]

The idea now is to relate the dissipative effect (power absorption) of the system when a perturbation is applied as above to fluctuations of the system when it is in thermal equilibrium and there is no applied perturbation. Consider, then, the mean–square fluctuation of $\dot{Q}$ when the system is in the energy eigenstate $|E_n\rangle$. Inserting unity in the form of the sum $\sum_m |E_m\rangle\langle E_m|$ over the complete set of energy eigenstates, and using the Heisenberg equation of motion $\dot{Q} = (i\hbar)^{-1}[Q, H_0]$ for the *unperturbed* evolution of Q, we can express this fluctuation as follows:

$$\langle E_n|\dot{Q}^2|E_n\rangle = \sum_m \langle E_n|\dot{Q}|E_m\rangle\langle E_m|\dot{Q}|E_n\rangle$$
$$= -\frac{1}{\hbar^2}\sum_m \langle E_n|(QH_0 - H_0Q)|E_m\rangle\langle E_m|(QH_0 - H_0Q)|E_n\rangle$$
$$= -\frac{1}{\hbar^2}\sum_m (E_m - E_n)\langle E_n|Q|E_m\rangle(E_n - E_m)\langle E_m|Q|E_n\rangle$$
$$= \frac{1}{\hbar^2}\sum_m (E_n - E_m)^2|\langle E_m|Q|E_n\rangle|^2. \quad (6.6.15)$$

We again use the assumption of a continuous distribution of energy, define a frequency $\omega = |E_n - E_m|/\hbar$, and replace the sum over E_m by two integrals, one for $E_m = E_n + \hbar\omega$ and one for $E_m = E_n - \hbar\omega$:

$$\langle \dot{Q}^2\rangle = \langle E_n|\dot{Q}^2|E_n\rangle = \frac{1}{\hbar^2}\int_0^\infty \hbar d\omega(\hbar\omega)^2 \sum_n \Big\{|\langle E_n + \hbar\omega|Q|E_n\rangle|^2 \rho(E_n + \hbar\omega)$$
$$+ |\langle E_n - \hbar\omega|Q|E_n\rangle|^2 \rho(E_n - \hbar\omega)\Big\}f(E_n). \quad (6.6.16)$$

The sum over n is the same as that in (6.6.8), except that it involves the sum instead of the difference of the two terms in curly brackets. It can be expressed as in (6.6.9) except that $1 + e^{-\hbar\omega/k_B T}$ appears instead of $1 - e^{-\hbar\omega/k_B T}$:

[42] When there is little chance of confusion, we frequently denote the Fourier transform of a function $Q(t)$ by $Q(\omega)$.

$$\langle \dot{Q}^2 \rangle = \hbar \int_0^\infty d\omega \omega^2 (1 + e^{-\hbar\omega/k_B T})$$

$$\times \int_0^\infty dE |\langle E + \hbar\omega |Q|E\rangle|^2 \rho(E + \hbar\omega)\rho(E)f(E). \tag{6.6.17}$$

Comparison with (6.6.13) gives

$$\langle \dot{Q}^2 \rangle = \hbar \int_0^\infty d\omega \omega^2 \frac{1}{\pi\omega} \frac{1 + e^{-\hbar\omega/k_B T}}{1 - e^{-\hbar\omega/k_B T}} \frac{R(\omega)}{|Z(\omega)|^2}$$

$$= \frac{1}{\pi} \int_0^\infty d\omega \hbar\omega \coth\left(\hbar\omega/2k_B T\right) \frac{R(\omega)}{|Z(\omega)|^2}. \tag{6.6.18}$$

Finally we can relate $\langle \dot{Q}^2 \rangle$ to $\langle V^2 \rangle$ using the relation (6.6.14) between their Fourier transforms:

$$\langle V^2 \rangle = \frac{1}{\pi} \int_0^\infty d\omega \hbar\omega \coth\left(\hbar\omega/2k_B T\right) R(\omega) = \frac{2}{\pi} \int_0^\infty d\omega R(\omega)E(\omega, T), \tag{6.6.19}$$

$$E(\omega, T) = \frac{1}{2}\hbar\omega + \frac{\hbar\omega}{e^{\hbar\omega/k_B T} - 1}. \tag{6.6.20}$$

This is the fluctuation–dissipation theorem of Callen and Welton.

The calculations leading to (6.6.19) started from the assumption of a linear dissipative system that can irreversibly absorb energy when it is subjected to a time-periodic perturbation. The rate of dissipation of energy was derived and associated with a generalized "resistance" $R(\omega)$, the real part of an "impedance" $Z(\omega)$ characterizing the linear response of the system to a force V. Then, an expression for fluctuations in thermal *equilibrium* of the unperturbed dissipative system was obtained and found to be related to the *irreversible* power dissipation characterized by $R(\omega)$. This relation, together with (6.6.14), implies a spontaneously fluctuating force.[43] The fluctuation–dissipation theorem relates the frequency spectrum of this fluctuating force to $R(\omega)$, as Callen and Welton illustrated with several examples.[44]

Callen and Welton obviously chose their notation to correspond directly to the example of Johnson noise. In that example, the fluctuation–dissipation theorem (6.6.19) implies a voltage fluctuation

$$\langle V^2 \rangle = 4R(\nu)\left(\frac{1}{2}h\nu + \frac{h\nu}{e^{h\nu/k_B T} - 1}\right)d\nu, \tag{6.6.21}$$

in a frequency band $d\nu$. This generalizes the Nyquist expression (6.6.4) to include the zero-point energy $\frac{1}{2}h\nu$. In other words, it allows for quantum fluctuations of the voltage at the absolute zero of temperature and relates the fluctuations at all temperatures to the resistance, a *macroscopic* property.

---

[43] It is perhaps worth stressing the simple point that the fluctuations are an inherent property of a thermal-equilibrium state and do not imply any sort of deviation from equilibrium!

[44] B. Callen and T. A. Welton, Phys. Rev. **83**, 34 (1951).

In Brownian motion, where there is a drag force $-\gamma v$, or $-\gamma \dot{Q}$, the equation describing the response to a periodic force is

$$m\frac{d\dot{Q}}{dt} + \gamma\dot{Q} = V(\omega)e^{-i\omega t}, \tag{6.6.22}$$

implying (see (6.6.14)) $Z(\omega) = -im\omega + \gamma$, a "resistance" $R(\omega) = \gamma$, and a fluctuating force $F\ (=V)$ such that

$$\langle F^2\rangle = \frac{2\gamma}{\pi}\int_0^\infty d\omega E(\omega, T), \tag{6.6.23}$$

in agreement with (6.5.1).

Consider now a slightly more complicated example in which the dissipation mechanism is the radiation reaction force on a point electric dipole (recall (1.11.14)):

$$F_{\mathrm{RR}} = \frac{e^2}{6\pi\epsilon_0 c^3}\,\dddot{x}\,(t). \tag{6.6.24}$$

For an applied perturbing force $F(t) = F(\omega)e^{-i\omega t}$,

$$m\frac{d\dot{x}}{dt} = -m\omega_0^2 x + \frac{e^2}{6\pi\epsilon_0 c^3}\frac{d^2\dot{x}}{dt^2} + F(\omega)e^{-i\omega t}, \tag{6.6.25}$$

implying

$$\dot{x}(\omega) = \frac{-i\omega F(\omega)}{m(\omega_0^2 - \omega^2) - ie^2\omega^3/6\pi\epsilon_0 c^3}, \tag{6.6.26}$$

$$Z(\omega) = \frac{i}{\omega}[m(\omega_0^2 - \omega^2) - ie^2\omega^3/6\pi\epsilon_0 c^3], \tag{6.6.27}$$

and $R(\omega) = e^2\omega^2/6\pi\epsilon_0 c^3$. The dissipative system in this example is the vacuum electric field that "absorbs" energy radiated by the dipole. According to the fluctuation–dissipation theorem, the dissipative effect ("radiative resistance") on the dipole means that the electric field in thermal equilibrium with it must have a mean-square fluctuation such that

$$\langle e^2 E_x^2\rangle = \frac{2}{\pi}\int_0^\infty d\omega R(\omega)E(\omega, T) = \frac{2}{\pi}\int_0^\infty d\omega\Big(\frac{e^2\omega^2}{6\pi\epsilon_0 c^3}\Big)E(\omega, T). \tag{6.6.28}$$

For the isotropic field there is, therefore, an energy density

$$\epsilon_0\langle\mathbf{E}^2\rangle = 3\epsilon_0\langle E_x^2\rangle = \frac{1}{\pi^2 c^3}\int_0^\infty d\omega\omega^2 E(\omega, T)$$
$$= \int_0^\infty d\omega\frac{\hbar\omega^3}{\pi^2 c^3}\Big[\frac{1}{2} + \frac{1}{e^{\hbar\omega/k_B T} - 1}\Big], \tag{6.6.29}$$

which, of course, corresponds to the Planck spectrum, including the zero-point energy.

We have already considered a model of a free, point electron in vacuum in Section 4.6. There, we saw that the fluctuating force $eE_x(t)$ was needed to prevent violation of the canonical commutation relation for the electron, and that the frequency spectrum of the zero-temperature fluctuating vacuum field is just what is needed to "enforce" that commutation law: dissipation necessitates fluctuations, and the spectrum of the fluctuations is determined by the form of the dissipation. In the case of ohmic dissipation, similarly, the zero-temperature spectrum of the fluctuating force specified by the fluctuation–dissipation theorem is such as to preserve the canonical commutation relation. Although satisfying, none of this is surprising. In both examples, and more generally, the dissipation and the fluctuating force both come from the coupling of the particle to the same reservoir. If no approximations are made, the commutation relations, which are independent of temperature, must be preserved, simply because the time evolution of the Heisenberg-picture operators is unitary. The fluctuating force comes from the unperturbed oscillations of the reservoir oscillators, whereas the dissipative force comes from the "back reaction" of the oscillators on the particle. It was essentially the fluctuation–dissipation relation between the vacuum field and the back reaction (radiation reaction) that allowed us to interpret spontaneous emission and the (non-relativistic) Lamb shift in terms of either vacuum-field fluctuations or radiation reaction.

Exercise 6.4: Consider a linear oscillator with natural oscillation frequency ω_0 and a dissipative force $-\gamma\dot{x}$. Derive the formula for the impedance $Z(\omega)$, and show that $R(\omega) = \gamma$.

6.6.1   Another Form of the Fluctuation–Dissipation Theorem

Equation (6.6.14) defines the impedance $Z(\omega)$ via the presumed linear response of $\dot{Q}(\omega) = -i\omega Q(\omega)$ to the force $V(\omega)$. It can also be written in terms of the *generalized susceptibility* $\alpha(\omega) = \alpha_R(\omega) + i\alpha_I(\omega)$:[45]

$$Q(\omega) = \frac{i}{\omega}\frac{V(\omega)}{Z(\omega)} = \frac{i}{\omega}\frac{V(\omega)}{R(\omega) + iX(\omega)} = \alpha(\omega)V(\omega). \tag{6.6.30}$$

Taking the real part, we have

$$R(\omega)/|Z(\omega)|^2 = \omega\alpha_I(\omega), \tag{6.6.31}$$

and, from (6.6.18),

$$\langle Q^2 \rangle = \frac{\hbar}{\pi}\int_0^\infty d\omega\,\alpha_I(\omega)\coth\left(\hbar\omega/2k_BT\right). \tag{6.6.32}$$

This is just another form of the fluctuation–dissipation theorem. The form (6.6.19) gives the mean–square fluctuation of the Langevin force acting on the Q system in

---

[45] In calling $\alpha(\omega)$ the generalized susceptibility, we follow the terminology of *Landau and Lifshitz (2004)*.

terms of the "resistance" $R(\omega)$ characterizing the dissipation, whereas (6.6.32) is the mean-square fluctuation of Q itself, expressed in terms of the imaginary (dissipative) part of the generalized susceptibility defined by (6.6.30).

Consider again the example of a linear oscillator with natural oscillation frequency ω_0 and a dissipative force $-\gamma \dot{x}$, where the operator x represents the electron coordinate:

$$\alpha(\omega) = (m\omega_0^2 - m\omega^2 - i\gamma\omega)^{-1}, \tag{6.6.33}$$

$$\alpha_I(\omega) = \frac{1}{m} \frac{\gamma\omega/m}{(\omega_0^2 - \omega^2)^2 + \gamma^2\omega^2/m^2}, \tag{6.6.34}$$

and, from (6.6.32) and (6.6.18),

$$\langle x^2 \rangle = \frac{\hbar\gamma}{\pi m^2} \int_0^\infty d\omega \frac{\omega \coth(\hbar\omega/2k_BT)}{(\omega_0^2 - \omega^2)^2 + \gamma^2\omega^2/m^2}, \tag{6.6.35}$$

$$\langle \dot{x}^2 \rangle = \frac{\hbar\gamma}{\pi m^2} \int_0^\infty d\omega \frac{\omega^3 \coth(\hbar\omega/2k_BT)}{(\omega_0^2 - \omega^2)^2 + \gamma^2\omega^2/m^2}. \tag{6.6.36}$$

The expectation value in the thermal equilibrium of the oscillator Hamiltonian is therefore

$$\left\langle \frac{1}{2}m\dot{x}^2 + \frac{1}{2}m\omega_0^2 x^2 \right\rangle = \frac{\hbar\gamma}{2\pi m} \int_0^\infty d\omega \frac{\omega(\omega^2 + \omega_0^2) \coth(\hbar\omega/2k_BT)}{(\omega_0^2 - \omega^2)^2 + \gamma^2\omega^2/m^2}, \tag{6.6.37}$$

in agreement with the expression (6.5.10) obtained from the quantum Langevin equation.

In the derivation of the quantum Langevin equation, it was assumed that the system of interest is linearly coupled to a reservoir of harmonic oscillators having a dense distribution of energy levels and acting as a passive "heat bath" in thermal equilibrium. These same assumptions were the basis of the derivation by Callen and Welton of the fluctuation–dissipation theorem. In fact, (6.5.57), relating the symmetrized correlation function of the Langevin force operator to the dissipation characterized by $\tilde{\mu}_R(\omega)$ in the quantum Langevin equation, has been called the "second" fluctuation–dissipation theorem.[46]

Introducing the Fourier transform of the symmetrized correlation function for x, and carrying out some algebra akin to that used in our discussion of the Wiener–Khintchine theorem (see Section 6.4.1), we obtain

$$\frac{1}{2}\langle x(t)x(t') + x(t')x(t) \rangle = \int_0^\infty d\omega \langle x^2(\omega) \rangle \cos\omega(t - t'). \tag{6.6.38}$$

$\langle x^2(\omega) \rangle$, the Fourier transform of $\langle x^2 \rangle$, can be read off from (6.6.32):

$$\langle x^2(\omega) \rangle = \frac{\hbar}{\pi}\alpha_I(\omega) \coth(\hbar\omega/2k_BT), \tag{6.6.39}$$

[46] R. Kubo, Rep. Prog. Phys. **29**, 255 (1966).

to give the fluctuation–dissipation theorem in the form relating the correlation function of x to the imaginary part of the generalized susceptibility:

$$\frac{1}{2}\langle x(t)x(t') + x(t')x(t)\rangle = \frac{\hbar}{\pi}\int_0^\infty d\omega\,\alpha_I(\omega)\coth\left(\hbar\omega/2k_BT\right)\cos\omega(t-t'). \quad (6.6.40)$$

Compare this to the correlation function for the Langevin force (cf. (6.5.57)), which depends only on the real part of the Fourier transform of the memory function.

From (6.6.40),

$$\langle x^2\rangle = \frac{\hbar}{\pi}\int_0^\infty d\omega\,\alpha_I(\omega)\coth\left(\hbar\omega/2k_BT\right), \quad (6.6.41)$$

$$\langle \dot{x}^2\rangle = \frac{\hbar}{\pi}\int_0^\infty d\omega\,\omega^2\alpha_I(\omega)\coth\left(\hbar\omega/2k_BT\right), \quad (6.6.42)$$

and

$$\left\langle\frac{1}{2}m\dot{x}^2 + \frac{1}{2}m\omega_0^2 x^2\right\rangle = \frac{\hbar m}{2\pi}\int_0^\infty d\omega(\omega^2 + \omega_0^2)\alpha_I(\omega)\coth\left(\hbar\omega/2k_BT\right), \quad (6.6.43)$$

which, of course, reduces to (6.6.37) in the case of ohmic dissipation.

Exercise 6.5: Using the quantum Langevin equation (6.5.39) with $V(x) = \frac{1}{2}m\omega_0^2$ and taking $t \to \infty$, show that

$$\alpha(\omega) = [m(\omega_0^2 - \omega^2) - i\omega\tilde{\mu}(\omega)]^{-1},$$

so that, in this oscillator model,

$$\alpha_I(\omega) = \omega|\alpha(\omega)|^2\tilde{\mu}_R(\omega).$$

The generalized susceptibility $\alpha(\omega)$ is essentially the Fourier transform of a Green function. We can express the "input–output" relation describing the response to a force $F_x(t)$,

$$x(\omega) = \alpha(\omega)F_x(\omega), \quad (6.6.44)$$

in the time domain using the Fourier transforms

$$x(t) = \frac{1}{2\pi}\int_{-\infty}^\infty d\omega\,x(\omega)e^{-i\omega t}, \quad (6.6.45)$$

$$F_x(t) = \frac{1}{2\pi}\int_{-\infty}^\infty d\omega\,F_x(\omega)e^{-i\omega t}, \quad (6.6.46)$$

from which

$$x(t) = \frac{1}{2\pi} \int_{-\infty}^{\infty} d\omega e^{-i\omega t} \alpha(\omega) \int_{-\infty}^{\infty} dt' F_x(t') e^{i\omega t'}$$

$$= \int_{-\infty}^{\infty} dt' F_x(t') \frac{1}{2\pi} \int_{-\infty}^{\infty} d\omega \alpha(\omega) e^{-i\omega(t-t')}$$

$$= \int_{-\infty}^{\infty} dt' G(t - t') F_x(t'), \tag{6.6.47}$$

$$G(t - t') = \frac{1}{2\pi} \int_{-\infty}^{\infty} d\omega \alpha(\omega) e^{-i\omega(t-t')}. \tag{6.6.48}$$

The causality condition that there can be "no output before any input" requires

$$G(t - t') = 0 \quad \text{for} \quad t < t' \tag{6.6.49}$$

and therefore

$$x(t) = \int_{-\infty}^{t} dt' G(t - t') F_x(t'). \tag{6.6.50}$$

We review below the fact that this causality condition requires that $\alpha(\omega)$ be an analytic function in the upper half of the complex ω plane.

As the notation suggests, $\alpha(\omega)$ in the example of an electric dipole in an electric field of frequency ω is just the polarizability (see Section 1.5), and like the polarizability it has some important properties that follow from very general considerations that are essentially, though not exclusively, classical in nature. We now briefly recall some of these properties.

The most obvious property is that, if we are dealing with real quantities $x(t)$ and $F_x(t)$, $\alpha(\omega)$ satisfies the "crossing relation"

$$\alpha^*(\omega) = \alpha(-\omega), \tag{6.6.51}$$

and therefore the real and imaginary parts of $\alpha(\omega)$ are even and off functions, respectively, of ω:

$$\alpha_R(-\omega) = \alpha_R(\omega) \quad \text{and} \quad \alpha_I(-\omega) = -\alpha_I(\omega). \tag{6.6.52}$$

This follows from (6.6.48) and the fact that $G(t)$ must be real.

Causality—in the sense that $G(t - t')$ must vanish for $t < t'$—requires that

$$\alpha(\omega) = \int_{-\infty}^{\infty} dt G(\tau) e^{i\omega\tau} = \int_{0}^{\infty} dt G(\tau) e^{i\omega\tau}, \tag{6.6.53}$$

so that $\alpha(\omega)$ is obviously an analytic function in the upper half of the complex ω plane. From Cauchy's theorem, therefore,

$$\alpha(\omega) = \frac{1}{2\pi i} \oint_C \frac{\alpha(\omega')}{\omega' - \omega} d\omega'. \tag{6.6.54}$$

For real values of ω the integration contour C is taken to be along the real axis from $-\infty$ to ∞, with a small semicircle over the pole at $\omega' = \omega$, and is closed by a large semicircle in the upper half of the complex plane; it is assumed that there is no contribution from the large semicircle, that is, that $\alpha(\omega) \to 0$ faster that $1/|\omega|$ as $|\omega| \to \infty$. Letting the radius of the small semicircle over ω go to zero, it follows that the real and imaginary parts of $\alpha(\omega)$ are related by Hilbert transforms:

$$\alpha_R(\omega) = \frac{1}{\pi} \fint_{-\infty}^{\infty} \frac{\alpha_I(\omega')}{\omega' - \omega} d\omega', \tag{6.6.55}$$

$$\alpha_I(\omega) = -\frac{1}{\pi} \fint_{-\infty}^{\infty} \frac{\alpha_R(\omega')}{\omega' - \omega} d\omega', \tag{6.6.56}$$

where $\fint$ denotes the Cauchy principal part of the integral.[47] The crossing relation can be used to re-express these relations:

$$\alpha_R(\omega) = \frac{1}{\pi} \fint_0^{\infty} \frac{\alpha_I(\omega')}{\omega' - \omega} d\omega' + \frac{1}{\pi} \fint_{-\infty}^{0} \frac{\alpha_I(\omega')}{\omega' - \omega} d\omega' = \frac{2}{\pi} \fint_0^{\infty} \frac{\omega' \alpha_I(\omega')}{\omega'^2 - \omega^2} d\omega', \tag{6.6.57}$$

and, likewise,

$$\alpha_I(\omega) = -\frac{2}{\pi} \fint_0^{\infty} \frac{\omega \alpha_R(\omega')}{\omega'^2 - \omega^2} d\omega'. \tag{6.6.58}$$

For imaginary frequencies, (6.6.54) gives

$$\alpha(i\omega) = \frac{1}{2\pi i} \oint_C \frac{\alpha(\omega')}{\omega' - i\omega} d\omega' = -\frac{i}{2\pi} \oint_C \frac{\omega' \alpha(\omega')}{\omega'^2 + \omega^2} d\omega' + \frac{\omega}{2\pi} \oint_C \frac{\alpha(\omega')}{\omega'^2 + \omega^2} d\omega'. \tag{6.6.59}$$

Assuming again that $\alpha(\omega)$ falls off fast enough as $|\omega| \to \infty$, the first integral in the second equality is

$$-\frac{i}{2\pi} \oint_C \frac{\omega' \alpha(\omega')}{\omega'^2 + \omega^2} d\omega' = \frac{1}{2\pi} \int_{-\infty}^{\infty} \frac{\omega' \alpha_I(\omega')}{\omega'^2 + \omega^2} d\omega' = \frac{1}{\pi} \int_0^{\infty} \frac{\omega' \alpha_I(\omega')}{\omega'^2 + \omega^2} d\omega', \tag{6.6.60}$$

since $\alpha_R(\omega)$ and $\alpha_I(\omega)$ are even and odd functions, respectively, of ω. The last integral in (6.6.59) can be evaluated using the residue theorem:

$$\frac{\omega}{2\pi} \oint_C \frac{\alpha(\omega')}{\omega'^2 + \omega^2} d\omega' = \frac{\omega}{2\pi} (2\pi i) \lim_{\omega' \to i\omega} \left[(\omega' - i\omega) \frac{\alpha(\omega')}{(\omega' + i\omega)(\omega' - i\omega)} \right]$$

$$= \frac{1}{2} \alpha(i\omega). \tag{6.6.61}$$

Therefore, (6.6.59) reduces to

$$\alpha(i\omega) = \frac{1}{\pi} \int_0^{\infty} \frac{d\omega' \omega' \alpha_I(\omega')}{\omega'^2 + \omega^2} + \frac{1}{2} \alpha(i\omega), \tag{6.6.62}$$

---

[47] It hardly seems necessary to go into these Kramers–Kronig relations in any detail, as this is done in standard texts such as *Jackson*. Readers wishing to study such dispersion relations in greater mathematical detail are referred to the authoritative treatment by H. M. Nussenzveig, *Causality and Dispersion Relations*, Academic Press, New York, 1972. For an interesting account of how Kramers and Kronig derived their dispersion relations in 1926 and 1927, see C. F. Bohren, Eur. J. Phys. **31**, 573 (2010).

or

$$\alpha(i\omega) = \frac{2}{\pi} \int_0^\infty \frac{\omega'\alpha_I(\omega')}{\omega'^2 + \omega^2} d\omega'. \qquad (6.6.63)$$

Similar relations hold for the complex dielectric permittivity $\epsilon(\omega)$. Assuming that $\epsilon(\omega) \to \epsilon_0$ as $\omega \to \infty$,[48] these relations are derived as above for the function $\epsilon(\omega) - \epsilon_0$. Thus, for example,

$$\epsilon_R(\omega) - \epsilon_0 = \frac{1}{\pi} \fint_{-\infty}^\infty \frac{\epsilon_I(\omega')}{\omega' - \omega} d\omega', \qquad (6.6.64)$$

$$\epsilon_I(\omega) = -\frac{1}{\pi} \fint_{-\infty}^\infty \frac{\omega'\epsilon_R(\omega')}{\omega' - \omega} d\omega', \qquad (6.6.65)$$

and

$$\epsilon(i\omega) - \epsilon_0 = \frac{2}{\pi} \int_0^\infty \frac{\omega'\epsilon_I(\omega')}{\omega'^2 + \omega^2} d\omega'. \qquad (6.6.66)$$

The relation (6.6.64) implies that, except for free space ($\epsilon(\omega) = \epsilon_0$), $\epsilon_I(\omega)$ cannot vanish at all frequencies. In other words, a medium cannot be non-absorbing at all frequencies.

Suppose the only force acting on a linear oscillator is the Langevin force $F(t)$. From (6.6.50) and (6.5.54),

$$\begin{aligned}
[x(t), x(t')] &= \int_{-\infty}^\infty dt_1 \int_{-\infty}^\infty dt_2 G(t - t_1) G(t' - t_2)[F(t_1), F(t_2)] \\
&= -\frac{2i\hbar}{\pi} \int_0^\infty d\omega\, \omega\tilde{\mu}_R(\omega) \int_{-\infty}^\infty dt_1 \int_{-\infty}^\infty dt_2 G(t - t_1) G(t' - t_2) \sin\omega(t_1 - t_2) \\
&= -\frac{2i\hbar}{\pi} \int_0^\infty d\omega\, \omega\tilde{\mu}_R(\omega)|\alpha(\omega)|^2 \sin\omega(t_1 - t_2), \qquad (6.6.67)
\end{aligned}$$

where we have used (6.6.48), (6.6.51), and

$$\delta(\omega \pm \omega') = \frac{1}{2\pi} \int_{-\infty}^\infty dt\, e^{i(\omega\pm\omega')t}. \qquad (6.6.68)$$

From Exercise 6.5,

$$[x(t), x(t')] = -\frac{2i\hbar}{\pi} \int_0^\infty d\omega\, \alpha_I(\omega) \sin\omega(t - t'), \qquad (6.6.69)$$

and, differentiating both sides with respect to t' and then setting $t' = t$,

$$[x(t), m\dot{x}(t)] = [x(t), p(t)] = \frac{2mi\hbar}{\pi} \int_0^\infty d\omega\, \omega\alpha_I(\omega). \qquad (6.6.70)$$

---

[48] This must be the case because the medium cannot respond sufficiently rapidly to fields at extremely high frequencies to affect their propagation. Polarizabilities, likewise, must vanish as $\omega \to \infty$.

We expect from physical considerations that $\alpha(\omega)$ varies as $-1/\omega^2$ at high frequencies; we assumed this in connection with the Thomas–Reiche–Kuhn sum rule in Section 2.3.3. In the present notation for the generalized susceptibility, this means

$$\alpha(\omega) \approx -1/m\omega^2, \quad \omega \to \infty \tag{6.6.71}$$

as exemplified by the susceptibility (6.6.33) in the case of ohmic damping. Assuming this high-frequency limit, we deduce from (6.6.57) that, in the high-frequency limit,

$$\alpha(\omega) = -\frac{1}{m\omega^2} = \alpha_R(\omega) = -\frac{2}{\pi\omega^2} \int_0^\infty d\omega' \omega' \alpha_I(\omega'), \tag{6.6.72}$$

or

$$\int_0^\infty d\omega \omega \alpha_I(\omega) = \frac{\pi}{2m} \tag{6.6.73}$$

and, therefore, $[x(t), p(t)] = i\hbar$. In other words, the property (6.6.71) ensures the preservation of the commutator.

In applying the fluctuation–dissipation theorem to nonlinear systems, the effect of the applied force is evaluated in linear response theory, as in (6.6.11), (6.6.30), and (6.6.44). In other words, *the system itself can be nonlinear, but the theorem only applies to its linear response*, described by the generalized susceptibility, to an applied force.

We have considered the fluctuation–dissipation theorem for only a single dynamical variable, the coordinate x of a linear oscillator. In Section 6.7, we apply the fluctuation–dissipation theorem to different operators of an interacting system, each characterized by a susceptibility in its (linear) response to an applied force.

6.6.2   The Fluctuation–Dissipation Theorem for the Electric Field

The fluctuation–dissipation theorem can also be applied to operators that depend on a continuous variable such as a spatial coordinate $\mathbf{r}$.[49] We consider as an example the electric field operator satisfying

$$\nabla \times \nabla \times \mathbf{E}(\mathbf{r}, t) + \frac{1}{c^2}\frac{\partial^2}{\partial t^2}\mathbf{E}(\mathbf{r}, t) = -\frac{1}{\epsilon_0 c^2}\frac{\partial^2}{\partial t^2}\mathbf{P}(\mathbf{r}, t) \tag{6.6.74}$$

when there is a polarization density (dipole moment per unit volume) $\mathbf{P}(\mathbf{r}, t)$. In terms of Fourier components $\mathbf{E}(\mathbf{r}, \omega)$ and $\mathbf{P}(\mathbf{r}, \omega)$ defined by

$$\mathbf{E}(\mathbf{r}, t) = \int_0^\infty d\omega \left[\mathbf{E}(\mathbf{r}, \omega)e^{-i\omega t} + \mathbf{E}^\dagger(\mathbf{r}, \omega)e^{i\omega t}\right] \tag{6.6.75}$$

and

$$\mathbf{P}(\mathbf{r}, t) = \int_0^\infty d\omega \left[\mathbf{P}(\mathbf{r}, \omega)e^{-i\omega t} + \mathbf{P}^\dagger(\mathbf{r}, \omega)e^{i\omega t}\right], \tag{6.6.76}$$

[49] See, for instance, G. W. Ford, Contemp. Phys. **58**, 244 (2017).

$$\nabla \times \nabla \times \mathbf{E}(\mathbf{r}, \omega) - \frac{\omega^2}{c^2}\mathbf{E}(\mathbf{r}, \omega) = \frac{\omega^2}{\epsilon_0 c^2}\mathbf{P}(\mathbf{r}, \omega). \qquad (6.6.77)$$

Each Cartesian component $E_i(\mathbf{r}, \omega)$ of the field $\mathbf{E}(\mathbf{r}, \omega)$ generated by $\mathbf{P}(\mathbf{r}, \omega)$ can be expressed as

$$E_i(\mathbf{r}, \omega) = \frac{\omega^2}{\epsilon_0 c^2} \int d^3 r' G_{ij}(\mathbf{r}, \mathbf{r}', \omega) P_j(\mathbf{r}', \omega). \qquad (6.6.78)$$

$G(\mathbf{r}, \mathbf{r}', \omega)$ is the dyadic Green tensor satisfying

$$\nabla \times \nabla \times G(\mathbf{r}, \mathbf{r}', \omega) - \frac{\omega^2}{c^2}G(\mathbf{r}, \mathbf{r}', \omega) = I\delta^3(\mathbf{r} - \mathbf{r}'). \qquad (6.6.79)$$

I is the unit tensor, and a summation over repeated indices is implied here and below. The fluctuation–dissipation theorem relates the thermal-equilibrium expectation value $\langle \mathbf{E}(\mathbf{r}, \omega) \cdot \mathbf{E}(\mathbf{r}, \omega') \rangle$ to the Green function. We can deduce this relation by analogy to (6.6.30) and (6.6.32): taking $\langle \mathbf{E}(\mathbf{r}, \omega) \cdot \mathbf{E}(\mathbf{r}, \omega') \rangle$, $\mathbf{P}(\mathbf{r}, \omega)$, and $G(\mathbf{r}, \mathbf{r}', \omega)$ to correspond to $\langle Q^2(\omega) \rangle$, $V(\omega)$, and $\alpha(\omega)$, respectively, the formula corresponding to (6.6.32) is[50]

$$\langle \mathbf{E}(\mathbf{r}, \omega) \cdot \mathbf{E}(\mathbf{r}, \omega') \rangle = \frac{\hbar}{\pi} \times \frac{\omega^2}{\epsilon_0 c^2}\mathrm{Im}[G_{ii}(\mathbf{r}, \mathbf{r}, \omega)]\coth(\hbar\omega/2k_BT)\delta(\omega - \omega')$$

$$= \frac{\hbar\omega^2}{\pi\epsilon_0 c^2}\mathrm{Tr}G_I(\mathbf{r}, \mathbf{r}, \omega)\coth(\hbar\omega/2k_BT)\delta(\omega - \omega'). \qquad (6.6.80)$$

The delta function $\delta(\omega - \omega')$ appears because different frequency components of thermal (and zero-point) fields are uncorrelated:

$$\langle \mathbf{E}^2(\mathbf{r}, t) \rangle = \int_0^\infty d\omega \int_0^\infty d\omega' \langle \mathbf{E}^\dagger(\mathbf{r}, \omega) \cdot \mathbf{E}(\mathbf{r}, \omega') + \langle \mathbf{E}(\mathbf{r}, \omega) \cdot \mathbf{E}^\dagger(\mathbf{r}, \omega') \rangle$$

$$= \frac{\hbar}{\pi\epsilon_0 c^2} \int_0^\infty d\omega G_I(\mathbf{r}, \mathbf{r}, \omega)\coth(\hbar\omega/2k_BT). \qquad (6.6.81)$$

At zero temperature, $\langle \mathbf{E}^\dagger(\mathbf{r}, \omega) \cdot \mathbf{E}(\mathbf{r}, \omega') \rangle = 0$, whereas

$$\langle \mathbf{E}(\mathbf{r}, \omega) \cdot \mathbf{E}^\dagger(\mathbf{r}, \omega') \rangle = \frac{\hbar\omega^2}{\pi\epsilon_0 c^2}\mathrm{Tr}G_I(\mathbf{r}, \mathbf{r}, \omega)\delta(\omega - \omega'). \qquad (6.6.82)$$

The Green dyadic $G_{ij}^0(\mathbf{r}, \mathbf{r}', \omega)$ for the electric field in free space can be calculated most easily by consideration of (6.6.78) for the special case of an electric dipole at $\mathbf{r} = \mathbf{r}'$, for which $\mathbf{P}(\mathbf{r}, \omega) = \mathbf{p}(\omega)\delta^3(\mathbf{r} - \mathbf{r}')$ and the positive-frequency part of the electric field is given by (4.3.11):

$$E_i(\mathbf{r}, \omega) = \frac{1}{4\pi\epsilon_0}\left[\frac{\omega^2}{c^2 R}(\delta_{ij} - \hat{R}_i\hat{R}_j) + (\delta_{ij} - 3\hat{R}_i\hat{R}_j)\left(\frac{i\omega}{cR^2} - \frac{1}{R^3}\right)\right]e^{ikR}p_j(\omega)$$

$$= \frac{\omega^2}{\epsilon_0 c^2}G_{ij}^0(\mathbf{r}, \mathbf{r}', \omega)p_j(\omega), \quad \mathbf{R} \equiv \mathbf{r} - \mathbf{r}', \qquad (6.6.83)$$

---

[50] This surmise is validated in Exercise 7.10.

and therefore

$$G_{ij}^0(\mathbf{r}, \mathbf{r}', \omega) = \frac{1}{4\pi} \left[\frac{\delta_{ij} - \hat{R}_i \hat{R}_j}{R} \right.$$
$$\left. + (\delta_{ij} - 3\hat{R}_i \hat{R}_j) \left(\frac{i}{kR^2} - \frac{1}{k^2 R^3} \right) \right] e^{ikR}, \tag{6.6.84}$$

where $k = \omega/c$. From (6.6.81) and

$$\lim_{\mathbf{r}' \to \mathbf{r}} \text{Im}[G_{ij}^0(\mathbf{r}, \mathbf{r}', \omega)] = \frac{\omega}{6\pi c} \delta_{ij}, \tag{6.6.85}$$

it follows, as expected, that the field energy density is

$$\left\langle \frac{1}{2} \epsilon_0 \mathbf{E}^2(\mathbf{r}, t) + \frac{1}{2} \mu_0 \mathbf{H}^2(\mathbf{r}, t) \right\rangle = \left\langle \epsilon_0 \mathbf{E}^2(\mathbf{r}, t) \right\rangle = \int_0^\infty d\omega \rho(\omega), \tag{6.6.86}$$

where

$$\rho(\omega) = \frac{\hbar \omega^3}{\pi^2 c^3} \left(\frac{1}{e^{\hbar\omega/k_B T} - 1} + \frac{1}{2} \right) \tag{6.6.87}$$

is the Planck spectrum, including the zero-point energy density.

6.7   The Energy and Free Energy of an Oscillator in a Heat Bath

Consider again the Hamiltonian (6.5.29),

$$p^2/2m + V(x) + \sum_j \frac{1}{2} \left[p_j^2 + \omega_j^2 \left(q_j + \frac{C_j x}{\omega_j^2} \right)^2 \right] = H_1 + H_2, \tag{6.7.1}$$

$$H_1 = p^2/2m + V(x), \quad H_2 = \sum_j \frac{1}{2} \left[p_j^2 + \omega_j^2 \left(q_j + \frac{C_j x}{\omega_j^2} \right)^2 \right], \tag{6.7.2}$$

which implies the Heisenberg equations of motion

$$\dot{x} = p/m, \tag{6.7.3}$$

$$\dot{p} = -m\omega_0^2 x - \sum_j C_j q_j - x \sum_j C_j^2/\omega_j^2, \tag{6.7.4}$$

$$\dot{q}_j = p_j \tag{6.7.5}$$

$$\dot{p}_j = -\omega_j^2 q_j - C_j x, \tag{6.7.6}$$

for a harmonic oscillator of natural frequency ω_0 coupled to a reservoir of harmonic oscillators. We have already calculated the thermal equilibrium average (see (6.6.43))

$$\langle H_1 \rangle = \left\langle \frac{p^2}{2m} + \frac{1}{2} m\omega_0^2 \right\rangle = \left\langle \frac{1}{2} m\dot{x}^2 + \frac{1}{2} m\omega_0^2 x^2 \right\rangle \tag{6.7.7}$$

in the frequency-continuum model of the reservoir. Using the fluctuation–dissipation theorem, we now calculate the thermal average

$$\langle H_2 \rangle = \sum_j \frac{1}{2} \langle p_j^2 + \omega_j^2 (q_j + \frac{C_j x}{\omega_j^2})^2 \rangle = \sum_j \frac{1}{2} \langle P_j^2 + \omega_j^2 Q_j^2 \rangle, \qquad (6.7.8)$$

where we have defined[51]

$$Q_j = q_j + (C_j/\omega_j^2)x \quad \text{and} \quad P_j = p_j. \qquad (6.7.9)$$

Obviously, $[Q_i, P_j] = i\hbar \delta_{ij}$, since Q_i and P_j commute with x. From the Heisenberg equations of motion, we obtain

$$m\ddot{x} + m\omega_0^2 x = -\sum_j C_j Q_j, \qquad (6.7.10)$$

$$\ddot{Q}_j + \omega_j^2 Q_j = (C_j/\omega_j^2)\ddot{x}. \qquad (6.7.11)$$

Now we can use the fluctuation–dissipation theorem to obtain the symmetrized auto-correlation functions for $x(t)$ and $Q_j(t)$ and therefore $\langle x^2 \rangle$ and $\langle Q_j^2 \rangle$. To this end, we introduce forces $F_x(\omega)e^{-i\omega t}$ and $F_j(\omega)e^{-i\omega t}$ in (6.7.10) and (6.7.11) to derive equations for the Fourier transforms $x(\omega)$ and $Q_j(\omega)$ that we use to deduce the generalized susceptibilities relating $x(\omega)$ to $F_x(\omega)$ and $Q_j(\omega)$ to $F_j(\omega)$:

$$m(\omega_0^2 - \omega^2)x(\omega) + \sum_j C_j Q_j(\omega) = F_x(\omega), \qquad (6.7.12)$$

$$(\omega_j^2 - \omega^2)Q_j(\omega) + (C_j\omega^2/\omega_j^2)x(\omega) = F_j(\omega). \qquad (6.7.13)$$

Consider first the expression for $x(\omega)$ obtained by putting

$$Q_j(\omega) = \frac{-C_j\omega^2/\omega_j^2}{\omega_j^2 - \omega^2}x(\omega) + \frac{F_j(\omega)}{\omega_j^2 - \omega^2} \qquad (6.7.14)$$

in (6.7.12):

$$\left[m(\omega_0^2 - \omega^2) - \sum_j \frac{C_j^2\omega^2/\omega_j^2}{\omega_j^2 - \omega^2}\right]x(\omega) = F_x(\omega) - \sum_j \frac{C_j}{\omega_j^2 - \omega^2}F_j(\omega). \qquad (6.7.15)$$

The term in square brackets is

$$m(\omega_0^2 - \omega^2) - i\omega\tilde{\mu}(\omega) = \alpha^{-1}(\omega), \qquad (6.7.16)$$

so that

$$x(\omega) = \alpha(\omega)F_x(\omega) - \alpha(\omega)\sum_j \frac{C_j F_j(\omega)}{\omega_j^2 - \omega^2}. \qquad (6.7.17)$$

---

[51] We employ here a simplified model and calculation based on G. W. Ford, J. T. Lewis, and R. F. O'Connell, Ann. Phys. (N. Y.) **185**, 270 (1988). See also G. W. Ford, J. T. Lewis, and R. F. O'Connell, Phys. Rev. Lett. **55**, 2273 (1985).

To verify (6.7.16), re-write (6.5.55) as

$$\tilde{\mu}(\omega + i\epsilon) = i \sum_j \frac{C_j^2}{2\omega_j^2} \left(\frac{1}{\omega + \omega_j + i\epsilon} + \frac{1}{\omega - \omega_j + i\epsilon} \right)$$

$$= i \sum_j \frac{C_j^2 \omega / \omega_j^2}{(\omega + i\epsilon)^2 - \omega_j^2}, \quad \text{where } \epsilon \to 0^+, \qquad (6.7.18)$$

and so

$$m(\omega_0^2 - \omega^2) - \sum_j \frac{C_j^2 \omega^2 / \omega_j^2}{\omega_j^2 - (\omega + i\epsilon)^2} = m(\omega_0^2 - \omega^2) - i\omega\tilde{\mu}(\omega) = \alpha^{-1}(\omega), \qquad (6.7.19)$$

from Exercise 6.5.

Now, use this result in the equation for $Q_j(\omega)$:

$$(\omega_j^2 - \omega^2)Q_j(\omega) = F_j(\omega) - \frac{C_j\omega^2}{\omega_j^2}\alpha(\omega)F_x(\omega) + \frac{C_j\omega^2}{\omega_j^2}\sum_k \frac{C_k F_k(\omega)}{\omega_k^2 - \omega^2}$$

$$= F_j(\omega) - \frac{C_j\omega^2}{\omega_j^2}\alpha(\omega)F_x(\omega) + \frac{C_j^2\omega^2}{\omega_j^2}\frac{F_j(\omega)}{\omega_j^2 - \omega^2}$$

$$+ \frac{C_j\omega^2}{\omega_j^2}\sum_{k \neq j} \frac{C_k F_k(\omega)}{\omega_k^2 - \omega^2}, \qquad (6.7.20)$$

from which we identify the generalized susceptibility

$$\alpha_j(\omega) = \frac{1}{\omega_j^2 - \omega^2} + \frac{C_j^2\omega^2/\omega_j^2}{(\omega_j^2 - \omega)^2}\alpha(\omega) \qquad (6.7.21)$$

relating $Q_j(\omega)$ to $F_j(\omega)$.[52]

From the fluctuation–dissipation theorem,

$$\langle x^2 \rangle = \frac{\hbar}{\pi} \int_0^\infty d\omega\, \alpha_I(\omega) \coth\left(\hbar\omega/2k_BT\right), \qquad (6.7.22)$$

$$\langle Q_j^2 \rangle = \frac{\hbar}{\pi} \int_0^\infty d\omega\, \alpha_{jI}(\omega) \coth\left(\hbar\omega/2k_BT\right), \qquad (6.7.23)$$

and

$$\langle H_1 \rangle = \frac{\hbar m}{2\pi} \int_0^\infty d\omega\, \alpha_I(\omega)(\omega^2 + \omega_0^2) \coth\left(\hbar\omega/2k_BT\right), \qquad (6.7.24)$$

---

[52] Since the generalized susceptibilities must be analytic in the upper half of the complex frequency plane, they must, of course, in these formulas be understood to be $\alpha(\omega + i\epsilon)$ and $\alpha_j(\omega + i\epsilon)$, $\epsilon \to 0^+$, when not explicitly indicated.

$$\langle H_2 \rangle = \frac{\hbar}{\pi} \sum_j \int_0^\infty d\omega (\omega^2 + \omega_j^2) \alpha_{jI}(\omega) \coth\left(\hbar\omega/2k_B T\right)$$

$$= -\frac{\hbar}{2\pi} \sum_j \int_0^\infty d\omega \mathrm{Im}\left(\frac{\omega^2 + \omega_j^2}{\omega^2 - \omega_j^2}\right) \coth\left(\hbar\omega/2k_B T\right)$$

$$+ \frac{\hbar}{2\pi} \sum_j \int_0^\infty d\omega \frac{C_j^2 \omega^2}{\omega_j^2} \mathrm{Im}\left[\frac{\omega^2 + \omega_j^2}{(\omega_j^2 - \omega^2)^2} \alpha(\omega)\right] \coth\left(\hbar\omega/2k_B T\right). \quad (6.7.25)$$

To evaluate the first contribution to $\langle H_2 \rangle$, we use (4.4.34) to write

$$\mathrm{Im}\frac{\omega^2 + \omega_j^2}{(\omega + i\epsilon)^2 - \omega_j^2} = \mathrm{Im}\frac{\omega^2 + \omega_j^2}{(\omega + i\epsilon - \omega_j)(\omega + i\epsilon + \omega_j)} \to \mathrm{Im}\frac{\omega^2 + \omega_j^2}{(\omega + \omega_j)} \frac{1}{\omega - \omega_j + i\epsilon}$$

$$= \mathrm{Im}\frac{\omega^2 + \omega_j^2}{(\omega + \omega_j)}\left[\mathrm{P}\frac{1}{\omega - \omega_j} - i\pi\delta(\omega - \omega_j)\right] = -i\pi\omega_j, \quad (6.7.26)$$

whence

$$-\frac{\hbar}{2\pi} \sum_j \int_0^\infty d\omega \mathrm{Im}\left(\frac{\omega^2 + \omega_j^2}{\omega^2 - \omega_j^2}\right) \coth\left(\hbar\omega/2k_B T\right) = \sum_j \frac{1}{2}\hbar\omega_j \coth(\hbar\omega_j/2k_B T)$$

$$= U_R(T), \quad (6.7.27)$$

which is just the reservoir energy *in the absence of the oscillator*.

To evaluate the second contribution to $\langle H_2 \rangle$, we use (6.7.18) and find, after differentiation, that

$$-i\sum_j \frac{C_j^2}{\omega_j^2} \frac{\omega^2 + \omega_j^2}{(\omega^2 - \omega_j^2)^2} = \frac{d\tilde{\mu}}{d\omega}, \quad (6.7.28)$$

and

$$\langle H_2 \rangle = U_R(T) + \frac{\hbar}{2\pi} \sum_j \int_0^\infty d\omega \omega^2 \mathrm{Im}\left[i\frac{d\tilde{\mu}}{d\omega}\alpha(\omega)\right] \coth\left(\hbar\omega/2k_B T\right), \quad (6.7.29)$$

so that, from (6.7.1) and (6.6.43),

$$\langle H_1 + H_2 \rangle = \left\langle p^2/2m + V(x) + \sum_j \frac{1}{2}\left[p_j^2 + \omega_j^2\left(q_j + \frac{C_j x}{\omega_j^2}\right)^2\right]\right\rangle$$

$$= U_R(T) + \frac{\hbar m}{2\pi} \int_0^\infty d\omega(\omega^2 + \omega_0^2)\alpha_I(\omega) \coth\left(\hbar\omega/2k_B T\right)$$

$$+ \frac{\hbar}{2\pi} \sum_j \int_0^\infty d\omega \omega^2 \mathrm{Im}\left[i\frac{d\tilde{\mu}}{d\omega}\alpha(\omega)\right] \coth\left(\hbar\omega/2k_B T\right). \quad (6.7.30)$$

From (6.7.16),

$$i\omega^2 \frac{d\tilde{\mu}}{d\omega} = -m(\omega_0^2 + \omega^2) + \frac{1}{\alpha(\omega)} + \frac{\omega}{\alpha^2(\omega)}\frac{d\alpha}{d\omega}, \quad (6.7.31)$$

which, after some simple algebra, leads finally to

$$\langle H_1 + H_2 \rangle = U_R(T) + \frac{\hbar}{2\pi} \int_0^\infty d\omega\, \omega \coth\left(\hbar\omega/2k_BT\right) \mathrm{Im}\left[\frac{d}{d\omega}\log\alpha(\omega)\right]. \qquad (6.7.32)$$

We now identify the mean oscillator energy U_{osc} as the difference between the mean energy $\langle H_1 + H_2 \rangle$ of the interacting oscillator-reservoir system and the mean energy of the reservoir in the absence of the oscillator:

$$U_{\mathrm{osc}}(T) = \langle H_1 + H_2 \rangle - U_R(T) = \frac{\hbar}{2\pi} \int_0^\infty d\omega\, \omega \coth\left(\hbar\omega/2k_BT\right) \mathrm{Im}\left[\frac{d}{d\omega}\log\alpha(\omega)\right]$$

$$= \frac{\hbar}{2\pi} \int_0^\infty d\omega\, \omega\, \mathrm{Im}\left[\frac{d}{d\omega}\log\alpha(\omega)\right]$$

$$+ \frac{\hbar}{\pi} \int_0^\infty d\omega\, \frac{\omega}{e^{\hbar\omega/k_BT}-1} \mathrm{Im}\left[\frac{d}{d\omega}\log\alpha(\omega)\right]. \qquad (6.7.33)$$

Ford et al. call this the "remarkable formula" for the energy of an oscillator interacting with a heat bath.[53] It is remarkable in part because it gives the exact expression for the energy of the oscillator interacting with a heat bath at all temperatures in terms of the oscillator susceptibility $\alpha(\omega)$ alone.[54]

Exercise 6.6: Assuming

$$\alpha(\omega) = [m(\omega_0^2 - \omega^2) - i\omega\gamma]^{-1},$$

show that (6.7.33) gives exactly the previously calculated oscillator energy (6.5.10) for the case of ohmic damping.

In general,

$$U_{\mathrm{osc}} \neq \langle H_1 \rangle = \langle p^2/2m + \tfrac{1}{2}m\omega_0^2 x^2 \rangle. \qquad (6.7.34)$$

The difference is found from (6.7.24) and (6.7.16) to be

$$U_{\mathrm{osc}} - \langle H_1 \rangle = \frac{\hbar}{2\pi} \int_0^\infty d\omega \coth\left(\hbar\omega/2k_BT\right) \mathrm{Im}\left[\omega\frac{d\alpha/d\omega}{\alpha(\omega)} - m(\omega^2 + \omega_0^2)\alpha(\omega)\right]$$

$$= \frac{\hbar}{2\pi} \int_0^\infty d\omega \coth\left(\hbar\omega/2k_BT\right) \mathrm{Re}\left[\omega^2\alpha(\omega)\frac{d\tilde{\mu}}{d\omega}\right], \qquad (6.7.35)$$

and vanishes only if $\tilde{\mu}(\omega)$ is independent of ω. This condition is satisfied in the special case of ohmic damping, but, in general, the mean oscillator energy defined by (6.7.33)

[53] G. W. Ford, J. T. Lewis, and R. F. O'Connell, Ann. Phys. (N. Y.) **185**, 270 (1988), p. 270.

[54] For another derivation of this formula, see P. R. Berman, G. W. Ford, and P. W. Milonni, J. Chem. Phys. **141**, 164105 (2014).

is not the same as the mean value of the oscillator Hamiltonian $p^2/2m + (1/2)m\omega_0^2$. When $\tilde{\mu}(\omega)$ is not constant, $\mu(t - t')$ is not proportional to the delta function $\delta(t - t')$, and then, according to the quantum Langevin equation (6.5.39), the interaction of the oscillator with the heat bath is not Markovian.

6.7.1 Free Energy

Recall that the Helmoltz free energy, energy available for reversible work at constant temperature according to classical thermodynamics, is defined as $F = U - TS$, where U, T, and S are, respectively, the internal energy, temperature, and entropy characterizing the state of a closed system. From the constant-volume relation $S = -\partial F/\partial T$, $U = F - T\partial F/\partial T$, and (6.7.33) implies the "remarkable formula" for the (Helmholtz) free energy of our oscillator interacting with the heat bath:[55]

$$
\begin{aligned}
F_{\mathrm{osc}}(T) &= \frac{k_B T}{\pi} \int_0^\infty d\omega \log\left[2\sinh(\hbar\omega/2k_B T)\right] \mathrm{Im}\left[\frac{d}{d\omega}\log\alpha(\omega)\right] \\
&= \frac{1}{\pi}\int_0^\infty d\omega \frac{1}{2}\hbar\omega \mathrm{Im}\left[\frac{d}{d\omega}\log\alpha(\omega)\right] \\
&\quad + \frac{k_B T}{\pi}\int_0^\infty d\omega \log(1 - e^{-\hbar\omega/k_B T})\mathrm{Im}\left[\frac{d}{d\omega}\log\alpha(\omega)\right].
\end{aligned}
\tag{6.7.36}
$$

An application of this formula appears in Section 6.8.1, and a slight generalization of it is used in Section 7.7.

6.8 Radiation Reaction Revisited

The quantum Langevin equation allows for any memory function consistent with basic physical principles such as causality, and, as such, can be applied to a wide variety of systems in which there is dissipation and, therefore, fluctuating forces. We now discuss a quantum Langevin equation for an electron interacting with its own field in addition to the fluctuating vacuum electromagnetic field and possibly externally applied forces. We begin by briefly recalling the main features of the classical theory as originally formulated by M. Abraham and H. A. Lorentz.[56]

The equation of motion that we derived for a Hertzian electric dipole, including its radiation reaction field, (4.3.26), applies as well to a non-relativistic free electron when we put $\nabla V = 0$. For motion along one dimension (x),

$$
m(\ddot{X} - \tau_e \dddot{X}) = f(t), \quad X = \langle x \rangle, \quad \tau_e = e^2/6\pi\epsilon_0 mc^3 \cong 6.3 \times 10^{-24} \text{ s}, \tag{6.8.1}
$$

with m the observed electron mass. Here, $f(t)$ is a mean applied force acting in addition to the radiation reaction force and the Langevin force $F(t)$, the latter having zero

---

[55] G. W. Ford, J. T. Lewis, and R. F. O'Connell, Ann. Phys. (N. Y.) **185**, 270 (1988); G. W. Ford, J. T. Lewis, and R. F. O'Connell, Phys. Rev. Lett. **55**, 2273 (1985).

[56] M. Abraham, Ann. Phys. **10**, 105 (1902); H. A. Lorentz, *The Theory of Electrons*, second edition, Teubner, Leipzig, 1916, Section 37, pp. 48–49, and Note 18, pp. 252–256.

expectation value. Equation (6.8.1) is the classical *Abraham–Lorentz equation* which, as is well known, has some peculiar features.

One peculiarity is the appearance of the *third* derivative of $X(t)$: the motion of the electron depends not only on its position and velocity at a starting time $t = 0$, but also on its acceleration $a(0)$ at $t = 0$. Even more troubling physically is that the Abraham–Lorentz equation admits a runaway solution for the acceleration. If $f(t) = 0$, for instance,

$$a(t) = a(0)e^{t/\tau_e}. \tag{6.8.2}$$

If $f(t) \neq 0$,

$$a(t) = \left[a(0) - \frac{1}{m\tau_e} \int_0^t dt' f(t') e^{-t'/\tau_e} \right] e^{t/\tau_e}, \tag{6.8.3}$$

and we again have a runaway unless

$$a(0) = \frac{1}{m\tau_e} \int_0^\infty dt' f(t') e^{-t'/\tau_e}, \tag{6.8.4}$$

in which case

$$a(t) = \frac{1}{m\tau_e} \int_t^\infty dt' f(t') e^{(t-t')/\tau_e} = \frac{1}{m\tau_e} \int_0^\infty dt' f(t' + t) e^{-t'/\tau_e}. \tag{6.8.5}$$

But now, the electron's motion at time t depends on the force $f(t)$ at times greater than t—the runaway dilemma has been replaced by that of "preacceleration." Although preacceleration, like runaway, is physically unpalatable, it is unobservable, because the time τ_e is so small. In fact, τ_e is on the order of the time r_0/c it takes for light to propagate a distance equal to the classical electron radius $r_0 = e^2/4\pi\epsilon_0 mc^2$, which is $1/137$ times the Compton radius $\hbar/mc$.

Radiation reaction typically amounts to a small perturbation. If it is ignored altogether by replacing the Abraham–Lorentz equation by $d^2X/dt^2 \cong (1/m)f(t)$ as a first approximation and then included in the next approximation using $d^3X/dt^3 \cong (\tau_e/m)df/dt$, we obtain the "Landau–Lifshitz equation"

$$m\ddot{X} = f(t) + \tau_e \dot{f}(t), \tag{6.8.6}$$

which has no third derivative and, unlike the Abraham–Lorentz equation, obviously has no runaway solution in the absence of any external forces. To get an idea of the range of validity of (6.8.6), consider a sinusoidal force $f(t) = f_0 \exp(-i\omega t)$, in which case

$$\left| \tau_e \dot{f}(t)/f(t) \right| = \omega\tau_e \approx \omega r_0/c \approx \frac{1}{137} \frac{\hbar\omega}{mc^2}. \tag{6.8.7}$$

If this is *not* very small, we must be dealing with frequencies ω for which the classical theory breaks down. If it is very small, (6.8.6) is at least compatible with classical non-relativistic theory.

Equation (6.8.6), which was proposed by Eliezer and by Landau and Lifshitz, can be put on a more secure foundation.[57] For this purpose, we first return to (4.3.18) for the radiation reaction field. Doing now only one partial integration, we can express (4.3.30) as

$$m_{\text{bare}}\ddot{\mathbf{x}} = -\nabla V + e\mathbf{E}_0(0,t) + \frac{e^2}{3\pi\epsilon_0 c^3}\int_{-\infty}^t dt'\,\dot{\mathbf{x}}(t')\ddot{\delta}(t'-t)$$

$$= -\nabla V + e\mathbf{E}_0(0,t) - \frac{e^2}{3\pi^2\epsilon_0 c^3}\int_{-\infty}^t dt'\,\dot{\mathbf{x}}(t')\int_{-\infty}^\infty d\omega\omega^2\cos\omega(t-t')..\quad(6.8.8)$$

This has the form of the quantum Langevin equation (6.5.39), with the memory function

$$\mu(t) = \frac{e^2}{3\pi^2\epsilon_0 c^3}\int_0^\infty d\omega\omega^2\cos\omega t,\qquad(6.8.9)$$

a Langevin force $e\mathbf{E}_0(0,t)$, and a mass equal to the *bare* mass m_{bare} (Section 4.3):

$$m_{\text{bare}}\ddot{\mathbf{x}} + \int_{-\infty}^t dt'\,\dot{\mathbf{x}}(t')\mu(t-t') + \nabla V = e\mathbf{E}_0(0,t).\qquad(6.8.10)$$

Of course, this operator equation of motion is equivalent to (4.3.30).

In this form of the equation of motion, the electron is point-like and interacts with arbitrarily large field frequencies. But if—as in Bethe's estimate of the Lamb shift (see Section 4.8.3) and its interpretation as a radiation reaction effect—the interaction is assumed to be primarily non-relativistic, it should be dominated by frequencies ω small compared to $mc^2/\hbar$. These correspond to wavelengths that are large compared to the Compton wavelength and, therefore, the classical electron radius, that is, wavelengths that are large compared to any measure of an effective electron radius. Assuming the interaction is dominated by such frequencies, and falls off rapidly at higher frequencies, we introduce an electron "form factor" $f(\omega)$ and replace the point-electron memory function (6.8.9) by[58]

$$\mu(t) = \frac{e^2}{3\pi^2\epsilon_0 c^3}\int_0^\infty d\omega f(\omega)\omega^2\cos\omega t.\qquad(6.8.11)$$

$f(\omega)$ should be slowly varying and near unity up to some "cutoff" frequency Ω but fall off rapidly thereafter, and the essential physical consequences of this modification should be insensitive to the specific form of $f(\omega)$. A convenient choice for $f(\omega)$ is

$$f(\omega) = \frac{\Omega^2}{\omega^2 + \Omega^2},\qquad(6.8.12)$$

[57] C. J. Eliezer, Proc. R. Soc. London A **194**, 543 (1948); *Landau and Lifshitz (2000)*; G. W. Ford and R. F. O'Connell, Phys. Lett. A **157**, 217 (1991); R. F. O'Connell, Contemp. Phys. **53**, 301 (2012), and references therein.

[58] G. W. Ford and R. F. O'Connell, Phys. Lett. A **157**, 217 (1991); R. F. O'Connell, Contemp. Phys. **53**, 301 (2012).

from which it follows that

$$\mu(t) = \frac{e^2\Omega^2}{3\pi\epsilon_0 c^3}\left[\delta(t) - \frac{1}{2}\Omega e^{-\Omega t}\right], \tag{6.8.13}$$

and the Fourier transform is[59]

$$\tilde{\mu}(\omega) = \frac{e^2\Omega^2}{6\pi\epsilon_0 c^3}\frac{\omega}{\omega + i\Omega}. \tag{6.8.14}$$

Putting (6.8.13) in (6.8.10) and then multiplying by $e^{\Omega t}$, differentiating with respect to t, and dividing by $\Omega e^{\Omega t}$, we obtain the equation of motion

$$\frac{m_{\text{bare}}}{\Omega}\dddot{x} + \left[m_{\text{bare}} + \frac{e^2\Omega}{6\pi\epsilon_0 c^3}\right]\ddot{x} + m_{\text{obs}}\omega_0^2 x + \frac{1}{\Omega}m_{\text{obs}}\omega_0^2\dot{x}$$
$$= eE_0(0,t) + \frac{1}{\Omega}\dot{E}_0(0,t), \tag{6.8.15}$$

for a harmonic potential $V = \frac{1}{2}m\omega_0^2 x^2$. For simplicity, we consider motion along one direction; $E_0(t)$ is the component of the vacuum electric field along this direction.

The mass multiplying $\ddot{x}$ is the observed mass

$$m_{\text{obs}} = m_{\text{bare}} + \frac{e^2\Omega}{6\pi\epsilon_0 c^3}, \tag{6.8.16}$$

in terms of which

$$m_{\text{bare}} = m_{\text{obs}}(1 - \Omega\tau_e) \tag{6.8.17}$$

and[60]

$$m_{\text{obs}}\left(\frac{1}{\Omega} - \tau_e\right)\dddot{x} + m_{\text{obs}}\ddot{x} + m_{\text{obs}}\omega_0^2 x + \frac{1}{\Omega}m_{\text{obs}}\omega_0^2\dot{x} = eE_0(0,t) + \frac{1}{\Omega}\dot{E}_0(0,t). \tag{6.8.18}$$

The generalization to include the effect of an externally applied force represented by an operator $f(t)$ is obviously

$$m_{\text{obs}}\left(\frac{1}{\Omega} - \tau_e\right)\dddot{x} + m_{\text{obs}}\ddot{x} + m_{\text{obs}}\omega_0^2 x + \frac{1}{\Omega}m_{\text{obs}}\omega_0^2\dot{x}$$
$$= eE_0(0,t) + \frac{1}{\Omega}\dot{E}_0(0,t) + f(t) + \frac{1}{\Omega}\dot{f}(t). \tag{6.8.19}$$

To compare with the classical Abraham–Lorentz equation, we take expectation values over the vacuum state of the field ($\langle E_0\rangle = \langle\dot{E}_0\rangle = 0$):

$$m\left(\frac{1}{\Omega} - \tau_e\right)\dddot{X} + m\ddot{X} + m\omega_0^2\left(X + \frac{1}{\Omega}\dot{X}\right) = f_{\text{ext}}(t) + \frac{1}{\Omega}\dot{f}_{\text{ext}}(t), \tag{6.8.20}$$

[59] Ibid.

[60] Recall the discussion in Section 4.3. The cutoff frequency Λ there is simply an upper limit on the frequency integration, whereas Ω here characterizes the form factor (6.8.12).

where we have written $m = m_{\text{obs}}$, $X = \langle x \rangle$, and $f_{\text{ext}} = \langle f \rangle$. In the point-electron model, $\Omega \to \infty$, we recover the Abraham–Lorentz equation for a free electron when we put $\omega_0 = 0$:

$$m(\ddot{X} - \tau_e \, \dddot{X}) = f_{\text{ext}}(t). \qquad (6.8.21)$$

The bare mass (6.8.17) in this model is negative and infinite.

Negative bare mass is related to runaway solutions of the Abraham–Lorentz equation; in fact it has been invoked to reconcile the runaway solutions with energy conservation: the total energy for a free electron and the field is the (negative) kinetic energy associated with the inertial bare mass plus the (positive) energy in the electromagnetic field.[61] If the electron acceleration continually increases, there is a continual increase in the field energy, but it is accompanied by a continual decrease in the kinetic energy. Analyses of different classical models for the electron charge distribution suggest that a positive bare mass precludes runaway solutions, and that runaway solutions are a consequence of treating the electron as a point charge.[62] These conclusions also follow from (6.8.20).

If we demand that the bare mass (6.8.17) not be negative—and therefore that there are no runaways—we must require that $\Omega \leq 1/\tau_e$. Such frequencies are consistent with the neglect of relativistic effects. The largest Ω satisfying this requirement, $\Omega = 1/\tau_e$, then gives $m_{\text{bare}} = 0$ and

$$m\ddot{X} + m\omega_0^2 (X + \tau_e \dot{X}) = f_{\text{ext}}(t) + \tau_e \dot{f}_{\text{ext}}(t). \qquad (6.8.22)$$

For a free electron we obtain the Landau–Lifshitz equation:

$$m\ddot{X} = f_{\text{ext}}(t) + \tau_e \dot{f}_{\text{ext}}(t). \qquad (6.8.23)$$

The radiation reaction theory based on the *quantum* Langevin equation and a physically acceptable electron form factor thus provides a consistent derivation of the classical (Eliezer–)Landau–Lifshitz equation. It is not difficult to see from the general expression (6.8.11) that the most physically significant results of this theory do not depend on the detailed form of the chosen form factor $f(\omega)$, provided, of course, that it has the properties stated earlier. In particular, if (6.8.12) is used with a cutoff $\Omega \approx 1/\tau_e$ but smaller (so that m_{bare} is positive), one obtains additional terms with higher powers of the very small time τ_e on the right side of (6.8.23).

For a sinusoidal applied force $f(t) = f_0 e^{-i\omega t}$, (6.8.22) takes the form

$$m\ddot{X} + m\Gamma_R \dot{X} + m\omega_0^2 X = (1 - i\omega\tau_e) f_0 e^{-i\omega t}, \quad \Gamma_R = e^2 \omega_0^2 / 6\pi\epsilon_0 mc^3 = \omega_0^2 \tau_e, \qquad (6.8.24)$$

implying the polarizability (generalized susceptibility)

---

[61] See, for instance, S. Coleman, in *Electromagnetism: Paths to Research*, ed. D. Teplitz, Plenum Press, New York, 1982, pp. 183–210.

[62] See, for instance, T. Erber, Fortsch. d. Phys. **9**, 343 (1961).

$$\alpha(\omega) = \frac{1}{m} \frac{1 - i\omega\tau_e}{(\omega_0^2 - \omega^2) - i\omega\Gamma_R}. \tag{6.8.25}$$

This satisfies the requirement of being an analytic function in the upper half of the complex ω plane (Section 6.6.1), and it is easily shown to be consistent with the optical theorem.[63] The polarizability that follows from the Abraham–Lorentz equation (indicated by the superscript ALE) for an oscillator,

$$\alpha^{\text{ALE}}(\omega) = \frac{1}{m} \frac{1}{(\omega_0^2 - \omega^2) - i\tau_e\omega^3}, \tag{6.8.26}$$

is *not* analytic in the upper half of the complex frequency plane. This is reflected in the runaway solutions of the Abraham–Lorentz equation.

6.8.1   Energy Shifts due to Blackbody Radiation

The susceptibility (6.8.25) characterizes an oscillator for which the only dissipative process is radiation reaction. It can therefore be used in the "remarkable formula" above to calculate the oscillator energy and free energy when it is in equilibrium with the interacting heat bath corresponding to blackbody radiation. For the free energy we obtain from (6.7.36),

$$F_{\text{osc}}(T) = F_{\text{osc}}(0) + \frac{\Gamma_R k_B T}{\pi} \int_0^\infty d\omega \, \log(1 - e^{-\hbar\omega/k_B T}) \frac{\omega_0^2 + \omega^2}{(\omega_0^2 - \omega^2)^2 + \Gamma_R^2\omega^2}$$

$$- \frac{\Gamma_R k_B T}{\pi\omega_0^2} \int_0^\infty d\omega \, \log(1 - e^{-\hbar\omega/k_B T}) \frac{1}{1 + \omega^2\tau_e^2}. \tag{6.8.27}$$

Let $F'_{\text{osc}}(T)$ denote the second term on the right side, and consider the corresponding energy

$$U'_{\text{osc}}(T) = F'_{\text{osc}}(T) - T\frac{\partial F'_{\text{osc}}}{\partial T}$$

$$= \frac{\hbar\Gamma_R}{2\pi} \int_0^\infty d\omega \frac{\omega(\omega_0^2 + \omega^2)}{(\omega_0^2 - \omega^2)^2 + \Gamma_R^2\omega^2} \big[\coth(\hbar\omega/2k_B T) - 1\big]. \tag{6.8.28}$$

For $\Gamma_R \to 0$,

$$U'_{\text{osc}}(T) \to \frac{\hbar\omega_0}{e^{\hbar\omega_0/k_B T} - 1}, \tag{6.8.29}$$

the mean energy of the oscillator in thermal equilibrium at temperature T. This implies that $F'_{\text{osc}}(T)$ is the corresponding free energy, allowing for the finite radiative linewidth

[63] F. Intravaia, R. Behunin, P. W. Milonni, G. W. Ford, and R. F. O'Connell, Phys. Rev. A **84**, 035801 (2011).

of the oscillator energy level.[64] Therefore, the last term on the right side of (6.8.27), $\Delta F_0(T)$, is a temperature-dependent *shift* in the free energy. Assuming $k_B T/\hbar \ll 1/\tau_e$, this shift is

$$\Delta F_{\text{osc}}(T) \cong -\frac{\Gamma_R k_B T}{\pi \omega_0^2} \int_0^\infty d\omega \, \log(1 - e^{-\hbar\omega/k_B T}) = \frac{1}{4\pi\epsilon_0} \frac{\pi e^2 (k_B T)^2}{9\hbar mc^3}, \qquad (6.8.30)$$

and the corresponding shift in energy of the oscillator in equilibrium with blackbody radiation is

$$\Delta U_{\text{osc}} = -\frac{1}{4\pi\epsilon_0} \frac{\pi e^2 (k_B T)^2}{9\hbar mc^3}. \qquad (6.8.31)$$

The Stark shift (2.3.57), which was obtained by assuming that the blackbody radiation is in effect an *externally applied* field, has a sign opposite to that of the energy shift as defined here. The derivation of the energy shift (6.8.31) and the free energy shift (6.8.30), in contrast, allows for the oscillator and the radiation with which it is in equilibrium to interact and exchange energy.[65]

6.8.2 The Commutator

Equation (6.8.20) implies the generalized susceptibility

$$\alpha(\omega) = \frac{1 - i\omega/\Omega}{m_{\text{obs}}(\omega_0^2 - \omega^2) + i(\omega/\Omega)(m_{\text{bare}}\omega^2 - m_{\text{obs}}\omega_0^2)} \qquad (6.8.32)$$

and

$$\alpha(\omega) \to -1/m_{\text{bare}}\omega^2 \quad \text{for } \omega \to \infty, \qquad (6.8.33)$$

as opposed to (6.6.71), where the observed mass appears. In fact, it is the *bare* mass that occurs in the canonical commutation relation, that is, $[x(t), m_{\text{bare}}\dot{x}(t)] = i\hbar$.[66] In the point-electron model, the observed mass does appear in the canonical commutation relation, which holds in spite of the fact that the generalized susceptibility in this model is not analytic in the upper half of the complex frequency plane (see Section 4.6).

$\cdots$

[64] G. W. Ford, J. T. Lewis, and R. F. O'Connell, J. Phys. B: At. Mol. Phys. **19**, L41–L46 (1986). We have used the cutoff frequency $\Omega = 1/\tau_e$, but the results (6.8.30) and (6.8.31) first derived in this paper require only that the cutoff frequency satisfies $\hbar\Omega \gg k_B T$. For the three-dimensional oscillator, these expressions for the energy and free energy should be multiplied by 3.

[65] For further discussion, see G. Barton, J. Phys. B: At. Mol. Phys. **20**, 879 (1987); G. W. Ford, J. T. Lewis, and R. F. O'Connell, J. Phys. B: At. Mol. Phys. **20**, 899 (1987).

[66] See G. W. Ford and R. F. O'Connell, J. Stat. Phys. **57**, 803 (1989), for details and further discussion of this point.

The Ford–O'Connell theory of radiation reaction described in this section has several attractive features. It provides a firm foundation for the equation of motion proposed by Eliezer and by Landau and Lifshitz, and negative bare mass and the runaway solutions of the Abraham–Lorentz equation are eliminated by a physically motivated choice for the form factor. The theory is formulated within the model of an electron bound by an elastic restoring force, and, as such, cannot be applied directly to the theory of spontaneous emission by an atom, although it can be applied as above in the case of the nearly free electron of a highly excited atom. A quantum Langevin equation can be derived for a two-state atom interacting with the vacuum and radiation reaction fields, including a form factor; closed-form exact solutions of the nonlinear operator equation, however, seem beyond reach. But since the dissipative effect of spontaneous emission can be attributed to radiation reaction (see Section 4.5), it is nevertheless natural to compare the dissipative effect of radiation reaction in the theory described in this section with that in the theory of spontaneous emission reviewed in Chapter 4. Returning to (4.4.75) and taking an expectation value over a vacuum state of the field, ignoring the radiative frequency shift Δ_{21}, we obtain for $\langle \sigma_x(t) \rangle = \langle \sigma(t) + \sigma^\dagger(t) \rangle$ the equation[67]

$$\langle \ddot{\sigma}_x(t) \rangle + \omega_0^2 \langle \sigma_x(t) \rangle + A_{21} \langle \dot{\sigma}_x(t) \rangle = 0. \qquad (6.8.34)$$

This has the same form as (6.8.24) when there is no applied force:

$$\ddot{X} + \omega_0^2 X + \Gamma_R \dot{X} = 0, \qquad (6.8.35)$$

which was derived for a linear oscillator rather than a two-state atom. If we introduce the oscillator strength f_{12} as discussed in Section 2.5.2,

$$\Gamma_R \to \Gamma_R f_{12} = \frac{1}{4\pi\epsilon_0} \frac{4d_{12}^2 \omega_0^3}{3\hbar c^3} = A_{21}, \quad d_{12}^2 = e^2 x_{12}^2 = e^2 |\mathbf{x}_{12}|^2 / 3, \qquad (6.8.36)$$

we see that (6.8.35) takes the same form as (6.8.34) for a two-state atom. In other words, the dissipative effect of radiation reaction in the Markovian approximation leading to (6.8.34) for a two-state atom's average dipole moment is essentially the same as in the theory of radiation reaction leading to (6.8.35) for a linear oscillator.

6.9   Spontaneous Emission Noise: The Laser Linewidth

The fluctuation–dissipation theorem relates the dissipative effect of the coupling of a system to a reservoir to fluctuations of the system when it is in thermal equilibrium with the reservoir. More generally, if there is a dissipative force on a system, there must also be a fluctuating force, as is clear when the system is described by a quantum

---

[67] If we choose not to simply ignore Δ_{21}, we could assume it can be made finite by an appropriate renormalization or other modification of the theory. Then, if the physical (finite) absolute frequency shift $|\Delta_{21}| \ll \omega_0$, so that $\omega_0^2 + 2\omega_0\Delta_{21} \cong (\omega_0 + \Delta_{21})^2$, we could just "absorb" Δ_{21} into the definition of ω_0.

Langevin equation: a fluctuating Langevin "noise" force must be present in order to preserve commutation relations and, therefore, Heisenberg uncertainty relations.

Amplification must likewise come with noise. In a laser, the amplification of light by stimulated emission from a population of excited states is accompanied by spontaneous emission of light from those excited states. We can write an equation for the electric field operator that includes a term accounting for amplification of the field due to stimulated emission, and a noise term accounting for spontaneous emission. As we now show, an effect of the noise is to limit the degree to which single-mode laser radiation can approach monochromaticity, that is, it puts a lower, quantum limit on the spectral width.

Consider the field in a laser resonator (or "cavity") filled with an amplifying medium. The medium and the resonator are characterized by photon gain and loss (per unit length) coefficients g and κ_c, respectively. Define $\mathcal{E}(t)$, the positive-frequency, photon annihilation part of the operator for a traveling-wave electric field propagating from one end of the resonator to the other. $\mathcal{E}(t)$ is assumed to be slowly varying compared to $\exp(-i\omega t)$. We "normalize" $\mathcal{E}(t)$ such that $\langle \mathcal{E}^\dagger(t)\mathcal{E}(t)\rangle$ is the average photon number of the field, whose time evolution we describe by the Langevin-type equation

$$\frac{1}{v_g}\frac{\partial \mathcal{E}}{\partial t} = \frac{1}{2}(g - \kappa_c)\mathcal{E}(t) + F_{sp}(t), \qquad (6.9.1)$$

where $v_g = d\omega/dk$ is the group velocity at the frequency ω (see Section 1.6) and F_{sp} is the spontaneous emission noise operator. We are assuming that the field is approximately uniform within the resonator. In particular, we assume that the resonator loss is not too large, say less than $\sim 50\%$ per round-trip pass of each traveling-wave component of the field, in which case $\partial \mathcal{E}/\partial z \cong 0$.[68] With this assumption, the two counter-propagating fields in the laser have approximately equal intensities, so that the total average photon number in the resonator is approximately $2\langle \mathcal{E}^\dagger(t)\mathcal{E}(t)\rangle$. We also assume for now that all field loss is due to transmission through the mirrors at the ends of the laser resonator, so that the *empty-resonator* photon loss rate is $\gamma_c = c\kappa_c$. The spontaneous emission noise operator satisfies $\langle F_{sp}(t)\rangle = 0$ and

$$\langle F_{sp}^\dagger(t')F_{sp}(t'')\rangle = \frac{\kappa_c n_{sp}}{2v_g}\delta(t'-t''), \qquad (6.9.2)$$

as shown below. Here, n_{sp} is the "spontaneous emission factor":

$$n_{sp} = \frac{P_2}{P_2 - P_1}, \qquad (6.9.3)$$

where P_2 and P_1 are, respectively, the occupation probabilities of the upper and lower states of the lasing transition. From (6.9.1) and (6.9.2), we obtain the steady-state correlation function

[68] W. W. Rigrod, J. Appl. Phys. **36**, 2487 (1965). See also *Milonni and Eberly*, Section 5.5.

$$\langle \mathcal{E}^\dagger(t)\mathcal{E}(t+\tau)\rangle = \frac{\kappa_c v_g n_{sp}}{4\gamma}e^{-\gamma\tau}, \quad \text{where } \gamma = \frac{1}{2}v_g(\kappa_c - g), \quad \tau > 0. \tag{6.9.4}$$

The steady-state average photon number for the field $\mathcal{E}(t)$ is

$$q = \langle \mathcal{E}^\dagger(t)\mathcal{E}(t)\rangle = \frac{\kappa_c v_g n_{sp}}{4\gamma}, \tag{6.9.5}$$

implying

$$\gamma = \frac{1}{2}v_g(\kappa_c - g) = \frac{\kappa_c n_{sp} v_g}{4q}. \tag{6.9.6}$$

This shows that the gain coefficient required for steady-state lasing is (typically very slightly) less than the loss coefficient (that is, $g < \kappa_c$) because spontaneously emitted photons contribute to q along with photons generated by stimulated emission. The laser spectrum, given according to the Wiener–Khintchine theorem by the Fourier transform of the correlation function (6.9.4), is a Lorentzian lineshape function with linewidth (full width at half-maximum)

$$\Delta\nu = 2 \times \gamma/2\pi = \frac{\kappa_c n_{sp} v_g}{4\pi q}. \tag{6.9.7}$$

The photon loss rate is

$$v_g\kappa_c = (v_g/c)\gamma_c, \tag{6.9.8}$$

implying that the total laser output power of the two counter-propagating fields in the resonator is $P_{\text{out}} = v_g\kappa_c(2q\hbar\omega)$ and[69]

$$\Delta\nu = \frac{\hbar\omega n_{sp}}{2\pi P_{\text{out}}}\frac{v_g^2}{c^2}\gamma_c^2. \tag{6.9.9}$$

In our simplified model, we have not allowed for saturation effects that suppress field amplitude fluctuations and reduce (6.9.9) by a factor $1/2$, as discussed below. We have also assumed, as discussed above, that the resonator loss is not too large and, consequently, that spatial variations of the field are negligible; this assumption is made in nearly all treatments of the fundamental laser linewidth.[70]

---

[69] The fact that the photon loss rate is determined by the group velocity $v_g = c/n_g$ rather than the phase velocity c/n has been confirmed experimentally by T. Lauprêtre, C. Proux, R. Ghosh, S. Schwartz, F. Goldfarb, and F. Bretenaker, Opt. Lett. **36**, 1551 (2011), who found that $v_g \approx 10^4$ m/s. Early experiments by F. R. Faxvog, C. N. Y. Chow, T. Bieber, and J. A. Carruthers, Appl. Phys. Lett. **17**, 192 (1970), and L. Casperson and A. Yariv, Phys. Rev. Lett. **26**, 293 (1971), demonstrated that the pulse repetition frequency of a mode-locked laser with resonator length L is $v_g/2L$, not $c/2L$. The fact that the longitudinal mode spacing is $v_g/2L$ provides one way of determining the group velocity of a laser medium. Implications of group velocities smaller or larger than c are discussed in *Milonni (2005)* and references therein. I made an error on p. 38 of that work and incorrectly concluded that the estimate of v_g by Faxvog et al. was too large by a factor of roughly 7.5.

[70] P. Goldberg, P. W. Milonni, and B. Sundaram, J. Mod. Opt. **38**, 1421 (1991); P. Goldberg, P. W. Milonni, and B. Sundaram, Phys. Rev. A **44**, 1969 (1991); P. Goldberg, P. W. Milonni and B. Sundaram, Phys. Rev. A **44**, 4556 (1991). This work includes effects on the laser linewidth of strong saturation, spatial variations of the intracavity field, and the possibility of spatial hole burning related to the standing-wave nature of the field.

The group velocity $v_g = c/[n + \omega dn/d\omega]_\omega$ has contributions from both lasing and non-lasing transitions. We denote the refractive index due to the lasing transition by $n_L(\omega)$, and that due to non-lasing transitions, including contributions from any additional (passive) medium in the resonator, by $n_A(\omega)$. Assuming without much loss of generality a homogeneously broadened gain medium, we can relate the refractive index $n_L(\omega)$ to the gain coefficient using (2.5.48):

$$n_L(\omega) \cong 1 + \frac{\lambda}{2\pi} \frac{\omega - \omega_0}{\gamma_g} g(\omega), \tag{6.9.10}$$

where the laser wavelength $\lambda = 2\pi c/\omega$, ω_0 is the resonance frequency of the gain medium, and γ_g is the width of the gain lineshape function. Then, for the laser frequency ω near the frequency ω_0 of maximum gain, in which case $(\partial g/\partial\omega)_{\omega_0} \cong 0$ and $n_L(\omega_0) \cong 1$, we have for the group index (Section 1.6)

$$n_g = \frac{d}{d\omega}[\omega(n_L + n_A)] \cong n_{Ag} + 1 + cg(\omega_0)/\gamma_g, \tag{6.9.11}$$

and, for the group velocity,

$$v_g = c/n_g = \frac{c}{n_{Ag} + 1 + cg(\omega_0)/\gamma_g}, \tag{6.9.12}$$

where $n_{Ag} = (d/d\omega)(\omega n_A)$ is the group index we associate with non-lasing transitions. The laser linewidth is then given by

$$\Delta\nu = \frac{\hbar\omega n_{sp}}{4\pi P_{out}}\gamma_c^2\left(\frac{1}{n_{Ag} + 1 + cg(\omega_0)/\gamma_g}\right)^2 = \frac{\hbar\omega n_{sp}}{4\pi P_{out}}\left(\frac{\gamma_c}{n_{Ag} + 1 + \gamma_c/\gamma_g}\right)^2$$
$$\cong \frac{\hbar\omega n_{sp}}{4\pi P_{out}}\left(\frac{\gamma_c}{1 + \gamma_c/\gamma_g}\right)^2, \tag{6.9.13}$$

where we use the steady-state lasing condition $g(\omega_0) \cong \kappa_c = \gamma_c/c$, assume that non-lasing transitions make no significant contribution, and include the aforementioned factor $1/2$.

In the case of a "bad cavity," where the photon loss rate γ_c is much larger than the gain bandwidth γ_g, we obtain the "Schawlow–Townes" linewidth:

$$\Delta\nu \cong \frac{\hbar\omega n_{sp}}{4\pi P_{out}}\gamma_g^2. \tag{6.9.14}$$

A similar expression in which the laser linewidth is proportional to the square of the gain linewidth γ_g was obtained by Schawlow and Townes in their seminal paper on the extension of maser concepts to the infrared and optical regions.[71] In the opposite, "good cavity" case in which $\gamma_c \ll \gamma_g$,

$$\Delta\nu \cong \frac{\hbar\omega n_{sp}}{4\pi P_{out}}\gamma_c^2. \tag{6.9.15}$$

[71] A. L. Schawlow and C. H. Townes, Phys. Rev. **112**, 1940 (1958).

6.9.1   Correlation Function for Spontaneous Emission Noise

We now derive the correlation function (6.9.2) characterizing the spontaneous emission noise, assuming for definiteness that the gain medium consists of two-state atoms. We begin with the equation for $\mathcal{E}$ when there is a single atom in the resonator:

$$\frac{\partial \mathcal{E}}{\partial t} = iDS(t) - \frac{1}{2}v_g\kappa_c\mathcal{E}(t), \qquad (6.9.16)$$

where $S(t)$ is the slowly varying part of the lowering operator $\sigma(t) = S(t)e^{-i\omega t}$ for a two-state atom, and D is a real coupling constant we need not specify. The fact that the slowly varying field operator $\mathcal{E}$ couples to S is easy to verify from the operator form of Maxwell's equations; it is made plausible by recalling the semiclassical equation (2.6.12), since $u(t) - iv(t)$ in the semiclassical approach is effectively just the expectation value of $S(t)$. The Heisenberg equation of motion for the operator $S(t)$ has the form

$$\dot{S}(t) = -\beta S(t) + F_s(t) - iD\sigma_z(t)\mathcal{E}(t). \qquad (6.9.17)$$

The damping rate β here can include collisions and other dissipative mechanisms in addition to spontaneous emission, and $F_s(t)$ is the Langevin noise operator associated with this dissipation. The coupling of S to $\mathcal{E}$ has the usual form for a two-state atom in an electric field; see, for instance, (5.6.9), with $E_0^{(+)}(t)e^{i\omega t}$ in that equation identified as the slowly varying electric field operator. We assume for simplicity, and without affecting our main results, that $\Delta = 0$, that is, that the laser frequency ω is equal to the transition frequency ω_0 of the two-state atom. Note that the same constant D appears in both (6.9.16) and (6.9.17) (see Exercise 6.7).

In the absence of coupling of the atom to the laser field ($D = 0$), the solution of (6.9.17) for $\beta t \gg 1$ is

$$\langle S^\dagger(t)S(t)\rangle = \int_0^t dt' \int_0^t dt'' \langle F_s^\dagger(t')F_s(t'')\rangle e^{\beta(t'+t''-2t)}. \qquad (6.9.18)$$

To be consistent with the operator identity $S^\dagger(t)S(t) = \frac{1}{2}[1 + \sigma_z(t)]$ for the two-state atom we take

$$\langle F_s^\dagger(t')F_s(t'')\rangle = 2\beta\delta(t'-t'')\langle 1 + \sigma_z(t)\rangle = 2\beta P_2\delta(t'-t''), \qquad (6.9.19)$$

where P_2 is the probability that the atom is in its excited state. Similarly, since $\langle S(t)S^\dagger(t)\rangle = \frac{1}{2}\langle 1 - \sigma_z(t)\rangle$ is the probability P_1 that the atom is in its unexcited state,

$$\langle F_s(t')F_s^\dagger(t'')\rangle = 2\beta P_1\delta(t'-t''). \qquad (6.9.20)$$

Exercise 6.7: (a) In the absence of any damping processes, the coupled atom–field equations of interest here have the general form

$$\dot{\mathcal{E}} = CS, \quad \dot{S} = -iD\sigma_z\mathcal{E}, \quad \dot{\sigma}_z = -2iD(\mathcal{E}^\dagger S - S^\dagger\mathcal{E}).$$

The last two equations follow from (5.6.9) and (5.6.10), for instance, but the coupling constant D here is different because the electric field operator is defined differently. Since we are assuming the field frequency equals the atom's transition frequency, energy conservation implies

$$\frac{\partial}{\partial t}\left[\mathcal{E}^\dagger(t)\mathcal{E}(t) + \frac{1}{2}\sigma_z(t)\right] = 0.$$

Show that this requires $C = iD$.
(b) Should we include in the field equation (6.9.16) a Langevin noise term related to the resonator loss?

These properties of the noise operator must also hold when the atom is coupled to the laser field. Then, in steady state,

$$S(t) = \frac{-iD}{\beta}\sigma_z(t)\mathcal{E}(t) + \frac{1}{\beta}F_s(t) \tag{6.9.21}$$

and, from (6.9.16),

$$\dot{\mathcal{E}}(t) \cong \frac{D^2}{\beta}\sigma_z(t)\mathcal{E}(t) + \frac{D}{\beta}F_s(t) - \frac{1}{2}v_g\kappa_c\mathcal{E}(t). \tag{6.9.22}$$

Comparison with (6.9.1)) shows that we can relate F_{sp} to F_s:

$$F_{sp}(t) = \frac{D}{\beta v_g}F_s(t); \tag{6.9.23}$$

therefore,

$$\langle F_{sp}^\dagger(t')F_{sp}(t'')\rangle \to \frac{1}{2}\frac{\mathcal{N}D^2}{\beta^2 v_g^2}\langle F_s^\dagger(t')F_s(t'')\rangle = \frac{\mathcal{N}D^2}{\beta v_g^2}P_2\delta(t'-t'') \tag{6.9.24}$$

when there are $\mathcal{N}$ atoms in the resonator. We have introduced the factor $1/2$ since each atom is equally likely to spontaneously emit a photon to the right or left along the cavity axis, whereas we require the spontaneous emission noise affecting a unidirectional, traveling-wave field $\mathcal{E}$.

Equation (6.9.2) follows when, based on (6.9.22), we associate $(2\mathcal{N}D^2/\beta)\langle\sigma_z\rangle = (2\mathcal{N}D^2/\beta)(P_2 - P_1)$ with the photon number amplification rate $v_g g \cong v_g \kappa_c$ in steady-state lasing.[72]

6.9.2   Amplitude-Stabilized Laser Linewidth

In a laser operating in a steady state and with the gain medium pumped sufficiently far above the (gain–loss) threshold for lasing, the gain coefficient is effectively "clamped" to the (constant) loss coefficient. Since the gain is saturated to some degree and therefore dependent on the field intensity (see Section 2.7), the field amplitude may be assumed to be approximately fixed at a constant value, so that the laser linewidth comes almost entirely from phase fluctuations. Of course, there are, in fact, amplitude fluctuations due to spontaneous emission, which causes the photon number in the cavity to fluctuate, but their effect on the above-threshold laser linewidth turns out to be relatively small compared to fluctuations in the field phase.[73] We now derive the fundamental laser linewidth for this operating regime, taking for simplification a semiclassical approach in which the complex field amplitude is written as

$$\tilde{\mathcal{E}}(t) = \mathcal{E}_0(t)e^{i\Phi(t)}, \tag{6.9.25}$$

with $\mathcal{E}_0(t)$ and $\phi(t)$ being real-valued functions of time. In similarity to the preceding quantum-operator approach, the amplitude $\mathcal{E}_0$ is defined such that, in steady state, $\mathcal{E}_0^2$ is the average number q of photons in the traveling-wave field, while the phase $\phi(t)$ undergoes random fluctuations due to spontaneous emission.

To obtain equations for $\mathcal{E}_0(t)$ and $\phi(t)$, we first recall that the loss and gain coefficients are related to the imaginary part, $n_I(\omega)$, of the refractive index. From (1.11.10), for instance, the gain coefficient (the negative of the absorption coefficient) can be expressed as $g(\omega) = (2\omega/c)n_I(\omega)$. In writing (6.9.1), we did not include the effect of the real part, $n_R(\omega)$, of the refractive index at the lasing transition, which would have added to the right side the term $[n_R(\omega)\omega/c]\mathcal{E}$, which would affect the phase of the field. If the pumping of the laser gain medium results in a relatively small change Δn_I in the imaginary part of the refractive index, the associated change in the real part of the index is $\alpha\Delta n_I$, where the *Henry α parameter*, also known for the reason discussed below as the *linewidth enhancement parameter*, is[74]

$$\alpha = \Delta n_R/\Delta n_I. \tag{6.9.26}$$

When the effect of the real part of the index is included, (6.9.1) is replaced in the semiclassical approach by

---

[72] In multiplying $\langle F_{sp}^\dagger(t')F_{sp}(t'')\rangle$ (but not $F_{sp}(t)$) by $\mathcal{N}$, we are assuming the spontaneous emission is not "superradiant," that is, that the different atoms radiate independently.

[73] R. D. Hempstead and M. Lax, Phys. Rev. **161**, 350 (1967).

[74] C. H. Henry, IEEE J. Quantum Electron. **QE-18**, 259 (1982); C. H. Henry, IEEE J. Quantum Electron. **QE-19**, 1391 (1983); C. H. Henry, J. Lightwave Tech. **LT-4**, 298 (1986), and references therein.

$$\frac{\partial \tilde{\mathcal{E}}}{\partial t} = -\gamma(1 + i\alpha)\tilde{\mathcal{E}}(t) + v_g \tilde{F}_{sp}(t), \tag{6.9.27}$$

where, as before, $\gamma = \frac{1}{2}v_g(\kappa_c - g)$ but now $\tilde{F}_{sp}(t)$ is a *classical* noise source. Then, from (6.9.25),

$$\dot{\mathcal{E}}_0 = -\gamma\mathcal{E}_0 + \frac{1}{2}v_g\big(\tilde{F}_{sp}e^{i\phi} + \tilde{F}_{sp}^*e^{-i\phi}\big), \tag{6.9.28}$$

$$\dot{\phi} = -\alpha\gamma + \frac{iv_g}{2\mathcal{E}_0}\big(\tilde{F}_{sp}e^{i\phi} - \tilde{F}_{sp}^*e^{-i\phi}\big). \tag{6.9.29}$$

Although we do not indicate it explicitly, the aforementioned saturability of the gain implies that γ in these equations depends on $\mathcal{E}_0$. If the gain is clamped to the loss, $\mathcal{E}_0$ will have the constant, steady-state value $\mathcal{E}_{oss}$ and

$$\langle\tilde{\mathcal{E}}^*(t)\tilde{\mathcal{E}}(t + \tau)\rangle = q\langle e^{-i[\phi(t+\tau)-\phi(t)]}\rangle, \quad q = \mathcal{E}_{0ss}^2. \tag{6.9.30}$$

To follow closely the fully quantum-theoretical approach, we assume

$$\langle\tilde{F}_{sp}^*(t')\tilde{F}_{sp}(t'')\rangle = \langle\tilde{F}_{sp}(t')\tilde{F}_{sp}^*(t'')\rangle = \frac{\kappa_c n_{sp}}{4v_g}\delta(t' - t'') \tag{6.9.31}$$

and

$$\langle\tilde{F}_{sp}(t')\tilde{F}_{sp}(t'')\rangle = \langle\tilde{F}_{sp}^*(t')\tilde{F}_{sp}^*(t'')\rangle = 0. \tag{6.9.32}$$

$\langle\tilde{F}_{sp}^*(t')\tilde{F}_{sp}(t'')\rangle$ is the semiclassical counterpart of the correlation function (6.9.2) appearing in the quantum-theoretical approach. There, we only required the noise correlation function $\langle F_{sp}^\dagger(t')F_{sp}(t'')\rangle$ because we dealt with the normally ordered field correlation function (6.9.4). To solve (6.9.29) for the phase fluctuations in the semiclassical approach, however, we require both $\langle\tilde{F}_{sp}^*(t')\tilde{F}_{sp}(t'')\rangle$ and $\langle\tilde{F}_{sp}(t')\tilde{F}_{sp}^*(t'')\rangle$: the mean-square fluctuation of the phase is determined by the sum of these two terms. The choice of (6.9.31) ensures that the average $\langle\tilde{F}_{sp}^*(t')\tilde{F}_{sp}(t'')\rangle + \langle\tilde{F}_{sp}(t')\tilde{F}_{sp}^*(t'')\rangle = 2\langle\tilde{F}_{sp}^*(t')\tilde{F}_{sp}(t'')\rangle$ for the classical random process has the same magnitude as the expectation value $\langle F_{sp}^\dagger(t')F_{sp}(t'')\rangle$. We similarly take

$$\langle\tilde{F}_{sp}(t')\tilde{F}_{sp}(t'')\rangle = 0, \tag{6.9.33}$$

since $\langle F_{sp}(t')F_{sp}(t'')\rangle = 0$.

The noise comes from a large number of independent spontaneous emission events, and, therefore, the moments of $[\phi(t + \tau) - \phi(t)]$ will be assumed, based on the central limit theorem, to be characteristic of a Gaussian random process. Then (see (6.4.6)),

$$\langle\tilde{\mathcal{E}}^*(t)\tilde{\mathcal{E}}(t + \tau)\rangle = qe^{-\frac{1}{2}\langle[\phi(t+\tau)-\phi(t)]^2\rangle}. \tag{6.9.34}$$

Suppose we set $\alpha = 0$ for now in (6.9.29) and integrate:

$$\phi(t + \tau) - \phi(t) = \frac{iv_g}{2\mathcal{E}_{0ss}}\int_t^{t+\tau} dt'\big[\tilde{F}_{sp}(t')e^{i\phi(t')} - \tilde{F}_{sp}^*(t')e^{-i\phi(t')}\big]. \tag{6.9.35}$$

It then follows from (6.9.31) and (6.9.32) that

$$\langle [\phi(t + \tau) - \phi(t)]^2 \rangle = \frac{\tau v_g^2}{4q} \frac{\kappa_c n_{sp}}{v_g}, \quad q = \mathcal{E}_{0ss}^2, \tag{6.9.36}$$

and

$$\langle \tilde{\mathcal{E}}^*(t)\tilde{\mathcal{E}}(t + \tau) \rangle = q e^{-\kappa_c n_{sp} v_g \tau/8q}, \tag{6.9.37}$$

implying the linewidth (full width at half-maximum)

$$\Delta \nu = 2 \times \frac{1}{2\pi} \frac{\kappa_c n_{sp} v_g}{8q} = \frac{\hbar \omega n_{sp}}{4\pi P_{\text{out}}} \frac{v_g^2}{c^2} \gamma_c^2. \tag{6.9.38}$$

This is half the linewidth (6.9.9) because we have assumed that the field amplitude is stabilized and therefore that the fundamental laser linewidth is attributable solely to the field phase fluctuations caused by spontaneous emission. The linewidth (6.9.7), which was derived under the implicit assumption that the gain coefficient does not depend on the field amplitude, evidently includes (equal) contributions from both amplitude and phase fluctuations.[75]

In terms of the output power and the phase excursion $\theta(\tau) = \phi(t + \tau) - \phi(t)$, (6.9.36) is

$$\langle \theta^2(\tau) \rangle = 2D\tau, \tag{6.9.39}$$

where the diffusion coefficient

$$D = \frac{\hbar \omega n_{sp}}{4 P_{\text{out}}} \frac{v_g^2}{c^2} \gamma_c^2 = \pi \Delta \nu. \tag{6.9.40}$$

In fact, an approach to the laser linewidth based on the Fokker–Planck equation leads to the diffusion equation

$$\frac{\partial P}{\partial \tau} = D \frac{\partial^2 P}{\partial \theta^2} \tag{6.9.41}$$

for the $P(\alpha)$ distribution (see Section 3.8) when, in the stable-amplitude approximation, it is assumed that $|\alpha| = |re^{i\theta}| = r$ is approximately constant.[76]

The fundamental linewidths of different lasers vary over many orders of magnitude, and therefore so do the fundamental phase diffusion times $\sim D^{-1}$, as will be clear from the numerical examples below. For a gas laser, the phase diffusion time due to spontaneous emission noise could be $\sim 10^2$ s or more, whereas, for a semiconductor laser, it might be more like a few nanoseconds.

---

[75] R. D. Hempstead and M. Lax, Phys. Rev. **161**, 350 (1967).
[76] See, for instance, *Scully and Zubairy*, Chapter 11.

6.9.3   Phase-Amplitude Coupling

We now include the coupling of the field phase and amplitude. This coupling stems from the parameter α in (6.9.29), which, in deriving (6.9.38), we took to be zero. Formally, using (6.9.28) in (6.9.29), we obtain an equation

$$\dot{\phi}(t) = \alpha\dot{\mathcal{E}}_0/\mathcal{E}_0 - \frac{\alpha v_g}{2\mathcal{E}_0}[\tilde{F}_{sp}(t')e^{i\phi(t')} + \tilde{F}_{sp}^*(t')e^{-i\phi(t')}]$$
$$+ \frac{iv_g}{2\mathcal{E}_0}[\tilde{F}_{sp}(t')e^{i\phi(t')} - \tilde{F}_{sp}^*(t')e^{-i\phi(t')}] \tag{6.9.42}$$

for $\phi(t)$, and therefore, in steady-state laser operation,

$$\phi(t+\tau) - \phi(t) \cong \frac{iv_g}{2\mathcal{E}_{0ss}} \int_t^{t+\tau} dt'[(1 + i\alpha)\tilde{F}_{sp}(t')e^{i\phi(t')}$$
$$- (1 - i\alpha)\tilde{F}_{sp}^*(t')e^{-i\phi(t')}], \tag{6.9.43}$$

when we take the field amplitude to be stabilized and constant and ignore the first term on the right side of (6.9.42). Now we can proceed exactly as we did in going from (6.9.35) to (6.9.38), with the result that the effect of the α parameter is to multiply (6.9.38) by $(1 + i\alpha)(1 - i\alpha) = 1 + \alpha^2$. In other words, the laser linewidth, including the effect of phase-amplitude coupling, is (see (6.9.9) and (6.9.38)),

$$\Delta\nu = (1 + \alpha^2)\frac{\hbar\omega n_{sp}}{4\pi P_{\text{out}}}\frac{v_g^2}{c^2}\gamma_c^2. \tag{6.9.44}$$

For an empty resonator of length L and with mirror reflectivities R_1, R_2, a field intensity I changes after a round-trip pass to

$$R_1 R_2 I = I e^{\log(R_1 R_2/2L)2L} = I e^{-2\kappa_c L}, \tag{6.9.45}$$

that is, the effective loss coefficient is

$$\kappa_c = -\frac{1}{2L}\log(R_1 R_2). \tag{6.9.46}$$

The total loss coefficient if there is a medium with the loss coefficient a filling the resonator is $\kappa_c + a$, and then κ_c in (6.9.38) is replaced by

$$\kappa_c + a = g_t, \tag{6.9.47}$$

the threshold gain coefficient for laser oscillation, assuming the gain medium occupies the entire length L of the resonator. Since the total output power as before is $P_{\text{out}} = v_g\kappa_c(2q\hbar\omega)$, the laser linewidth (6.9.38) can be expressed as

$$\Delta\nu = \frac{\hbar\omega n_{sp}g_t\kappa_c v_g^2}{4\pi P_{\text{out}}}, \tag{6.9.48}$$

or, more generally,

$$\Delta\nu = \frac{\hbar\omega n_{sp}g_t\kappa_c v_g^2}{4\pi P_{\text{out}}}(1+\alpha^2) \tag{6.9.49}$$

when the linewidth enhancement parameter is included. Early studies of the fundamental laser linewidth dealt with gas lasers, where $\Delta\nu$ is typically extremely small. For example, for a 1 mW, 632.8 nm helium–neon laser with $R_1 \cong 1$, $R_2 \cong 0.97$, $L \cong 30$ cm, $n_{sp} \approx 1$, and $\gamma_c = c\kappa_c$, (6.9.15) gives $\Delta\nu \approx 10^{-2}$ Hz. In such lasers, the linewidth is dominated by "technical noise" attributable to effects such as mechanical vibrations of the laser mirrors. Moreover, α is negligibly small in such lasers because the real part of the refractive index is not much different from unity near the peak of the gain spectrum, as is seen from (6.9.10), for instance. The α parameter attracted interest only with the development of semiconductor lasers in which the peak of the gain spectrum actually occurs at frequencies well above the laser frequency ω and $\Delta n_R(\omega)$ is relatively large and such that α^2 can be ~ 25 or larger. Mainly because of their small dimensions and large threshold gains, the linewidths $\Delta\nu$ of semiconductor lasers can be ten orders of magnitude larger than the fundamental linewidths of other lasers. In early work[77] demonstrating semiconductor laser linewidths much larger than predicted by the Schawlow–Townes formula, for example, the experimental parameters were estimated to be $a \approx 45$ cm^{-1}, $\kappa_c \approx 39$ cm^{-1}, $g_t = a + \kappa_c \approx 84$ cm^{-1}, $\hbar\omega \approx 1.5$ eV, $n_{sp} \approx 2.6$, $v_g \approx c/4.33$, and $\alpha \approx 6$.[78] For output powers $P_{\text{out}} \approx 1$ mW, (6.9.49) gives $\Delta\nu \approx 100$ MHz, consistent with the measured linewidths.

6.9.4 Amplification of Spontaneous Emission Noise

Spontaneous emission puts a fundamental limit on the spectral purity of laser radiation, as the preceding calculations make clear. One of the assumptions implicit in these calculations is that spontaneous emission noise is not amplified. But spontaneously emitted radiation can be amplified in the gain medium before it propagates out of the resonator. This can affect the linewidth in lasers with lossy resonators, since large amplification is required to overcome large loss. To account rather simply for this noise amplification, we first recall that spontaneous emission can be attributed physically to the source field (radiation reaction) of the emitter or to the vacuum field, or to some combination of these two fields, depending on how we choose to order the commuting source and field operators (see Section 4.5). The fundamental laser linewidth can similarly be interpreted in different ways, depending on our choice of operator orderings. Our linewidth calculations thus far have proceeded from the normally ordered field correlation function $\langle\mathcal{E}^\dagger(t)\mathcal{E}(t+\tau)\rangle$, but we could just as well have worked, for instance, with $\langle\mathcal{E}(t)\mathcal{E}^\dagger(t+\tau)\rangle$. With this anti-normally ordered correlation function, the fundamental laser linewidth naturally has a contribution from the vacuum field,

[77] M. W. Fleming and A. Mooradian, Appl. Phys. Lett. **38**, 511 (1981).

[78] C. H. Henry, IEEE J. Quantum Electron. **QE-18**, 259 (1982); C. H. Henry, IEEE J. Quantum Electron. **QE-19**, 1391 (1983); C. H. Henry, J. Lightwave Tech. **LT-4**, 298 (1986) and references therein.

and the amplification of the spontaneous emission noise can be related simply to the amplification of the vacuum field "leaking" into the resonator, as indicated in Figure 6.3.[79]

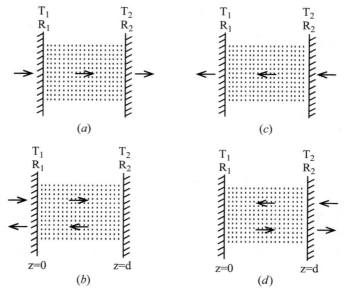

Fig. 6.3 The four contributions to the amplification of vacuum field fluctuations by a laser. T_i and R_i, $i = 1, 2$, denote the mirror transmissivities and reflectivities, respectively. From P. Goldberg, P. W. Milonni, and B. Sundaram, J. Mod. Opt. **38**, 1421 (1991), with permission.

Figure 6.3a depicts a vacuum field incident on the mirror at $z = 0$, passing through the gain medium, and emerging after amplification from the mirror at $z = d$. Transmission through the mirrors at $z = 0$ and $z = d$ reduces the intensity by factors T_1 and T_2, respectively, whereas a single pass through the gain medium increases the intensity by a factor G, which satisfies the steady-state (gain equals loss) condition $R_1 G R_2 G = 1$, or $G = 1/\sqrt{R_1 R_2}$. The net effect of the transmission corresponding to Figure 6.3a is therefore to give an effective vacuum-field noise proportional to $T_1 G T_2$. The processes indicated in Figures 6.3b, 6.3c, and 6.3d likewise contribute effective noise proportional to $T_2 G T_1$, $T_1 G R_2 G T_1$, and $T_2 G R_1 G T_2$, respectively. The four pro-

[79] P. Goldberg, P. W. Milonni and B. Sundaram, J. Mod. Opt. **38**, 1421 (1991); P. Goldberg, P. W. Milonni and B. Sundaram, Phys. Rev. A **44**, 1969 (1991); P. Goldberg, P. W. Milonni and B. Sundaram, Phys. Rev. A **44**, 4556 (1991).

cesses indicated in Figure 6.3 therefore result in spontaneous emission noise and a laser linewidth proportional to[80]

$$T_1 G T_2 + T_2 G T_1 + T_1 G R_2 G T_1 + T_2 G R_1 G T_2 = \left(\frac{T_1}{\sqrt{R_1}} + \frac{T_2}{\sqrt{R_2}} \right)^2 . \tag{6.9.50}$$

To see how this affects our previous results for the laser linewidth, suppose that transmission of light through the resonator mirrors is the only loss mechanism, so that $a = 0$ and therefore the threshold gain $g_t = \kappa_c$. Then, according to (6.9.49), for instance, $\Delta\nu \propto \kappa_c^2$ and, for a given output power, this depends on R_1 and R_2 only through the factor $[\log(R_1 R_2)]^2$. In contrast, the dependence on R_1 and R_2 of the linewidth according to (6.9.50) is

$$\left(\frac{T_1}{\sqrt{R_1}} + \frac{T_2}{\sqrt{R_2}} \right)^2 = \left[\frac{(\sqrt{R_1} + \sqrt{R_2})(1 - \sqrt{R_1 R_2})}{\sqrt{R_1 R_2}} \right]^2 , \tag{6.9.51}$$

where we have used $T_i = 1 - R_i$, $i = 1, 2$. The effect of the amplification of spontaneous emission noise is therefore to multiply our previous expressions for the fundamental laser linewidth by the K factor:

$$K = \left[\frac{(\sqrt{R_1} + \sqrt{R_2})(1 - \sqrt{R_1 R_2})}{\sqrt{R_1 R_2} \log(R_1 R_2)} \right]^2 > 1, \tag{6.9.52}$$

which in the literature is also called the *excess spontaneous emission factor* or the *Petermann factor*.[81] For most lasers, K represents only a very small correction; for $R_1 = R_2 = 0.90$, for example, $K = 1.001$. In a semiconductor laser, however, the reflectivities of the cleaved facets might be only $R_1 \approx R_2 \approx 0.30$, in which case $K \approx 1.13$.[82]

Exercise 6.8: Why does it suffice for our derivation of the K factor to consider only the four processes shown in Figure 6.3, and not processes involving more than two passes through the gain medium?

Our derivation of (6.9.52) pertains only to longitudinal resonator modes. However, the physical explanation of the K factor as a consequence of the amplification of spontaneous emission noise holds also for transverse field modes, and more generally for

---

[80] Ibid.

[81] The "excess spontaneous emission factor" K is not the result of any enhancement of the spontaneous emission rate of each emitter in the gain medium. See I. H. Deutsch, J. C. Garrison, and E. M. Wright, J. Opt. Soc. Am. B **8**, 1244 (1991); P. Goldberg, P. W. Milonni, and B. Sundaram, Phys. Rev. A **44**, 1969 (1991); K. Petermann, IEEE J. Quantum Electron. **15**, 566 (1979), considered "gain-guided" semiconductor lasers and introduced K as an "astigmatism parameter" characterizing a transverse laser mode.

[82] Early experimental evidence for the increase of the linewidth of a semiconductor laser by the K factor (6.9.52) was reported by W. A. Hamel and J. P. Woerdman, Phys. Rev. Lett. **64**, 1506 (1990).

any lossy resonator: greater loss requires more gain and therefore more noise ampli-fication.[83] Different derivations of the K factor have associated the K factor more mathematically with the fact that the longitudinal and transverse modes of a lossy resonator are not orthogonal.[84]

6.10   Amplification and Attenuation: The Noise Figure

We have emphasized the fact that the preservation of commutation relations in quantum theory requires that there be quantum noise when an electromagnetic field is amplified or attenuated. In other words, there must be a Langevin noise operator in the Heisenberg equations of motion for the field. In Section 6.9, we showed how the noise puts a fundamental limit on the spectral purity of laser radiation. We now consider the effect of noise on the photon number fluctuations of an amplified or attenuated field. It will be shown that amplification and attenuation both degrade the signal-to-noise ratio.

6.10.1   Amplification

Our model for this purpose is one in which the output field is linearly related to the input field. In terms of the photon annihilation operators b and a for single-mode output and input fields, respectively, we can write the input–output relation in the case of amplification as

$$b = \sqrt{G}a + L^\dagger, \tag{6.10.1}$$

where G is the power gain factor due to stimulated emission, and L is a Langevin noise operator associated with spontaneous emission.[85] The assumption of linearity means that G is independent of a and therefore that the amplifier is not saturable. The condition that the bosonic commutation relations are satisfied by both the input and the output fields, that is, that $[a, a^\dagger] = [b, b^\dagger] = 1$, requires that the Langevin operator satisfy

$$[L, L^\dagger] = G - 1 > 0. \tag{6.10.2}$$

This follows trivially from (6.10.1) when we assume also that the Langevin noise source is uncorrelated with the input field, that is, that $[L, a] = [L^\dagger, a] = 0$. The noise source is regarded, as usual, as a reservoir of harmonic oscillators, and we therefore assume that $L = \sqrt{G-1}c$, $[c, c^\dagger] = 1$. We will also assume that the reservoir is at

[83] P. Goldberg, P. W. Milonni and B. Sundaram, J. Mod. Opt. **38**, 1421 (1991).

[84] A. E. Siegman, Phys. Rev. A **39**, 1253 (1989); Phys. Rev. A **39**, 1264 (1989). The very large diffractive losses of transverse modes in unstable laser resonators can result in K factors $\sim 10^2$. See Y.-J. Cheng, C. G. Fanning, and A. E. Siegman, Phys. Rev. Lett. **77**, 627 (1996). For a discussion of different interpretations of the K factor, and a review of various other aspects of noise in photonics, see C. H. Henry and R. F. Kazarinov, Rev. Mod. Phys. **68**, 801 (1996).

[85] See, for instance, C. M. Caves, Phys. Rev. D **26**, 1817 (1982); *Desurvire*, Chapter 2; *Haus*, Chapter 9. We write $L^\dagger$ rather that L in (6.10.1) in order to have the right side of (6.10.2) be positive, as required for a reservoir of harmonic oscillators.

zero temperature and therefore that $\langle c^\dagger c \rangle = 0$ and $\langle c c^\dagger \rangle = 1$. The reservoir need not actually be at zero temperature for this assumption to be a good approximation, but only at a temperature T small enough that $\hbar\omega/k_B T \gg 1$, where ω is the field frequency. It follows from (6.10.1) that the output photon number expectation value $\langle n_b \rangle = \langle b^\dagger b \rangle$ is related to the input photon number expectation value $\langle n_a \rangle = \langle a^\dagger a \rangle$ by

$$\langle n_b \rangle = G\langle n_a \rangle + G - 1. \tag{6.10.3}$$

The first term on the right obviously represents the amplification of the input signal field, whereas the second term is added noise, corresponding to a noise power proportional to $(G - 1)\hbar\omega$ for radiation of frequency ω. Implicit in this simplified model is that only noise within the same bandwidth as the signal is relevant, as is the case in practice when the detection includes a spectral filtering. It is perhaps worth emphasizing that the noise and fluctuations under consideration here are inherent to the output field and independent of the photodetector.

It follows easily, based on our assumptions for the Langevin noise operator L, that the variance in the photon number of the output field is

$$\langle \Delta n_b^2 \rangle = G^2 \langle \Delta n_a^2 \rangle + G(G - 1)(\langle n_a \rangle + 1). \tag{6.10.4}$$

The first term corresponds to amplification of the variance of the input photon number, whereas the second term is the noise contribution to the variance of the output photon number. If there were no noise, and therefore no amplification ($G = 1$), this term would vanish, as would the second term on the right side of (6.10.3).

Different contributions to the variance of the output photon number are often identified by writing (6.10.4) equivalently as

$$\begin{aligned}\langle \Delta n_b^2 \rangle &= G^2(\langle \Delta n_a^2 \rangle - \langle n_a \rangle) + G\langle n_a \rangle + N + 2GN\langle n_a \rangle + N^2 \\ &= G^2(\langle \Delta n_a^2 \rangle - \langle n_a \rangle) + \langle n_b \rangle + 2GN\langle n_a \rangle + N^2,\end{aligned} \tag{6.10.5}$$

where $N = G - 1$ is the noise contribution to $\langle n_b \rangle$ in (6.10.3). The first term on the right is said to be *excess noise*. It depends on the photon statistics of the input field and vanishes when the statistics is Poissonian ($\langle \Delta n_a^2 \rangle = \langle n_a \rangle$); any deviation from Poisson statistics of the input field increases the photon number variance of the output field. The second term in (6.10.5), $\langle n_b \rangle$, represents *shot noise*. Finally, the last term, $2GN\langle n_a \rangle + N^2$, is usually referred to, somewhat loosely, as *beat noise*.

If there is no input field (6.10.5) reduces to

$$\langle \Delta n_b^2 \rangle = N + N^2, \tag{6.10.6}$$

which is the variance for a Bose–Einstein distribution with average photon number $N = G - 1$ equal to the noise contribution to (6.10.3). We show below that the photon number distribution of the amplified field when there is no input is, in fact, a Bose–Einstein distribution with $\langle n_b \rangle = N$.

The signal-to-noise ratios of the input and output fields are conventionally defined as

$$\text{SNR}_{\text{in}} = \frac{\langle n_a \rangle^2}{\langle \Delta n_a^2 \rangle} \tag{6.10.7}$$

and

$$\text{SNR}_{\text{out}} = \frac{G^2 \langle n_a \rangle^2}{\langle \Delta n_b^2 \rangle}, \tag{6.10.8}$$

respectively. The *noise figure* of the amplifier is then defined as

$$F = \frac{\text{SNR}_{\text{in}}}{\text{SNR}_{\text{out}}}, \tag{6.10.9}$$

a larger value of F corresponding to a smaller signal-to-noise ratio of the output field. It follows from (6.10.4) that

$$F = 1 + \left(1 - \frac{1}{G}\right) \frac{\langle n_a \rangle + 1}{\langle \Delta n_a^2 \rangle}. \tag{6.10.10}$$

Obviously, $F > 1$ for $G > 1$: *the amplified field always has a smaller signal-to-noise ratio than the input field*. For an input field having a Poissonian photon number distribution, $\langle \Delta n_a^2 \rangle = \langle n_a \rangle$, and

$$F = 1 + \left(1 - \frac{1}{G}\right)\left(1 + \frac{1}{\langle n_a \rangle}\right). \tag{6.10.11}$$

For $\langle n_a \rangle \gg 1$,

$$F \cong 2 - \frac{1}{G}. \tag{6.10.12}$$

In the high-gain limit $(G \gg 1)$, $F \cong 2$ or, in decibels, $F \cong 3$ dB, the so-called quantum limit for a high-gain amplifier.[86] Smaller gain implies less degradation of the signal-to-noise ratio.

If an amplifier with gain G_1 is followed by an amplifier with gain G_2, the net gain factor is $G_1 G_2$, and the noise figure for the two-amplifier system is $F \cong 2 - 1/G_1 G_2$ in the approximation (6.10.12). In terms of the noise figures $F_1 \cong 2 - 1/G_1$ and $F_2 \cong 2 - 1/G_2$ for the individual amplifiers when they act alone,

$$F \cong F_1 + \frac{F_2 - 1}{G_1}. \tag{6.10.13}$$

Obviously, the noise figure F for the two amplifiers acting together depends on the order in which the amplifiers are placed. The lowest noise figure for the two-amplifier system is obtained when the amplifier with the lowest noise figure acts first.

The equation $F = F_1 + (F_2 - 1)/G_1$ is well known in the theory of electronic amplifiers, as is its generalization for a chain of more than two electronic amplifiers. It should be noted, however, that (6.10.13) in the present context is obtained in the *approximation* $\langle n_a \rangle \gg 1$. In particular, the definition (6.10.9) of the noise figure used here and in much of the literature

---

[86] $F(\text{dB}) = 10 \log_{10} F = 10 \log_{10} 2 = 3.01$.

on optical fiber amplifiers depends on the input field as well as the amplifier gain, as is seen from (6.10.11), for example. This is a consequence of defining the noise figure using the *square* of $\langle n_a \rangle$ and the variance $\langle \Delta n_a^2 \rangle$. Haus has suggested a different definition of the noise figure that is independent of the input field and satisfies exactly the formula for the noise figure of a chain of amplifiers that has historically been used for electronic amplifiers.[87]

These results assume an "ideal amplifier" in which the gain medium is fully inverted and therefore the spontaneous emission factor (see (6.9.3)), $n_{sp} = 1$. More generally, $n_{sp} > 1$, and the noise figure (6.10.12), for example, is replaced by

$$F = 2n_{sp} + \frac{1}{G}(1 - 2n_{sp}) \tag{6.10.14}$$

for a non-ideal amplifier, as discussed below. In the high-gain limit, $F \cong 2n_{sp} > 2$ for an input field with Poissonian photon statistics; all else being equal, the non-ideal amplifier results in a greater noise figure than the ideal amplifier.

6.10.2   Attenuation

We proceed similarly in the case of a linear attenuator. Instead of (6.10.1), we write

$$b = ta + L, \tag{6.10.15}$$

where t is the amplitude transmission factor of the attenuator and is assumed to be independent of a. The condition $[b, b^\dagger] = [a, a^\dagger] = 1$ requires that

$$[L, L^\dagger] = 1 - T > 0, \quad T = |t|^2, \tag{6.10.16}$$

when we again assume that the Langevin noise is uncorrelated with the input field ($[L, a] = [L^\dagger, a] = 0$). We again relate the noise operator to a harmonic-oscillator lowering operator c: $L = \sqrt{1-T}c$, $[c, c^\dagger] = 1$, and assume the oscillator reservoir is in its zero-temperature ground state, $\langle c^\dagger c \rangle = 0$, $\langle cc^\dagger \rangle = 1$. Then we obtain straightforwardly the noise figure

$$F = \frac{\text{SNR}_{\text{in}}}{\text{SNR}_{\text{out}}} = 1 + \left(\frac{1}{T} - 1\right)\frac{\langle n_a \rangle}{\langle \Delta n_a^2 \rangle} \tag{6.10.17}$$

for the attenuator. For Poissonian photon statistics of the input field,

$$F = 1/T, \tag{6.10.18}$$

which, unlike the noise figure for an ideal amplifier, can be arbitrarily large.

---

[87] H. A. Haus, IEEE Photonics Tech. Lett. **10**, 1602 (1998). See also *Haus*, Chapter 9.

Exercise 6.9: (a) The Langevin operator for amplification by stimulated emission in a medium with an inverted population is associated with spontaneous emission. What physical process is responsible for the quantum noise in an absorbing medium? (b) Explain (or show) how the input–output relations (6.10.1) and (6.10.15) can be derived from Heisenberg equations of motion describing amplification and attenuation. What assumptions are made in obtaining these relations?

6.11   Photon Statistics of Amplification and Attenuation

We now consider the more general problem of calculating not just the photon number variance but the photon probability distribution p_n^b of the output field, given p_n^a for the input field. The algebra is a little simpler for attenuation, which we will consider first.

6.11.1   Attenuation: Photon Statistics

As in Section 3.7.2, we define

$$F_N^b(\xi) = \sum_{r=0}^{\infty} \frac{\xi^r}{r!} \langle b^{\dagger r} b^r \rangle \tag{6.11.1}$$

for the output field, and use (6.10.15) and the assumption that the reservoir is in its ground state. As a consequence of normal ordering, the Langevin operator L makes no contribution to $\langle b^{\dagger r} b^r \rangle$:

$$\langle b^{\dagger r} b^r \rangle = (t^* t) \langle a^{\dagger r} a^r \rangle = T^r \langle a^{\dagger r} a^r \rangle \tag{6.11.2}$$

and

$$F_N^b(\xi) = \sum_{r=0}^{\infty} \frac{(T\xi)^r}{r!} \langle a^{\dagger r} a^r \rangle = F_N^a(T\xi). \tag{6.11.3}$$

Therefore, the generating function for the photon number probability distribution of the attenuated field is (see (3.7.31))

$$p^b(\xi) = F_N^b(\xi - 1) = F_N^a((\xi - 1)T). \tag{6.11.4}$$

If the photon probability distribution of the input field is Poissonian with average photon number $\langle n_a \rangle$, (6.11.4) and (3.7.32) imply

$$p^b(\xi) = e^{(\xi - 1)T \langle n_a \rangle}, \tag{6.11.5}$$

which is just the generating function for a Poisson distribution with average photon number $T\langle n_a \rangle$. If the input field has a Bose–Einstein distribution with average photon number $\langle n_a \rangle$, (6.11.4) and (3.7.33) imply

$$p^b(\xi) = \frac{1}{1 - (\xi - 1)T\langle n_a \rangle}, \tag{6.11.6}$$

which is the generating function for a Bose–Einstein distribution with average photon number $T\langle n_a \rangle$. So, in both cases, the only change in the photon statistics as a result of attenuation is simply that the average photon number has the diminished value $T\langle n_a \rangle$. More generally, if $T\langle n_a \rangle \ll 1$,

$$p^b(\xi) \cong 1 + (\xi - 1)T\langle n_a \rangle \cong e^{(\xi - 1)T\langle n_a \rangle}, \tag{6.11.7}$$

implying that a highly attenuated field will have an approximately Poissonian photon probability distribution.

6.11.2 Amplification: Photon Statistics

We proceed similarly in the case of amplification. Since $L^\dagger$ appears in (6.10.1), it is convenient now to employ the anti-normally ordered expectation value from (3.7.34), since $\langle b^r b^{\dagger r} \rangle$ involves expectation values of normally ordered products of L and $L^\dagger$, which vanish under our assumption that the oscillator reservoir is in its ground state. From (6.10.1),

$$F_A^b(\xi) = \sum_{r=0}^{\infty} \frac{\xi^r}{r!} \langle b^r b^{\dagger r} \rangle = 1 + \xi G \langle aa^\dagger \rangle + \frac{\xi^2}{2!} G^2 \langle aaa^\dagger a^\dagger \rangle + \cdots$$

$$= 1 + \xi G(n_a + 1) + \frac{\xi^2}{2!} G^2 (n_a + 1)(n_a + 2) + \cdots = \left(\frac{1}{1 - \xi G} \right)^{n_a + 1} \tag{6.11.8}$$

for a number (Fock) state $|n_a\rangle$ of the field input to an ideal amplifier, and from (3.7.36),

$$p^b(\xi) = \frac{1}{\xi} F_A^b \left(1 - \frac{1}{\xi} \right)$$

$$= \frac{1}{\xi} \left[\frac{1}{1 - (1 - 1/\xi)G} \right]^{n_a + 1} = \frac{1}{G - \xi(G - 1)} \left[\frac{\xi}{G - \xi(G - 1)} \right]^{n_a} \tag{6.11.9}$$

for this case. If the input field has a photon number probability distribution p_n^a, the photon number probability generating function for the amplified field is

$$p^b(\xi) = \frac{1}{G - \xi(G - 1)} \sum_{n=0}^{\infty} p_n^a \left[\frac{\xi}{G - \xi(G - 1)} \right]^n. \tag{6.11.10}$$

Suppose first that there is no signal field input to the amplifier. Then, $p_n^a = \delta_{n,0}$ and

$$p^b(\xi) = \frac{1}{G - \xi(G - 1)}. \tag{6.11.11}$$

Comparison with (3.7.33) shows that this is the moment generating function for a Bose–Einstein distribution with average photon number $G - 1$, which is the average photon number given by (6.10.3) with $\langle n_a \rangle = 0$. The fact that an optical amplifier with no input signal generates noise power proportional to $G - 1$ is a consequence of amplified spontaneous emission, and can also be explained by accounting for the amplification of the vacuum field at the amplifier input port, as discussed below.

If the input field has a Poissonian photon number distribution,

$$p^b(\xi) = \frac{1}{G - \xi(G-1)} \sum_{n=0}^{\infty} \frac{\langle n_a \rangle^n e^{-\langle n_a \rangle}}{n!} \left[\frac{\xi}{G - \xi(G-1)} \right]^n$$

$$= \frac{e^{-\langle n_a \rangle}}{G - \xi(G-1)} \sum_{n=0}^{\infty} \frac{1}{n!} \left[\frac{\xi \langle n_a \rangle}{G - \xi(G-1)} \right]^n$$

$$= \frac{1}{G - \xi(G-1)} \exp \left[\frac{G \langle n_a \rangle (\xi - 1)}{G - \xi(G-1)} \right], \qquad (6.11.12)$$

which leads to a rather complicated, non-Poissonian photon number distribution involving generalized Laguerre polynomials.[88] But the lowest-order moments follow easily from the definition of the generating function:

$$\langle n_b \rangle = \frac{d}{d\xi} p^b(\xi) \Big|_{\xi=1}, \qquad \langle n_b(n_b - 1) \rangle = \frac{d^2}{d\xi^2} p^b(\xi) \Big|_{\xi=1}, \qquad (6.11.13)$$

and, of course, the previously derived results (6.10.3) and (6.10.4) are reproduced. Those results already show that the photon statistics of the amplified field is not Poissonian.

If the input field has a Bose–Einstein photon number distribution,[89]

$$p^b(\xi) = \frac{1}{G - \xi(G-1)} \frac{1}{\langle n_a \rangle + 1} \sum_{n=0}^{\infty} \left\{ \frac{\langle n_a \rangle \xi}{(\langle n_a \rangle + 1)[G - \xi(G-1)]} \right\}^n$$

$$= \frac{1}{G - \xi(G-1)} \frac{1}{\langle n_a \rangle + 1} \left\{ 1 - \frac{\langle n_a \rangle \xi}{(\langle n_a \rangle + 1)[G - \xi(G-1)]} \right\}^{-1}$$

$$= \frac{1}{G(\langle n_a \rangle + 1)(1 - \xi) + \xi} = \frac{1}{1 - \langle n_b \rangle (\xi - 1)}, \qquad (6.11.14)$$

which implies, for the amplified field, a Bose–Einstein distribution with an average photon number $\langle n_b \rangle$ given by (6.10.3).

6.11.3 Remarks

The term $N = G - 1$ in (6.10.3) is the noise contribution to $\langle n_b \rangle$ for an ideal amplifier in which the gain medium is fully inverted and therefore the spontaneous emission

---

[88] See, for instance, *Desurvire*, Chapter 2, or *Haus*, Chapter 9.
[89] H. A. Haus, IEEE Photonics Tech. Lett. **10**, 1602 (1998). See also *Haus*, Chapter 9.

factor $n_{sp} = 1$. It is shown below that, for a non-ideal amplifier, for which $n_{sp} > 1$, N is replaced by $n_{sp}N = n_{sp}(G-1)$ and (6.10.3) by

$$\langle n_b \rangle = G\langle n_a \rangle + n_{sp}(G-1). \tag{6.11.15}$$

Similarly, it may be shown by introducing an appropriate Langevin operator that N in (6.10.5) is replaced by $n_{sp}(G-1)$ and

$$\langle \Delta n_b^2 \rangle = G^2(\langle \Delta n_a^2 \rangle - \langle n_a \rangle) + G\langle n_a \rangle + n_{sp}(G-1)$$
$$+ 2n_{sp}G(G-1)\langle n_a \rangle + n_{sp}^2(G-1)^2 \tag{6.11.16}$$

for a non-ideal amplifier. For $\langle \Delta n_a^2 \rangle = \langle n_a \rangle$ and $\langle n_a \rangle \gg 1$, for example, (6.11.15) and (6.11.16) imply the noise figure (6.10.14).

We have assumed that the input and output fields belong to one and the same mode in order to focus on the most basic physics of spontaneous emission noise in amplification and attenuation. This restriction is removed below. For amplification in a fiber supporting M modes, for example, (6.11.15) is replaced by

$$\langle n_b \rangle = G\langle n_a \rangle + Mn_{sp}(G-1) \tag{6.11.17}$$

and (6.11.16) by

$$\langle \Delta n_b^2 \rangle = G^2(\langle \Delta n_a^2 \rangle - \langle n_a \rangle) + G\langle n_a \rangle + Mn_{sp}(G-1)$$
$$+ 2n_{sp}G(G-1)\langle n_a \rangle + Mn_{sp}^2(G-1)^2. \tag{6.11.18}$$

In these expressions, the photon numbers are the sums over all M modes, and the amplification factor G, as well as the spontaneous emission rate, is assumed to be approximately the same for all the modes. If, for example, there are two differently polarized modes with the same spatial profile, amplification factor, and spontaneous emission rate in a doped optical fiber, $M = 2$. Photon probability distributions allowing for $n_{sp} > 1$ and $M > 1$ tend to be rather complicated and may be found in the more specialized literature.[90]

6.12 Amplified Spontaneous Emission

The degradation of the signal-to-noise ratio in an amplifier is a consequence of amplified spontaneous emission noise. If we assume a single field mode, as for example in the case of amplification in a single-mode fiber, we can describe the growth with propagation distance z of the average number of photons by the simple rate equation

$$\frac{d\langle n_b \rangle}{dz} = R(N_2 - N_1)\langle n_b \rangle + RN_2. \tag{6.12.1}$$

The first and second terms on the right describe stimulated and spontaneous emission, respectively. N_2 and N_1 are the population densities of the upper- and lower-energy

[90] Ibid.

states, respectively, of the atoms of the gain medium, R is proportional to the rate per unit volume of spontaneous emission into the single mode, and we have used the fact that for a single mode the stimulated emission rate is $\langle n_a \rangle$ times the spontaneous emission rate. Assuming N_2 and N_1 to be constant over the propagation path, the average photon number at the output port of the amplifier of length L is found from this rate equation to be given by (6.11.15):

$$\langle n_b(L) \rangle = G\langle n_b(0)\rangle + \frac{N_2}{N_2 - N_1}(G-1) = G\langle n_b(0)\rangle + n_{sp}(G-1), \qquad (6.12.2)$$

where now G is identified as $\exp[R(N_2 - N_1)L]$. We have assumed non-degenerate energy levels. If the upper and lower levels of the amplifying transition have degeneracies g_2 and g_1, respectively, and the degenerate states of each level are equally populated, the amplification rate is proportional to $N_2 - (g_2/g_1)N_1$. In that case,

$$n_{sp} = \frac{N_2}{N_2 - (g_2/g_1)N_1}. \qquad (6.12.3)$$

If the amplifier supports M modes, and the rates of spontaneous emission into different modes are approximately equal, the total average photon number at L is

$$\langle n_b(L) \rangle = G\sum_{i=1}^{M}\langle n_{bi}(0)\rangle + \sum_{i=1}^{M} n_{sp}(G-1) = G\langle n_b(0)\rangle + Mn_{sp}(G-1), \qquad (6.12.4)$$

which is (6.11.17), and, likewise, $\langle \Delta n_b^2(L) \rangle$ is given by (6.11.18). Note that, in (6.11.18), the term $2n_{sp}G(G-1)\langle n_a \rangle$, which represents a beat noise involving the signal and spontaneous emission, is independent of M, whereas the beat noise term $Mn_{sp}^2(G-1)^2$ representing spontaneous emission beat noise is M times (not M^2 times) the corresponding single-mode contribution. These results simply reflect the fact that the zero-mean fluctuations of the different modes are uncorrelated. If there is no input field, (6.11.17) and (6.11.18) give the photon number average and variance for amplified spontaneous emission:

$$\langle n_b \rangle = Mn_{sp}(G-1) = MN \qquad (6.12.5)$$

and

$$\langle \Delta n_b^2 \rangle = Mn_{sp}(G-1) + Mn_{sp}^2(G-1)^2 = M(N + N^2). \qquad (6.12.6)$$

The photon number distribution in this case is found to be

$$p_n = \frac{(n+M-1)!}{n!(M-1)!}\frac{N^n}{(N+1)^{n+M}}, \qquad (6.12.7)$$

which reduces to the Bose–Einstein distribution when $M = 1$.[91]

[91] Ibid.

In a semiclassical approach to amplified spontaneous emission, we consider a traveling plane wave in an amplifier with gain coefficient g, which is assumed to be unsaturable and constant over the length of the amplifier. For the variation of the average field intensity I with propagation distance, we write

$$\frac{dI}{dz} = gI + (A_{21}N_2\hbar\omega)\frac{\Omega}{4\pi}. \tag{6.12.8}$$

A_{21} is the spontaneous emission rate for the emission by each atom of a photon of energy $\hbar\omega$ into free space, and Ω is a solid angle chosen in accord with the geometry of the amplifier. If N_2 is independent of I and z,

$$I(L) = I(0)e^{gL} + \frac{A_{21}\hbar\omega\Omega}{4\pi}\frac{N_2}{g}(e^{gL} - 1) = GI(0) + \frac{A_{21}h\nu\Omega}{4\pi}\frac{N_2}{g}(G - 1). \tag{6.12.9}$$

Now, from (2.7.12) and (2.7.13), the gain coefficient at the resonance frequency ν_0 ($= \omega_0/2\pi$) of the amplifying medium, which we assume for definiteness to be homogeneously broadened, is

$$g = \frac{\lambda^2 A_{21}}{8\pi}(N_2 - N_1)S(\nu_0) = \frac{\lambda^2 A_{21}}{8\pi^2}(N_2 - N_1)\frac{1}{\Delta\nu}, \quad \lambda = c/\nu_0. \tag{6.12.10}$$

We can therefore express (6.12.9) as

$$I(L) = GI(0) + \frac{\pi h\nu^3\Omega}{c^2}\Delta\nu_f n_{sp}(G - 1) = I_N n_{sp}(G - 1), \tag{6.12.11}$$

where $\Delta\nu_f = 2\Delta\nu$ is the full width at half-maximum of the gain lineshape function and

$$I_N = \frac{\pi h\nu^3\Omega}{c^2}\Delta\nu_f. \tag{6.12.12}$$

If there is no field input at $z = 0$, the amplified spontaneous emission intensity at L is

$$I(L) = I_N n_{sp}(G - 1). \tag{6.12.13}$$

This has the same form as (6.12.5), which was derived for a discrete number M of field modes.

The intensity (6.12.13) is obviously the result of amplification of spontaneously emitted radiation over the length of the amplifier. We can interpret it alternatively by assuming there is an *effective noise intensity* $I(0) = I_N$ into the amplifier input port at $z = 0$ and that this intensity is amplified to $GI(0) = GI_N$ at $z = L$. I_N can be related to the fluctuating vacuum electromagnetic field, as discussed below, and as such is not directly observable with only a photodetector in a vacuum. But assuming that the *difference* $GI_N - I_N$ with the amplifier in place is observable, we obtain, aside from the factor n_{sp}, the amplified spontaneous emission intensity (6.12.13).[92] So,

---

[92] The appearance of n_{sp} in amplified spontaneous emission simply accounts for the fact that unexcited atoms also increase the noise intensity because absorption removes photons from the field in a random fashion. This is why the noise figure of an attenuator increases with decreasing transmission (see (6.10.17)) and can be arbitrarily large.

amplified spontaneous emission can be interpreted as amplified, "internal" spontaneous emission noise or as the amplification of an "external" noise intensity I_N leaking into the amplifier. This is the same sort of interpretive flexibility that occurs in the theory of the fundamental laser linewidth.[93] By attributing physical significance to the difference $GI_N - I_N$, we are following along the lines of the interpretation of the Lamb shift or Casimir forces in terms of the *change* in the zero-point vacuum-field energy resulting from the presence of an atom or material media (see Chapter 7).

It is a simple matter to relate I_N to the zero-point, vacuum electromagnetic field. Given the fact that the energy density in free space of the zero-point field in a small frequency interval $[\omega, \omega + \Delta\omega]$ is $\rho_0(\omega)\Delta\omega$, where $\rho_0(\omega)$ is given by (4.5.17), we infer that an effective vacuum-field intensity

$$I_0 \approx c(\hbar\omega^3/2\pi^2 c^3)\Delta\omega\frac{\Omega}{4\pi} = \frac{h\nu^3\Omega}{c^2}\Delta\nu \qquad (6.12.14)$$

is input to our amplifier for which the gain bandwidth is $\Delta\nu$. We have not accounted for the fact that I_N above refers to a single polarization component of the field, and we have assumed that a fraction $\Omega/4\pi$ of the modes in free space gets amplified, but, nevertheless, I_0 differs from I_N by only a factor of π.

For an amplifier of cross-sectional area S, we define a noise power

$$P_N \approx I_N n_{sp}(G-1)S. \qquad (6.12.15)$$

Assuming a pencil-shaped amplifier of diameter D, we estimate a divergence angle $\sim \lambda/D$, owing to diffraction for the field in the amplifier, and approximate the solid angle Ω by $\lambda^2/D^2 \approx c^2/\nu^2 S$:

$$P_N \approx \frac{\pi h\nu^3}{c^2}\frac{c^2}{\nu^2 S}\Delta\nu_f n_{sp}(G-1)S = h\nu n_{sp}\Delta\nu_f(G-1). \qquad (6.12.16)$$

More generally, the *noise power per mode* is defined as

$$P_N = n_{sp}h\nu B(G-1). \qquad (6.12.17)$$

B is the amplifier bandwidth, or the bandwidth of a detector that responds to some portion of the amplifier bandwidth.

6.12.1 Quantum Noise in Absorbers

The formula (6.12.17) applies as well to an absorbing medium: we only have to replace $G = \exp(gL)$ by $\exp(-aL)$, where a (> 0) is the absorption coefficient:

---

[93] P. Goldberg, P. W. Milonni, and B. Sundaram, J. Mod. Opt. **38**, 1421 (1991); P. Goldberg, P. W. Milonni, and B. Sundaram, Phys. Rev. A **44**, 1969 (1991); P. Goldberg, P. W. Milonni, and B. Sundaram, Phys. Rev. A **44**, 4556 (1991).

$$P_N = h\nu \frac{N_2}{N_2 - N_1} B(e^{-aL} - 1) = h\nu \frac{N_2}{N_1 - N_2} B(1 - e^{-aL})$$

$$= h\nu \frac{N_1 e^{-h\nu/k_B T}}{N_1 - N_1 e^{-h\nu/k_B T}} B(1 - e^{-aL}) \qquad (6.12.18)$$

for an absorbing medium composed of two-state atoms in thermal equilibrium at temperature T. If $aL \to \infty$,

$$P_N = \frac{h\nu}{e^{h\nu/k_B T} - 1} B \qquad (6.12.19)$$

and the absorber just acts as a source of traveling-wave thermal radiation. In the classical limit, $k_B T/h\nu \to \infty$,

$$P_N = k_B T B. \qquad (6.12.20)$$

Equation (6.12.20) follows from thermodynamic considerations and, as such, has a much broader significance and applicability than might be suggested from our derivation based on the spontaneous emission noise of an absorbing medium of two-state atoms. Consider, for example, an electric circuit with Johnson noise due to a resistance R. The power per unit bandwidth B that could theoretically be extracted from the resistor by a purely resistive load R_L is $P = \langle I^2 \rangle R_L = \langle V^2 \rangle R_L/(R + R_L)^2$, which is found from the Nyquist formula (6.6.1) to have the maximum value (6.12.20) when $R_L = R$. The maximum extractable noise power per unit bandwidth is independent of R and depends only on the temperature T.

6.13   The Beam Splitter

Consider a lossless beam splitter followed in one "output" arm by a nearly perfect photodetector (see Figure 6.4). Let a and b denote single-mode photon annihilation operators for "input" fields at the beam splitter, and c and d denote single-mode photon annihilation operators for output fields. The amplitude reflection and transmission coefficients r and t for a symmetric beam splitter satisfy the "unitarity" condition $|r + t|^2 = 1$, or $|r|^2 + |t|^2 = R + T = 1$ (by conservation of energy) and $r^*t + rt^* = 0$. We can satisfy these relations using the *Stokes relations* $t = \sqrt{T}$ and $r = i\sqrt{1 - T}$. Then the input–output relations take the form[94]

$$c = \sqrt{T}b + i\sqrt{1 - T}a, \qquad (6.13.1)$$

$$d = i\sqrt{1 - T}b + \sqrt{T}a. \qquad (6.13.2)$$

Since $[a, a^\dagger] = [b, b^\dagger] = 1$, $[a, b] = [a, b^\dagger] = 0$, these input–output relations imply $[c, c^\dagger] = [d, d^\dagger] = 1$, as required. The b and c modes might be unoccupied, vacuum-field modes, but we must nevertheless include them in the input–output relations in order to consistently maintain commutation relations.

---

[94] Since only a single frequency ω is considered, we do not indicate time dependence, $\exp(\pm i\omega t)$, of the fields. We can also ignore the spatial dependence of the fields, which are presumed for simplicity to be plane waves.

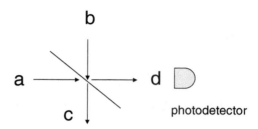

Fig. 6.4 A beam splitter with inputs a and b and outputs c and d.

Suppose the state $|\psi\rangle$ of the field is such that the b mode is unoccupied. For the expectation value of the photon number at the detector we obtain

$$\langle n_d \rangle = \langle d^\dagger d \rangle = \langle T a^\dagger a + (1-T)b^\dagger b + i\sqrt{T(1-T)}(a^\dagger b - b^\dagger b)\rangle$$
$$= T\langle a^\dagger a \rangle = T\langle n_a \rangle, \tag{6.13.3}$$

and for the variance, similarly,

$$\langle \Delta n_d^2 \rangle = \langle d^\dagger d d^\dagger d \rangle - \langle d^\dagger d \rangle^2 = T^2 \langle \Delta n_a^2 \rangle + T(1-T)\langle a^\dagger a b b^\dagger \rangle$$
$$= T^2 \langle \Delta n_a^2 \rangle + T(1-T)\langle n_a \rangle. \tag{6.13.4}$$

Since $bb^\dagger |\psi\rangle = |\psi\rangle$, the vacuum field at the unused port of the beam splitter contributes to the photon number variance at the detector.[95]

If, for example, the signal field a is in a coherent state $|\alpha\rangle$ with photon number expectation value $\langle n_a \rangle = |\alpha|^2$,

$$\langle d^{\dagger r} d^r \rangle = T^r |\alpha|^{2r} = T^r \langle n_a \rangle^r = \langle n_d \rangle^r, \tag{6.13.5}$$

and the function $F_N(\xi)$ defined by (3.7.27) is

$$F_N(\xi) = \sum_{r=0}^{\infty} \frac{\xi^r}{r!} \langle n_d \rangle^r = e^{\xi \langle n_d \rangle}. \tag{6.13.6}$$

The generating function for the photon number probability distribution at the detector is then (see equation (3.7.31))

$$p(\xi) = F_N(\xi - 1) = e^{(\xi - 1)\langle n_d \rangle}, \tag{6.13.7}$$

which is the generating function for a Poisson distribution with average $\langle n_d \rangle$ (see (3.7.32)). The variance in the photon number at the detector is therefore

---

[95]This effect of vacuum fields in interferometers must be included in analyses of photon shot noise in the Laser Interferometer Gravitational-Wave Observatory (LIGO). See C. M. Caves, Phys. Rev. D **23**, 1693 (1981).

$$\langle \Delta n_d^2 \rangle = \langle n_d \rangle = T \langle n_a \rangle, \tag{6.13.8}$$

and the normalized variance

$$V_d = \frac{\langle \Delta n_d^2 \rangle}{\langle n_d \rangle^2} = \frac{T \langle n_a \rangle}{T^2 \langle n_a \rangle^2} = \frac{1}{T \langle n_a \rangle} \tag{6.13.9}$$

is larger than the normalized photon number variance

$$V_a = \frac{\langle \Delta n_a^2 \rangle}{\langle n_a \rangle^2} = \frac{1}{\langle n_a \rangle} \tag{6.13.10}$$

of the input signal field.

Things get more interesting when photodetectors are placed in both the c and the d arms (see Figure 6.5). Consider again a state $|\psi\rangle$ for which the b field is in its vacuum state. Since $[a, b] = [a, b^\dagger] = [c, d] = [c, d^\dagger] = 0$,

$$\langle n_c \rangle = (1 - T) \langle n_a \rangle, \quad \langle n_d \rangle = T \langle n_a \rangle, \tag{6.13.11}$$

and

$$\langle d^\dagger c^\dagger cd \rangle = \langle c^\dagger cd^\dagger d \rangle = \langle n_c n_d \rangle, \quad \text{where } n_c = c^\dagger c, n_d = d^\dagger d. \tag{6.13.12}$$

From the input–output relations (6.13.1) and (6.13.2),

$$\langle n_c n_d \rangle = T(1 - T) \left[\langle (a^\dagger a)^2 \rangle - \langle a^\dagger abb^\dagger \rangle \right] = T(1 - T) \left[\langle (a^\dagger a)^2 \rangle - \langle a^\dagger a \rangle \right]$$
$$= T(1 - T) \left(\langle n_a^2 \rangle - \langle n_a \rangle \right). \tag{6.13.13}$$

If the signal field is in a one-photon Fock state, $\langle n_a^2 \rangle = \langle n_a \rangle$ and, therefore, $\langle n_c n_d \rangle = 0$. The probability of finding a photon in both the c and d arms is therefore zero: a single photon cannot be split (see Sections 2.9 and 5.7). Equation (6.13.13) shows that this follows formally from the fact that $bb^\dagger |\psi\rangle = 1$; once again we require the quantum nature of the vacuum b field.

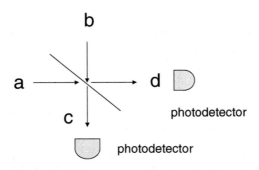

Fig. 6.5 As in Figure 6.4, but with an additional detector.

Suppose now that both the a and b input fields are in one-photon states. Another easy calculation using (6.13.1) and (6.13.2) gives in this case (see Exercise 6.10)

$$\langle n_c n_d \rangle = (|t|^2 - |r|^2)^2 = (T - R)^2 = (2T - 1)^2, \qquad (6.13.14)$$

which vanishes for a 50/50 beam splitter ($T = 1/2$). The probability of finding one photon in each output arm of the beam splitter is zero. Evidently, either both photons are found in the c arm, or both photons are found in the d arm. This result can be understood as a consequence of interfering probability amplitudes. There are two *indistinguishable* ways by which a single photon could be found at both the c and the d detectors in Figure 6.5: (1) both photons result from reflection of the fields at the beam splitter, or (2) they both result from transmission by the beam splitter. The probability amplitude for the first process is r^2, and that for the second process is t^2. The total probability amplitude for finding a c photon *and* a d photon is therefore $r^2 + t^2$, and the probability is

$$|r^2 + t^2|^2 = |-(1 - T) + T|^2 = (2T - 1)^2, \qquad (6.13.15)$$

as obtained from the calculation using photon annihilation and creation operators. In both ways of obtaining this result, we have assumed that the fields at the two photodetectors are mutually coherent (perfectly correlated) and have therefore ignored any difference in the path lengths of the fields incident on the two photodetectors. If, however, the path difference is large compared to the distance over which the fields are mutually coherent, there is a probability of 1/2 that one photon is (ideally) counted at each of the two photodetectors *and* a probability 1/2 that two photons are counted at one or the other of the two detectors. The simplest way to obtain this result is to assume that the two processes (1) and (2) above are distinguishable, and therefore that we should add probabilities, not probability amplitudes (see Exercise 6.10).

The dip in the coincidence counting rate for a 50/50 beam splitter with one photon in each of the input a and b fields is the basis of the *Hong–Ou–Mandel interferometer* sketched in Figure 6.6.[96] The two photons (a and b) are produced by the parametric downconversion of a pump laser field in a crystal and are incident on the beam splitter as shown. The coincidence counting rate depends on the path delay between the two fields, and allows temporal resolution on the order of femtoseconds between two optical paths.

Exercise 6.10: (a) Verify (6.13.13) and (6.13.14). (b) Show that, if the two processes (1) and (2) identified before (6.13.15) are regarded as distinguishable, then the probability of counting one photon at each of the two detectors in Figure 6.6 is found to be 1/2 for the 50/50 beam splitter, the same as the probability of counting two photons at one or the other of the two detectors.

[96] C. K. Hong, Z. Y. Ou, and L. Mandel, Phys. Rev. Lett. **59**, 2044 (1987).

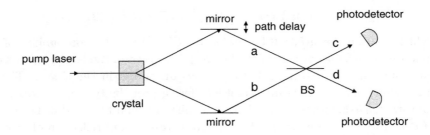

Fig. 6.6 The Hong–Ou–Mandel interferometer.

6.14   Homodyne Detection

In photodetection an incident field generates charge carriers and an electric current proportional to the average photon flux within some detection bandwidth. Such "direct" detection gives no phase information about the field, which is required, for example, in studies of squeezed light (see Section 3.6.4). Phase-sensitive detection schemes work by mixing a signal field with another, "local oscillator" field, as indicated in Figure 6.7. Such techniques were originally developed (and are widely used) in radio receivers, where the local oscillator and signal fields have different frequencies (heterodyne detection). We conclude this chapter with a brief introduction to *homodyne* detection in quantum optics, in which the local oscillator and signal frequencies have the same frequency. The main purpose is to show how the most elementary quantum properties of the fields come into play.

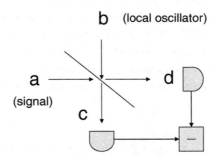

Fig. 6.7 Basic setup for heterodyne or homodyne detection of light.

Suppose first that only a single detector is used, as in Figure 6.4, and that the local oscillator field b is in a coherent state $|\beta\rangle$, $b|\beta\rangle = \beta|\beta\rangle$, $\beta = |\beta|e^{i\Phi_\beta}$. Then, from (6.13.2), the current measured by the d detector is proportional to

$$\langle n_d \rangle = \langle d^\dagger d \rangle = T\langle a^\dagger a \rangle + (1-T)\langle b^\dagger b \rangle + i\sqrt{T(1-T)}\langle a^\dagger b \rangle - i\sqrt{T(1-T)}\langle b^\dagger a \rangle$$
$$= T\langle a^\dagger a \rangle + (1-T)\langle b^\dagger b \rangle - i|\beta|\sqrt{T(1-T)}\langle ae^{-i\Phi_\beta} - a^\dagger e^{i\Phi_\beta} \rangle. \tag{6.14.1}$$

The interference term provides phase information about the signal field a. This field is typically very weak, in which case the beam splitter is chosen such that $T \gg R$ so that a substantial part of the signal field reaches the detector. If the local oscillator field is sufficiently strong compared to the signal field that $(1-T)|\beta|^2 \gg T\langle a^\dagger a \rangle$, then

$$\langle d^\dagger d \rangle \cong (1-T)|\beta|^2 - 2|\beta|\sqrt{T(1-T)}\langle X \rangle, \tag{6.14.2}$$

$$X \equiv \frac{i}{2}\left(ae^{-i\Phi_\beta} - a^\dagger e^{i\Phi_\beta}\right). \tag{6.14.3}$$

It also follows in this approximation that

$$\langle \Delta n_d^2 \rangle \cong \langle (d^\dagger d)^2 \rangle - \langle d^\dagger d \rangle^2 = (1-T)^2|\beta|^2 + 4T(1-T)|\beta|^2\langle (\Delta X)^2 \rangle, \tag{6.14.4}$$

in which contributions of third and higher order in $|\beta|$ are assumed to be negligible. The variance (6.14.4) provides a measure of squeezing of the signal field. If $\Phi_\beta = 0$, for example, X is the signal field quadrature component X_1 defined in Section 3.6.4.

Consider now the arrangement of Figure 6.7 in which the difference in the photodetector currents, proportional in our simplified model to

$$\langle n_d \rangle - \langle n_c \rangle = (2T-1)\langle n_a \rangle + (1-2T)|\beta|^2 - 4|\beta|\sqrt{T(1-T)}\langle X \rangle$$
$$\cong (1-2T)|\beta|^2 - 4|\beta|\sqrt{T(1-T)}\langle X \rangle, \tag{6.14.5}$$

is measured. In *balanced homodyne detection*, a 50/50 beam splitter ($T = 1/2$) is used. This eliminates the term depending solely on the local oscillator field:

$$\langle n_d \rangle - \langle n_c \rangle = -2|\beta|\langle X \rangle. \tag{6.14.6}$$

The variance measured in balanced homodyne detection is found similarly to be proportional to

$$\langle (n_d - n_c)^2 \rangle - \langle (n_d - n_c) \rangle^2 = 4|\beta|^2\langle \Delta X^2 \rangle. \tag{6.14.7}$$

The elimination of the local oscillator contribution, which in practice might undergo random fluctuations, is an important advantage of balanced homodyne detection.

If the signal and local oscillator fields are in coherent states $|\alpha\rangle$ and $|\beta\rangle$ ($|\psi\rangle = |\alpha\rangle|\beta\rangle, a|\psi\rangle = \alpha|\psi\rangle, b|\psi\rangle = \beta|\psi\rangle$),

$$\langle (n_d - n_c)^2 \rangle - \langle (n_d - n_c) \rangle^2 = \langle a^\dagger abb^\dagger \rangle + \langle b^\dagger baa^\dagger \rangle - 2|\alpha|^2|\beta|^2$$
$$= \langle a^\dagger a \rangle + \langle b^\dagger b \rangle = \langle n_a \rangle + \langle n_b \rangle. \tag{6.14.8}$$

The second equality follows from $aa^\dagger = a^\dagger a + 1$ and $bb^\dagger = b^\dagger b + 1$. The first commutation relation results in the term $\langle n_b \rangle$ in the variance $\langle (n_d - n_c)^2 \rangle - \langle (n_d - n_c) \rangle^2$, while the second commutation relation results in $\langle n_a \rangle$. In other words, the shot noise of the signal and local oscillator fields can be attributed to the quantum fluctuations of the local oscillator and signal fields, respectively. If there were no signal field ($\langle n_a \rangle = 0$), the variance (6.14.8) reduces to $\langle b^\dagger b a a^\dagger \rangle = \langle b^\dagger b \rangle = \langle n_b \rangle$, and might be thought of as a result of homodyning the local oscillator field with the vacuum field entering the unused port of the beam splitter.

References and Suggested Additional Reading

The literature on subjects addressed in this chapter is large, and the recommendations here only represent the small segment with which I can claim some familiarity.

MacDonald is a very readable, concise, and elementary introduction to the classical theory of Brownian motion and other fluctuation phenomena. *Wax* collects some of the classic foundational papers on stochastic processes in physics and chemistry. More recent books include those by *Van Kampen* and *Gardiner and Zoller*, the latter focusing especially on applications to quantum optics. *Risken* is a clearly written introduction to the Fokker–Planck equation and its applications. The fluctuation–dissipation theorem is treated in textbooks such as *Landau and Lifshitz (2004)* and *Van Kampen* and has been reviewed more recently by G. W. Ford (Contemp. Phys. **58**, 244 (2017)).

For recent reviews of radiation reaction theory, see R. F. O'Connell, Contemp. Phys. **53**, 301 (2012), and A. M. Steane, Am. J. Phys. **83**, 256 (2015). Limitations of the classical theory are discussed by K. T. McDonald, arXiv:physics/0003062, accessed September 24, 2018. This article includes a large number of references on the subject of radiation reaction.

Desurvire and *Haus* include discussions of noise and fluctuations in applications of amplifiers and attenuators in optical communications. *Haus* and references therein should also be consulted for more discussion of optical heterodyning and homodyning.

7

Dipole Interactions and Fluctuation-Induced Forces

Atoms and molecules interact through their coupling to the electromagnetic field. In this chapter, we consider dispersion interactions, so called because they depend on electric dipole matrix elements and polarizabilities that determine refractive indices. The field at each atom in these interactions is a fluctuating field affected by other atoms, host media, or material boundaries, as in the universal van der Waals interaction and the closely related Casimir effects we consider in the first part of the chapter. We then discuss the "fluctuation-induced" character of dispersion forces, their derivations based on zero-point energy, and other effects attributable to quantum fluctuations and zero-point energy. The essential features of the influential Lifshitz theory of "molecular attractive forces between solids" are reviewed and related through the fluctuation–dissipation theorem to the theory of quantized fields and zero-point energy in dispersive and dissipative media. Interactions of excited molecules include the resonance dipole–dipole interaction and Förster resonance energy transfer, aspects of which we address in Sections 7.9 and 7.10. Finally, we consider some effects of cavities and dielectric media on some basic radiative processes.

7.1 Van der Waals Interactions

As far back as in my treatise of 1873 I came to the conclusion that the attraction of the molecules decreases extremely quickly with distance, indeed that the attraction only has an appreciable value at distances close to the size of the molecules.

— Johannes D. van der Waals[1]

Atoms and molecules interact even when they are in their ground states and even when there is no possibility of chemical bonding. For example, the atoms of inert gases do not form chemical bonds, but the fact that they have liquid and solid phases under suitable conditions means there must be an attractive force between them, as was presumed by van der Waals in his work on corrections to the ideal-gas equation of state.[2]

---

[1] J. D. van der Waals, "The Equation of State for Gases and Liquids," http://www.nobelprize.org/prizes/physics/1910/waals/lecture, accessed October 25, 2018.

[2] J. D. van der Waals, *Over de Continuiteit van den Gas-en Vloeistoftoestand [On the Continuity of the Gas and Liquid State]*, A. W. Sijhoff, Leiden, 1873.

An Introduction to Quantum Optics and Quantum Fluctuations. Peter W. Milonni.
© Peter W. Milonni 2019. Published in 2019 by Oxford University Press.
DOI:10.1093/oso/9780199215614.001.0001

Van der Waals assumed that each molecule has a finite volume that cannot be penetrated by other molecules. The volume available to any one molecule is then $V - b$, where V is the total volume of the gas, and b depends on the number $\mathcal{N}$ of molecules and the volume of each molecule. The attraction between the molecules decreases the pressure of the gas on container walls by an amount approximately proportional to the number of pairs of molecules and therefore to $\mathcal{N}^2/V^2$. The pressure exerted by the gas is then $P_{\text{ideal}} - a/V^2$, where a characterizes the intermolecular attractive forces, P_{ideal} is the pressure that would be exerted if there were no attractive forces, and the ideal-gas equation of state $PV = RT$ for one mole of the gas should be replaced according to van der Waals by

$$(P + \frac{a}{V^2})(V - b) = RT. \tag{7.1.1}$$

This is "the simplest and most convenient" of "an infinity of possible interpolation formulae" for the equation of state of a non-ideal gas.[3]

One type of van der Waals interaction occurs between (polar) molecules with permanent electric dipole moments $\mathbf{p}_A$ and $\mathbf{p}_B$.[4] Their interaction energy as a function of their separation $\mathbf{r} = r\hat{\mathbf{r}}$ depends on the orientations of their dipole moments:

$$V = -\frac{1}{4\pi\epsilon_0}\left[\frac{3(\mathbf{p}_A \cdot \hat{\mathbf{r}})(\mathbf{p}_B \cdot \hat{\mathbf{r}}) - \mathbf{p}_A \cdot \mathbf{p}_B)}{r^3}\right]. \tag{7.1.2}$$

In thermal equilibrium, the lower-energy, attractive orientations ($V < 0$) are statistically favored over repulsive ones. Averaging $V \exp(-V/k_B T)$ over all orientations gives a net attractive interaction

$$V = -\left(\frac{1}{4\pi\epsilon_0}\right)^2 \frac{p_A^2 p_B^2}{3k_B T r^6} \tag{7.1.3}$$

for $k_B T \gg V$. For most polar molecules, this is an excellent approximation for temperatures T above around 273 K.

An attractive interaction also occurs when the electric field of a polar molecule A with electric dipole moment $\mathbf{p}_A$,

$$\mathbf{E} = \frac{1}{4\pi\epsilon_0}\frac{1}{r^3}\left[3(\mathbf{p}_A \cdot \hat{\mathbf{r}})\hat{\mathbf{r}} - \mathbf{p}_A\right], \tag{7.1.4}$$

induces an electric dipole moment in a molecule B. If α_B is the static polarizability of B, the interaction energy $-(1/2)\alpha_B \mathbf{E}^2$ (cf. (2.3.41)), averaged over orientations of $\mathbf{p}_A$, is

$$V = -\left(\frac{1}{4\pi\epsilon_0}\right)^2 \frac{p_A^2 \alpha_B}{r^6}. \tag{7.1.5}$$

Both the (Keesom) *dipole orientation interaction* (7.1.3) and the (Debye) *induction interaction* (7.1.5) require a permanent dipole moment. Both were derived prior to

---

[3] *Landau and Lifshitz (2004)*, p. 234.

[4] Polar molecules have a "permanent" electric dipole moment in the same sense that the linear Stark effect in the hydrogen atom implies, in effect, a permanent dipole moment (Section 2.3.4).

the development of quantum theory, and neither is sufficiently general to account in all cases for corrections to the ideal gas law. A "universal" type of van der Waals interaction between ground-state atoms was derived from quantum theory by London, who found that this interaction between any two ground-state atoms could be roughly approximated by

$$V = -\left(\frac{1}{4\pi\epsilon_0}\right)^2 \frac{3\alpha_A\alpha_B}{2r^6} \frac{I_A I_B}{I_A + I_B}, \tag{7.1.6}$$

where α_A and α_B are the static polarizabilities of the two atoms, and I_A and I_B are their ionization potentials.[5] This (London) *dispersion interaction*, which we derive below, is typically stronger than the dipole orientation and induction forces, although, for strongly polar molecules like H_2O, the orientation and induction forces can dominate.

More generally, the attractive interaction between ground-state atoms is expressed as $V = -C_6/r^6 - C_8/r^8 - C_{10}/r^{10} - \ldots$ when contributions to the interaction from higher multipole moments are included. There is a large literature on the computation of the coefficients in this series. For two He atoms, for example, $C_6 \cong 0.88$ eV-$\mathring{A}^6$; as for all rare-gas atoms, this is due entirely to the dispersion interaction. The static polarizability of He is $\alpha/4\pi\epsilon_0 \cong 0.17$ $\mathring{A}^3$, and the ionization potential is $I = 24.6$ eV.[6] In the approximation (7.1.6), therefore, $C_6 = 0.53$ eV-$\mathring{A}^6$.

Exercise 7.1: (a) Verify the form of the orientation interaction (7.1.3). (b) How does the pairwise van der Waals dispersion energy compare to the thermal kinetic energy for helium at standard temperature and pressure? (c) The van der Waals interaction energy between two hydrogen atoms has been computed by a variational approach to be (see L. Pauling and G. Y. Beach, Phys. Rev. **47**, 686 (1935))

$$V = -\left(\frac{1}{4\pi\epsilon_0}\right)^2 \times 6.5\left(\frac{a_0}{r}\right)^6 \frac{e^2}{a_0},$$

where a_0 is the Bohr radius. The hydrogen polarizability is $\alpha/4\pi\epsilon_0 = 9a_0^3/2$. Compare V with the London approximation (7.1.6). (d) The H_2O molecule has an electronic polarizability $\alpha/4\pi\epsilon_0 = 1.48$ $\mathring{A}^3$, a permanent dipole moment $p = 6.2 \times 10^{-30}$ C-m, and an ionization potential $I = 12.6$ eV. Estimate the orientation, induction, and dispersion interaction energies between two water vapor molecules at standard conditions for temperature and pressure.

The London dispersion interaction can be derived in different ways, most directly from standard perturbation theory, as in London's original approach. It can, of course,

---

[5] F. London, Z. Phys. **63**, 245 (1930); F. London, Trans. Faraday Soc. **33**, 8 (1937).

[6] See, for instance, Z. C. Yan, J. F. Babb, A. Dalgarno, and G. W. F. Drake, Phys. Rev. A **54**, 2824 (1996).

also be derived in the Heisenberg-picture approach taken in the preceding chapters. We will instead take a different route based on the formula (2.3.42) for the shift δE_i of an atomic energy level when an electric field induces in the atom an electric dipole moment. For an atom in free space, this energy shift is

$$\delta E_i = -\frac{1}{2} \sum_{\mathbf{k}\lambda} \alpha_i(\omega_k) \langle \mathbf{E}_{\mathbf{k}\lambda}^2(\mathbf{r}_A, t) \rangle, \tag{7.1.7}$$

where $\alpha_i(\omega_k)$ is the polarizability at frequency ω_k of the atom in its ground state i, and $\mathbf{E}_{\mathbf{k}\lambda}(\mathbf{r}_A, t)$ is the electric field operator, at the position $\mathbf{r}_A$ of the atom, for the field mode with wave vector $\mathbf{k}$, frequency $\omega_k = kc$, and polarization λ. The expectation value here is over the zero-temperature, vacuum state of the field, in which case the different field modes are uncorrelated, as we have implicitly assumed in writing (7.1.7); the energy δE_i is then time independent. For the polarizability $\alpha_i(\omega_k)$, we use the Kramers–Heisenberg formula (2.3.32), including an imaginary part $i\epsilon$, $\epsilon \to 0^+$, in the denominator:

$$\alpha_i(\omega_k) = \frac{1}{3\hbar} \sum_j |\mathbf{d}_{ij}|^2 \left(\frac{1}{\omega_{ji} - \omega_k - i\epsilon} + \frac{1}{\omega_{ji} + \omega_k} \right). \tag{7.1.8}$$

This ensures that $\alpha_i(\omega_k)$ has no singularities for real frequencies and is analytic in the upper half of the complex frequency plane, as required by causality (see Section 6.6.1). This form of the polarizability is consistent with our goal of calculating the van der Waals interaction to lowest (fourth) order in the atom–field coupling, since ϵ is independent of this coupling. In particular, the polarizability (7.1.8) does not account for radiative damping.

The photon annihilation and creation parts of the electric field in (7.1.7) are symmetrically ordered and, as discussed in Chapter 4, δE_i as expressed by (7.1.7) is just a "Lamb shift" in the case of an isolated atom in free space. But we are interested now in two atoms, A and B, in otherwise free space. The field in each mode at Atom A is

$$\mathbf{E}_{\mathbf{k}\lambda}(\mathbf{r}_A, t) = \mathbf{E}_{0,\mathbf{k}\lambda}(\mathbf{r}_A, t) + \mathbf{E}_{B,\mathbf{k}\lambda}(\mathbf{r}_A, t). \tag{7.1.9}$$

$\mathbf{E}_{0,\mathbf{k}\lambda}(\mathbf{r}_A, t)$ is the source-free, "vacuum" field of mode $\mathbf{k}, \lambda$ at A, and $\mathbf{E}_{B,\mathbf{k}\lambda}(\mathbf{r}_A, t)$ is the electric field operator, for this same mode, for the field at $\mathbf{r}_A$ from Atom B. The part of (7.1.7) that depends on the distance between A and B, that is, the part identifiable as an interaction energy, is

$$V_{AB} = -\mathrm{Re} \sum_{\mathbf{k}\lambda} \alpha_A(\omega_k) \Big\langle \mathbf{E}_{0,\mathbf{k}\lambda}(\mathbf{r}_A, t) \cdot \mathbf{E}_{B,\mathbf{k}\lambda}(\mathbf{r}_A, t) \Big\rangle. \tag{7.1.10}$$

We have now written $\alpha_A(\omega_k)$ for the polarizability at frequency ω_k of the ground-state Atom A.

Atom B has no permanent dipole moment. Whatever dipole moment $\mathbf{p}_B$ it has is induced by an electric field. To the lowest (zeroth) order in the atom–atom coupling, the inducing field is the source-free, vacuum field at B:

$$\mathbf{p}_B(t) = \hat{\mathbf{d}}_B p_B(t) = \sum_{\mathbf{k}\lambda} \alpha_B(\omega_k) \mathbf{E}_{0,\mathbf{k}\lambda}(\mathbf{r}_B, t)$$

$$= i \sum_{\mathbf{k}\lambda} \alpha_B(\omega_k) \sqrt{\frac{\hbar\omega_k}{2\epsilon_0 V}} \Big[a_{\mathbf{k}\lambda}(0) e^{-i\omega_k t} e^{i\mathbf{k}\cdot\mathbf{r}_B}$$

$$- a_{\mathbf{k}\lambda}^\dagger(0) e^{i\omega_k t} e^{-i\mathbf{k}\cdot\mathbf{r}_B} \Big] \mathbf{e}_{\mathbf{k}\lambda}, \qquad (7.1.11)$$

and the field from this dipole moment is (recall (4.3.11))

$$\mathbf{E}_B(\mathbf{r}, t) = -\frac{1}{4\pi\epsilon_0} \Big\{ \frac{1}{c^2 r} \big[\hat{\mathbf{d}}_B - (\hat{\mathbf{d}}_B \cdot \hat{\mathbf{r}})\hat{\mathbf{r}} \big] \ddot{p}_B(t - r/c)$$

$$- \big[3(\hat{\mathbf{d}}_B \cdot \hat{\mathbf{r}})\hat{\mathbf{r}} - \hat{\mathbf{d}}_B \big] \Big[\frac{1}{cr^2} \dot{p}_B(t - r/c) + \frac{1}{r^3} p_B(t - r/c) \Big] \Big\}. \quad (7.1.12)$$

Since $\langle\mathrm{vac}| a_{\mathbf{k}\lambda}^\dagger(0) = 0$, only the part

$$\mathbf{E}_{0,\mathbf{k}\lambda}^{(+)}(\mathbf{r}_A, t) = i \sqrt{\frac{\hbar\omega_k}{2\epsilon_0 V}} a_{\mathbf{k}\lambda}(0) e^{-i\omega_k t} e^{i\mathbf{k}\cdot\mathbf{r}_A} \mathbf{e}_{\mathbf{k}\lambda} \qquad (7.1.13)$$

of $\mathbf{E}_{0,\mathbf{k}\lambda}(\mathbf{r}_A, t)$ will contribute to the mode summation in (7.1.10). Similarly, only the part $\mathbf{E}_{B,\mathbf{k}\lambda}^{(-)}(\mathbf{r}_A, t)$ of (7.1.12) that involves $a_{\mathbf{k}\lambda}^\dagger(0)$ will contribute; from (7.1.11), (7.1.12), and

$$\mathbf{E}_{0,\mathbf{k}\lambda}^{(-)}(\mathbf{r}_B, t) = -i \sum_{\mathbf{k}\lambda} \sqrt{\frac{\hbar\omega_k}{2\epsilon_0 V}} a_{\mathbf{k}\lambda}^\dagger(0) e^{i\omega_k t} e^{-i\mathbf{k}\cdot\mathbf{r}_B} \mathbf{e}_{\mathbf{k}\lambda}, \qquad (7.1.14)$$

this part is

$$\mathbf{E}_{B,\mathbf{k}\lambda}^{(-)}(\mathbf{r}_A, t) = \frac{-i}{4\pi\epsilon_0} \alpha_B(\omega_k) \sqrt{\frac{\hbar\omega_k}{2\epsilon_0 V}} a_{\mathbf{k}\lambda}^\dagger(0) \Big\{ \frac{k^2}{r} \big[\mathbf{e}_{\mathbf{k}\lambda} - (\mathbf{e}_{\mathbf{k}\lambda} \cdot \hat{\mathbf{r}})\hat{\mathbf{r}} \big]$$

$$+ \big[3(\mathbf{e}_{\mathbf{k}\lambda} \cdot \hat{\mathbf{r}})\hat{\mathbf{r}} - \mathbf{e}_{\mathbf{k}\lambda} \big] \Big(\frac{ik}{r^2} + \frac{1}{r^3} \Big) \Big\} e^{-i\mathbf{k}\cdot\mathbf{r}_B} e^{-ikr}, \qquad (7.1.15)$$

$r = |\mathbf{r}_A - \mathbf{r}_B|$. Since the different modes are uncorrelated ($\langle\mathrm{vac}| a_{\mathbf{k}'\lambda'}(0) a_{\mathbf{k}\lambda}^\dagger(0)|\mathrm{vac}\rangle = \delta_{\mathbf{k},\mathbf{k}'}^3 \delta_{\lambda\lambda'}$),

$$V_{AB} = -\frac{1}{4\pi\epsilon_0} \mathrm{Re} \sum_{\mathbf{k}\lambda} \alpha_A(\omega_k) \Big\{ \frac{\hbar\omega_k}{2\epsilon_0 V} \alpha_B(\omega_k) \langle a_{\mathbf{k}\lambda}(0) a_{\mathbf{k}\lambda}^\dagger(0) \rangle e^{i\mathbf{k}\cdot\mathbf{r}} e^{-ikr}$$

$$\times \Big\{ \frac{k^2}{r} \big[1 - (\mathbf{e}_{\mathbf{k}\lambda} \cdot \hat{\mathbf{r}})^2 \big] + \big[3(\mathbf{e}_{\mathbf{k}\lambda} \cdot \hat{\mathbf{r}})^2 - 1 \big] \Big(\frac{ik}{r^2} + \frac{1}{r^3} \Big) \Big\}$$

$$\to -\Big(\frac{1}{4\pi\epsilon_0} \Big)^2 \frac{\hbar}{4\pi^2 c^6} \mathrm{Re} \int_0^\infty d\omega \, \omega^6 \alpha_A(\omega) \alpha_B(\omega) e^{-ikr}$$

$$\times \int d\Omega_{\mathbf{k}} \Big\{ \frac{1}{kr} \big[1 + (\hat{\mathbf{k}} \cdot \hat{\mathbf{r}})^2 \big] + \Big(\frac{i}{k^2 r^2} + \frac{1}{k^3 r^3} \Big) \big[1 - 3(\hat{\mathbf{k}} \cdot \hat{\mathbf{r}})^2 \big] \Big\} e^{i\mathbf{k}\cdot\mathbf{r}}. \quad (7.1.16)$$

We have used

$$\sum_{\mathbf{k}\lambda}(\ldots) \to \frac{V}{8\pi^3}\int d^3k(\ldots) = \frac{V}{8\pi^3}\int dk\, k^2 \int d\Omega_\mathbf{k}(\ldots)$$

$$= \frac{V}{8\pi^3 c^3}\int d\omega\,\omega^2 \int d\Omega_\mathbf{k}(\ldots) \tag{7.1.17}$$

and

$$\sum_{\lambda=1}^{2}[1-(\mathbf{e}_{\mathbf{k}\lambda}\cdot\hat{\mathbf{r}})^2] = 2-\sum_{\lambda=1}^{2}(\mathbf{e}_{\mathbf{k}\lambda}\cdot\hat{\mathbf{r}})^2 = 1+(\hat{\mathbf{k}}\cdot\hat{\mathbf{r}})^2,$$

$$\sum_{\lambda=1}^{2}[3(\mathbf{e}_{\mathbf{k}\lambda}\cdot\hat{\mathbf{r}})^2-1] = 1-3(\hat{\mathbf{k}}\cdot\hat{\mathbf{r}})^2. \tag{7.1.18}$$

The integration $\int d\Omega_\mathbf{k}(\ldots)$ over solid angles in k space is easily performed:

$$\int d\Omega_\mathbf{k}(\ldots) = 2\pi \int_0^\pi d\theta \sin\theta \Big[\frac{1}{x}(1+\cos^2\theta)$$

$$+ (\frac{i}{x^2}+\frac{1}{x^3})(1-3\cos^2\theta)\Big]e^{ix\cos\theta}$$

$$= 8\pi \Big[\Big(\frac{1}{x^2}-\frac{i}{x^3}-\frac{2}{x^4}+\frac{3i}{x^5}+\frac{3}{x^6}\Big)\sin x + \Big(\frac{1}{x^3}-\frac{3i}{x^4}-\frac{3}{x^5}\Big)\cos x\Big], \tag{7.1.19}$$

and, using elementary trigonometric identities, we obtain

$$V_{AB} = -\Big(\frac{1}{4\pi\epsilon_0}\Big)^2\frac{\hbar}{\pi c^6}\int_0^\infty d\omega\,\omega^6\alpha_A(\omega)\alpha_B(\omega)G(\omega r/c), \tag{7.1.20}$$

$$G(x) \equiv \frac{\sin 2x}{x^2}+\frac{2\cos 2x}{x^3}-\frac{5\sin 2x}{x^4}-\frac{6\cos 2x}{x^5}+\frac{3\sin 2x}{x^6}. \tag{7.1.21}$$

The integral

$$\int_0^\infty d\omega\,\omega^6\alpha_A(\omega)\alpha_B(\omega)G(\omega r/c) = \mathrm{Re}\int_0^\infty d\omega\,\omega^6\alpha_A(\omega)\alpha_B(\omega)\Big(\frac{-i}{k^2 r^2}+\frac{2}{k^3 r^3}+\frac{5i}{k^4 r^4}$$

$$-\frac{6}{k^5 r^5}-\frac{3i}{k^6 r^6}\Big)e^{2ikr} \tag{7.1.22}$$

can be written, using Cauchy's theorem and the integration contour shown in Figure 7.1, as an integral along the positive imaginary axis, since the integrand in (7.1.22) is analytic in the upper half of the complex frequency plane and goes to zero as the radius of the quarter-circle extends to infinity.[7] Then,

$$V_{AB} = -\Big(\frac{1}{4\pi\epsilon_0}\Big)^2\frac{\hbar c}{\pi r^2}\int_0^\infty du\, u^4\alpha_A(icu)\alpha_B(icu)$$

$$\times \Big(1+\frac{2}{ur}+\frac{5}{u^2 r^2}+\frac{6}{u^3 r^3}+\frac{3}{u^4 r^4}\Big)e^{-2ur}. \tag{7.1.23}$$

---

[7] This analytic continuation of the integral along the positive real axis to an integral along the positive imaginary axis, effectively a rotation of the integration line by $\pi/2$, is a so-called *Wick rotation*. It will be used several times in this chapter. See, for example, (7.3.41).

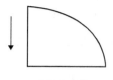

Fig. 7.1 Integration path used in obtaining (7.1.23).

7.1.1 Nonretarded (London) van der Waals Interaction

If r is small compared to transition wavelengths that contribute significantly to the polarizabilities, V_{AB} varies as $1/r^6$. Suppose, for example, that the two atoms are identical and that the transition from the ground state to a single excited energy level, with transition dipole moment $\mathbf{d}$ and frequency ω_0, accounts almost entirely for the polarizability:

$$\alpha_A(i\xi) = \alpha_B(i\xi) \cong \frac{2|\mathbf{d}|^2\omega_0/3\hbar}{\omega_0^2 + \xi^2}. \tag{7.1.24}$$

This, of course, is only a crude approximation, but often not a terribly bad one, since transitions from the ground level to the first excited level typically have the largest oscillator strength. For the hydrogen atom, for example, this oscillator strength is 0.42, and, for the alkali atoms, it is nearly 1.0. In this approximation,

$$V_{AB} \cong -\left(\frac{1}{4\pi\epsilon_0}\right)^2 \frac{\hbar\omega_0}{\pi c^4 r^2} \left(\frac{2|\mathbf{d}|^2\omega_0}{3\hbar}\right)^2$$
$$\int_0^\infty dx \frac{e^{-2k_0rx}}{(1+x^2)^2}\left(x^4 + \frac{2x^3}{k_0r} + \frac{5x^2}{k_0^2r^2} + \frac{6x}{k_0^3r^3} + \frac{3}{k_0^4r^4}\right), \tag{7.1.25}$$

and, if $k_0r \ll 1$, the integral is approximately

$$\frac{3}{k_0^4r^4}\int_0^\infty \frac{dx}{(1+x^2)^2} = \frac{3\pi}{4k_0^4r^4}, \tag{7.1.26}$$

and therefore

$$V_{AB} \cong -\left(\frac{1}{4\pi\epsilon_0}\right)^2 \frac{3\hbar\omega_0\alpha_A^2}{4r^6}, \tag{7.1.27}$$

where $\alpha_A = 2|\mathbf{d}|^2/3\hbar\omega_0$ is the static (zero-frequency) polarizability. If the two atoms are different but we can still make the dominant-transition approximation for each atom, that is,

$$\alpha_A(i\xi) \cong \frac{2|\mathbf{d}_A|^2\omega_A}{\omega_A^2 + \xi^2}, \qquad \alpha_B(i\xi) \cong \frac{2|\mathbf{d}_B|^2\omega_B}{\omega_B^2 + \xi^2}, \tag{7.1.28}$$

we find from (7.1.23) that, again for small separations ($k_Ar, k_Br \ll 1$),

$$V_{AB} \cong -\left(\frac{1}{4\pi\epsilon_0}\right)^2 \frac{3\hbar}{\pi r^6}(\omega_A^2\alpha_A)(\omega_B^2\alpha_B)\int_0^\infty \frac{dy}{(\omega_A^2 + y^2)(\omega_B^2 + y^2)}$$
$$= -\left(\frac{1}{4\pi\epsilon_0}\right)^2 \frac{3\hbar\alpha_A\alpha_B}{2r^6}\frac{\omega_A\omega_B}{\omega_A + \omega_B}, \tag{7.1.29}$$

where, again, α_A and α_B are static polarizabilities. If, furthermore, $\hbar\omega_A$ and $\hbar\omega_B$ are *roughly* the atoms' ionization potentials I_A and I_B,

$$V_{AB} \cong -\left(\frac{1}{4\pi\epsilon_0}\right)^2 \frac{3\alpha_A\alpha_B}{2r^6} \frac{I_A I_B}{I_A + I_B}, \tag{7.1.30}$$

which is London's approximation (7.1.6).

The more general London dispersion energy that follows from the short-range limit of (7.1.23) is

$$V_{AB} = -\left(\frac{1}{4\pi\epsilon_0}\right)^2 \frac{3\hbar}{\pi r^6} \int_0^\infty d\xi \alpha_A(i\xi)\alpha_B(i\xi). \tag{7.1.31}$$

For the ground state of Atom A,

$$\alpha_A(i\xi) = \frac{2}{3\hbar} \sum_m \frac{|\mathbf{d}_{Am}|^2 \omega_{Am}}{\omega_{Am}^2 + \xi^2}, \tag{7.1.32}$$

where $\mathbf{d}_{Am}$ is the transition dipole moment between the ground state and the excited state m, and ω_{Am} is the corresponding transition (angular) frequency. Using the corresponding expression for Atom B, we can write (7.1.31) in the form obtained by London:[8]

$$V_{AB} = -\left(\frac{1}{4\pi\epsilon_0}\right)^2 \frac{3\hbar}{\pi r^6} \left(\frac{2}{3\hbar}\right)^2 \sum_m \sum_n |\mathbf{d}_{Am}|^2 |\mathbf{d}_{Bn}|^2 \omega_{Am}\omega_{Bn}$$

$$\times \int_0^\infty \frac{d\xi}{(\omega_{Am}^2 + \xi^2)(\omega_{Bn}^2 + \xi^2)}$$

$$= -\left(\frac{1}{4\pi\epsilon_0}\right)^2 \frac{2}{3\hbar r^6} \sum_m \sum_n \frac{|\mathbf{d}_{Am}|^2 |\mathbf{d}_{Bn}|^2}{\omega_{Am} + \omega_{Bn}}. \tag{7.1.33}$$

In the derivation of this dispersion energy, the factor e^{-2ur}, which comes from the factor e^{2ikr} in (7.1.22), is neglected. In other words, the London energy applies when *retardation* is negligible.

7.1.2 Retarded (Casimir–Polder) van der Waals Interaction

From their experiments on quartz suspensions, and their theoretical considerations on the theory of colloidal stability, Verwey and Overbeek were led to suggest that the van der Waals interaction potential must fall off faster than $1/r^6$ at large separations.[9] They suggested, in particular, that the London theory must be modified to account for the finite velocity of light. This led Casimir and Polder to include retardation in the theory of the van der Waals dispersion interaction, and to derive the formula (7.1.23).[10]

---

[8] F. London, Z. Phys. **63**, 245 (1930); F. London, Trans. Faraday Soc. **33**, 8 (1937).

[9] E. J. W. Verwey and J. T. G. Overbeek *Theory of the Stability of Lyophobic Colloids*, Elsevier, Amsterdam, 1948.

[10] H. B. G. Casimir and D. Polder, Nature **158**, 787 (1946); H. B. G. Casimir and D. Polder, Phys. Rev. **73**, 360 (1948).

Suppose we again make the approximation (7.1.25) for identical atoms, but now assuming $k_0 r \gg 1$. In this limit, only values of $x \ll 1$ contribute significantly to the integral, and we can replace $1/(1 + x^2)^2$ in the integrand by 1:

$$
\begin{aligned}
V_{AB} &\cong -\Big(\frac{1}{4\pi\epsilon_0}\Big)^2 \frac{\hbar\omega_0}{\pi c^4 r^2} \Big(\frac{2|\mathbf{d}|^2\omega_0}{3\hbar}\Big)^2 \\
&\quad \times \int_0^\infty dx\, e^{-2k_0 r x}\Big(x^4 + \frac{2x^3}{k_0 r} + \frac{5x^2}{k_0^2 r^2} + \frac{6x}{k_0^3 r^3} + \frac{3}{k_0^4 r^4}\Big) \\
&= -\Big(\frac{1}{4\pi\epsilon_0}\Big)^2 \frac{\hbar\omega_0}{\pi c^4 r^2} \Big(\frac{2|\mathbf{d}|^2\omega_0}{3\hbar}\Big)^2 \times \frac{23}{4k_0^5 r^5} \\
&= -\Big(\frac{1}{4\pi\epsilon_0}\Big)^2 \frac{23\hbar c \alpha_A^2}{4\pi r^7}.
\end{aligned}
\tag{7.1.34}
$$

In fact, this $1/r^7$ form of the van der Waals interaction in the large-separation limit does not depend on the approximation (7.1.25): from (7.1.23),

$$
\begin{aligned}
V_{AB} &\cong -\Big(\frac{1}{4\pi\epsilon_0}\Big)^2 \frac{\hbar c}{\pi r^2} \alpha_A \alpha_B \int_0^\infty du\,\Big(u^4 + \frac{2u^3}{r} + \frac{5u^2}{r^2} + \frac{6u}{r^3} + \frac{3}{r^4}\Big)e^{-2ur} \\
&= -\Big(\frac{1}{4\pi\epsilon_0}\Big)^2 \frac{23\hbar c \alpha_A \alpha_B}{4\pi r^7}
\end{aligned}
\tag{7.1.35}
$$

for r much greater than the wavelengths of any transitions that contribute significantly to the polarizabilities. This is the famous Casimir–Polder formula for the retarded van der Waals interaction.

Exercise 7.2: Show that r should typically be greater than at least about 137 Bohr radii before the retarded form of the van der Waals interaction (7.1.35) applies.

7.1.3 Physics behind the Dispersion Interaction: Fluctuations

London interpreted the dispersion interaction as follows:

If one were to take an instantaneous photograph of a molecule at any time, one would find various configurations of nuclei and electrons, showing in general dipole moments. In a spherically symmetrical . . . molecule . . . the average over very many of such snapshots would of course give no preference for any direction. These very quickly varying dipoles, represented by the zero-point motion of a molecule, produce an electric field and act upon the polarisability of the other molecule and produce there induced dipoles, which are in phase and in interaction with the instantaneous dipoles producing them. The zero-point motion is, so to speak, accompanied by a synchronised electric alternating field . . . [11]

Using the model of two coupled oscillators, London showed that the dispersion energy he derived could be regarded as a change in the zero-point energy resulting from the

[11] F. London, Trans. Faraday Soc. **33**, 8 (1937), p. 13.

coupling of the oscillators. A slightly modified version of his model is described by the equations

$$\ddot{x}_A + \omega_A^2 x_A = K x_B,$$
$$\ddot{x}_B + \omega_B^2 x_B = K x_A, \qquad (7.1.36)$$

with

$$K = -\frac{1}{4\pi\epsilon_0}\frac{e^2}{mr^3}[3(\hat{d}_A \cdot \hat{r})(\hat{d}_B \cdot \hat{r}) - \hat{\mathbf{d}}_A \cdot \hat{\mathbf{d}}_B] \equiv -\frac{1}{4\pi\epsilon_0}\frac{e^2}{mr^3}q, \qquad (7.1.37)$$

and $\hat{\mathbf{r}}$ the unit vector pointing from one oscillator to the other. Particle j $(j = A, B)$ has mass m and charge e, is subject to a restoring force $-m\omega_j^2 x_j$, and results in an electric dipole moment $ex_j\hat{\mathbf{d}}_j$ when it is displaced by x_j from its equilibrium position. In response to an applied electric field of frequency ω, the oscillators have polarizabilities

$$\alpha_j(\omega) = \frac{e^2/m}{\omega_j^2 - \omega^2}. \qquad (7.1.38)$$

Normal-mode oscillation frequencies are found from (7.1.36) to be

$$\Omega_1 \cong \omega_A + \frac{K^2 q^2}{2\omega_A}\frac{1}{\omega_A^2 - \omega_B^2}, \quad \text{and} \quad \Omega_2 \cong \omega_B - \frac{K^2 q^2}{2\omega_B}\frac{1}{\omega_A^2 - \omega_B^2}, \qquad (7.1.39)$$

for $|K^2|q^2/\omega_j^2 \ll |\omega_A^2 - \omega_B^2|$. Without any coupling, the total zero-point energy of the two oscillators is

$$E_{ZP} = \frac{1}{2}\hbar(\omega_A + \omega_B), \qquad (7.1.40)$$

whereas it is

$$E_{ZP} = \frac{1}{2}\hbar(\Omega_1 + \Omega_2) \qquad (7.1.41)$$

when they interact. The difference is the interaction energy:

$$\Delta E_{ZP} = \frac{1}{2}\hbar(\Omega_1 + \Omega_2) - \frac{1}{2}\hbar(\omega_A + \omega_B) = \frac{\hbar K^2 q^2/4}{\omega_A^2 - \omega_B^2}\left(\frac{1}{\omega_A} - \frac{1}{\omega_B}\right)$$
$$= -\frac{\hbar K^2 q^2}{4\omega_A\omega_B}\frac{1}{\omega_A + \omega_B} \qquad (7.1.42)$$

for one-dimensional oscillators. To obtain the interaction energy allowing for oscillations in three dimensions, we note from the definition of q in (7.1.37) that $q^2 = 0$ if the dipole moments are perpendicular, $q^2 = 4$ if they are parallel and lie along the line joining the molecules, and $q^2 = 1$ if they are parallel to each other but perpendicular to that line. Since there are two independent ways for them to be parallel to each other and perpendicular to the line, and one way for them to be parallel to it, we multiply

(7.1.42) by $2(1) + 1(4) = 6$, and express the resulting interaction energy in terms of the static polarizabilities $\alpha_j = e^2/m\omega_j^2$ to obtain

$$\Delta E_{ZP} = -\Big(\frac{1}{4\pi\epsilon_0}\Big)^2 \frac{3\hbar\alpha_A\alpha_B}{2r^6}\frac{\omega_A\omega_B}{\omega_A + \omega_B}. \tag{7.1.43}$$

The right-hand side is London's result (7.1.29) for the interaction energy V_{AB}. The London dispersion interaction is thus seen to be closely related to zero-point energy, or, more precisely, to the change in zero-point energy resulting from the electric dipole coupling of two oscillators.

The dispersion interaction can be attributed to the zero-point energy of the coupled oscillators, or to correlated quantum fluctuations of the dipoles resulting from the correlated quantum fluctuations of the electric fields at the positions of the dipoles. Equation (7.1.10) expresses the interaction in terms of the expectation value of the product of the vacuum electric field at Atom A and the electric field at A due to Atom B. The latter is the field from the dipole moment of Atom B that is induced by the vacuum field at B. So, the non-vanishing interaction comes from the fact that the vacuum electric fields at $\mathbf{r}_A$ and $\mathbf{r}_B$ are correlated and, therefore, the induced dipole moments at $\mathbf{r}_A$ and $\mathbf{r}_B$ are correlated. The interaction falls off with the dipole separation $r = |\mathbf{r}_A - \mathbf{r}_B|$ because both the dipole–dipole interaction and the electric field correlation fall off with r. The vacuum-field correlations fall off rapidly with decreasing field wavelengths, and, at large separations, the dominant contribution to the interaction comes from frequencies near zero. We can then replace the polarizabilities in (7.1.23) by their static values $\alpha_A(0)$ and $\alpha_B(0)$, and then it is clear from simple dimensional considerations that the interaction must vary as $1/r^7$.

In concluding their 1948 paper on retarded van der Waals interactions, Casimir and Polder remarked that the very simple forms of the interaction they derived "suggest that it might be possible to derive these expressions, perhaps apart from the numerical factors, by more elementary considerations. This would be desirable since it would also give a more physical background to our result, a result which in our opinion is rather remarkable. So far we have not been able to find such a simple argument." [12] Already in the same year their paper was published, however, Casimir was "well aware that the Casimir Polder formula can be derived more simply using the methods of zero-point energy," and in a short paper he derived (7.1.35), and the energy-level shift of a ground-state atom near a conducting plate (see Section 7.3), by a method followed closely here in our derivation of (7.1.23). [13]

The van der Waals dispersion interaction can be calculated in other ways, besides those already discussed. Casimir and Polder derived the basic formula (7.1.23)

---

[12] H. B. G. Casimir and D. Polder, Phys. Rev. **73**, 360 (1948), p. 372.

[13] H. B. G. Casimir, private communication, 15 January 1993; H. B. G. Casimir, J. Chim. Phys. **46** 407 (1949). Casimir does not cite London's earlier ideas about zero-point energy and the dispersion interaction in this paper or in his 1948 paper with Polder, although both papers refer to "London–van der Waals forces."

using Schrödinger-picture perturbation theory and the minimal coupling interaction Hamiltonian, but this result can also be obtained from the electric dipole form of the Hamiltonian and approximate solutions of Heisenberg equations of motion.[14] We have taken what is perhaps a more intuitive approach starting from the Stark-shift formula (7.1.7). The interaction can also be obtained—nonperturbatively and including temperature dependence (see Section 7.1.5)—using the remarkable formula (6.7.36), for the Helmholtz free energy of oscillators coupled to a heat bath.[15]

Exercise 7.3: The sum of the zeros minus the sum of the poles inside a simple closed curve C of a function $f(z)$ that is analytic on and inside C follows from the argument theorem of complex variable theory:

$$\frac{1}{2\pi i} \oint_C z \frac{f'(z)}{f(z)} dz, \quad \text{where } f'(z) = df/dz.$$

Consider the generalization of (7.1.36) to include retardation, and determine the function whose zeros are the normal-mode frequencies of the coupled dipole oscillators. Show that, to fourth order in the dipole-field coupling, the van der Waals dispersion interaction (7.1.23) is the difference between the zero-point energies of the oscillators when they are at a finite separation r and when they are infinitely far apart. (Note: The poles of the relevant function $f(z)$ are independent of the dipole separation r.)

Unlike the Coulomb interaction in electrostatics or the forces on masses in Newtonian gravitation, the dispersion interaction is not pairwise additive: the interaction of an Atom A with Atoms B and C is not simply the sum of the A–B and A–C interactions.[16] This can be understood, for instance, from a Heisenberg-picture calculation with normal ordering of field operators.[17] In such an approach, the dispersion interaction is interpretable as a consequence of radiation reaction, just as the normal ordering of field operators in Section 4.5 leads us to think of spontaneous emission as a consequence of radiation reaction. In the case of the dispersion interaction, the physical picture is that the field of Atom A scatters off Atom B, so that the total source (radiation reaction) field acting on Atom A is modified by B, and depends on the polarizability of B and the distance separating it from A. The dispersion interaction between A and B is attributable to this modification of radiation reaction. In the presence of a third atom, C, the source field at A depends not only on the field at A that results from single scattering of the field from A by Atoms B and C separately

---

[14] See, for instance, P. W. Milonni, Phys. Rev. A **25**, 1315 (1982), and references therein.

[15] P. R. Berman, G. W. Ford, and P. W. Milonni, Phys. Rev. **89**, 022127 (2014).

[16] B. M. Axilrod and E. Teller, J. Chem. Phys. **11**, 299 (1943).

[17] See, for instance, P. W. Milonni, Phys. Rev. A **25**, 1315 (1982), and references therein.

but also on the field at A that has been scattered by B and then by C, and vice versa. From this perspective, it is because of multiple scattering that the dispersion interaction is not pairwise additive.[18]

7.1.4 The Hellmann–Feynman Theorem

What is usually called the Hellmann–Feynman theorem was derived by Pauli and, independently, by Hellmann and Feynman.[19] The proof is straightforward: suppose the Hamiltonian H depends on some parameter λ, and consider the time-independent Schrödinger equation,

$$H(\lambda)|\psi(\lambda)\rangle = E(\lambda)|\psi(\lambda)\rangle, \quad \text{where } \langle\psi(\lambda)|\psi(\lambda)\rangle = 1. \quad (7.1.44)$$

Then,

$$E(\lambda) = \langle\psi(\lambda)|H(\lambda)|\psi(\lambda)\rangle, \quad (7.1.45)$$

$$\begin{aligned}
\frac{d}{d\lambda}E(\lambda) &= \langle\frac{d\psi}{d\lambda}|H(\lambda)|\psi(\lambda)\rangle + \langle\psi(\lambda)|H(\lambda)|\frac{d\psi}{d\lambda}\rangle + \langle\psi(\lambda)|\frac{dH}{d\lambda}|\psi(\lambda)\rangle \\
&= E(\lambda)\frac{d}{d\lambda}\langle\psi(\lambda)|\psi(\lambda)\rangle + \langle\psi(\lambda)|\frac{dH}{d\lambda}|\psi(\lambda)\rangle \\
&= \langle\psi(\lambda)|\frac{dH}{d\lambda}|\psi(\lambda)\rangle.
\end{aligned} \quad (7.1.46)$$

This is one form of the Hellmann–Feynman theorem. It is often used to describe "forces in molecules," in which case the parametrization is with respect to a nuclear coordinate $\mathbf{R}$, and $-\partial E/\partial(\lambda\mathbf{R})$ is then a force.[20] The force between molecular nuclei is then expressed as the Coulomb repulsion of the nuclei together with the attractive forces between the nuclei and the electron charge distribution. This suggests yet another interpretation of the van der Waals interaction, as described at the end of Feynman's paper:

Van der Waals forces can also be interpreted as arising from charge distributions with higher concentration between the nuclei. The Schrödinger perturbation theory for two interacting atoms at a separation R, large compared to the radii of the atoms, leads to the result that the charge distribution of each is distorted from central symmetry, a dipole moment of order $1/R^7$ being induced in each atom. The negative charge distribution of each atom has its center of gravity moved slightly toward the other. It is not the interaction of these dipoles which leads to van der Waals' force, but rather the attraction of each nucleus for the distorted charge distribution of it *own* electrons that gives the attractive $1/R^7$ force.[21]

---

[18] For an analysis of non-additivity along the lines of the coupled-oscillator model of Section 7.1.3, see C. Farina, F. C. Santos, and A. C. Tort, Am. J. Phys. **67**, 344 (1999).

[19] W. Pauli, *Handbuch der Physik* **24**, 162 (1933); H. Hellmann, *Einführung in die Quantechimie* (Deuticke, Leipzig, 1937), p. 285; R. P. Feynman, Phys. Rev. **56**, 340 (1939). This paper was based on Feynman's MIT senior thesis, supervised by J. C. Slater. For some history of the theorem, see J. Musher, Am. J. Phys. **34**, 267 (1966). For some applications in basic quantum mechanics, see D. J. Griffiths, *Introduction to Quantum Mechanics*, second edition, Cambridge University Press, Cambridge, 2017.

[20] "Forces in Molecules" was the title of Feynman's 1939 paper.

[21] R. P. Feynman, Phys. Rev. **56**, 340 (1939), p. 343.

In an integral form, the Hellmann–Feynman theorem is also known as the "coupling constant integration algorithm." Let H_1 in a Hamiltonian $H = H_0 + H_1$ depend on some parameter (e.g., charge). We again scale by a dimensionless parameter λ and write

$$H(\lambda) = H_0 + \lambda H_1. \tag{7.1.47}$$

It follows from (7.1.46) that

$$E(1) = E(0) + \int_0^1 \frac{d\lambda}{\lambda} \langle \psi(\lambda)|\lambda H_1|\psi(\lambda)\rangle \equiv \int_0^1 \frac{d\lambda}{\lambda} f(\lambda). \tag{7.1.48}$$

In this form, the Hellmann–Feynman theorem yields, exactly, the shift $E(1) - E(0)$ in the ground-state energy of the system, completely in terms of an expectation value involving only the part λH_1 of the Hamiltonian.

Of course, things in reality are not that easy because exact expressions for the $|\psi(\lambda)\rangle$ are not usually available. Overhauser and Tsai, for instance, remark that (7.1.48) "is exact provided exact $|\psi(\lambda)\rangle$ are employed. In practice this is not the case," and they give examples for which (7.1.48) is "rather unreliable."[22] Fetter and Walecka likewise note that "unfortunately . . . $[\langle \psi(\lambda)|\lambda H_1|\psi(\lambda)\rangle]$ is required for *all* values of the coupling constant $0 \leq \lambda \leq 1$."[23]

7.1.5 Temperature Dependence of the van der Waals Interaction

In deriving the general (fourth-order) expression (7.1.23) for the van der Waals dispersion interaction, we assumed that the two atoms are at zero temperature. Expressions for finite temperature are not difficult to derive. The necessary modification of the calculation leading to (7.1.23) amounts simply to the replacement of $\langle a_{\mathbf{k}\lambda}(0)a^\dagger_{\mathbf{k}\lambda}(0)\rangle$ in the first line of (7.1.16) by

$$\langle a_{\mathbf{k}\lambda}(0)a^\dagger_{\mathbf{k}\lambda}(0) + a^\dagger_{\mathbf{k}\lambda}(0)a_{\mathbf{k}\lambda}(0)\rangle = 2\left[\langle a^\dagger_{\mathbf{k}\lambda}(0)a_{\mathbf{k}\lambda}(0)\rangle + \frac{1}{2}\right] = 2\left(N_{\mathbf{k}\lambda} + \frac{1}{2}\right)$$

$$= \frac{2}{e^{\hbar\omega/k_B T} - 1} + 1, \tag{7.1.49}$$

and the calculation leading from (7.1.16) to (7.1.23) goes through in essentially the same way as the zero-temperature calculation. In the approximation that a single transition of frequency ω_0 makes the dominant contribution to the polarizability, one obtains, for small separations ($\omega_0 r/c \ll 1$) and low temperatures ($\hbar\omega_0/k_B T \gg 1$),

$$V_{AB}(r) \cong -\left(\frac{1}{4\pi\epsilon_0}\right)^2 \frac{3\hbar\omega_0\alpha^2}{4r^6}\left[1 + 2\left(1 + \frac{\hbar\omega_0}{k_B T}\right)e^{-\hbar\omega_0/k_B T}\right] \tag{7.1.50}$$

[22] A. W. Overhauser and J. T. Tsai, Phys. Rev. B **13**, 607 (1976), p. 607.

[23] A. L. Fetter and J. D. Walecka, *Quantum Theory of Many-Particle Systems*, McGraw-Hill, New York, 1971, p. 69.

for identical atoms having a static polarizability α.[24] In the high-temperature limit $(\hbar\omega_0/k_B T \ll 1)$, the free energy of interaction is

$$V_{AB} \cong -\Big(\frac{1}{4\pi\epsilon_0}\Big)^2 \frac{3\alpha^2 k_B T}{r^6} \qquad (7.1.51)$$

for all separations r, as shown by McLachlan and Boyer,[24] among others.[25] Since atomic and molecular excitation frequencies ω_0 are much larger than $k_B T/\hbar$ for temperatures of practical interest, it is usually an excellent approximation to use the zero-temperature expression (7.1.23) for the dispersion interaction.

The dispersion interaction is also modified by quasi-monochromatic fields, and, for nearly unidirectional fields, can very as $1/r^3$ rather than $1/r^6$.[26] A sufficiently intense applied field can similarly produce an *optical binding* of dielectric particles.[27]

7.2   Van der Waals Interaction in Dielectric Media

The dispersion interaction between dielectric materials is best described by the Lifshitz theory, which we consider in Section 7.5. Here, we consider briefly the simpler example of two polarizable particles embedded in a dielectric medium, treated (as in the Lifsihtz theory) as a continuum. Since it involves electric fields at the two particles, the dispersion interaction is changed from its vacuum form when the particles are in a host dielectric medium. The simplest way to derive the interaction energy in this case is to modify (7.1.20) by replacing $(1/4\pi\epsilon_0)^2$ by $[1/4\pi\epsilon(\omega)]^2 = [1/4\pi\epsilon_0 n^2(\omega)]^2$, and ω/c by $n(\omega)\omega/c$ in the integrand:

$$V_{AB} = -\Big(\frac{1}{4\pi\epsilon_0}\Big)^2 \frac{\hbar}{\pi c^6} \int_0^\infty d\omega\, \omega^6 \frac{\alpha_A(\omega)\alpha_B(\omega)}{n^4(\omega)} G\Big(\frac{n(\omega)\omega r}{c}\Big). \qquad (7.2.1)$$

As in the step from (7.1.20) to (7.1.23), we can express (7.2.1) as

$$V_{AB} = -\Big(\frac{1}{4\pi\epsilon_0}\Big)^2 \frac{\hbar c}{\pi r^2} \int_0^\infty du\, u^4 \alpha_A(icu)\alpha_B(icu)\Big[1 + \frac{2}{n(iuc)ur} + \frac{5}{n^2(iuc)u^2 r^2}$$
$$+ \frac{6}{n^3(iuc)u^3 r^3} + \frac{3}{n^4(iuc)u^4 r^4}\Big] e^{-2n(iuc)ur}. \qquad (7.2.2)$$

This expression remains valid when the refractive index $n(\omega)$ is complex, that is, when the dielectric medium is dissipative.[28] For small separations,

---

[24] See, for instance, T. H. Boyer, Phys. Rev. A **11**, 1650 (1975).

[25] A. D. McLachlan, Proc. R. Soc. Lond. A **274**, 80 (1963); T. H. Boyer, Phys. Rev. A **11**, 1650 (1975); see, for instance, P. R. Berman, G. W. Ford, and P. W. Milonni, Phys. Rev. **89**, 022127 (2014).

[26] T. Thirunamachandran, Mol. Phys. **40**, 393 (1980); P. W. Milonni and A. Smith, Phys. Rev. A **53**, 3484 (1996).

[27] Such optical binding was described and demonstrated by M. M. Burns, J.-M. Fournier, and J. A. Golovchenko, Phys. Rev. Lett. **63**, 1233 (1989).

[28] S. Spagnolo, D. A. R. Dalvit, and P. W. Milonni, Phys. Rev. A **75**, 052117 (2007), and references therein. This paper considers more generally the dispersion interaction of electrically or magnetically polarizable particles in a magnetodielectric medium. Note that, for dissipative $(\epsilon_I(\omega) \geq 0)$ dielectric media, $n = (\epsilon/\epsilon_0)^{1/2}$ is real (and positive) for frequencies on the positive imaginary axis (see (6.6.66)).

$$V_{AB} = -\left(\frac{1}{4\pi\epsilon_0}\right)^2 \frac{3\hbar}{\pi r^6} \int_0^\infty d\xi \frac{\alpha_A(i\xi)\alpha_B(i\xi)}{n^4(i\xi)}, \tag{7.2.3}$$

indicating that the dispersion interaction energy can be significantly diminished in a dielectric medium. For two polarizable particles in a vacuum ($n = 1$), (7.2.3) obviously reduces to the London expression (7.1.31).

Suppose, for example, that A and B are small dielectric spheres with radii a_A, a_B and permittivities ϵ_A, ϵ_B. Their polarizabilities are given by (1.5.59), and their dispersion interaction according to (7.2.3) is therefore

$$V_{AB} = -\frac{3\hbar}{\pi r^6} a_A^3 a_B^3 \int_0^\infty d\xi \left[\frac{\epsilon_A(i\xi) - \epsilon(i\xi)}{\epsilon_A(i\xi) + 2\epsilon(i\xi)}\right] \left[\frac{\epsilon_B(i\xi) - \epsilon(i\xi)}{\epsilon_B(i\xi) + 2\epsilon(i\xi)}\right], \tag{7.2.4}$$

where $\epsilon(\omega)$ is the permittivity of the host dielectric medium. For identical spheres,

$$V_{AA} = -\frac{3\hbar}{\pi r^6} a_A^6 \int_0^\infty d\xi \left[\frac{\epsilon_A(i\xi) - \epsilon(i\xi)}{\epsilon_A(i\xi) + 2\epsilon(i\xi)}\right]^2, \tag{7.2.5}$$

and the force between them is always attractive. For example, the dispersion interaction of two bubbles in a liquid will always be an effective attraction.[29]

The dispersion force implied by (7.2.5) is smallest when $\epsilon_A(i\xi) \approx \epsilon(i\xi)$, that is, when the guest particles are similar to the host medium in their dielectric properties. In this case, there is less of a tendency for the guest particles to congregate. This is related to the "rule of thumb" that "like dissolves like": in a liquid solution, the attraction between the solute and the solvent molecules is not too different from the attraction between the solute molecules themselves, and so the solute molecules do not congregate, and solute and solvent are miscible.[30]

The fact that the dispersion interaction is weak when $\epsilon_A(i\xi) \approx \epsilon(i\xi)$ is easy to understand if we think of the interaction in terms of the scattering of the field of one particle by the other, resulting in an r-dependent change in the (radiation reaction) field acting on each particle: if $\epsilon_A(i\xi) - \epsilon(i\xi)$ is small, the dielectric inhomogeneities associated with the guest particles, and the scattering by them, are weak, and therefore so must be their dispersion interaction.

The van der Waals dispersion interaction between *unlike* particles can be attractive or repulsive. If $\epsilon_A(i\xi) - \epsilon(i\xi)$ and $\epsilon_B(i\xi) - \epsilon(i\xi)$ have different signs over the range of frequencies that contribute significantly to (7.2.3), the dispersion force between A and B will be repulsive; otherwise, it will be attractive. Note also that the general relation

$$\left[\frac{\epsilon_A(i\xi) - \epsilon(i\xi)}{\epsilon_A(i\xi) + 2\epsilon(i\xi)}\right]^2 + \left[\frac{\epsilon_B(i\xi) - \epsilon(i\xi)}{\epsilon_B(i\xi) + 2\epsilon(i\xi)}\right]^2$$
$$\geq 2\left[\frac{\epsilon_A(i\xi) - \epsilon(i\xi)}{\epsilon_A(i\xi) + 2\epsilon(i\xi)}\right]\left[\frac{\epsilon_B(i\xi) - \epsilon(i\xi)}{\epsilon_B(i\xi) + 2\epsilon(i\xi)}\right] \tag{7.2.6}$$

[29] A. D. McLachlan, Discuss. Faraday Soc. **40**, 239 (1965).
[30] *Israelachvili*, p. 81.

suggests that, in a mixture of A and B particles, the dispersion interaction will tend to make A's congregate near A's, and B's near B's.[31] Birds of a feather flock together.[32]

7.3   The Casimir Force

As remarked earlier, Casimir was soon "well aware" that the formulas he and Polder obtained for the long-range van der Waals interaction "can be derived more simply using the methods of zero-point energy." He was led to such a derivation following a discussion of the formulas with Niels Bohr, who "mumbled something about zero-point energy." From his applied work at the Philips Laboratories in Eindhoven, Casimir was "somewhat familiar with the theory of modes of resonant cavities and their perturbations," and this led him to consider the effect of zero-point energy on two parallel, uncharged, perfectly conducting plates (see Figure 7.2).[33] What he found was that, because of the change in the zero-point energy of the electromagnetic field in the presence of the plates, the plates should attract each other with a force per unit plate area that is inversely proportional to the fourth power of the plate separation.[34] This bold prediction was the inspiration for some remarkable experimental work during the next decade, but definitive experimental confirmations came much later, half a century after Casimir's work. *Casimir effects* and related phenomena involving zero-point energy are now of much interest in areas ranging from the wetting of surfaces to cosmology, and Casimir's formula for the force between conducting plates has been rederived and discussed in quite a few papers and books.[35] We will briefly review Casimir's calculation.

Consider a box in the form of a parallelepiped with perfectly conducting walls of length $L_x = L_y = L$ and $L_z = d$. The condition that the tangential components of the electric field vanish at the walls requires that the electric fields of the allowed modes have the spatial dependence

$$
\begin{aligned}
\mathbf{A}_{\ell mn}(x,y,z) = \left(\frac{8}{V}\right)^{1/2}\Big[&\mathbf{e}_x \cos(\ell\pi x/L)\sin(m\pi y/L)\sin(n\pi z/d)\\
+ &\mathbf{e}_y \sin(\ell\pi x/L)\cos(m\pi y/L)\sin(n\pi z/d)\\
+ &\mathbf{e}_z \sin(\ell\pi x/L)\sin(m\pi y/L)\cos(n\pi z/L)\Big],
\end{aligned}
\tag{7.3.1}
$$

with $\mathbf{e}_x^2 + \mathbf{e}_y^2 + \mathbf{e}_z^2 = 1$, and ℓ, m, and n positive integers, including zero. $(8/V)^{1/2} = (8/L^2 d)^{1/2}$ is a mode normalization factor. The mode frequencies are

$$
\omega_{\ell mn} = c\left(\frac{\pi^2\ell^2}{L^2} + \frac{\pi^2 m^2}{L^2} + \frac{\pi^2 n^2}{d^2}\right)^{1/2},
\tag{7.3.2}
$$

---

[31] See, for instance, *Israelachvili* or *Parsegian*.

[32] G. F. Chew, Phys. Rev. A **46**, 1194 (1992).

[33] H. B. G. Casimir, private communication, March 12, 1992.

[34] H. B. G. Casimir, Proc. K. Ned. Akad. Wet. **51**, 793 (1948).

[35] See, for instance, *Milonni (1994)*, *Milton*, and *Bordag et al.*

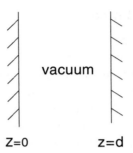

$$Z=0 \qquad\qquad z=d$$

Fig. 7.2 Two parallel, perfectly conducting plates experience an attractive force per unit area proportional to $1/d^4$.

and the zero-point field energy in the box is

$$E_0 = 2 \times \sideset{}{'}\sum_{\ell,m,n} \frac{1}{2}\hbar\omega_{\ell mn} = \hbar c \sideset{}{'}\sum_{\ell,m,n} \Big(\frac{\pi^2\ell^2}{L^2} + \frac{\pi^2 m^2}{L^2} + \frac{\pi^2 n^2}{d^2} \Big)^{1/2}. \qquad (7.3.3)$$

The factor 2 accounts for two independent polarizations when $\ell, m, n \neq 0$, and the prime indicates that a factor $1/2$ should be included when one of these integers is zero, since, in that case, there is only one independent polarization. For the geometry of Figure 7.2, we take $L \gg d$. Then, (7.3.3) can be approximated by

$$E_0(d) = \hbar c \Big(\frac{L}{\pi}\Big)^2 \sideset{}{'}\sum_{n=0}^{\infty} \int_0^\infty dq_1 \int_0^\infty dq_2 \Big(q_1^2 + q_2^2 + \frac{\pi^2 n^2}{d^2} \Big)^{1/2}$$

$$= \frac{\hbar c L^2}{\pi^2} \frac{\pi}{2} \sideset{}{'}\sum_{n=0}^{\infty} \int_0^\infty du u \Big(u^2 + \frac{n^2\pi^2}{d^2} \Big)^{1/2}. \qquad (7.3.4)$$

If d were arbitrarily large, the zero-point field energy would similarly be

$$E_0(\infty) = \frac{\hbar c L^2}{\pi^2} \frac{\pi}{2} \frac{d}{\pi} \int_0^\infty du u \int_0^\infty dq \Big(u^2 + q^2 \Big)^{1/2}. \qquad (7.3.5)$$

Following Casimir, we define a potential energy $U(d)$ as the difference between $E_0(d)$ and $E_0(\infty)$, this being the energy required to bring the plates from a very large separation to a finite separation d:

$$U(d) = \frac{\hbar c L^2}{\pi^2} \frac{\pi}{2} \Big[\sideset{}{'}\sum_{n=0}^{\infty} \int_0^\infty du u \Big(u^2 + \frac{n^2\pi^2}{d^2} \Big)^{1/2}$$

$$- \frac{d}{\pi} \int_0^\infty dq \int_0^\infty du u \Big(u^2 + q^2 \Big)^{1/2} \Big]. \qquad (7.3.6)$$

$U(d)$ is a difference between two infinite energies, the infinities being a result of our having allowed contributions from arbitrarily large frequencies. In the actual physical situation of interest, the assumption of perfectly conducting walls at all frequencies is unrealistic: fields of sufficiently high frequencies will not be confined to a resonator and will therefore not contribute to the zero-point field energy inside the resonator. To account for this physical reality, we introduce in the integrands in (7.3.6) a cutoff function $f(k) = f((u^2 + q^2)^{1/2})$ such that $f(k) = 1$ for $k \ll k_m$, and $f(k) = 0$ for $k \gg k_m$. k_m might be taken to be on the order of the inverse of the average distance between the atoms of the wall, but a particular numerical value need not be specified. $f(k)$ itself, similarly, need not be specified, but it will be assumed to be sufficiently flat for small k that all its derivatives vanish at $k = 0$. Thus, we redefine $U(d)$ as

$$
\begin{aligned}
U(d) &= \frac{\hbar c L^2}{\pi^2} \frac{\pi}{2} \Big\{ \sum_{n=0}^{\infty}{}' \int_0^\infty duu \Big(u^2 + \frac{n^2\pi^2}{d^2}\Big)^{1/2} f[(u^2 + n^2\pi^2/d^2)^{1/2}] \\
&\quad - \frac{d}{\pi} \int_0^\infty dq \int_0^\infty duu \Big(u^2 + q^2\Big)^{1/2} f[(u^2 + q^2)^{1/2}] \Big\} \\
&= \frac{\pi^2\hbar c}{4d^3} L^2 \Big\{ \sum_{n=0}^{\infty}{}' \int_0^\infty dx(x + n^2)^{1/2} f\Big[\frac{\pi}{d}(x + n^2)^{1/2}\Big] \\
&\quad - \int_0^\infty d\kappa \int_0^\infty dx(x + \kappa^2)^{1/2} f\Big[\frac{\pi}{d}(x + \kappa^2)^{1/2}\Big] \Big\} \\
&= \frac{\pi^2\hbar c}{4d^3} L^2 \Big[\frac{1}{2}F(0) + \sum_{n=1}^{\infty} F(n) - \int_0^\infty d\kappa F(\kappa)\Big],
\end{aligned}
\tag{7.3.7}
$$

where, in the second equality, we have defined the (dimensionless) integration variables $x = u^2 d^2/\pi^2$ and $\kappa = qd/\pi$, and, in the last equality,

$$
F(\kappa) \equiv \int_0^\infty dx(x + \kappa^2)^{1/2} f\Big[\frac{\pi}{d}(x + \kappa^2)^{1/2}\Big] = -\int_\infty^{\kappa^2} duu^{1/2} f\Big[\frac{\pi}{d}u^{1/2}\Big].
\tag{7.3.8}
$$

We next use the Euler–Maclaurin formula for the difference between a sum and an integral:

$$
\sum_{n=1}^{\infty} F(n) - \int_0^\infty dq F(q) = -\frac{1}{2}F(0) - \frac{1}{12}F'(0) + \frac{1}{720}F'''(0) + \cdots,
\tag{7.3.9}
$$

the primes denoting derivatives. Now,

$$
F'(\kappa) = -\frac{d\kappa^2}{d\kappa} f\Big(\frac{\pi}{d}\kappa\Big) = -2\kappa f\Big(\frac{\pi}{d}\kappa\Big),
\tag{7.3.10}
$$

and therefore $F'(0) = 0$, $F'''(0) = -4$, and all higher derivatives $F^{(n)}(0)$ vanish, since all derivatives of the cutoff function vanish at $\kappa = 0$. Therefore,

$$U(d) = \frac{\pi^2 \hbar c}{4d^3} L^2 \left(-\frac{4}{720} \right) = -\frac{\pi^2 \hbar c}{720 d^3} L^2, \tag{7.3.11}$$

and the (attractive) force between the plates is

$$\mathcal{F}(d) = -U'(d) = -\frac{\pi^2 \hbar c}{240 d^4} \tag{7.3.12}$$

per unit area.

Casimir remarked that the force (7.3.12) per unit area "may be interpreted as a zero-point pressure of electromagnetic waves."[36] Indeed, (7.3.12) is obtained when we calculate the radiation pressure of the vacuum field on the plates.[37] An intuitive but superficial explanation for the Casimir force is suggested by such a calculation: fields outside the resonator reflect off the plates and act to push them together, whereas field reflections between the plates act to push them apart. Since all field frequencies are allowed outside the resonator, whereas, between the plates, the possible frequencies are restricted by the boundary conditions on the fields, the pressure pushing the plates together exceeds the pressure pushing them apart.[38] (In Section 7.5, we consider the zero-point field pressure for the case of dielectric media.)

It was only after nearly half a century that Casimir's predicted force was accurately measured experimentally.[39]

7.3.1 A Simpler, Approximate Derivation of the Casimir Force

The fact that the Casimir force is a consequence of the different field modes in the spaces between and outside the plates suggests a simpler, albeit approximate, derivation.[40] The zero-point field energy in a volume $V = L^2 d$ of free space is (see (4.5.21))

$$E_0 = L^2 d \left(\frac{\hbar}{2\pi^2 c^3} \right) \int_{\omega_1}^{\omega_2} d\omega \omega^3 \tag{7.3.13}$$

for frequencies between ω_1 and ω_2. In the space between the plates, the field modes are restricted to wavelengths smaller than $2d$, or frequencies greater than $\omega_1 = 2\pi c/2d$, and the zero-point field energy is

$$E_0(d) = L^2 d \left(\frac{\hbar}{2\pi^2 c^3} \right) \int_{\pi c/d}^{\omega_2} d\omega \omega^3. \tag{7.3.14}$$

---

[36] H. B. G. Casimir, Proc. K. Ned. Akad. Wet. **51**, 793 (1948).

[37] L. S. Brown and G. J. Maclay, Phys. Rev. **184**, 1272 (1969); P. W. Milonni, R. J. Cook, and M. E. Goggin, Phys. Rev. A **38**, 1621 (1988).

[38] The superficiality of this argument becomes evident when one considers the Casimir force on a conducting spherical shell. There are more modes allowed outside the sphere than inside, and yet the force is repulsive (see Section 7.3.2).

[39] For a brief survey of experiments up to the early 1990s, see, for instance, *Milonni (1994)*, and references therein. Discussion of the more recent experimental work, beginning with that of S. K. Lamoreaux in 1996 and U. Mohideen in 1998, may be found in S. K. Lamoreaux, Rep. Prog. Phys. **68**, 201 (2005), Physics Today **60**, 40 (February, 2007), and *Bordag et al.* See also the chapters from various experimental groups in *Dalvit et al.*

[40] See, for instance, *Power*.

Outside the plates, there is no upper limit on the possible wavelengths. If d is very large, the zero-point energy is approximately

$$E_0(\infty) = L^2 d\left(\frac{\hbar}{2\pi^2 c^3}\right) \int_0^{\omega_2} d\omega\,\omega^3, \qquad (7.3.15)$$

and

$$E_0(d) - E_0(\infty) = L^2 d\left(\frac{\hbar}{2\pi^2 c^3}\right) \int_{\pi c/d}^0 d\omega\,\omega^3 = -L^2 \frac{\pi^2 \hbar c}{8d^3}. \qquad (7.3.16)$$

This is 90 times the Casimir energy (7.3.11). However, our choice for ω_1 ignored the fact that the restriction on wavelengths between the plates applies to wave vector components k_z but not to k_x or k_y. A better estimate is expected, therefore, if we choose a smaller value for ω_1. If we take $\omega_1 = (\pi c/d)/(90)^{1/4} \cong 0.32(\pi c/d)$, we recover exactly the Casimir energy and force.

7.3.2 Casimir Effects and the van der Waals Interaction

The Casimir force between conducting plates may be, and often is, considered counter-intuitive, but a force between two *dielectric* materials is to be expected, since the atoms of the two materials experience van der Waals forces. A single molecule near a dielectric wall consisting of molecules of the same kind, for example, is expected to be attracted to the wall as a result of its van der Waals attraction to the molecules constituting the wall. Assuming an attractive pairwise interaction $V(r) = -B/r^P$ between the molecules of the wall and the molecule at a distance d from the wall, and that the wall occupies the half-space $z \geq d$, we calculate an interaction energy

$$
\begin{aligned}
U_1(d) &= -N_1 B \int_{-\infty}^{\infty} dx \int_{-\infty}^{\infty} dy \int_d^{\infty} dz (x^2 + y^2 + z^2)^{-P/2} \\
&= -\frac{2\pi N_1 B}{(P-2)(P-3)}\left(\frac{1}{d}\right)^{P-3},
\end{aligned}
\qquad (7.3.17)
$$

where N_1 is the number density of wall molecules. Now replace the single molecule outside the wall by a dielectric consisting of N_2 identical molecules per unit volume occupying the half-space $z \leq 0$. The interaction energy between the two dielectrics, assuming again pairwise van der Waals interactions, is

$$
\begin{aligned}
U(d) &= -\frac{2\pi N_1 N_2 B}{(P-2)(P-3)} \int_0^{\infty} dR(R+d)^{3-P} \\
&= -\frac{2\pi N_1 N_2 B}{(P-2)(P-3)(P-4)}\left(\frac{1}{d}\right)^{P-4}
\end{aligned}
\qquad (7.3.18)
$$

per unit area, and the (attractive) force per unit area between the dielectrics is therefore

$$\mathcal{F}(d) = -\frac{2\pi N_1 N_2 B}{(P-2)(P-3)}\left(\frac{1}{d}\right)^{P-3}. \qquad (7.3.19)$$

If the molecules of the two dielectric experience nonretarded van der Waals interactions ($P = 6$), then

$$U(d) = -\frac{\pi N_1 N_2 B}{12d^2},$$ (7.3.20)

and $\mathcal{F}(d) \propto 1/d^3$, whereas, for retarded intermolecular van der Waals interactions, ($P = 7$) $\mathcal{F}(d) \propto 1/d^4$. In the latter case, $B = (23\hbar c \alpha^2/4\pi)/(4\pi\epsilon_0)^2$ (see (7.1.35)) and

$$\mathcal{F}(d) = -\left(\frac{1}{4\pi\epsilon_0}\right)^2 \frac{N^2 \alpha^2}{40d^4}$$ (7.3.21)

for identical dielectric media.

As discussed in Section 7.1.3, the van der Waals dispersion interaction is not pairwise additive, and formulas such as (7.3.21) that derive from the assumption of pairwise additivity are only approximations. Hamaker, who assumed pairwise interactions, was among the first to calculate van der Waals interaction energies between macroscopic objects, and these energies for various geometries are to this day characterized by "Hamaker constants" A_{Ham}, even if they are calculated without the pairwise approximation.[41] The energy (7.3.20), for instance, is conventionally expressed as $U(d) = -A_{\text{Ham}}/12\pi d^2$.

The force (7.3.21), although it is obtained for dielectrics and by pairwise summation, has the $1/d^4$ dependence on distance that characterizes the Casimir force for perfectly conducting plates. A way of relating (7.3.21) to the Casimir force is to use the Clausius–Mossotti relation

$$N\alpha = (4\pi\epsilon_0)\frac{3}{4\pi}\frac{\epsilon - 1}{\epsilon + 2}$$ (7.3.22)

in (7.3.21) and then go to the limit $\epsilon \to \infty$ of a perfect conductor. This gives about 80% of the Casimir force:

$$\mathcal{F}(d) \to -\left(\frac{3}{4\pi}\right)^2 \frac{(23)(240)}{40\pi^2} \times \frac{\pi^2\hbar c}{240d^4} \cong -0.80 \times \frac{\pi^2\hbar c}{240d^4}.$$ (7.3.23)

In these heuristic considerations, we require the pairwise van der Waals interactions to have their *retarded* form in order to obtain the $1/d^4$ dependence of the Casimir force between perfectly conducting plates. Independent of these considerations, the distance d must be large compared with the skin depth of the plates for all wavelengths that contribute significantly to the force, and, in particular, for wavelengths comparable to d, in order for the formula (7.3.12) to apply to real conductors. For dielectric materials, the force varies as $1/d^4$ at large separations but as $1/d^3$ at small separations, "large" and "small" being defined with respect to wavelengths at which there is significant absorption (see Section 7.5). In the pairwise approximation, the $1/d^4$ and $1/d^3$ forces are seen from (7.3.20) and (7.3.21) to follow from retarded and nonretarded van der Waals interactions, respectively.

[41] H. C. Hamaker, Physica **4**, 1058 (1937); see, for instance, *Israelachvili* or *Parsegian*.

Casimir effects are associated more generally with changes in the zero-point energies of quantum fields and appear when constraints (boundary conditions) are imposed on the fields. The force (7.3.12) is the best-known such effect. For more complicated geometries, calculations of Casimir forces are generally quite difficult, and one typically lacks intuition beforehand even as to whether the forces are attractive or repulsive.[42] For example, the Casimir force in the case of a conducting spherical shell turns out to be repulsive, in contrast to the expectation of Casimir himself.[43]

7.3.3 Casimir–Polder Shift of an Atom near a Conducting Plate

Consider first a ground-state atom at the point $\mathbf{r}_A = (L/2, L/2, d)$ in an otherwise empty box for which the mode functions are given by (7.3.1). The quantized electric field at the atom is (see (7.3.1) and (3.3.4))

$$
\begin{aligned}
\mathbf{E}(\mathbf{r}_A, t) &= i \sum_{\ell m n} \sqrt{\frac{\hbar \omega_{\ell m n}}{2 \epsilon_0}} \left[\mathbf{A}_{\ell m n}(\mathbf{r}) a_{\ell m n}(t) - \mathbf{A}_{\ell m n}^*(\mathbf{r}) a_{\ell m n}^\dagger(t) \right] \\
&= \sum_{\ell m n} \mathbf{E}_{\ell m n}(\mathbf{r}_A, t).
\end{aligned}
\tag{7.3.24}
$$

The fluctuations of the zero-point electric field at the atom will cause a shift $\delta E_g(d)$ in its energy level. To derive it, we again start with the Stark-shift formula (7.1.7), which in this case takes the form

$$
U(\mathbf{r}_A) = -\frac{1}{2} \sum_{\ell m n} \alpha(\omega_{\ell m n}) \langle \mathbf{E}_{\ell m n}^2(\mathbf{r}_A, t) \rangle.
\tag{7.3.25}
$$

The vacuum-field expectation value $\langle \mathbf{E}_{\ell m n}^2(\mathbf{r}_A, t) \rangle$ is easily obtained from (7.3.24), using $a_{\ell m n}(t) = a_{\ell m n}(0) \exp(-i \omega_{\ell m n} t)$, $a_{\ell m n}(0)|\text{vac}\rangle = 0$, and $\langle a_{\ell m n}(0) a_{\ell' m' n'}^\dagger(0) \rangle = \delta_{\ell \ell'} \delta_{m m'} \delta_{n n'}$, with the result that

$$
U(\mathbf{r}_A) = -\frac{1}{2} \frac{8}{V} \sum_{\ell m n} \alpha(\omega_{\ell m n}) \frac{\hbar \omega_{\ell m n}}{2 \epsilon_0} \mathbf{A}_{\ell m n}^2(\mathbf{r}_a).
\tag{7.3.26}
$$

Suppose now that the atom is not in a box but is at a distance d from a perfectly conducting plate in the plane $z = 0$. Then we replace (7.3.26) by

---

[42] Different analytical and numerical techniques for the computation of Casimir forces are reviewed by several practitioners in the volume edited by *Dalvit et al.*

[43] This was discovered by T. H. Boyer, Phys. Rev. **174**, 1764 (1968). His result for the Casimir energy of a conducting spherical shell of diameter D, $E(a) \cong +0.09 \hbar c/D$, was subsequently confirmed by several others.

$$U(d) = -\frac{4}{V} \sum_{\mathbf{k}\lambda} \alpha(\omega_k) \frac{\hbar\omega_k}{2\epsilon_0} \Big[e_x^2 \cos^2 \frac{1}{2} k_x L \sin^2 \frac{1}{2} k_y L \sin^2 k_z d$$

$$+ e_y^2 \sin^2 \frac{1}{2} k_x L \cos^2 \frac{1}{2} k_y L \sin^2 k_z d$$

$$+ e_z^2 \sin^2 \frac{1}{2} k_x L \sin^2 \frac{1}{2} k_y L \cos^2 k_z d \Big]$$

$$\rightarrow -\frac{\hbar}{2\epsilon_0 V} \sum_{\mathbf{k}\lambda} \omega_k \alpha(\omega_k) \Big[(e_{\mathbf{k}\lambda x}^2 + e_{\mathbf{k}\lambda y}^2) \sin^2 k_z d + e_{\mathbf{k}\lambda z}^2 \cos^2 k_x d \Big], \quad (7.3.27)$$

where we have replaced $\sin^2 \frac{1}{2} k_x L$, $\sin^2 \frac{1}{2} k_y L$, $\cos^2 \frac{1}{2} k_x L$, and $\cos^2 \frac{1}{2} k_y L$ by their average values $(1/2)$. The mode summation is over all wave vectors $\mathbf{k}$, with two independent polarizations $(\lambda = 1, 2)$ for each $\mathbf{k}$. In the limit $d \rightarrow \infty$ in which the atom is far removed from the wall, we can likewise replace $\sin^2 k_z d$ and $\cos^2 k_z d$ by $1/2$:

$$U(\infty) = -\frac{\hbar}{2\epsilon_0 V} \sum_{\mathbf{k}\lambda} \omega_k \alpha(\omega_k) \frac{1}{2} \left(e_{\mathbf{k}\lambda x}^2 + e_{\mathbf{k}\lambda y}^2 + e_{\mathbf{k}\lambda z}^2 \right) = -\frac{\hbar}{2\epsilon_0 V} \sum_{\mathbf{k}\lambda} \omega_k \alpha(\omega_k) \frac{1}{2}. \quad (7.3.28)$$

We define the observable shift in the atom's ground-state energy level, due to the presence of the conducting plate, as the difference

$$\delta E_g(d) = U(d) - U(\infty)$$

$$= -\frac{\hbar}{2\epsilon_0 V} \sum_{\mathbf{k}\lambda} \omega_k \alpha(\omega_k) \left(e_{\mathbf{k}\lambda x}^2 + e_{\mathbf{k}\lambda y}^2 - e_{\mathbf{k}\lambda z}^2 \right) \left(\sin^2 k_z d - \frac{1}{2} \right)$$

$$= \frac{\hbar}{4\epsilon_0 V} \sum_{\mathbf{k}} \omega_k \alpha(\omega_k) \sum_{\mathbf{k}} \omega_k \cos 2k_z d \sum_{\lambda=1}^{2} \left(e_{\mathbf{k}\lambda x}^2 + e_{\mathbf{k}\lambda y}^2 - e_{\mathbf{k}\lambda z}^2 \right)$$

$$= \frac{\hbar}{4\epsilon_0 V} \sum_{\mathbf{k}} \omega_k \alpha(\omega_k) \cos 2k_z d \frac{2k_z^2}{k^2}, \quad (7.3.29)$$

where we have used $e_{\mathbf{k}\lambda x}^2 + e_{\mathbf{k}\lambda y}^2 + e_{\mathbf{k}\lambda z}^2 = 1$, and $\sum_{\lambda=1}^{2} e_{\mathbf{k}\lambda z}^2 = 1 - k_z^2/k^2$.[44] Thus, in the large-volume limit,

$$\delta E_g(d) = \frac{\hbar}{2\epsilon_0 V} \frac{V}{8\pi^3} \int d^3 k \, \omega \alpha(\omega) \frac{k_z^2}{k^2} \cos 2k_z d$$

$$= \frac{\hbar c}{8\pi^2 \epsilon_0} \int_0^\infty dk k^3 \alpha(\omega) \int_0^\pi d\theta \sin \theta \cos^2 \theta \cos(2kd \cos \theta)$$

$$= \frac{\hbar c}{4\pi^2 \epsilon_0} \int_0^\infty dk k^3 \alpha(\omega) \left(\frac{\sin 2kd}{2kd} + \frac{2 \cos 2kd}{4k^2 d^2} - \frac{2 \sin 2kd}{8k^3 d^3} \right). \quad (7.3.30)$$

By the same application of Cauchy's theorem that was used to obtain (7.1.23) in the calculation of the van der Waals interaction, we can express (7.3.30) as

$$\delta E_g(d) = -\frac{\hbar c}{4\pi^2 \epsilon_0} \int_0^\infty du u^3 \alpha(icu) \left(\frac{1}{2ud} + \frac{2}{4u^2 d^2} + \frac{2}{8u^3 d^3} \right) e^{-2ud}. \quad (7.3.31)$$

---

[44] The latter relation follows from (4.4.16), when we take $\mathbf{d}_{12}$ to be the unit vector $\hat{\mathbf{z}}$.

Small Atom–Plate Separations.   In the limit of very small d,

$$\delta E_g(d) \cong -\frac{\hbar c}{4\pi^2 \epsilon_0} \frac{1}{4d^3} \int_0^\infty du\, \alpha(icu)$$

$$= -\frac{1}{4\pi\epsilon_0} \frac{c}{6\pi d^3} \sum_n |\mathbf{d}_{ng}|^2 \omega_{ng} \int_0^\infty \frac{du}{\omega_{ng}^2 + c^2 u^2}$$

$$= -\frac{1}{4\pi\epsilon_0} \frac{1}{12d^3} \sum_n |\mathbf{d}_{ng}|^2, \tag{7.3.32}$$

where we have used the Kramers–Heisenberg formula (2.3.32),

$$\alpha(icu) = \frac{2}{3\hbar} \sum_n \frac{\omega_{ng}|\mathbf{d}_{ng}|^2}{\omega_{ng}^2 + c^2 u^2}, \tag{7.3.33}$$

for the ground-state polarizability, at imaginary frequencies, of the spherically symmetric atom. Since

$$|\mathbf{d}_{ng}^\parallel|^2 + 2|\mathbf{d}_{ng}^\perp|^2 = 2|\mathbf{d}_{ng}|^2/3 + 2|\mathbf{d}_{ng}|^2/3, \tag{7.3.34}$$

the small-separation energy (7.3.32) can also be expressed as

$$\delta E_g(d) = -\frac{1}{4\pi\epsilon_0} \frac{1}{16d^3} \sum_n \left(|\mathbf{d}_{ng}^\parallel|^2 + 2|\mathbf{d}_{ng}^\perp|^2\right). \tag{7.3.35}$$

This has the same form as the classical electrostatic interaction energy between an electric dipole $\mathbf{p}$ and its mirror image $\mathbf{p}_i$: if the dipole is perpendicular to the mirror, $\mathbf{p}_i = \mathbf{p} = p\hat{\mathbf{z}}$, and this interaction is

$$V^\perp(d) = -\frac{1}{2} \times \frac{1}{4\pi\epsilon_0} \frac{1}{(2d)^3} \left[3\mathbf{p}\cdot\hat{\mathbf{z}})(\mathbf{p}_i\cdot\hat{\mathbf{z}}) - \mathbf{p}\cdot\mathbf{p}_i\right] = -\frac{1}{2} \times \frac{1}{4\pi\epsilon_0} \frac{2p^2}{(2d)^3}, \tag{7.3.36}$$

whereas if the dipole is parallel to the mirror, $\mathbf{p}_i = -\mathbf{p} = -p\hat{\mathbf{z}}$, and[45]

$$V^\parallel(d) = -\frac{1}{2} \times \frac{1}{4\pi\epsilon_0} \frac{p^2}{(2d)^3}. \tag{7.3.37}$$

The factor $1/2$ appears because a force acting to bring the dipole from infinity to $z = d$ acts only on the dipole, not on its image.[46]

---

[45] J. Bardeen, Phys. Rev. **58**, 727 (1940).

[46] The method of images as applied to the Casimir–Polder interaction for all separations is discussed by S. M. Barnett, A. Aspect, and P. W. Milonni, J. Phys. B: At. Mol. Opt. Phys. **33**, L143 (2000).

Large Atom–Plate Separation. In the large-separation limit,

$$
\begin{aligned}
\delta E_g(d) &= -\frac{\hbar c}{4\pi^2\epsilon_0}\left(\frac{1}{2d}\right)^4\int_0^\infty dx\, x^3 \alpha\!\left(\frac{icx}{2d}\right)\left(\frac{1}{x}+\frac{2}{x^2}+\frac{2}{x^3}\right)e^{-x} \\
&\cong -\frac{\hbar c}{4\pi^2\epsilon_0}\frac{\alpha(0)}{16d^4}\int_0^\infty dx\,(x^2+2x+2)e^{-x} \\
&= -\frac{1}{4\pi\epsilon_0}\frac{3\hbar c\alpha(0)}{8\pi d^4},
\end{aligned}
\tag{7.3.38}
$$

as originally obtained by Casimir and Polder and measured many years later.[47] The effect of retardation is again to make an interaction weaken faster with distance than would be expected from the short-distance, nonretarded form of the interaction.

The correct distance dependence of the atom–plate interaction is obtained in the pairwise summation approximation (7.3.17): the nonretarded ($p = 6$) and retarded ($p = 7$) van der Waals interactions give atom–plate interaction energies varying as $1/d^3$ and $1/d^4$, respectively, consistent with (7.3.32) and (7.3.38).

7.3.4   Polarizability and the Electrostatic Interaction

Our derivation of the Casimir–Polder energy (7.3.31) was based on the Stark-shift formula (7.3.25) involving the ground-state polarizability $\alpha(\omega)$. This intuitive approach can be justified in a more formal way based on standard second-order perturbation theory in which the shift in the ground-state energy of an atom at $\mathbf{r}$ is given by (4.4.39). The spontaneous emission rate A_{ng} for an atom at a distance d from a perfect planar reflector is derived in Section 7.11; the d-dependent part of $A_{ng}(\omega, d)$ for a spherically symmetric atom follows from (7.11.7), with the state labels 2 and 1 replaced by n and g, respectively:

$$
A_{ng}(\omega, d) = -\frac{\omega^3|\mathbf{d}_{ng}|^2}{3\pi\hbar\epsilon_0 c^3}\left(\frac{\sin 2kd}{2kd}+\frac{2\cos 2kd}{(2kd)^2}-\frac{2\sin 2kd}{(2kd)^3}\right).
\tag{7.3.39}
$$

Equation (4.4.39) then gives the d-dependent shift in the ground-state energy:

$$
\delta E_g(d) = \frac{c}{6\pi^2\epsilon_0}\sum_n|\mathbf{d}_{ng}|^2\int_0^\infty\frac{dk\,k^3}{kc+\omega_{ng}}\left[\frac{\sin 2kd}{2kd}+\frac{2\cos 2kd}{(2kd)^2}-\frac{2\sin 2kd}{(2kd)^3}\right].
\tag{7.3.40}
$$

With Cauchy's theorem and the integration contour of Figure 7.1, we can express this as an integral along the positive imaginary axis in the complex k plane. For example,

[47] H. B. G. Casimir and D. Polder, Nature **158**, 787 (1946); H. B. G. Casimir and D. Polder, Phys. Rev. **73**, 360 (1948). Both the small- and large-separation limits have been measured and carefully analyzed. See V. Sandoghdar, C. I. Sukenik, E. A. Hinds, and S. Haroche, Phys. Rev. Lett. **68**, 3432 (1992); C. I. Sukenik, M. G. Boshier, D. Cho, V. Sandoghdar, and E. A. Hinds, Phys. Rev. Lett. **70**, 560 (1993); A. Landragin, J.-Y. Courtois, G. Labeyrie, N. Vansteenkiste, C. I. Westbrook, and A. Aspect, Phys. Rev. Lett. **77**, 1464 (1996).

$$\int_0^\infty \frac{dk k^3}{kc + \omega_{ng}} \frac{\sin 2kd}{2kd} = \text{Im} \int_0^\infty \frac{dk k^3}{kc + \omega_{ng}} \frac{e^{2kd}}{2kd}$$

$$= -\text{Im} \int_0^\infty \frac{i d u u^3}{i u c + \omega_{ng}} \frac{e^{-2ud}}{2ud}$$

$$= -\omega_{ng} \int_0^\infty \frac{d u u^3}{\omega_{ng}^2 + u^2 c^2} \frac{e^{-2ud}}{2ud}. \tag{7.3.41}$$

When the remaining terms in (7.3.40) are evaluated in the same way, and the result is expressed in terms of the ground-state polarizability $\alpha(icu)$, it is found that (7.3.40) is identical to (7.3.31). Thus, (4.4.39) leads to the Casimir–Polder energy involving the polarizability, as in the derivation starting from (7.3.25).

But a more interesting point is brought out by this simple exercise in perturbation theory. In their derivation of (7.3.31), Casimir and Polder used the Coulomb-gauge, minimal coupling form of the Hamiltonian.[48] As discussed in Section 4.2, a nonretarded, electrostatic dipole–dipole interaction, in addition to the contribution from the $\mathbf{A} \cdot \mathbf{p}$ interaction, must be included in the minimal coupling Hamiltonian in order to obtain the total interaction. Part of the $\mathbf{A} \cdot \mathbf{p}$ interaction results in a nonretarded electrostatic interaction that cancels the contribution from the nonretarded electrostatic interaction in the Hamiltonian, resulting in a properly retarded total interaction. This is exactly what Casimir and Polder found.

The electric dipole Hamiltonian we have used, in contrast, does not require the addition of a nonretarded dipole–dipole interaction. In the limit of small atom–plate (or atom–image) separations, we recover the "electrostatic" form (7.3.36) of the interaction as an *approximation* to the energy (7.3.31) that obviously includes effects of retardation. In this small-separation approximation, the field acting on the atom is the near field of its image dipole $\mathbf{p}_i(t) = \hat{\mathbf{p}}_i p_i(t - 2d/c)$:

$$\mathbf{E}(\mathbf{r}, t) = -\frac{1}{4\pi\epsilon_0} \frac{1}{(2d)^3} \left[\hat{\mathbf{p}}_i - 3(\hat{\mathbf{p}}_i \cdot \hat{\mathbf{z}})\hat{\mathbf{z}} \right] p_i(t - 2d/c)$$

$$\cong -\frac{1}{4\pi\epsilon_0} \frac{1}{(2d)^3} \left[\hat{\mathbf{p}}_i - 3(\hat{\mathbf{p}}_i \cdot \hat{\mathbf{z}})\hat{\mathbf{z}} \right] p_i(t) \tag{7.3.42}$$

7.3.5 Cavity Perturbation and Stark Shift in a Vacuum

The atom–wall energy, like the intermolecular van der Waals interaction, is derivable as a consequence of zero-point, quantum fluctuations of the electromagnetic field, as we have shown assuming the validity of the intuitively appealing formula (7.1.7). The fluctuations responsible for the interaction in both cases are characterized by $\langle \Delta \mathbf{E}^2(\mathbf{r}, t) \rangle = \langle \mathbf{E}^2(\mathbf{r}, t) \rangle - \langle \mathbf{E}(\mathbf{r}, t) \rangle^2 = \langle \mathbf{E}^2(\mathbf{r}, t) \rangle \neq 0$, which, of course, reflects the fact that the field has zero-point energy. We now show how (7.1.7) can be derived and related more directly to changes in zero-point field energy.

---

[48] H. B. G. Casimir and D. Polder, Phys. Rev. **73**, 360 (1948).

Consider a single-mode field in a cavity filled with a non-magnetic, non-absorbing medium with permittivity $\epsilon(\omega)$. The electric field operator is given by (3.4.13). If dispersion is negligible, $\mu c^2/2nn_g = \mu_0 c^2/2n^2 = \mu_0 \epsilon_0 c^2/2\epsilon = 1/2\epsilon$, and

$$\mathbf{E}(\mathbf{r}, t) = i\left(\frac{\hbar\omega}{2\epsilon}\right)^{1/2}\left[a(t)\mathbf{A}(\mathbf{r}) - a^\dagger(t)\mathbf{A}^*(\mathbf{r})\right]. \tag{7.3.43}$$

Now suppose a particle with polarizability $\alpha(\omega)$ is placed at point $\mathbf{r}$ in the cavity. The electric dipole interaction of the field with this particle, $-\mathbf{d}(t)\cdot\mathbf{E}(\mathbf{r}, t)$, implies, for the photon annihilation operator $a(t)$, the Heisenberg equation of motion

$$\dot{a} = -i\omega a + \frac{1}{\hbar}\left(\frac{\hbar\omega}{2\epsilon}\right)^{1/2}\mathbf{d}(t)\cdot\mathbf{A}^*(\mathbf{r}). \tag{7.3.44}$$

$\mathbf{A}(\mathbf{r})$ is the mode function for the cavity without the particle; we assume here and below that the perturbation of the cavity due to any added particles is sufficiently weak that $\mathbf{A}(\mathbf{r})$ is approximately unchanged. We now use in (7.3.44) the expression

$$\mathbf{d}(t) = \alpha(\omega)\mathbf{E}(\mathbf{r}, t) = i\alpha(\omega)\left(\frac{\hbar\omega}{2\epsilon}\right)^{1/2}\left[a(t)\mathbf{A}(\mathbf{r}) - a^\dagger(t)\mathbf{A}^*(\mathbf{r})\right] \tag{7.3.45}$$

for the dipole moment operator, and make the approximation of neglecting the counter-rotating term involving $a^\dagger(t)$, since it varies approximately as $\exp(i\omega t)$ whereas $a(t)$ varies approximately as $\exp(-i\omega t)$:

$$\dot{a} \cong -i\omega a + i\alpha(\omega)\frac{\omega}{2\epsilon}|\mathbf{A}(\mathbf{r})|^2 a = -i(\omega + \delta\omega)a, \tag{7.3.46}$$

$$\delta\omega \equiv -\frac{\omega}{2\epsilon}\alpha(\omega)|\mathbf{A}(\mathbf{r})|^2. \tag{7.3.47}$$

The change in the zero-point energy of the field due to the polarizable particle is

$$U = \frac{1}{2}\hbar(\omega + \delta\omega) - \frac{1}{2}\hbar\omega = -\frac{1}{2}\frac{\hbar\omega}{2\epsilon}\alpha(\omega)|\mathbf{A}(\mathbf{r})|^2. \tag{7.3.48}$$

The vacuum expectation value of $\mathbf{E}^2(\mathbf{r}, t)$ that follows from (7.3.43) is

$$\langle\mathbf{E}^2(\mathbf{r}, t)\rangle = \frac{\hbar\omega}{2\epsilon}\langle\text{vac}|a(0)a^\dagger(0)|\text{vac}\rangle|\mathbf{A}(\mathbf{r})|^2 = \frac{\hbar\omega}{2\epsilon}|\mathbf{A}(\mathbf{r})|^2, \tag{7.3.49}$$

and so (7.3.48) is equivalent to

$$U = -\frac{1}{2}\alpha(\omega)\langle\mathbf{E}^2(\mathbf{r}, t)\rangle. \tag{7.3.50}$$

Thus, the Stark-shift energy (7.1.7) for the interaction of an electrically polarizable particle and a vacuum field is the change in the zero-point field energy caused by the particle. We have assumed a single-mode field, but the multimode generalization is straightforward and leads to (7.1.7).

If the single-mode cavity is perturbed by the addition at points $\mathbf{r}_j$ of $\mathcal{N}$ identical particles, (7.3.47) is replaced by

$$\delta\omega = -\frac{\omega}{2\epsilon} \sum_{j-1}^{\mathcal{N}} \alpha_j(\omega)|\mathbf{A}(\mathbf{r}_j)|^2, \tag{7.3.51}$$

and if there is a continuous distribution with number density $N(\mathbf{r})$ of added particles,

$$\delta\omega = -\frac{\omega}{2\epsilon} \int d^3r N(\mathbf{r})\alpha(\omega)|\mathbf{A}(\mathbf{r})|^2. \tag{7.3.52}$$

The permittivity of the medium in this case is approximately

$$\epsilon'(\mathbf{r}, \omega) = \epsilon(\omega) + N(\mathbf{r})\alpha(\omega) \equiv \epsilon + \Delta\epsilon(\mathbf{r}, \omega), \tag{7.3.53}$$

and

$$\delta\omega = -\frac{\omega}{2\epsilon(\omega)} \int d^3r \Delta\epsilon(\mathbf{r}, \omega)|\mathbf{A}(\mathbf{r})|^2. \tag{7.3.54}$$

As a simple application, suppose that a dielectric material with real permittivity $\epsilon_r(\omega) \approx \epsilon_0$ fills an otherwise empty, single-mode cavity. Then $\epsilon(\omega) = \epsilon_0$, $\Delta\epsilon(\mathbf{r}, \omega) = \epsilon_r(\omega) - \epsilon_0$, and the resonance frequency ω of the weakly perturbed cavity is shifted by

$$\delta\omega = -\frac{\omega}{2\epsilon_0}[\epsilon_r(\omega) - \epsilon_0] \int d^3r |\mathbf{A}(\mathbf{r})|^2 = -\frac{\omega}{2}[\epsilon_r(\omega)/\epsilon_0 - 1]. \tag{7.3.55}$$

Expressions such as (7.3.54) can also be obtained by a perturbative analysis of the classical Maxwell equations. More general expressions allowing for a spatially varying ϵ (and μ) have been derived by that approach.[49] Casimir used such a cavity perturbation formula in his derivations of the interaction energies (7.1.35) and (7.3.38) from the perspective of zero-point field energy.[50] Such formulas are used to infer the dielectric and magnetic properties of objects placed in microwave resonators by observing changes in cavity resonance frequencies.

Exercise 7.4: Use the cavity perturbation formula (7.3.47) with $\epsilon = \epsilon_0$ to derive under appropriate approximations the frequency shifts from (5.5.19) and (5.5.20), which were obtained using dressed states.

---

[49] H. A. Bethe and J. Schwinger, Report MIT-RL-D1-117, Massachusetts Institute of Technology, Radiation Laboratory, Cambridge, Massachusetts, 1943; H. B. G. Casimir, Philips Res. Rep. **6**, 162 (1951); D. M. Pozar, *Microwave Engineering*, fourth edition, John Wiley & Sons, New York, 2012, Section 6.7.

[50] H. B. G. Casimir, J. Chim. Phys. **46** 407 (1949).

7.4   Zero-Point Energy and Fluctuations

In Chapter 4, we showed that, depending on how we choose to order commuting atom and field operators, radiative frequency shifts could be ascribed to radiation reaction, to vacuum, zero-point field fluctuations, or to some combination of the two. The vacuum-field fluctuations are reflected in the fact that the vacuum average $\langle \text{vac}|\Delta \mathbf{E}^2(\mathbf{r}, t)|\text{vac}\rangle = \langle \text{vac}|\mathbf{E}^2(\mathbf{r}, t)|\text{vac}\rangle \neq 0$, which, in turn, reflects a non-vanishing zero-point energy of the electromagnetic field.

The cavity perturbation theory given in Section 7.3 shows a direct connection between radiative level shifts and zero-point field energy: the change in the zero-point energy of the field in the presence of a single atom is exactly the vacuum-field Stark shift of the atom's energy level. The generalization of (7.3.48) allowing for all field modes of free space ($\epsilon = \epsilon_0$, $\mathbf{A}(\mathbf{r}) \to \mathbf{A}_{\mathbf{k}\lambda}(\mathbf{r}) = (1/\sqrt{V})e^{i\mathbf{k}\cdot\mathbf{r}}\mathbf{e}_{\mathbf{k}\lambda}$) is

$$U_j = -\frac{\hbar}{4\epsilon_0}\frac{1}{V}\sum_{\mathbf{k}\lambda}\omega_k \alpha_j(\omega_k) \to -\frac{\hbar}{4\epsilon_0}\frac{1}{V}\frac{V}{8\pi^3}\sum_{\lambda=1}^{2}\int d^3k\,\omega_k \alpha_j(\omega_k)$$

$$= -\frac{\hbar}{4\epsilon_0}\frac{2}{8\pi^3}4\pi\int dk\,k^2\omega_k \alpha_j(\omega_k) = -\frac{\hbar}{4\pi^2\epsilon_0 c^3}\int d\omega\,\omega^3 \alpha_j(\omega)$$

$$= -\frac{1}{6\pi^2\epsilon_0 c^3}\sum_m \omega_{mj}|\mathbf{d}_{mj}|^2 \int \frac{d\omega\,\omega^3}{\omega_{mj}^2 - \omega^2} \tag{7.4.1}$$

for an atom in a state for which the polarizability is $\alpha_j(\omega)$.[51] We have used the Kramers–Heisenberg dispersion formula, (2.3.32), for the polarizability $\alpha_j(\omega)$. As in the derivation of the Casimir force between conducting plates, we associate the change in zero-point energy with a change in the energy of the material system—in this case, a shift in the energy level E_j of the atom. And as in Section 4.8.3, we argue that the energy

$$\delta E_{\text{free}} = \frac{1}{6\pi^2\epsilon_0 c^3}\sum_m |\mathbf{d}_{mj}|^2 \omega_{mj} \int d\omega\,\omega \tag{7.4.2}$$

of the free electron, the energy when the electron–nucleus separation $\to \infty$, must be subtracted from (7.4.1) in order to obtain the *observable* shift in the energy level E_j:

$$\delta E_j^{\text{obs}} \equiv U_j - \delta E_{\text{free}} = -\frac{1}{6\pi^2\epsilon_0 c^3}\sum_m |\mathbf{d}_{mj}|^2 \omega_{mj} \int_0^{\Lambda} d\omega \left(\frac{\omega^3}{\omega_{mj}^2 - \omega^2} + \omega\right)$$

$$= \frac{e^2}{6\pi^2\epsilon_0\hbar c}\left(\frac{1}{mc}\right)^2 \sum_m (E_m - E_j)|\mathbf{p}_{mj}|^2 \log\frac{\Lambda}{|\hbar\omega_{mj}|}, \tag{7.4.3}$$

where, again as in Section 4.8.3, we have introduced a high-frequency cutoff Λ. The resulting level shift is seen to be identical to (4.8.31). In other words, we have shown

---

[51] For a treatment of the Lamb shift of an atom in a dielectric medium, see P. W. Milonni, M. Schaden, and L. Spruch, Phys. Rev. A **59**, 4259 (1999).

that the Lamb shift (or, more precisely, Bethe's non-relativistic approximation to the Lamb shift in the hydrogen atom) can be regarded as a consequence of the change in the zero-point energy of the field due to the mere presence of the atom. As noted in Section 5.5, the atom acts in effect as a non-absorbing dielectric "medium." It is effectively non-absorbing because, as discussed in Section 4.5, there is no absorption of energy from the zero-point, vacuum field.

The role of zero-point field *fluctuations* in the Lamb shift was clearly established in a frequently cited argument of Welton, which we now summarize.[52] From $m\ddot{\mathbf{r}} = e\mathbf{E}$, an electron in a zero-point field $\mathbf{E}_{0,\omega}$, oscillating at frequency ω, will have no average displacement because $\langle \mathbf{E}_{0,\omega} \rangle = 0$, whereas

$$\langle \Delta r_\omega^2 \rangle = \frac{e^2}{m^2\omega^4}\langle \mathbf{E}_{0,\omega}^2 \rangle = \frac{e^2}{m^2\omega^4}\langle \text{vac}|\mathbf{E}_{0,\omega}^2|\text{vac}\rangle, \tag{7.4.4}$$

with $\langle \mathbf{E}_{0,\omega}^2 \rangle$ defined such that

$$\frac{1}{2}\epsilon_0\langle \mathbf{E}_{0,\omega}^2 + \mathbf{H}_{0,\omega}^2\rangle d\omega = \epsilon_0\langle \mathbf{E}_{0,\omega}^2\rangle d\omega \tag{7.4.5}$$

is the energy density of the vacuum field in the frequency interval $[\omega, \omega + d\omega]$. This energy density is $\rho_0(\omega)d\omega = (\hbar\omega^3/2\pi^2c^3)d\omega$ (cf. (4.5.17)), and so the mean-square electron displacement is

$$\langle \Delta \mathbf{r}^2 \rangle = \int d\omega \langle \Delta r_\omega^2 \rangle = \int d\omega \frac{e^2}{m^2\omega^4}\frac{\hbar\omega^3}{2\pi\epsilon_0 c^3} = \frac{e^2\hbar}{2\pi^2\epsilon_0 m^2c^3}\int \frac{dE}{E}. \tag{7.4.6}$$

Now, if the electron is in the Coulomb field of the nucleus, its position fluctuations due to the vacuum field will cause a fluctuation in the Coulomb energy:

$$V(\mathbf{r} + \Delta\mathbf{r}) = V(\mathbf{r}) + \Delta\mathbf{r}\cdot\nabla V(\mathbf{r}) + \frac{1}{6}\Delta r^2\nabla^2 V(\mathbf{r}) + \cdots \tag{7.4.7}$$

The fluctuations in $\mathbf{r}$ due to the vacuum field result in a change in the energy levels of the electron. For the state j, the expectation value of this change is

$$\begin{aligned}
\delta E_j &= \langle j; \text{vac}|V(\mathbf{r} + \Delta\mathbf{r}) - V(\mathbf{r})|j; \text{vac}\rangle \\
&= \frac{1}{6}\langle \Delta \mathbf{r}^2 \rangle\langle j|\nabla^2 V|j\rangle = \frac{1}{6}\langle \Delta \mathbf{r}^2 \rangle\int d^3r|\psi_j(\mathbf{r})|^2\nabla^2\left(-\frac{e^2}{4\pi\epsilon_0 r}\right) \\
&= \frac{1}{6}\langle \Delta \mathbf{r}^2 \rangle\int d^3r|\psi_j(\mathbf{r})|^2 e^2\delta^3(\mathbf{r})/\epsilon_0 \\
&= \frac{\hbar e^4}{12\pi^2\epsilon_0^2 m^2c^3}|\psi_j(0)|^2\int \frac{dE}{E}.
\end{aligned} \tag{7.4.8}$$

---

[52] T. A. Welton, Phys. Rev. **74**, 1157 (1948).

If, as in Bethe's computation (see Section 4.8.3), we take the lower limit of integration to be the average $|E_n - E_j|_{\text{avg}}$, and the upper limit to be mc^2, we obtain exactly Bethe's estimate (4.8.36) for the Lamb shift in hydrogen.[53]

Exercise 7.5: Why is mass renormalization unnecessary in this approach to the Lamb shift or in the approach based on the change in the zero-point field energy?

Are the Lamb shift, van der Waals interactions, and Casimir forces incontrovertible evidence for the reality of vacuum-field fluctuations and zero-point field energy? The preceding arguments certainly lend strong support to that viewpoint, whereas the calculations in Chapter 4 suggest that the interpretation of the Lamb shift (and spontaneous emission) based on vacuum-field fluctuations is only one side of the same fluctuation–dissipation coin. To quote Jaynes again, the fact that both source-field and vacuum-fluctuation interpretations are possible "does not show that vacuum fluctuations are 'real.' It shows that source-field effects are the same <u>as if</u> vacuum fluctuations were present."[54]

Like the Lamb shift, van der Waals and Casimir interactions are ascribed to source fields, without explicit reference to vacuum fields or zero-point energy, when they are calculated using normally ordered field operators in the Heisenberg picture.[55] And like spontaneous emission (see Section 4.5), these effects have been attributed to either source fields or vacuum fields. For instance, Feynman wrote in a letter to John Wheeler in 1951 that the Lamb shift "is supposedly due to the self-action of the electron," but ten years later suggested that it could be understood as a change in zero-point field energy.[56] In connection with the cosmological constant problem, Weinberg wrote, "Perhaps surprisingly, it was a long time before particle physicists began seriously to worry about this problem despite the demonstration in the Casimir effect of the reality of zero-point energies."[57]

One argument *against* the reality of zero-point energy is based on the obvious fact that the *classical* Hamiltonian $H = \frac{1}{2}(p^2 + \omega^2 q^2)$ for the harmonic oscillator can be written as

$$H = \frac{1}{2}(p + i\omega q)(p - i\omega q). \tag{7.4.9}$$

---

[53] The neglect of higher derivatives of V is justified by the relatively small variation of $|\psi_j(\mathbf{r})|^2$. The cutoff at the lower limit of integration is justified in part by a more careful calculation in which $\langle \Delta \mathbf{r}^2 \rangle$ is calculated for a bound state rather than for a free electron, as done here.

[54] E. T. Jaynes, "Electrodynamics Today," http://bayes.wustl.edu/etj/articles/electrodynamics. today.pdf, p. 12, accessed September 24, 2018. See also Section 4.6.

[55] *Milonni (1994)* and references therein. See also Section 7.8.

[56] See *Schweber*, p. 462; R. P. Feynman, in *The Quantum Theory of Fields. Proceedings of the 1961 Solvay Conference*, Interscience Publishers, New York, 1962, p. 76. See also E. A. Power, Am. J. Phys. **34**, 516 (1966) or *Milonni (1994)*.

[57] S. Weinberg, Rev. Mod. Phys. **61**, 1 (1989), p. 3.

If we now treat the oscillator quantum mechanically by taking q and p to be operators satisfying $[q, p] = i\hbar$, and define $a = \sqrt{1/2\hbar\omega}(p - i\omega q)$, $a^\dagger = \sqrt{1/2\hbar\omega}(p + i\omega q)$ and $[a, a^\dagger] = 1$, we can write (7.4.9) as

$$H = \hbar\omega a^\dagger a, \qquad (7.4.10)$$

for which, unlike (3.1.9), the lowest energy eigenvalue is 0, not $\frac{1}{2}\hbar\omega$. The Hamiltonian (7.4.10), in which there is no zero-point energy, is just as valid a Hamiltonian for the harmonic oscillator as (3.1.9), suggesting that the zero-point energy of the oscillator is not physically meaningful. Only *differences* in the oscillator energy levels are observable, and these are unaffected by zero-point energy.

An even simpler argument against the physical significance of the zero-point energy in the Hamiltonian (3.1.9) is that it commutes with all operators and therefore does not affect Heisenberg equations of motion for any observables. It can simply be subtracted from the Hamiltonian (3.1.9) without affecting those equations. The same argument has been used to reject the infinite zero-point energy of the electromagnetic field in free space; the physical significance of this infinite field energy was rejected by Pauli and others (see Section 4.6). Regarding divergences, Feynman in the early stages of his work in quantum electrodynamics "believed that the difficulty arose somehow from a combination of the electron acting on itself and the infinite number of degrees of freedom of the field."[58] The latter presented the difficulty that

if you quantized the harmonic oscillators of the field (say in a box), each oscillator has a ground state energy of $\hbar\omega/2$, and there is an infinite number of modes in a box of ever increasing frequency ω, and therefore there is an infinite energy in the box. I now realize that that wasn't a completely correct statement of the central problem; it can be removed simply by changing the zero from which energy is measured.[59]

What cannot be removed by simply changing the zero from which energy is measured are *changes* in zero-point energy. In the case of the Lamb shift, for example, the change due to the atom in the zero-point field energy depends on the state of the atom; in the Casimir force between conducting plates, it is similarly the change in zero-point field energy due to the plates that depends on the plate separation.

A compelling reason to accept the physical significance of the zero-point electromagnetic field is found in the non-relativistic model of a free electron considered in Section 4.6: the source-free, vacuum field provides the Langevin force without which the commutator $[x, p_x] = i\hbar$ cannot be maintained. More generally, the quantum theory of the interaction of light with matter requires the vacuum field for its formal consistency. The vacuum electric and magnetic field operators in the Heisenberg picture are the homogeneous, source-free solutions of Maxwell's equations and, as such, couple to charged particles regardless of whether we include zero-point field energy in

---

[58] R. P. Feynman, "The Development of the Space-Time View of Quantum Electrodynamics," http://www.nobelprize.org/prizes/physics/1965/feynman/lecture, accessed May 11, 2018.
[59] Ibid.

the matter–field Hamiltonian. These vacuum fields, quantized with appropriate mode functions determined by boundary conditions and the distribution of matter in the surrounding space, appear as Langevin force operators in equations of motion for particles interacting with the field. The boundary conditions and the distribution of matter in the space surrounding the particles are what make the zero-point fields that the particles experience different from the vacuum fields in the absence of any matter, and that result in physically significant *changes* in zero-point field energy compared to the (infinite) zero-point energy of free space. The possibility of interpreting observable effects of these changes in terms of source fields arises because the source and source-free fields are (and must be) quantized with the same mode functions.

A simple point about zero-point field fluctuations is worth emphasizing. Dropping the zero-point field energy from the Hamiltonian does not eliminate it or the field fluctuations once and for all! In the free-space vacuum, the Heisenberg equation of motion $\dot{a} = [a, H]/i\hbar = -i\omega a$ that follows from the Hamiltonian (7.4.10) and $[a, a^\dagger] = 1$ for a single mode of the field implies the electric field operator (see (3.3.11))

$$E(\mathbf{r}, t) = i\Big(\frac{\hbar\omega}{2\epsilon_0 V}\Big)^{1/2}\Big[a(0)e^{-i(\omega t - \mathbf{k}\cdot\mathbf{r})} - a^\dagger(0)e^{i(\omega t - \mathbf{k}\cdot\mathbf{r})}\Big], \tag{7.4.11}$$

a vacuum-field variance

$$\langle\text{vac}|\Delta\mathbf{E}^2(\mathbf{r}, t)|\text{vac}\rangle = \Big(\frac{\hbar\omega}{2\epsilon_0 V}\Big)\langle\text{vac}|a(0)a^\dagger(0)|\text{vac}\rangle = \frac{\hbar\omega}{2\epsilon_0 V}, \tag{7.4.12}$$

and a zero-point energy $\epsilon_0 V\langle\text{vac}|\mathbf{E}^2(\mathbf{r}, t)|\text{vac}\rangle = \hbar\omega/2$.

The concept of zero-point energy is, of course, a general and essential feature of quantum theory, tied to the uncertainty relation $\Delta x \Delta p \geq \hbar/2$ (see Section 3.1.1). A particle of mass m, confined to a spatial region $\approx a$, will experience zero-point momentum fluctuations $\Delta p^2 \approx \hbar^2/4a^2$ and a resulting uncertainty $\approx \hbar^2/8ma^2$ in kinetic energy, roughly its lowest possible (zero-point) energy.

As a general feature of quantum theory, zero-point energy, of course, has implications beyond the quantum theory of radiation and the examples we have considered. A frequently cited example is the failure of liquid He^4 to solidify at normal pressures as $T \to 0$: the small mass of the helium atom results in a zero-point energy that is too large to allow solidification via the van der Waals interaction. From x-ray diffraction data, it can be inferred that each atom in liquid He^4 is separated by $r \cong 3.16$ Å from each of its six nearest neighbors. The attractive energy between two He atoms at this separation has been computed to be approximately $C_6/r^6 = 1.4 \times 10^{-22}$ J.[60] In the pairwise approximation, the cohesive energy is, therefore, roughly $6\frac{1}{2}(6.022 \times 10^{23})(1.4 \times 10^{-22}) \approx 250$ J/mole or ~ 60 J/g, three times the experimental value of ~ 20 J/g for the latent heat of liquid helium. However, the zero-point energy of each helium atom confined to the "cage" formed by its nearest

---

[60] Z. C. Yan, J. F. Babb, A. Dalgarno, and G. W. F. Drake, Phys. Rev. A **54**, 2824 (1996).

neighbors acts as an effective repulsive energy, which London estimated to be ~ 35 J/g.[61] This would suggest a latent heat of $\sim 60 - 35 = 25$ J/g, which, if nothing else, indicates that zero-point energy has a significant effect on the latent heat of liquid helium.[62] And so zero-point energy may be said to play two roles in determining the latent heat of liquid helium: the zero-point *field* energy and fluctuations cause the attractive van der Waals forces, while the zero-point oscillations of the helium *atoms* themselves result in a significantly smaller net attraction.

Laser cooling techniques have made it possible to trap ions predominantly in the zero-point, ground state of their confining potential well.[63] Similar "sideband-cooling" techniques have made it possible to observe effects of zero-point motion in mechanical systems.[64]

The concept of zero-point energy first appeared in 1912, in Planck's "second theory" of blackbody radiation; according to that theory, the average energy of an oscillator of frequency ω, in thermal equilibrium at temperature T, was found to be[65]

$$U = \frac{\hbar\omega}{e^{\hbar\omega/k_B T} - 1} + \frac{1}{2}\hbar\omega. \tag{7.4.13}$$

Planck's path to this formula included the assumption that the oscillator absorbs energy from the field continuously, but emits radiation discontinuously in discrete quanta of energy.[66] In 1913, Einstein and Otto Stern invoked the zero-point energy of dipole oscillators in a derivation of the Planck spectrum based on considerations of the kinetic energy of particles as they absorb and emit radiation and experience recoil and a drag force; the Planck law follows from the assumption that, in equilibrium, the average kinetic energy satisfies the equipartition theorem. Assuming a rotational zero-point energy of a diatomic molecule, they also derived a formula for the specific heat of H_2 that agreed remarkably well with experiment, and concluded that zero-point energy is probably real; Einstein, however, soon came to regard zero-point energy as untenable.[67] Around the same time, Debye sought evidence for zero-point motion of the atoms in a crystal from its effect on x-ray diffraction. Such evidence came later in work by James, Waller, and Hartree, who found that theory and data on x-ray diffraction by rock salt "agree very closely . . . on the assumption that the crystal possesses zero-point energy of amount half a quantum per degree of freedom, as proposed by Planck."[68]

[61] F. London, Proc. Roy. Soc. Lond. A**153**, 576 (1936).

[62] Similar rough estimates can be made for surface tension. See, for instance, P. W. Milonni and P. B. Lerner, Phys. Rev. **46**, 1185 (1992).

[63] F. Diedrich, J. C. Berquist, W. M. Itano, and D. J. Wineland, Phys. Rev. Lett. **62**, 403 (1989).

[64] See, for instance, J. D. Teufel, T. Donner, D. Li, J. W. Harlow, M. S. Allman, K. Cicak, A. J. Sirois, J. D. Whitaker, K. W. Lehnert, and R. W. Simmons, Nature **475**, 359 (2011), and A. H. Safavi-Naeini, J. Chan, J. T. Hill, T. P. Mayer Alegre, A. Krause, and O. Painter, Phys. Rev. Lett. **108**, 033602 (2012). See also the review by Yu. M. Tsipenyuk, Sov. Phys. Usp. **55**, 796 (2012).

[65] M. Planck, Ann. Physik. **37**, 642 (1912).

[66] For some early history of zero-point energy, see, for instance, *Milonni (1994)*, Chapter 1, and references therein.

[67] A. Einstein and O. Stern, Ann. Physik. **40**, 551 (1913).

[68] R. W. James, I. Waller, and D. R. Hartree, Proc. R. Soc. Lond. A **118**, 334 (1928), p. 350. The fluctuations in the positions of the atoms diminish the intensity of the Bragg-diffracted field by the Debye–Waller factor. There is ample experimental evidence that this diminution persists as $T \to 0$ and that the atoms in the crystal undergo zero-point motion. See, for instance, C. A. Burns and E. D. Isaacs, Phys. Rev. B **55**, 5767 (1997).

Experimental evidence for zero-point energy was apparently first found by Mulliken. The energy levels of a diatomic molecule in an electronic level E_e were well known to fit the formula $E_e + an - bn^2$, where n is a vibrational quantum number and the constants a and b ($0 < b \ll a$) characterizing the molecule. What Mulliken found was that the different spectra he measured for $B^{10}O^{16}$ and $B^{11}O^{16}$ required, barring seemingly less acceptable possibilities, that n have half-integral values, "in line" with earlier work by Heisenberg, Tolman, and others in which half-integral quantum numbers were used to fit other spectroscopic phenomena. In particular, Mulliken's experiments on the "vibrational isotope effect" implied that $n = 1/2$ for the lowest vibrational level of BO "and doubtless in many or perhaps most molecules," and he noted that the assumption that there is no zero-point energy "definitely does not fit the facts."[69] In 1925 Heisenberg, and then Born and Jordan, obtained the formula $E_n = (n + 1/2)\hbar\omega$, $n = 0, 1, 2, \ldots$, from the new quantum mechanics, but did not comment on the zero-point energy.[70]

Zero-point energy of a quantum *field* appeared a few months later in the paper "On Quantum Mechanics II" by Born, Heisenberg, and Jordan.[71] The concept of a vacuum *state* (what Dirac called the "zero state") was an important part of Dirac's 1927 paper on the quantization of the electromagnetic field and spontaneous emission (see Section 4.4).[72]

7.4.1 Stochastic Electrodynamics

We have emphasized that, in the quantum theory of the electromagnetic field and its interaction with matter, we must include the fluctuating, zero-point field that appears even in the absence of any sources.[73] This "vacuum" field is required for the formal consistency of the theory: without it, canonical commutation rules would be broken, and *operators* would decay to zero in dissipative systems—the whole formal edifice of quantum electrodynamics would collapse! In classical electrodynamics, in contrast, it is generally taken for granted that there are simply no fields in the absence of any sources. This does not lead to any formal inconsistencies within classical theory, but effects such as the Casimir force indicate that it is inconsistent with experiment: it fails to account for effects that can be ascribed to zero-point field energy and fluctuations.

Such phenomena suggest that we might adopt a different "boundary condition" in classical electrodynamics. Instead of taking it for granted that there are no fields when there no sources, we could, based on what we know from quantum electrodynamics, modify the classical theory to include a fluctuating zero-point field with non-vanishing average energy. Provided that this *classical* vacuum field satisfies Maxwell's equations, there is nothing obviously wrong with this alteration. In addition to satisfying Maxwell's equations, a classical zero-point field should be consistent with special relativity, and in particular there should be no net force acting on a particle moving with

[69] R. S. Mulliken, Nature **114**, 349 (1924); Phys. Rev. **25**, 259 (1925), p. 281.

[70] W. Heisenberg, Z. Phys. **33**, 879 (1925); M. Born and P. Jordan, Z. Phys. **34**, 858 (1926).

[71] M. Born, W. Heisenberg, and P. Jordan, Z. Phys. **35**, 557 (1926). This paper has already been cited, in Section 2.8.2, in connection with the Einstein fluctuation formula. The section of this paper on field quantization was largely the work of Jordan; see, for instance, *Schweber*.

[72] Walter Nernst (Verhandlungen DPG **4**, 83 (1916)) was evidently the first to suggest that the electromagnetic field has zero-point energy. For an interesting discussion of Nernst's speculations and other matters, including the skepticism about zero-point energy expressed by Einstein, Pauli, and others, see H. Kragh, Arch. Hist. Exact Sci. **66**, 199 (2012).

[73] Recall especially Section 4.6.

constant velocity in this field. In Section 2.8.1, we showed that a two-state atom with transition frequency ω experiences a drag force

$$F \propto \left[\rho(\omega) - \frac{\omega}{3}\frac{d\rho}{d\omega}\right]v \tag{7.4.14}$$

as it moves with constant velocity v through an isotropic, unpolarized field having a spectral energy density $\rho(\omega)$. The same dependence on $\rho(\omega)$ is found for the drag force on a classical dipole oscillator with resonance frequency ω, as is clear from the analysis in Section 2.8.1. Since the force exerted on a classical dipole oscillator moving with constant velocity must be zero, therefore, (7.4.14) must vanish. Thus, a zero-point field in classical theory must have a spectral energy density $\rho_0(\omega)$ proportional to ω^3, just as in quantum electrodynamics. With the appropriate choice of multiplicative constants, the classical zero-point spectrum can therefore be chosen to have exactly the same form as the zero-point spectrum of quantum electrodynamics,

$$\rho_0(\omega) = \frac{\hbar\omega^3}{2\pi^2 c^3}, \tag{7.4.15}$$

implying a zero-point energy $\frac{1}{2}\hbar\omega$ for each field mode of frequency ω (see (4.5.17)).

The free-space, zero-point electric field $\mathbf{E}_0(\mathbf{r}, t)$ in this classical *stochastic electrodynamics* can be expanded in plane waves:

$$\mathbf{E}_0(\mathbf{r}, t) = i \sum_{\mathbf{k}\lambda} \left(C_{\mathbf{k}\lambda}\mathbf{e}_{\mathbf{k}\lambda}e^{i(\mathbf{k}\cdot\mathbf{r}-\omega_k t)}e^{i\theta_{\mathbf{k}\lambda}} - C^*_{\mathbf{k}\lambda}\mathbf{e}^*_{\mathbf{k}\lambda}e^{-i(\mathbf{k}\cdot\mathbf{r}-\omega_k t)}e^{-i\theta_{\mathbf{k}\lambda}}\right); \tag{7.4.16}$$

the zero-point magnetic induction field $\mathbf{B}_0(\mathbf{r}, t)$ follows from the Maxwell equation $\partial\mathbf{B}_0/\partial t = -\nabla \times \mathbf{E}_0$. The $\mathbf{e}_{\mathbf{k}\lambda}$ are as usual polarization unit vectors, $\mathbf{k} \cdot \mathbf{e}_{\mathbf{k}\lambda} = 0$. The $\theta_{\mathbf{k}\lambda}$ are assumed to be independent random variables, uniformly distributed over $[0, 2\pi]$, such that the phase averages

$$\langle e^{i\theta_{\mathbf{k}\lambda}}\rangle_\theta = 0, \quad \langle e^{i\theta_{\mathbf{k}\lambda}}e^{-i\theta_{\mathbf{k}'\lambda'}}\rangle_\theta = \delta^3_{\mathbf{k}\mathbf{k}'}\delta_{\lambda\lambda'}. \tag{7.4.17}$$

Such stochastic fields were assumed by Planck, Einstein, Hopf, and others to describe *thermal* fields, as we did in Section 3.6.5. The average energy density of the zero-point field is

$$\frac{1}{2}\epsilon_0\langle\mathbf{E}_0^2 + \mathbf{B}_0^2\rangle_\theta = \epsilon_0\langle\mathbf{E}_0^2\rangle_\theta = 2\epsilon_0\sum_{\mathbf{k}\lambda}|C_{\mathbf{k}\lambda}|^2, \tag{7.4.18}$$

which, if we assume a contribution $\frac{1}{2}\hbar\omega_k$ from each mode $(\mathbf{k}, \lambda)$ in order to be consistent with (7.4.15), requires

$$2\epsilon_0\sum_{\mathbf{k}\lambda}|C_{\mathbf{k}\lambda}|^2 = \frac{1}{V}\sum_{\mathbf{k}\lambda}\frac{1}{2}\hbar\omega_k, \tag{7.4.19}$$

or $|C_{\mathbf{k}\lambda}|^2 = \hbar\omega_k/4\epsilon_0 V$, where V is a quantization volume. We choose the $C_{\mathbf{k}\lambda}$ and $\mathbf{e}_{\mathbf{k}\lambda}$ to be real and take the zero-point electric field in free space to be

$$\mathbf{E}_0(\mathbf{r}, t) = i \sum_{\mathbf{k}\lambda} \left(\frac{\hbar\omega_k}{4\epsilon_0 V}\right)^{1/2} \left(e^{i(\mathbf{k}\cdot\mathbf{r} - \omega_k t + \theta_{\mathbf{k}\lambda})} - e^{-i(\mathbf{k}\cdot\mathbf{r} - \omega_k t + \theta_{\mathbf{k}\lambda})}\right) \mathbf{e}_{\mathbf{k}\lambda}. \qquad (7.4.20)$$

This has a form similar to the quantum-electrodynamical expression (3.3.11):

$$\mathbf{E}(\mathbf{r}, t) = i \sum_{\mathbf{k}\lambda} \sqrt{\frac{\hbar\omega_k}{2\epsilon_0 V}} \left[a_{\mathbf{k}\lambda}(0)e^{i(\mathbf{k}\cdot\mathbf{r} - \omega_k t)} - a_{\mathbf{k}\lambda}^\dagger(0)e^{-i(\mathbf{k}\cdot\mathbf{r} - \omega_k t)}\right] \mathbf{e}_{\mathbf{k}\lambda}, \qquad (7.4.21)$$

except, of course, that there are no photon annihilation and creation operators in the classical field (7.4.20). Another difference is that the factor $(\hbar\omega_k/4\epsilon_0 V)$ appears in (7.4.20) as opposed to $(\hbar\omega_k/2\epsilon_0 V)$ in (7.4.21). This derives from the fact that both

$$\left\langle e^{i(\mathbf{k}\cdot\mathbf{r} - \omega_k t + \theta_{\mathbf{k}\lambda})} e^{-i(\mathbf{k}\cdot\mathbf{r} - \omega_k t + \theta_{\mathbf{k}\lambda})} \right\rangle_\theta \quad \text{and} \quad \left\langle e^{-i(\mathbf{k}\cdot\mathbf{r} - \omega_k t + \theta_{\mathbf{k}\lambda})} e^{i(\mathbf{k}\cdot\mathbf{r} - \omega_k t + \theta_{\mathbf{k}\lambda})} \right\rangle_\theta$$

contribute to the energy density in stochastic electrodynamics, whereas in quantum electrodynamics only $\langle \text{vac}|a_{\mathbf{k}\lambda}a_{\mathbf{k}\lambda}^\dagger|\text{vac}\rangle = 1$ contributes, since $\langle \text{vac}|a_{\mathbf{k}\lambda}^\dagger a_{\mathbf{k}\lambda}|\text{vac}\rangle = 0$. In other words, an "extra" factor $1/\sqrt{2}$ appears in stochastic electrodynamics, so that the energy density is the same as in quantum electrodynamics. Of course, these differences appear not only for plane-wave mode functions, but also when the fields of stochastic electrodynamics are expressed in terms of whatever mode functions are appropriate in order to satisfy Maxwell's equations and boundary conditions. These mode functions will be the same in stochastic electrodynamics and quantum electrodynamics.

Stochastic electrodynamics is a classical field theory. Planck's constant appears in the theory as a multiplicative constant (see 7.4.15)) whose numerical value is fixed by comparison with experiment (or, perhaps more accurately, by comparison with quantum electrodynamics); this is not very different from how $\hbar$ was introduced by Planck. Combining stochastic electrodynamics with the quantum theory of atoms and molecules would result in just another version of semiclassical radiation theory that, like any theory based on the interaction of a system described quantum mechanically with one described classically, would be fundamentally inconsistent, regardless of whether the classical system itself were stochastic or deterministic.

On the other hand, stochastic electrodynamics can account, for instance, for the Casimir force between conducting plates. As is clear from the derivation in Section 7.3, all that is really required is that there be a zero-point energy $\frac{1}{2}\hbar\omega$ per mode of frequency ω. It doesn't matter whether we associate this energy with quantum electrodynamics or stochastic electrodynamics. Similarly, we can derive the Casimir–Polder interaction using stochastic electrodynamics—but only up to a point: since it is a purely classical theory, there is no way, as far as we know, to compute the polarizability α of an atom in stochastic electrodynamics. (In quantum electrodynamics, α is given by the Kramers–Heisenberg formula.) Likewise, the van der Waals interaction between two ground-state atoms can be derived in stochastic electrodynamics, once we assume that the induced dipole moments of the atoms are related through a polarizability to the field at the atom, but stochastic electrodynamics cannot provide an expression for the polarizability.

Can stochastic electrodynamics account for any properties of matter that are ordinarily explained only by quantum mechanics? In the simple case of a harmonically bound charge, the charge's position and momentum have the same steady-state probability distributions as x and p for the ground state of the harmonic oscillator in quantum mechanics, independently of the charge.[74] This is not surprising, as the (linear) equation of motion for the oscillator in stochastic electrodynamics is formally the same as the Heisenberg equation of motion of the oscillator in quantum electrodynamics, and the zero-point fields of stochastic electrodynamics and quantum electrodynamics both satisfy Gaussian statistics. More intriguing are numerically intensive computations of electron trajectories in the Coulomb field of a nucleus, including the effect on the electron of the zero-point stochastic-electrodynamical field.[75] These computations indicate that the stochastic-electrodynamical field prevents the electron from spiraling into the nucleus and that the radial position and energy distributions of the electron are, very roughly, in agreement with the quantum-mechanical results for the ground state of the hydrogen atom. However, the electron in the stochastic-electrodynamical field is also found to "self-ionize." In stochastic electrodynamics, "the classical problem of atomic collapse has been replaced by the problem of the zero-point radiation delivering too much energy to the orbiting electron so as to cause self-ionization of the atom."[76]

7.5 The Lifshitz Theory

Lifshitz derived a force per unit area between *dielectric* materials (see Figure 7.3), assuming that the space between the dielectrics is a vacuum ($\epsilon_3 = \epsilon_0$).[77] The more general force per unit area is

$$F(d) = -\frac{\hbar}{2\pi^2} \int_0^\infty dk\, k \int_0^\infty d\xi\, K_3 \left[\left(\frac{\epsilon_3 K_1 + \epsilon_1 K_3}{\epsilon_3 K_1 - \epsilon_1 K_3} \frac{\epsilon_3 K_2 + \epsilon_2 K_3}{\epsilon_3 K_2 - \epsilon_2 K_3} e^{2K_3 d} - 1 \right)^{-1} \right.$$
$$\left. + \left(\frac{K_1 + K_3}{K_1 - K_3} \frac{K_2 + K_3}{K_2 - K_3} e^{2K_3 d} - 1 \right)^{-1} \right]. \tag{7.5.1}$$

Here, d is the distance between two half-spaces with permittivities ϵ_1 and ϵ_2, with ϵ_3 the permittivity of the material separating them; these permittivities are evaluated at imaginary frequences: $\epsilon_j = \epsilon_j(i\xi)$. The K_j's are defined by

$$K_j = \sqrt{k^2 + \epsilon_j(i\xi)\xi^2/c^2\epsilon_0}. \tag{7.5.2}$$

[74] T. W. Marshall, Proc. Roy. Soc. Lond. A **276**, 475 (1963).

[75] D. C. Cole and Y. Zou, Phys. Lett. A **317**, 14 (2003); T. M. Nieuwenhuizen and M. T. P. Liska, Phys. Scr. T **165**, 014006 (2015), and T. M. Nieuwenhuizen and M. T. P. Liska, Found. Phys. **45**, 1190 (2015).

[76] T. H. Boyer, Found. Phys. **46**, 880 (2016), p. 890. The status of these computations is reviewed in this paper.

[77] E. M. Lifshitz, Sov. Phys. JETP **2**, 73 (1956).

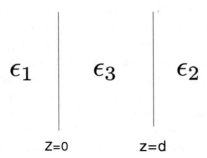

$$\epsilon_1 \qquad\qquad \epsilon_3 \qquad\qquad \epsilon_2$$

$$\text{Z=0} \qquad\qquad \text{z=d}$$

Fig. 7.3 Geometry assumed in the Lifshitz theory: two dielectric half-spaces with permittivities ϵ_1 and ϵ_2, between which is a dielectric of width d and permittivity ϵ_3.

As functions of real frequencies ω, the ϵ_j are complex, so that (7.5.1) allows for absorption. As functions of imaginary frequencies $i\xi$, on the positive imaginary axis, the ϵ_j in (7.5.1) are real, as is clear from (6.6.66). To express (7.5.1) in the more convenient form obtained by Lifshitz, we define p and $s_{1,2}$ by writing

$$k^2 = \frac{\epsilon_3}{\epsilon_0}(p^2-1)\xi^2/c^2,$$

$$K_3 = \sqrt{\frac{\epsilon_3}{\epsilon_0}}p\xi/c,$$

$$K_{1,2}^2 = \frac{\epsilon_3}{\epsilon_0}(p^2-1+\epsilon_{1,2}/\epsilon_3)\xi^2/c^2 \equiv \frac{\epsilon_3}{\epsilon_0}s_{1,2}^2\xi^2/c^2,$$

$$dkk = (\xi^2/c^2)\frac{\epsilon_3}{\epsilon_0}dpp. \tag{7.5.3}$$

In terms of p and $s_{1,2}$,

$$F(d) = -\frac{\hbar}{2\pi^2 c^3}\int_1^\infty dpp^2\int_0^\infty d\xi\xi^3\left(\frac{\epsilon_3}{\epsilon_0}\right)^{3/2}$$
$$\times\left[\left(\frac{\epsilon_3 s_1+\epsilon_1 p}{\epsilon_3 s_1-\epsilon_1 p}\frac{\epsilon_3 s_2+\epsilon_2 p}{\epsilon_3 s_2-\epsilon_2 p}e^{2\xi p\sqrt{\epsilon_3/\epsilon_0}d/c}-1\right)^{-1}\right.$$
$$\left.+\left(\frac{s_1+p}{s_1-p}\frac{s_2+p}{s_2-p}e^{2\xi p\sqrt{\epsilon_3/\epsilon_0}d/c}-1\right)^{-1}\right]. \tag{7.5.4}$$

7.5.1   A One-Dimensional Model

Lifshitz's calculation, which allows for absorption, is complicated. It is based on a fluctuation–dissipation relation for a "random field" (what we refer to in Section 7.6 as a noise polarization **K**) and the Maxwell stress tensor. It has since been derived in different ways, but the physics behind most other approaches follows closely that of the Lifshitz theory. Some aspects of it are discussed below. Here, we present a heuristic derivation for a one-dimensional model in order to highlight *some* of the salient physics.[78] In particular, we show how, as in the case of perfectly conducting plates, the (zero-temperature) force is attributable to zero-point field pressure.[79] We consider two *non-absorbing* dielectric media with (real) refractive indices n_1 and n_2 separated by a vacuum region ($n_3 = 1$) of width d (see Figure 7.4) and restrict ourselves to fields propagating along the z axis.[80] For $z < -a$, and $z > d + a$, we assume there is, again, vacuum.

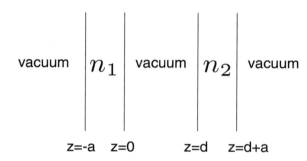

Fig. 7.4 Two dielectric media with refractive indices n_1 and n_2, separated by a distance d by a vacuum.

Consider a field of frequency ω, in the vacuum Region 3 between the two dielectrics, resulting from propagation into that space of vacuum fields from $z = \pm\infty$. The field at $z = 0^+$ and propagating to the left, for instance, has contributions from: (1) the field incident from the vacuum region $z < -a$, propagated through Dielectric 1, reflected off Dielectric 2, and propagated to $z = 0^+$; it has amplitude $t_1 r_2 e^{2i\omega d/c} E_L$, where t_1 is the transmission coefficient for Dielectric 1, r_2 is the reflection coefficient for Dielectric 2, and E_L is the amplitude of the field incident from the left at $z = -a$; and (2) the field incident from the vacuum region $z > d + a$ and propagated to $z = 0^+$; it has amplitude $t_2 e^{i\omega d/c} E_R$, where t_2 is the transmission coefficient for Dielectric 2, and E_R is the amplitude of the field incident at $z = d + a$ from vacuum. The field at $z = 0^+$ from these two contributions therefore has amplitude $t_1 r_2 e^{2i\omega d/c} E_L + t_2 e^{i\omega d/c} E_R$. It reflects off the interface at $z = 0$, picks up a phase factor $\exp(i\omega d/c)$ as it propagates to $z = d$, reflects there with reflection coefficient r_2, and picks up another phase factor

[78] For a more formal and detailed analysis of this model, see D. Kupiszewska and J. Mostowski, Phys. Rev. A **41**, 4636 (1990), and D. Kupiszewska, Phys. Rev. A **46**, 2286 (1992).

[79] L. S. Brown and G. J. Maclay, Phys. Rev. **184**, 1272 (1969); P. W. Milonni, R. J. Cook, and M. E. Goggin, Phys. Rev. A **38**, 1621 (1988).

[80] Of course, the assumption that there is no absorption at any frequency cannot be valid for real dielectric media. But the force calculated under this assumption turns out to be valid (see Section 7.6).

$\exp(iwd/c)$ as it propagates back to $z = 0^+$. After this round-trip propagation, the field incident from the right at $z = 0^+$ is multiplied by $r_1 r_2 e^{2iwd/c}$ and so forth for further round-trip propagations. Since

$$\left(1 + r_1 r_2 e^{2iwd/c} + r_1^2 r_2^2 e^{4iwd/c} + \ldots\right) = \frac{1}{1 - r_1 r_2 e^{2iwd/c}}, \tag{7.5.5}$$

the amplitude of the field propagating to the left in Region 3 is

$$E_\leftarrow = \frac{t_1 r_2 e^{2iwd/c} E_L + t_2 E_R}{1 - r_1 r_2 e^{2iwd/c}}. \tag{7.5.6}$$

The field propagating to the right in Region 3 is derived similarly:

$$E_\rightarrow = \frac{t_2 r_1 e^{2iwd/c} E_R + t_1 E_L}{1 - r_1 r_2 e^{2iwd/c}}. \tag{7.5.7}$$

The corresponding intensities are

$$I_\leftarrow = \frac{|t_1 r_2|^2 + |t_2|^2}{|1 - r_1 r_2 e^{2iwd/c}|^2} I_0 \tag{7.5.8}$$

and

$$I_\rightarrow = \frac{|t_2 r_1|^2 + |t_1|^2}{|1 - r_1 r_2 e^{2iwd/c}|^2} I_0, \tag{7.5.9}$$

assuming the fields E_L and E_R do not interfere and have the same intensity I_0. Since there are no absorption losses, $|t_j|^2 = 1 - |r_j|^2$, $j = 1, 2$, and

$$I_\leftarrow + I_\rightarrow = \frac{2(1 - |r_1|^2|r_2|^2}{|1 - r_1 r_2 e^{2iwd/c}|^2} I_0. \tag{7.5.10}$$

Zero-Point Radiation Pressure. Now consider the force between Dielectrics 1 and 2. We suppose the vacuum fields have energy $\frac{1}{2}\hbar\omega$ and momentum $\frac{1}{2}\hbar\omega/c$. The force per unit area on Dielectric 1 (c times momentum density) due to the $\leftarrow$ and $\rightarrow$ fields in Region 3 is then

$$\mathbf{F}_1/L^2 = \frac{c}{L^3}\frac{\hbar\omega}{2c}\left(-\hat{\mathbf{z}}I_\leftarrow - \hat{\mathbf{z}}I_\rightarrow\right)\frac{1}{I_0} = -\frac{c}{L^3}\frac{\hbar\omega}{2c} \times \frac{2(1 - |r_1|^2|r_2|^2}{|1 - r_1 r_2 e^{2iwd/c}|^2}\hat{\mathbf{z}}, \tag{7.5.11}$$

where $V = L^3$ is a quantization volume, assuming, as usual, periodic boundary conditions for normalization of the field modes, and $\hat{\mathbf{z}}$ is the unit vector pointing in the positive z direction. This is the pressure exerted on Dielectric 1 due to the vacuum field in the space *between* the two media. The minus sign indicates that the force $\mathbf{F}_1$ acts to push it away from Dielectric 2.

There is also a force on Dielectric 1 due to fields *outside* Regions 1, 2, and 3. This comes from the vacuum field incident from $z < -a$ being reflected by both dielectrics, and the field incident from $z > d + a$ being transmitted by them. Denote the reflection and transmission coefficients by r and t (where $|r|^2 + |t|^2 = 1$). The force on Dielectric 1 due to this reflection and transmission is in the positive z direction:

$$\mathbf{F}_2/L^2 = \frac{c}{L^3}\frac{\hbar\omega}{2c}\left(\hat{\mathbf{z}} + |r|^2\hat{\mathbf{z}} - |t|^2(-\hat{\mathbf{z}})\right) = 2 \times \frac{c}{L^3}\frac{\hbar\omega}{2c}\hat{\mathbf{z}}. \tag{7.5.12}$$

The net force per unit area on Dielectric 1 is therefore

$$\mathbf{F}/L^2 = \mathbf{F}_1/L^2 + \mathbf{F}_2/L^2 = \frac{c}{L^3}\frac{\hbar\omega}{2c} \times 2\left(1 - \frac{1 - |r_1|^2|r_2|^2}{|1 - r_1r_2e^{2i\omega d/c}|^2}\right)\hat{\mathbf{z}}. \tag{7.5.13}$$

The *total* pressure on Dielectric 1 in this one-dimensional model is obtained by adding the contributions from all modes:

$$\sum(\ldots) \to 2 \times \frac{L}{2\pi}\int_0^\infty dk(\ldots) = 2 \times \frac{L}{2\pi c}\int_0^\infty d\omega(\ldots). \tag{7.5.14}$$

The factor 2 accounts for two independent polarizations for each frequency ω.[81] Then, the total force acting to push Dielectric 1 in the z direction, toward Dielectric 2, is

$$\begin{aligned}
F &= \frac{\hbar}{\pi c}\int_0^\infty d\omega\omega\left(1 - \frac{1 - |r_1|^2|r_2|^2}{|1 - r_1r_2e^{2i\omega d/c}|^2}\right) \\
&= -\frac{2\hbar}{\pi c}\mathrm{Re}\int_0^\infty d\omega\omega\frac{r_1r_2e^{2i\omega d/c}}{1 - r_1r_2e^{2i\omega d/c}},
\end{aligned} \tag{7.5.15}$$

and, of course, the same force acts to push Dielectric 2 toward Dielectric 1. The force is the difference between two infinite forces, but it is finite, since $r_1, r_2 \to 0$ as $\omega \to \infty$.

The reflection coefficient r_1 for Dielectric 1 is given by the well-known "Fabry–Pérot" formula (see Exercise 7.6):

$$r_1 = -\frac{1 - e^{2in_1\omega a/c}}{1 - r_{31}^2 e^{2in_1\omega a/c}}r_{31}, \tag{7.5.16}$$

where

$$r_{31} = -\frac{n_1 - 1}{n_1 + 1} \tag{7.5.17}$$

is the reflection coefficient for normal incidence of a field from vacuum onto a medium with refractive index n_1. To apply our model to the case of two dielectric half-spaces ($a \to \infty$), we replace $\exp(2in_1\omega a/c)$ by its average (zero) and, therefore, r_1 by $-r_{31}$. Similarly, we replace r_2 by $-r_{32} = (n_2 - 1)/(n_2 + 1)$ and, therefore, (7.5.15) by

$$F = -\frac{2\hbar}{\pi c}\mathrm{Re}\int_0^\infty d\omega\omega\frac{r_{31}r_{32}e^{2i\omega d/c}}{1 - r_{31}r_{32}e^{2i\omega d/c}}. \tag{7.5.18}$$

---

[81] The two polarizations have the same reflection and transmission coefficients for the normal incidence in the one-dimensional model. Note also that the mode summation generally involves all positive and all negative values of k, but we need only sum over k's of one sign because we are working specifically with traveling-wave fields.

We once again make a Wick rotation (cf. (7.1.23)) to replace the integral along the positive real axis by an integral along the positive imaginary axis:

$$F_{1D}(d) = -\frac{2\hbar}{\pi c} \int_0^\infty d\xi \xi \frac{r_{31} r_{32} e^{-2\xi d/c}}{1 - r_{31} r_{32} e^{-2\xi d/c}}. \tag{7.5.19}$$

We have now included a minus sign to indicate that the force between the dielectrics is attractive. r_{31} and r_{32} are, of course, to be evaluated at $\omega = i\xi$, and are real since, as noted above, the $\epsilon_j(i\xi)$ are real. The transformation from expression (7.5.18) to (7.5.19) is useful because (7.5.18) involves rapid oscillations of the integrand; this is why Lifshitz performed a similar transformation to write the force in the form (7.5.4).

Exercise 7.6: (a) Derive the reflection coefficient (7.5.16). (b) What is the justification for assuming there is no interference of the fields E_R and E_L? (c) Why was it not necessary in the derivation of the force (7.5.19) to distinguish between the Abraham and Minkowski forms of the photon momentum (Section 1.7)? (d) Show that the divergent

$$\frac{\hbar}{\pi c} \int_0^\infty d\omega \omega$$

appearing as part of (7.5.15) is the total energy density (per unit length) of the free-space vacuum field in the one-dimensional model.

Now compare $F_{1D}(d)$ with the one-dimensional version of the Lifshitz force (7.5.1). The integration over k in the Lifshitz formula for the force per unit area comes from a summation over wave vectors parallel to the two interfaces:

$$\left(\frac{1}{2\pi}\right)^2 \int d^2k(\ldots) = \left(\frac{1}{2\pi}\right)^2 2\pi \int dk k(\ldots). \tag{7.5.20}$$

No such summation appears in the one-dimensional model, so to obtain $F_{1D}(d)$ from the Lifshitz theory, we just ignore (7.5.20), in effect replacing it by unity. In our one-dimensional model, furthermore, $\epsilon_3 = \epsilon_0$, $K_1 = n_1 \xi/c$, $K_2 = n_2 \xi/c$, and $K_3 = \xi/c$. With these substitutions, (7.5.1) reduces to the force obtained in the one-dimensional model:

$$F(d) = -\frac{\hbar c}{\pi} \int_0^\infty d\xi \xi \left[2 \times \left(\frac{n_1 + 1}{n_1 - 1} \frac{n_2 + 1}{n_2 - 1} e^{2\xi d/c} - 1 \right)^{-1} \right]$$

$$= -\frac{2\hbar}{\pi c} \int_0^\infty d\xi \xi \frac{r_{31} r_{32} e^{-2\xi d/c}}{1 - r_{31} r_{32} e^{-2\xi d/c}} \equiv F_{1D}(d). \tag{7.5.21}$$

The Lifshitz force (7.5.1) is not written explicitly in terms of reflection coefficients at the interfaces, but it can be done so using the definitions (7.5.2) of the K_j's. In terms of real frequencies ω, $K_j = i\sqrt{n_j^2 \omega^2/c^2 - k^2} = i k_\perp^{(j)}$, or

$$K_j = ik_\perp^{(j)} = i\left(n_j\frac{\omega}{c}\right)\cos\theta_j, \tag{7.5.22}$$

where $k_\perp^{(j)}$ is the component of a wave vector in Region j that is perpendicular to the two interfaces, and θ_j is the angle of incidence defined with respect to propagation in a direction into Region 3. Thus, for the interface between Regions 3 and j ($j = 1, 2$), $K_j = in_j\cos\theta_t$ and $K_3 = i\cos\theta_i$, where θ_i and θ_t are, respectively, the angles of incidence and transmission for plane waves of the same frequency and polarization. Therefore, in the Lifshitz formula (7.5.1),

$$\frac{K_3 - K_j}{K_3 + K_j} = \frac{\cos\theta_i - n_j\cos\theta_t}{\cos\theta_i - n_j\cos\theta_t} = r_s^{(1)}, \tag{7.5.23}$$

and

$$\frac{\epsilon_j K_3 - K_j}{\epsilon_j K_3 + K_j} = \frac{n_j\cos\theta_i - \cos\theta_t}{n_j\cos\theta_i - \cos\theta_t} = r_p^{(1)}, \tag{7.5.24}$$

which are, respectively, the Fresnel reflection coefficients for polarization perpendicular (s) and parallel (p) to the plane of incidence. Thus, the Lifshitz force per unit area (7.5.1) when Region 3 is a vacuum ($n_3 = 1, K_3 = \xi/c$) can be written as

$$F(d) = -\frac{\hbar}{2\pi^2 c}\int_0^\infty dk\,k \int_0^\infty d\xi\,\xi\left(\frac{r_p^{(1)}r_p^{(2)}e^{-2\xi d/c}}{1 - r_p^{(1)}r_p^{(2)}e^{-2\xi d/c}} + \frac{r_s^{(1)}r_s^{(2)}e^{-2\xi d/c}}{1 - r_s^{(1)}r_s^{(2)}e^{-2\xi d/c}}\right). \tag{7.5.25}$$

Zero-Point Energy. The zero-point electromagnetic energy for the configuration of Figure 7.3 is found to be

$$U(d) = \frac{\hbar}{4\pi^2}\int_0^\infty dk\,k \int_0^\infty d\xi\left[\log\left(\frac{e^{2K_3 d}}{r_p^{(1)}r_p^{(2)}} - 1\right) + \log\left(\frac{e^{2K_3 d}}{r_s^{(1)}r_s^{(2)}} - 1\right)\right] \tag{7.5.26}$$

per unit interface area. The easiest way to arrive at this result is to write the general form of the electric and magnetic fields that follow from the geometry of Figure 7.3 and to impose the required boundary conditions on these fields. This restricts the physically allowed frequencies ω of the fields, and the total zero-point energy is then the sum $\sum \frac{1}{2}\hbar\omega$ over all these allowed frequencies, which may be evaluated using the argument theorem stated in Exercise 7.3.[82]

As in Casimir's calculation for conducting plates, the zero-point energy of physical interest is not $U(d)$ as defined by (7.5.26), but rather the difference $E(d)$ between this energy and the zero-point energy $U_0(d)$ when the two dielectrics are very far apart. From (7.5.26),

$$U_0(d) = \frac{\hbar}{4\pi^2}\int_0^\infty dk\,k \int_0^\infty d\xi\left[\log\left(\frac{e^{2K_3 d}}{r_p^{(1)}r_p^{(2)}}\right) + \log\left(\frac{e^{2K_3 d}}{r_s^{(1)}r_s^{(2)}}\right)\right], \tag{7.5.27}$$

---

[82] This approach was evidently first taken by N. G. van Kampen, B. R. A. Nijboer, and K. Schram, Phys. Lett. **26A**, 307 (1968), for the case in which d is sufficiently small that retardation is negligible.

$$E(d) = U(d) - U_0(d)$$

$$= \frac{\hbar}{4\pi^2} \int_0^\infty dk\, k \int_0^\infty d\xi \Big[\log \Big(1 - r_p^{(1)} r_p^{(2)} e^{-2K_3 d} \Big)$$

$$+ \log \Big(1 - r_s^{(1)} r_s^{(2)} e^{-2K_3 d} \Big) \Big], \qquad (7.5.28)$$

and the Lifshitz force (7.5.1) is $F(d) = -\partial E/\partial d$.[83]

Exercise 7.7: What is the zero-point energy corresponding to $U_0(d)$ in the one-dimensional model? Does it affect our calculation of the force (7.5.19)?

The derivation of the energy (7.5.26) using the argument theorem is not entirely straightforward, as there are branch points associated with the square roots in (7.5.2). However, when the calculation is done with due attention to that "unpleasant" feature, one obtains exactly (7.5.26).[84] As in our calculation for the one-dimensional model, the derivation based on the argument theorem also assumes that the permittivities are real or, in other words, that there is no absorption. But, again, the formulas obtained for the energy and the force turn out to be valid when there is absorption. Zero-point energy in the case of absorbing media is discussed in Section 7.7.

The formula for the Lifshitz energy in terms of real frequencies ω, before the Wick rotation, involves an integration over ω from 0 to ∞ (cf. (7.5.15)). If, for example, $\epsilon(\omega)$ is assumed to have the form (1.6.4), it is negative for frequencies

$$\omega < \sqrt{\omega_p^2 + \omega_0^2} \qquad (7.5.29)$$

and positive otherwise. Thus, the Lifshitz energy has contributions from "photonic" ($\epsilon(\omega) > 0$) as well as "surface" and "waveguide" ($\epsilon(\omega) < 0$) modes.[85]

Comparison of the Lifshitz theory to measured forces requires permittivities (and more generally permeabilities) or reflectivities at all frequencies, or at least over a broad range of frequencies.[86] Careful measurements of Casimir forces in more recent years have revealed some disagreements between theory and experiment with regard to finite temperatures and conductivities.[87] These appear to stem from uncertainties in the characterization of real materials.

---

[83] See also, for instance, *Milton*.

[84] Yu. S. Barash and V. L. Ginzburg, Sov. Phys. Usp. **18**, 306 (1975); V. V. Nesterenko and I. G. Pirozhenko, Phys. Rev. A **86**, 052503 (2012).

[85] For an analysis of the different field modes and their contributions to the energy for the geometry of Figure 7.3, see M. Bordag, Phys. Rev. D **85**, 025005 (2012). See also G. Barton, Rep. Prog. Phys. **42**, 65 (1979), for a detailed analysis of "surface" and "bulk" modes in the hydrodynamic model for a metal.

[86] See E. S. Sabisky and C. H. Anderson, Phys. Rev. A **7**, 790 (1973), for an early example of a numerical evaluation of the Lifshitz force and its agreement with experiment.

[87] See Footnote 39.

More general expressions can be derived for the Lifshitz force and energy. For anisotropic and generally magnetodielectric media, for example, the physically relevant zero-point energy (7.5.28) is replaced by

$$E(d) = \frac{\hbar}{2\pi} \sum_{\text{pol}} \int \frac{d^2k}{(2\pi)^2} \int_0^\infty d\xi K_3 \text{Tr} \log \left(1 - R_1 \cdot R_2 e^{2K_3 d} \right), \tag{7.5.30}$$

where $\sum_{\text{pol}}$ denotes a sum over polarizations and the 2×2 matrices

$$R_j \equiv \begin{bmatrix} r_j^{ss}(i\xi, k) & r_j^{sp}(i\xi, k) \\ r_j^{ps}(i\xi, k) & r_j^{pp}(i\xi, k) \end{bmatrix}, \tag{7.5.31}$$

with r_j^{ab} the ratio between a reflected field with b-polarization and an incoming field with a-polarization.[88] The indices s, p again denote perpendicular and parallel polarizations, respectively, with respect to the plane of incidence. For isotropic dielectric media, the off-diagonal elements vanish, and the diagonal elements are the usual Fresnel coefficients. Then, (7.5.30) reduces to the Lifshitz formula (7.5.1).

The trace of the logarithm of a square matrix $1 + X$ such as appears in (7.5.30) can be expressed in terms of the determinant (det) of the matrix:

$$\text{Tr} \log(1 + X) = \log \det(1 + X). \tag{7.5.32}$$

This identity appears frequently in the theory of van der Waals interactions and more generally in many-body theory. A proof for a diagonalizable matrix Y with eigenvalues λ_j goes as follows.[89] Let U be the matrix that diagonalizes Y:

$$Y = U^{-1}\Lambda U, \tag{7.5.33}$$

where Λ is the (diagonal) matrix with elements $\Lambda_{ij} = \lambda_j \delta_{ij}$. Then, $Y^n = U^{-1}\Lambda^n U$, $n = 1, 2, \ldots$, and

$$f(Y) = U^{-1} f(\Lambda) U \tag{7.5.34}$$

for any function f of Y that can be expressed as a power series. $f(\Lambda)$ is diagonal, $\det[U^{-1}] = 1/\det[U]$, and therefore, since the determinant of a product of matrices is equal to the product of the determinants,

$$\det[f(Y)] = \det[f(\Lambda)] = \prod_j f(\lambda_j). \tag{7.5.35}$$

Now, let $f(Y) = \exp(Y)$:

$$\det(e^Y) = \prod_j e^{\lambda_j},$$

$$\log \det(e^Y) = \log \prod_j e^{\lambda_j} = \sum_j \lambda_j = \text{Tr}(Y). \tag{7.5.36}$$

Equation (7.5.32) then follows by putting $Y = \log(1 + X)$.

---

[88] See, for instance, A. Lambrecht, P. A. M. Neto, and S. Reynaud, New J. Phys. **8**, 243 (2006).
[89] For a more general proof, see J. Schwinger, Phys. Rev. **93**, 615 (1954).

As will be discussed in Section 7.8, Casimir and Lifshitz forces can, in fact, be derived by different approaches in which there is no explicit reference to zero-point field energy or pressure. But first, we derive some of the simpler results that follow from the Lifshitz theory.

7.5.2 Casimir Force between Perfectly Conducting Plates

For two perfect conductors separated by a vacuum, we take $\epsilon_{1,2} \to \infty$, and $\epsilon_3 = \epsilon_0$. The Lifshitz force (7.5.4) then reduces to the Casimir force (7.3.12):

$$
\begin{aligned}
F(d) &= -\frac{\hbar}{2\pi^2 c^3} \int_1^\infty dp p^2 \int_0^\infty d\xi \xi^3 \left[2 \times \left(e^{2\xi pd/c} - 1 \right)^{-1} \right] \\
&= -\frac{\hbar c}{16\pi^2 d^4} \int_1^\infty dp p^{-2} \int_0^\infty \frac{dx x^3}{e^x - 1} \\
&= -\frac{\pi^2 \hbar c}{240 d^4}.
\end{aligned}
\tag{7.5.37}
$$

In the one-dimensional model, the force in the case of perfect conductors ($r_{31}, r_{32} \to -1$) is

$$
F(d) = -\frac{\pi \hbar c}{12 d^2}.
\tag{7.5.38}
$$

7.5.3 Rarefied Dielectrics and van der Waals Interactions

Consider two identical dielectric half-spaces separated by a vacuum: $\epsilon_3 = \epsilon_0$, $\epsilon_1 = \epsilon_2 = n^2 \epsilon_0$. The main contribution to (7.5.4) comes from values of p such that $p\xi d/c \approx 1$, and it follows from the form of the integrand that the main contribution for d small compared with absorption wavelengths $2\pi c/\xi_0$ of the media comes from large values of p ($p \gg c/d\xi_0$), for which $s_1 = s_2 \cong p$ and

$$
\begin{aligned}
F(d) &\cong -\frac{\hbar}{2\pi^2 c^3} \int_0^\infty dp p^2 \int_0^\infty d\xi \xi^3 \left[\left(\frac{n^2 + 1}{n^2 - 1} \right)^2 e^{2\xi pd/c} - 1 \right]^{-1} \\
&= -\frac{\hbar}{16\pi^2 c^3} \int_0^\infty dx x^2 \int_0^\infty d\xi \left[\left(\frac{n^2 + 1}{n^2 - 1} \right)^2 e^x - 1 \right]^{-1}.
\end{aligned}
\tag{7.5.39}
$$

For rarefied media in which there are N atoms per unit volume, $n^2(i\xi) + 1 \cong 2$, $n^2(i\xi) - 1 = N\alpha(i\xi)/\epsilon_0 \ll 1$, and

$$
\begin{aligned}
F(d) &\cong -\frac{N^2 \hbar}{64\pi^2 \epsilon_0^2 d^3} \int_0^\infty dx x^2 e^{-x} \int_0^\infty d\xi \alpha^2(i\xi) \\
&= -\left(\frac{1}{4\pi\epsilon_0} \right)^2 \frac{N^2 \hbar}{2 d^3} \int_0^\infty d\xi \alpha^2(i\xi).
\end{aligned}
\tag{7.5.40}
$$

Equating this to (7.3.19) for the force obtained by pairwise addition of interactions implies $P = 6$ in that formula, and $\pi N^2 B/6 = (1/4\pi\epsilon_0)^2 N^2 \hbar/2$, and therefore a pairwise interaction energy

$$V(r) = -\frac{B}{r^6} = -\left(\frac{1}{4\pi\epsilon_0}\right)^2 \frac{3\hbar}{\pi r^6} \int_0^\infty d\xi \alpha^2(i\xi), \tag{7.5.41}$$

exactly the London dispersion energy (7.1.31) between two identical electrically polarizable particles.

For *d large* compared with any significant absorption wavelengths, it is convenient to first write (7.5.4) for $\epsilon_3 = \epsilon_0$, and $\epsilon_1 = \epsilon_2 \equiv \epsilon$, as

$$F(d) = -\frac{\hbar}{2\pi^2 c^3}\left(\frac{c}{2d}\right)^4 \int_1^\infty dp\, p^2 \int_0^\infty dx \frac{x^3}{p^4}\left\{\left[\left(\frac{s + \epsilon p/\epsilon_0}{s - \epsilon p/\epsilon_0}\right)^2 e^x - 1\right]^{-1} \right.$$
$$\left. + \left[\left(\frac{s + p}{s - p}\right)^2 e^x - 1\right]^{-1}\right\}, \tag{7.5.42}$$

$$\epsilon = \epsilon(icx/2pd), \quad s = \sqrt{p^2 - 1 + \epsilon/\epsilon_0}, \tag{7.5.43}$$

after a change of integration variable to $x = (2pd/c)\xi$. The main contribution here comes from values of $x \approx 1$, and so, for *d* large compared with any significant absorption wavelength c/ξ_0, we can replace ϵ by the electrostatic (zero-frequency) permittivity $\epsilon_e = \epsilon(0)$:

$$F(d) \cong -\frac{\hbar}{2\pi^2 c^3}\left(\frac{c}{2d}\right)^4 \int_1^\infty dp\, p^2 \int_0^\infty dx \frac{x^3}{p^4}\left\{\left[\left(\frac{s_e + \epsilon_e p/\epsilon_0}{s_e - \epsilon_e p/\epsilon_0}\right)^2 e^x - 1\right]^{-1}\right.$$
$$\left. + \left[\left(\frac{s_e + p}{s_e - p}\right)^2 e^x - 1\right]^{-1}\right\}, \tag{7.5.44}$$

$s_e \equiv \sqrt{p^2 - 1 + \epsilon_e/\epsilon_0}$. Then, for rarefied media ($\epsilon_e \cong \epsilon_0$, $s_e \cong p + (\epsilon_e/\epsilon_0 - 1)/2p$),

$$F(d) \cong -\frac{\hbar c}{32\pi^2 d^4} \int_1^\infty dp\, p^{-2}\left[\left(\frac{s_e - \epsilon_e p/\epsilon_0}{s_e + \epsilon_e p/\epsilon_0}\right)^2 + \left(\frac{s_e - p}{s_e + p}\right)^2\right] \int_0^\infty dx\, x^3 e^{-x}$$
$$\cong -\frac{12\hbar c(\epsilon_e/\epsilon_0 - 1)^2}{(16)(32\pi^2 d^4)} \int_1^\infty dp\, p^{-6}(1 - 2p^2 + 2p^4)$$
$$= -\frac{23\hbar c}{640\pi^2 d^4}(\epsilon_e/\epsilon_0 - 1)^2 = \frac{23\hbar c}{640\pi^2 d^4} N^2/\epsilon_0^2. \tag{7.5.45}$$

Comparison now with (7.3.19) for the force obtained by pairwise addition of particle interactions implies $P = 7$, $2\pi N^2 B/20 = (2\hbar c/640\pi^2 d^4)N^2\alpha^2/\epsilon_0^2$, and therefore a pairwise interaction energy

$$V(r) = -\frac{B}{r^7} = -\left(\frac{1}{4\pi\epsilon_0}\right)^2 \frac{23\hbar c\alpha^2}{4\pi r^7}, \tag{7.5.46}$$

which is exactly the retarded van der Waals interaction (7.1.35).

The forces given by the Lifshitz formula in the limits of large and small separations therefore imply the interatomic retarded and nonretarded van der Waals interactions, respectively, when we assume particle densities sufficiently small that the forces can be derived by simple pairwise additions of the interaction energies. This is, in one sense, remarkable when we consider that the Lifshitz force can be calculated, as we have shown with a simple one-dimensional model, essentially just from boundary conditions on classical electromagnetic fields, together with the assumption that these fields have zero-point energy and momentum. Since the solution for the electric field in continuous media includes effects of multiple scatterings among the electrically polarizable particles (see Section 1.5.1), the Lifshitz theory includes all effects of non-additive, multiparticle van der Waals interactions.

7.5.4 Finite Temperature

We have considered the Lifshitz force in the limit of zero temperature. The simplest way to allow for a finite equilibrium temperature T is to replace zero-point field energies $\frac{1}{2}\hbar\omega$ by (see (7.4.13))

$$\frac{1}{2}\hbar\omega + \frac{\hbar\omega}{e^{\hbar\omega/k_B T} - 1} = \frac{1}{2}\hbar\omega \coth(\hbar\omega/2k_B T). \tag{7.5.47}$$

In applying Cauchy's theorem to transform the integral over real frequencies in the Lifshitz theory to one over imaginary frequencies (cf. (7.5.19)), one must account for the poles of $\coth(\hbar\omega/2k_B T)$ at the *Matsubara frequencies*

$$\omega_n = (2\pi i k_B T)/\hbar)n \equiv i\xi_n \tag{7.5.48}$$

for integers n.[90] One obtains, for the case $\epsilon_3 = \epsilon_0$ treated by Lifshitz,

$$F(d) = -\frac{\hbar}{2\pi^2 c^3}\left(\frac{2\pi k_B T}{\hbar}\right)\sum_{n=0}^{\infty}{}' \xi_n^3 \int_1^{\infty} dp\, p^2$$
$$\times \left[\left(\frac{\epsilon_0 s_{1n} + \epsilon_{1n}p}{\epsilon_0 s_{1n} - \epsilon_{1n}p}\frac{\epsilon_0 s_{2n} + \epsilon_{2n}p}{\epsilon_0 s_{2n} - \epsilon_{2n}p}e^{2p\xi_n d/c} - 1\right)^{-1}\right.$$
$$\left.+ \frac{s_{1n} + p}{s_{1n} - p}\frac{s_{2n} + p}{s_{2n} - p}e^{2\xi_n pd/c} - 1\right)^{-1}\right], \tag{7.5.49}$$

$$s_{jn} = \sqrt{p^2 - 1 + \epsilon_{jn}/\epsilon_0}, \quad \epsilon_{jn} = \epsilon_j(i\xi_n), \quad j = 1, 2.[91] \tag{7.5.50}$$

The prime indicates that a factor $1/2$ should be included in the $n = 0$ term.

For perfect conductors ($\epsilon_1, \epsilon_2 \to \infty$),

$$F(d) = -\frac{\hbar}{2\pi^2 c^3}\left(\frac{2\pi k_B T}{\hbar}\right)\sum_{n=0}^{\infty}{}' 2\xi_n^3 \int_1^{\infty} \frac{dp\, p^2}{e^{2p\xi_n d/c} - 1}$$
$$= -\frac{2k_B T}{\pi c^3}\left(\frac{c}{2d}\right)^2 \sum_{n=0}^{\infty}{}' \int_{nx}^{\infty} \frac{dy\, y^2}{e^y - 1}, \quad x \equiv 4\pi k_B T d/\hbar c. \tag{7.5.51}$$

---

[90] See, for instance, *Bordag et al.*
[91] E. M. Lifshitz, Sov. Phys. JETP **2**, 73 (1956).

For high temperatures, such that $x \gg 1$, the $n = 0$ term dominates:

$$F(d) \cong -\frac{k_B T}{8\pi d^3} \int_0^\infty \frac{dy\, y^2}{e^y - 1} = -\frac{2.4 k_B T}{8\pi d^3}. \tag{7.5.52}$$

Like (7.1.51), this result is independent of $\hbar$, which of course is not surprising, since

$$\frac{1}{2}\hbar\omega + \frac{\hbar\omega}{e^{\hbar\omega/k_B T} - 1} \cong k_B T \tag{7.5.53}$$

for $k_B T \gg \hbar\omega$. In the low-temperature limit, $k_B T/\hbar \ll c/4\pi d$

$$F(d) \cong -\frac{\pi^2 \hbar c}{240 d^4}\left[1 + \frac{16}{3}\left(\frac{k_B T d}{\hbar c}\right)^4 - \frac{240}{\pi}\left(\frac{k_B T d}{\hbar c}\right)e^{-\pi\hbar c/k_B T d}\right].92 \tag{7.5.54}$$

These results indicate that finite-temperature corrections are determined by $k_B T d/\hbar c$ and take on greater importance as the separation d becomes larger. For $T = 293$ K, $k_B T d/\hbar c = 1$ when $d \cong 8$ μm.

Exercise 7.8: Show that, for high temperatures, the Lifshitz theory for rarefied media implies the van der Waals interaction (7.1.51) for all separations r.

7.6   Quantized Fields in Dissipative Dielectric Media

The original derivation of the force (7.5.3) by Lifshitz allows for $\epsilon_j(\omega)$ to be complex and thereby accounts for absorption. Although, as discussed below, Lifshitz's theory might be said to be essentially classical, the same force is obtained when the field is quantized. In fact a slight modification of Lifshitz's approach leads to expressions for quantized electric and magnetic fields in dissipative media.

We will employ the simplest model for an absorbing medium, treating it as a collection of N two-state atoms per unit volume that with, high probability, remain in their ground states.[93] Then, as discussed in Section 2.5.2, the atoms act for most purposes as harmonic oscillators. To allow for dissipation by the atoms and therefore absorption from electromagnetic fields in the medium, we include a damping force $-\gamma\dot{\mathbf{x}}_j$ in the Heisenberg equation of motion for each oscillator j:

$$m\ddot{\mathbf{x}}_j + \gamma\dot{\mathbf{x}}_j + m\omega_0^2 \mathbf{x}_j = \mathbf{F}_{Lj}(t). \tag{7.6.1}$$

$\mathbf{F}_{Lj}(t)$ is a Langevin operator that enforces the commutation relation $[x_{j\mu}(t), p_{k\nu}(t)] = i\hbar\delta_{jk}\delta_{\mu\nu}$, where $x_{j\mu}$ and $p_{j\mu} = m\dot{x}_{j\mu}$ are, respectively, the Cartesian components

---

[92] See, for instance, J. Schwinger, L. L. DeRaad, Jr., and K. A. Milton, Ann. Phys. (N. Y.) **115**, 1 (1978).

[93] We follow parts of F. S. S. Rosa, D. A. R. Dalvit, and P. W. Milonni, Phys. Rev. A **81**, 033812 (2010), and F. S. S. Rosa, D. A. R. Dalvit, and P. W. Milonni, Phys. A **84**, 053813 (2011).

($\mu = 1, 2, 3$) of the coordinate and momentum operators for the atom oscillator j. From Sections 4.6 and 6.5,

$$\mathbf{F}_{Lj}(t) = \int_0^\infty d\omega \sqrt{\frac{\gamma \hbar \omega}{\pi}} [\mathbf{a}_j(\omega)e^{-i\omega t} + \mathbf{a}_j^\dagger(\omega)e^{i\omega t}], \tag{7.6.2}$$

$$[a_{j\mu}(\omega), a_{k\nu}^\dagger(\omega')] = \delta_{jk}\delta_{\mu\nu}\delta(\omega - \omega'), \quad [a_{j\mu}(\omega), a_{k\nu}(\omega')] = 0, \tag{7.6.3}$$

$$[F_{Lj}(t), F_{Lk}(t')] = 2i\gamma\hbar\delta_{jk}\frac{\partial}{\partial t}\delta(t - t'). \tag{7.6.4}$$

Some properties of an oscillator described by (7.6.1) were discussed in Section 6.5. Here, we are interested in the coupling of each oscillator j of our model medium to an electric field $\mathbf{E}(\mathbf{r}_j, t)$ at the position $\mathbf{r}_j$ of the oscillator, as described by the Heisenberg equation

$$\ddot{\mathbf{x}}_j + \frac{\gamma}{m}\dot{\mathbf{x}}_j + \omega_0^2 \mathbf{x}_j = \frac{1}{m}\mathbf{F}_{Lj}(t) + \frac{e}{m}\mathbf{E}(\mathbf{r}_j, t). \tag{7.6.5}$$

This implies a polarization density operator $\mathbf{P}(\mathbf{r}, t)$ (dipole moment per unit volume) that depends not only on the electric fields $\mathbf{E}(\mathbf{r}_j, t)$ but also on the Langevin noise operators $\mathbf{F}_{Lj}(t)$. For $\mathbf{E}(\mathbf{r}, t)$ and $\mathbf{P}(\mathbf{r}, t)$, we write

$$\mathbf{E}(\mathbf{r}, t) = \int_0^\infty d\omega [\mathbf{E}(\mathbf{r}, \omega)e^{-i\omega t} + \mathbf{E}^\dagger(\mathbf{r}, \omega)e^{i\omega t}], \tag{7.6.6}$$

$$\mathbf{P}(\mathbf{r}, t) = \int_0^\infty d\omega [\mathbf{P}(\mathbf{r}, \omega)e^{-i\omega t} + \mathbf{P}^\dagger(\mathbf{r}, \omega)e^{i\omega t}], \tag{7.6.7}$$

with

$$\mathbf{P}(\mathbf{r}, \omega) = e \sum_j \mathbf{x}_j(\omega)\delta^3(\mathbf{r} - \mathbf{r}_j) \tag{7.6.8}$$

and $\mathbf{x}_j(\omega)$ defined by

$$\mathbf{x}_j(t) = \int_0^\infty d\omega [\mathbf{x}_j(\omega)e^{-i\omega t} + \mathbf{x}_j^\dagger(\omega)e^{i\omega t}]. \tag{7.6.9}$$

From (7.6.2), (7.6.5), and (7.6.8),

$$\mathbf{P}(\mathbf{r}, \omega) = \frac{1}{\omega_0^2 - \omega^2 - i\gamma\omega/m}\left[\frac{e^2}{m}\sum_j \mathbf{E}(\mathbf{r}_j, \omega)\delta^3(\mathbf{r} - \mathbf{r}_j)\right.$$

$$\left. + \frac{e}{m}\sqrt{\frac{\gamma\hbar\omega}{\pi}}\sum_j \mathbf{a}_j(\omega)\delta^3(\mathbf{r} - \mathbf{r}_j)\right]. \tag{7.6.10}$$

For a continuous distribution of N atoms per unit volume, the part of $\mathbf{P}(\mathbf{r}, \omega)$ proportional to $\mathbf{E}(\mathbf{r}, \omega)$ is obtained by the replacement

$$\frac{e^2/m}{\omega_0^2 - \omega^2 - i\gamma\omega/m}\sum_j \mathbf{E}(\mathbf{r}_j, \omega)\delta^3(\mathbf{r} - \mathbf{r}_j) \to \frac{Ne^2/m}{\omega_0^2 - \omega^2 - i\gamma\omega/m}\mathbf{E}(\mathbf{r}, \omega), \tag{7.6.11}$$

and, for a homogeneous medium with a spatially uniform permittivity,

$$\nabla^2 \mathbf{E}(\mathbf{r}, \omega) + \frac{\omega^2}{c^2} \mathbf{E}(\mathbf{r}, \omega) = -\frac{\omega^2}{\epsilon_0 c^2} \mathbf{P}(\mathbf{r}, \omega) = -\frac{Ne^2/m}{\omega_0^2 - \omega^2 - i\gamma\omega/m} \frac{\omega^2}{\epsilon_0 c^2} \mathbf{E}(\mathbf{r}, \omega)$$
$$- \frac{e/m}{\omega_0^2 - \omega^2 - i\gamma\omega/m} \sqrt{\frac{\gamma\hbar\omega}{\pi}} \frac{\omega^2}{c^2\epsilon_0} \sum_j \mathbf{a}_j(\omega)\delta^3(\mathbf{r} - \mathbf{r}_j), \tag{7.6.12}$$

or

$$\nabla^2 \mathbf{E}(\mathbf{r}, \omega) + \frac{\epsilon(\omega)}{\epsilon_0} \frac{\omega^2}{c^2} \mathbf{E}(\mathbf{r}, \omega) = -\frac{\omega^2}{\epsilon_0 c^2} \mathbf{K}(\mathbf{r}, \omega). \tag{7.6.13}$$

Here, the complex permittivity is

$$\epsilon(\omega) = \epsilon_0 + \frac{Ne^2/m}{\omega_0^2 - \omega^2 - i\gamma\omega/m} = \epsilon_R(\omega) + i\epsilon_I(\omega), \tag{7.6.14}$$

and

$$\mathbf{K}(\mathbf{r}, \omega) \equiv \frac{e/m}{\omega_0^2 - \omega^2 - i\gamma\omega/m} \sqrt{\frac{\gamma\hbar\omega}{\pi}} \sum_j \mathbf{a}_j(\omega)\delta^3(\mathbf{r} - \mathbf{r}_j). \tag{7.6.15}$$

From the properties of the operators $\mathbf{a}_j(\omega)$, it follows that $[K_\mu(\mathbf{r}, \omega), K_\nu(\mathbf{r}', \omega')] = 0$, and

$$[K_\mu(\mathbf{r}, \omega), K_\nu^\dagger(\mathbf{r}', \omega')] = \frac{e^2/m^2}{(\omega_0^2 - \omega^2)^2 + \gamma^2\omega^2/m^2} \frac{\gamma\hbar\omega}{\pi} \sum_j \sum_k [a_{j\mu}(\omega), a_{k\nu}^\dagger(\omega')]$$
$$\times \delta^3(\mathbf{r} - \mathbf{r}_j)\delta^3(\mathbf{r}' - \mathbf{r}_k)$$
$$= \frac{e^2/m^2}{(\omega_0^2 - \omega^2)^2 + \gamma^2\omega^2/m^2} \frac{\gamma\hbar\omega}{\pi} \delta_{\mu\nu}\delta(\omega - \omega') \sum_j \delta^3(\mathbf{r} - \mathbf{r}_j)\delta^3(\mathbf{r}' - \mathbf{r}_j). \tag{7.6.16}$$

For a continuous distribution of N atoms per unit volume,

$$\sum_j \delta^3(\mathbf{r} - \mathbf{r}_j)\delta^3(\mathbf{r}' - \mathbf{r}_j) \to N \int d^3r'' \delta^3(\mathbf{r} - \mathbf{r}'')\delta^3(\mathbf{r}' - \mathbf{r}'') = N\delta^3(\mathbf{r} - \mathbf{r}'), \tag{7.6.17}$$

and therefore

$$[K_\mu(\mathbf{r}, \omega), K_\nu^\dagger(\mathbf{r}', \omega')] = \frac{Ne^2/m^2}{(\omega_0^2 - \omega^2)^2 + \gamma^2\omega^2/m^2} \frac{\gamma\hbar\omega}{\pi} \delta_{\mu\nu}\delta(\omega - \omega')\delta^3(\mathbf{r} - \mathbf{r}')$$
$$= \frac{\hbar}{\pi}\epsilon_I(\omega)\delta_{\mu\nu}\delta(\omega - \omega')\delta^3(\mathbf{r} - \mathbf{r}'), \tag{7.6.18}$$

where $\epsilon_I(\omega)$, the imaginary part of $\epsilon(\omega)$, is defined by (7.6.14):

$$\epsilon_I(\omega) = \frac{Ne^2\gamma\omega/m^2}{(\omega_0^2 - \omega^2)^2 + \gamma^2\omega^2/m^2}. \tag{7.6.19}$$

For thermal equilibrium at temperature T,

$$\langle a_{j\mu}^{\dagger}(\omega)a_{k\nu}(\omega')\rangle = \delta_{jk}\delta_{\mu\nu}\delta(\omega-\omega')\frac{1}{e^{\hbar\omega/k_BT}-1}, \tag{7.6.20}$$

$$\langle a_{j\mu}(\omega)a_{k\nu}^{\dagger}(\omega')\rangle = \delta_{jk}\delta_{\mu\nu}\delta(\omega-\omega')\left(\frac{1}{e^{\hbar\omega/k_BT}-1}+1\right), \tag{7.6.21}$$

and, proceeding as above from (7.6.15), we derive the correlation functions

$$\langle K_{\mu}^{\dagger}(\mathbf{r},\omega)K_{\nu}(\mathbf{r}',\omega')\rangle = \frac{\hbar}{\pi}\epsilon_I(\omega)\delta_{\mu\nu}\delta(\omega-\omega')\delta^3(\mathbf{r}-\mathbf{r}')\frac{1}{e^{\hbar\omega/k_BT}-1}, \tag{7.6.22}$$

and

$$\langle K_{\mu}(\mathbf{r},\omega)K_{\nu}^{\dagger}(\mathbf{r}',\omega')\rangle = \frac{\hbar}{\pi}\epsilon_I(\omega)\delta_{\mu\nu}\delta(\omega-\omega')\delta^3(\mathbf{r}-\mathbf{r}')\left(\frac{1}{e^{\hbar\omega/k_BT}-1}+1\right). \tag{7.6.23}$$

$\mathbf{K}(\mathbf{r},\omega)$ is a polarization density due to the Langevin noise forces acting on the atoms of the medium. These Langevin forces induce fluctuating dipole moments in the atoms, giving rise to the fluctuating *noise polarization* $\mathbf{K}(\mathbf{r},\omega)$ at position $\mathbf{r}$ and frequency ω. To get the fluctuating electric field due to this noise polarization, we first write the noise polarization in a form similar to that used by Lifshitz in his calculation of the force between dielectric half-spaces:[94]

$$K_{\lambda}(\mathbf{r},\omega) = \int d^3k g_{\lambda}(\mathbf{k},\omega)e^{i\mathbf{k}\cdot\mathbf{r}}. \tag{7.6.24}$$

We are assuming, for the purpose of deriving an expression for the quantized electric field in a homogeneous, isotropic, dissipative medium, that the medium is unbounded. The commutation relation (7.6.18) implies a commutation relation for the operators $g_{\lambda}(\mathbf{k},\omega)$: from

$$g_{\lambda}(\mathbf{k},\omega) = \left(\frac{1}{2\pi}\right)^3 \int d^3r K_{\lambda}(\mathbf{r},\omega)e^{-i\mathbf{k}\cdot\mathbf{r}}, \tag{7.6.25}$$

it follows from (7.6.18) that

$$[g_{\lambda}(\mathbf{k},\omega),g_{\lambda'}^{\dagger}(\mathbf{k}',\omega')] = \left(\frac{1}{2\pi}\right)^6 \int d^3r \int d^3r' [K_{\lambda}(\mathbf{r},\omega),K_{\lambda'}^{\dagger}(\mathbf{r}',\omega')]e^{-i\mathbf{k}\cdot\mathbf{r}}e^{i\mathbf{k}'\cdot\mathbf{r}'}$$

$$= \left(\frac{1}{2\pi}\right)^3 \frac{\hbar}{\pi}\epsilon_I(\omega)\delta_{\lambda\lambda'}\delta(\omega-\omega')\delta^3(\mathbf{k}-\mathbf{k}'). \tag{7.6.26}$$

It is also convenient to define

$$c_{\lambda}(\mathbf{k},\omega) = \left[\frac{\hbar\epsilon_I(\omega)}{8\pi^4}\right]^{-1/2} g_{\lambda}(\mathbf{k},\omega), \tag{7.6.27}$$

which satisfies the commutation relation

$$[c_{\lambda}(\mathbf{k},\omega),c_{\lambda}^{\dagger}(\mathbf{k}',\omega')] = \delta_{\lambda\lambda'}\delta(\omega-\omega')\delta^3(\mathbf{k}-\mathbf{k}'). \tag{7.6.28}$$

[94] E. M. Lifshitz, Sov. Phys. JETP **2**, 73 (1956).

Thus, $c_\lambda(\mathbf{k},\omega)$ and $c_\lambda^\dagger(\mathbf{k},\omega)$ are bosonic lowering and raising operators, respectively; they are the lowering and raising operators for damped polaritons.[95] In terms of $c_\lambda(\mathbf{k},\omega)$,

$$K_\lambda(\mathbf{r},\omega) = \left[\frac{\hbar\epsilon_I(\omega)}{8\pi^4}\right]^{1/2} \int d^3k\, c_\lambda(\mathbf{k},\omega)e^{i\mathbf{k}\cdot\mathbf{r}}, \tag{7.6.29}$$

$$K_\lambda(\mathbf{r},t) = \int d^3k \int_0^\infty d\omega \left[\frac{\hbar\epsilon_I(\omega)}{8\pi^4}\right]^{1/2} c_\lambda(\mathbf{k},\omega)e^{-i(\omega t - \mathbf{k}\cdot\mathbf{r})}. \tag{7.6.30}$$

Finally, we express the electric and magnetic field operators in terms of the "quasi-particle" lowering and raising operators. From (7.6.6) and (7.6.13),

$$E_\lambda(\mathbf{r},t) = E_\lambda^{(+)}(\mathbf{r},t) + E_\lambda^{(-)}(\mathbf{r},t), \quad E^{(-)}(\mathbf{r},t) = [E_\lambda^{(+)}(\mathbf{r},t)]^\dagger, \tag{7.6.31}$$

$$
\begin{aligned}
E_\lambda^{(+)}(\mathbf{r},t) &= \int d^3k \int_0^\infty d\omega \left[\frac{\hbar\epsilon_I(\omega)}{8\pi^4}\right]^{1/2} \frac{\omega^2/\epsilon_0 c^2}{k^2 - \epsilon(\omega)\omega^2/\epsilon_0 c^2} c_\lambda(\mathbf{k},\omega)e^{-i(\omega t - \mathbf{k}\cdot\mathbf{r})} \\
&= \int_0^\infty d\omega\, E_\lambda^{(+)}(\mathbf{r},\omega)e^{-i\omega t},
\end{aligned}
\tag{7.6.32}
$$

$$E_\lambda^{(+)}(\mathbf{r},\omega) = \left[\frac{\hbar\epsilon_I(\omega)}{8\pi^4}\right]^{1/2} \frac{\omega^2}{\epsilon_0 c^2} \int d^3k\, \frac{c_\lambda(\mathbf{k},\omega)e^{i\mathbf{k}\cdot\mathbf{r}}}{k^2 - \epsilon(\omega)\omega^2/\epsilon_0 c^2}. \tag{7.6.33}$$

Since we have assumed in writing (7.6.13) for a homogeneous medium that $\nabla \cdot \mathbf{E} = 0$,

$$
\begin{aligned}
\mathbf{E}^{(+)}(\mathbf{r},t) = \sum_{\lambda=1,2} &\int d^3k \int_0^\infty d\omega \left[\frac{\hbar\epsilon_I(\omega)}{8\pi^4}\right]^{1/2} \\
&\times \frac{\omega^2/\epsilon_0 c^2}{k^2 - \epsilon(\omega)\omega^2/\epsilon_0 c^2} c_\lambda(\mathbf{k},\omega)\mathbf{e}_{\mathbf{k}\lambda}e^{-i(\omega t - \mathbf{k}\cdot\mathbf{r})},
\end{aligned}
\tag{7.6.34}
$$

where the two unit vectors $\mathbf{e}_{\mathbf{k}\lambda}$, $\lambda = 1, 2$, are orthogonal to $\mathbf{k}$ and to each other, and are chosen here to be real. The magnetic field operator follows from the Maxwell equation $\nabla \times \mathbf{E} = -\partial\mathbf{B}/\partial t$. The field operators have the same form as those derived by Huttner and Barnett in their elegant approach based on Fano diagonalization of the entire system of atom, field, and Langevin "bath" oscillators.[96] Unlike the field operators for a vacuum or a non-absorbing dielectric medium (Section 3.4), the $\mathbf{k}$'s and ω's when there is dissipation are not related by $|\mathbf{k}| = \omega/c$ or $|\mathbf{k}| = n_R\omega/c$ and must both be summed as in (7.6.34).

---

[95] The case of *inhomogeneous* dissipative media has been treated in the Fano diagonalization approach by L. G. Suttorp and A. J. van Wonderen, Europhys. Lett. **67**, 766 (2004). See also A. Drezet, Phys. Rev. A **95**, 023832 (2017), and A. Drezet, Phys. Rev. A **96**, 033849 (2017).

[96] B. Huttner and S. M. Barnett, Phys. Rev. A **46**, 4306 (1992).

7.6.1 Zero-Point Energy in Dissipative Media

In derivations of the Lifshitz force, a huge simplification is achieved by pretending that there is no absorption at any frequency. Absorption is ignored in the one-dimensional model of Section 7.5.1, the approach based on the argument theorem, and in a significant part of the large literature on zero-point electromagnetic energy and the theory of the Lifshitz force. And yet, as remarked earlier, the force obtained without accounting for absorption is the same as that derived in more complicated theories that allow for absorption. This has led to the persistent question of how the concept of zero-point electromagnetic energy can be applied to dissipative media. To address this question, we consider first the expectation value

$$\langle \mathbf{E}^2(\mathbf{r}, t) \rangle = \frac{\hbar}{8\pi^4 \epsilon_0^2 c^4} \int d^3k \sum_{\lambda=1,2} \int_0^\infty \frac{d\omega \omega^4 \epsilon_I(\omega)}{|k^2 - \epsilon(\omega)\omega^2/\epsilon_0 c^2|^2}$$

$$= \frac{\hbar}{\pi^3 \epsilon_0^2 c^4} \int_0^\infty d\omega \omega^4 \epsilon_I(\omega) \int_0^\infty \frac{dk k^2}{|k^2 - \epsilon(\omega)\omega^2/\epsilon_0 c^2|^2} \qquad (7.6.35)$$

for the zero-temperature state of the medium, for which

$$\langle c_\lambda(\mathbf{k}, \omega) c_{\lambda'}(\mathbf{k}', \omega') \rangle = 0,$$
$$\langle c_\lambda^\dagger(\mathbf{k}, \omega) c_{\lambda'}(\mathbf{k}', \omega') \rangle = 0, \quad \text{and}$$
$$\langle c_\lambda(\mathbf{k}, \omega) c_{\lambda'}^\dagger(\mathbf{k}', \omega') \rangle = \delta_{\lambda\lambda'} \delta(\omega - \omega') \delta^3(\mathbf{k} - \mathbf{k}'). \qquad (7.6.36)$$

These properties describe the zero-temperature "bath" oscillators associated with the Langevin forces: $\langle a_j^\dagger(\omega) a_k(\omega') \rangle = 0$, $\langle a_j(\omega) a_k^\dagger(\omega') \rangle = \delta_{jk} \delta(\omega - \omega')$, and so on.

For the integration over k in (7.6.35), make a change of integration variable, from k to $x = kc/\omega$:

$$\int_0^\infty \frac{dk k^2}{|k^2 - \epsilon(\omega)\omega^2/\epsilon_0 c^2|^2} = \frac{c}{\omega} \int_0^\infty \frac{x^2}{|x^2 - \epsilon(\omega)/\epsilon_0|^2}$$

$$= \frac{c}{2\omega} \frac{1}{\epsilon_I(\omega)} \text{Im}\left[\epsilon(\omega) \int_{-\infty}^\infty \frac{dx}{[x + n(\omega)][x - n(\omega)]}\right]$$

$$= \frac{c}{2\omega} \frac{1}{\epsilon_I(\omega)} \text{Im}\left[\epsilon(\omega) \frac{i\pi}{n(\omega)}\right] = \frac{\pi c \epsilon_0}{2\omega} \frac{n_R(\omega)}{\epsilon_I(\omega)}, \qquad (7.6.37)$$

in which $n(\omega) = \sqrt{\epsilon(\omega)/\epsilon_0} = n_R(\omega) + i n_I(\omega)$ is the complex refractive index at frequency ω. Then,

$$\langle \mathbf{E}^2(\mathbf{r}, t) \rangle = \frac{\hbar}{\pi^3 \epsilon_0^2 c^4} \int_0^\infty d\omega \omega^4 \epsilon_I(\omega) \frac{\pi c \epsilon_0}{2\omega} \frac{n_R(\omega)}{\epsilon_I(\omega)} = \frac{\hbar}{2\pi^2 \epsilon_0 c^3} \int_0^\infty d\omega \omega^3 n_R(\omega).$$
$$(7.6.38)$$

It follows similarly from (7.6.6) and (7.6.33) that the zero-temperature expectation values are

$$\langle \mathbf{E}(\mathbf{r}, \omega) \cdot \mathbf{E}(\mathbf{r}, \omega') \rangle = \langle \mathbf{E}^\dagger(\mathbf{r}, \omega) \cdot \mathbf{E}(\mathbf{r}, \omega') \rangle = 0, \qquad (7.6.39)$$

$$\langle \mathbf{E}(\mathbf{r}, \omega) \cdot \mathbf{E}^{\dagger}(\mathbf{r}, \omega') \rangle = \frac{\hbar \omega^3 n_R(\omega)}{2\pi^2 \epsilon_0 c^3} \delta(\omega - \omega'). \tag{7.6.40}$$

The formula (7.6.38) is identical to (3.4.19), which was obtained without allowing for absorption. So this is another example where identical results are obtained with or without absorption. But this example relates more directly to energy and suggests, as in fact we will now show, that the expression for zero-point energy has the same form with or without absorption.

We proceed as in Section 1.6, but now with quantized fields and allowance for dissipation ($\epsilon_I(\omega) \neq 0$). The Poynting-vector operator, symmetrized to make it manifestly Hermitian, is $\mathbf{S} = (\mathbf{E} \times \mathbf{H} - \mathbf{H} \times \mathbf{E})$, and Poynting's theorem for the zero-temperature expectation value of interest is

$$\oint \langle \mathbf{S} \rangle \cdot \hat{n} da = -\frac{1}{2} \int_V \left\langle \mathbf{E} \cdot \frac{\partial \mathbf{D}}{\partial t} + \frac{\partial \mathbf{D}}{\partial t} \cdot \mathbf{E} + \mu_0 \frac{\partial}{\partial t} \mathbf{H}^2 \right\rangle dV, \tag{7.6.41}$$

assuming $\mu = \mu_0$. As usual,

$$\frac{\partial u}{\partial t} = \frac{1}{2} \left\langle \mathbf{E} \cdot \frac{\partial \mathbf{D}}{\partial t} + \frac{\partial \mathbf{D}}{\partial t} \cdot \mathbf{E} + \mu_0 \frac{\partial}{\partial t} \mathbf{H}^2 \right\rangle \tag{7.6.42}$$

is identified as the rate of change of the energy density u. The electric displacement vector $\mathbf{D}$ here is

$$\mathbf{D} = \epsilon_0 \mathbf{E} + \mathbf{P} = \epsilon \mathbf{E} + \mathbf{K} \equiv \mathbf{D}_\epsilon + \mathbf{K}; \tag{7.6.43}$$

$\mathbf{D}_\epsilon$ is the part of $\mathbf{D}$ that determines the permittivity ϵ of the medium, and $\mathbf{K}$ is, again, the noise polarization. It is convenient to deal separately with the three contributions to the energy density:

$$\frac{\partial u}{\partial t} = \frac{\partial}{\partial t}(u_1 + u_2 + u_3), \tag{7.6.44}$$

$$\frac{\partial u_1}{\partial t} = \frac{1}{2} \left\langle \mathbf{E} \cdot \frac{\partial \mathbf{D}_\epsilon}{\partial t} + \frac{\partial \mathbf{D}_\epsilon}{\partial t} \cdot \mathbf{E} \right\rangle, \tag{7.6.45}$$

$$\frac{\partial u_2}{\partial t} = \frac{1}{2} \left\langle \mathbf{E} \cdot \frac{\partial \mathbf{K}}{\partial t} + \frac{\partial \mathbf{K}}{\partial t} \cdot \mathbf{E} \right\rangle, \tag{7.6.46}$$

$$\frac{\partial u_3}{\partial t} = \frac{1}{2} \left\langle \mu_0 \frac{\partial \mathbf{H}^2}{\partial t} \right\rangle. \tag{7.6.47}$$

The positive-frequency ($e^{-i\omega t}$) part of each component λ of $\mathbf{D}_\epsilon(\mathbf{r}, t)$ is

$$D_{\epsilon\lambda}(\mathbf{r}, t) = \int_0^\infty d\omega \epsilon(\omega) E_\lambda(\mathbf{r}, \omega) e^{-i\omega t}, \tag{7.6.48}$$

and so

$$\frac{\partial D_{\epsilon\lambda}}{\partial t} = -i \int_0^\infty d\omega \omega \epsilon(\omega) E_\lambda(\mathbf{r}, \omega) e^{-i\omega t}, \tag{7.6.49}$$

and, for $T = 0$,

$$\left\langle \mathbf{E} \cdot \frac{\partial \mathbf{D}_\epsilon}{\partial t} \right\rangle = \sum_{\lambda=1,2} \left\langle E_\lambda \frac{\partial D_{\epsilon\lambda}^\dagger}{\partial t} \right\rangle$$

$$= i \sum_{\lambda=1,2} \int_0^\infty d\omega' \int_0^\infty d\omega \, \omega \epsilon(\omega) \langle E_\lambda(\mathbf{r},\omega') E_\lambda^\dagger(\mathbf{r},\omega) \rangle e^{-i(\omega'-\omega)t}$$

$$= \frac{i\hbar}{2\pi^2 \epsilon_0 c^3} \int_0^\infty d\omega' \int_0^\infty d\omega \, \omega^4 \epsilon(\omega) n_R(\omega) e^{-i(\omega'-\omega)t} \delta(\omega-\omega'). \quad (7.6.50)$$

$\partial u_1/\partial t$ is the real part of (7.6.50). Integrating, and ignoring a physically irrelevant integration constant, we obtain

$$u_1 = -\frac{\hbar}{4\pi^2 \epsilon_0 c^3} \int_0^\infty d\omega' \int_0^\infty d\omega \, n_R(\omega) \omega^3 \left[\frac{\omega \epsilon(\omega) e^{-i(\omega'-\omega)t} + \omega \epsilon^*(\omega) e^{i(\omega'-\omega)t}}{\omega'-\omega} \right],$$
$$\times \, \delta(\omega - \omega'), \quad (7.6.51)$$

or

$$u_1 = \frac{\hbar}{4\pi^2 \epsilon_0 c^3} \int_0^\infty d\omega' \int_0^\infty d\omega \, n_R(\omega) \omega^3 \left[\frac{\omega' \epsilon_R(\omega') - \omega \epsilon_R(\omega)}{\omega'-\omega} \right] e^{-i(\omega'-\omega)t} \delta(\omega-\omega')$$
$$- \frac{i\hbar}{4\pi^2 \epsilon_0 c^3} \int_0^\infty d\omega' \int_0^\infty d\omega \, n_R(\omega) \epsilon_I(\omega) \omega^4$$
$$\times \left(\frac{e^{i(\omega'-\omega)t} - e^{-i(\omega'-\omega)t}}{\omega'-\omega} \right) \delta(\omega-\omega'). \quad (7.6.52)$$

Then, from

$$\lim_{\omega'\to\omega} \frac{\omega' \epsilon_R(\omega' - \omega \epsilon_R(\omega)}{\omega'-\omega} = \frac{d}{d\omega}[\omega \epsilon_R(\omega)] \quad (7.6.53)$$

and

$$\lim_{\omega'\to\omega} \frac{e^{i(\omega'-\omega)t} - e^{-i(\omega'-\omega)t}}{\omega'-\omega} = 2it, \quad (7.6.54)$$

$$u_1 = \frac{\hbar}{4\pi^2 \epsilon_0 c^3} \int_0^\infty d\omega \, n_R(\omega) \omega^3 \frac{d}{d\omega}[\omega \epsilon_R(\omega)]$$
$$+ \frac{\hbar t}{2\pi^2 \epsilon_0 c^3} \int_0^\infty d\omega \, \omega^4 n_R(\omega) \epsilon_I(\omega). \quad (7.6.55)$$

The evaluation of u_2 and u_3 proceeds in the same fashion, and is not particularly interesting. What is found is that part of u_3 cancels the part of (7.6.55) proportional to time t, and that, from the remaining part of u_3, together with u_2,

$$u = u_1 + u_2 + u_3 = \frac{\hbar}{4\pi^2 c^3} \int_0^\infty d\omega \left\{ \mathrm{Re}\left[n_R \frac{d}{d\omega}(\omega n^2) \right] + n^3 + \frac{2}{\omega} n_R n_I \frac{d}{d\omega}(\omega^2 n_I) \right\}. \quad (7.6.56)$$

This simplifies remarkably to

$$u = \frac{\hbar}{2\pi^2 c^3} \int_0^\infty d\omega\, \omega^3 n_R^2 \frac{d}{d\omega}(\omega n_R), \tag{7.6.57}$$

and is just the zero-point energy density u_0 derived when absorption is ignored:

$$u_0 = \sum_{\mathbf{k}\lambda} \frac{1}{2}\hbar\omega_k = \sum_{\lambda=1,2} \int d^3k \frac{1}{2}\hbar\omega_k = \frac{\hbar}{2\pi^2} \int_0^\infty dk\, k^2 \omega_k = \frac{\hbar}{2\pi^2 c^3} \int_0^\infty d\omega \frac{dk}{d\omega} k^2 \omega$$

$$= \frac{\hbar}{2\pi^2 c^3} \int_0^\infty d\omega\, \omega^3 n_R^2 \frac{d}{d\omega}(\omega n_R). \tag{7.6.58}$$

So the total zero-point energy has the same form (cf. (3.4.19)), with or without accounting for absorption. Of course, the numerical value of $n_R(\omega)$ for a dissipative medium is related to the imaginary (dissipative) parts of $\epsilon(\omega)$ and $n_I(\omega)$: $\epsilon/\epsilon_0 = (n_R + in_I)^2$ implies $n_R^2 = (\epsilon_R/\epsilon_0 + n_I^2)^{1/2}$. The point is that we can calculate the zero-point energy by first ignoring dissipation and taking the refractive index $n(\omega)$ to be purely real, and then just replacing $n(\omega)$ by $n_R(\omega)$. This makes it less surprising that the Lifshitz force (7.5.1) can be derived under the pretense of no absorption. Of course, the derivation of the Lifshitz force must include the spatial variation of the permittivity and, therefore, the zero-point energy (see Section 7.7).

7.6.2 Langevin Noise and Fano Diagonalization

The noise polarization $\mathbf{K}$ is the source of electric and magnetic fields of the form obtained by Huttner and Barnett in their approach in which the coupled system of atom, bath, and field oscillators is diagonalized by a method known as Fano diagonalization.[97] Since this approach is taken in other contexts, we briefly outline it here and compare it with the "Langevin approach" we have taken. The model consists of an oscillator A of frequency ω_0 coupled to a reservoir ("bath") R of other oscillators in the rotating-wave approximation. The coupling is chosen such that the Hamiltonian is

$$H = \hbar\omega_0 a^\dagger a + \int_0^\infty d\omega\, \hbar\omega b^\dagger(\omega)b(\omega) + \hbar\sqrt{\frac{\gamma}{\pi}} \int_0^\infty d\omega[a^\dagger b(\omega) + b^\dagger(\omega)a], \tag{7.6.59}$$

with $[a, a^\dagger] = 1$, $[b(\omega), b(\omega')] = 0$, $[b(\omega), b^\dagger(\omega')] = \delta(\omega - \omega')$. We first summarize some results of this model, and show that the chosen coupling results in a frictional damping of $a(t)$ at the rate γ.

In the Heisenberg picture, the a and b operators are time dependent, and their equations of motion are

$$\dot{a}(t) = -i\omega_0 a(t) - i\sqrt{\frac{\gamma}{\pi}} \int_0^\infty d\omega\, b(\omega, t), \tag{7.6.60}$$

---

[97] U. Fano, Phys. Rev. **103**, 1202 (1956).

$$\dot{b}(\omega, t) = -i\omega b(\omega, t) - i\sqrt{\frac{\gamma}{\pi}} a(t). \tag{7.6.61}$$

Using the formal solution of (7.6.61) in (7.6.60), and defining $b_0(\omega) = b(\omega, 0)$, we obtain in familiar fashion

$$\dot{a}(t) + i\omega_0 a(t) + \frac{\gamma}{\pi} \int_0^\infty d\omega \int_0^t dt' a(t') e^{i\omega(t'-t)} = -i\sqrt{\frac{\gamma}{\pi}} \int_0^\infty d\omega b_0(\omega) e^{-i\omega t}, \tag{7.6.62}$$

the operator on the right representing a Langevin "force." We solve this equation for "steady state" ($\gamma t \gg 1$) by first writing

$$a(t) = \int_0^\infty d\Omega A(\Omega) b_0(\Omega) e^{-i\Omega t}. \tag{7.6.63}$$

Then, from (7.6.62) and the approximation (4.4.31),

$$A(\Omega) = \frac{\sqrt{\gamma/\pi}}{\Omega - \omega_0 + \Delta(\Omega) + i\gamma}, \tag{7.6.64}$$

where

$$\Delta(\Omega) = \frac{\gamma}{\pi} \fint_0^\infty \frac{d\omega}{\omega - \Omega} \tag{7.6.65}$$

and f denotes the Cauchy principal value. Assuming weak coupling between the A and R systems, in which case mainly frequencies $\Omega \cong \omega_0$ contribute to the time evolution of A, we approximate $\Delta(\Omega)$ by $\Delta(\omega_0)$. Of course, $\Delta(\omega_0)$ diverges unless the upper limit of integration is cut off; for our purposes, there is no need to explicitly indicate any cutoff. We will simply include $\Delta(\omega_0)$ in the definition of ω_0 and write the solution for $a(t)$ as

$$a(t) = \sqrt{\frac{\gamma}{\pi}} \int_0^\infty d\Omega \frac{b_0(\Omega) e^{-i\Omega t}}{\Omega - \omega_0 + i\gamma}. \tag{7.6.66}$$

Assuming $\omega_0 \gg \gamma$, consistent with the rotating-wave approximation, we can approximate the integral by extending the lower limit of integration to $-\infty$. Then

$$[a(t), a^\dagger(t)] \cong \frac{\gamma}{\pi} \int_{-\infty}^\infty \frac{d\Omega}{(\Omega - \omega_0)^2 + \gamma^2} = 1, \tag{7.6.67}$$

as required for the consistency of the model.

Turning now to Fano diagonalization, we define an operator

$$B(\Omega, t) = \alpha(\Omega) a(t) + \int_0^\infty d\omega \beta(\Omega, \omega) b(\omega, t) \tag{7.6.68}$$

such that

$$[B(\Omega, t), B^\dagger(\Omega', t)] = \delta(\Omega - \Omega'), \quad [B(\Omega, t), B(\Omega', t)] = 0, \tag{7.6.69}$$

and in terms of which the Hamiltonian (7.6.59) takes the *diagonal* form

$$H = \int_0^\infty d\Omega \hbar \Omega B^\dagger(\Omega) B(\Omega). \tag{7.6.70}$$

The commutation relations (7.6.69) imply $[B(\Omega), H] = \hbar \Omega B(\Omega)$ and the Heisenberg equation of motion $\dot{B}(\Omega, t) = -i\Omega B(\Omega, t)$, and, from (7.6.68) and the equations of motion (7.6.60) and (7.6.61) for $a(t)$ and $b(\omega, t)$, we deduce equations for the coefficients $\alpha(\Omega)$ and $\beta(\Omega, \omega)$:

$$(\omega_0 - \Omega)\alpha(\Omega) = -\sqrt{\frac{\gamma}{\pi}} \int_0^\infty d\omega \beta(\Omega, \omega), \tag{7.6.71}$$

$$\int_0^\infty d\omega \left[(\omega - \Omega)\beta(\Omega, \omega) + \sqrt{\frac{\gamma}{\pi}} \alpha(\Omega) \right] b(\omega, t) = 0. \tag{7.6.72}$$

The last equation is satisfied if

$$\beta(\Omega, \omega) = \alpha(\Omega) f(\Omega) \delta(\omega - \Omega) - \sqrt{\frac{\gamma}{\pi}} \frac{\alpha(\Omega)}{\omega - \Omega}, \tag{7.6.73}$$

and the first equation then requires

$$f(\Omega) = \sqrt{\frac{\pi}{\gamma}} (\Omega - \omega_0) \tag{7.6.74}$$

and therefore

$$\beta(\Omega, \omega) = \sqrt{\frac{\pi}{\gamma}} \alpha(\Omega)(\Omega - \omega_0)\delta(\omega - \Omega) - \sqrt{\frac{\gamma}{\pi}} \frac{\alpha(\Omega)}{\omega - \Omega}. \tag{7.6.75}$$

As explained earlier, we have approximated $\Delta(\Omega)$ by $\Delta(\omega_0)$ and will continue to include, here and below, this frequency shift in the definition of ω_0.

It remains to determine $\alpha(\Omega)$. To do so, we impose the requirement that the commutation relations (7.6.69) be satisfied. From the commutation relations for the a and b operators,

$$[B(\Omega, t), B^\dagger(\Omega', t)] = \alpha(\Omega)\alpha^*(\Omega')$$
$$+ \int_0^\infty d\omega \int_0^\infty d\omega' \beta(\Omega, \omega)\beta^*(\Omega', \omega')\delta(\omega - \omega'), \tag{7.6.76}$$

or, from (7.6.75) and some algebra,

$$[B(\Omega), B^\dagger(\Omega')] = \frac{\pi}{\gamma}\alpha(\Omega)\alpha^*(\Omega') \left[(\Omega - \omega_0)^2 \delta(\Omega - \Omega') + \frac{\gamma}{\pi} \frac{\Delta(\Omega') - \Delta(\Omega)}{\Omega' - \Omega} \right.$$
$$\left. + \frac{\gamma^2}{\pi^2} \int_0^\infty \frac{d\omega}{\omega - \Omega} \int_0^\infty \frac{d\omega'}{\omega' - \Omega'} \delta(\omega - \omega') \right]. \tag{7.6.77}$$

Using now (recall (4.4.34))

$$P\frac{1}{\omega - \Omega} = \frac{1}{\omega - \Omega + i\epsilon} + i\pi\delta(\omega - \Omega), \quad \text{where } \epsilon \to 0^+, \tag{7.6.78}$$

and some manipulations with partial fractions, we find that (7.6.77) reduces to

$$[B(\Omega), B^\dagger(\Omega')] = \frac{\pi}{\gamma}\alpha(\Omega)\alpha^*(\Omega')[\gamma^2 + (\Omega - \omega_0)^2]\delta(\Omega - \Omega'), \qquad (7.6.79)$$

and therefore we can satisfy (7.6.69) with

$$\alpha(\Omega) = \frac{\sqrt{\gamma/\pi}}{\Omega - \omega_0 - i\gamma}. \qquad (7.6.80)$$

Together with (7.6.75) for $\beta(\Omega, \omega)$, this specifies $B(\Omega, t)$, and so the diagonalization of the Hamiltonian to the form (7.6.70) has been accomplished.

For comparison with the Langevin approach leading to (7.6.66), we express $a(t)$ in terms of the operators $B(\Omega, t)$. Multiplication of both sides of (7.6.68) by $\alpha^*(\Omega)$ and integration over Ω yields

$$a(t)\int_0^\infty d\Omega|\alpha(\Omega)|^2 = \int_0^\infty d\Omega\alpha^*(\Omega)B(\Omega, t)$$
$$+ \int_0^\infty d\omega b(\omega, t)\int_0^\infty d\Omega\beta(\Omega, \omega)\alpha^*(\Omega). \qquad (7.6.81)$$

The integral on the left is

$$\frac{\gamma}{\pi}\int_0^\infty \frac{d\Omega}{(\Omega - \omega_0)^2 + \gamma^2} \cong \frac{\gamma}{\pi}\int_{-\infty}^\infty \frac{d\Omega}{(\Omega - \omega_0)^2 + \gamma^2} = 1. \qquad (7.6.82)$$

The last last integral on the right-hand side of (7.6.81) can be evaluated by multiplying both sides of (7.6.71) by $\alpha^*(\Omega)$ and integrating:

$$-\frac{\gamma}{\pi}\int_0^\infty d\omega\int_0^\infty d\Omega\beta(\Omega, \omega)\alpha^*(\Omega) = -\int_0^\infty d\Omega(\omega_0 - \Omega)|\alpha(\Omega)|^2$$
$$= \frac{\gamma}{\pi}\int_0^\infty \frac{(\Omega - \omega_0)d\Omega}{(\Omega - \omega_0)^2 + \gamma^2}$$
$$\cong \frac{\gamma}{\pi}\int_{-\infty}^\infty \frac{(\Omega - \omega_0)d\Omega}{(\Omega - \omega_0)^2 + \gamma^2}. \qquad (7.6.83)$$

The Cauchy principal value of the last integral vanishes, implying

$$\int_0^\infty d\omega\int_0^\infty d\Omega\beta(\Omega, \omega)\alpha^*(\Omega) = 0 \qquad (7.6.84)$$

and

$$a(t) = \int_0^\infty d\Omega\alpha^*(\Omega)B(\Omega, t) = \sqrt{\frac{\gamma}{\pi}}\int_0^\infty \frac{d\Omega B(\Omega, 0)e^{-i\Omega t}}{\Omega - \omega_0 + i\gamma}. \qquad (7.6.85)$$

The expressions (7.6.66) and (7.6.85) obtained, respectively, in the Langevin and Fano approaches look formally the same in the sense that

$$[b_0(\Omega), b_0^\dagger(\Omega')] = [B_0(\Omega), B_0^\dagger(\Omega')] = \delta(\Omega - \Omega') \tag{7.6.86}$$

and

$$[b_0(\Omega), b_0(\Omega')] = [B_0(\Omega), B_0(\Omega')] = 0. \tag{7.6.87}$$

They differ in that $B(\Omega, 0)$ is a linear combination of A and R operators, whereas only bath (R) operators appear in the formula (7.6.66) for $a(t)$. In the ground state $|\psi\rangle$ of the coupled rotating-wave-approximation system,

$$B^\dagger(\Omega, t)B(\Omega, t)|\psi\rangle = 0, \tag{7.6.88}$$

all the oscillators are in their ground states, and the properties of Oscillator A derived from (7.6.66) are equivalent to those derived from (7.6.85). In a state of thermal equilibrium, similarly, $\langle B^\dagger(\Omega, t)B(\Omega, t)\rangle = [\exp(\hbar\Omega/k_B T) - 1]^{-1}\delta(\Omega - \Omega')$, and it follows from either (7.6.66) or (7.6.85) that $\langle a^\dagger(t)a(t)\rangle = [\exp(\hbar\Omega/b_B T) - 1]^{-1}$. But the two formulas for $a(t)$ have different interpretations. In the Langevin approach, there are fluctuations and dissipation: $a(t)$ experiences a Langevin noise force in addition to a damping force, the noise coming from unperturbed bath oscillators $[b_0(\Omega)e^{-i\Omega t}]$. In the Fano approach, the complete (A + R) system of coupled oscillators is described by its normal modes of oscillation, and $a(t)$ is just a linear combination of the lowering operators $B(\Omega, 0)e^{-i\Omega t}$ for these normal modes. Similarly, the quantized field (7.6.34) was derived as a consequence of the damping and Langevin noise forces on the atoms (oscillators) of the medium, and has the same form when expressed in terms of the bosonic operators for the normal modes (polaritons) of the coupled system of atom, bath, and field oscillators.

7.6.3 Lifshitz Theory and Stochastic Electrodynamics

In Lifshitz's paper the noise polarization is assumed to satisfy, in our units and notation, the fluctuation–dissipation relation[98]

---

[98] E. M. Lifshitz, Sov. Phys. JETP **2**, 73 (1956); to compare with Lifshitz's formulas, which are expressed in the Gaussian system of units, we multiply e^2 in our noise polarization correlation function by $4\pi\epsilon_0$, and $\epsilon_I(\omega)$ in (7.6.23) by ϵ_0, resulting in an overall factor of 4π. Thus, $\hbar\epsilon_I(\omega)/\pi$ in (7.6.92) translates to the factor $4\hbar\epsilon_I(\omega)$ in equation (1.2) of Lifshitz's paper.

$$\langle\langle K_\mu^*(\mathbf{r},\omega)K_\nu(\mathbf{r}',\omega')\rangle\rangle = \langle\langle K_\mu(\mathbf{r},\omega)K_\nu^*(\mathbf{r}',\omega')\rangle\rangle$$
$$= \frac{\hbar}{\pi}\epsilon_I(\omega)\delta_{\mu\nu}\delta(\omega-\omega')\delta^3(\mathbf{r}-\mathbf{r}')$$
$$\times\left(\frac{1}{e^{\hbar\omega/k_BT}-1}+\frac{1}{2}\right). \tag{7.6.89}$$

$K_\nu(\mathbf{r}',\omega')$ is a *classical* stochastic process, and $\langle\langle\ldots\rangle\rangle$ denotes a classical ensemble average. The appearance of $\hbar$ on the right-hand side allows classical thermal averages to reproduce quantum-electrodynamical expectation values. But, for $T \to 0$, the classical average

$$\langle\langle K_\mu^*(\mathbf{r},\omega)K_\nu(\mathbf{r}',\omega')\rangle\rangle = \langle\langle K_\mu(\mathbf{r},\omega)K_\nu^*(\mathbf{r}',\omega')\rangle\rangle$$
$$= \frac{\hbar}{2\pi}\epsilon_I(\omega)\delta_{\mu\nu}\delta(\omega-\omega')\delta^3(\mathbf{r}-\mathbf{r}'), \tag{7.6.90}$$

as opposed to the quantum-mechanical expectation values (7.6.22) and (7.6.23):

$$\langle K_\mu^\dagger(\mathbf{r},\omega)K_\nu(\mathbf{r}',\omega')\rangle = 0 \tag{7.6.91}$$

and

$$\langle K_\mu(\mathbf{r},\omega)K_\nu^\dagger(\mathbf{r}',\omega')\rangle = \frac{\hbar}{\pi}\epsilon_I(\omega)\delta_{\mu\nu}\delta(\omega-\omega')\delta^3(\mathbf{r}-\mathbf{r}'). \tag{7.6.92}$$

This fluctuation–dissipation relation derives in the quantum-electrodynamical theory from the bosonic properties of the bath oscillators associated with the Langevin force. Only (7.6.92) contributes to the mean square of the noise polarization in quantum electrodynamics, whereas, in the Lifshitz theory,

$$\langle\langle K_\mu^*(\mathbf{r},\omega)K_\nu(\mathbf{r}',\omega')\rangle\rangle = \langle\langle K_\mu(\mathbf{r},\omega)K_\nu^*(\mathbf{r}',\omega')\rangle\rangle, \tag{7.6.93}$$

and these *both* contribute to the mean squares of the noise polarization and the electric and magnetic fields. The end results of the two theories are then the same. The situation here is much like that discussed in connection with the form (7.4.20) of the electric field in zero-temperature stochastic electrodynamics, and, in this sense, Lifshitz's derivation of the force (7.5.1) might be said to be an example of the application of that theory.

7.7 Green Functions and Many-Body Theory of Dispersion Forces

The quantized fields in Section 7.6 were obtained for dissipative but homogeneous media and therefore are not directly relevant to the calculation of Casimir–Lifshitz forces. The theory of these forces is often formulated with dyadic Green functions. For dielectric media in which $\mu(\omega) \cong \mu_0$, the positive-frequency part $\mathbf{E}(\mathbf{r},\omega)$ of the electric field operator (see (7.6.6)) satisfies

$$\nabla\times\nabla\times\mathbf{E}(\mathbf{r},\omega)-\frac{\omega^2}{\epsilon_0c^2}\epsilon(\mathbf{r},\omega)\mathbf{E}(\mathbf{r},\omega)=\frac{\omega^2}{\epsilon_0c^2}\mathbf{K}(\mathbf{r},\omega), \tag{7.7.1}$$

and the dyadic Green function is the solution of

$$\nabla \times \nabla \times G(\mathbf{r}, \mathbf{r}', \omega) - \frac{\omega^2}{\epsilon_0 c^2} \epsilon(\mathbf{r}, \omega) G(\mathbf{r}, \mathbf{r}', \omega) = \delta^3(\mathbf{r} - \mathbf{r}') \qquad (7.7.2)$$

that satisfies the boundary conditions required by Maxwell's equations. The components of the electric field operator due to the noise polarization $\mathbf{K}(\mathbf{r}, \omega)$ are then

$$E_i(\mathbf{r}, \omega) = \frac{\omega^2}{\epsilon_0 c^2} \int d^3 r' G_{ij}(\mathbf{r}, \mathbf{r}', \omega) K_j(\mathbf{r}', \omega). \qquad (7.7.3)$$

(Summation over repeated indices is implied here and in most of what follows in this section.)

To derive a useful relation between an electric field correlation function and the Green function, we consider first, for simplicity, a homogeneous medium: from $\mathbf{D}(\mathbf{r}, \omega) = \epsilon(\omega)\mathbf{E}(\mathbf{r}, \omega)$, together with $\nabla \cdot \mathbf{D}(\mathbf{r}, \omega) = 0$ for a charge-neutral medium, $\nabla \cdot \mathbf{E}(\mathbf{r}, \omega) = 0$ and therefore $\nabla \times \nabla \times \mathbf{E} = -\nabla^2 \mathbf{E}$ and

$$\nabla^2 \mathbf{E}(\mathbf{r}, \omega) + \frac{\omega^2}{\epsilon_0 c^2} \epsilon(\omega) \mathbf{E}(\mathbf{r}, \omega) = -\frac{\omega^2}{\epsilon_0 c^2} \mathbf{K}(\mathbf{r}, \omega), \qquad (7.7.4)$$

$$\nabla^2 G(\mathbf{r}, \mathbf{r}', \omega) + \frac{\omega^2}{\epsilon_0 c^2} \epsilon(\omega) G(\mathbf{r}, \mathbf{r}', \omega) = -\delta^3(\mathbf{r} - \mathbf{r}'), \qquad (7.7.5)$$

$$\mathbf{E}(\mathbf{r}, \omega) = \frac{\omega^2}{\epsilon_0 c^2} \int d^3 r' G(\mathbf{r}, \mathbf{r}', \omega) \mathbf{K}(\mathbf{r}', \omega). \qquad (7.7.6)$$

The correlation function for the electric field is

$$\langle E_i(\mathbf{r}, \omega) E_j^\dagger(\mathbf{r}, \omega') \rangle = \frac{\omega^4}{\epsilon_0^2 c^4} \int d^3 r' \int d^3 r'' G(\mathbf{r}, \mathbf{r}', \omega) G^*(\mathbf{r}, \mathbf{r}'', \omega')$$
$$\times \langle K_i(\mathbf{r}', \omega) K_j^\dagger(\mathbf{r}'', \omega') \rangle, \qquad (7.7.7)$$

or, from (7.6.23),

$$\langle E_i(\mathbf{r}, \omega) E_j^\dagger(\mathbf{r}, \omega') \rangle = \frac{\hbar}{\pi} \frac{\omega^4}{\epsilon_0^2 c^4} \epsilon_I(\omega) \delta_{ij} \delta(\omega - \omega') \int d^3 r' G(\mathbf{r}, \mathbf{r}', \omega) G^*(\mathbf{r}, \mathbf{r}', \omega). \qquad (7.7.8)$$

The integral can be simplified: multiply both sides of (7.7.5) by $G^*(\mathbf{r}, \mathbf{r}', \omega)$ and subtract from the resulting equation its complex conjugate:

$$G^*(\mathbf{r}, \mathbf{r}', \omega) \nabla^2 G(\mathbf{r}, \mathbf{r}', \omega) - G(\mathbf{r}, \mathbf{r}', \omega) \nabla^2 G^*(\mathbf{r}, \mathbf{r}', \omega)$$
$$+ \frac{\omega^2}{\epsilon_0 c^2} [\epsilon(\omega) - \epsilon^*(\omega)] G^*(\mathbf{r}, \mathbf{r}', \omega) G(\mathbf{r}, \mathbf{r}', \omega)$$
$$= -[G^*(\mathbf{r}, \mathbf{r}', \omega) - G(\mathbf{r}, \mathbf{r}', \omega)] \delta^3(\mathbf{r} - \mathbf{r}'). \qquad (7.7.9)$$

Then, integrate both sides over all $\mathbf{r}'$:

$$\frac{\omega^2}{\epsilon_0 c^2} \epsilon_I(\omega) \int d^3 r' G(\mathbf{r}, \mathbf{r}', \omega) G^*(\mathbf{r}, \mathbf{r}', \omega) = \text{Im}[G(\mathbf{r}, \mathbf{r}, \omega)] \equiv G_I(\mathbf{r}, \mathbf{r}, \omega), \qquad (7.7.10)$$

and therefore (see Exercise 7.10)

$$\langle E_i(\mathbf{r}, \omega) E_j^\dagger(\mathbf{r}, \omega') \rangle = \frac{\hbar \omega^2}{\pi \epsilon_0 c^2} \delta_{ij} G_I(\mathbf{r}, \mathbf{r}, \omega) \delta(\omega - \omega'). \qquad (7.7.11)$$

The more general relation, applicable to inhomogeneous media, is (see Exercise 7.10)

$$\langle E_i(\mathbf{r}, \omega) E_j^\dagger(\mathbf{r}', \omega') \rangle = \frac{\hbar \omega^2}{\pi \epsilon_0 c^2} \text{Im}[G_{ij}(\mathbf{r}, \mathbf{r}', \omega)] \delta(\omega - \omega'), \qquad (7.7.12)$$

from which

$$\langle \mathbf{E}(\mathbf{r}, \omega) \cdot \mathbf{E}^\dagger(\mathbf{r}', \omega') \rangle = \frac{\hbar \omega^2}{\pi \epsilon_0 c^2} \text{Tr} G_I(\mathbf{r}, \mathbf{r}', \omega) \delta(\omega - \omega'), \qquad (7.7.13)$$

which generalizes (6.6.82).

Exercise 7.10: (a) What assumptions about the Green function are needed to obtain (7.7.11)? (b) Derive (7.7.12) and its extension to finite temperatures.

The application of dyadic Green functions to the theory of fluctuation-induced dispersion forces is formally straightforward. Consider a collection of $\mathcal{N}$ atoms at positions $\mathbf{r}_n$ in otherwise free space. The nth atom has an induced electric dipole moment $\mathbf{p}_n(t) = e\mathbf{x}_n(t)$ with Cartesian components

$$p_{ni}(t) = \int_0^\infty d\omega \left[p_{ni}(\omega) e^{-i\omega t} + p_{ni}^\dagger(\omega) e^{i\omega t} \right]. \qquad (7.7.14)$$

From (7.6.5),

$$(\omega_0^2 - \omega^2 - i\gamma\omega/m) p_{ni}(\omega) = \frac{e}{m} F_{Lni}(\omega) + \frac{e^2}{m} E_i(\mathbf{r}_n, \omega), \qquad (7.7.15)$$

where each component of the electric field operator is again written as a Fourier integral:

$$E_i(\mathbf{r}, t) = \int_0^\infty d\omega \left[E_i(\mathbf{r}, \omega) e^{-i\omega t} + E_i^\dagger(\mathbf{r}, \omega) e^{i\omega t} \right], \qquad (7.7.16)$$

and likewise for the Langevin force. According to the superposition principle, the electric field acting on the nth dipole is the sum of a source-free ("vacuum") field $E_{0i}(\mathbf{r}_n, \omega)$ and a field generated by all the other ($\mathcal{N}$) dipoles:

$$E_i(\mathbf{r}_n, \omega) = E_{0i}(\mathbf{r}_n, \omega) + \frac{\omega^2}{\epsilon_0 c^2} G_{ij}^0(\mathbf{r}_n, \mathbf{r}_m, \omega) p_{mj}(\omega). \qquad (7.7.17)$$

The dyadic Green function G^0 for *free space* satisfies (7.7.2) with $\epsilon(\mathbf{r}, \omega) = \epsilon_0$:

$$\nabla \times \nabla \times G^0(\mathbf{r}, \mathbf{r}', \omega) - \frac{\omega^2}{c^2} G^0(\mathbf{r}, \mathbf{r}', \omega) = \delta^3(\mathbf{r} - \mathbf{r}').\tag{7.7.18}$$

In terms of this Green function,

$$(\omega_0^2 - \omega^2 - i\gamma/m)p_{ni}(\omega) = \frac{e}{m}F_{Lni}(\omega) + \frac{e^2}{m}E_{0i}(\mathbf{r}_n, \omega)$$

$$+ \frac{e^2}{m}\frac{\omega^2}{\epsilon_0 c^2}G_{ij}^0(\mathbf{r}_n, \mathbf{r}_m, \omega)p_{mj}(\omega).\tag{7.7.19}$$

These coupled equations for the dipole moments can be written in matrix form as

$$p(\omega) = \alpha_0 \frac{\omega^2}{\epsilon_0 c^2}G^0 p(\omega) + \alpha_0 E_0 + \frac{\alpha_0}{e}F_L,\tag{7.7.20}$$

where $p(\omega)$, $F_L(\omega)$, and $E_0(\omega)$ are $3\mathcal{N}$-dimensional column vectors, G^0 is a $3\mathcal{N} \times 3\mathcal{N}$ matrix, and

$$\alpha_0(\omega) = \frac{e^2/m}{\omega_0^2 - \omega^2 - i\gamma\omega/m}\tag{7.7.21}$$

is the polarizability of each atom. The solution of the set of $3\mathcal{N}$ equations in (7.7.20) is

$$p(\omega) = \left(1 - \alpha_0 G_0\right)^{-1}\left(\alpha_0 E_0 + \frac{\alpha_0}{e}F_L\right),\tag{7.7.22}$$

where "1" here is the $3\mathcal{N} \times 3\mathcal{N}$ unit matrix and to simplify the notation slightly, we have defined

$$G_0 = \frac{\omega^2}{\epsilon_0 c^2}G^0.\tag{7.7.23}$$

We now calculate the free energy of the equilibrium system of atoms in the presence of the vacuum electric field E_0. One way to do this is to use the "remarkable formula" (6.7.36). For the system of interest here, in which we are dealing not with a single, one-dimensional oscillator coupled to a heat bath but a set of $\mathcal{N}$ three-dimensional oscillators, the generalized susceptibility $\alpha(\omega)$ in the remarkable formula is replaced by the matrix $\alpha_0(1-\alpha_0 G_0)^{-1}$ that characterizes the linear response of $p(\omega)$ to the "applied force" E_0 acting on the collection of oscillators. And since the different oscillators as well as the different components of each oscillator are uncorrelated in thermal equilibrium, we sum (trace) over the $3\mathcal{N}$ contributions to the free energy. For simplicity, we focus on the case of zero temperature; the generalization to finite temperature is easy. Thus, the energy of interest becomes

$$U = \frac{\hbar}{2\pi}\text{Im}\int_0^\infty d\omega\,\omega\,\text{Tr}\frac{d}{d\omega}\log\left[\alpha_0\left(1 - \alpha_0 G_0\right)^{-1}\right]$$

$$= -\frac{\hbar}{2\pi}\text{Im}\int_0^\infty d\omega\,\text{Tr}\log\left[\alpha_0\left(1 - \alpha_0 G_0\right)^{-1}\right]$$

$$= -\frac{\hbar}{2\pi}\text{Im}\int_0^\infty d\omega\,\text{Tr}\log\alpha_0 + \frac{\hbar}{2\pi}\text{Im}\int_0^\infty d\omega\,\text{Tr}\log\left(1 - \alpha_0 G_0\right)\tag{7.7.24}$$

after a partial integration. The part of U that depends on G_0 and therefore accounts for the interaction of the dipoles with the electric field is

$$W = \frac{\hbar}{2\pi}\mathrm{Im}\int_0^\infty d\omega\,\mathrm{Tr}\log\left(1 - \alpha_0 G_0\right). \tag{7.7.25}$$

This formula can also be derived using the Hellmann–Feynman theorem.[99] Note that, in the trace operation here, the spatial coordinates $\mathbf{r}_n, \mathbf{r}_m$ are regarded along with Cartesian components i, j as matrix indices. Using the identity (7.5.32), we can write W alternatively as

$$W = \frac{\hbar}{2\pi}\mathrm{Im}\int_0^\infty d\omega\,\log\det\left(1 - \alpha_0 G_0\right). \tag{7.7.26}$$

Writing (7.7.25) more explicitly shows its many-body character:

$$W = -\frac{\hbar}{2\pi}\mathrm{Im}\int_0^\infty d\omega\,\mathrm{Tr}\left(\alpha_0 G_0 + \frac{1}{2}\alpha_0^2 G_0 G_0 + \frac{1}{3}G_0 G_0 G_0 + \ldots\right). \tag{7.7.27}$$

The contribution from the first term in brackets is a sum of (divergent) single-particle self-energies:

$$W_1 = \frac{\hbar}{2\pi}\int_0^\infty d\omega\left[\alpha_{0R}(G_{0I})_{ii}(\mathbf{r}_n, \mathbf{r}_n, \omega) + \alpha_{0I}(\omega)(G_{0R})_{ii}(\mathbf{r}_n, \mathbf{r}_n, \omega)\right]. \tag{7.7.28}$$

The second term in (7.7.27) contributes

$$W_2 = -\frac{\hbar}{2\pi}\mathrm{Im}\int_0^\infty d\omega\,\frac{1}{2}\alpha_0^2(\omega)G_{0ij}(\mathbf{r}_n, \mathbf{r}_m, \omega)G_{0ji}(\mathbf{r}_m, \mathbf{r}_n, \omega), \tag{7.7.29}$$

and is found to be just the sum of pairwise van der Waals interactions of $\mathcal{N}$ atoms when self-energy terms with $m = n$ are excluded (see Exercise 7.11). In the continuum approximation, we assume a density $N(\mathbf{r})$ of atoms and replace the summation by

$$\int d^3r \int d^3r'\, N(\mathbf{r})N(\mathbf{r}')\alpha_0^2(\omega)G_{0ij}(\mathbf{r}, \mathbf{r}', \omega)G_{0ji}(\mathbf{r}', \mathbf{r}, \omega) \equiv \mathrm{Tr}\left[(\epsilon - \epsilon_0)G_0\right]^2, \tag{7.7.30}$$

where we have used the relation $N(\mathbf{r})\alpha_0(\omega) = \epsilon(\mathbf{r}, \omega) - \epsilon_0$. Similarly, we replace (7.7.27) by

$$W = -\frac{\hbar}{2\pi}\mathrm{Im}\int_0^\infty d\omega\,\mathrm{Tr}\sum_{n=1}^\infty \frac{1}{n}\left(\epsilon - \epsilon_0\right)^n (G_0)^n \tag{7.7.31}$$

and define an energy density

$$u(\mathbf{r}) = \frac{\hbar}{2\pi}\mathrm{Im}\,\mathrm{Tr}\int_0^\infty d\omega\,\log\left[1 - (\epsilon - \epsilon_0)G_0\right]. \tag{7.7.32}$$

[99] F. S. S. Rosa, D. A. R. Dalvit, and P. W. Milonni, Phys. Rev. A **81**, 033812 (2010), and F. S. S. Rosa, D. A. R. Dalvit, and P. W. Milonni, Phys. A **84**, 053813 (2011).

Exercise 7.11: Using the dyadic Green function, (6.6.84), show that the terms with $m \neq n$ in W_2 are van der Waals interaction energies, given by (7.1.20), between atoms m and n. Note:

$$\sum_{i,j=1}^{3} \delta_{ij}\delta_{ij} = 3, \quad \sum_{i,j=1}^{3} \delta_{ij}\hat{\mathbf{R}}_i\hat{\mathbf{R}}_j = 1, \quad \sum_{i,j=1}^{3} \hat{\mathbf{R}}_i^2\hat{\mathbf{R}}_j^2 = 1.$$

The many-body formula (7.7.24) for the energy can be expressed differently using the Green function G defined by (7.7.2). Subtraction of (7.7.18) from (7.7.2) gives

$$\nabla \times \nabla \times (G - G^0) - \frac{\omega^2}{c^2}(G - G^0) = \frac{\omega^2}{\epsilon_0 c^2}(\epsilon - \epsilon_0)G \tag{7.7.33}$$

and therefore

$$G = G^0 + \frac{\omega^2}{\epsilon_0 c^2}G^0(\epsilon - \epsilon_0)G, \tag{7.7.34}$$

or, more explicitly, a "Lippmann–Schwinger" equation

$$G_{ij}(\mathbf{r}, \mathbf{r}, \omega) = G_{ij}^0(\mathbf{r}, \mathbf{r}, \omega) + \frac{\omega^2}{\epsilon_0 c^2} \int d^3 r' G_{ik}^0(\mathbf{r}, \mathbf{r}', \omega)\left[\epsilon(\mathbf{r}', \omega) - \epsilon_0\right]G_{kj}(\mathbf{r}, \mathbf{r}', \omega). \tag{7.7.35}$$

From (7.7.23), (7.7.32), and (7.7.34),

$$u(\mathbf{r}) = \frac{\hbar}{2\pi}\mathrm{Im}\,\mathrm{Tr}\int_0^\infty d\omega \log\left(G^0 G^{-1}\right). \tag{7.7.36}$$

We wish to express $u(\mathbf{r})$ in terms of the permittivity $\epsilon(\mathbf{r}, \omega)$, whose spatial variation will result in a variation in field energy density and give rise to forces. G^0 is independent of any properties of the medium, and its contribution can therefore be ignored, leaving the interaction energy

$$u(\mathbf{r}) = -\frac{\hbar}{2\pi}\mathrm{Im}\,\mathrm{Tr}\int_0^\infty d\omega \log G. \tag{7.7.37}$$

Now, by partial integration,

$$u(\mathbf{r}) = \frac{\hbar}{2\pi}\mathrm{Im}\,\mathrm{Tr}\int_0^\infty d\omega\,\omega\frac{\partial}{\partial\omega}\log G = \frac{\hbar}{2\pi}\mathrm{Im}\,\mathrm{Tr}\int_0^\infty d\omega\,\omega G^{-1}G', \tag{7.7.38}$$

the prime denoting differentiation with respect to ω.[100] From (7.7.2),

$$\nabla \times \nabla \times G' - \frac{\omega^2}{\epsilon_0 c^2}\epsilon G' = \frac{\omega}{\epsilon_0 c^2}\left(2\epsilon + \omega\epsilon'\right)G, \tag{7.7.39}$$

---

[100] K. A. Milton, J. Wagner, P. Parashar, and I. Brevik, Phys. Rev. D **81**, 065007 (2010).

$$G^{-1}G' = \frac{\omega}{\epsilon_0 c^2}(2\epsilon + \omega\epsilon')G, \tag{7.7.40}$$

and, from (7.7.38), finally,

$$\begin{aligned} u(\mathbf{r}) &= \frac{\hbar}{2\pi\epsilon_0 c^2}\mathrm{Im}\mathrm{Tr}\int_0^\infty d\omega\omega^2\big[2\epsilon(\mathbf{r},\omega) + \omega\epsilon'(\mathbf{r},\omega)\big]G(\mathbf{r},\mathbf{r},\omega) \\ &= \frac{\hbar}{2\pi\epsilon_0 c^2}\mathrm{Im}\int_0^\infty d\omega\omega^2\big[2\epsilon(\mathbf{r},\omega) + \omega\epsilon'(\mathbf{r},\omega)\big]G_{ii}(\mathbf{r},\mathbf{r},\omega). \end{aligned} \tag{7.7.41}$$

This energy density has a familiar form. For a homogeneous, non-absorbing medium, for example, $\epsilon(\omega)$ is real, and

$$\begin{aligned} u(\mathbf{r}) &= \frac{\hbar}{2\pi\epsilon_0 c^2}\int_0^\infty d\omega\omega^2\big[2\epsilon(\omega) + \omega\epsilon'(\omega)\big]\mathrm{Tr}G_I(\mathbf{r},\mathbf{r},\omega) \\ &= \frac{\hbar}{2\pi\epsilon_0 c^2}\int_0^\infty d\omega\int_0^\infty d\omega'\omega^2\big[2\epsilon(\omega) + \omega\epsilon'(\omega)\big]\mathrm{Tr}G_I(\mathbf{r},\mathbf{r},\omega)\delta(\omega-\omega') \\ &= \frac{1}{2}\int_0^\infty d\omega\int_0^\infty d\omega'\big[2\epsilon(\omega) + \omega\epsilon'(\omega)\big]\langle\mathbf{E}(\mathbf{r},\omega)\cdot\mathbf{E}^\dagger(\mathbf{r},\omega')\rangle, \end{aligned} \tag{7.7.42}$$

where we have used (7.7.13). This has an obvious similarity to the average energy density (1.6.27), which was derived in classical electromagnetic theory for fields in which different frequencies are uncorrelated.

More generally, (7.7.41) involves the real as well as the imaginary part of the Green function:

$$\begin{aligned} u(\mathbf{r}) = \frac{\hbar}{2\pi\epsilon_0 c^2}\mathrm{Tr}\int_0^\infty d\omega\omega^2\Big\{&\big[2\epsilon_R(\mathbf{r},\omega) + \omega\epsilon'_R(\mathbf{r},\omega)\big]G_I(\mathbf{r},\mathbf{r},\omega) \\ &+ \big[2\epsilon_I(\mathbf{r},\omega) + \omega\epsilon'_I(\mathbf{r},\omega)\big]G_R(\mathbf{r},\mathbf{r},\omega)\Big\}. \end{aligned} \tag{7.7.43}$$

Whereas $\mathrm{Tr}G_I(\mathbf{r},\mathbf{r},\omega) = \omega/2\pi c$ (see (6.6.85)), $\mathrm{Tr}G_R(\mathbf{r},\mathbf{r},\omega)$ diverges, a consequence of the continuum model in which there are no spaces between the atoms. If we retain only the finite part of $u(\mathbf{r})$, the result for a dissipative homogeneous medium, for example, is found to be just (7.6.58).

To calculate forces we consider, following the seminal work of Dzyaloshinski et al.,[101] a variation $\delta\epsilon$ in ϵ leading to a variation $\delta u(\mathbf{r})$ in the energy density:

$$\delta u(\mathbf{r}) = \frac{\hbar}{2\pi\epsilon_0 c^2}\mathrm{Im}\int_0^\infty d\omega\omega^2\big[(2\delta\epsilon + \omega\delta\epsilon')G_{ii} + (2\epsilon + \omega\epsilon')\delta G_{ii}\big]. \tag{7.7.44}$$

To put this in a more convenient form, we note first that, from (7.7.2),

$$\nabla\times\nabla\times\delta G - \frac{\omega^2}{\epsilon_0 c^2}\epsilon\delta G = \frac{\omega^2}{\epsilon_0 c^2}\delta\epsilon G, \tag{7.7.45}$$

[101] I. E. Dzyaloshinski, E. M. Lifshitz, and L. P. Pitaevskii, Sov. Phys. JETP **9**, 1282 (1959); I. E. Dzyaloshinski, E. M. Lifshitz, and L. P. Pitaevskii, Adv. Phys. **10**, 165 (1965); see also *Abrikosov, Gorkov, and Dzyaloshinski*.

and therefore

$$\delta G = \frac{\omega^2}{\epsilon_0 c^2} \delta \epsilon G G,$$ (7.7.46)

and

$$
\begin{aligned}
\left(2\delta\epsilon + \omega\delta\epsilon'\right)G + \left(2\epsilon + \omega\epsilon'\right)\delta G &= \left(2\delta\epsilon + \omega\delta\epsilon'\right)G + \left(2\epsilon + \omega\epsilon'\right)\frac{\omega^2}{\epsilon_0 c^2}\delta\epsilon G G \\
&= \left(2\delta\epsilon + \omega\delta\epsilon'\right)G + \omega\delta\epsilon G' \\
&= 2\delta\epsilon G + \omega\frac{\partial}{\partial\omega}\left(\delta\epsilon G\right),
\end{aligned}
$$ (7.7.47)

where we have used (7.7.46) and (7.7.40).[102] Partial integration then yields

$$\delta u(\mathbf{r}) = -\frac{\hbar}{2\pi\epsilon_0 c^2}\mathrm{Im}\int_0^\infty d\omega\omega^2\delta\epsilon(\mathbf{r},\omega)G_{ii}(\mathbf{r},\mathbf{r},\omega).$$ (7.7.48)

This result for $\delta u(\mathbf{r})$ can also be derived starting from the formula (7.7.32) expressing $u(\mathbf{r})$ in terms of ϵ and the free-space Green function G_0. Since G_0 is independent of any properties of the medium, any variation of $u(\mathbf{r})$ can only result from a variation $\delta\epsilon$ in ϵ:

$$\delta u(\mathbf{r}) = -\frac{\hbar}{2\pi}\mathrm{Im}\mathrm{Tr}\int_0^\infty d\omega\sum_{n=0}^\infty (\epsilon - \epsilon_0)^n\delta\epsilon)G_0)^{n+1} = -\frac{\hbar}{2\pi}\mathrm{Im}\mathrm{Tr}\int_0^\infty d\omega\delta\epsilon$$
$$\times\left[G_0 + (\epsilon - \epsilon_0)G_0 G_0 + (\epsilon - \epsilon_0)^2 G_0 G_0 G_0 + \ldots\right].$$ (7.7.49)

The term in brackets is seen from (7.7.49) to be just $(\omega^2/\epsilon_0 c^2)G$; thus, we obtain again (7.7.48).

Now, a *local* infinitesimal displacement $\mathbf{R}(\mathbf{r})$ of the medium results in a variation in $\epsilon(\mathbf{r},\omega)$ at $\mathbf{r}$: $\epsilon(\mathbf{r} - \mathbf{R},\omega) = \epsilon(\mathbf{r},\omega) + \delta\epsilon(\mathbf{r},\omega)$, $\delta\epsilon(\mathbf{r},\omega) = -\nabla\epsilon\cdot\mathbf{R}$. Then,

$$\delta u(\mathbf{r}) = -\frac{\hbar}{2\pi\epsilon_0 c^2}\mathrm{Im}\int_0^\infty d\omega\omega^2\left(-\nabla\epsilon\cdot\mathbf{R}\right)G_{ii}(\mathbf{r},\mathbf{r},\omega),$$ (7.7.50)

implying a force density[103]

$$\mathbf{f}(\mathbf{r}) = -\frac{\hbar}{2\pi\epsilon_0 c^2}\mathrm{Im}\int_0^\infty d\omega\omega^2\nabla\epsilon(\mathbf{r},\omega)G_{ii}(\mathbf{r},\mathbf{r},\omega).$$ (7.7.51)

The generalization for finite temperatures T is

$$\mathbf{f}(\mathbf{r}) = -\frac{\hbar}{2\pi\epsilon_0 c^2}\mathrm{Im}\int_0^\infty d\omega\omega^2\nabla\epsilon(\mathbf{r},\omega)G_{ii}(\mathbf{r},\mathbf{r},\omega)\coth(\hbar\omega/2k_B T).$$ (7.7.52)

---

[102] K. A. Milton, J. Wagner, P. Parashar, and I. Brevik, Phys. Rev. D **81**, 065007 (2010).

[103] I. E. Dzyaloshinski, E. M. Lifshitz, and L. P. Pitaevskii, Sov. Phys. JETP **9**, 1282 (1959); I. E. Dzyaloshinski, E. M. Lifshitz, and L. P. Pitaevskii, Adv. Phys. **10**, 165 (1965); see also *Abrikosov, Gorkov, and Dzyaloshinski*.

These formulas for the force density are compact and general for the force density in dissipative and dispersive dielectric media. As is always the case with fluctuation-induced dispersion forces, they can be evaluated analytically only for simple geometries. The simplest, not surprisingly, is the most studied one (see Figure 7.3), for which

$$\epsilon(\mathbf{r}, \omega) = \epsilon(z, \omega) = \epsilon_1(\omega)\theta(-z) + \epsilon_2(\omega)\theta(z - d) + \epsilon_3(\omega)[\theta(z) - \theta(z - d)]. \quad (7.7.53)$$

The calculation is made easy by the fact that the derivatives of the step functions θ result in delta functions in z. Once the Green function is determined by solution of the classical boundary-value problem for this geometry, the finite, d-dependent force per unit area follows. The Green function involves reflection coefficients at the two interfaces, and the force obtained is, of course, just that given by the Lifshitz formula: minus the derivative with respect to d of the energy (7.5.28).[104]

7.8   Do Casimir Forces Imply the Reality of Zero-Point Energy?

Van der Waals and Casimir interactions are subjects of entire books. We have focused only on some of the basic physics involved and have emphasized the role of zero-point energy and fluctuations of fields and polarizations. Casimir effects in particular are almost universally regarded as consequences, and even proof, of zero-point field energy. This is hardly surprising in light of Casimir's original calculation of the force between conducting plates. While zero-point energy per se appears less explicitly in his theory for dielectric media, Lifshitz clearly attributes the force to the fluctuating electromagnetic field, including its zero-point fluctuations:

The interaction of the objects is regarded as occurring through the medium of the fluctuating electromagnetic field which is always present in the interior of any absorbing medium, and also extends beyond its boundaries—partially in the form of travelling waves radiated by the body, partially in the form of standing waves which are damped exponentially as we move away from the surface of the body. It must be emphasized that this field does not vanish even at absolute zero, at which point it is associated with the zero point vibrations of the radiation field.[105]

Casimir's original calculation of the force between perfectly conducting plates—and the subsequent measurements of this force—leave little doubt as to the physical significance of zero-point energy. The Lifshitz force for dielectric media, similarly, is attributable, at $T = 0$, to zero-point field energy and momentum (see Section 7.5.1). Of course, there is no doubt that zero-point energy is a general feature of quantum theory: as shown in Section 3.1.1, it is required by uncertainty relations. Experimental evidence for the reality of zero-point energy is found in the spectra of diatomic molecules, x-ray diffraction at low temperatures, and liquid helium, among many other phenomena (see Section 7.4). But can Casimir forces be said to "prove" to the same extent the reality of zero-point energy?

---

[104] See, for instance, *Milton* or F. S. S. Rosa, D. A. R. Dalvit, and P. W. Milonni, Phys. A **84**, 053813 (2011).

[105] E. M. Lifshitz, Sov. Phys. JETP **2**, 73 (1956), p. 73.

In Schwinger's "source theory," there are no zero-point field fluctuations or zero-point energy: "the vacuum is regarded as truly a state with all physical properties equal to zero."[106] Yet, Schwinger et al. were able to derive the Casimir force within source theory, and later work showed that, in more conventional formulations of quantum electrodynamics, "Casimir effects can be formulated and Casimir forces can be computed without reference to zero-point energies."[107] In other words, Casimir forces can be calculated in different ways, and not only as direct consequences of zero-point energies. We now present a brief, if facile, argument for why this is so.

The matter–field interaction leading to Casimir forces is the energy of dipoles *induced* by an electric field:

$$E = -\frac{1}{2} \int d^3r \mathbf{P} \cdot \mathbf{E}. \tag{7.8.1}$$

The electric field (operator) is a vacuum field $\mathbf{E}_0$ plus a source field $\mathbf{E}_s$ due to all the dipoles:

$$\mathbf{E}(\mathbf{r}, t) = \mathbf{E}_0(\mathbf{r}, t) + \mathbf{E}_s(\mathbf{r}, t). \tag{7.8.2}$$

Each of these fields can be expressed as a positive-frequency (+) part and a negative-frequency (−) part:

$$\mathbf{E}_0(\mathbf{r}, t) = \mathbf{E}_0^{(+)}(\mathbf{r}, t) + \mathbf{E}_0^{(-)}(\mathbf{r}, t), \tag{7.8.3}$$

$$\mathbf{E}_s(\mathbf{r}, t) = \mathbf{E}_s^{(+)}(\mathbf{r}, t) + \mathbf{E}_s^{(-)}(\mathbf{r}, t), \tag{7.8.4}$$

$$\mathbf{E}(\mathbf{r}, t) = \mathbf{E}^{(+)}(\mathbf{r}, t) + \mathbf{E}^{(-)}(\mathbf{r}, t), \tag{7.8.5}$$

$$\mathbf{E}^{(\pm)}(\mathbf{r}, t) = \mathbf{E}_0^{(\pm)}(\mathbf{r}, t) + \mathbf{E}_s^{(\pm)}(\mathbf{r}, t). \tag{7.8.6}$$

The operators $\mathbf{E}^{(\pm)}(\mathbf{r}, t)$ commute with the (equal-time) matter operator $\mathbf{P}(\mathbf{r}, t)$, and so we can write (7.8.1) in different, equivalent ways. In a normal ordering of field operators, for example, there is no explicit contribution from the vacuum field, and $\langle E \rangle$ appears as just the result of dipole–dipole interactions:

$$\langle E \rangle = -\frac{1}{2} \int d^3r \langle \mathbf{P}(\mathbf{r}, t) \cdot \mathbf{E}_s^{(+)}(\mathbf{r}, t) + \mathbf{E}_s^{(-)}(\mathbf{r}, t) \cdot \mathbf{P}(\mathbf{r}, t) \rangle,$$

$$= -\mathrm{Re} \int d^3r \langle \mathbf{P}(\mathbf{r}, t) \cdot \mathbf{E}_s^{(+)}(\mathbf{r}, t) \rangle$$

$$= -\mathrm{Re} \int d^3r \int d^3r' \langle P_i(\mathbf{r}, t) G_{ij}(\mathbf{r}, \mathbf{r}', t) P_j(\mathbf{r}', t) \rangle. \tag{7.8.7}$$

Except for a sign, this has essentially the same form as the action W in the work of Schwinger et al., from which they derive (7.7.48), and eventually the Lifshitz and Casimir forces, by introducing an infinitesimal variation $\delta\epsilon$ in the permittivity.[108]

[106] J. Schwinger, L. L. DeRaad, Jr., and K. A. Milton, Ann. Phys. (N. Y.) **115**, 1 (1978), p. 3.

[107] P. W. Milonni, Phys. Rev. A **25**, 1315 (1982); P. W. Milonni and M.-L. Shih, Phys. Rev. A **45**, 4241 (1992); R. L. Jaffe, Phys. Rev. D **72**, 021301(R) (2005); quote from p. 1.

[108] J. Schwinger, L. L. DeRaad, Jr., and K. A. Milton, Ann. Phys. (N. Y.) **115**, 1 (1978).

By following their analysis after that point, one likewise derives these forces from (7.8.7).[109]

From the present perspective of conventional quantum electrodynamics, upon which (7.8.7) is of course based, the fact that the Casimir force (and Casimir effects more generally) can be derived without explicit reference to the vacuum field implies that the reality of Casimir effects does not prove the reality of zero-point field energy. But, of course, zero-point energy and zero-point fluctuations in quantum theory are real phenomena. The situation here is very much like that discussed in Chapter 4 with regard to spontaneous emission and radiative frequency shifts: how we choose to order commuting matter and field operators leads us to different interpretations of these effects, but both source fields and vacuum fields are required for the internal consistency of the theory. As discussed in connection with the fluctuation–dissipation theorem, both fields are required to maintain canonical commutation relations (see Section 4.6). In the present context, the fluctuation–dissipation relation (7.7.13) between the electric field correlation function characterizing field fluctuations, and the Green function that determines source fields, exemplifies the fundamental inseparability of vacuum and source fields.

Exercise 7.12: (a) How, if at all, do source fields appear in the approach to dispersion and Casimir forces, based on (7.7.32)? (b) Can the effects of zero-point energy cited in Section 7.4 be attributed alternatively to source fields?

7.9   The Dipole–Dipole Resonance Interaction

The interactions considered thus far in this chapter do not involve excited states of atoms. We now consider some basic interactions involving excited states. We begin with an example of long-standing interest—the interaction energy and the radiation rate of two atoms, one of which (Atom A) is initially excited. For the purpose of calculating the interaction energy and the radiation rate, the atoms can be accurately modeled as two-state systems (see Figure 7.5) separated by a distance r.

Fig. 7.5 Model for the dipole–dipole resonance interaction of two atoms.

---

[109] P. W. Milonni, Phys. Rev. A **25**, 1315 (1982); P. W. Milonni and M.-L. Shih, Phys. Rev. A **45**, 4241 (1992).

For identical atoms ($\omega_A = \omega_B \equiv \omega_0$) in free space, the probability amplitudes for the states $|\phi_A\rangle = |+\rangle_A|-\rangle_B|\text{vac}\rangle$ and $|\phi_B\rangle = |-\rangle_A|+\rangle_B|\text{vac}\rangle$ are found to be

$$b_A(t) = \frac{1}{2}\exp\{-\beta[1+g(r)]t\} + \frac{1}{2}\exp\{-\beta[1-g(r)]t\} \tag{7.9.1}$$

and

$$b_B(t) = \frac{1}{2}\exp\{-\beta[1+g(r)]t\} - \frac{1}{2}\exp\{-\beta[1-g(r)]t\}, \tag{7.9.2}$$

respectively.[110] $|\phi_A\rangle$ is the state in which Atom A is excited, Atom B is unexcited, and the field is in the vacuum state $|\text{vac}\rangle$ of no photons; $|\phi_B\rangle$ is the state in which Atom A is unexcited, Atom B is excited, and the field is in the vacuum state. β is half the Einstein A coefficient (spontaneous emission rate) of each atom alone (recall (4.4.21)):

$$\beta = \frac{1}{2}A_{21} = \frac{1}{4\pi\epsilon_0}\frac{2|d|^2\omega_0^3}{3\hbar c^3}, \tag{7.9.3}$$

and

$$g(r) = \frac{3}{2}\left(p\frac{\sin k_0 r}{k_0 r} + q\frac{\sin k_0 r}{k_0^3 r^3} - q\frac{\cos k_0 r}{k_0^2 r^2}\right) - \frac{3i}{2}\left(p\frac{\cos k_0 r}{k_0 r} + q\frac{\cos k_0 r}{k_0^3 r^3} + q\frac{\sin k_0 r}{k_0^2 r^2}\right), \tag{7.9.4}$$

$$p = \hat{\mathbf{d}}_A \cdot \hat{\mathbf{d}}_B - (\hat{\mathbf{d}}_A \cdot \hat{\mathbf{r}})(\hat{\mathbf{d}}_B \cdot \hat{\mathbf{r}}),$$
$$q = 3(\hat{\mathbf{d}}_A \cdot \hat{\mathbf{r}})(\hat{\mathbf{d}}_B \cdot \hat{\mathbf{r}}) - \hat{\mathbf{d}}_A \cdot \hat{\mathbf{d}}_B, \tag{7.9.5}$$

where $k_0 = \omega_0/c$, $\hat{\mathbf{d}}_j$ is the unit vector in the direction of the transition dipole moment $\mathbf{d}_j$ of Atom j ($j = A, B$), $\hat{\mathbf{r}}$ is a unit vector pointing from one atom to the other, and the transition dipole moments are taken, without loss of generality, to be real. As indicated by the terms varying as $\sin k_0 r$ and $\cos k_0 r$ in $g(r)$, the probability amplitudes (7.9.1) and (7.9.2) include the effect of retardation in the dipole–dipole interaction.

These probability amplitudes apply when it is known for certain which atom is excited at a time $t = 0$. If the atoms are very close together, however, it is more likely that the excitation mechanism will leave both atoms with a finite excitation probability. Excitation by a resonant field, for instance, will excite the atoms with approximately equal probabilities if the atoms are very close together compared with the wavelength, since the two atoms then see the same field. It is more convenient then to work with the symmetric (+) and antisymmetric (−) entangled states

$$|\phi_\pm\rangle = \frac{1}{\sqrt{2}}\left(|\phi_A\rangle \pm |\phi_B\rangle\right) \tag{7.9.6}$$

rather than $|\phi_A\rangle$ and $|\phi_B\rangle$. If the state of the atom–field system at time $t = 0$ is $|\phi_+\rangle$, for example, the probability amplitude for this state at a time $t \geq 0$ is

---

[110] See, for instance, M. J. Stephen, J. Chem. Phys. **40**, 669 (1964).

$$c_+(t) = e^{-\beta(1+\mathrm{Re}[g(r)])t}e^{-\beta\mathrm{Im}[g(r)]t} = e^{-\beta_+(r)t}e^{-i\delta E_+(r)t/\hbar}, \qquad (7.9.7)$$

and its probability is $|c_+(t)|^2 = \exp[-2\beta_+(r)t]$, where the radiative damping rate is

$$2\beta_+(r) = 2\beta\{1 + \mathrm{Re}[g(r)]\} = 2\beta\left[1 + \frac{3}{2}\left(p\frac{\sin k_0 r}{k_0 r} - q\frac{\cos k_0 r}{k_0^2 r^2} + q\frac{\sin k_0 r}{k_0^3 r^3}\right)\right] \qquad (7.9.8)$$

and the energy shift is

$$\delta E_+(r) = \hbar\beta\mathrm{Im}[g(r)] = -\frac{3\hbar\beta}{2}\left(p\frac{\cos k_0 r}{k_0 r} + q\frac{\sin k_0 r}{k_0^2 r^2} + q\frac{\cos k_0 r}{k_0^3 r^3}\right). \qquad (7.9.9)$$

For $k_0 r \ll 1$,

$$2\beta_+(r) \cong 2\beta\left(1 + \frac{3}{2}p + \frac{1}{2}q\right) = 4\beta = 2A_{21}. \qquad (7.9.10)$$

The radiation rate by the two atoms in this case is twice the single-atom spontaneous emission rate A_{21}. This is the simplest example of *superradiance*, in which $\mathcal{N}$ atoms radiate at a rate $\mathcal{N}$ times the spontaneous emission rate of a single atom.[111] If the initial state is $|\phi_-\rangle$, in contrast, its probability amplitude at time t is

$$c_-(t) = e^{-\beta(1-\mathrm{Re}[g(r)])t}e^{\beta\mathrm{Im}[g(r)]t} = e^{-\beta_-(r)t}e^{-i\delta E_-(r)t/\hbar}, \qquad (7.9.11)$$

with a damping rate

$$2\beta_-(r) = 2\beta\{1 - \mathrm{Re}[g(r)]\} = 2\beta\left[1 - \frac{3}{2}\left(p\frac{\sin k_0 r}{k_0 r} - q\frac{\cos k_0 r}{k_0^2 r^2} + q\frac{\sin k_0 r}{k_0^3 r^3}\right)\right] \qquad (7.9.12)$$

and an energy shift

$$\delta E_-(r) = -\hbar\beta\mathrm{Im}[g(r)] = \frac{3\hbar\beta}{2}\left(p\frac{\cos k_0 r}{k_0 r} + q\frac{\sin k_0 r}{k_0^2 r^2} + q\frac{\cos k_0 r}{k_0^3 r^3}\right). \qquad (7.9.13)$$

For $k_0 r \ll 1$,

$$\beta_-(r) \cong \beta\left(1 - \frac{3}{2}p - \frac{1}{2}q\right) = 0. \qquad (7.9.14)$$

In this small-separation limit, there is no radiation when the atoms are in the antisymmetric state. The excitation in this "subradiant" case is trapped in the atoms.

The state $|\phi_A\rangle$ is a superposition of the states $|\phi_\pm\rangle$: $|\phi_A\rangle = (|\phi_+\rangle + |\phi_-\rangle)/\sqrt{2}$. If the atoms are very close together ($k_0 r \ll 1$), and the initial state is $|\phi_A\rangle$, therefore, the system at times $t \gg 1/\beta$ will be left in the nonradiative, stationary state $|\phi_-\rangle$, with an interaction energy

$$\delta E_-(r) \cong \frac{3\hbar\beta}{2}q\frac{\cos k_0 r}{k_0^3 r^3} = \frac{1}{4\pi\epsilon_0}\left[3(\hat{\mathbf{d}}_A \cdot \hat{\mathbf{r}})(\hat{\mathbf{d}}_B \cdot \hat{\mathbf{r}}) - \hat{\mathbf{d}}_A \cdot \hat{\mathbf{d}}_B\right]\frac{|\mathbf{d}|^2}{r^3}. \qquad (7.9.15)$$

Except for having a minus sign because it refers to the antisymmetric state, $\delta E_-(r)$ has the same form as the classical, near-field dipole–dipole interaction. In fact, the r

[111] R. H. Dicke, Phys. Rev. **93**, 99 (1954).

dependence of the expressions $\beta_\pm(r)$ and $\delta E_\pm(r)$ can all be obtained from a purely classical model of two electric dipoles oscillating in phase $(+)$ or π out of phase $(-)$.[112] For $k_0 r > 1$, there is no stationary state, and the energies $\delta E_\pm(r)$ are not (time-independent) interaction energies in the usual sense.

Single-atom radiative damping in the results summarized above is treated in the Weisskopf–Wigner approximation of exponential decay, as can be seen from the large-separation limit $(k_0 r \to \infty)$ of (7.9.1) and (7.9.2), the limit in which there is no influence of one atom on the other:

$$b_A(t) \cong e^{-\beta t}, \quad \text{and} \quad b_B(t) \cong 0. \tag{7.9.16}$$

Other approximations are made in obtaining the amplitudes (7.9.1) and (7.9.2). It is assumed that the atoms are far enough apart that there is effectively no overlap of the electron wave functions, and that the time r/c it takes for the field to propagate from one atom to the other is so small as to be completely negligible. In an analysis that improves on the latter approximation, it is found that

$$b_A(t) = \sum_{n=0,2,4,\ldots}^{\infty} \frac{1}{n!} \left[\beta g(r)\left(t - \frac{nr}{c}\right) \right]^n \theta\left(\left(t - \frac{nr}{c}\right) e^{-\beta(t - nr/c)}, \tag{7.9.17}$$

$$b_B(t) = \sum_{n=1,3,5,\ldots}^{\infty} \frac{1}{n!} \left[-\beta g(r)\left(t - \frac{nr}{c}\right) \right]^n \theta\left(\left(t - \frac{nr}{c}\right) e^{-\beta(t - nr/c)}, \tag{7.9.18}$$

where $\theta(t)$ is the unit step function.[113] The appearance of the step functions $\theta(t - nr/c)$ has an obvious physical interpretation: the initially unexcited Atom B has zero probability of excitation until $t > r/c$, the time it takes for light to propagate from A to B, and the initially excited Atom A is not affected by the presence of Atom B until time $t > 2r/c$, the time it takes for light to propagate from A to B and back to A, and so forth. For times $t \gg r/c$, the amplitudes (7.9.1) and (7.9.2) are recovered. Thus,

$$b_A(t) \cong e^{-\beta t} \sum_{n=0,2,4,\ldots}^{\infty} \frac{1}{n!} \left[\beta g(r)t \right]^n, \quad t \gg r/c, \tag{7.9.19}$$

which is the same as (7.9.1), and (7.9.18) likewise reduces to (7.9.2).

The dipole orientation factors (7.9.5) are

$$p = (\hat{d}_A)_x (\hat{d}_B)_x + (\hat{d}_A)_y (\hat{d}_B)_y,$$
$$q = 2(\hat{d}_A)_z (\hat{d}_B)_z - (\hat{d}_A)_x (\hat{d}_B)_x - (\hat{d}_A)_y (\hat{d}_B)_y, \tag{7.9.20}$$

$(\hat{d}_A)_x^2 + (\hat{d}_A)_y^2 + (\hat{d}_A)_z^2 = 1$, when we take $\hat{\mathbf{r}}$ to be along the z axis. Suppose, for example, that the upper and lower states of our two-state atoms have magnetic quantum numbers $m = 1$ and $m = 0$, respectively, defined with $\hat{\mathbf{r}} = \hat{z}$ as the quantization axis, and

---

[112] P. W. Milonni and P. L. Knight, Am. J. Phys. **44**, 741 (1976).

[113] P. W. Milonni and P. L. Knight, Phys. Rev. A **10**, 1096 (1974).

that excitation is exchanged simply by spontaneous emission by Atom A followed by absorption of this radiation by Atom B. Then, in emitting a photon, Atom A makes a $\Delta m = -1$ transition and, by conservation of angular momentum, Atom B, after absorbing the photon from A, can only emit via a $\Delta m = -1$ transition, implying that we can take the transition moments of A and B to be parallel. This same conclusion applies if $m = -1$ and $m = 0$ for the upper and lower states, respectively, or if $m = 0$ for both states. For the $\Delta m = \pm 1$ transitions, where the emitted fields are polarized orthogonally to the $\hat{z}$ axis (and therefore $(\hat{d}_A)_z = (\hat{d}_B)_z = 0$), $p = 1$, $q = -1$, and

$$2\beta_{\pm}(r) = 2\beta\left[1 \pm \frac{3}{2}\left(\frac{\sin k_0 r}{k_0 r} + \frac{\cos k_0 r}{k_0^2 r^2} - \frac{\sin k_0 r}{k_0^3 r^3}\right)\right], \qquad (7.9.21)$$

$$\delta E_{\pm}(r) = \mp \frac{3\hbar\beta}{2}\left(\frac{\cos k_0 r}{k_0 r} - \frac{\sin k_0 r}{k_0^2 r^2} - \frac{\cos k_0 r}{k_0^3 r^3}\right). \qquad (7.9.22)$$

For $\Delta m = 0$, the x and y components of the dipole moments vanish, and therefore $p = 0$, $q = 2$, and

$$2\beta_{\pm}(r) = 2\beta\left[1 \mp 3\left(\frac{\cos k_0 r}{k_0^2 r^2} - \frac{\sin k_0 r}{k_0^3 r^3}\right)\right], \qquad (7.9.23)$$

$$\delta E_{\pm} = \mp 3\hbar\beta\left(\frac{\sin k_0 r}{k_0^2 r^2} + \frac{\cos k_0 r}{k_0^3 r^3}\right). \qquad (7.9.24)$$

In this case, there are no "far-field" contributions varying as $\sin k_0 r / k_0 r$ or $\cos k_0 r / k_0 r$, since the transition dipole moments lie along the line joining the atoms.

The two-state model for the atoms is less restrictive than might appear at first thought, since it can be applied as just described to the interaction of two atoms in which the upper state is triply degenerate. It can also be adapted, for instance, to derive the Förster rate of excitation transfer from one molecule to another (see Section 7.10).

The (approximate) expressions (7.9.1) and (7.9.2) are derived in many papers. Their derivation is typically based on the time-dependent Schrödinger equation with the state of the atom–field system at time t approximated by

$$|\psi(t)\rangle = b_A(t)|\phi_A\rangle + b_B(t)|\phi_B\rangle + \Sigma_{\mathbf{k}\lambda}b_{\mathbf{k}\lambda}(t)|-\rangle_A|-\rangle_B|1_{\mathbf{k}\lambda}\rangle$$
$$+ \Sigma_{\mathbf{k}\lambda}c_{\mathbf{k}\lambda}(t)|+\rangle_A|+\rangle_B|1_{\mathbf{k}\lambda}\rangle. \qquad (7.9.25)$$

$|1_{\mathbf{k}\lambda}\rangle$ is the field state with one photon in the free-space, plane-wave field mode with wave vector $\mathbf{k}$ and polarization index λ. The state $|+\rangle_A|+\rangle_B|1_{\mathbf{k}\lambda}\rangle$ is coupled by the electric dipole Hamiltonian to the states $|+\rangle_A|-\rangle_B|\text{vac}\rangle$ and $|-\rangle_A|+\rangle_B|\text{vac}\rangle$ by the "energy nonconserving" process in which a photon is created as an atom becomes excited. Without this process, the terms involving $1/(\omega + \omega_0)$ do not appear in the expression

$$g(r) = -\frac{3i\beta}{2\pi\omega_0^3} \int_0^\infty d\omega \left[p\frac{\sin\omega r/c}{\omega r/c} - q\frac{\cos\omega r/c}{(\omega r/c)^2} + q\frac{\sin\omega r/c}{(\omega r/c)^3} \right] \left(\frac{1}{\omega - \omega_0 - i\epsilon} \right.$$
$$\left. + \frac{1}{\omega + \omega_0 + i\epsilon} \right)$$
$$= -\frac{3i\beta}{2\pi\omega_0^3} \int_{-\infty}^\infty d\omega \left(p\frac{\sin\omega r/c}{\omega r/c} - q\frac{\cos\omega r/c}{(\omega r/c)^2} \right.$$
$$\left. + q\frac{\sin\omega r/c}{(\omega r/c)^3} \right) \left(\frac{1}{\omega - \omega_0 - i\epsilon} \right), \qquad \epsilon \to 0^+, \tag{7.9.26}$$

obtained for $g(r)$; this is found by a simple application of the residue theorem to be equivalent to (7.9.4). Thus, the state $|+\rangle_A|+\rangle_B|1_{\mathbf{k}\lambda}\rangle$ must be included in order to obtain the interaction described by $g(r)$. This point is discussed further below, in the Heisenberg-picture approach to the dipole–dipole resonance interaction.

Two-atom superradiant and subradiant spontaneous emission was observed in an experiment in which one of two well-separated trapped ions was excited by a laser pulse.[114] The laser field was sufficiently weak that excitation of more than one ion was unlikely, and $k_o r$ was sufficiently large that the radiation rate could be well approximated by

$$2\beta(r) = 2\beta\left(1 + \alpha\cos\Phi\frac{\sin k_0 r}{k_0 r}\right), \tag{7.9.27}$$

where Φ is the difference in the phase of the field at the two ions. In the two-state model, $\alpha = 3/2$ (see (7.9.21)) but, for the more complicated level structure of the Ba ions in the experiment, an averaging over all the possible $\Delta m = \pm 1$ and $\Delta m = 0$ transitions gives $\alpha = 1/2$.[115] The separation distance r was determined from the oscillation frequency of one of the trapped ions. Figure 7.6 shows the measured spontaneous emission lifetime for three different ion separations r, along with the variation expected from (7.9.27). The deviations from the measured single-ion lifetime were small, since $k_o r \approx 10$, but statistically significant. For example, the superradiance observed at the separation $r = 1380$ nm was $1.5\% \pm 0.8\%$ below the single-ion lifetime.

Superradiant and subradiant emission has also been observed in oriented polymer nanostructures with dipole moments of different emitters parallel to each other and perpendicular to the plane of a substrate.[116] The radiative decay rate of pairs of nanostructures excited by 100 ps pulses was measured as a function of their separation r_{12}. The emitters experienced the same excitation field, and so their cooperative emission rate γ, normalized with respect to the single-emitter rate γ_0, should be given by the emission rate (7.9.21) of the symmetric state of the two-"atom" system with dipole moments parallel to each other and perpendicular to the line joining them $((p = 1, q = -1)$:

$$\gamma/\gamma_0 = 1 + \frac{3}{2}\left(\frac{\sin\Phi}{\Phi} + \frac{\cos\Phi}{\Phi^2} - \frac{\sin\Phi}{\Phi^3}\right), \tag{7.9.28}$$

where $\Phi = 2\pi r_{12}/\lambda$. The agreement of the experimental data with the theoretical prediction (7.9.28) (see Figure 7.7) was "quite good."[117]

[114] R. G. DeVoe and R. G. Brewer, Phys. Rev. Lett. **76**, 2049 (1996).

[115] Ibid.

[116] M. D. Barnes, P. S. Krstic, P. Kumar, A. Mehta, and J. C. Wells, Phys. Rev. B **71**, 241303(R) (2005).

[117] Ibid.

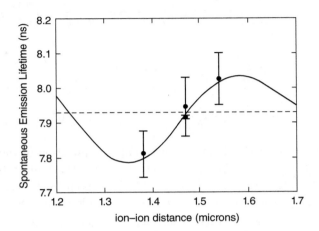

Fig. 7.6 Observed spontaneous emission lifetime of two trapped ions for three different ion separations (1380, 1470, and 1540 nm). The solid curve is obtained from the formula (7.9.27) without any adjustable parameters. From R. G. DeVoe and R. G. Brewer, Phys. Rev. Lett. **76**, 2049 (1996), with permission.

Not surprisingly, cooperative damping phenomena analogous to superradiance and subradiance can be observed in classical systems of coupled oscillators such as tuning forks and piano strings.[118]

7.9.1 Heisenberg Picture

We now consider, in the Heisenberg picture, a few other aspects of the dipole–dipole interaction, allowing for the two atoms to have different transition frequencies and dipole moments. The Heisenberg equations of motion for the operators $\sigma_{A,B}(t)$ and $(\sigma_z)_{A,B}(t)$ for the two atoms A and B are straightforward extensions of those for a single two-state atom (see (4.5.1) and (4.5.31)). For Atom A,

$$\dot{\sigma}_A(t) = -i\omega_A\sigma_A(t) - i\big[\sigma_{zA}(t)\mathcal{E}_{TA}^{(+)}(t) + \mathcal{E}_{TA}^{(-)}(t)\sigma_{zA}(t)\big], \qquad (7.9.29)$$

and

$$\dot{\sigma}_{zA}(t) = -2i\big[\sigma_A(t)\mathcal{E}_{TA}^{(+)}(t) \\ + \mathcal{E}_{TA}^{(-)}(t)\sigma_A(t)\big] + 2i\big[\sigma_A^\dagger(t)\mathcal{E}_{TA}^{(+)}(t) + \mathcal{E}_{TA}^{(-)}(t)\sigma_A^\dagger(t)\big], \qquad (7.9.30)$$

$$\mathcal{E}_{TA}^{(\pm)}(t) \equiv (\mathbf{d}_A/\hbar) \cdot \mathbf{E}_{TA}^{(\pm)}(t). \qquad (7.9.31)$$

$\mathbf{E}_{TA}^{(+)}(t)$ and $\mathbf{E}_{TA}^{(-)}(t)$ are the parts of the total electric field operator at Atom A that involve the photon annihilation and creation operators, respectively. We have chosen

[118] See *Mandel and Wolf (1995)*, Section 16.6.6.

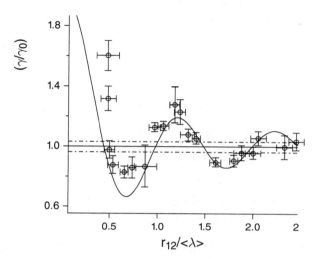

Fig. 7.7 Measured fluorescence decay rate, normalized to the single-emitter rate, as a function of emitter spacing r_{12} divided by the most probable emission wavelength $\lambda = 607$ nm. The solid curve is the theoretical prediction (7.9.28) with no adjustable parameters. The dashed horizontal lines represent ± 1 standard deviations in the distribution of single-emitter decay rates. From M. D. Barnes et al., Phys. Rev. B **71**, 241303(R) (2005), with permission.

to order the commuting, equal-time atom and field operators such that $\mathbf{E}_{TA}^{(+)}(t)$ appears to the right of the operators $\sigma_A(t)$ and $\sigma_{zA}(t)$, whereas $\mathbf{E}_{TA}^{(-)}(t)$ appears to the left. As discussed in Chapter 4, with this ordering there will be no explicit contributions from the source-free, "vacuum" field when expectation values are taken over a state in which there are initially no photons.

The total field acting on Atom A consists of its radiation reaction field, the source-free field at A, and, of course, the field at A from other sources. In the Markovian approximation (Chapter 4), the main effects of the radiation reaction field in the Heisenberg equations of motion are to add the terms $-\beta_A \sigma_A(t)$ and $-2\beta_A[\sigma_{zA}(t) + 1]$ to the right-hand sides of (7.9.29) and (7.9.30), respectively, where $2\beta_A$ is the rate in free space of radiative decay of the excited state of Atom A; we assume that radiative frequency (Lamb) shifts are included in the definition of the transition frequency ω_A. We denote the vacuum field acting on A by $\mathbf{E}_{A0}(t)$ and define

$$\mathcal{E}_{A0}^{(\pm)} = (\mathbf{d}_A/\hbar) \cdot \mathbf{E}_{A0}^{(\pm)}(t). \tag{7.9.32}$$

The remaining field acting on A is the field from Atom B. We denote it by $\mathbf{E}_{AB}(t)$ and define

$$\mathcal{E}_{AB}^{(\pm)}(t) = (\mathbf{d}_A/\hbar) \cdot \mathbf{E}_{AB}^{(\pm)}(t). \tag{7.9.33}$$

To simplify the calculations, we will make a rotating-wave approximation in which we ignore "counter-rotating" terms on the right-hand sides of (7.9.29) and (7.9.30). Then we replace these equations by

$$\dot{\sigma}_A(t) = -i(\omega_A - i\beta_A)\sigma_A(t) - i\sigma_{zA}(t)\mathcal{E}_A^{(+)}(t) \tag{7.9.34}$$

and

$$\dot{\sigma}_{zA}(t) = -2\beta_A[\sigma_{zA}(t) + 1] + 2i\sigma_A^\dagger(t)\mathcal{E}_A^{(+)}(t) - 2i\mathcal{E}_A^{(-)}(t)\sigma_A(t), \tag{7.9.35}$$

where

$$\mathcal{E}_A^{(\pm)}(t) = \mathcal{E}_{A0}^{(\pm)}(t) + \mathcal{E}_{AB}^{(\pm)}(t). \tag{7.9.36}$$

The rotating-wave approximation here is based on the assumptions that $\mathcal{E}_A^{(+)}(t)$ and $\mathcal{E}_A^{(-)}(t)$ are primarily positive-frequency ($e^{-i\omega t}$) and negative-frequency ($e^{i\omega t}$) fields, respectively, $\sigma_A(t)$ varies predominantly as $e^{-i\omega_A t}$, and $\sigma_{zA}(t)$ varies slowly compared with $e^{\pm i\omega_A t}$.

The electric field of an Atom is given by (4.3.11). For the field from the two-state Atom B, we replace $ex(t)$ in that formula by $d_B\sigma_{xB}(t) = d_B[\sigma_B(t) + \sigma_B^\dagger(t)]$; in the rotating-wave approximation,

$$\begin{aligned}
\mathcal{E}_{AB}^{(+)}(t) &\cong -\frac{1}{4\pi\epsilon_0}\frac{d_A d_B}{\hbar}\left[\frac{p}{c^2 r}\ddot{\sigma}_B(t - r/c)\right. \\
&\quad \left. -\frac{q}{cr^2}\dot{\sigma}_B(t - r/c) - \frac{q}{r^3}\sigma_B(t - r/c)\right]\theta(t - r/c) \\
&\cong F_{AB}(r)\sigma_B(t - r/c)\theta(t - r/c), \tag{7.9.37}
\end{aligned}$$

$$F_{AB}(r) \equiv \frac{1}{4\pi\epsilon_0}\frac{d_A d_B}{\hbar}k_B^3\left(\frac{p}{k_B r} - \frac{iq}{k_B^2 r^2} + \frac{q}{k_B^3 r^3}\right), \qquad k_B = \omega_B/c. \tag{7.9.38}$$

We have made the weak-interaction approximation that $\ddot{\sigma}_B(t) \cong -\omega_B^2 \sigma_B(t)$, and $\dot{\sigma}_B(t) \cong -i\omega_B\sigma_B(t)$.

With these approximations and definitions, we replace (7.9.34), for instance, by

$$\begin{aligned}
\dot{\sigma}_A(t) &= -i(\omega_A - i\beta_A)\sigma_A(t) - i\sigma_{zA}\mathcal{E}_{A0}^{(+)}(t) \\
&\quad - iF_{AB}(r)\sigma_{zA}(t)\sigma_B(t - r/c)\theta(t - r/c). \tag{7.9.39}
\end{aligned}$$

Similarly, for Atom B,

$$\begin{aligned}
\dot{\sigma}_B(t) &= -i(\omega_B - i\beta_B)\sigma_B(t) - i\sigma_{zB}\mathcal{E}_{B0}^{(+)}(t) \\
&\quad - iF_{BA}(r)\sigma_{zB}(t)\sigma_A(t - r/c)\theta(t - r/c), \tag{7.9.40}
\end{aligned}$$

and

$$\dot{\sigma}_{zB}(t) = -2\beta_B[\sigma_{zB}(t) + 1] + 2i\sigma_B^\dagger(t)\mathcal{E}_B^{(+)}(t) - 2i\mathcal{E}_B^{(-)}(t)\sigma_B(t), \tag{7.9.41}$$

$$\mathcal{E}_B^{(\pm)}(t) = \mathcal{E}_{B0}^{(\pm)}(t) + \mathcal{E}_{BA}^{(\pm)}(t), \tag{7.9.42}$$

$$
\begin{aligned}
\mathcal{E}_{BA}^{(+)}(t) &\cong -\frac{1}{4\pi\epsilon_0}\frac{d_A d_B}{\hbar}\left[\frac{p}{c^2 r}\ddot{\sigma}_A(t-r/c)\right.\\
&\qquad \left. -\frac{q}{cr^2}\dot{\sigma}_A(t-r/c) - \frac{q}{r^3}\sigma_A(t-r/c)\right]\theta(t-r/c)\\
&\cong F_{BA}(r)\sigma_A(t-r/c)\theta(t-r/c), \tag{7.9.43}
\end{aligned}
$$

$$F_{BA}(r) \equiv \frac{1}{4\pi\epsilon_0}\frac{d_A d_B}{\hbar}k_A^3\left(\frac{p}{k_A r} - \frac{iq}{k_A^2 r^2} + \frac{q}{k_A^3 r^3}\right), \quad k_A = \omega_A/c. \tag{7.9.44}$$

The expressions (7.9.37) and (7.9.43) entail not only the "weak-interaction" approximation for $\sigma_A(t)$ and $\sigma_B(t)$ but also the approximation that $\mathcal{E}_{BA}^{(+)}(t)$, for example, depends only on the two-state lowering operator $\sigma_A(t)$. Thus, we have ignored a part of $\mathcal{E}_{BA}^{(+)}(t)$ that depends on the two-state raising operator $\sigma_A^\dagger(t)$, since $\mathcal{E}_{BA}^{(+)}(t)$ is the part of the electric field operator that derives from photon annihilation operators, which, in a rotating-wave approximation, depend only on the two-state lowering operators. In this approximation, $\mathcal{E}_{BA}^{(\pm)}(t)$ and $\mathcal{E}_{AB}^{(\pm)}(t)$ involve the step function $\theta(t-r/c)$, whereas, in fact, they are not strictly "causal" in this sense when evaluated exactly: only the full electric field operators $\mathcal{E}_{BA}^{(+)}(t) + \mathcal{E}_{BA}^{(-)}(t)$ and $\mathcal{E}_{AB}^{(+)}(t) + \mathcal{E}_{AB}^{(-)}(t)$ are truly causal. This is discussed further in Appendix C.

7.9.2 Identical Atoms

To illustrate the Heisenberg-picture approach, we now calculate the excitation probabilities

$$P_A(t) = \langle \sigma_A^\dagger(t)\sigma_A(t)\rangle, \quad \text{and} \quad P_B(t) = \langle \sigma_B^\dagger(t)\sigma_B(t)\rangle \tag{7.9.45}$$

of Atoms A and B, assuming they are identical ($\omega_A = \omega_B \equiv \omega_0$, $d_A = d_B \equiv d$, $\beta_A = \beta_B \equiv \beta$, and $F_{AB}(r) = F_{BA}(r)$). The (approximate) equations for $\sigma_A(t)$ and $\sigma_B(t)$ in this case are

$$\dot{\sigma}_A(t) = -i(\omega_0 - i\beta)\sigma_A(t) - i\sigma_{zA}(t)\mathcal{E}_{A0}^{(+)}(t) + \beta g(r)\sigma_{zA}(t)\sigma_B(t), \tag{7.9.46}$$

and

$$\dot{\sigma}_B(t) = -i(\omega_0 - i\beta)\sigma_B(t) - i\sigma_{zB}(t)\mathcal{E}_{B0}^{(+)}(t) + \beta g(r)\sigma_{zB}(t)\sigma_A(t) \tag{7.9.47}$$

for $t \gg r/c$; we have made the approximations

$$\sigma_A(t-r/c) \cong \sigma_A(t)e^{i\omega_0 r/c} = \sigma_A(t)e^{ik_0 r}, \tag{7.9.48}$$

and

$$\sigma_B(t-r/c) \cong \sigma_B(t)e^{i\omega_0 r/c} = \sigma_B(t)e^{ik_0 r}, \tag{7.9.49}$$

which are accurate under the physically realistic assumption that $\omega_0 \gg \beta, |F_{AB}(r)|$, and have used the relation

$$F_{AB}(r)e^{ik_0 r} = i\beta g(r), \tag{7.9.50}$$

with $g(r)$ defined by (7.9.4). From (7.9.46),

$$\frac{dP_A}{dt} = -2\beta P_A(t) + \beta g(r)\langle \sigma_A^\dagger(t)\sigma_{zA}(t)\sigma_B(t)\rangle + \beta g^*(r)\langle \sigma_{zA}(t)\sigma_A(t)\sigma_B^\dagger(t)\rangle. \tag{7.9.51}$$

We have used the fact that equal-time A and B operators commute and have taken expectation values over a state of the atom–field system in which the field is in its vacuum state; the normal ordering we have chosen has eliminated any contribution from the vacuum-field operator $\mathcal{E}_{A0}^{(+)}(t)$. Now, since the two-state operators satisfy

$$\sigma_A^\dagger(t)\sigma_{zA}(t) = -\sigma_A^\dagger(t), \quad \text{and} \quad \sigma_{zA}(t)\sigma_A(t) = -\sigma_A(t), \tag{7.9.52}$$

(7.9.51) reduces to

$$\frac{dP_A}{dt} = -2\beta P_A(t) - \beta g(r)X(t) - \beta g^*(r)X^*(t), \tag{7.9.53}$$

$$X(t) \equiv \langle \sigma_A^\dagger(t)\sigma_B(t)\rangle. \tag{7.9.54}$$

Similarly, for the excitation probability of Atom B,

$$\frac{dP_B}{dt} = -2\beta P_B(t) - \beta g(r)X^*(t) - \beta g^*(r)X(t). \tag{7.9.55}$$

It follows from (7.9.46) and (7.9.47) that

$$\frac{dX}{dt} = -2\beta X(t) + \beta g(r)\langle \sigma_A^\dagger(t)\sigma_A(t)\sigma_{zB}(t)\rangle + \beta g^*(r)\langle \sigma_B^\dagger(t)\sigma_B(t)\sigma_{zA}(t)\rangle, \tag{7.9.56}$$

and, from the operator identities

$$\sigma_A^\dagger(t)\sigma_A(t)\sigma_{zB}(t) = \sigma_A^\dagger(t)\sigma_A(t)\big[\sigma_B^\dagger(t)\sigma_B(t) - 1\big], \tag{7.9.57}$$

$$\sigma_B^\dagger(t)\sigma_B(t)\sigma_{zA}(t) = \sigma_B^\dagger(t)\sigma_B(t)\big[\sigma_A^\dagger(t)\sigma_A(t) - 1\big], \tag{7.9.58}$$

we see that the right-hand side of (7.9.56) involves the expectation value of the product $\sigma_A^\dagger(t)\sigma_A(t)\sigma_B^\dagger(t)\sigma_B(t)$ of the excitation operators of the two atoms. Since there is initially only one excitation in the atom–field system, the contribution of this expectation value to $P_A(t)$ and $P_B(t)$ is negligible. We therefore take

$$\langle \sigma_A^\dagger(t)\sigma_A(t)\sigma_{zB}(t)\rangle \cong -\langle \sigma_A^\dagger(t)\sigma_A(t)\rangle = -P_A(t), \tag{7.9.59}$$

$$\langle \sigma_B^\dagger(t)\sigma_B(t)\sigma_{zA}(t)\rangle \cong -\langle \sigma_B^\dagger(t)\sigma_B(t)\rangle = -P_B(t),) \tag{7.9.60}$$

and replace (7.9.56) by

$$\frac{dX}{dt} = -2\beta X(t) - \beta g(r)P_A(t) - \beta g^*(r)P_B(t). \tag{7.9.61}$$

For the initial conditions $P_A(0) = 1$, $P_B(0) = 0$, and $X(0) = 0$, the solution of (7.9.53), (7.9.55), and (7.9.61) for $P_B(t)$, for example, is ($g_R \equiv \mathrm{Re}(g)$, $g_I \equiv \mathrm{Im}(g)$)

$$P_B(t) = \frac{1}{4}\big(e^{-2\beta[1+g_R(r)]t} + e^{-2\beta[1-g_R(r)]t} - 2e^{-2\beta t}\cos(2\beta g_I t)\big), \tag{7.9.62}$$

which is the excitation probability $|b_B(t)|^2$ given by (7.9.2).

Although our approach using the Heisenberg picture just reproduces known expressions for the excitation probabilities, it helps us to better understand why *these expressions could be derived from the simpler model of two coupled dipole oscillators*: the operator identity (7.9.52) and the approximations (7.9.59) and (7.9.60) allow us, in effect, to replace both $\sigma_{zA}(t)$ and $\sigma_{zB}(t)$ by -1, which, as discussed in Section 2.5.2, amounts to replacing two-state atoms by harmonic oscillators.

7.9.3 Non-Identical Atoms

An approach similar to the preceding one can be taken when the atoms are different, but it will suffice for our purposes to take a perturbative approach starting from (7.9.40), which implies

$$P_B(t) \cong |F_{BA}(r)|^2 \int_0^t dt' \int_0^t dt'' \langle \sigma_A^\dagger(t'-r/c)\sigma_A(t''-r/c)\rangle \theta(t'-r/c)\theta(t''-r/c)$$
$$\times e^{i\omega_B(t'-t'')}e^{\beta_B(t'+t''-2t)}$$
$$= |F_{BA}(r)|^2 \int_{r/c}^t dt' \int_{r/c}^t dt'' \langle \sigma_A^\dagger(t'-r/c)\sigma_A(t''-r/c)\rangle e^{i\omega_B(t'-t'')}e^{\beta_B(t'+t''-2t)}$$
$$\times \theta(t-r/c) \tag{7.9.63}$$

for the excitation probability $P_B(t) = \langle \sigma_B^\dagger(t)\sigma_B(t)\rangle$ of Atom B. We have assumed again that Atom B is initially unexcited, the field is initially in its vacuum state, and therefore that $\langle \sigma_B^\dagger(0)\sigma_B(0)\rangle = 0$, $\langle \sigma_B^\dagger(0)\sigma_{zB}(0)\rangle = -\langle \sigma_B^\dagger(0)\rangle = 0$ and $\mathcal{E}_{B0}^{(+)}(t)|\psi(0)\rangle = 0$. We have also used the identities $\sigma_B^\dagger(0)\sigma_{zB}(0) = -\sigma_B^\dagger(0) = 0$ and $\sigma_{zB}^2(0) = 1$ for the Pauli two-state operators. To calculate $P_B(t)$ to lowest order in the atom–field coupling, we make the weak-interaction approximation:

$$\sigma_A(t-r/c) \cong \sigma_A(0)e^{-i(\omega_A - i\beta_A)(t-r/c)}, \tag{7.9.64}$$

so that, assuming Atom A is initially excited ($\langle \sigma_A^\dagger(0)\sigma_A(0)\rangle = 1$),

$$P_B(t) \cong \frac{|F_{BA}(r)|^2}{\Delta^2 + \gamma^2}\theta(t-r/c)\Big[e^{-2\beta_A(t-r/c)} + e^{-2\beta_B(t-r/c)} - 2e^{-(\beta_A+\beta_B)(t-r/c)}$$
$$\times \cos\Delta(t-r/c)\Big], \tag{7.9.65}$$

$$\Delta \equiv \omega_A - \omega_B, \qquad \gamma \equiv \beta_A - \beta_B. \tag{7.9.66}$$

For $\Delta^2 \gg \gamma^2$ and times much longer than r/c but much shorter than the radiative lifetimes,

$$P_B(t) \cong \frac{4|F_{BA}(r)|^2}{\Delta^2}\sin^2\frac{1}{2}\Delta t. \tag{7.9.67}$$

This result, for $k_A r \gg 1$, was obtained by Breit.[119] For $\beta_A \gg \beta_B$, $\beta_A(t-r/c) \gg 1$, and $\beta_B(t-r/c) \ll 1$,

[119] G. Breit, Rev. Mod. Phys. **5**, 91 (1933).

$$P_B(t) \cong \frac{|F_{BA}(r)|^2}{\Delta^2 + \beta_A^2} \theta(t - r/c), \tag{7.9.68}$$

which is essentially the result obtained by Fermi.[120] In the special case of identical atoms ($k_A = k_B = k_0$, $d_A = d_B = d$, $\beta_A = \beta_B = \beta$, and $|F_{AB}(r)|^2 = |F_{BA}(r)|^2 = |\beta g(r)|^2$), (7.9.65) reduces to

$$P_B(t) \cong \left| \beta g(r)(t - r/c) e^{-\beta(t - r/c)} \right|^2 \theta(t - r/c), \tag{7.9.69}$$

which is the first in the series of terms involving $\theta(t - nr/c)$, n odd, for the probability $|b_B(t)|^2$ that follows from (7.9.18). Successive excitation-exchange probabilities for nonidentical atoms fall off with increasingly higher powers of $|F_{BA}(r)/\Delta|$ and $|F_{AB}(r)/\Delta|$.

We remarked earlier on the values of the dipole orientation factors p and q for $\Delta m = 0$ and $\Delta m = \pm 1$ transitions. It is sometimes appropriate to average p^2 and q^2 over random and independent orientations of $\hat{\mathbf{d}}_A$ and $\hat{\mathbf{d}}_B$; this is often done in the theory of Förster resonance energy transfer, as discussed in Section 7.10. Consider, for example, this averaging of

$$q^2 = [3(\hat{\mathbf{d}}_A \cdot \hat{\mathbf{r}})(\hat{\mathbf{d}}_B \cdot \hat{\mathbf{r}}) - \hat{\mathbf{d}}_A \cdot \hat{\mathbf{d}}_B]^2. \tag{7.9.70}$$

For orientations with $\hat{\mathbf{d}}_A \parallel \hat{\mathbf{d}}_B$, $q^2 = 1$ if the dipoles are orthogonal to $\hat{\mathbf{r}}$, whereas $q^2 = 4$ if they lie along the $\hat{r}$ axis. For orientations such that $\hat{\mathbf{d}}_A$ and $\hat{\mathbf{d}}_B$ are orthogonal to each other and to $\hat{\mathbf{r}}$, $q^2 = 0$. There are 3×3 independent orientations in all, one for which $q^2 = 4$, and two for which $q^2 = 1$, and the orientational average of q^2 is therefore $(4 \times 1 + 1 \times 2)/9 = 2/3$.

Exercise 7.13: Show that

$$|F_{BA}(r)|^2 \rightarrow \left(\frac{1}{4\pi\epsilon_0} \right)^2 \frac{2d_A^2 d_B^2}{9\hbar^2} k_A^6 \left(\frac{1}{k_A^2 r^2} + \frac{1}{k_A^4 r^4} + \frac{3}{k_A^6 r^6} \right)$$

when we average over random, independent orientations of $\hat{\mathbf{d}}_A$ and $\hat{\mathbf{d}}_B$.

The point of Fermi's calculation was to show how the propagation of light at the velocity c in vacuum is accounted for by the quantum theory of radiation: the initially unexcited Atom B cannot be excited before the time r/c it takes for light to propagate from A to B.[121] In his calculation, Fermi neglected the possibility that both atoms could be excited while there is one photon in the field. As discussed in connection with (7.9.26), the correct, retarded dipole interaction is not obtained in this approximation.

[120] E. Fermi, Rev. Mod. Phys. **4**, 87 (1932).
[121] Ibid.

But Fermi could recover the correct interaction within this approximation by making an additional approximation of replacing an integral over all positive frequencies by one including all negative frequencies. The consequence of this approximation can be seen by keeping only the part of the integrand involving $1/(\omega - \omega_0 - i\epsilon)$ in (7.9.26), while extending the integration to $-\infty$ on the grounds that the contributions from $1/(\omega - \omega_0 - i\epsilon)$ with $\omega < 0$ should be small compared to those from $\omega > 0$:

$$-\frac{3i\beta}{2\pi\omega_0^3}\int_0^\infty d\omega \left(p\frac{\sin \omega r/c}{\omega r/c} - q\frac{\cos \omega r/c}{(\omega r/c)^2} + q\frac{\sin \omega r/c}{(\omega r/c)^3}\right)\frac{1}{\omega - \omega_0 - i\epsilon}$$

$$\rightarrow -\frac{3i\beta}{2\pi\omega_0^3}\int_{-\infty}^\infty d\omega \left(p\frac{\sin \omega r/c}{\omega r/c} - q\frac{\cos \omega r/c}{(\omega r/c)^2} + q\frac{\sin \omega r/c}{(\omega r/c)^3}\right)\frac{1}{\omega - \omega_0 - i\epsilon}$$

$$= -\frac{3i\beta}{2\pi\omega_0^3}\int_0^\infty d\omega \left(p\frac{\sin \omega r/c}{\omega r/c} - q\frac{\cos \omega r/c}{(\omega r/c)^2} + q\frac{\sin \omega r/c}{(\omega r/c)^3}\right)\left(\frac{1}{\omega - \omega_0 - i\epsilon}\right.$$

$$\left. + \frac{1}{\omega + \omega_0 + i\epsilon}\right) = g(r). \tag{7.9.71}$$

In other words, extending the integration to $-\infty$ in effect corrects for the neglect of the "energy-non-conserving" states in which both atoms are excited and there is one photon in the field. In fact, our approximations (7.9.43) and (7.9.37) are really just the Heisenberg-picture version of Fermi's approximation, as discussed briefly in a broader context in Appendix C.

The field from Atom A also causes a shift in the transition frequency of Atom B. To calculate this shift most simply, we return again to (7.9.40) and take expectation values on both sides after inserting the formal solution of (7.9.41):

$$\langle \dot\sigma_B(t)\rangle \cong -i(\omega_B - i\beta_B)\langle \sigma_B(t)\rangle + (\ldots)$$

$$- 2\int_0^t dt' \Big\langle \mathcal{E}_{BA}(t')^{(-)}\sigma_B(t')\mathcal{E}_{BA}^{(+)}(t)\Big\rangle e^{2\beta_B(t'-t)}, \tag{7.9.72}$$

where $(\ldots)$ denotes terms not depending directly on $\sigma_B(t)$. Next, we make the approximation

$$\sigma_B(t') \cong \sigma_B(t)e^{-i(\omega_B - i\beta_B)(t'-t)}, \tag{7.9.73}$$

use (7.9.43) for $\mathcal{E}_{BA}^{(+)}(t)$ and $\mathcal{E}_{BA}^{(-)}(t)$, and set $\langle \sigma_A^\dagger(0)\sigma_A(0)\rangle = 1$, since Atom A is initially excited:

$$\langle \dot\sigma_B(t)\rangle \cong -i(\omega_B - i\beta_B)\langle \sigma_B(t)\rangle + (\ldots) - 2|F_{BA}(r)|^2\theta(t - r/c)e^{-2\beta_A(t-r/c)}\langle \sigma_B(t)\rangle$$

$$\times \int_{r/c}^t dt' e^{i(\Delta + i\gamma)(t'-t)}$$

$$= -i(\omega_B - i\beta_B)\langle \sigma_B(t)\rangle + (\ldots) + \frac{2i|F_{BA}(r)|^2}{\Delta + i\gamma}\theta(t - r/c)\langle \sigma_B(t)\rangle$$

$$\times \left(e^{-2\beta_A(t-r/c)} - e^{-i\Delta(t-r/c)}e^{-(\beta_A+\beta_B)(t-r/c)}\right). \tag{7.9.74}$$

The factor proportional to $|F_{BA}(r)|^2$ and multiplying $\langle \sigma_B(t) \rangle$ accounts for the modification of the resonance frequency and radiative decay rate of Atom B due to Atom A.

Suppose, for example, that $\beta_A(t - r/c) \ll 1 \ll \beta_B(t - r/c)$, so that de-excitation of Atom A is negligible but Atom B is rapidly de-excited. Then, for $t > r/c$,

$$\langle \dot{\sigma}_B(t) \rangle \cong -i[\omega_B + \delta\omega_B - i(\beta_B + \delta\beta_B)]\langle \sigma_B(t) \rangle + (\ldots), \tag{7.9.75}$$

where

$$\delta\omega_B = -\frac{2\Delta |F_{BA}(r)|^2}{\Delta^2 + \beta_B^2}, \tag{7.9.76}$$

$$\delta\beta_B = \frac{2\beta_B |F_{BA}(r)|^2}{\Delta^2 + \beta_B^2}. \tag{7.9.77}$$

$\delta\omega_B$ is a shift in the transition frequency of Atom B—the difference between upper- and lower-state energy shifts. It has the form of an AC Stark shift (Section 2.5.3) caused by the field from Atom A, and it vanishes in the case of identical two-state atoms ($\Delta = 0$). Since the real parts of the upper- and lower-state polarizabilities of a two-state atom have opposite signs (see Section 2.5.3), and the level shifts are proportional to the real parts of the polarizabilities, the level shift of the upper state is $-\delta E_B$, where δE_B is the level shift of the lower state of Atom B. In other words,

$$\delta E_B = -\frac{1}{2}\hbar\delta\omega_B = \frac{\hbar\Delta}{\Delta^2 + \beta_B^2}|F_{BA}(r)|^2 \cong \frac{\hbar}{\Delta}|F_{BA}(r)|^2. \tag{7.9.78}$$

We are assuming $|\Delta| \gg \beta_B$ as part of the condition that the two atoms are "nonidentical." Now, recall that we have made a rotating-wave approximation. Without this approximation, we obtain a term involving $1/(\omega_A + \omega_B)$ in addition to $1/\Delta = 1/(\omega_A - \omega_B)$, with the result that $1/\Delta$ is replaced by

$$\frac{1}{\Delta} + \frac{1}{\omega_A + \omega_B} = \frac{2\omega_A}{\omega_A^2 - \omega_B^2} \tag{7.9.79}$$

and (7.9.78) by

$$\delta E_B \cong \left(\frac{1}{4\pi\epsilon_0}\right)^2 \frac{4}{9} \frac{d_A^2 d_B^2}{\hbar c} \frac{k_B k_A^6}{k_A^2 - k_B^2} \left(\frac{1}{k_A^2 r^2} + \frac{1}{k_A^4 r^4} + \frac{3}{k_A^6 r^6}\right) \tag{7.9.80}$$

when we average over dipole orientations, as in Exercise 7.13. Without the rotating-wave approximation, δE_B is seen to depend on the real part of the polarizability at frequency ω_A of Atom B,

$$\alpha_B(\omega_A) = \frac{2\omega_B d_B^2/\hbar}{\omega_B^2 - \omega_A^2} = \frac{2k_B d_B^2/\hbar c}{k_B^2 - k_A^2}, \tag{7.9.81}$$

and, as already noted, is an AC Stark shift of the lower state of Atom B. The energy (7.9.80) was obtained in fourth-order perturbation theory by Power and Thirunamachandran as the interaction energy of the atom–field state in which A is excited, B is unexcited, and the field is in the vacuum state.[122]

These changes in the transition frequency and radiative decay rate of Atom B only apply, of course, for times greater than the time r/c it takes for the field from A to reach B. Assuming as we have that

$$|F_{BA}(r)|^2 \ll \Delta^2 + \beta_B^2, \qquad (7.9.82)$$

they are good approximations for times greater than r/c but shorter than the radiative lifetimes.

The effect of Atom B back on Atom A can be determined in similar fashion.[123] For $|\Delta/\gamma| \ll 1$ and times longer than $2r/c$ but shorter than the radiative lifetimes, the shift in Atom A's transition frequency and the radiative decay rate are found to be

$$\delta\omega_A \cong \left(\frac{1}{4\pi\epsilon_0}\right)^2 \frac{d_A^2 d_B^2 p^2 k_A^6}{\hbar^2 \Delta (k_A r)^2} \cos(2k_A r) \qquad (7.9.83)$$

and

$$\beta_A' \cong \beta_A + \left(\frac{1}{4\pi\epsilon_0}\right)^2 \frac{d_A^2 d_B^2 p^2 k_A^6}{\hbar^2 \Delta (k_A r)^2} \sin(2k_A r) \qquad (7.9.84)$$

for $k_0 r \gg 1$. Unlike the monotonic r dependence found above for the effects of Atom A on Atom B, these shifts have a sinusoidal dependence on r. This difference has been discussed from different approaches and physical perspectives.[124]

Our treatment of the dipole–dipole interaction is based on the electric dipole form of the interaction. If we had used instead the Coulomb-gauge, minimal coupling interaction involving the vector potential, we would have had to include, as discussed in Section 4.2, the nonretarded dipole–dipole interaction that varies as $1/r^3$ in order to obtain the correct retarded and causal interaction between the atoms.[125]

7.10 Förster Resonance Energy Transfer

The rate of excitation transfer from A to B is the rate of change of the excited-state probability of Atom B:

$$r_T = \frac{d}{dt}\langle \sigma_B^\dagger(t)\sigma_B(t)\rangle = \langle \sigma_B^\dagger(t)\dot\sigma_B(t)\rangle + \langle \dot\sigma_B^\dagger(t)\sigma_B(t)\rangle. \qquad (7.10.1)$$

---

[122] E. A. Power and T. Thirunamachandran, Phys. Rev. A **47**, 2539 (1993), and E. A. Power and T. Thirunamachandran, Phys. Rev. A **51**, 3660 (1995).

[123] P. W. Milonni and S. M. H. Rafsanjani, Phys. Rev. A **92**, 062711 (2015).

[124] Ibid; P. R. Berman, Phys. Rev. A **91**, 042127 (2015); M. Donaire, Phys. Rev. A **93**, 052706 (2016).

[125] P. W. Milonni and P. L. Knight, Phys. Rev. A **10**, 1096 (1974).

We assume the excitation probability of B is small and approximate (7.9.40) by replacing $\sigma_{zB}(t)$ by -1, since the lower state of Atom B is an eigenstate of σ_{zB} with eigenvalue -1:

$$\dot{\sigma}_B(t) \cong -i(\omega_B - i\beta_B)\sigma_B(t) + i\mathcal{E}_{B0}^{(+)}(t) + iF_{BA}(r)\sigma_A(t - r/c)\theta(t - r/c). \quad (7.10.2)$$

Since $\sigma_B(0)|\psi(0)\rangle = \mathcal{E}_{B0}^{(+)}(t)|\psi(0)\rangle = 0$, it follows from (7.10.1) that, for times $t > r/c$,

$$r_T \cong 2|F_{BA}(r)|^2 \text{Re} \int_0^t dt' \langle \sigma^{\dagger}_A(t' - r/c)\sigma_A(t - r/c)\rangle e^{-i(\omega_B + i\beta_B)(t'-t)}. \quad (7.10.3)$$

We now use the approximation (7.9.64) for $\sigma_A(t)$ and assume, as in the derivation of (7.9.76) and (7.9.77), that β_A is very small compared to β_B and effectively negligible over times of interest. It then follows from (7.10.3) that

$$r_T \cong \frac{2\beta_B}{\Delta^2 + \beta_B^2}|F_{BA}(r)|^2 = 2\pi|F_{BA}(r)|^2 S_B(\omega_A)$$

$$= \left(\frac{1}{4\pi\epsilon_0}\right)^2 \frac{2\pi d_A^2 d_B^2}{\hbar^2} k_A^6 \left|\frac{p}{k_A r} - \frac{iq}{k_A^2 r^2} + \frac{q}{k_A^3 r^3}\right|^2 S_B(\omega_A), \quad (7.10.4)$$

which is the same as (7.9.77), that is, the excitation transfer rate is equal to the damping rate $\delta\beta_B$ of the dipole moment induced in Atom B by the field of Atom A. Here,

$$S_B(\omega_A) = \frac{\beta_B/\pi}{\Delta^2 + \beta_B^2} = \frac{\beta_B/\pi}{(\omega_B - \omega_A)^2 + \beta_B^2} \quad (7.10.5)$$

is the absorption lineshape function for Atom B, evaluated at the transition frequency ω_A of Atom A. r_T is largest at distances small compared to the transition wavelength $\lambda_A \, (= 2\pi/k_A)$ of Atom A:

$$r_T \cong \left(\frac{1}{4\pi\epsilon_0}\right)^2 \frac{d_A^2 d_B^2}{\hbar^2} \frac{q^2}{r^6} S_B(\omega_A), \quad (k_A r \ll 1). \quad (7.10.6)$$

We next express this result for the excitation transfer rate in a different form, and also modify it by allowing for the two atoms to be embedded in a dielectric medium of refractive index $n(\omega)$. For this purpose, we use the formulas (see Section 7.12)

$$A_D = \frac{1}{4\pi\epsilon_0} n(\omega_A) \frac{4d_A^2 \omega_A^3}{3\hbar c^3} \quad (7.10.7)$$

and (cf. (2.7.7))

$$\sigma_{ac}(\omega) = \frac{1}{4\pi\epsilon_0} \frac{1}{n(\omega)} \frac{4\pi^2 \omega d_B^2/3}{\hbar c} S_B(\omega) \quad (7.10.8)$$

for the spontaneous emission rate A_D of Atom A and the absorption cross section $\sigma_{ac}(\omega)$ of Atom B in the medium. Then, in terms of A_D and $\sigma_{ac}(\omega_A)$,

$$r_T = \frac{9q^2c^4}{8\pi\omega_A^4 r^6}\sigma_{ac}(\omega_A)A_D. \tag{7.10.9}$$

But, in addition, since we are now allowing for the two atoms to be guests in a host dielectric medium, we must account for the fact that the electric field acting on Atom B depends on $n(\omega_A)$. The simplest way to do this is to replace $1/4\pi\epsilon_0$ by

$$\frac{1}{4\pi\epsilon} = \frac{1}{4\pi n^2 \epsilon_0}, \tag{7.10.10}$$

which amounts to recognizing that the square of the electric field acting on B is proportional to $1/n^4$, and therefore to modify the algebra leading from (7.10.3) to (7.10.9) to include the factor $1/n^4(\omega_A)$:

$$r_T = \frac{9q^2c^4}{8\pi n^4(\omega_A)\omega_A^4 r^6}\sigma_{ac}(\omega_A)A_D. \tag{7.10.11}$$

One more, essential modification is necessary to realistically describe the excitation transfer of interest—the transfer of excitation between electronic energy levels of "donor" and "acceptor" molecules. Instead of assuming as we have that A has a sharply defined transition frequency ω_A, we allow for A to have an emission spectrum $S_D(\omega)$, $\int_0^\infty d\omega S_D(\omega) = 1$, and replace (7.10.11) by

$$R_T = \frac{9q^2c^4A_D}{8\pi r^6}\int_0^\infty d\omega\, \frac{\sigma_{ac}(\omega)S_D(\omega)}{n^4(\omega)\omega^4}, \tag{7.10.12}$$

which is one form of the excitation transfer rate derived by Förster.[126] Equivalently,

$$R_T = \frac{1}{\tau_D}\left(\frac{R_0}{R}\right)^6, \tag{7.10.13}$$

with $\tau_D = 1/A_D$ the radiative lifetime of the donor molecule, R now denoting the distance separating the centers of the two molecules ("chromophores"), and

$$R_0^6 = \frac{9q^2c^4}{8\pi}\int_0^\infty d\omega\, \frac{\sigma_{ac}(\omega)S_D(\omega)}{n^4(\omega)\omega^4}. \tag{7.10.14}$$

R_T is often cast in an equivalent form involving the "molar decadic extinction coefficient" of the acceptor rather than its absorption cross section, and q^2, which is usually denoted κ^2, is often replaced by $2/3$, the orientational average calculated in Section 7.9.

---

[126] T. Förster, Ann. Phys. (Germany) **2**, 55 (1948). See R. M. Clegg, in *Reviews in Fluorescence*, Vol. **2006**, eds. J. R. Lakowicz and C. D. Geddes, Springer, New York, 2007, pp. 1-45, for a detailed history of Förster excitation transfer. See also R. S. Knox, Photosynthesis Res. **48**, 35 (1996), and J. Biomed. Opt. **17**, 01103 (2012), for additional historical perspectives.

The efficiency of the transfer of excitation from the donor to the acceptor is defined as the ratio of the excitation rate of the acceptor to the rate at which the donor de-excites via fluorescence and excitation transfer, assuming there are no other ways for the donor to be de-excited:

$$e = \frac{R_T}{A_D + R_T} = \frac{1}{1 + (R/R_0)^6}. \tag{7.10.15}$$

The formulas (7.10.13)–(7.10.15) are the primary results of the theory of *Förster resonance energy transfer*—also called *fluorescence resonance energy transfer*, or *FRET*—in the form originally developed by Förster.[127] The transfer rate depends on the overlap of the emission spectrum of the donor with the absorption spectrum of the acceptor, the radiative lifetime (in the medium) of the donor, and the distance between the molecules, which could be identical. There is a vast literature on fluorescence resonance energy transfer and its applications in photochemistry, biology, and other areas.[128] Excitation transfer causes a decrease in the fluorescence of the donor after absorption of light and an increase in the fluorescence of the acceptor. Because of its strong distance dependence, this makes Förster resonance energy transfer very useful as a "molecular ruler" for the determination of inter- and intra-molecular separations in typically a 1–10 nm range, or to determine the radiative lifetime of the donor by measuring its fluorescence lifetime with and without the presence of the acceptor.

The derivation of (7.10.13) did not include the process of excitation of the donor. We simply asserted that the donor, however it is excited, has an excitation loss rate that is small compared to that of the acceptor, and, in particular, that the transfer rate R_T is small compared to the acceptor de-excitation rate. The de-excitation at the acceptor in our model was assumed to be radiative, but actually β_B in (7.10.2) could be the decay rate for *any* dipole damping process that results in a lineshape function (7.10.5). In fact, the rate (7.10.4) for $k_A R \ll 1$ follows from Fermi's golden rule,

$$R_T = \frac{2\pi}{\hbar} |V|^2 \rho, \tag{7.10.16}$$

with the dipole–dipole interaction matrix element

$$V = -\frac{1}{4\pi\epsilon_0} \left[\frac{3(\mathbf{d}_A \cdot \hat{\mathbf{r}})(\mathbf{d}_B \cdot \hat{\mathbf{r}}) - \mathbf{d}_A \cdot \mathbf{d}_B}{r^3} \right] \tag{7.10.17}$$

between the states $|\phi_A\rangle$ and $|\phi_B\rangle$ defined in Section 7.9 and a density per unit energy of final states $\rho = S_B(\omega_A)/\hbar$. In other words, R_T describes an *irreversible* transfer of energy. This irreversibility was built into our model when it was assumed that $\beta_A(t - r/c) \ll 1 \ll \beta_B(t - r/c)$, as in (7.9.77) and (7.10.4).

[127] T. Förster, Ann. Phys. (Germany) **2**, 55 (1948).

[128] See, for instance, *FRET: Förster Resonance Energy Transfer*, eds. I. Medintz and N. Hildebrandt, Wiley-VCH, Weinheim, 2014 or the many online reviews of Förster resonance energy transfer and its applications.

Based as it is on perturbation theory (Fermi's golden rule), Förster's theory of electronic excitation transfer between molecules requires that there be a dense distribution of final states and rapid de-excitation at the acceptor, so that transferred excitation does not return to the donor. The assumption in our simple model that β_B is large is consistent with these conditions: the lineshape function $S_B(\omega)$ is a continuous distribution, although, in reality, the distribution of final states might be a quasi-continuum associated with molecular vibrational states, and β_B would characterize a rate of vibrational relaxation.

Despite being called "fluorescence" resonance energy transfer, Förster energy transfer is not simply absorption by the acceptor of radiation (fluorescence) from the donor. The intensity of radiation from the donor falls off with distance as $1/R^2$, so that the excitation rate of the acceptor via absorption of fluorescence radiation from the donor would fall off as $1/R^2$, not $1/R^6$. This rate follows from the far-field limit ($k_A r \gg 1$) of (7.9.44); as noted many years ago by W. Arnold and J. R. Oppenheimer in the context of photosynthesis, it is too slow and inefficient to explain observed photochemical effects requiring energy transfer.[129] Förster transfer occurs in the *near* field of the donor.

Some additional remarks about the nature of the Förster resonance energy transfer process may be worthwhile. Let us return to the case of two identical atoms, and consider the excitation probabilities for times much smaller than the radiative lifetimes $\{2\beta[1 \pm g(r)]\}^{-1}$ of the symmetric and antisymmetric states. From (7.9.62),

$$P_B(t) \cong \sin^2(Vt/\hbar), \tag{7.10.18}$$

and, similarly,

$$P_A(t) \cong \cos^2(Vt/\hbar) \tag{7.10.19}$$

for $k_0 r \ll 1$, where V is defined by (7.10.17). For such times, the two atoms simply exchange excitation at a rate proportional to $1/R^3$. This does not, of course, correspond to an *irreversible* transfer of excitation from a donor molecule to an acceptor, which requires rapid de-excitation at the donor and a continuum (or quasi-continuum) of final states. This is all very much like the difference discussed in Chapter 2 between Rabi oscillations and excitation described by Fermi's golden rule for atoms in near-resonance fields, the former being analogous to the two-atom population oscillations (7.10.18) and (7.10.19), and the latter to Förster resonance energy transfer. In fact, for times short compared to the radiative lifetimes and for $k_0 r \ll 1$, (7.9.55) and (7.9.61) for identical atoms take the forms

$$\dot{v} = \frac{2V}{\hbar} w(t), \tag{7.10.20}$$

and

[129] See Footnote 126.

$$\dot{w} = -\frac{2V}{\hbar}v(t) \tag{7.10.21}$$

when written in terms of $w(t) \equiv P_B(t) - P_A(t)$ and $v(t) \equiv i[X(t) - X^*(t)]$ and with V again defined by (7.10.17). These have exactly the form of the optical Bloch equations, (2.6.7) and (2.6.8), with $\Delta = 0$. But we emphasize again that Förster resonance energy transfer occurs in the near field of the donor, as opposed to the far (radiation) field, and therefore is not absorption by the acceptor of a photon emitted by the donor. Recall the remarks in Section 1.4.2, about storage fields and far fields of antennas.

For second-order perturbation theory to be accurate—and therefore for the Förster theory to be applicable—the transfer rate (7.10.6) must be small compared to the acceptor de-excitation rate, so that the acceptor excitation probability is small. In our simple model, this requires that $V \ll \hbar\beta_B$, or, more generally, $|V| \ll \Delta E$, ΔE being a measure of the absorption width of the acceptor. But this weak-coupling approximation is not always valid; if there is strong coupling, excitation can oscillate between the two molecules instead of flowing irreversibly from one to the other. In our simple model, this requires a description based on the coupled equations (7.9.53), (7.9.55), and (7.9.61) in the case of identical molecules, or more general equations of this type for non-identical molecules. In other words, we must take account of the "coherence" represented by $X(t)$. The situation is then again analogous to that for atoms in near-resonance fields, the coherence in the excitation exchange corresponding to the Rabi oscillations of atoms driven by optical pulses of duration short compared to dipole relaxation times. There is clear evidence of short-time coherence effects in biological and chemical systems.[130]

7.11   Spontaneous Emission near Reflectors

In Section 7.3.3, we calculated the change in energy of an atom due to its proximity to a conducting surface. The spontaneous emission rate of an atomic transition is also changed from its free-space value and depends on the distance of the atom from the conductor. The rate $A_{21}(\mathbf{r})$ for an excited two-state atom with transition electric dipole matrix element $\mathbf{d}_{12}$ and transition frequency $\omega_0 = k_0 c$ is given by (4.4.37) for an atom at position $\mathbf{r}$.

For an atom at a distance d from a conducting plate, the calculation of $A_{21}(d)$ requires the mode functions $\mathbf{A}_\beta(\mathbf{r})$ for the half-space $z > 0$ bounded by the conductor. These mode functions can be expressed as combinations of plane waves with wave vectors $\mathbf{k} = k_x\hat{\mathbf{x}} + k_y\hat{\mathbf{y}} + k_z\hat{\mathbf{z}} = \mathbf{k}_\parallel + k_z\hat{\mathbf{z}}$ and $\mathbf{k}_r = \mathbf{k}_\parallel - k_z\hat{\mathbf{z}}$. (The carets denote unit vectors.) Thus, we define

$$\mathbf{A}_{\mathbf{k}1}(\mathbf{r}) = \left(\frac{2}{V}\right)^{1/2}\left(\hat{\mathbf{k}}_\parallel \times \hat{\mathbf{z}}\right)\sin k_z z\, e^{i\mathbf{k}_\parallel \cdot \mathbf{r}}, \tag{7.11.1}$$

---

[130] For a review of research in this area, see G. D. Scholes, G. R. Fleming, L. X. Chen, et al., Nature **543**, 647 (2017).

which satisfies $\nabla \cdot \mathbf{A}_{\mathbf{k}1} = 0$ and the Helmholtz equation $\nabla^2 \mathbf{A}_{\mathbf{k}1} + k^2 \mathbf{A}_{\mathbf{k}1} = 0$, and $\hat{\mathbf{z}} \times \mathbf{A}_{\mathbf{k}1}(\mathbf{r})$ vanishes on the conductor at $z = 0$, as required. $V = L^3$ ($L \to \infty$) is a quantization volume for the normalization

$$\int_V d^3r \, |\mathbf{A}_{\mathbf{k}1}(\mathbf{r})|^2 = \frac{2}{V} \int_V d^3r \, \sin^2 k_z z = 1, \tag{7.11.2}$$

so that the vector functions $\mathbf{A}_{\mathbf{k}1}(\mathbf{r})$ satisfy all the conditions required for them to be mode functions for the transverse electric field. We also define the vector functions

$$\mathbf{A}_{\mathbf{k}2}(\mathbf{r}) = \frac{1}{k}\left(\frac{2}{V}\right)^{1/2}\left(k_\| \hat{\mathbf{z}} \cos k_z z - i k_z \hat{\mathbf{k}}_\| \sin k_z z\right) e^{i\mathbf{k}_\| \cdot \mathbf{r}}, \tag{7.11.3}$$

which are orthogonal to $\mathbf{A}_{\mathbf{k}1}(\mathbf{r})$ and satisfy the same aforementioned conditions for mode functions. The set of all $\mathbf{A}_{\mathbf{k}1}(\mathbf{r})$ and $\mathbf{A}_{\mathbf{k}2}(\mathbf{r})$ form a complete orthonormal set of mode functions for the half-space bounded by the conductor at $z = 0$.

Suppose that $\mathbf{d}_{12} = d_{12}\hat{\mathbf{z}}$. Then $\mathbf{d}_{12} \cdot \mathbf{A}_{\mathbf{k}1}(\mathbf{r}) = 0$,

$$\mathbf{d}_{12} \cdot \mathbf{A}_{\mathbf{k}2}(\mathbf{r}) = \frac{1}{k}\left(\frac{2}{V}\right)^{1/2} d_{12} k_\| \cos k_z z, \tag{7.11.4}$$

and, from (4.4.37),[131]

$$\begin{aligned}
A_{21}(d) \equiv A_{21}^\perp(d) &= \frac{2}{V}\sum_{\mathbf{k}}\left(\frac{\pi\omega_k}{\hbar\epsilon_0}\frac{1}{k^2}k_\|^2\cos^2 k_z d\right)\delta(\omega_k - \omega) \\
&= \frac{2\pi\omega_0 d_{12}^2}{\hbar\epsilon_0}\frac{1}{k_0^2}\frac{1}{V}\frac{V}{8\pi^3}\int_0^\infty dk\,k^2\int_0^\pi d\theta\,\sin\theta\left[k_0^2\sin^2\theta\cos^2(k_0 d\cos\theta)\right]\delta(\omega - \omega_0) \\
&= \frac{3}{2}A_{21}\int_0^\pi d\theta\,\sin^3\theta\cos^2(k_0 d\cos\theta) \\
&= 3A_{21}\left[\frac{1}{3} + \frac{\sin 2k_0 d}{(2k_0 d)^3} - \frac{\cos 2k_0 d}{(2k_0 d)^2}\right], \tag{7.11.5}
\end{aligned}$$

where $A_{21} = \omega_0^3 d_{12}^2/3\pi\hbar\epsilon_0 c^3$ is the spontaneous emission rate when the atom is in free, unbounded space ($d \to \infty$). For $\mathbf{d}_{12}$ parallel to the plate, we obtain in similar fashion

$$A_{21}^\|(d) = \frac{3}{2}A_{21}\left[\frac{2}{3} - \frac{\sin 2k_0 d}{2k_0 d} - \frac{\cos 2k_0 d}{(2k_0 d)^2} + \frac{\sin 2k_0 d}{(2k_0 d)^3}\right]. \tag{7.11.6}$$

The emission rate averaged over dipole orientations is

$$A_{21}^{\text{avg}}(d) = \frac{1}{3}A_{21}^\perp(d) + \frac{2}{3}A_{21}^\|(d) = A_{21}\left[1 - \frac{\sin 2k_0 d}{2k_0 d} - \frac{2\cos 2k_0 d}{(2k_0 d)^2} + \frac{2\sin 2k_0 d}{(2k_0 d)^3}\right]. \tag{7.11.7}$$

$A_{21}^\perp(d)$, $A_{21}^\|(d)$, and $A_{21}^{\text{avg}}(d)$ reduce, of course, to A_{21} for distances d large compared to c/ω_0. For $k_0 d \to 0$,

---

[131] The substitutions $k_z = k_0\cos\theta$ and $k_\|^2 = k_0^2\sin^2\theta$ are made for the k-space integration $\int d^3k(\ldots) = 2\pi\int_0^\infty dk\,k^2\int_0^\pi d\theta\,\sin\theta(\ldots)$.

$$A_{21}^{\perp}(d) \to 2A_{21}, \qquad (7.11.8)$$

whereas

$$A_{21}^{\parallel}(d) \to 0. \qquad (7.11.9)$$

In this small-separation limit, the emission is "superradiant" when $\mathbf{d}_{12}$ is perpendicular to the conductor, and "subradiant" when it is parallel. In fact, $A_{21}^{\perp}(d)$ and $A_{21}^{\parallel}(d)$ can be interpreted as cooperative spontaneous emission rates for two atoms, one of which is initially excited: $A_{21}^{\perp}(d)$ is identical to the emission rate $2\beta_{+}(2d)$ (see (7.9.23)) for $\Delta m = 0$ transitions when the two atoms are initially in the symmetric state $|\phi_{+}\rangle$, and $A_{21}^{\parallel}(d)$ is identical to the rate $2\beta_{-}(2d)$ (see (7.9.22)) when the two atoms are initially in the antisymmetric state $|\phi_{-}\rangle$. Figure 7.8 shows why there is this connection between the spontaneous emission for a single atom near a "mirror" and the cooperative emission of two identical atoms in free space. Spontaneous emission can result in a photon in a plane-wave mode k in two ways: (a) by direct emission into that mode, or (b) after a reflection off the mirror. The probability amplitude for a photon to be found in the mode is the sum of the probability amplitudes for the indistinguishable processes (a) and (b), since the photon provides no information as to which of the two alternatives has occurred. We can therefore describe the emission by an atom near a mirror as emission by the two-atom system in a state for which there is no information as to whether the atom or its image is initially excited. In other words, the atom and its image are described by entangled states $|\phi_{\pm}\rangle$. If the transition dipole moment of the atom is perpendicular to the mirror, the dipole and its image oscillate in phase, and so the atom-image pair is described by the $|\phi_{+}\rangle$ state; if the dipole moment is parallel to the mirror, the dipole and image oscillations are π out of phase, and then $|\phi_{-}\rangle$ is the appropriate atom-image state.

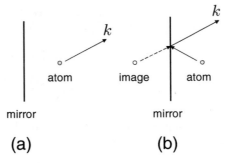

Fig. 7.8 Two ways by which spontaneous emission by an atom near a reflector can result in a photon in a plane-wave mode k.

Early, elegant experiments with molecular monolayers demonstrated the oscillatory dependence of the fluorescence rate on the distance from a reflecting plate.[132] The experimental results were well described by extending the theory above to allow for imperfect conductivity. Experiments with a "Chinese gong" near a wall demonstrate analogous effects for sound waves.[133]

As in the case of spontaneous emission in free space, we can derive the emission rates here in different, equivalent ways. Consider the radiation reaction field $\mathbf{E}_{RR}(d,t)$ of an electric dipole $p(t)\hat{\mathbf{z}}$ at a distance d from a mirror in the plane $z = 0$. The component of this field in the z direction is found in a straightforward calculation to be

$$E_{RR}(d,t) = E_{RR}(\infty,t) - \frac{1}{4\pi\epsilon_0}(-2)\left[\frac{1}{c(2d)^2}\dot{p}(t-2d/c) + \frac{1}{(2d)^3}p(t-2d/c)\right] \quad (7.11.10)$$

for $t > 2d/c$.[134] The second term is the field from an image dipole at $z = -d$ (cf. equation (4.3.11)). The positive-frequency part of $E_{RR}(d,t)$ for a dipole oscillation $p(t) = pe^{-i\omega_0 t}$ is

$$E_{RR}^{(+)}(d,t) = E_{RR}^{(+)}(\infty,t) + \frac{1}{4\pi\epsilon_0}(2k_0^3)\left[\frac{-i}{(2k_0d)^2} + \frac{1}{(2k_0d)^3}\right]p(t)e^{2ik_0d}. \quad (7.11.11)$$

The d-dependent imaginary part of $E_{RR}^{(+)}(d,t)$—the part that determines the emission rate—is proportional to

$$\frac{\sin 2k_0d}{(2k_0d)^3} - \frac{\cos 2k_0d}{(2k_0d)^2}, \quad (7.11.12)$$

which is exactly the d-dependence of $A_{12}^{\perp}(d)$. If, instead, the dipole at $z = d$ is parallel to the conductor, we require the component of the radiation reaction field parallel to the conducting plate:

$$E_{RR}(d,t) = E_{RR}(\infty,t) + \frac{1}{4\pi\epsilon_0}\left[\frac{1}{c^2(2d)}\ddot{p}(t-2d/c) + \frac{1}{c(2d)^2}\dot{p}(t-2d/c)\right.$$
$$\left. + \frac{1}{(2d)^3}p(t-2d/c)\right], \quad (7.11.13)$$

and, for $p(t) = pe^{-i\omega_0 t}$,

$$E_{RR}^{(+)}(d,t) = E_{RR}^{(+)}(\infty,t) - \frac{1}{4\pi\epsilon_0}\frac{1}{2}(2k_0^3)\left[\frac{1}{2k_0d} + \frac{i}{(2k_0d)^2} - \frac{1}{(2k_0d)^3}\right]p(t)e^{2ik_0d}. \quad (7.11.14)$$

The d-dependent imaginary part of this expression is proportional to

[132] K. H. Drexhage, Sci. Am. **222**, 108 (March, 1970); K. H. Drexhage, "Interaction of light with molecular dye lasers," *Progress in Optics* **12**, ed. E. Wolf, North-Holland, Amsterdam, 1974, pp. 163–232.

[133] L. Langguth, R. Fleury, A. Alú, and A. F. Koenderink, Phys. Rev. Lett. **116**, 224301 (2016).

[134] P. W. Milonni, Phys. Rev. A **25**, 1315 (1982).

$$-\frac{1}{2}\left[\frac{\sin 2k_0 d}{2k_0 d} + \frac{\cos 2k_0 d}{(2k_0 d)^2} - \frac{\sin 2k_0 d}{(2k_0 d)^3}\right], \tag{7.11.15}$$

which has the d-dependence of $A_{21}^{\parallel}(d)$. Thus, the spontaneous emission rate of an atom near a reflector can be understood as a consequence of the radiation reaction field acting back on the atom. This radiation reaction field is just the sum of the radiation reaction field of the atom in free space and the field from the image of the atomic dipole at $z = -d$. Just as in the case of an atom in free space (see Section 4.5), this interpretation based on radiation reaction follows naturally when we normally order atom and field operators in the Heisenberg picture, and if we choose different operator orderings, we can identify contributions to $A_{21}^{\perp}(d)$ and $A_{21}^{\parallel}(d)$ from the zero-point, "vacuum" field in the half-space $z > 0$.

These different interpretations do not depend on the simplifying assumption of a perfecting conducting plate, nor do they depend on the specific geometry we have considered. As in the case of Casimir forces, the calculations for different geometries require the determination of appropriate mode functions for the electromagnetic field, a *classical* boundary-value problem that is generally difficult to solve analytically.

The modification of a spontaneous emission rate can also be calculated analytically for the case of an atom between two perfectly conducting parallel plates at $z = \pm L/2$. If the transition dipole moment is perpendicular to the plates, the emission rate is found to be

$$A_{21}^{\perp}(d) = A_{21}\left(\frac{3\pi}{k_0 L}\right)\left[\frac{1}{2} + \sum_{n=1}^{N}\left(1 - \frac{n^2\pi^2}{k_0^2 L^2}\right)\cos^2\left(\frac{n\pi d}{L} - \frac{n\pi}{L}\right)\right], \tag{7.11.16}$$

where d is the distance of the atom from one of the mirrors, and N is the greatest integer part of $k_0 L/\pi$. If the dipole moment is parallel to the plates,

$$A_{21}^{\parallel}(d) = A_{21}\left(\frac{3\pi}{2k_0 L}\right)\sum_{n=1}^{N}\left(1 + \frac{n^2\pi^2}{k_0^2 L^2}\right)\sin^2\left(\frac{n\pi d}{L} - \frac{n\pi}{L}\right). \tag{7.11.17}$$

Like the rates for an atom near a single mirror, these results can be obtained by an image method.[135] For $k_0 L/\pi = 2L/\lambda < 1$, that is, when L is less than half a transition wavelength λ, $A_{21}^{\parallel}(d) = 0$. Such inhibited spontaneous emission by atoms in cavities has been observed.[136]

In the derivation of these formulas, it is implicitly assumed that the time $2L/c$ it takes for light to make a round trip in the cavity is sufficiently small compared to the radiative lifetime that the emission rate can, in fact, be modified by the cavity. The situation here is just like that in the system of two atoms, where the formulas (7.9.22)

[135] See, for instance, P. W. Milonni and P. L. Knight, Opt. Commun. **9**, 119 (1973), and references therein.

[136] For a review of early work on such effects in cavity quantum electrodynamics, see S. Haroche and D. Kleppner, Physics Today (January, 1989).

and (7.9.23) for cooperative emission apply only for times sufficiently long that the atoms can "communicate" with each another. It is also assumed that the cavity is lossless and, in particular, that the plates are perfectly reflecting.

To allow for both the propagation time $T = 2L/c$ and imperfect reflectivities, it is convenient to consider a simplified model in which only field modes propagating along the (z) axis perpendicular to the parallel plates are included. The probability amplitude $b(t)$ for the state in which the atom is excited and the field in the cavity is in its vacuum state, assuming $b(0) = 1$, is found in the rotating-wave approximation to satisfy the delay-differential equation[137]

$$\dot{b}(t) = -\frac{1}{2}\Omega^2 T \Big[\frac{1}{2}b(t) + \sum_{n=1}^{\infty} \left(r^2 e^{i\Delta_0 T}\right)^n b(t - nT)\theta(t - nT)$$

$$+ \frac{1}{2}re^{2ik_0 d} \sum_{n=0}^{\infty} \left(r^2 e^{i\Delta_0 T}\right)^n b(t - T_0 - nT)\theta(t - T_0 - nT)$$

$$+ \frac{1}{2}r^{-1}e^{-2ik_0 d} \sum_{n=1}^{\infty} \left(r^2 e^{i\Delta_0 T}\right)^n b(t + T_0 - nT)\theta(t + T_0 - nT)\Big]. \quad (7.11.18)$$

Here, $\Omega = (d_{12}^2\omega_0/\hbar\epsilon_0 SL)^{1/2}$, where, again, ω_0 is the transition frequency, and d_{12} is the transition electric dipole moment, which is assumed to be real and perpendicular to the z axis; S is a cross-sectional area used in defining a field quantization volume $V = SL$; $\Delta_0 = \omega_0 - \omega$, with ω the cavity-mode frequency closest to the transition frequency; $T_0 = 2d/c$, where d is, again, the distance of the atom from one of the mirrors; r is the field reflection coefficient of the (identical) plates for frequencies $\omega \approx \omega_0$; and $\theta(t)$ is the unit step function.

It follows from (7.11.18) that, for times before the time it takes for light to propagate from the atom to the nearest plate and back,

$$\dot{b}(t) = -(\frac{1}{4}\Omega^2 T)b(t) = -(d_{12}^2\omega_0/2\hbar\epsilon_0 cS)b(t). \quad (7.11.19)$$

This also describes the spontaneous emission when the atom is infinitely far from either plate. The emission rate is proportional to ω_0, not to ω_0^3 as in three-dimensional free space, because, in this model, we allow for field propagation only along one axis, and therefore the density of field modes into which the atom can radiate is proportional to k_0 rather than k_0^3.

Suppose instead that T is very small and $\Omega T \ll 1$. If the plates are perfectly reflecting ($r = -1$),

---

[137] R. J. Cook and P. W. Milonni, Phys. Rev. A **35**, 5081 (1987).

$$\dot{b}(t) = -\frac{1}{2}\Omega^2 e^{i\Delta_0 t}\Big[T\sum_{n=1}^{\infty} e^{-i\Delta_0(t-nT)}b(t-nT)\theta(t-nT)$$

$$-\frac{1}{2}e^{2ik_0 d}T\sum_{n=0}^{\infty} e^{-i\Delta_0(t-nT)}b(t-T_0-nT)\theta(t-T_0-nT)$$

$$-\frac{1}{2}e^{-2ik_0 d}T\sum_{n=1}^{\infty} e^{-i\Delta_0(t-nT)}b(t+T_0-nT)\theta(t+T_0-nT)\theta(t+T_0-nT)\Big]$$

$$\cong -\frac{1}{2}\Omega^2 e^{i\Delta_0 t}\Big(1-\frac{1}{2}e^{2ik_0 d}-\frac{1}{2}e^{2ik_0 d}\Big)\int_0^t dt'\, b(t')e^{-i\Delta_0 t'}$$

$$= -\Omega^2 \sin^2(k_0 d)e^{i\Delta_0 t}\int_0^t dt'\, b(t')e^{-i\Delta_0 t'}, \tag{7.11.20}$$

or, equivalently,

$$\ddot{b}(t) - i\Delta_0\dot{b}(t) + \frac{1}{4}\Lambda^2 b(t) \cong 0, \quad \text{where } \Lambda \equiv 2\Omega\sin(k_0 d), \tag{7.11.21}$$

and therefore

$$b(t) \cong e^{i\Delta_0 t}\Big\{\cos\big[\frac{1}{2}(\Delta_0^2+\Lambda^2)^{1/2}t\big] - \frac{i\Delta_0}{(\Delta_0^2+\Lambda^2)^{1/2}}\sin\big[\frac{1}{2}(\Delta_0^2+\Lambda^2)^{1/2}t\big]\Big\}. \tag{7.11.22}$$

In this limit, we obtain the familiar "vacuum Rabi oscillations" for a two-state atom in a single-mode resonator (see Section 5.3). We recover single-mode behavior because the assumption $\Omega T^2 \ll 1$ implies that the photon bounce frequency $T^{-1} = c/2L$—which is also the spacing of the cavity-mode frequencies—is large compared to the spontaneous emission rate and therefore the radiative linewidth. The atom therefore effectively interacts with only a single mode of the (Fabry–Pérot) resonator in this limit.

Consider now the same single-mode limit when $r \neq -1$. Define a damping rate γ by writing

$$r^2 e^{i\Delta_0 T} = e^{i(\Delta_0 T - iT^{-1}\log r^2)} = e^{i(\Delta_0+i\gamma)T} \tag{7.11.23}$$

in (7.11.18), with $\gamma = -(c/2L)\log r^2$ the field damping rate due to imperfectly reflecting walls. The same approximations that led to (7.11.21) then lead to

$$\ddot{b}(t) + (\gamma - i\Delta_0)\dot{b}(t) + \frac{1}{4}\Lambda^2 b(t) \cong 0. \tag{7.11.24}$$

For $\Delta_0 = 0$ and $\Lambda \gg \gamma$, the upper-state probability undergoes damped Rabi oscillations:

$$|b(t)|^2 \cong e^{-\gamma t}\cos^2(\frac{1}{2}\Lambda t). \tag{7.11.25}$$

In the opposite, "overdamped" case, where $\gamma \gg \Lambda$ and $\Delta_0 = 0$,

$$|b(t)|^2 \cong e^{-\Lambda^2 t/2\gamma} = e^{-\Gamma t}. \tag{7.11.26}$$

The spontaneous emission rate is then

$$\Gamma = \frac{\Lambda^2}{2\gamma} = \frac{1}{2\gamma} 4\Omega^2 \sin^2(k_0 d) = \frac{\omega_0}{2\gamma} \frac{4d_{12}^2}{\hbar\epsilon_0 SL} \sin^2(k_0 d). \tag{7.11.27}$$

Expressed in terms of the free-space spontaneous emission rate A_{21}, the transition wavelength $\lambda = 2\pi c/\omega_0$, the volume $V = SL$, and the cavity Q factor ($Q \equiv \omega_0/2\gamma$),

$$\Gamma = \frac{3Q\lambda^3}{2\pi^2 V} A_{21} \sin^2(k_0 d), \tag{7.11.28}$$

which is $2\sin^2(k_0 d)$ times the oft-cited result obtained by Purcell for the spontaneous emission rate of an atom in a lossy single-mode resonator.[138] The *Purcell effect* has been observed, for instance, in the spontaneous emission rate of Rydberg sodium atoms in a Fabry–Pérot resonator; spontaneous emission rates as large as ~ 500 times the free-space emission rate A_{21} in the mm region were observed.[139] Both enhanced and inhibited spontaneous emission rates by the Purcell effect have been observed in optical transitions.[140]

7.12 Spontaneous Emission in Dielectric Media

Rates for spontaneous emission and other radiative processes are modified from their free-space values not only by reflectors and cavities but also, for instance, by bulk, passive dielectric media in which the emitters are embedded. In particular, radiative processes depend on the refractive index $n(\omega)$ of the surrounding medium. We now consider briefly the dependence of the spontaneous rate on the refractive index.[141]

One effect of the (real) refractive index in a transparent, homogeneous, and isotropic dielectric medium is to change the k-space volume element $d^3 k$ from its free-space form, $4\pi k^2 dk = 4\pi \omega^2 d\omega/c^3$, to

$$4\pi k^2 \frac{dk}{d\omega} d\omega = [4\pi n^2(\omega)\omega^2 n_g(\omega)/c^3]d\omega, \tag{7.12.1}$$

where $n_g(\omega) = d(\omega n)/d\omega$ is the group index.[142] This implies, for example, that the spectral energy density $\rho(\omega)$ of radiation in thermal equilibrium at temperature T in a dielectric medium that is transparent at frequency ω is

$$\rho(\omega)d\omega = \frac{2 \times \hbar\omega d^3 k/(2\pi)^3}{e^{\hbar\omega/k_B T} - 1}, \tag{7.12.2}$$

---

[138] E. M. Purcell, Phys. Rev. **69**, 681 (1946).

[139] P. Goy, J. M. Raimond, M. Gross, and S. Haroche, Phys. Rev. Lett. **50**, 1903 (1983).

[140] D. J. Heinzen, J. J. Childs, J. F. Thomas, and M. S. Feld, Phys. Rev. Lett. **58**, 1320 (1987); F. DeMartini, G. Innocenti, G. R. Jacobovitz, and P. Mataloni, Phys. Rev. Lett. **59**, 2955 (1987).

[141] For a treatment of absorption and stimulated emission by atoms in a dielectric medium based, as in the treatment here, on field quantization in a dielectric medium, see P. W. Milonni, J. Mod. Opt. **42**, 1991 (1995), and references therein.

[142] Recall from Section 1.6 that the group index n_g is positive for a passive medium.

or

$$\rho(\omega) = \frac{n^2(\omega)n_g(\omega)\hbar\omega^3/\pi^2 c^3}{e^{\hbar\omega/k_B T} - 1}. \tag{7.12.3}$$

This in turn implies that the ratio of the Einstein A and B coefficients for a transition at frequency ω depends on $n^2(\omega)n_g(\omega)$:

$$A'/B' = n^2(\omega)n_g(\omega)\hbar\omega^3/\pi^2 c^3, \tag{7.12.4}$$

since, from (2.8.2), this ratio is equal to the numerator of (7.12.3). A' and B' denote the Einstein A and B coefficients as modified by the dielectric medium, respectively.

A' or B' can be calculated directly from the expression (3.4.15) for the electric field in the dielectric medium. For $\mu \cong \mu_0$, this field differs from the free-space electric field (3.3.11) at the frequency ω_k by the factor $1/\sqrt{n(\omega_k)n_g(\omega_k)}$. The rate of spontaneous emission for a transition with electric dipole matrix element $\mathbf{d}_{12}$ and frequency ω_0 is therefore changed from its free-space value

$$A_{21} = \frac{\pi|\mathbf{d}_{12}|^2}{3\epsilon_0\hbar}\frac{2}{(2\pi)^3}\int d^3k\,\omega_k\delta(\omega_k - \omega_0) = \frac{\pi|\mathbf{d}_{12}|^2}{3\epsilon_0\hbar}\frac{2}{(2\pi)^3}\int 4\pi dk\,k^2\omega\delta(\omega - \omega_0)$$

$$= \frac{\pi|\mathbf{d}_{12}|^2}{3\epsilon_0\hbar}\frac{2}{(2\pi)^3}\int 4\pi d(\frac{\omega}{c})(\frac{\omega}{c})^2\omega\delta(\omega - \omega_0) = \frac{1}{4\pi\epsilon_0}\frac{4|\mathbf{d}_{12}|^2\omega_0^3}{3\hbar c^3} \tag{7.12.5}$$

to

$$A'_{12} = \frac{\pi|\mathbf{d}_{12}|^2}{3\epsilon_0\hbar}\frac{2}{(2\pi)^3}\int 4\pi d\Big[\frac{n(\omega)\omega}{c}\Big]\Big[\frac{n(\omega)\omega}{c}\Big]^2\omega \times \frac{1}{n(\omega)n_g(\omega)}\delta(\omega - \omega_0)$$

$$= \frac{1}{4\pi\epsilon_0}n(\omega_0)\frac{4|\mathbf{d}_{12}|^2\omega_0^3}{3\hbar c^3} = n(\omega_0)A_{21}. \tag{7.12.6}$$

Note that we have allowed for dispersion ($n_g(\omega_0) \neq 0$) at the transition frequency, but that A'_{12} turns out to be independent of $n_g(\omega_0)$.[143]

This derivation of the index dependence of the spontaneous emission rate ignores the fact that the local field acting on an atom is, in general, different from the average, "macroscopic" field in the medium. The effect of this difference is to replace (7.12.6) with

$$A'_{12} = L(\omega_0)n(\omega_0)A_{21}, \tag{7.12.7}$$

where $L(\omega_0)$ is the "local field correction." In the "real cavity model" for $L(\omega_0)$, it is assumed that the emitters are inclusions or dopants and radiate as if they are within an otherwise empty microscopic cavity; the local field correction in this model is

$$L = \frac{3n^2}{2n^2 + 1}. \tag{7.12.8}$$

The more familiar Lorentz or "virtual cavity" model derived in many textbooks applies for a homogeneous system of identical atoms, or when the emitters are "substitutional" rather than "interstitial."[144] Experiments have clearly demonstrated not only

---

[143] For a microscopic approach to spontaneous emission by an atom in a dielectric medium, see, for instance, P. R. Berman and P. W. Milonni, Phys. Rev. Lett. **92**, 053601 (2004).

[144] P. de Vries and A. Lagendijk, Phys. Rev. Lett. **81**, 1381 (1998).

that spontaneous emission rates depend on the refractive index of the host medium, but also that they exhibit local field corrections.[145]

References and Suggested Additional Reading

Van der Waals forces are treated in considerable detail by *Israelachvili* and *Parsegian*, the latter taking a more pedagogical approach. Many-body approaches to van der Waals and Casimir interactions are presented by M. J. Renne (in Physica **53**, 193 (1971), and **56**, 125 (1971)) and *Abrikosov, Gorkov, and Dzyaloshinski*. Books focusing specifically on the general theory of Casimir interactions include *Milton* and *Bordag et al. Milonni (1994)* and *Compagno, Passante, and Persico* treat various aspects of vacuum-field fluctuations in electrodynamics, and some discussion of Casimir and related effects appears also in *Novotny and Hecht*. An entire chapter is devoted to fluctuations and van der Waals forces in *Ginzburg*.

Some experiments to detect zero-point energy are reviewed by Yu. M. Tsipenyuk, Sov. Phys. Usp. **55**, 796 (2012).

The theory of stochastic electrodynamics is described in detail in papers by T. H. Boyer and in *de la Peña and Cetto*.

Among early papers on the theory of the dipole–dipole resonance interaction including retardation are D. A. Hutchinson and H. F. Hameka, J. Chem. Phys. **41**, 2006 (1964); A. D. McLachlan, Mol. Phys. **8**, 409 (1964); R. R. McClone and E. A. Power, Proc. R. Soc. Lond. A **286**, 573 (1965); and M. R. Philpott, Proc. Phys. Soc. Lond. **87**, 619 (1966). For an analysis employing the argument theorem, see P. R. Berman, G. W. Ford, and P. W. Milonni, J. Chem. Phys. **141**, 164105 (2014). The nonretarded, $1/r^3$ interaction is discussed in much detail for specific systems by G. W. King and J. H. van Vleck, Phys. Rev. **55**, 1169 (1939), and R. S. Mulliken, Phys. Rev. **120**, 1674 (1960).

For a careful analysis of causality in the two-atom interaction see P. R. Berman and B. Dubetsky, Phys. Rev A **55**, 4060 (1997). See also P. W. Milonni, D. F. V. James, and H. Fearn, Phys. Rev. A **52**, 1525 (1995), and references therein.

The edited volumes by *Berman (1994)* and *Dalvit et al.* include theoretical and experimental papers by experts in cavity quantum electrodynamics and Casimir forces.

---

[145] See, for instance, G. L. J. A. Rikken and Y. A. R. R. Kessener, Phys. Rev. Lett. **74**, 880 (1995); K. Dolgaleva, R. W. Boyd, and P. W. Milonni, J. Opt. Soc. Am. B **24**, 516 (2007).

Appendices

A. Retarded Electric Field in the Coulomb Gauge

The expression (1.3.46) for the electric field derived in the Coulomb gauge is identical to the manifestly retarded expression obtained in the Lorentz gauge, in spite of the appearance of the static, nonretarded Coulomb field. To show this, it is convenient to write

$$\mathbf{E}(\mathbf{r}, t) = \int_{-\infty}^{\infty} d\omega \tilde{\mathbf{E}}(\mathbf{r}, \omega) e^{-i\omega t} \tag{A.1}$$

and, likewise, for $\rho(\mathbf{r}, t)$ and $\mathbf{J}(\mathbf{r}, t)$. Then, (1.3.45) and (1.3.46) imply

$$\tilde{\mathbf{E}}(\mathbf{r}, \omega) = -\frac{1}{4\pi\epsilon_0} \nabla \int d^3 r' \frac{\tilde{\rho}(\mathbf{r}', \omega) e^{i\omega|\mathbf{r}-\mathbf{r}'|/c}}{|\mathbf{r}-\mathbf{r}'|} + \frac{i\mu_0\omega}{4\pi} \int d^3 r' \frac{\tilde{\mathbf{J}}(\mathbf{r}', \omega) e^{i\omega|\mathbf{r}-\mathbf{r}'|/c}}{|\mathbf{r}-\mathbf{r}'|} \tag{A.2}$$

and

$$\tilde{\mathbf{E}}(\mathbf{r}, \omega) = -\frac{1}{4\pi\epsilon_0} \nabla \int d^3 r' \frac{\tilde{\rho}(\mathbf{r}', \omega)}{|\mathbf{r}-\mathbf{r}'|} + \frac{i\mu_0\omega}{4\pi} \int d^3 r' \frac{\tilde{\mathbf{J}}^{\perp}(\mathbf{r}', \omega) e^{i\omega|\mathbf{r}-\mathbf{r}'|/c}}{|\mathbf{r}-\mathbf{r}'|} \tag{A.3}$$

in the Lorentz and Coulomb gauges, respectively. We use the identity $\tilde{\mathbf{J}}^{\perp} = \tilde{\mathbf{J}} - \tilde{\mathbf{J}}^{\parallel}$ to express (A.3) in the form

$$\begin{aligned} \mathbf{E}(\mathbf{r}, \omega) = &-\frac{1}{4\pi\epsilon_0} \nabla \int d^3 r' \frac{\tilde{\rho}(\mathbf{r}', \omega)}{|\mathbf{r}-\mathbf{r}'|} + \frac{i\mu_0\omega}{4\pi} \int d^3 r' \frac{\tilde{\mathbf{J}}(\mathbf{r}', \omega) e^{i\omega|\mathbf{r}-\mathbf{r}'|/c}}{|\mathbf{r}-\mathbf{r}'|} \\ &- \frac{i\mu_0\omega}{4\pi} \int d^3 r' \frac{\tilde{\mathbf{J}}^{\parallel}(\mathbf{r}', \omega) e^{i\omega|\mathbf{r}-\mathbf{r}'|/c}}{|\mathbf{r}-\mathbf{r}'|}. \end{aligned} \tag{A.4}$$

Denote the third term by X and use (1.3.41) and the continuity equation (1.1.6) to express it as

$$\begin{aligned} X &= -\frac{i\mu_0\omega}{4\pi} \int d^3 r' \frac{e^{i\omega|\mathbf{r}-\mathbf{r}'|/c}}{|\mathbf{r}-\mathbf{r}'|} \left(\frac{-1}{4\pi}\right) \nabla' \int d^3 r'' \frac{\nabla'' \cdot \tilde{\mathbf{J}}(\mathbf{r}'', \omega)}{|\mathbf{r}'-\mathbf{r}''|} \\ &= -\frac{\mu_0\omega^2}{(4\pi)^2} \int d^3 r' \frac{e^{i\omega|\mathbf{r}-\mathbf{r}'|/c}}{|\mathbf{r}-\mathbf{r}'|} \nabla' \int d^3 r'' \frac{\tilde{\rho}(\mathbf{r}'', \omega)}{|\mathbf{r}'-\mathbf{r}''|}. \end{aligned} \tag{A.5}$$

Next, we use the identity

$$\left(\nabla'^2 + \frac{\omega^2}{c^2}\right) \frac{e^{i\omega|\mathbf{r}-\mathbf{r}'|/c}}{|\mathbf{r}-\mathbf{r}'|} = -4\pi\delta^3(\mathbf{r}-\mathbf{r}') \tag{A.6}$$

to write

$$X = \frac{\mu_0 c^2}{(4\pi)^2} \int d^3 r' \left[\nabla'^2 \frac{e^{i\omega|\mathbf{r}-\mathbf{r}'|/c}}{|\mathbf{r}-\mathbf{r}'|} + 4\pi \delta^3 (\mathbf{r}-\mathbf{r}') \right] \nabla' \int d^3 r'' \frac{\tilde{\rho}(\mathbf{r}'',\omega)}{|\mathbf{r}'-\mathbf{r}''|}$$

$$= \frac{1}{4\pi} \frac{1}{4\pi\epsilon_0} \int d^3 r' \nabla'^2 \left(\frac{e^{i\omega|\mathbf{r}-\mathbf{r}'|/c}}{|\mathbf{r}-\mathbf{r}'|} \right) \nabla' \int d^3 r'' \frac{\tilde{\rho}(\mathbf{r}'',\omega)}{|\mathbf{r}'-\mathbf{r}''|}$$

$$+ \frac{1}{4\pi\epsilon_0} \int d^3 r' \frac{\tilde{\rho}(\mathbf{r}',\omega)}{|\mathbf{r}-\mathbf{r}'|}. \tag{A.7}$$

Integration by parts, together with

$$\nabla'^2 \left(\frac{1}{|\mathbf{r}'-\mathbf{r}''|} \right) = -4\pi \delta^3 (\mathbf{r}'-\mathbf{r}''), \tag{A.8}$$

then yields

$$X = \frac{1}{4\pi} \frac{1}{4\pi\epsilon_0} \int d^3 r' \nabla \frac{e^{i\omega|\mathbf{r}-\mathbf{r}'|/c}}{|\mathbf{r}-\mathbf{r}'|} \nabla'^2 \int d^3 r'' \frac{\tilde{\rho}(\mathbf{r}'',\omega)}{|\mathbf{r}'-\mathbf{r}''|} + \frac{1}{4\pi\epsilon_0} \int d^3 r' \frac{\tilde{\rho}(\mathbf{r}',\omega)}{|\mathbf{r}-\mathbf{r}'|}.$$

$$= -\frac{1}{4\pi\epsilon_0} \nabla \int d^3 r' \frac{\tilde{\rho}(\mathbf{r}',\omega)e^{i\omega|\mathbf{r}-\mathbf{r}'|/c}}{|\mathbf{r}-\mathbf{r}'|} + \frac{1}{4\pi\epsilon_0} \int d^3 r' \frac{\tilde{\rho}(\mathbf{r}',\omega)}{|\mathbf{r}-\mathbf{r}'|}. \tag{A.9}$$

The (instantaneous) second term cancels the (instantaneous) first term in (A.4), which therefore reduces to

$$\mathbf{E}(\mathbf{r},\omega) = \frac{i\mu_0\omega}{4\pi} \int d^3 r' \frac{\tilde{\mathbf{J}}(\mathbf{r}',\omega)e^{i\omega|\mathbf{r}-\mathbf{r}'|/c}}{|\mathbf{r}-\mathbf{r}'|} - \frac{1}{4\pi\epsilon_0} \nabla \int d^3 r' \frac{\tilde{\rho}(\mathbf{r}',\omega)e^{i\omega|\mathbf{r}-\mathbf{r}'|/c}}{|\mathbf{r}-\mathbf{r}'|}, \tag{A.10}$$

which is identical to the electric field (A.2) derived in the Lorentz gauge.

B. Transverse and Longitudinal Delta Functions

Consider the vector field

$$\mathbf{G}(\mathbf{r}) \equiv \frac{1}{4\pi} \nabla \times \nabla \times \int d^3 r' \frac{\mathbf{F}(\mathbf{r}')}{|\mathbf{r}-\mathbf{r}'|} \tag{B.1}$$

obtained from another vector field $\mathbf{F}(\mathbf{r})$. Using the identities $\nabla \times \nabla \times \mathbf{C} = \nabla(\nabla \cdot \mathbf{C}) - \nabla^2\mathbf{C}$, and $\nabla^2(1/|\mathbf{r}-\mathbf{r}'|) = -4\pi\delta^3(\mathbf{r}-\mathbf{r}')$, we obtain, for any point $\mathbf{r}$,

$$4\pi\mathbf{G}(\mathbf{r}) = \nabla \int d^3 r' \mathbf{F}(\mathbf{r}') \cdot \nabla \frac{1}{|\mathbf{r}-\mathbf{r}'|} + 4\pi\mathbf{F}(\mathbf{r})$$

$$= -\nabla \int d^3 r' \mathbf{F}(\mathbf{r}') \cdot \nabla' \frac{1}{|\mathbf{r}-\mathbf{r}'|} + 4\pi\mathbf{F}(\mathbf{r})$$

$$= \nabla \int d^3 r' \frac{\nabla' \cdot \mathbf{F}(\mathbf{r}')}{|\mathbf{r}-\mathbf{r}'|} + 4\pi\mathbf{F}(\mathbf{r}), \tag{B.2}$$

after integrating by parts and assuming $\mathbf{F}(\mathbf{r})$ vanishes sufficiently rapidly at infinity that the "boundary" term is zero. Therefore,

$$
\begin{aligned}
\mathbf{F}(\mathbf{r}) &= \mathbf{G}(\mathbf{r}) - \frac{1}{4\pi}\nabla \int d^3r' \frac{\nabla' \cdot \mathbf{F}(\mathbf{r}')}{|\mathbf{r} - \mathbf{r}'|} \\
&= \frac{1}{4\pi}\nabla \times \nabla \times \int d^3r' \frac{\mathbf{F}(\mathbf{r}')}{|\mathbf{r} - \mathbf{r}'|} - \frac{1}{4\pi}\nabla \int d^3r' \frac{\nabla' \cdot \mathbf{F}(\mathbf{r}')}{|\mathbf{r} - \mathbf{r}'|} \\
&= \mathbf{F}^{\perp}(\mathbf{r}) + \mathbf{F}^{\parallel}(\mathbf{r}),
\end{aligned}
\tag{B.3}
$$

which is the Helmholtz theorem cited in Section 1.3.2. Here,

$$
\mathbf{F}^{\perp}(\mathbf{r}) \equiv \frac{1}{4\pi}\nabla \times \nabla \times \int d^3r' \frac{\mathbf{F}(\mathbf{r}')}{|\mathbf{r} - \mathbf{r}'|} \,,
\tag{B.4}
$$

$$
\mathbf{F}^{\parallel}(\mathbf{r}) \equiv -\frac{1}{4\pi}\nabla \int d^3r' \frac{\nabla' \cdot \mathbf{F}(\mathbf{r}')}{|\mathbf{r} - \mathbf{r}'|},
\tag{B.5}
$$

and, obviously, $\nabla \cdot \mathbf{F}^{\perp} = 0$, and $\nabla \times \mathbf{F}^{\parallel} = 0$, since the divergence of a curl and the curl of a gradient are both zero. These equations uniquely identify the transverse and longitudinal parts of an arbitrary vector field $\mathbf{F}(\mathbf{r})$.

The transverse and longitudinal parts of $\mathbf{F}$ can also be expressed as follows, using the summation convention for repeated indices:

$$
F_i^{\perp}(\mathbf{r}) = \int d^3r' \delta_{ij}^{\perp}(\mathbf{r} - \mathbf{r}')F_j(\mathbf{r}'),
\tag{B.6}
$$

and

$$
F_i^{\parallel}(\mathbf{r}) = \int d^3r' \delta_{ij}^{\parallel}(\mathbf{r} - \mathbf{r}')F_j(\mathbf{r}'),
\tag{B.7}
$$

where the transverse and longitudinal delta functions (tensors) are defined by

$$
\delta_{ij}^{\perp}(\mathbf{r}) = \left(\frac{1}{2\pi}\right)^3 \int d^3k \left(\delta_{ij} - \frac{k_i k_j}{k^2}\right) e^{i\mathbf{k}\cdot\mathbf{r}},
\tag{B.8}
$$

$$
\delta_{ij}^{\parallel}(\mathbf{r}) \equiv \left(\frac{1}{2\pi}\right)^3 \int d^3k \frac{k_i k_j}{k^2} e^{i\mathbf{k}\cdot\mathbf{r}},
\tag{B.9}
$$

from which it follows by carrying out the integrations that

$$
\delta_{ij}^{\perp}(\mathbf{r}) + \delta_{ij}^{\parallel}(\mathbf{r}) = \delta_{ij}\delta^3(\mathbf{r}),
\tag{B.10}
$$

$$
\delta_{ij}^{\perp}(\mathbf{r}) = \frac{2}{3}\delta_{ij}\delta^3(\mathbf{r}) - \frac{1}{4\pi r^3}\left(\delta_{ij} - \frac{3r_i r_j}{r^2}\right),
\tag{B.11}
$$

$$
\delta_{ij}^{\parallel}(\mathbf{r}) = \frac{1}{3}\delta_{ij}\delta^3(\mathbf{r}) + \frac{1}{4\pi r^3}\left(\delta_{ij} - \frac{3r_i r_j}{r^2}\right).
\tag{B.12}
$$

We can prove (B.6), for instance, by showing that $\nabla \cdot \mathbf{F}^{\perp}(\mathbf{r}) = 0$, that is, $\partial F_i^{\perp}(\mathbf{r})/\partial x_i = 0$:

$$\frac{\partial F_i^\perp(\mathbf{r})}{\partial x_i} = \frac{\partial}{\partial x^i} \int d^3r'\, \delta_{ij}^\perp(\mathbf{r} - \mathbf{r}') F_j(\mathbf{r}')$$

$$= \left(\frac{1}{2\pi}\right)^3 \frac{\partial}{\partial x_i} \int d^3r' \int d^3k \left(\delta_{ij} - \frac{k_i k_j}{k^2}\right) e^{i\mathbf{k}\cdot(\mathbf{r}-\mathbf{r}')} F_j(\mathbf{r}')$$

$$= \left(\frac{1}{2\pi}\right)^3 \int d^3r' \int d^3k \left(\delta_{ij} - \frac{k_i k_j}{k^2}\right) i k_i e^{i\mathbf{k}\cdot(\mathbf{r}-\mathbf{r}')} F_j(\mathbf{r}')$$

$$= i \left(\frac{1}{2\pi}\right)^3 \int d^3r' \int d^3k [\mathbf{k}\cdot\mathbf{F}(\mathbf{r}') - \mathbf{k}\cdot\mathbf{F}(\mathbf{r}')] e^{i\mathbf{k}\cdot(\mathbf{r}-\mathbf{r}')} = 0. \qquad \text{(B.13)}$$

Note that $\delta_{ij}^\perp(\mathbf{r})$ and $\delta_{ij}^\parallel(\mathbf{r})$, unlike $\delta^3(\mathbf{r})$, do not vanish for $\mathbf{r} \neq 0$.

Exercise B.1: (a) Let $\mathbf{A}(\mathbf{r})$ and $\mathbf{B}(\mathbf{r})$ be two vector fields. Show that

$$\int d^3r\, \mathbf{A}^\perp(\mathbf{r}) \cdot \mathbf{B}^\parallel(\mathbf{r}) = 0,$$

$$\int d^3r\, \mathbf{A}^\perp(\mathbf{r}) \cdot \mathbf{B}^\perp(\mathbf{r}) = \int d^3r\, \mathbf{A}^\perp(\mathbf{r}) \cdot \mathbf{B}(\mathbf{r}),$$

$$\int d^3r\, \mathbf{A}^\parallel(\mathbf{r}) \cdot \mathbf{B}^\parallel(\mathbf{r}) = \int d^3r\, \mathbf{A}^\parallel(\mathbf{r}) \cdot \mathbf{B}(\mathbf{r}).$$

(b) Using the general expression above for the transverse part of a vector field, show that the transverse part of the vector potential $\mathbf{A}$ is gauge-invariant.

C.   Photodetection, Normal Ordering, and Causality

Consider first a two-state atom, initially in its lower eigenstate $|g\rangle$ of energy E_g, interacting with a field initially in a state $|I\rangle$. The probability that, at time t, the atom is in its upper eigenstate of energy E_e and the field is in a state $|F\rangle$ is given in second-order perturbation theory by (2.1.17) with the (interaction-picture) interaction Hamiltonian

$$h_I(t) = -\mathbf{d}(t) \cdot \mathbf{E}(\mathbf{r}, t), \qquad \text{(C.1)}$$

$$\mathbf{d}(t) = e^{-i\omega_{eg}t} \mathbf{d}_{ge}\sigma + e^{i\omega_{eg}t} \mathbf{d}_{eg}\sigma^\dagger, \qquad \text{(C.2)}$$

$$\mathbf{E}(\mathbf{r}, t) = \mathbf{E}^{(+)}(\mathbf{r}, t) + \mathbf{E}^{(-)}(\mathbf{r}, t), \qquad \text{(C.3)}$$

$$\mathbf{E}^{(+)}(\mathbf{r}, t) = \sum_\beta C_\beta a_\beta e^{-i\omega_\beta t} \mathbf{A}_\beta(\mathbf{r}), \qquad \text{(C.4)}$$

$$\mathbf{E}^{(-)}(\mathbf{r},t) = \sum_\beta C_\beta^* a_\beta^\dagger e^{i\omega_\beta t} \mathbf{A}_\beta^*(\mathbf{r}). \tag{C.5}$$

Here, $\omega_{eg} = (E_e - E_g)/\hbar > 0$, σ and $\sigma^\dagger$ are, as usual, the two-state lowering and raising operators, and $\mathbf{r}$ denotes the position of the atom. $\mathbf{E}^{(+)}(\mathbf{r},t)$ and $\mathbf{E}^{(-)}(\mathbf{r},t)$ are, respectively, the positive- and negative-frequency parts of the electric field operator, which has been expressed as an expansion in mode functions $\mathbf{A}_\beta$. Since $\langle e|\sigma|g\rangle = 0$, $\langle e|\sigma^\dagger|g\rangle = \langle e|e\rangle = 1$, $|i\rangle = |g\rangle|I\rangle$ and $|f\rangle = |e\rangle|F\rangle$,

$$\int_0^t dt' \langle f|h_I(t')|i\rangle = -\int_0^t dt' e^{i\omega_{eg}t'} \mathbf{d}_{eg} \cdot \langle F|\mathbf{E}(\mathbf{r},t')|I\rangle \tag{C.6}$$

and[1]

$$p_{i\to f}(t) = \frac{1}{\hbar^2} \int_0^t dt' \int_0^t dt'' d_{ge,\mu} d_{eg,\nu} \langle I|E_\mu(\mathbf{r},t'')|F\rangle\langle F|E_\nu(\mathbf{r},t')|I\rangle e^{i\omega_{eg}(t'-t'')}. \tag{C.7}$$

Summing over all possible final field states $|F\rangle$, assuming there is no discrimination among them in a measurement, and using the completeness relation $|F\rangle\langle F| = 1$, we replace (C.7) by

$$p^{(1)}(t) \equiv \frac{1}{\hbar^2} \int_0^t dt' \int_0^t dt'' d_{ge,\mu} d_{eg,\nu} \langle E_\mu(\mathbf{r},t'')E_\nu(\mathbf{r},t')\rangle e^{i\omega_{eg}(t'-t'')}, \tag{C.8}$$

in which the expectation value $\langle E_\mu(t'')E_\nu(t')\rangle = \langle I|E_\mu(\mathbf{r},t'')E_\nu(\mathbf{r},t')|I\rangle$ refers to the initial state of the field.

We now modify this result of second-order perturbation theory to allow for a *continuum* of possible (photo-)electron states $|e\rangle$, assuming an energy distribution function $P(E_e)$ for these states:

$$p^{(1)}(t) = \frac{1}{\hbar^2} \int_0^\infty dE_e P(E_e) d_{eg,\mu}^* d_{eg,\nu}$$
$$\times \int_0^t dt' \int_0^t dt'' \langle E_\mu(\mathbf{r},t'')E_\nu(\mathbf{r},t')\rangle e^{i\omega_{eg}(t'-t'')}. \tag{C.9}$$

This expression does not allow for any possibility that the electron can make a transition from a state $|e\rangle$ back to the initial bound state $|g\rangle$. It therefore serves to model, for example, a photodetector based on the photoelectric effect.

If the electric field acting on the atom were monochromatic,

$$E_\mu(\mathbf{r},t) = \frac{1}{2}\left[E_\mu^{(+)}(\mathbf{r})e^{-i\omega t} + E_\mu^{(-)}(\mathbf{r})e^{i\omega t}\right], \tag{C.10}$$

we would replace (C.9) by

---

[1] The summation convention for repeated indices (μ,ν) is used in writing (C.7).

$$p^{(1)}(t) = \frac{1}{\hbar^2} \int_0^\infty dE_e P(E_e) d_{eg,\mu}^* d_{eg,\nu} \langle E_\mu^{(-)}(\mathbf{r}) E_\nu^{(+)}(\mathbf{r}) \rangle$$

$$\times \int_0^t dt' \int_0^t dt'' e^{i(\omega_{eg}-\omega)(t'-t'')}. \tag{C.11}$$

Then, $p^{(1)}(t)$ varies in time as

$$\int_0^t dt' \int_0^t dt'' e^{i(\omega_{eg}-\omega)(t'-t'')} = \frac{\sin^2[(\omega_{eg}-\omega)t/2]}{[(\omega_{eg}-\omega)/2]^2}, \tag{C.12}$$

which leads to a rate of excitation $dp^{(1)}/dt$ given by Fermi's golden rule (see Section 2.3).

More generally, if the electric field is slowly varying in time compared to $\exp(i\omega_{eg}t)$ for frequencies ω_{eg} that contribute significantly to (C.9), we replace (C.9) by

$$p^{(1)}(t) = \frac{1}{\hbar^2} \int_0^\infty dE_e P(E_e) d_{eg,\mu}^* d_{eg,\nu} \int_0^t dt' \int_0^t dt'' \langle E_\mu^{(-)}(\mathbf{r},t'') E_\nu^{(+)}(\mathbf{r},t') \rangle e^{i\omega_{eg}(t'-t'')}$$

$$= \int_0^t dt'' \int_0^t dt' S_{\mu\nu}(t'-t'') \langle E_\mu^{(-)}(\mathbf{r},t'') E_\nu^{(+)}(\mathbf{r},t') \rangle, \tag{C.13}$$

$$S_{\mu\nu}(t) \equiv \frac{1}{\hbar^2} \int_0^\infty dE_e P(E_e) d_{eg,\mu}^* d_{eg,\nu} e^{i\omega_{eg}t}. \tag{C.14}$$

In this approximation the photodetection probability depends on the *normally ordered* field correlation function $\langle E_\mu^{(-)}(t'') E_\nu^{(+)}(t') \rangle$. Another approximation, often made in theoretical discussions, is based on the assumption that the photodetection is insensitive to $d_{eg,\mu}^* d_{eg,\nu}$ and to the frequencies ω_{eg} within the bandwidth of the applied field; actual detectors at optical frequencies can approximate fairly well such an "ideal broadband detector." In this approximation,

$$S_{\mu\nu}(t) \propto \frac{1}{\hbar^2} d_{eg,\mu}^* d_{eg,\nu} \int_{-\infty}^\infty d\omega_{eg} e^{i\omega_{eg}t} \equiv s_{\mu\nu}\delta(t), \tag{C.15}$$

$$p^{(1)}(t) \cong s_{\mu\nu} \int_0^t dt' \langle E_\mu^{(-)}(\mathbf{r},t') E_\mu^{(+)}(\mathbf{r},t') \rangle, \tag{C.16}$$

and we define a rate

$$R^{(1)}(t) = \frac{d}{dt} p^{(1)}(t) = s_{\mu\nu} \langle E_\mu^{(-)}(\mathbf{r},t) E_\nu^{(+)}(\mathbf{r},t) \rangle \equiv s_{\mu\nu} G_{\mu\nu}^{(1)}(\mathbf{r},t;\mathbf{r},t). \tag{C.17}$$

The first-order field correlation function

$$G_{\mu\nu}^{(1)}(\mathbf{r}_1,t_1;\mathbf{r}_2,t_2) \equiv \langle E_\mu^{(-)}(\mathbf{r}_1,t_1) E_\nu^{(+)}(\mathbf{r}_2,t_2) \rangle, \tag{C.18}$$

as defined in Section 3.9 for a single vector component of the electric field. The main assumption in the derivation of (C.13) and (C.17) is the applicability of second-order perturbation theory, which implies an irreversible transition of electrons from a ground state to a continuum of possible final states. The replacement of the electric field correlation function $\langle E_\mu(\mathbf{r}, t'')E_\nu(\mathbf{r}, t')\rangle$ by the normally ordered correlation function $\langle E_\mu^{(-)}(\mathbf{r}, t'')E_\nu^{(+)}(\mathbf{r}, t')\rangle$ is an approximation of exactly the type used in the derivation of Fermi's golden rule for a transition rate (see Section 2.3).

Exercise C.1: Show that a hypothetical photodetection system based on stimulated emission rather than absorption responds to an *anti-normal-ordered* field correlation function.

Application of (C.13) and (C.17) to an actual photodetector requires integration over the positions $\mathbf{r}$ of the detector atoms, and account of practical effects such as reflections at the detector surface. For a nearly ideal broadband detector, the rate obtained from (C.17) would be proportional to the rate at which electrons are irreversibly removed from their initial ground states; in a photomultiplier, for example, this would be proportional to the electric current generated by the applied field. We will conclude our simplified treatment with some remarks relating to normal ordering, causality, and the vacuum field.

The electric field in (C.9), of course, satisfies the condition that a source of radiation turned on at a time $t = 0$ at a distance R from a detector atom cannot affect the atom before a time $t = R/c$. The fields $\mathbf{E}^{(\pm)}(\mathbf{r}, t)$, however, are not retarded in this sense. The appearance of the normally ordered correlation $\langle E_\mu^{(-)}(\mathbf{r}, t'')E_\nu^{(+)}(\mathbf{r}, t')\rangle$ in photodetection theory might therefore imply, at first glance, a violation of causality. However, the replacement of $\langle E_\mu(\mathbf{r}, t'')E_\nu(\mathbf{r}, t')\rangle$ by the normally ordered correlation $\langle E_\mu^{(-)}(\mathbf{r}, t'')E_\nu^{(+)}(\mathbf{r}, t')\rangle$ is clearly an *approximation*. The situation here is essentially the same as in the discussion in Section 7.9.3 of causality in the dipole interaction of two atoms. It is only when "energy-non-conserving," negative-frequency contributions to the interaction are included in the calculation that the interaction is found to be causal in the sense that the initially unexcited atom can only be excited after the time it takes for the field from the initially excited atom to reach it; in the approximation made by Fermi, these negative-frequency contributions were included by extending an integration to include all negative as well as all positive field frequencies. In the Heisenberg-picture calculation given in Section 7.9.1, similarly, we began with the fully retarded electric field from an atom and then identified a fully retarded expression (7.9.37) for its positive-frequency part in a rotating-wave approximation. The same approach can be employed in! the present context to modify (C.13) and (C.17) to explicitly include retardation, although, as a practical matter, this is hardly necessary.[2]

---

[2] For further discussion of causality in photodetection theory, see, for instance, P. W. Milonni, D. F. V. James, and H. Fearn, Phys. Rev. A **52**, 1525 (1995).

Because of the normal ordering of field operators in (C.13) and (C.17), there is no contribution from the vacuum field. The same cannot be said for (C.9), which appears to include unphysical vacuum-field contributions $\langle a_\beta a_\beta^\dagger \rangle$. However, the *total* field acting on an atom at $\mathbf{r}$ in our model is

$$\mathbf{E}(\mathbf{r}, t) = \mathbf{E}_0(\mathbf{r}, t) + \mathbf{E}_{\mathrm{RR}}(\mathbf{r}, t) + \mathbf{E}_{\mathrm{ext}}(\mathbf{r}, t), \tag{C.19}$$

in which $\mathbf{E}_0(\mathbf{r}, t)$ is the source-free, vacuum field, $\mathbf{E}_{\mathrm{RR}}(\mathbf{r}, t)$ is the radiation reaction field, and $\mathbf{E}_{\mathrm{ext}}(\mathbf{r}, t)$ is the field from sources external to the detector atom. As discussed in Chapter 4, the only effect of $\mathbf{E}_0(\mathbf{r}, t) + \mathbf{E}_{\mathrm{RR}}(\mathbf{r}, t)$ on a ground-state atom is a radiative frequency shift. The only field that can cause an atomic electron to make a transition out of the ground state is therefore $\mathbf{E}_{\mathrm{ext}}(\mathbf{r}, t)$, and it is only this field that contributes to (C.13) and (C.17). In other words, the field appearing in these expressions should be understood to be $\mathbf{E}_{\mathrm{ext}}(\mathbf{r}, t)$.

Photodetection theory for two or more detectors at different positions and times can be modeled in essentially the same way, leading to higher-order field correlation functions such as the second-order correlation function (see Section 3.9)[3]

$$\begin{aligned} G^{(2)}_{ijk\ell})(\mathbf{r}_1, t_1; \mathbf{r}_2, t_2; \mathbf{r}_3, t_3; \mathbf{r}_4, t_4) = \langle E_i^{(-)}(\mathbf{r}_1, t_1) E_j^{(-)}(\mathbf{r}_2, t_2) \\ \times\; E_k^{(+)}(\mathbf{r}_3, t_3) E_\ell^{(+)}(\mathbf{r}_4, t_4) \rangle, \end{aligned} \tag{C.20}$$

which we used in the theory of photon bunching (Section 3.9) and anti-bunching (see Section 5.7) and in the calculation of photon polarization correlations (see Section 5.8).

---

[3] R. J. Glauber, in *Quantum Optics and Electronics*, C. DeWitt, A. Blandin, and C. Cohen-Tannoudji, eds., Gordon and Breach, New York, 1965, pp. 65–185.

Bibliography

Abrikosov, A. A., L. P. Gorkov, and I. E. Dzyaloshinski, *Methods of Quantum Field Theory in Statistical Physics*, Dover Publications, Mineola, New York, 1975.

Allen, L. and J. H. Eberly, *Optical Resonance and Two-Level Atoms*, Dover Publications, Mineola, New York, 1987.

Ballentine, L. E., *Quantum Mechanics*, Prentice-Hall, Englewood Cliffs, New Jersey, 1990.

Berman, P. R., ed., *Cavity Quantum Electrodynamics*, Academic Press, Boston, 1994.

Berman, P. R., *Introductory Quantum Mechanics*, Springer-Verlag, Berlin, 2018.

Berman, P. R. and V. S. Malinovsky, *Principles of Laser Spectroscopy and Quantum Optics*, Princeton University Press, Princeton, 2011.

Bohren, C. F. and D. R. Huffman, *Absorption and Scattering of Light by Small Particles*, Wiley-Interscience, New York, 1998.

Bordag, M., G. L. Klimchitskaya, U. Mohideen, and V. M. Mostepanenko, *Advances in the Casimir Effect*, Oxford University Press, Oxford, 2009.

Born, M. and E. Wolf, *Principles of Optics*, seventh edition, Cambridge University Press, Cambridge, 1999.

Budker, D., D. F. Kimball, and D. P. DeMille, *Atomic Physics: An Exploration through Problems and Solutions*, second edition, Oxford University Press, Oxford, 2008.

Carmichael, H. J., *An Open Systems Approach to Quantum Optics*, Springer-Verlag, Berlin, 1993.

Compagno, G., R. Passante, and F. Persico *Atom-Field Interactions and Dressed Atoms,*, Cambridge University Press, Cambridge, 1995.

Cowan, R. D., *The Theory of Atomic Structure and Spectra*, University of California Press, Berkeley, 1981.

Dalvit, D., P. Milonni, D. Roberts, and F. da Rosa, eds., *Casimir Physics*, Springer-Verlag, Berlin, 2011.

De la Peña, L. and A. M. Cetto, *The Quantum Dice: An Introduction to Stochastic Electrodynamics*, Kluwer Academic Publishers, Dordrecht, 2010.

Desurvire, E., *Erbium-Doped Fiber Amplifiers: Principles and Applications*, Wiley-Interscience, New York, 1994.

Feynman, R. P., R. B. Leighton, and M. Sands, *The Feynman Lectures on Physics*, Volumes I–III, Addison-Wesley, Reading, Massachusetts, 1964.

Fürth, R. F., ed., *Investigations on the Theory of the Brownian Movement by Albert Einstein*, Dover Publications, New York, 1956.

Gardiner, C. W. and P. Zoller, *Quantum Noise: A Handbook of Markovian and Non-Markovian Quantum Stochastic Methods with Applications to Quantum Optics*, Springer-Verlag, Berlin, 2010.

Garrison, J. C. and R. Y. Chiao, *Quantum Optics*, Oxford University Press, Oxford, 2008.

Ginzburg, V. L., *Theoretical Physics and Astrophysics*, Pergamon Press, Oxford, 1979.

Gottfried, K. and T.-M. Yan, *Quantum Mechanics: Fundamentals*, second edition, Springer, New York, 2004.

Haroche, S. and J.-M. Raimond, *Exporing the Quantum: Atoms, Cavities, and Photons*, Oxford University Press, Oxford, 2006.

Haus, H. A., *Electromagnetic Noise and Quantum Optical Measurements*, Springer-Verlag, Berlin, 2000.

Heisenberg, W., *The Physical Principles of the Quantum Theory*, Dover Publications, New York, 1949.

Heitler, W., *The Quantum Theory of Radiation*, third edition, Clarendon Press, Oxford, 1954.

Israelachvili, J. N., *Intermolecular and Surface Forces, with Applications to Colloidal and Biological Systems*, Academic Press, San Diego, 1985.

Jackson, J. D., *Classical Electrodynamics*, third edition, John Wiley & Sons, New York, 1999.

Klauder, J. R. and E. C. G. Sudarshan, *Fundamentals of Quantum Optics*, W. A. Benjamin, Inc., New York, 1968.

Knight, P. L. and L. Allen, *Concepts of Quantum Optics*, Pergamon Press, Oxford, 1983.

Landau, L. D. and E. M. Lifshitz, *Statistical Physics*, third edition, Part 1, Elsevier Butterworth-Heinemann, Oxford, 2004.

Landau, L. D. and E. M. Lifshitz, *The Classical Theory of Fields*, fourth edition, Butterworth–Heinmann, Oxford, 2000.

Landau, L. D., E. M. Lifshitz, and L. P. Pitaevskii, *Electrodynamics of Continuous Media*, second edition, Elsevier Butterworth–Heinemann, Oxford, 2004.

Loudon, R., *The Quantum Theory of Light*, third edition, Oxford University Press, Oxford, 2000.

Louisell, W. H., *Quantum Statistical Properties of Radiation*, John Wiley & Sons, New York, 1973.

MacDonald, D. K. C., *Noise and Fluctuations: An Introduction*, Dover Publications, Mineola, New York, 2006.

Mandel, L. and E. Wolf, *Optical Coherence and Quantum Optics*, Cambridge University Press, Cambridge, 1995.

Mandel, L. and E. Wolf, eds., *Selected Papers on Coherence and Fluctuations of Light*, Volumes 1 and 2, Dover Publications, New York, 1970.

Menzel, R., *Photonics. Linear and Nonlinear Interactions of Laser Light and Matter*, second edition, Springer-Verlag, Berlin, 2007.

Meystre, P. and M. Sargent, *Elements of Quantum Optics*, Springer-Verlag, Berlin, 1990.

Milonni, P. W., *Fast Light, Slow Light, and Left-Handed Light*, Taylor & Francis, New York, 2005.

Milonni, P. W., *The Quantum Vacuum: An Introduction to Quantum Electrodynamics*, Academic Press, San Diego, 1994.

Milonni, P. W. and J. H. Eberly, *Laser Physics*, John Wiley & Sons, New York, 2010.

Milton, K. A., *The Casimir Effect. Physical Manifestations of Zero-Point Energy*, World-Scientific Publishing, Singapore, 2001.

Novotny, L. and B. Hecht, *Principles of Nano-Optics*, second edition, Cambridge University Press, Cambridge, 2012.

Pais, A., *Niel's Bohr's Times, in Physics, Philosophy, and Polity*, Oxford University Press, Oxford, 1991.

Pais, A., *"Subtle is the Lord ..." The Science and the Life of Albert Einstein*, Oxford University Press, Oxford, 1982.

Parsegian, V. A., *Van der Waals Forces. A Handbook for Biologists, Chemists, Engineers, and Physicists*, Cambridge University Press, New York, 2006.

Peres, A., *Quantum Theory: Concepts and Methods*, Kluwer Academic Publishers, Dordrecht, 1995.

Power, E. A., *Introductory Quantum Electrodynamics*, Longmans, London, 1964.

Risken, H., *The Fokker-Planck Equation: Methods of Solution and Applications*, second edition, Springer-Verlag, Berlin, 1996.

Saleh, B. A. and M. C. Teich, *Fundamentals of Photonics*, John Wiley & Sons, New York, 1991.

Sargent, M., M. O. Scully, and W. E. Lamb, Jr., *Laser Physics*, Addison-Wesley. Reading, Massachusetts, 1974.

Schiff, L. I., *Quantum Mechanics*, third edition, McGraw-Hill, New York, 1968.

Schweber, S. S., *QED and the Men Who Made It: Dyson, Feynman, Schwinger, and Tomonaga*, Princeton University Press, Princeton, 1994.

Scully, M. O. and S. Suhail Zubairy, *Quantum Optics*, Cambridge University Press, Cambridge, 1997.

Shore, B. W., *The Theory of Coherent Atomic Excitation: Volume 1, Simple Atoms and Fields*, Wiley-Interscience, New York, 1990.

Shore, B. W, *The Theory of Coherent Atomic Excitation: Volume 2, Multilevel Atoms and Incoherence*, Wiley-Interscience, New York, 1990.

Van der Waerden, B. L., ed., *Sources of Quantum Mechanics*, Dover Publications,

New York, 1968.

Van Kampen, N. G., *Stochastic Processes in Physics and Chemistry*, third edition,
Elsevier, Amsterdam, 2007.

Vogel, W. and D.-K. Welsch, *Quantum Optics: An Introduction*, third edition,
Wiley-VCH, Weinheim, 2006.

Wax, N., ed., *Selected Papers on Noise and Stochastic Processes*, Dover, London, 1954.

Index

Index